International Conference PhysicA.SPb/2016

AA006260

Journal of Physics: Conference Series Volume 929

Saint-Petersburg, Russia
1 - 3 November 2016

Editors:

Nikita S. Averkiev
Sergey A. Poniaev
Grigorri S. Sokolovskii

ISBN: 978-1-5108-5313-3
ISSN: 1742-6588

Printed from e-media with permission by:

Curran Associates, Inc.
57 Morehouse Lane
Red Hook, NY 12571

Printed by Curran Associates, Inc. (2018)

For permission requests, please contact the Institute of Physics
at the address below.

Institute of Physics
Dirac House, Temple Back
Bristol BS1 6BE UK

Phone: 44 1 17 929 7481
Fax: 44 1 17 920 0979

techtracking@iop.org

Additional copies of this publication are available from:

Curran Associates, Inc.
57 Morehouse Lane
Red Hook, NY 12571 USA
Phone: 845-758-0400
Fax: 845-758-2633
Email: curran@proceedings.com
Web: www.proceedings.com

Table of contents

Volume 929

International Conference PhysicA.SPb/2016
(previous edition: PhysicASPb-2015)

1–3 November 2016, Saint-Petersburg, Russian Federation

Accepted papers received: 9 November 2017
Published online: 27 November 2017

Preface

International Conference PhysicA.SPb/2016

Peer review statement

Papers

Astronomy and Astrophysics

Investigation of radiation pressure influence on the asteroids that subsequently have fallen to the Earth

A Martyusheva.....1

The fractal analysis of the gravitational field and topography of the Mars

S A Demin, A O Andreev, N Y Demina and Y A Nefedyev.....5

D Na I absorption lines formation in the RZ Psc spectrum

S G Shulman, V P Grinin, I S Potravnov and D E Mkrtichian.....11

Modelling non-dissociative shock waves in interstellar medium: first results

A V Nesterenok.....16

What can we learn about the stellar wind of massive stars from studying spin evolution of the X-ray pulsar OAO 1657-415

V.Yu. Kim and N.R. Ikhsanov.....22

Mirror instability in plasma with relativistic component

S M Osipov, A M Bykov, D C Ellison and P E Gladilin.....28

Glitches and precession damping in the framework of 3-component model of neutron star

K Y Kraav, A I Tsygan, D P Barsukov and A N Yurkin.....34

The angular and energy distributions of neutrons, gamma- and x-rays produced in solar flares. Related changes in the isotopic composition of the photosphere and the solar wind

G I Vasiliev and M E Chakchurina.....40

Numerical studies of instability of generalized polytropic models of stellar disks

A. A. Smirnov and N. Ya. Sotnikova.....44

The impact of heat blanketing envelopes on neutron stars cooling

M V Beznogov, D G Yakovlev, M Fortin, P Haensel and J L Zdunik.....51

The γ-ray pulsar J1932+1916 in X-rays

A Karpova, P Shternin, D Zyuzin, A Danilenko and Yu Shibanov.....56

Thermal evolution of old white dwarfs

Andrew Kozhberov.....60

The method for celestial bodies' center of mass position relative to their figures determination on the basis of harmonic analysis of the expansion in spherical functions in order to refine the physical libration parameters

N Y Demina, A O Andreev, S A Demin and Y A Nefedyev.....66

Particle-in-cell simulation of shock waves with different kinds of ions.

V I Romanky, A M Bykov, S M Osipov and P E Gladilin.....72

Modeling of relativistic pulsar wind interaction with the interstellar medium

A E Petrov, A M Bykov and S M Osipov.....78

Biophysics

Method of recording bioelectrical signals using a capacitive coupling

V A Simon, V A Gerasimov, D K Kostrin, L M Selivanov and A A Uhov.....84

The development of the experimental setup for measuring the cell membrane electrical potential by Sucrose-Gap Technique

AD Yuzhakov, AV Nosarev and AN Aleinik.....88

Specialized nanowhisker probes for high-precision investigation of native bio-objects in liquid by means of atomic force microscopy

MV Zhukov, IV Kukhtevich, FE Komissarenko and VI Chubinskiy-Nadezhdin.....94

Temperature sensitivity of ligand-gated ion channels: ryanodine receptor case

B.I. Iaparov, A.S. Moskvin and O.E. Solovyova.....98

Interaction of indole-papaverine with DNA in solutions of various ionic strength

V I Travkina, E B Moroshkina and D N Osinnikova.....104

Fluorescent spectroscopy features in the study of green leaves of plants

I S Kudryashova, V Yu Rud, Yu V Rud, V Ch Shpunt, A P Glinushkin and N N Bykova.....109

Devices and materials of the THz and microwave ranges

Design of microwave antenna system on planar Yagi-Uda elements and microstrip coupler

K V Petrovnin and R R Latypov.....115

Investigation of self-generation of broadband microwave chaotic and noise signals in microwave photonic ring oscillator

A.B. Ustinov, A.V. Kondrashov and B.A. Kalinikos.....120

A device for water content measuring of oil

A Gorbunov and N Evstegneev.....125

Investigation of frequency-selective properties of microwave wideband bandpass filters

S A Khodenkov and N M Boev.....129

Microstrip filters based on 2D electromagnetic crystal

S A Khodenkov and B A Belyaev.....134

Investigation of spin waves in magnonic crystal based on periodically metallized YIG film

V.V. Vitko, A.A. Nikitin, A.B. Ustinov, A.V. Drozdovskii, A.A. Semenov and E. Lähderanta.....139

Magnetically tunable high-Q spin-wave optoelectronic single-loop resonator

V V Vitko, A A Nikitin, A V Kondrashov, M I Martynov and A B Ustinov.....143

Impurities and defects in solids

Positron annihilation spectroscopy studies of bronze exposed to sandblasting at different pressure

S Kurdyumov, K Siemek and P Horodek.....149

X-ray diffraction investigations of the thermal expansion of iron borate $FeBO_3$ crystals

A Zamkovskaya, E Maksimova, I Nauhatsky and M Shapoval.....155

Boson peak in two-dimensional random matrix models

D A Conyuh, Y M Beltukov and D A Parshin.....161

Mathematical physics and numerical methods

Stability investigation of implicit parametrical schemes for the systems of kinetic equations

Elizaveta A Prokhorova and Gerasim V Krivovichev.....167

Analysis of parametric finite-difference schemes for the system of linear advection equations

Elena S Marnopolskaya and Gerasim V Krivovichev.....173

Functional differential equations of neutral type with integrable weak singularity: hydrogen thermal desorption model

Yury V Zaika and Ekaterina K Kostikova.....179

The arrowhead decomposition method for a block-tridiagonal system of linear equations

P A Belov, E R Nugumanov and S L Yakovlev.....185

The coefficient smoothing method application to the problem of gas pipeline glaciation

S A Mikheev, N N Ermolaeva and G V Krivovichev.....192

Numerical determination of the CFT central charge in the site-diluted Ising model

P A Belov, A A Nazarov and A O Sorokin.....198

Reduction of the dimension of neural network models in problems of pattern recognition and forecasting

A D Nasertdinova and V V Bochkarev.....205

Modelling of hydrogen permeability of membranes for high-purity hydrogen production

Yury V Zaika and Natalia I Rodchenkova.....211

Equation of state for computer simulation of metal ablation by femtosecond laser pulses

R V Davydov and V I Antonov.....217

Algorithm for detection the QRS complexes based on support vector machine

G.V. Van and K.V. Podmasteryev.....222

Nano-structured and thin film materials

Electrical and optical properties of iodide nanoparticles embedded into zeolite porous matrices

A V Cvetkov, S E Gango, M S Ivanova, S D Khanin, A E Lukin, V G Solovyev, S V Trifonov and V L Veisman.....227

Effect of tungsten implantation on the switching parameters in V_2O_5 films

S.V. Burdyukh, O.Ya. Berezina and A.L. Pergament.....233

Effect of temperature on the value of the surface potential and the phase state of 4'-n-octyl-4-p-cyanobiphenyl liquid crystal

A S Chumakov, Ammar J K Al-Alwani, I A Gorbachev, E G Glukhovskoy and N V Usoltseva.....238

Synchronization in the system of coupled oscillators based on VO_2 switches

V V Perminov, V V Putrolaynen, M A Belyaev and A A Velichko.....243

Dependence of the photoconductivity of nanofibers based on indium-zinc oxide on the concentration of components

N.P. Markova, O.Ya. Berezina and A.L. Pergament.....248

The dependence of the wavelength on MBE growth parameters of GaAs quantum dot in AlGaAs NWs on Si (111) substrate

R R Reznik, I V Shtrom, Yu B Samsonenko, A I Khrebtov, I P Soshnikov and G E Cirlin.....252

Thin-film fractal nanostructures formed by electrical breakdown

P O Tadtaev, A A Bobkov, V F Borodzyulya, I A Lamkin, I I Mihailov, V A Moshnikov, N V Permyakov, A V Solomonov, N T Sudar and S A Tarasov.....258

Study of inverse magnetostrictive effect in metallic glasses $Fe_{80-x}Co_xP_{14}B_6$

V S Severikov, A M Grishin and V S Ignahin.....264

Study of functional properties of gas-sensitive cobalt-containing polyacrylonitrile films

T.V. Semenistaya and A.A. Voronova.....269

Investigation of porous silicon obtained under different conditions by the contact angle method

A O Belorus, Y V Bukina, A I Pastukhov, D S Stebko, Yu M Spivak and V A Moshnikov.....275

Investigation of helium charged beams passing through the porous alumina membranes

E N Muratova and A A Shemukhin.....280

Laser deposition of resonant silicon nanoparticles on perovskite for photoluminescence enhancement

E Y Tiguntseva, A S Zalogina, V A Milichko, D A Zuev, M M Omelyanovich, A Ishteev, A Cerdan Pasaran, R Haroldson, S V Makarov and A A Zakhidov.....287

The electron tunneling from fullerene molecule in a weak electric field

P A Golovinski, A A Drobyshev, E A Mikhin and A V Tuchin.....292

The formation of multi-axis micromechanical gyroscopes and accelerometers using surface micromachining

E Yu Gusev, J Y Jityaeva and Al V Bykov.....298

Molecular dynamics simulation of metal nanoislands growth

N V Kapralov, E S Babich and A V Redkov.....303

The technique of visualization and evaluation of the emission site distribution for large area field emitters

S V Filippov, E O Popov, A G Kolosko and A.V. Shchegolkov.....309

Investigation of porous silicon nanopowders functionalized by antibiotic Kanamycin, fluorophore Indocyanine Green

K Bespalova, P A Somov and Yu M Spivak.....315

Screening effect in matrix graphene / SiC planar field emmiters

I L Jityaev, A M Svetlichnyi, A S Kolomiytsev and O A Ageev.....319

Nuclear and elementary particle physics

Coulomb explosion of nanosize cylindrical target

P A Golovinski and E A Mikhin.....325

Neutral meson production in Cu+Au collisions at 200 GeV

A. Berdnikov, Ya. Berdnikov, D. Kotov, P. Radzevich, V. Riabov and Sergei Zharko.....331

Optics and Spectroscopy

Determination of the composition of liquids using spectral analysis of the electric discharge radiation

A N Ramazanov, D K Kostrin and A A Lisenkov.....336

Investigation of Doppler spectra of laser radiation scattered inside hand skin during occlusion test

I.O. Kozlov, E.A. Zherebtsov, A.I. Zherebtsova, V.V. Dremin and A.V. Dunaev.....340

New method for determining concentrations of the mixture components during rapid control

N S Myazin, S E Logunov, V V Davydov, V Yu Rud', N M Grebenikova and V V Yushkova.....345

Destructive interference between electric and toroidal dipole moments in TiO_2 cylinders and frustums with coaxial voids

P D Terekhov, K V Baryshnikova, A B Evlyukhin and A S Shalin.....352

Core-shell Yagi-Uda nanoantenna for highly efficient and directive emission

Olga N. Sergaeva, Roman S. Savelev, Denis G. Baranov and Alexander E. Krasnok.....358

Influence of the dynamic Stark effect on long-term frequency stability of a self-oscillating magnetometer with laser-pumped alkali atoms

A A Baranov, S V Ermak, N K Kulachenkov, M V Petrenko, E A Sagitov and V V Semenov.....366

The luminescence properties of nanocrystalline phosphors $Mg_2SiO_4:Eu^{3+}$

A Y Kolomytsev, D V Mamonova, A A Manshina and I E Kolesnikov.....371

Application of optical non-invasive methods to diagnose the state of the lower limb tissues in patients with diabetes mellitus

E.V. Zharkikh, V.V. Dremin, M.A. Filina, I.N. Makovik, E.V. Potapova, E.A. Zherebtsov, A.I. Zherebtsova and A.V. Dunaev.....378

Investigation of lasers based on coupled waveguides by near-field scanning optical microscopy

Yu S Polubavkina, N Yu Gordeev, A S Payusov, N V Kryzhanovskaya, E I Moiseev, F I Zubov, S A Mintairov, N A Kalyuzhnyy, M M Kulagina, Yu M Shernyakov, M V Maximov and A E Zhukov.....384

Plasmonic reflectance anisotropy spectroscopy of metal nanoclusters in a dielectric multilayer. Theory

A V Korotchenkov and V A Kosobukin.....388

Coating of Au nanoparticle by Si shell for enhanced local heating

G P Zograf, M I Petrov and S V Makarov.....394

Investigation of the methods for optical wavefront parameter manipulation

Tatiana A Vovk and Nikolay V Petrov.....398

Optical second harmonic generation and two-photon luminescence in organic microstructures

K. D. Zhdanova, N. V. Mitetelo, E.A. Mamonov and T.V. Murzina.....404

Optoelectronic devices

Compensation of the nonuniformity of sensitivity for the multielement infrared photodetector

D K Kostrin, V A Simon, A A Uhov, V A Gerasimov and L M Selivanov.....409

Study of electroluminescence of InAs(Sb,P) LED heterostructures

A A Semakova, N L Bazhenov and K D Mynbaev.....413

Edge-emitting lasers based on coupled large optical cavity with high beam stability

A Serin, N Gordeev, A Payusov, Y Shernyakov, y Kalyuzhnyy, S Mintairov and M Maximov.....417

The luminescent structures based on colloidal quantum dots CdZnSeS/ZnS

I I Mikhailov, L I Kozlovich, P O Tadtaev, T D Lebedeva, A V Solomonov and S A Tarasov.....421

Calculation of Quantum Efficiency of Photodetectors Based on A3B5 Metal-Solid Solutions Contacts

V S Zemtsovskaya, S A Tarasov, I A Lamkin and A V Solomonov.....426

Spectral characteristics of InP photocathode with a surface grid electrode

N S Myazin, K J Smirnov, V V Davydov and S E Logunov.....434

Molecular-beam epitaxy of 7-8 μm range quantum-cascade laser heterostructures

A V Babichev, D V Denisov, A V Filimonov, V N Nevedomsky, A S Kurochkin, A G Gladyshev, L Ya Karachinsky, G S Sokolovskii, I I Novikov, A Bousseksou and A Yu Egorov.....439

The concept for realization of quantum-cascade lasers emitting at 7.5 μm wavelength

I I Novikov, A V Babichev, V E Bugrov, A G Gladyshev, L Ya Karachinsky, E S Kolodeznyi, A S Kurochkin, A V Savelyev, G S Sokolovskii and A Yu Egorov.....445

Physics and technology of energy conversion

Simulating of solid-state electrocaloric cooler based on multi-layered ferroelectric capacitor structures

A V Es'kov, P Yu Belyavskiy, A S Anokhin, O V Pakhomov, A A Semenov, A A Nikitin, V A Krylov and M T Bui.....451

Spectrum tuning in multi-junction solar cells measurements

S.A. Bogomolova, M.Z. Shvarts and N. Kh. Timoshina.....457

Physics of ferroics

Mechanism of domain wall interaction with electric field in iron garnet films

A. S. Sergeev.....463

Magnonic crystals based on ferrite and multiferroic periodic structures

A V Drozdovskiy, I A Ustinova, A A Nikitin, V V Vitko and A B Ustinov.....468

Physics of quantum structures

Entanglement between two Rydberg atoms induced by a thermal field

T S Mastyugina and E K Bashkirov.....472

Quantum size effect in single-crystalline bismuth nanorods

E A Sedov, K –P Riikonen and K Yu Arutyunov.....478

Effect of carrier spectra nonsphericity and subband mixing on CHHS Auger process rate in deep quantum wells

N V Pavlov and G G Zegrya.....483

Bleaching compensation in GaAs/AlGaAs quantum wells by above-barrier illumination

I. A. Solovev, Yu. V. Kapitonov, V. G. Davydov, Yu. P. Efimov, S. A. Eliseev, V. V. Petrov and V. V. Ovsyankin.....488

Plasmonic enhancement of photoluminescence from cadmium sulfide and lead sulfide quantum dots

L Matyushkin and A Percova.....492

Plasma physics, hydrodynamics and aerodynamics

Assessing the impact of the discharge conditions on the accommodation coefficient of electrons in plasma of helium and neon in narrow extended tubes

A S Kiselev and E A Smirnov.....498

Plasma of a vacuum-arc discharge for obtaining carbon-based coatings

M I Pikus, A A Lisenkov, D K Kostrin and S A Trifonov.....502

On prediction of turbulent heat transfer in the framework of Improved Delayed Detached Eddy Simulation with Wall Functions

M S Gritskevich and A V Garbaruk.....507

The influence of interatomic collisions on the angular distribution of Cs atomic flow from microchannel

A Yu Rumyantsev, M V Petrenko, S A Poniaev, Yu A Shustrov, V P Kochegarov and D L Schennikov.....513

Investigation of electrode erosion products of alternating current plasma torch operating on mixture of steam, carbon dioxide, methane and carbon tetrachloride

J. A. Kuchina, V.E. Kuznetsov, D.I. Subbotin, V.E. Popov, E.O. Serba, J.D. Dudnik, A.I. Litvyakova, I.A. Cherepkova and A.V. Surov.....518

Numerical simulation of natural convection in wedge-shaped domain with isothermal free surface

N A Yaroslavtseva and N G Ivanov.....523

Numerical study of MHD supersonic flow control

A.I. Ryakhovskiy and A.A. Schmidt.....529

An automatic hybrid numerical scheme for global RANS-LES approaches

E K Guseva, A V Garbaruk and M Kh Strelets.....535

Laboratory simulation of Hot Jupiters atmosphere expansion

P A Prokopov, I F Shaikhislamov, Yu P Zakharov, V G Posukh, A V Melekhov, E L Boyarintsev and A G Ponomarenko.....541

Comparative analysis of transition models at different farfield turbulence intensities

A S Stabnikov and A V Garbaruk.....548

Non-linear correction for the k-ω SST turbulence model

A A Matyushenko and A V Garbaruk.....554

Analytical and numerical solutions to Stefan problem in model of the glaciation dynamics of the multilayer cylinder in sea water

V V Mikova, G I Kurbatova and N N Ermolaeva.....560

The influence of thermal properties of anisotropic thermoelement substrate to the temperature distribution and the value of thermal EMF

P.A. Popov.....566

Photoionization microplasma sensor

A S Mustafaev, I V Rastvorova, S S Podenko and A B Tsyganov.....571

The axisymmetric oscillations of a cylindrical bubble in a liquid bounded volume with free deformable interface

A A Alabuzhev and M I Kaysina.....576

The oscillations of oblate drop under the influence of a alternating electric field

A A Alabuzhev and M A Kashina.....582

A study of stationary states of a plasma diode in the presence of a transverse magnetic field.

V.I. Kuznetsov, A.B. Gerasimenko, S. Pramanik and N. Chakrabarti.....587

Numerical simulation of multicellular natural convection in air-filled vertical cavities

A I Kunaeva and N G Ivanov.....594

Surface phenomena

Evaluation of dispersion interaction in glyoxal/silica organomineral system

Y V Sokolova, A M Ayzenshtadt and V V Strokova.....600

Sorption-desorption properties of saponite-containing material

M V Morozova, M A Frolova and T A Makhova.....605

IOP Conf. Series: Journal of Physics: Conf. Series **929** (2017) 011001 doi:10.1088/1742-6596/929/1/011001

International Conference PhysicA.SPb/2016

The International Conference PhysicA.SPb was held 1-3 November 2016 in Saint-Petersburg, Russia. The Conference continues the tradition of Saint-Petersburg Seminars on Physics and Astronomy originating from mid-90s. Since then PhysicA.SPb maintains both scientific and educational quality of contributions delivered to the young audience. This is the main feature of the Conference that makes it possible to combine the whole spectrum of modern Physics and Astronomy within one event.

PhysicA.SPb/2016 has brought together about 300 students, young scientists and their colleague professors from many universities and research institutes across whole Russia as well as from Belarus, Ukraine, Germany, Finland, Netherlands, and France. Oral and poster presentations were combined into well-defined sections among which one should name Astronomy and Astrophysics, Optics and spectroscopy, Physics of ferroics, Nanostructured and thin-film materials, Mathematical physics and numerical methods, Devices and materials for the THz and microwave ranges, Biophysics, Optoelectronic devices, Surface phenomena, Physics and technology of energy conversion, Plasma physics, hydrodynamics and aerodynamics, Nuclear and elementary particle physics, Impurities and defects in solids, and Physics of quantum structures.

This issue of the *Journal of Physics: Conference Series* presents the extended contributions from participants of PhysicA.SPb/2016 that were peer-reviewed by expert referees through processes administered by the Presiders of the Organising and Programme Committees to the best professional and scientific standards. This was made possible by the efforts of the Sectional Editors of this Issue: Prof. Petr Arseev (Lebedev Physical Institute), Prof. Yuri Kusraev (Ioffe Institute), Prof. Alexander Ivanchik (Ioffe Institute), Prof. Polina Ryabochkina (Ogarev Mordova State University), Dr. Karina Litvinova (Aston University), Dr. Nikolay Bert (Ioffe Institute), Dr. Yana Kuznetsova (Ioffe Institute), Dr. Prokhor Alekseev (Ioffe Institute), Dr. Mikhail Dunaevskii (Ioffe Institute), Dr. Alexandra Kalashnikova (Ioffe Institute) and Prof. Dmitry Khokhlov (Moscow State University).

The Editors: **Nikita S. Averkiev, Sergey A. Poniaev and Grigorii S. Sokolovskii**

IOP Conf. Series: Journal of Physics: Conf. Series **929** (2017) 011002 doi:10.1088/1742-6596/929/1/011002

Peer review statement

All papers published in this volume of *Journal of Physics: Conference Series* have been peer reviewed through processes administered by the proceedings Editors. Reviews were conducted by expert referees to the professional and scientific standards expected of a proceedings journal published by IOP Publishing.

Investigation of radiation pressure influence on the asteroids that subsequently have fallen to the Earth

A Martyusheva[1]

[1]The Central Astronomical Observatory of the Russian Academy of Sciences, Pulkovskoe chaussee 65, Saint-Petersburg, 196140, Russia

E-mail: alex.mart13@gmail.com

Abstract. The deviations in orbital motion of the asteroid-meteoroid 2008 TC3 and the asteroid, which subsequently became the Chelyabinsk meteorite, under radiation pressure influence were calculated. These celestial objects are of special interest for investigation, since due to the close encounters with the Earth their orbits had dramatically changed, which resulted in their downfall. Calculations were made for several values of albedo of the asteroids in order to compare the extent of deviations. Obtained results allowed estimating the effect of solar radiation pressure depending on changes of surface reflectance of the objects.

1. Introduction

Identification of potentially hazardous Near-Earth asteroids remains an urgent problem. As part of the model problem for the asteroid-meteoroid 2008 TC3 and the asteroid, which subsequently became the Chelyabinsk meteorite, the values of deviations in their orbital motion under the influence of solar radiation pressure were calculated. These asteroids are of particular interest for the study, since due to the close encounters with the Earth their orbits had changed significantly, which in turn, led to their downfall.

The asteroid 2008 TC3 met with the Earth on the 7th of October 2008. This asteroid was discovered in Mount Lemmon Observatory (Arizona, USA) on the day before the collision with the Earth. Observation data, passed to the Minor Planet Center, allowed a calculation of preliminary orbit the asteroid before the anticipated collision. During the fall it collapsed in the atmosphere above the Nubian Desert in Sudan, some fragments reached the Earth's surface. All fragments were collectively named after the railway station Almahata Sitta where the bolide was noticed. Therefore, the fall of this asteroid was the first predicted fall of a celestial body.

The Chelyabinsk meteorite fell to the Earth on the 15th of February 2013 as a result of braking in the atmosphere of a small asteroid. The meteorite fall was accompanied by shock waves spread over the territory of the Chelyabinsk region and other regions of Russia and Kazakhstan. Fragments of the asteroid were found in the Chelyabinsk region. The major fragment was reached out of Chebarkul Lake.

Estimation of radiation pressure influence on the motion of the asteroid 2008 TC3 and the Chelyabinsk meteorite was produced by means of specially developed model [1] based on numerical integration of motion equations by Everhart method. The three-body system was chosen for this

IOP Conf. Series: Journal of Physics: Conf. Series **929** (2017) 012001 doi:10.1088/1742-6596/929/1/012001

problem which splits into two different models: the restricted and the photogravitational restricted planar circular three-body problem. In first model only gravity affects on an asteroid. In second one, an asteroid is considered as an extended spherical body (of finite radius and definite density) so that gravity and radiation pressure influence on it. Comparison of these two problems gives us the deviations in motions of asteroids, caused by solar radiation pressure. The motion is defined in heliocentric rectangular coordinates. The initial orbits of the asteroids are elliptical. Calculations for each of the asteroids were made with three different values of albedo: the real albedo, albedo of a bright body (increased reflectance) and of an absolutely white body (the maximal reflectance). Thus, one can compare the deviations due to the radiation pressure influence.

The displacement along the heliocentric radius-vector Δr, the displacement along the orbit Δl and the total displacement Δd were also calculated for each object.

2. Results

It is important to note that calculations were made for the model asteroids with the parameters of real objects (2008 TC3 and the Chelyabinsk meteorite) on the assumption as if these asteroids did not fall to the Earth and continued their movement in space.

The orbital and physical parameters of the asteroid 2008 TC3, used for taking into account the solar radiation pressure, were taken on the epoch 2454746.5 (2008-Oct-07.0):

$e = 0.312065$ — eccentricity [2],
$a = 1.308201$ AU — semi-major axis [2],
$H_V = 30.4$ — absolute magnitude [2],
$D = 4.8$ m — diameter [3],
$\rho = 2330$ kg/m^3 — density [4],
$\delta = 0.05$ — albedo in the band V, derived from the asteroid's absolute magnitude and its diameter by using the formula $\lg D = 3.122 - 0.5\lg\delta - 0.2H_V$ [5],
$k = 1.02$ — optical coefficient for an extended spherical diffusely-reflecting body, derived from [6]:

$$k = \alpha + \rho + 13\delta/9, \quad (1)$$

where ρ — is the mirror reflection ($\rho = 0$ for a natural body) and $\alpha = 1 - \delta$ — is the absorption coefficient.

Within the next 20 years, the maximum deviations of the asteroid 2008 TC3, caused by the radiation pressure, would have amounted:

Table 1. The maximum deviations of 2008 TC3 in the next 20 years.

Deviation	$\delta = 0.05$, k = 1.02	$\delta = 0.80$ (bright body), k = 1.36	$\delta = 1.0$ (absolutely white body), k = 1.44
Δr	782.2 ± 0.1 km	1042.9 ± 0.1 km	1104.3 ± 0.1 km
Δl	3293.3 ± 0.1 km	4391.0 ± 0.1 km	4649.3 ± 0.1 km
Δd	3293.4 ± 0.1 km	4391.2 ± 0.1 km	4649.5 ± 0.1 km

The orbital and physical parameters of the Chelyabinsk meteorite, used for taking into account the solar radiation pressure, were taken on the epoch 2456338.5 (2013-Feb-15.0):

$e = 0.51$ — eccentricity [7],
$a = 1.69$ AU — semi-major axis [7],
$D = 18$ m — diameter [8],
$\rho = 3600$ kg/m^3 — density [8],

$\delta = 0.21$ — albedo [9],
$k = 1.09$ — optical coefficient, derived from (1).

Within the next 20 years, the maximum deviations of the asteroid, which subsequently became the Chelyabinsk meteorite, would have amounted:

Table 2. The maximum deviations of the Chelyabinsk meteorite in the next 20 years.

Deviation	$\delta = 0.21$, k = 1.09	$\delta = 0.80$ (bright body), k = 1.36	$\delta = 1.0$ (absolutely white body), k = 1.44
Δr	379.6 ± 0.1 km	473.7 ± 0.1 km	501.5 ± 0.1 km
Δl	1104.9 ± 0.1 km	1378.6 ± 0.1 km	1459.7 ± 0.1 km
Δd	1109.1 ± 0.1 km	1383.8 ± 0.1 km	1465.2 ± 0.1 km

3. Conclusion

The studies revealed that the asteroid 2008 TC3, which became the meteorite, was formed in a collision of three different asteroids [9]. Nysa-Polana asteroid family consists of three types of asteroids: the relatively rare B-type, the siliceous S-type and the transitional X-type. The fragments of Sudanese meteorite have spectral characteristics that are common to all three types of asteroids. This fact could mean that 2008 TC3 was presumably formed in a collision of an asteroid of S-type with an object of B-type, followed by another collision with an asteroid of X-type.

In case of the Chelyabinsk meteorite, it has been found by the group of scientists that about 10 million years ago a parent asteroid survived the collision in outer space, whereupon its trajectory had intercrossed with the Earth [10].

During the evolution small asteroids come under the impact of such non-gravitational effects, which can significantly change their orbits in a relatively short period of time by virtue of low mass of objects. Therefore, the effect of radiation pressure on the orbital motion of an asteroid should be taken into account in calculation of close encounters with the Earth. These calculations allowed estimating the possible effect of radiation pressure on asteroids by changing their albedo. It should be noted that these deviations are quite significant and should be considered at future calculations. Solar radiation pressure is the important disturbing factor especially in case of closeness of asteroid orbits to resonance zones with Jupiter and the Earth at large time intervals.

References

[1] Martyusheva A. A., Petrov N. A., Polyakhova E. N. Vestnik of SPbU, Vol.2 (60), Issue 1, pp. 135-147, 2015.
[2] http://ssd.jpl.nasa.gov/sbdb.cgi?sstr=2008%20TC3;old=0;orb=0;cov=0;log=0;cad=0#elem
[3] Aleshkina E. Y., Kupriyanov V. V., Devyatkin A. V. Vereshchagina I. A., Slesarenko V. Yu., L'vov V. N., Tsekmeyster S. D. Solar System Research, Vol. 45, No. 1, pp. 36-44, 2011.
[4] Jenniskens P., Shaddad M. H., Numan D., Elsir S., Kudoda A. M., Zolensky M. E. et al. Nature. Vol. 458, pp. 485-488, 2009.
[5] Vinogradova, T. A., Zheleznov N. B., Kuznetsov V. B., Chernetenko Yu. A., Shor V. A. Tr. Inst. Prikl. Astron. Ross. Akad. Nauk, Issue 9, 2003.
[6] Polyakhova E. N., Shmyrov A. S. Vestnik of SPbU, Vol. 2, Issue 8, pp. 87–104, 1994.
[7] Zuluaga J. I., Ferrin I. arXiv:1302.5377, 2013.

IOP Conf. Series: Journal of Physics: Conf. Series **929** (2017) 012001 doi:10.1088/1742-6596/929/1/012001

[8] http://neo.jpl.nasa.gov/news/fireball_130301.html
[9] Gayon-Markt J., Delbo M., Morbidelli A., Marchi S. arXiv:1206.3042, 2012.
[10] Antipin N. A. (Ed.) Materials All-Russian Scientific Conference «The Chelyabinsk meteorite — one year on the Earth», p. 410, 2014.
[11] Ozawa S., Miyahara M., Ohtani E., Koroleva O. N., Ito Y., Litasov K. D., Pokhilenko N. P. Scientific Reports, Vol. 4, 2014.

The fractal analysis of the gravitational field and topography of the Mars

S A Demin[1,2], A O Andreev[2], N Y Demina[1,2], Y A Nefedyev[1,2]
[1] Kazan Federal University, Institute of Physics, Kazan, 420008 Russia
[2] Kazan Federal University, Engelhardt Astronomical observatory, Zelenodolski region, st. Observatory, 422526 Russia

E-mail: serge_demin@mail.ru

Abstract. The aim of this paper is to represent the results of a structured fractal analysis of gravitational and topographical parameters of Mars on the basis of data obtained from the space missions. To analyze Martian fractal structures the observations from the data received from boards of the space missions including «Mars Global Surveyor» has been used. The models of relief and Mars' gravitational field have been constructed on the basis of harmonic analysis of the expansion in spherical functions of the satellite observations data. As a result, fractal dimensions of Martian relief's anomalies and Mars' gravitational potential by longitude and latitude have been determined. Mean values of the fractal dimensions $\overline{D}$ have been obtained as well: mean fractal dimensions of Martian topographic model by latitude $\overline{D} = 0.86$, by longitude $\overline{D} = 0.88$; mean fractal dimensions of Martian gravitational potential anomalies by latitude $\overline{D} = 1.06$, by longitude $\overline{D} = 1.092$.

1. Introduction

For evolution of complex celestial systems investigation the methods of multiparametric statistical physics [1, 2] are used. Particularly, during the analysis of non-linear processes and structures of natural objects, including determination of its properties, fractal geometry plays the key role, this also applies both to gravimetric and topographical parameters of Mars [3, 4]. Dilated symmetry and spatio-temporal heterogeneity are common properties of various physical system. These properties in, in case of a scale change, lead to basic geometrical parameters invariance [5, 6]. Fractal dimension is a quantitative measure characterizing structure of an object. The connection between geometrical parameters of an object and processes of its formation may be investigated by an analysis of fractal dimensions. This is why development of methods for fractal structure of celestial systems determination is a relevant issue. During the space missions to the bodies of Solar system [7, 8] a significant amount of data on geophysics, morphology, gravitational fields and other planet's atmosphere has been accumulated. As a result of these investigations comparative planetology has been significantly developed. This field focuses on the problems related not just to cosmogony and the theory of Solar system's bodies evolution, but to the analysis of processes of physical nature as well. The purpose of this work is to conduct the fractal structured analysis of gravimetric and topological parameters of Mars on the basis of the space missions' data.

2. The method of Martial structures fractal analysis

The models of gravitational field and the physical surface of Mars variations may be considered as fractal structures. To investigate them the corresponding approaches of statistical physics and multi-parameter analysis are required. The formulation diversity of «fractal dimension», ways to determine that by the results of experiments allow to consider fractal analysis as a separate field of the dynamical systems theory. The fact, that fractal dimension of an object's structure is different from topological one, indicates there is some significant complexity in structural organization of an object being investigated (distribution in space). As a model of a fractal object the scale-invariant fractal curve – the fractal Weierstrass-Mandelbrot function is used. From the one hand the model of fractal is a recursive one, i.e. each part of it repeats the whole model and reproduces that at different scales without any change. From the other hand, real natural systems can not be described as one fractal and

IOP Conf. Series: Journal of Physics: Conf. Series **929** (2017) 012002 doi:10.1088/1742-6596/929/1/012002

present interdependent multifractals where each fractal included in a set has its dimension. Multifractal analysis estimates a system in a form of fractal dimensions spectrum. Thus, such an approach provides sufficient accuracy of global fractal structures description on the basis of certain local areas investigation. In this work as a mathematical algorithm the Minkovsky method on determination and analysis of dimension has been used which is a simplified option of Hausdorff–Besicovitch dimension but at the same time provides sufficient accuracy and reliability.

Let M be a limited set in the plane and we define $N(\sigma)$ as a minimal number of an area being investigated covers whose combination covers M [9]. In the general case the set M has a dimension of $D=\dim M$, $0 \leq D \leq 2$, if at $\sigma \to 0$ the number of covers $N(\sigma)$ increases as R/σ^D, where R is a positive constant called D-measure of the set M. For R in the general case we may write:

$$R = \lim_{\sigma \to 0} \sigma^D N(\sigma). \tag{1}$$

Expression (1) is not convenient when calculating since it contains the value of D-measure. The fractal dimension D may be determined according to the following expression [9]:

$$D = \dim M = \lim_{\sigma \to 0} \frac{\ln N(\sigma)}{\ln \frac{1}{\sigma}}, \tag{2}$$

as

$$\lim_{\sigma \to 0} \frac{\ln N(\sigma)}{\ln \frac{1}{\sigma}} = \lim_{\sigma \to 0} \frac{\ln \frac{R}{\sigma^D}}{\ln \frac{1}{\sigma}} = \lim_{\sigma \to 0} \frac{D \ln \frac{1}{\sigma} + \ln R}{\ln \frac{1}{\sigma}} = D. \tag{3}$$

In order to analyze Martian structures the data received from the boards of the spaceships, particularly "Mars Global Surveyor" [10–12], has been used. The model of topographical anomalies is based on height measurements from the board of the spaceship, while the model of gravitational potentials includes investigations of Martian artificial satellites' orbits and gravitational observations made by other methods. The constructed geoid of Mars (as it is well known, names of gravimetric and topographical characteristics for the Solar system bodies coincide with terrestrial ones and only different for the Moon) defines both gravity potential and difference between normal and real gravity potentials. With this model one can investigate the gravitational field near the surface of Mars and mass distribution in bowels of Mars. Determination of the basic level surface on Mars is set by a point on its surface that coincides with the geoid's surface or a certain value of the potential. Thus, depending the gravimetric potential the geoid may be located above the physical surface of Mars as well as under that which is similar to Earth's geoid [3]. The obtained models of variations of the gravitational field and the topographical surface of Mars have been studied on the basis of the Minkovsky fractal dimension's values (see equation (2)). The analysis of the obtained dimension has allowed to conduct a comparative analysis of various geophysical Martian parameters.

3. Harmonic models of topography and gravitational field of Mars

Using harmonic analysis of expansion in spherical functions of the topographical data received from boards of «Mars Global Surveyor» a topographical model of Mars has been constructed [13]:

$$h(\varphi, \lambda) = \sum_{n=0}^{N} \sum_{m=0}^{n} (\bar{C}_{nm} \cos m\lambda + \bar{S}_{nm} \sin m\lambda) \cdot \bar{P}_{nm}(\cos \varphi) + \varepsilon, \tag{4}$$

where $h(\varphi, \lambda)$ – function of altitude;
φ, λ – latitude, longitude (known parameters);
$\bar{C}_{nm}$, $\bar{S}_{nm}$ – normalized harmonic amplitudes;
$\bar{P}_{nm}$ – normalized associated Legendre functions;
ε – random regression error.

This equation has also been used for an analysis of positional observations made in Engelhardt Astronomical Observatory [14].

On the basis of gravimetric measurements a harmonic model of Mars' gravitational potential has been constructed [15]:

$$U(\varphi,\lambda,r) = \frac{GM}{r}\left[1 + \sum_{n=0}^{\infty}\left(\frac{R_0}{r}\right)\sum_{m=0}^{n}(\bar{C}_{nm}\cos m\lambda + \bar{S}_{nm}\sin m\lambda)\cdot \bar{P}_{nm}(\sin\varphi)\right] + \varepsilon, \quad (5)$$

where G – gravitational constant;
GM – gravitational field of Mars (42 828.31 km^3/sec^2);
φ, λ, r – spherical coordinate of reference point;
R_0 – equatorial radius of planet.

4. Study of Martian fractal structures and the results of analysis

The topographical model of Mars has been cut by planes in order to pick out macroprofiles of the planet's surface on parallels with latitudes of the following values: -80, -60, -40, -20, 0, 20, 40, 60, 80 degrees. As an example for the latitude of 20 degrees this may be seen on the figure 1. The units systems on both x and y axis correspond to the measured data and are not critical, since the profiles are considered as self-similar structures.

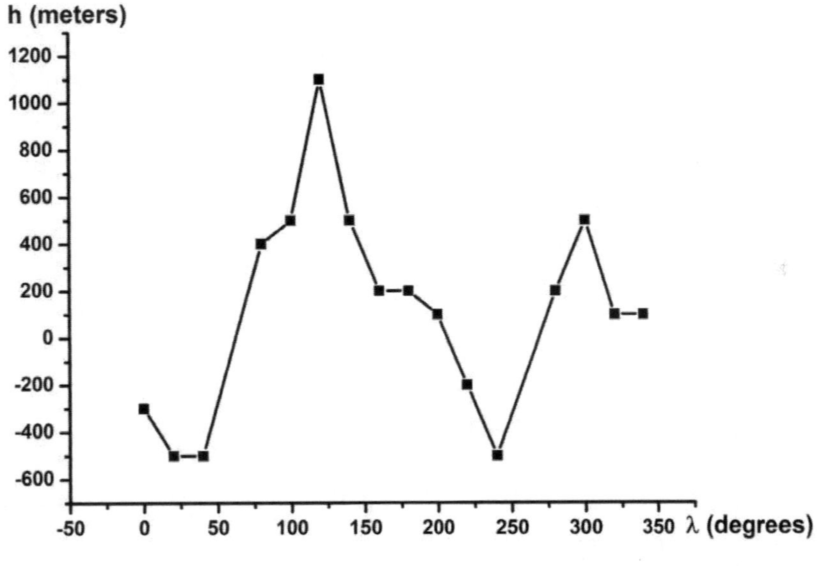

Figure 1. Macroprofile of Mars topography on the latitude of 20 degrees. *X*–axis corresponds to longitude (λ) in degrees, *Y*–axis corresponds to altitude in meters (h).

Similarly to this the models of gravity's anomalies on the parallels with the following latitudes: -80, -60, -40, -20, 0, 20, 40, 60, 80 degrees. As an example for the latitude of 0 degrees this may be seen on the figure 2.

For each cut fractal dimensions according to equation (2) have been calculated, where N – number of covers in the investigation area. The division into covers starts with value of 24 for number of covers to be integer. The obtained values for the topographical model and the model of gravity field's anomalies are represented in figure 3.

Analysis of figure 3 shows that for the topographical data the variations of fractal dimensions vary within 0.8–1.0. Considering the fact that the more significantly topological dimensions of a structure differs from fractal one, the more complex formation of a system is, so the similarity of topographical formations on Mars is sufficiently defined. The variations of gravity's anomalies fractal dimensions change within 0.6–1.2. Thus, the gravity field of Mars may be described as a topological dissimilar system with a complex structure in latitude measurements.

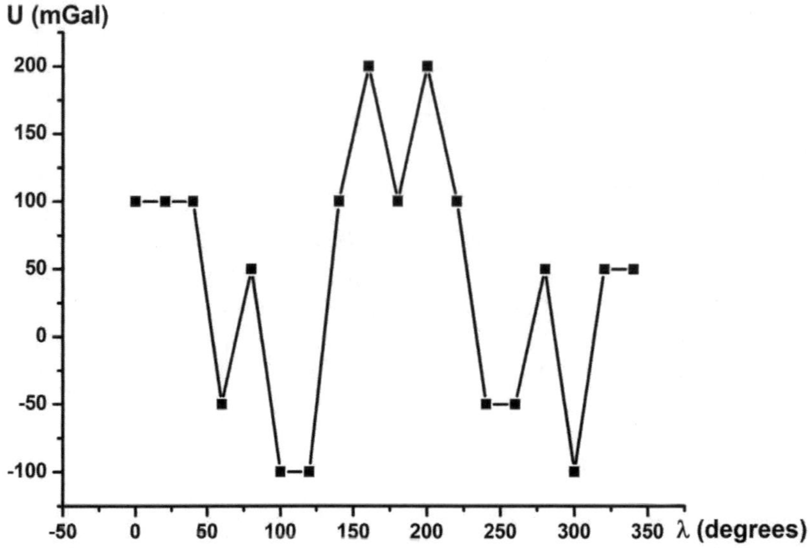

Figure 2. Macroprofile of Mars gravity anomalies on the latitude of 0 degrees. *X*–axis corresponds to latitude in degrees (λ), *Y*–axis corresponds to unit of acceleration mGal (0.001 cm/sec^2) (U).

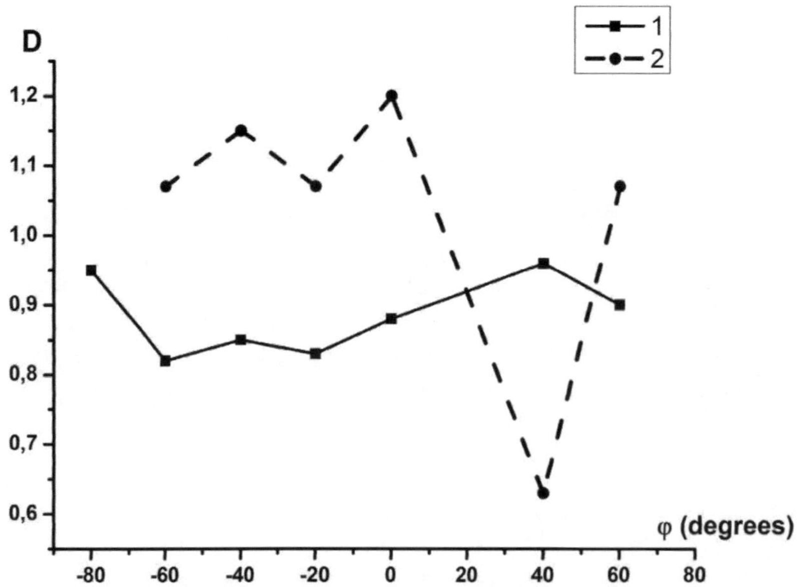

Figure 3. Fractal dependence of topographical model and gravity anomalies by Mars latitudes. *X*–axis corresponds to latitudes in degrees (φ), *Y*–axis corresponds to fractal dimension values *D*. Line 1 corresponds to fractal dimensions of Mars topographical model; line 2 corresponds to fractal dimensions of gravity anomalies on Mars.

The next stage is determination of fractal dimensions by longitude. To conduct this, cuts of Mars models by Martian meridians in certain Martian longitudes within 0–320 degrees with a step of 25 degrees have been made. As a result, the curve on figure4 has been plotted.

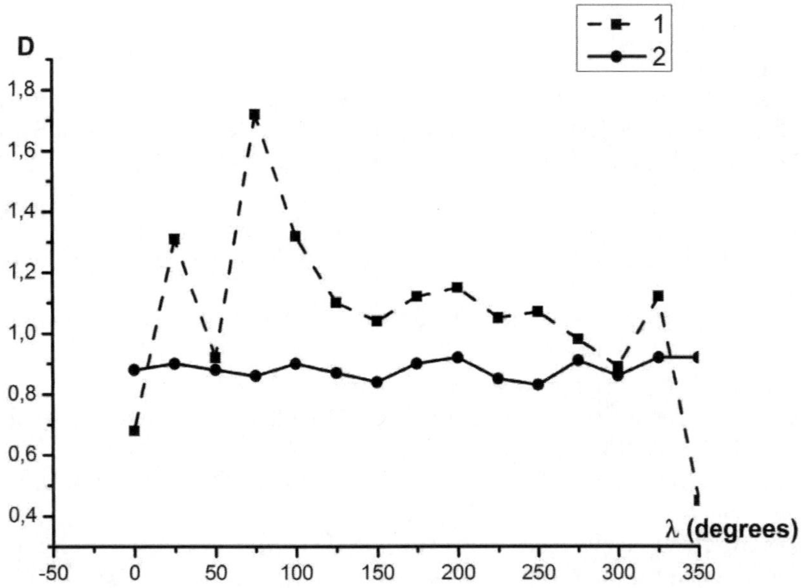

Figure 4. Dependence of fractal dimension of topographical model and gravity anomalies on longitude of Mars. *X*–axis corresponds to longitude in degrees (λ), *Y*–axis corresponds to values of fractal dimension *D*. Line 1 corresponds to fractal dimensions of gravity anomalies on Mars; line 2 corresponds to fractal dimensions of Mars topographical model.

Analysis of figure 4 well is in well agreement with the conclusions drawn for the analysis of figure 3. It only should be said that topographical fractal dimensions behave more evenly while gravitational fractal dimensions show more anomalies than in case of figure 3.

5. Summary and conclusions

In this work the fractal dependences for Martian topographic model and anomalies of gravitational potential by longitude as well as by latitude have been defined; the values of mean fractal dimensions have been obtained: mean fractal dimensions of Martian topographic model by latitude $\overline{D} = 0.86$, by longitude $\overline{D} = 0.88$; mean fractal dimensions of Martian gravitational potential anomalies by latitude $\overline{D} = 1.06$, by longitude $\overline{D} = 1.092$.

It is also found that use of fractal method for topological and gravimetric structures variations allows to estimate them by topological inhomogeneity and to draw conclusion about their structural similarity.

Thus, investigation of fractal structures is a very important and interesting direction in the studies of physical objects as well as natural systems [16]. A lot of materials are heterogeneous. Structures of disordered materials or connections between their structures and formation processes or properties are still not enough studied. The reason for it is that the methods of study of such structures have been developed recently. The qualitative feature of fractal objects is their inherent invariance when rescaling. Fractal geometry allows to estimate inhomogeneity of complex physical systems quantitatively.

We should also note that use of fractal analysis in this work allows to introduce independent estimates of Martian macrostrucutre which leads to new approaches to interpret physical process taking place on Mars. The further use of fractal comparative analysis at reduce of space measurements data will surely bring some interesting results which will allow to solve certain problems of space astrometry.

Acknowledgments
The work is performed according to the Russian Government Program of Competitive Growth of Kazan Federal University. This work was supported by the Russian Foundation for Basic Research, project nos. 15-02-01638 a, 16-02-00496 a, 16-32-60071 mol_a_dk (N.D.).

References
[1] Demin S A, Panischev O Y and Nefedyev Y A 2014 *Kinematics and Physics of Celestial Bodies* **30** 63
[2] Demin S A, Panischev O Y and Nefedyev Y A 2014 *Nonlinear Phenomena in Complex Systems* **17** 177
[3] Turcotte D L 1987 *Journal of Geophysical Research* **92** 597
[4] Stepinski T F, Collier M L, McGovern P J and Clifford S M 2004 *Journal of Geophysical Research: Planets* **109** E02005-1
[5] Nefedyev Y A, Bezmenov V M, Demin S A, Andreev A O and Demina N Y 2016 *Nonlinear Phenomena in Complex Systems* **19** 102
[6] Sokolova M, Kondratyeva E and Nefedyev Y 2013 *Advances in Space Research* **52** 1217
[7] Nefedjev Y A and Rizvanov N G 2002 *Astronomische Nachrichten* **323** 135
[8] Nefedyev Y, Valeev S, Mikeev R, Varaksina N and Andreev A 2012 *Advances in Space Research* **50** 1564
[9] Zhikov VV 1996 *Soros Educational Journal* **12** 109
[10] Genova A, Goossens S, Lemoine F G, Mazarico E, Neumann G A, Smith D E and Zuber M T 2016 *Icarus* **272** 228
[11] Shan J, Yoon J S, Lee D S, Kirk R L, Neumann G A and Acton C H 2005 *Photogrammetric Engineering and Remote Sensing* **71** 97
[12] Ferguson B B, Cain J C, Crider D H, Brain D A, Harnett E M 2005 *Geophysical Research Letters* **32** L16105-1
[13] Andreev A O, Demina N Y, Demin S A, Nefedyev Y A and Churkin K O 2016 *Nonlinear Phenomena in Complex Systems* **19** 271
[14] Rizvanov N G, Nefed'ev Y A, Kibardina M I 2007 *Solar System Research* **41** 140
[15] Busarev V V, Shevchenko V V and Surdin V G 2007 The physical conditions near the Moon and planets of the Solar System *Model of the Cosmos* ed M I Panasyuk (Moscow: KDU) pp 794–861
[16] Panischev O Y, Demin S A and Rusanova I A 2015 *Biomedical Engineering* **49** 236

IOP Conf. Series: Journal of Physics: Conf. Series **929** (2017) 012003 doi:10.1088/1742-6596/929/1/012003

D Na I absorption lines formation in the RZ Psc spectrum

S G Shulman[1,2], V P Grinin[1,2], I S Potravnov[2] and D E Mkrtichian[3]

[1] St.-Petersburg State University, 7-9 Universitetskaya Nab., St.-Petersburg, 199034, Russia
[2] Pulkovo Astronomical Observatory of RAS, 65/1 Pulkovskoye chaussee, St.-Petersburg, 196140, Russia
[3] National Astronomical Research Institute of Thailand, Chiang Mai, 50200, Thailand

E-mail: sgshulman@gmail.com

Abstract. Discrete variable absorption components in the RZ Psc spectrum are studied. Two hypothesis of their formation are considered. The effective star radiation scattering computation method is presented. The resonance scattering in both cases is modeled. The model spectra are reviewed and analyzed. The conclusion about the possible reasons of the observed spectra formation is made.

1. Introduction

Discrete absorption components with night to night variations are observed in spectra of several young stars. RZ Psc holds a special place among them: absorption details without any emission in the sodium resonance lines presents in its spectra, while some emission presents in the spectra of the other stars with this type of spectral variability [1]. As a result RZ Psc requires a close study. We summarized the data about it and obtained eleven new high-quality spectra during the two observation runs [2]. The first one took place in September 2014 at the 2.6-m Shajn telescope of Crimean Astrophysical Observatory with the ESPL echelle spectrograph. The second set of observations was hold in December 2014 at the 2.4-m telescope of Thai National Observatory with MRES echelle spectrograph. These spectra are presented on the Fig. 1. Absorption features were also detected in the calcium and potassium lines.

All observed discrete absorptions are placed in the blue wind of the line. This means that they are formed in the matter out-flowing against the star. The outflow must have a complicate structure to produce a few absorption lines at the same time and do not lead to notable emission presence.

We developed two hypothesis of the outflow form. They are based on the existing theories of the interaction of an accretion disc with a magnetosphere. The first one is based on the magnetic propeller regime of the interaction when the spiral out-flowing believed to be possible [3]. The out-flow shape for this case was already analyzed [4, 5]. The second one considers low-speed magnetosphere wind.

To analyze the absorptions formation the resonance scattering of the star radiation in the hypothesized matter outflows is calculated with a designed method. The method considers photon packages propagation through the circumstellar matter using directions enumeration to speed up the modeling process.

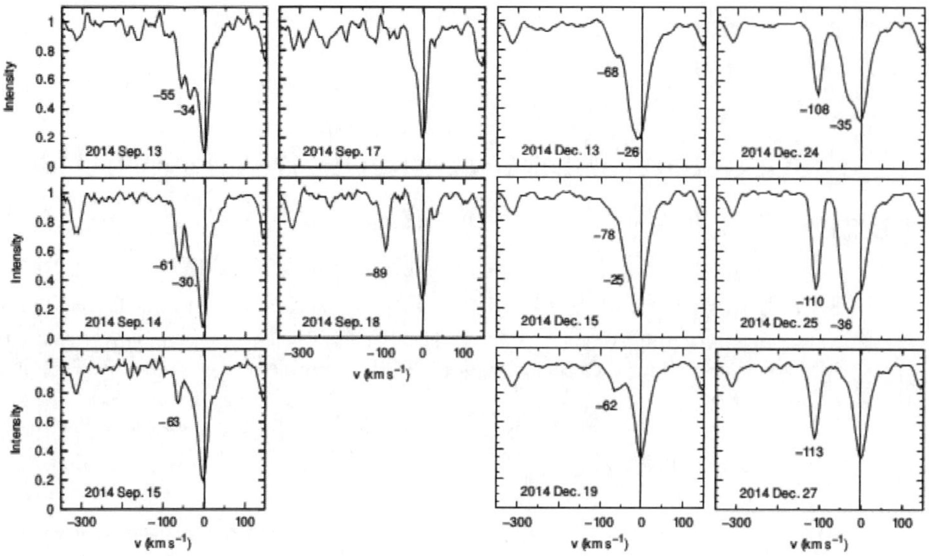

Figure 1. Na I 5889Å lines in the RZ Psc spectra obtained during two observations series [2].

2. Calculation method

In this work we calculate photon packages propagation through the matter. For this purpose we introduce a uniform photon wavelengths grid and a sampling of the photon propagation directions. Scattering matter is approximated with small areas with precalculated positions, matter motion velocities and matter densities. The directions grid consists of the vectors pointing from the unit sphere center to the barycenters of nearly equal triangles covering the sphere.

We emit photon packages from the star with all possible wavelengths and propagation directions. The initial package weights are proportional to the star photosphere intensity on the same wavelength. When the photon package reaches the scattering area we model the scattering dividing the package into non-scattered package and a pack of scattered packages propagating in all directions in the grid. The weights of the new packages depend on the optical depth of the matter and a ratio between the photon wavelength and the matter motion velocity. After the scatterings the photon packages are counted with the model spectrographs which build model spectra for various observer positions.

The described technique allows analyzing propagation directions before the photons are emitted. As a result we do not model photons which will never be scattered or registered by the observers. This makes the method significantly faster than Monte Carlo method for our problems. For the Monte Carlo realization and some common reception see the review [6].

3. Magnetic propeller regime

We assume that the magnetic propeller regime of the star rotating magnetosphere interaction with an accretion disk is possible in RZ Psc case [3]. In this regime strong magnetic field may reject the matter from the star. After the rejection the matter leaves the Alfven radios and moves without a magnetosphere influence. As a result it is possible to use ballistic approximation for the out-flow [4, 5]. Fig. 2 demonstrates the out-flow shape which is optimal for the absorption components formation.

For this scattering matter geometry model spectra were calculated for different phases of the star rotation period which coincides the variations period. Fig. 3 shows some of this spectra [7].

IOP Conf. Series: Journal of Physics: Conf. Series **929** (2017) 012003 doi:10.1088/1742-6596/929/1/012003

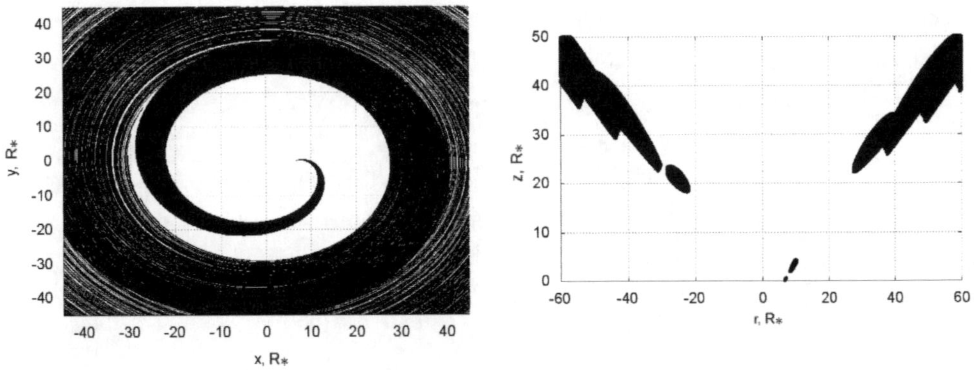

Figure 2. The optimal outflow form for the multiple absorption components formation. The top view is on the left panel and the crossection is on the right panel.

Model absorption components are close to the observed ones but have a period, which is not observed yet. The reason may be the out-flow instability which is likely in RZ Psc case. Modeled Spectra demonstrate no marked scattering light. Its intensity is very low (see Fig. 4).

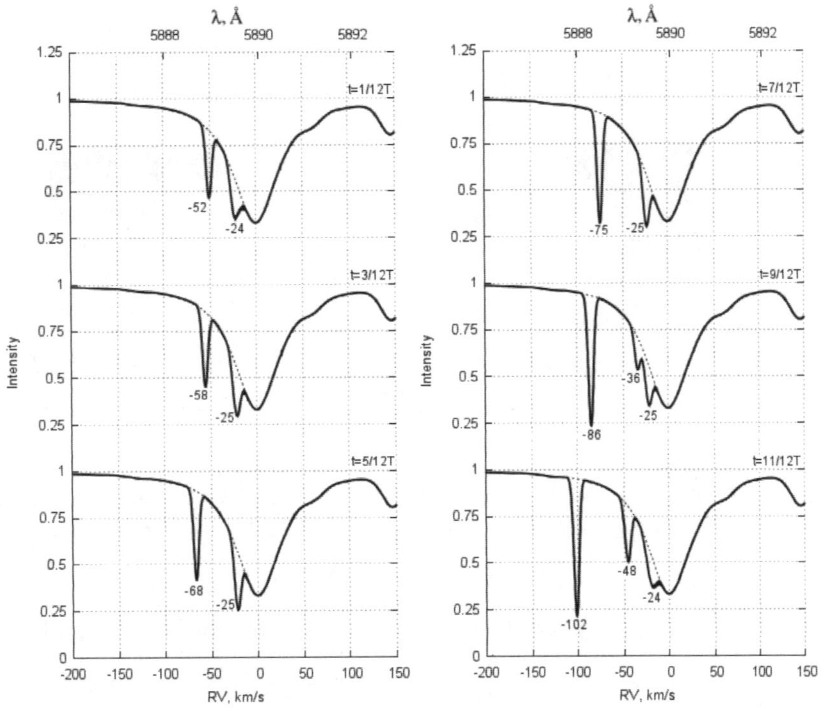

Figure 3. Model spectra with discrete absorption components, corresponding to different phases of the star rotation period. The synthetic RZ Psc atmosphere spectrum is shown with a dashed line.

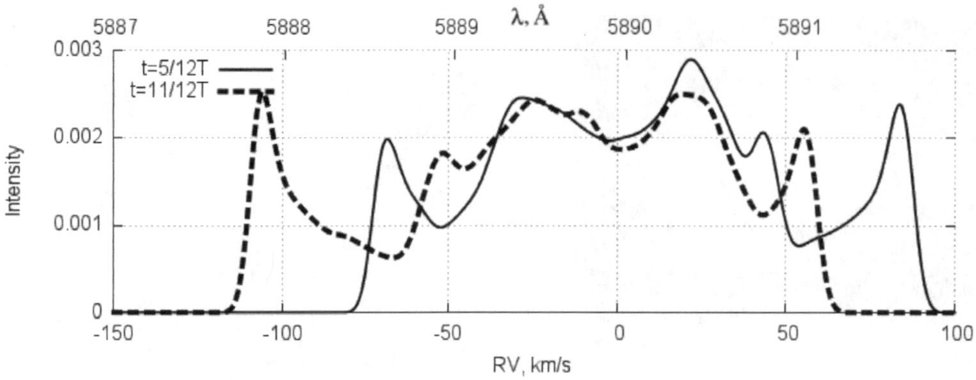

Figure 4. The resonance scattering intensities for two phases of the rotation period. The intensity is measured in the continuum units.

4. Low-speed magnetosphere wind

Another theory is based on the low-speed magnetosphere wind. An interaction of the star magnetosphere and an accretion disk can form this wind. When the interaction is not strong, the matter speed is low. In this case the out-flowing matter forms a symmetric stream which returns backwards to the disk. Hence, the line of sight may cross the stream twice in the points with different velocities of the matter. As a result two absorption components might be observed. A sample cross-section of this out-flow is shown in the Fig. 5.

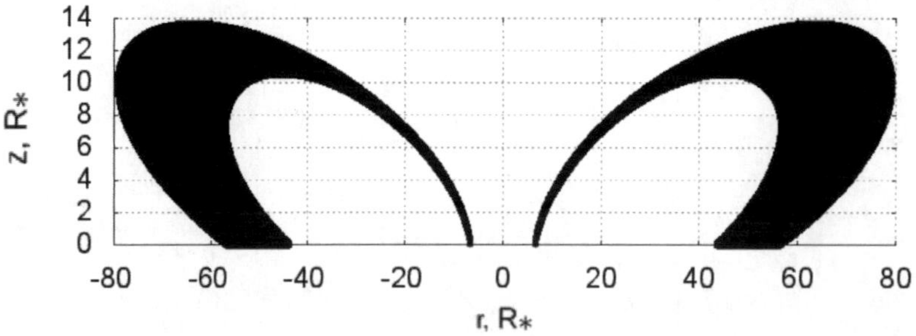

Figure 5. An example of the outflow crossection in the case of the low-speed magnetosphere wind

In this model when the wind formation process is stationary the absorption components might be seen at the same positions at any time. In the observation absorption components appear on different offsets. As a result we have to take into account wind instability. It vary the form of the out-flow and the velocities of the matter on the sight line. Possible spectra are presented in the Fig. 6. The scattering light intensity is larger than in the previous model but is still under the sensibility threshold.

These model spectra have no periodicity but should have some prevailing discrete absorption positions. The statistic of the velocities of the additional Na I D components is small now but will be vaster in the future. It will allow us to analyze the statistic properties of the absorption lines.

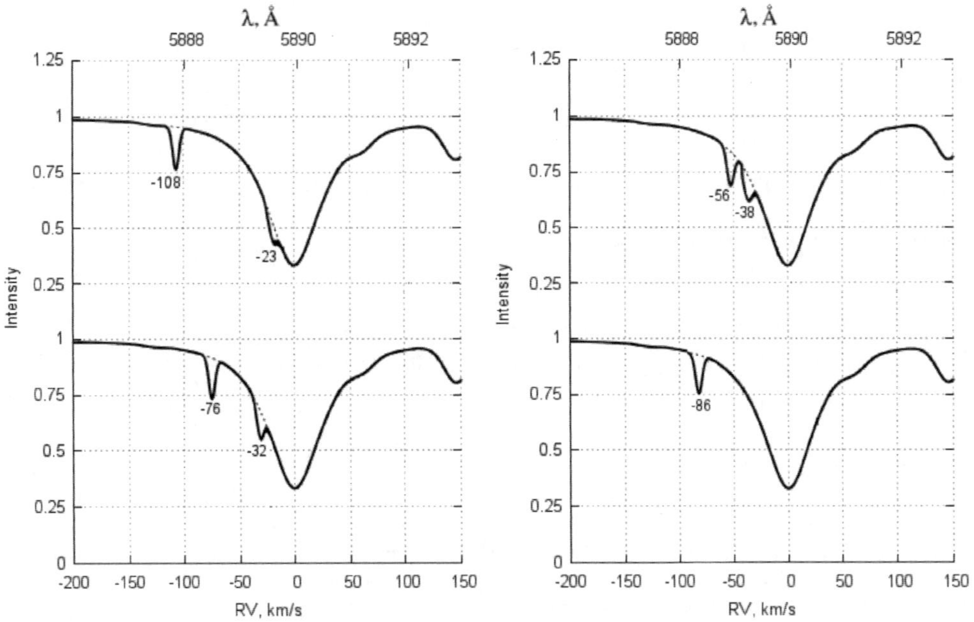

Figure 6. Model spectra in the case of the low-speed magnetosphere wind. The spectral variations are non-cyclic in this model. The synthetic RZ Psc atmosphere spectrum is shown with a dashed line.

5. Conclusion

We considered two hypothesis explaining the formation of the Na I D additional components in spectra of RZ Psc and obtained model spectra for them. The both hypothesis could explain the observations with taking into account the out-flow instability.

The magnetic propeller hypothesis predicts some periodicity in the spectra while the low-velocity magnetosphere wind one predicts some statistically preferential components positions. The additional observations may clarify the situation.

The suggested method shows its effectiveness. We are going to develop it and make applicable for more problems.

Acknowledgments

This work was partially supported by the Russian Foundation for Basic Research (RFFI), grant 15-02-05399, the Programs of Fundamental Research of the Russian Academy of Sciences 7P Experimental and theoretical researches of the objects of Solar system and planetary systems of stars and the Program for Support of Leading Scientific Schools (NSh-7241.2016.2).

References

[1] Petrov P P 2003, *Astrophysics* **46**, 506–29
[2] Potravnov I S, Mkrtichian D E, Grinin V P, Ilyin I V and Shakhovskoy D N 2016 *Astron. Astrophys.* in press
[3] Grinin V P, Potravnov I S, Ilyin I V and Shulman S G 2015 *Astron. Lett.* **41** 407–16
[4] Shulman S G 2015 *Astrophysics* **58** 258–75
[5] Shulman S G, Grinin V P and Potravnov I S 2015 *J. Phys.: Conf. Ser.* **661** 012012
[6] Whitney B 2011 *Bull. Astr. Soc. India* **39** 101–27
[7] Shulman S G 2017 *Astrophysics* in preparation

IOP Conf. Series: Journal of Physics: Conf. Series **929** (2017) 012004 doi:10.1088/1742-6596/929/1/012004

Modelling non-dissociative shock waves in interstellar medium: first results

A V Nesterenok

Ioffe Institute, 26 Polytechnicheskaya St., 194021, Saint Petersburg, Russia

E-mail: alex-n10@yandex.ru

Abstract.
The structure of steady one-dimensional shock waves in partially ionized interstellar gas is studied. The model is constructed of so called C-type shocks where the physical parameters characterizing the flow change continuously at the shock front. We present our first results of the modelling of C-type shocks. The model will be employed to study molecular line emission from interstellar shocks, in particular, maser emission.

1. Introduction

Interstellar shock waves play a major role in the chemical and physical evolution of the interstellar medium. These shocks result from the expansion of HII regions, cloud-cloud collisions, stellar winds, supernova explosions [1].

The nature and structure of interstellar shocks are strongly dependent upon the strength and orientation of the magnetic field and the local ionization fraction in the pre-shock gas [2, 3]. For strong transverse magnetic fields or low shock velocities, magnetosonic waves can propagate faster than the shock front forming a magnetic precursor to the shock. The ionized fluid is accelerated and heated by the magnetic precursor. The resulting drift between the ionized and neutral fluids heats and accelerates the neutral component of the gas. Due to ion-neutral scattering the discontinuity in the neutral fluid can then disappear – a C-type, or 'continuous', shock forms. If the temperature of the neutral gas is low enough due to molecular and atomic cooling, the neutral gas remains supersonic everywhere within the shock. In C-type shocks, molecules are at most partially dissociated. Therefore, a characteristic feature of C-type shocks is strong molecular emission.

The intent of the present paper is to discuss the general properties of C-type magnetohydrodynamic shock waves.

2. Shock wave model

2.1. Magnetohydrodynamic equations

A steady, one-dimensional flow along the z-direction is considered. The magnetic field B_0 is adopted to be transverse to the flow direction. The flow components considered are neutrals, ions, and electrons, denoted by subscripts n, i, and e, respectively. Each component is assumed to have velocity $v(z)$, density $\rho(z)$, and temperature $T(z)$, with the condition of stationarity $\partial v/\partial t = \partial\rho/\partial t = \partial T/\partial t = 0$. Electrons and ions are assumed to move at the same velocity [2].

The fluid equations that express conservation of particle number density, mass, momentum, and energy are [3]:

$$\frac{d}{dz}\left[\frac{\rho_n}{\mu_n}v_n\right] = \mathcal{N}_n \tag{1}$$

$$\frac{d}{dz}\left[\frac{\rho_i}{\mu_i}v_i\right] = \mathcal{N}_i \tag{2}$$

$$\frac{d}{dz}\left[\rho_n v_n\right] = \mathcal{S}_n \tag{3}$$

$$\frac{d}{dz}\left[\rho_i v_i\right] = -\mathcal{S}_n \tag{4}$$

$$\frac{d}{dz}\left[\rho_n v_n^2 + kT_n\frac{\rho_n}{\mu_n}\right] = \mathcal{F}_n \tag{5}$$

$$\frac{d}{dz}\left[\rho_i v_i^2 + kT_i\frac{\rho_i}{\mu_i} + kT_e n_e + \frac{B_0^2 v_s^2}{8\pi v_i^2}\right] = -\mathcal{F}_n \tag{6}$$

$$\frac{d}{dz}\left[\frac{1}{2}\rho_n v_n^3 + \left(\frac{5}{2}kT_n + U_n\right)\frac{\rho_n}{\mu_n}v_n\right] = \mathcal{G}_n + \mathcal{F}_n v_n - \frac{1}{2}\mathcal{S}_n v_n^2 \tag{7}$$

$$\frac{d}{dz}\left[\frac{1}{2}\rho_i v_i^3 + \frac{5}{2}kT_i\frac{\rho_i}{\mu_i}v_i + \frac{5}{2}kT_e n_e v_i + \frac{B_0^2 v_s^2}{4\pi v_i}\right] = \mathcal{G}_i + \mathcal{G}_e - \mathcal{F}_n v_i + \frac{1}{2}\mathcal{S}_n v_i^2 \tag{8}$$

$$\frac{d}{dz}\left[\frac{3}{2}kT_i\frac{\rho_i}{\mu_i}v_i - \frac{3}{2}kT_e n_e v_i\right] + \left[kT_i\frac{\rho_i}{\mu_i} - kT_e n_e\right]\frac{dv_i}{dz} = \mathcal{G}_i - \mathcal{G}_e \tag{9}$$

where v_s is the shock speed, μ_n is the mean mass per neutral particle and μ_i is the mean mass per ion particle; U_n is the mean internal energy per neutral particle. The source terms $\mathcal{N}_n$, $\mathcal{N}_i$, $\mathcal{S}_n$, $\mathcal{F}_n$, $\mathcal{G}_n$, $\mathcal{G}_i$ and $\mathcal{G}_e$ have the following definitions: $\mathcal{N}_n$ and $\mathcal{N}_i$ are the numbers of neutral particles and ions created per unit volume and time as a result of chemical reactions; $\mathcal{S}_n$ is the rate per volume at which ion-electron mass is converted into neutral mass; $\mathcal{F}_n$ is the rate per volume at which momentum is transferred from the charged fluid to the neutral fluid as a result of elastic scattering and chemical reactions; $\mathcal{G}_n$, $\mathcal{G}_i$ and $\mathcal{G}_e$ are the rates, per volume, at which thermal energy is added to the neutral, ion and electron fluids, respectively.

2.2. Ion-neutral elastic scattering

Momentum exchange between neutral and charged fluids occurs through ion-neutral elastic scattering and by collisions of neutral particles with charged dust grains. The momentum transfer cross sections for the scattering of ions and neutral particles (H, He, H_2) were calculated using the results by [4]. For the elastic scattering between electrons and neutral particles, the momentum transfer cross sections were adopted from [5–7]. The rates of momentum transfer and gas heating due to elastic scattering were calculated according to [8] using the given dependence of the cross section on velocity.

The expression for energy transfer between ions and electrons was taken from [9].

2.3. Chemistry

The UMIST database for astrochemistry [10] was used in our simulations. The chemical species containing atoms H, He, C, N, O, Na, Mg, Si, S, Cl, Fe were taken into account.

The total number of gas species is 426. We took into account gas-phase chemical reactions, direct cosmic-ray ionization, cosmic-ray-induced photoreactions, photoreactions induced by interstellar background ultraviolet (UV) field, adsorption of neutral species on grains and various desorption mechanisms [11, 12]. The rates of endothermic reactions between ions and neutrals were calculated by the method of 'effective' temperature [2].

The differential equations for the concentrations of chemical species were solved together with the magnetohydrodynamic equations.

2.4. Calculation of molecule level populations

The energy level populations of H_2, CO and H_2O molecules were computed in parallel with dynamical and chemical rate equations. We took into account 150 ro-vibrational levels for each of molecules H_2, ortho- and para-H_2O and 41 levels of ground vibrational state of CO molecule. Spectroscopic data were taken from [13, 14] for H_2 molecule and from HITRAN 2012 database [15] for CO and H_2O molecules. Rate coefficients for collisions of H_2 molecule with He, H_2 and H were taken from [16–21]. For collisions of H_2 with thermal electrons, rate coefficients were calculated using the recipes given by [22, 23]. We took into account the excitation of ro-vibrational levels of H_2 molecule in collisions with fast electrons produced by cosmic ray ionization of the gas [24]. Rate coefficients for collisions of CO with He, H_2, H were taken from [25–27]. Information on H_2O collision data used in calculations can be found in [28].

Level populations of ions CI, CII and OI were computed. The excitation of atomic levels by H_2, He, H and thermal electrons was taken into account [29]. The large velocity gradient approximation was used in calculations of radiative transition probabilities of molecular and atomic species [30]. The dust radiation was neglected in these calculations.

The system of differential equations was solved using the computer program CVODE v2.8.2, which is available in the web[1] [31].

2.5. Gas cooling

We took into account cooling of gas via excitation of fine structure transitions of ions (OI, CI and CII) and ro-vibrational lines of molecules (H_2, CO and H_2O).

2.6. Dust model

A simple single-size grain model was adopted in the calculations. The grain radius was assumed to equal 0.1 μm and the dust-gas mass ratio is equal to 0.01. The grains were assumed to compose from carbonaceous material [32]. We considered the heating of dust by diluted star light and UV interstellar background, by collisions with gas species, by gas species adsorption, and H_2 molecule formation. The photoelectric emission from dust grains, the electron attachment and ion recombination were taken into account. Reaction rates were calculated according to [33]. Reactions of neutralization and dissociative neutralization of ions on grain surface were considered [34]. Dust grain dynamics was treated according to [2]. The momentum transfer between neutral particles and charged grains was taken into account.

2.7. UV field

We used the average local interstellar UV field approximated by [35]. The interstellar radiation field is reduced in accordance with visual extinction. We took into account the UV field generated internally via excitation and fluorescence of electronic states of H_2 molecules in the process of cosmic ray propagation through the gas [36].

[1] URL:http://computation.llnl.gov/projects/sundials

IOP Conf. Series: Journal of Physics: Conf. Series **929** (2017) 012004 doi:10.1088/1742-6596/929/1/012004

Table 1. Parameters of the molecular cloud

Total H concentration, n_{H}	$2{\times}10^4$ cm^{-3}
Visual extinction, A_{V}	10
Cosmic ray ionization rate, ζ	10^{-16} s^{-1}
The ratio of visual extinction to reddening, R_{V}	5.5
Transverse magnetic field, B_0	0.1 mG
Grain surface area, S_{g}	7.8×10^{-22} cm^2/H

3. Results

3.1. Chemical evolution of stationary molecular cloud

Parameters of a molecular cloud are given in the table 1. The chemical and thermal evolution of the stationary cloud was calculated, starting with some arbitrary initial values of specimen concentrations and gas temperature. We took low metal elemental abundances that are expected to reproduce the gas-phase chemistry of dense cold clouds [34]. All metal atoms with the exception of O and N, are in ionic state in the adopted initial chemical abundances.

The calculation results are shown in the figure 1. It is seen that the gas reaches thermal equilibrium after about 10^4 years. A small decrease of the gas temperature at about several 10^3 years comes from the increase of H_2O and CO concentrations and, hence, increase of gas cooling. The temperature of the gas remains at relatively constant level close to 20 K until approximately 10^6 years. After this time, the effect of molecule depletion from the gas phase becomes significant – main coolants H_2O and CO become adsorbed on dust grains and the temperature of the gas rises. According to our results, the dominant O-bearing specimen at the end of the chemical evolution is CO molecule. Carbon monoxide is effectively produced in the gas, while the production of H_2O is not effective at low gas temperatures. The ionization fraction of the gas is low, of the order 10^{-7}. The total charge carried by dust grains is negligibly small compared to the electron concentration.

For simulations of shock waves, the chemical composition and temperature of the gas at 5×10^5 years were taken.

3.2. C-type shock wave

Figure 2 presents simulation results for the C-type shock of speed 20 km/s. The results are consistent in general with previous studies of C-type shocks [3]. The scattering of gas molecules on charged grains is the dominant mechanism of the momentum transfer from the charged fluid to the neutral gas at gas densities in question. Consequently, simulation results are sensitive to the assumed dust properties such as specific surface area. Two velocity dependences of cross section for H_2-ion elastic scattering were considered: (i) the cross section is inversely proportional to the relative velocity and (ii) the cross section dependence found by [4]. The shock width is almost insensitive to the form of the cross section dependence on velocity of ion-neutral scattering. According to our calculations, the electron temperature is much lower than the ion temperature. The excitation of ro-vibrational levels of H_2 is very effective cooling mechanism of electron fluid [3].

Some important physical effects such as grain mantle sputtering are not yet included in our model. Grain mantle sputtering releases molecules to the gas phase which can significantly modify cooling rate of the gas.

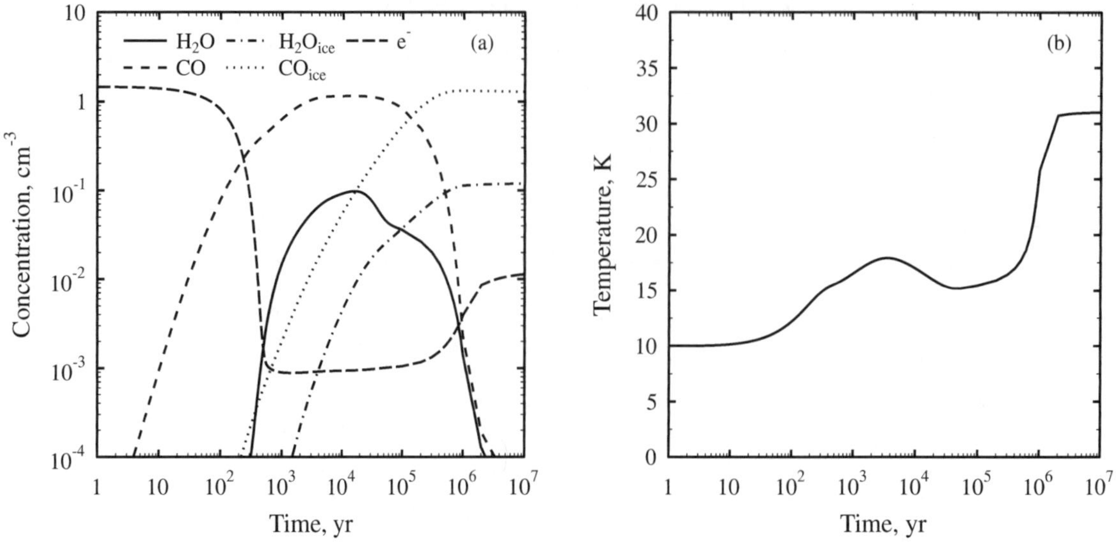

Figure 1. Chemical evolution of a dark cloud; (a) concentrations of chemical species; (b) gas temperature.

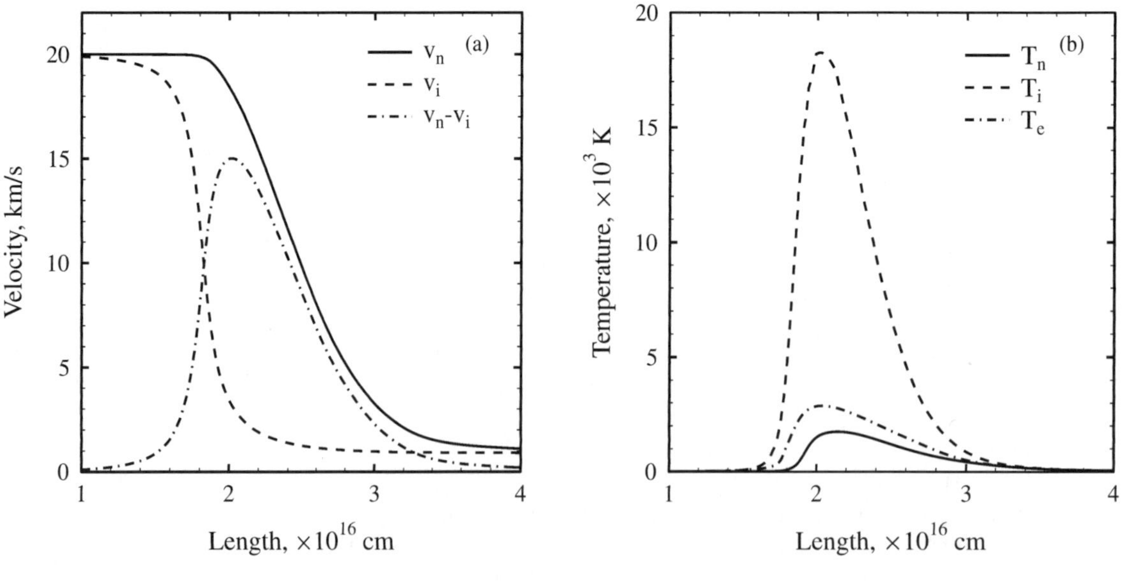

Figure 2. Results of C-type shock wave calculations. The steady-state profiles of (a) velocity, (b) temperature.

4. Conclusions

The model of the C-type shock in interstellar cloud is presented. Simulation results are consistent in general with previous studies. Important physical effects such as sputtering of grain mantles are not yet included in our model.

Acknowledgments
The author was supported by the RSF grant 16-12-10225.

References

[1] Draine B T and McKee C F 1993 *Annu. Rev. Astron. Astr.* **31** 373–432
[2] Draine B T 1980 *Astrophys. J.* **241** 1021–1038
[3] Draine B T, Roberge W G and Dalgarno A 1983 *Astrophys. J.* **264** 485–507
[4] Flower D R 2000 *MNRAS* **313** L19–L21
[5] Crompton R W, Elford M T and Robertson A G 1970 *Aust. J. Phys.* **23** 667
[6] Dalgarno A, Yan M and Liu W 1999 *Astrophys. J. Suppl. S.* **125** 237–256
[7] Yoon J S, Song M Y, Han J M, Hwang S H, Chang W S, Lee B and Itikawa Y 2008 *Journal of Physical and Chemical Reference Data* **37** 913–931
[8] Draine B T 1986 *MNRAS* **220** 133–148
[9] Spitzer L 1956 *Physics of Fully Ionized Gases* (New York: Interscience Publishers)
[10] McElroy D, Walsh C, Markwick A J, Cordiner M A, Smith K and Millar T J 2013 *Astron. Astrophys.* **550** A36 (*Preprint* 1212.6362)
[11] Chaparro Molano G and Kamp I 2012 *Astron. Astrophys.* **537** A138 (*Preprint* 1111.1149)
[12] Roberts J F, Rawlings J M C, Viti S and Williams D A 2007 *MNRAS* **382** 733–742 (*Preprint* 0708.3374)
[13] Dabrowski I 1984 *Can. J. Phys.* **62** 1639–1664
[14] Wolniewicz L, Simbotin I and Dalgarno A 1998 *Astrophys. J. Suppl. S.* **115** 293–313
[15] Rothman L S *et al.* 2013 *J. Quant. Spectrosc. Radiat. Transfer* **130** 4–50
[16] Flower D R, Roueff E and Zeippen C J 1998 *J. Phys. B: At. Mol. Opt. Phys.* **31** 1105–1113
[17] Flower D R and Roueff E 1998 *J. Phys. B: At. Mol. Opt. Phys.* **31** 2935–2947
[18] Flower D R and Roueff E 1999 *J. Phys. B: At. Mol. Opt. Phys.* **32** 3399–3407
[19] Martin P G and Mandy M E 1995 *Astrophys. J. Lett.* **455** L89
[20] Wrathmall S A, Gusdorf A and Flower D R 2007 *MNRAS* **382** 133–138
[21] Lique F 2015 *MNRAS* **453** 810–818
[22] Gredel R and Dalgarno A 1995 *Astrophys. J.* **446** 852
[23] Faure A and Josselin E 2008 *Astron. Astrophys.* **492** 257–264
[24] Tiné S, Lepp S, Gredel R and Dalgarno A 1997 *Astrophys. J.* **481** 282–295
[25] Cecchi-Pestellini C, Bodo E, Balakrishnan N and Dalgarno A 2002 *Astrophys. J.* **571** 1015–1020
[26] Yang B, Stancil P C, Balakrishnan N and Forrey R C 2010 *Astrophys. J.* **718** 1062–1069 (*Preprint* 1004.3923)
[27] Walker K M, Song L, Yang B H, Groenenboom G C, van der Avoird A, Balakrishnan N, Forrey R C and Stancil P C 2015 *Astrophys. J.* **811** 27 (*Preprint* 1511.04981)
[28] Nesterenok A V 2015 *MNRAS* **449** 2875–2882 (*Preprint* 1503.07041)
[29] Draine B T 2011 *Physics of the Interstellar and Intergalactic Medium* (Princeton University Press)
[30] Hummer D G and Rybicki G B 1985 *Astrophys. J.* **293** 258–267
[31] Cohen S D, Hindmarsh A C and Dubois P F 1996 *Comput. Phys.* **10** 138
[32] Laor A and Draine B T 1993 *Astrophys. J.* **402** 441–468
[33] Weingartner J C and Draine B T 2001 *Astrophys. J. Suppl. S.* **134** 263–281 (*Preprint* astro-ph/9907251)
[34] Wakelam V and Herbst E 2008 *Astrophys. J.* **680** 371-383 (*Preprint* 0802.3757)
[35] Draine B T 1978 *Astrophys. J. Suppl. S.* **36** 595–619
[36] Cecchi-Pestellini C and Aiello S 1992 *MNRAS* **258** 125–133

IOP Conf. Series: Journal of Physics: Conf. Series **929** (2017) 012005 doi:10.1088/1742-6596/929/1/012005

What can we learn about the stellar wind of massive stars from studying spin evolution of the X-ray pulsar OAO 1657-415

V.Yu. Kim[1] **and N.R. Ikhsanov**[1,2,3]

[1]Pulkovo Observatory, St. Petersburg, 196140 Russia
[2]Saint Petersburg State University, St. Petersburg, 198504 Russia
[3]Special Astrophysical Observatory, Nizhny Arkhyz, 369167 Russia

E-mail: kim@gao.spb.ru, ikhsanov@gao.spb.ru

Abstract. The persistent X-ray pulsar OAO 1657-415 is associated with a wind-fed High Mass X-ray Binary system and shows a peculiar spin evolution. On a timescale of more than 30 years it is observed to experience a regular spin-up which is superposed on episodic and chaotic variations of the period at a very high rate. We explore a possibility to explain the spin evolution of the pulsar in terms of Magnetic Levitation Accretion scenario in which the neutron star captures material from a magnetized stellar wind of its massive companion and accretes matter from a non-Keplerian magnetized disk (Magnetic Levitating disk or ML-disk). We show this scenario to be applicable if the magnetic field in the wind at a distance of binary separation lies in the interval $20 - 70$ mG.

1. Introduction

Persistent X-ray pulsars represent a subclass of pulsars whose X-ray luminosity has not been observed to vary by more than an order of magnitude [1]. A majority of these sources are associated with High Mass X-ray Binaries (HMXBs) composed of a massive O/B-type star and a magnetized rotating Neutron Star (NS). The X-ray emission of these systems is usually powered by a wind-fed accretion onto the NS. The appearance of the X-ray sources depends on physical conditions in the matter captured by the NS and, hence, these pulsars can be considered as "laboratories" for studying parameters of stellar wind ejected by their massive companions. This is a complimentary method to the currently used spectral analysis of UV radiation, which allows to evaluate terminal wind velocity and the mass-loss rate by the massive stars [2, 3]. An application of the method to determination of magnetic field in the wind of the massive component of the HMXB OAO 1657-415 is discussed in this paper.

OAO 1657-415 has initially been discovered as a point-like X-Ray source by [4] and recognized to be $P_s \simeq 38.22$ s pulsar by [5]. It was later associated with the $P_{orb} \simeq 10.44$ d HMXB [6] composed of a NS and a B0 − 6 Iab supergiant and located at the distance of 6.4 ± 1.5 kpc [7]. The X-ray luminosity of the source is smoothly varying by a factor of 8 around its average value $L_x \simeq 3 \times 10^{36}\,\mathrm{erg\,s^{-1}}$, [8]. A presence of the ~ 36 keV cyclotron line in its X-ray spectrum suggests the surface magnetic field of the neutron star to be $B_{\rm ns} \simeq 3 \times 10^{12}$ G [9].

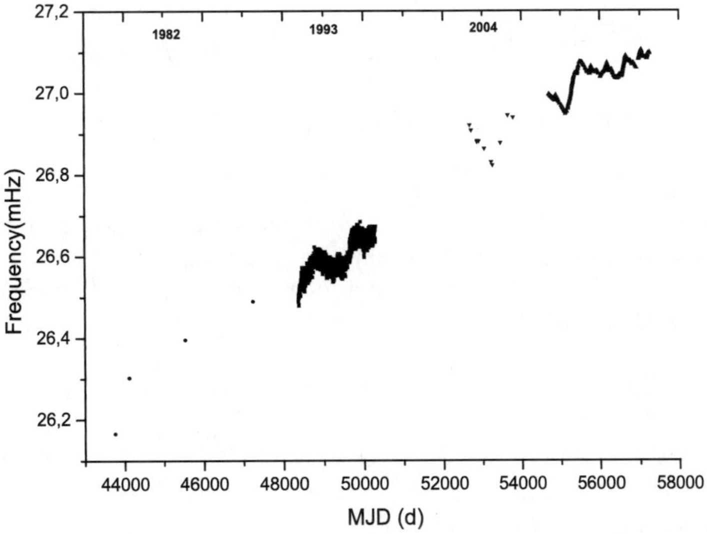

Figure 1. Evolution of spin period OAO 1657 from 1979 to 2015 received by "HEAO-1", "Temna", "Ginga", "Einstein" [10], "Integral" [11], "Fermi GMB" [12]

2. Observed spin evolution

The pulsar is observed to experience a regular spin-up (see fig. 1) at the average rate $\dot{\nu}_{\rm su}^{(\rm g)} \simeq (7.7-8.9)\times10^{-13}\,{\rm Hz\,s^{-1}}$ [11, 8]. The spin period of the pulsar since its discovery decreased by $\Delta P \simeq 1.312\,{\rm s}$ to its currently observed value 36.9 s.

The regular spin-up trend is superimposed with rapid spin-up/spin-down episodes. These spin variations occur at the spin-up, $\dot{\nu}_{\rm su}^{(1)} \simeq 6.07\times10^{-12}\,{\rm Hz\,s^{-1}}$, and spin-down, $\dot{\nu}_{\rm sd}^{(1)} \simeq -3.92\times10^{-12}\,{\rm Hz\,s^{-1}}$ rates and show no correlation with intensity variations of the system X-ray emission [13].

3. Regular spin-up

The observed regular spin-up of the pulsar implies that the average spin-up torque exerted on the NS by the accreting matter is $< K_{\rm su} > \geq 2\pi I \dot{\nu}_{\rm su}^{(\rm g)}$, where I is the moment of inertia of the NS. On the other hand, the average rate of transfer of angular momentum into the Roche lobe of a NS by the material which the NS of mass $M_{\rm ns}$ captures at the Bondi radius, $r_{\rm G} = 2GM_{ns}/\upsilon_{\rm rel}^2$, is limited to[1] $< K_{\rm su} > \leq \xi \dot{M} \Omega_{\rm orb} r_{\rm G}^2$. Here $\upsilon_{\rm rel}$ is the relative velocity of the NS to the surrounding material, $\dot{M}$ is the mass-transfer rate between the system components, $\Omega_{\rm orb} = 2\pi/P_{\rm orb}$ is the orbital angular velocity and ξ is the efficiency parameter accounting for dissipation of the angular momentum due to density and velocity gradients in the accretion flow inside the Bondi radius.

Solving inequality $2\pi I \dot{\nu}_{su}^{(\rm g)} \leq \xi \dot{M} \Omega_{\rm orb} r_{\rm G}^2$ for $\upsilon_{\rm rel}$ one finds that the observed regular spin-up

[1] It is a product of the maximum possible specific angular momentum of the material captured by the NS at the Bondi radius, $\Omega_{\rm orb} r_{\rm G}^2$, with the mass captured by the NS in a unit time, $\dot{M}$, corrected by the factor $\xi \leq 1$

of the pulsar implies that the relative velocity of the NS satisfies the condition $v_{\rm rel} \leq v_0$, where

$$v_0 \simeq 273\,{\rm km\,s^{-1}} \times \xi_{0.2}^{1/4}\, m^{1/2}\, I_{45}^{-1/4} \left(\frac{\dot{M}}{2\times10^{16}\,{\rm g\,s^{-1}}}\right)^{1/4} \left(\frac{P_{\rm orb}}{10.44\,{\rm d}}\right)^{-1/4} \left(\frac{\dot{\nu}_{su}^{(g)}}{8\times10^{-13}\,{\rm Hz\,s^{-1}}}\right)^{-1/4} \quad (1)$$

Here $\xi_{0.2} = \xi/0.2$ is normalized according to [14], $m = M_{\rm ns}/1.4\,M_{\odot}$ and $I_{45} = I/10^{45}\,{\rm g\,cm^2}$.

4. Accretion scenario

Traditional quasi-spherical and Keplerian disk accretion scenarios encounter major difficulties explaining the observed rates of spin evolution of the pulsar during the spin-up/spin-down episodes [13]. Indeed, the maximum possible spin-down rate of the NS which can be expected for the parameters of OAO 1657-415 in these scenarios,

$$\dot{\nu}_{sd}^{(0)} = \frac{k_{\rm t}\dot{M}\omega_{\rm s}r_{\rm A}^2}{2\pi I} \simeq -10^{-13}\,{\rm Hz\,s^{-1}} \times k_{\rm t}\,\mu_{30}^{8/7}\,\dot{M}_{16}^{3/7}\,m^{-2/7}\,I_{45}^{-1}\,P_{37}^{-1}, \quad (2)$$

is almost an order of magnitude smaller than the spin-down rate inferred from observations of the spin-down episodes. Here $k_{\rm t} \leq 1$ is the efficiency parameter, $\omega_{\rm s} = 2\pi/P_{\rm s}$ is the angular velocity of the NS and $r_{\rm A} = \left(\mu^2/\dot{M}\sqrt{2GM_{\rm ns}}\right)^{2/7}$ is the Alfvén radius, which is assumed to be the radius of the magnetosphere of a NS in the traditional accretion scenarios. $\mu = (1/2)B_{\rm ns}R_{\rm ns}^3$ is the dipole magnetic moment and $R_{\rm ns}$ is the radius of the NS and $P_{37} = P_{\rm s}/37\,{\rm s}$.

A situation in which the observed rates of spin evolution exceed the maximum possible rate predicted by the traditional accretion scenarios is rather common among pulsars in HMXBs [15]. It can be improved by incorporating the magnetic field of the accreting material, $B_{\rm f}$, into the model. The magnetic energy associated with this field, $B_{\rm f}^2/8\pi$, increases rapidly as the initially quasi-spherical accretion flow is approaching the NS and reaches the dynamical pressure of the flow at a distance

$$R_{\rm sh} = \beta_0^{-2/3}\,r_{G}\left(\frac{c_{\rm s0}}{v_{\rm rel}}\right)^{4/3}, \quad (3)$$

which is referred to as Shvartsman radius [16, 17]. Here $\beta_0 = \beta(r_G) = 8\pi\rho_0 c_{\rm s0}^2/B_{\rm f0}^2$ is the ratio of a thermal to magnetic pressure at the Bondi radius, $\rho_0 = \rho(r_G)$ is the density, $c_{\rm s0} = c_{\rm s}(r_G)$ is the sound speed and $B_{\rm f0} = B_f(r_G)$ is the magnetic field in the stellar wind of the massive companion at the Bondi radius.

Being controlled by its own magnetic field the accretion flow inside the Shvartsman radius changes its geometry from quasi-spherical to a non-Keplerian magnetized disk [18, 19] (which in later studies is referred to as Magnetic Levitating disk, or ML-disk [20]). The torque exerted on the NS from the ML-disk in this case [20],

$$K_{\rm ml} = \frac{k_{\rm t}\,\mu^2}{(r_{\rm ma}\,r_{\rm cor})^{3/2}}\left(\frac{\Omega_{\rm f}(r_{\rm ma})}{\omega_{\rm s}} - 1\right), \quad (4)$$

is large enough for the expected spin-down rate of the NS in OAO 1657-415,

$$\dot{\nu}_{sd}^{(ml)} = \frac{K_{\rm ml}}{2\pi I} \simeq -8\times10^{-12}\,{\rm Hz\,s^{-1}} \times m^{-8/13}\,\alpha_{0.1}^{-3/13}\,P_{37}^{-1}\,T_6^{3/13}\,\dot{M}_{16}^{6/13}\,\mu_{30}^{17/13}, \quad (5)$$

to be in excess of the observed value by more than a factor of 2. Here $k_{\rm t}$ is the efficiency parameter, $r_{\rm cor} = \left(GM_{\rm ns}/\omega_{\rm s}^2\right)^{1/3}$ is the corotation radius, $\Omega_{\rm f}$ is the angular velocity of matter in the ML-disk and [15]

$$r_{\rm ma} = \left(\frac{c\,m_{\rm p}^2}{16\sqrt{2}\,e\,k_{\rm B}}\right)^{2/13} \frac{\alpha_{\rm B}^{2/13}\mu^{6/13}(GM_{\rm ns})^{1/13}}{T_0^{2/13}\dot{M}^{4/13}}, \quad (6)$$

is the inner radius of the disk which plays a role of the magnetospheric radius of the NS within this so called Magnetic Levitation Accretion scenario (MLA-scenario). Here c is the speed of light, $m_{\rm p}$ is the proton mass, e is the electric charge of an electron, $k_{\rm B}$ is the Boltzmann constant and T_0 is the temperature of matter at the inner radius of the ML-disk ($T_6 = T_0/10^6$ K). The parameter $\alpha_{\rm B}$ is the ratio of the effective diffusion coefficient at the magnetospheric boundary to the Bohm diffusion coefficient and $\alpha_{0.1} = \alpha_{\rm B}/0.1$ (for discussion see [20] and references therein).

Hence, the observed spin evolution of OAO 1657-415 can be explained within the MLA-scenario provided the relative velocity of a NS is limited as $v_{\rm rel} \leq v_0$ and stellar wind of the massive star in this system is magnetized. Constrains for the validity of this explanation are discussed in the next sections.

5. Magnetization of the stellar wind

As recently shown by Ikhsanov and Mereghetti [20] the MLA-scenario can be realized if $R_{\rm sh} \geq \max\{r_{\rm A}, r_{\rm circ}\}$, where $r_{\rm cir}$ is the circularization radius. Under these conditions the magnetic field of the accreting material prevents formation of a Keplerian disk in the system and changes the flow geometry from a quasi-spherical to ML-disk. Solving this inequality for $v_{\rm rel}$ yields

$$v_{\rm ca} < v_{\rm rel} < v_{\rm ma}, \tag{7}$$

where

$$v_{\rm ca} \simeq 160\,{\rm km\,s^{-1}} \times \xi_{0.2}^{3/7}\,\beta_0^{1/7}\,m^{3/7} \left(\frac{P_{\rm orb}}{10.44\,{\rm d}}\right)^{-3/7} \left(\frac{c_{\rm s0}}{10\,{\rm km\,s^{-1}}}\right)^{-2/7} \tag{8}$$

and

$$v_{\rm ma} \simeq 540\,{\rm km\,s^{-1}} \times \beta_0^{-1/5} m^{12/35} \mu_{30}^{-6/35}\,\dot{M}_{16}^{3/35} \left(\frac{c_{\rm s0}}{10\,{\rm km\,s^{-1}}}\right)^{2/5}. \tag{9}$$

As we have seen in Sect. 3, the observed regular spin-up of OAO 1657-415 implies that the relative velocity of a NS is limited as $v_{\rm rel} \leq v_0$, where v_0 is expressed by Eq. (1). This allows us to set an upper limit to β_0. Indeed, according to condition (7), the MLA-scenario can be realized in the system only if $v_{\rm ca} \leq v_0$. Otherwise the magnetic field of the accreting material would not be able to prevent formation of a Keplerian disk. Combining Eqs. (1) and (9) and solving inequality $v_{\rm ca}(\beta_0) \leq v_0$ for β_0 one finds that the MLA-scenario in OAO 1657-415 can be realized if $\beta_0 \leq \beta_0^{\rm max}$

$$\beta_0^{\rm max} \simeq 50 \times m^{1/2}\,\xi_{0.2}^{-5/4}\,I_{45}^{-7/4}\,\dot{M}_{16}^{7/4} \left(\frac{c_{s0}}{10\,{\rm km/s}}\right)^2 \left(\frac{P_{orb}}{10.44\,{\rm d}}\right)^{5/4}. \tag{10}$$

A lower limit to β_0 can be derived by taking into account that the regular spin-up of a NS within the MLA-scenario can be expected only if the angular velocity of material at the inner radius of the ML-disk (see Eq. 22 in [20]), $\Omega_{\rm f}(r_{\rm ma}) \simeq \xi\Omega_{\rm orb}\,(r_G/R_{\rm sh})^2$, exceeds the angular velocity of the NS itself, $\omega_{\rm s}$. Solving inequality $\Omega_{\rm f}(r_{\rm ma}) \geq \omega_{\rm s}$ for β_0 and using Eq. (3) one finds $\beta_0 \geq \beta_0^{\rm min}$, where

$$\beta_0^{\rm min} \simeq 4 \times m^{-8/7}\,\xi_{0.2}^{-5/4}\,P_{37}^{-5/4}\,\dot{M}_{16}^{-2/7} \left(\frac{c_{s0}}{10\,{\rm km/s}}\right)^2 \left(\frac{P_{orb}}{10.44\,{\rm d}}\right)^{5/4}. \tag{11}$$

The velocities estimated above are plotted in Fig. 2 versus β_0. As seen from this figure the relative velocity for all points within the interval $\beta_0^{\rm min} \leq \beta_0 \leq \beta_0^{\rm max}$ is smaller than $v_{\rm ma}$. This indicates that the modeling of the accretion process in OAO 1657-415 within the MLA-scenario is basically applicable.

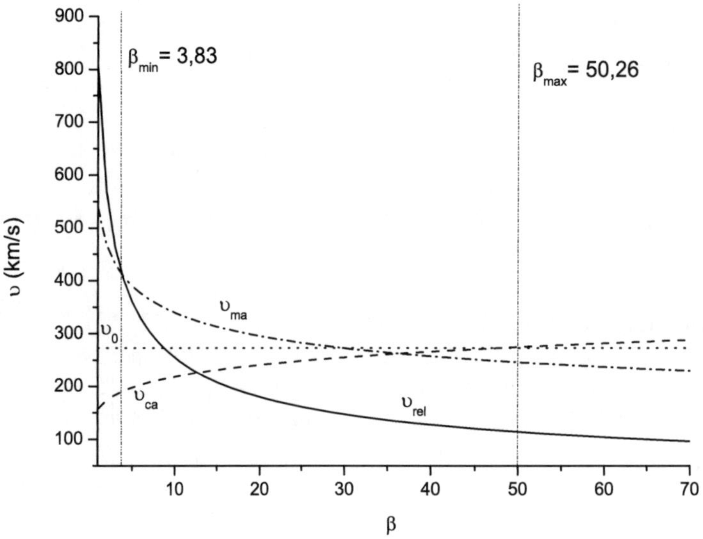

Figure 2. The characteristic velocities versus β_0 -parameter for parameters of OAO 1657-415

Finally, using the range of possible values of the parameter β_0 one can also evaluate the magnetic field strength in the stellar wind of the massive companion at the distance of binary separation. Using the above given definition of β_0 one finds

$$B_{\mathrm{f0}} = \left(\frac{2\dot{M}\, v_0^3\, c_{\mathrm{s0}}^2}{(GM_{\mathrm{ns}})^2\, \beta_0} \right)^{1/2} . \tag{12}$$

Then, putting $\beta_0 = \beta_0^{\max}$ and $\beta_0 = \beta_0^{\min}$ to this equation one gets $B_{\mathrm{f0}}^{\min} \le B_{\mathrm{f0}} \le B_{\mathrm{f0}}^{\max}$, where

$$\begin{aligned} B_{\mathrm{f0}}^{\min} &\simeq 19\,\mathrm{mG} \times \dot{M}_{16}^{1/2} \left(\frac{\beta_0^{\max}}{50}\right)^{1/2} \left(\frac{c_{\mathrm{s0}}}{10\,\mathrm{km\,s^{-1}}}\right) \left(\frac{v_0}{273\,\mathrm{km\,s^{-1}}}\right)^{3/2} \\ B_{\mathrm{f0}}^{\max} &\simeq 70\,\mathrm{mG} \times \dot{M}_{16}^{1/2} \left(\frac{\beta_0^{\min}}{4}\right)^{1/2} \left(\frac{c_{\mathrm{s0}}}{10\,\mathrm{km/s}}\right) \left(\frac{v_0}{273\,\mathrm{km/s}}\right)^{3/2} \end{aligned} \tag{13}$$

Thus, the peculiar spin evolution of OAO 1657-415 can be explained within the MLA-scenario provided the relative velocity of the NS in this system does not exceed v_0 given by Eq. (1), and the magnetic field in the stellar wind of the massive component at the distance of orbital separation lies in the range 20 − 70 mG.

Acknowledgements

VYK acknowledges support of the National Scientific School program under the grant No. NSh-7241-2016.2. NRI acknowledges support of the Russian Science Foundation under the grant No. 14-50-00043.

References

[1] Reig, P. and Roche, P. 1999, *Mon. Not. Roy. Astron. Soc.*, **306**, 100

[2] Kudritzki, R. and Puls, J. 2000, *Annu. Rev. of Astron. Astrophys.*, **38**, 613
[3] Markova, N. and Puls, J. 2008, *Astron. and Astrophys.*, **478**, 823
[4] Polidan, R.S., Sanford, P.W., White, N.E., Pollard, G.S.G., Locke, M.C., Peters, G.J. and Dobias, J. 1978, *IAU Circ.*, **3234**, 1
[5] White, N.E. and Pravdo, S.H. 1979, *Astrophys. J.*, **233**, L121
[6] Chakrabarty, D., Grunsfeld, J.M., Prince, T.A., Bildsten, L., Finger, M.H., Wilson, R.B., Fishman, G.J., Meegan, C.A. and Paciesas, W.S. 1993, *Astrophys. J.*, **403**, L33
[7] Chakrabarty, D., Wang, Z., Juett, A.M., Lee, J.C. and Roche, P. 2002, *Astrophys. J.*, **573**, 789
[8] Jenke, P.A., Finger, M.H., Wilson-Hodge, C.A. and Camero-Arranz, A. 2012, *Astrophys. J.*, **759**, 124
[9] Orlandini, M., Fiume, D., Sordo, S., Frontera, F., Parmar, A.N., Santangelo, A. and Segreto. A. 1999, *Astron. and Astrophys.*, **349**, L9
[10] Nagase, F. 1989, *Pub. Astr. Soc. Japen*, **41**, 1
[11] Barnstedt, J., Staubert, R., Santangelo, A., Ferrigno, C., Horns, D., Klochkov, D., Kretschmar, P., Kreykenbohm, I., Segreto, A. and Wilms, J. 2008, *Astron. and Astrophys.*, **486**, 293
[12] http://gammaray.msfc.nasa.gov/gbm/science/pulsars/lightcurves/oao1657.fits.gz
[13] Inam, S. and Baykal, A. 2000, *Astron. and Astrophys.*, **353**, 617
[14] Ruffert, M. 1999, *Astron. and Astrophys.*, **346**, 861
[15] Ikhsanov, N.R., Likh, Yu.S., Beskrovnaya, N.G., 2014, *Astron. Rep.*, **58**, 376
[16] Ikhsanov, N.R., Beskrovnaya, N.G. 2012, *Astron. Rep.*, **56**, 589
[17] Ikhsanov, N. and Finger, M. 2012, *Astrophys.J.*, **753**, 8
[18] Bisnovatyi-Kogan, G.S. and Ruzmaikin, A.A. 1976, *Astrophys. and Space Sci.*, **42**, 401
[19] Igumenshchev, I.V., Narayan, R., Abramowicz, M.A. 2003, *Astrophys.J.*, **592**, 1042
[20] Ikhsanov, N.R. and Mereghetti, S. 2015, *Monthly. Not. Roy. Astron. Soc.*, **454**, 3760
[21] Gosling, J.T., Thomsen, M.F., Bame, S.J., Elphic, R.C. and Russell, C.T. 1991, *Journal Geophys. Res.*, **96**, 14097

Mirror instability in plasma with relativistic component

S M Osipov[1], A M Bykov[1], D C Ellison[2] and P E Gladilin[1]
[1]Ioffe Institute, 194021, Saint-Petersburg, Russia,
[2]Physics Department, North Carolina State University, Box 8202, Raleigh, NC 27695, USA
E-mail: osm2004@mail.ru

Abstract. A kinetic model of the mirror instability in highly non-equilibrium anisotropic relativistic plasma is presented. The growth rate is derived for long-wavelength compressible magnetic perturbations in a magnetized collisionless plasma with an anisotropic pressure of the relativistic component. The mirror-type mode perturbations of the magnetic field are growing for the wave vectors preferentially directed transverse to the mean magnetic field. The mirror instability may occur in the vicinity of relativistic particle accelerators in the interstellar and intergalactic media.

1. Introduction
Instabilities in cosmic plasmas are responsible for relaxation of non-equilibrium states in various objects starting from the planetary magnetospheres to early cosmology. Instabilities of the relativistic component (RC) has long been recognized to play a role in the dynamics of the interstellar gas [1]. The firehose and mirror instabilities occur in the case of anisotropic thermal background plasma pressure in the presence of magnetic fields [2]. The mirror instability in the anisotropic magnetized interstellar and intergalactic plasma was studied in detail in [3, 4, 5]. A pressure anisotropy of RC may result in the growth of a firehorse or mirror instability depending on the ratio of the partial pressures of relativistic particles along and transverse to the mean magnetic field. The firehose instability of the RC was studied recently in [6, 7, 8], while the mirror instability case requires a special kinetic treatment, which is out lined below.

2. Mirror instability of anisotropic RC distributions
The derivation given below treats kinetically the anisotropic RC which is the source of free energy for the mirror instability, while the background plasma is described with a magnetohydrodynamic (MHD) approximation as it was earlier done in [8].

The background plasma obey the continuity and the momentum conservation equations

$$\frac{\partial \rho}{\partial t} + \nabla(\rho \mathbf{u}) = 0 \ , \tag{1}$$

and

$$\rho\left(\frac{\partial \mathbf{u}}{\partial t} + (\mathbf{u}\nabla)\mathbf{u}\right) = -\nabla p_g + \frac{1}{4\pi}(\nabla \times \mathbf{B}) \times \mathbf{B} - \frac{1}{c}(\mathbf{j}^{rc} - e n_{\rm rc}\mathbf{u}) \times \mathbf{B} - \int \mathbf{p} I[\phi^{rc}] d^3p \ , \tag{2}$$

where $\mathbf{u}$, ρ, and p_g are the plasma bulk velocity, density, and the thermal gas pressure respectively. The electromagnetic fields $\mathbf{E}$ and $\mathbf{B}$ obey

$$\frac{\partial \mathbf{B}}{\partial t} = \nabla \times (\mathbf{u} \times \mathbf{B}) \ , \tag{3}$$

and the kinetic equation for the RC is

$$\frac{\partial \phi^{rc}}{\partial t} + \mathbf{v} \cdot \frac{\partial \phi^{rc}}{\partial \mathbf{r}} + e\mathbf{E} \cdot \frac{\partial \phi^{rc}}{\partial \mathbf{p}} - \frac{ec}{\mathcal{E}} \mathbf{B} \cdot \widehat{\mathcal{O}} \phi^{rc} = I[\phi^{rc}] \ , \tag{4}$$

where ϕ^{rc}, n_{rc}, and j_{rc} are the RC distribution function, concentration, and electric current, $\mathcal{E}$, $\mathbf{v}$, and $\mathbf{p}$ are the RC particle energy, velocity, and momentum, respectively, $\widehat{\mathcal{O}} = \mathbf{p} \times \frac{\partial}{\partial \mathbf{p}}$ is the momentum rotation operator, c is the speed of light, e is the particle charge, and $\Omega = eB_0c/\mathcal{E}$ is the RC particle gyro-frequency.

The collision operator $I[\phi^{rc}]$ describes RC interactions with magnetic fluctuations carried by the thermal plasma. In the thermal background plasma frame

$$I[\phi^{rc}] = -\nu \left(\phi^{rc} - \phi^{rc}_{iso}\right) \ , \tag{5}$$

where ϕ^{rc}_{iso} is the isotropic part of the distribution function, $\nu = \epsilon\Omega$ is the RC scattering frequency by magnetic fluctuations with $\epsilon \leq 1$, and $\int \mathbf{p} I[\phi^{rc}] d^3p = -\epsilon B_0 \mathbf{j}^{rc}/c$ in Eq. (2).

The small variations of the background plasma parameters in the plasma rest frame Eqs. (2)–(4) are $\xi = \xi_0 + \delta\xi$, where $\delta\xi \propto \exp(i\mathbf{kr} - i\omega t)$. Then in the case of the adiabatic process $\nabla \delta p_g = c_s^2 \nabla \delta\rho$, where $c_s = \sqrt{\gamma_g p_0/\rho_0}$, and γ_g is the adiabatic index. The wavevector is $\mathbf{k} = k_\parallel \mathbf{e}_z + k_\perp \mathbf{e}_x$, where the parallel direction is $\mathbf{e}_z = \mathbf{B}_0/B_0$, while the transverse direction is $\mathbf{e}_x$.

We will consider below instabilities initiated by the quadrupole anisotropy of the relativistic component. Such anisotropy of the relativistic particle distribution can be parameterized as

$$\phi_0^{rc}(p, \mu) = \frac{n_{rc} F(p)}{4\pi} \left[1 + \frac{\zeta}{2}\left(3\mu^2 - 1\right)\right] \ , \tag{6}$$

where $\int_0^\infty F(p) p^2 dp = 1$ and ζ is a quadrupole anisotropy parameter with $|\zeta| < 1$.

Keeping only linear responses to the perturbations in Eqs. (1)–(2), we obtain

$$\left(\omega^4 - \omega^2 \left(c_a^2 + c_s^2\right) k^2 + c_a^2 c_s^2 k^2 k_\parallel^2\right) \delta B_x = \left(\omega^2 - c_s^2 k_\parallel^2\right) i \frac{B_0^2 k_\parallel}{c\rho_0} \left(\delta j_y^{rc} - \epsilon \delta j_x^{rc}\right)$$
$$-i c_s^2 k_\parallel^2 \frac{B_0^2 k_\perp}{c\rho_0} \epsilon \delta j_z^{rc} \ , \tag{7}$$

and

$$\left(\omega^2 - c_a^2 k_\parallel^2\right) \delta B_y = -i \frac{B_0^2 k_\parallel}{c\rho_0} \left(\delta j_x^{rc} + \epsilon \delta j_y^{rc}\right) \ , \tag{8}$$

where $c_a = B_0/\sqrt{4\pi\rho_0}$ is the Alfvén velocity. The linearized Eq. (4) has the form

$$\left[\nu + i\left(-\omega + k_\parallel v \cos\theta\right)\right] \delta\phi^{rc} + i k_\perp v \sin\theta \cos\varphi \, \delta\phi^{rc} - \Omega \frac{\partial \delta\phi^{rc}}{\partial \varphi} = -e\delta\mathbf{E} \cdot \frac{\partial \phi_0^{rc}}{\partial \mathbf{p}} + \frac{ec}{\mathcal{E}} \delta\mathbf{B} \cdot \widehat{\mathcal{O}} \phi_0^{rc} \ , \tag{9}$$

where θ is the pitch angle and φ is the azimuthal angle between the field direction and the particle velocity.

In the weakly collisional case we have $\epsilon \ll 1$, and therefore one can omit the frequency ν compared to ω in Eq. (9). The response of relativistic particles' current $\delta \mathbf{j}^{rc}$ on the imposed magnetic perturbation $\delta \mathbf{B}$ can be derived from Eq. (9) and has the form

$$\delta j_x^{rc} = \frac{\pi e}{2B_0} \int dpp^2 \int_{-1}^{1} d\mu v S(\mu) \frac{\partial \phi_0^{rc}(p,\mu)}{\partial \mu}$$
$$\left\{ -\delta B_y \sum_{n=-\infty}^{\infty} \left[J_n(\beta)\left(G_n^+ + G_n^-\right) + 2J_{n-2}(\beta) G_n^- \right] + \right.$$
$$\left. + i\delta B_x \sum_{n=-\infty}^{\infty} J_n(\beta)\left(G_n^+ - G_n^-\right) \right\}, \tag{10}$$

and

$$\delta j_y^{rc} = \frac{\pi e}{2B_0} \int dpp^2 \int_{-1}^{1} d\mu v S(\mu) \frac{\partial \phi_0^{rc}(p,\mu)}{\partial \mu}$$
$$\left\{ \delta B_x \sum_{n=-\infty}^{\infty} \left[J_n(\beta)\left(G_n^+ + G_n^-\right) - 2J_{n-2}(\beta) G_n^- \right] + \right.$$
$$\left. + i\delta B_y \sum_{n=-\infty}^{\infty} J_n(\beta)\left(G_n^+ - G_n^-\right) \right\}, \tag{11}$$

$$\delta j_z^{rc} = \frac{\pi e}{B_0} \int dpp^2 \int_{-1}^{1} d\mu v \mu \sqrt{S(\mu)} \frac{\partial \phi_0^{rc}(p,\mu)}{\partial \mu}$$
$$\left\{ i\delta B_x \sum_{n=-\infty}^{\infty} \left[J_{n+1}(\beta) G_n^+ - J_{n-1}(\beta) G_n^- \right] - \right.$$
$$\left. -\delta B_y \sum_{n=-\infty}^{\infty} \left[J_{n+1}(\beta) G_n^+ + J_{n-1}(\beta) G_n^- \right] \right\}, \tag{12}$$

where

$$G_n^{\pm} = \frac{J_n(\beta)}{a + i(n \pm 1)}, \tag{13}$$

$S(\mu) = 1 - \mu^2$, $J_n(\beta)$ is the Bessel function of integer n, $\beta = k_\perp v \sin\theta/\Omega$, and $a = [\nu + i(k_\parallel v \cos\theta - \omega)]/\Omega$ (see, e.g., [9]). The equations were obtained by neglecting the contributions proportional to $\delta \mathbf{E}$ in (9), since they $\sim \dfrac{|\omega|}{ck} \ll 1$ for $v \approx c$.

The mirror instability appears mostly in the regime where $k_\parallel v/\Omega \ll 1$, $k_\perp v/\Omega \ll 1$, and $\omega/\Omega \ll 1$. In Eq. (13) we assumed $\epsilon \ll 1$ as well as $\beta \ll 1$ and $a \ll 1$. Therefore, in the long-wavelength regime from Eqs. (10) and (11) one can obtain the relativistic current response as

$$\delta j_x^{rc} = \frac{\pi e}{2B_0} \int dpp^2 \int_{-1}^{1} d\mu v S(\mu) \left\{ 2\delta B_x - 2i \frac{k_\parallel v \mu}{\Omega} \delta B_y \right\} \frac{\partial \phi_0^{rc}(p,\mu)}{\partial \mu}, \tag{14}$$

and

$$\delta j_y^{rc} = \frac{\pi e}{2B_0} \int dpp^2 \int_{-1}^{1} d\mu v S(\mu) \left\{ 2\delta B_y + \delta B_x \left[2i \frac{k_\parallel v \mu}{\Omega} - i \frac{k_\perp^2 v^2 S(\mu)}{\Omega\left(k_\parallel v \mu - \omega\right)} \right] \right\} \frac{\partial \phi_0^{rc}}{\partial \mu}. \tag{15}$$

The second term in the brace in Eq. (15) results in the firehorse instability which grows if the quadrupole anisotropy parameter $\zeta > 0$. On the contrary the mirror instability appears from the third term in the brace in Eq. (15) if $\zeta < 0$. Note that if the RC scattering rate by magnetic turbulence $\nu \sim \Omega$, which may occur if $\epsilon \sim 1$, then the mirror instability is suppressed.

Integrating Eq. (15) over μ, with an account of Eq. (6), one obtains the asymptotic form of the third term in the brace in Eq. (15) for $\psi \ll 1$ as

$$\int_{-1}^{1} \frac{S(\mu)^2 \mu d\mu}{1 - \psi\mu} \to \frac{16}{105}\psi + O\left(\psi^3\right) , \tag{16}$$

where $\psi = k_{\|} v/\omega$. For $\psi \gg 1$

$$\int_{-1}^{1} \frac{S(\mu)^2 \mu d\mu}{1 - \psi\mu} \to -\frac{16}{15\psi} + O\left(\frac{1}{\psi^2}\right) . \tag{17}$$

The response of the RC electric current in the last case is

$$\delta j_x^{rc} = -i\frac{\delta B_y \zeta n_{rc} k_{\|} c}{5B_0^2} \int_0^{\infty} v p^3 F(p)\, dp , \tag{18}$$

and

$$\delta j_y^{rc} = i\frac{\delta B_x \zeta n_{rc} c}{5B_0^2} \left(k_{\|} - \frac{2k_{\perp}^2}{k_{\|}}\right) \int_0^{\infty} v p^3 F(p)\, dp . \tag{19}$$

Let us define the unperturbed pressure

$$P_0 = \frac{n_{rc}}{3} \int_0^{\infty} v p^3 F(p)\, dp , \tag{20}$$

and note that for the RC distribution function of Eq. (6)

$$P_{\|} = \int v p \mu^2 \phi_0^{rc}(p, \mu)\, d^3p = P_0 \left(1 + \frac{2\zeta}{5}\right) , \tag{21}$$

$$P_{\perp} = \int v p S(\mu) \cos^2\varphi \phi_0^{rc}(p, \mu)\, d^3p = \int v p S(\mu) \sin^2\varphi \phi_0^{rc}(p, \mu)\, d^3p$$
$$= P_0 \left(1 - \frac{\zeta}{5}\right) , \tag{22}$$

and

$$\delta P = P_{\|} - P_{\perp} = \frac{3\zeta}{5} P_0 . \tag{23}$$

Now, subsituting Eqs. (18) and (19) into (7) and (8) we obtain the dispersion relations

$$\omega^4 - \omega^2 \left(c_a^2 + c_s^2\right) k^2 + c_a^2 c_s^2 k^2 k_{\|}^2 = -\left(\omega^2 - c_s^2 k_{\|}^2\right) \frac{3\zeta P_0}{5\rho_0} \left(k_{\|}^2 - 2k_{\perp}^2\right) , \tag{24}$$

and

$$\omega^2 - c_a^2 k_{\|}^2 = -\frac{3\zeta P_0}{5\rho_0} k_{\|}^2 . \tag{25}$$

The dispersion equation splits into two independent equations where the mirror instability is determined by Eq. (24) under conditions $2k_\perp^2 > k_\parallel^2$ and $\zeta < 0$. For $k_\perp^2 \gg k_\parallel^2$ Eq. (24) simplifies to

$$\omega^2 - \left(c_a^2 + c_s^2\right) k^2 = \frac{3\zeta P_0}{5\rho_0} 2k_\perp^2 \ . \tag{26}$$

Here the perturbation of the magnetic field is mostly δB_z which is connected to the background plasma density variations by $\delta\rho \approx \rho_0 \dfrac{\delta B_z}{B_0}$.

Then from Eq. (23) with account of $k \approx k_\perp$, one finally obtains

$$\omega^2 = \left(c_a^2 + c_s^2 + 2\frac{\delta P}{\rho_0}\right) k^2 \ . \tag{27}$$

Eq. (27) shows that growing modes occur if $\delta P \propto \zeta < 0$ and $2\dfrac{|\delta P|}{\rho_0} > (c_a^2 + c_s^2)$, i.e., the mirror instability is driven by the anisotropic RC pressure. Thus, the growth rate of the mirror instability rate from (27) is

$$\Gamma_{mir} = k\sqrt{2\frac{|\delta P|}{\rho_0} - c_a^2 - c_s^2} \ , \tag{28}$$

for $\delta P \propto \zeta < 0$ and $2\dfrac{|\delta P|}{\rho_0} > (c_a^2 + c_s^2)$.

If $k_\perp = 0$, $k_\parallel = k$ then Eq. (24) is reduced to the dispersion relation

$$\left(\omega^2 - c_s^2 k^2\right)\left(\omega^2 - c_a^2 k^2 + \frac{3\zeta P_0}{5\rho_0} k^2\right) = 0 \ . \tag{29}$$

While the first bracket in Eq. (29) is the longitudinal acoustic mode, the second bracket and Eq. (25) gives

$$\omega^2 - c_a^2 k^2 + \frac{3\zeta P_0}{5\rho_0} k^2 = 0 \ , \tag{30}$$

which may be unstable if $\delta P \propto \zeta > 0$ and $\dfrac{\delta P}{\rho_0} > c_a^2$. In this case the anisotropy of the RC pressure results in the growth of the long-wavelength firehose instability

$$\Gamma_{fh} = k\sqrt{\frac{\delta P}{\rho_0} - c_a^2} \ , \tag{31}$$

which was studied earlier in [8].

3. Discussion

Let us estimate the growth rate for the long-wavelength compressible mirror mode perturbation in the warm interstellar plasma with a typical number density about 0.1 cm^{-3}. The cosmic ray pressure in the interstellar medium is of the order of 1 eV cm^{-3}. Therefore, the anisotropy of the cosmic ray pressure of about $|\zeta| \sim 10^{-3}$ would provide the growth rate $\Gamma \sim 10^5 k\ \text{s}^{-1}$ if the wavenumber k is in cm^{-1}. Depending on the character of the pressure anisotropy which is determined by the sign of ζ, either the mirror-type or the firehose mode perturbations of the magnetic field are growing on the time scale above some hundred years. The mirror instability should occur in the vicinity of relativistic particle accelerators. In the case of supernova shocks with super-diffusive particle transport the magnetic field amplification due to the mirror instability may be efficient [10].

4. Acknowledgements
The results of the work were obtained using computational resources of Peter the Great Saint-Petersburg Polytechnic University Supercomputing Center (http://www.spbstu.ru).

References
[1] Parker E N 1969 *Space Sci. Rev.* **9** 651–712
[2] Baumjohann W and Treumann R A 1996 *Basic space plasma physics*
[3] Hall A N 1980 *MNRAS* **190** 353–369
[4] Achterberg A 2013 *MNRAS* **436** 705–717
[5] Komarov S V, Churazov E M, Kunz M W and Schekochihin A A 2016 *MNRAS* **460** 467–477 (*Preprint* 1603.00524)
[6] Bykov A M, Gladilin P E and Osipov S M 2011 *Mem. Societa Astronom. Italiana* **82** 800 (*Preprint* 1111.2587)
[7] Schure K M, Bell A R, O'C Drury L and Bykov A M 2012 *Space Sci. Rev.* **173** 491–519 (*Preprint* 1203.1637)
[8] Bykov A M, Brandenburg A, Malkov M A and Osipov S M 2013 *Space Sci. Rev.* **178** 201–232 (*Preprint* 1304.7081)
[9] Akhiezer A I 1975 *Plasma electrodynamics - Vol.1: Linear theory; Vol.2: Non-linear theory and fluctuations*
[10] Bykov A M, Ellison D C and Osipov S M 2017 *Phys. Rev. E* **95** 033207 (*Preprint* 1703.01160)

Glitches and precession damping in the framework of 3-component model of neutron star

K Y Kraav[1], A I Tsygan[2], D P Barsukov[2,1], A N Yurkin[1]
[1] SPbPU, Saint-Petersburg, Russia
[2] Ioffe Institute, Saint-Petersburg, Russia
E-mail: kirill-kraav@yandex.ru

Abstract. The evolution of the inclination angle and precession damping is considered. It is assumed that the neutron star consists of 3 freely rotating components: the crust and two core components, one of them contains pinned superfluid vortices. We suppose that each component rotates as a rigid body. Also the influence of the small-scale magnetic field on the star's breaking process is examined. Within the framework of this model the star simultaneously can have glitch-like events combined with long-period precession (with periods $10 - 10^4$ years). It is shown that the case of the small quantity of pinned superfluid vortices seems to be more consistent with observations.

1. Introduction

Radio pulsars can be considered as exceptionally stable clock. But sometimes besides gradually braking its rotation suffers some irregularities called glitches [1, 2]. Usually it is related to superfluid pinning. Pinned vortices do not allow superfluid to participate in slow pulsar braking and in some moments it leads to the unpinning and to the transfer of the angular momentum from superfluid to the crust [2]. That is observed as a glitch. There are some evidences that isolated neutron stars may precess with enough long period T_p. Pulsar B1821-11 precesses with period $T_p \sim 500$ days [3]. Some other pulsars show periodic variations which may be explained by precession with $T_p \sim 100 - 500$ days [4]. The changing of pulse profile of Crab pulsar also may be related to precession with $T_p \sim 10^2$ years [5]. The increase of radio luminosity of Geminga pulsar [6] may be related to precession with $T_p > 10$ years too. The nonregular pulse variations known as "red noise" [1, 2] with timescales $T \sim 1$ month $- 10^4$ years also can be related to precession [7]. But presence of pinned vortices leads to precession with periods $T_p \sim P \cdot (I_{tot}/L_g) \sim 10^2 - 10^6 P$ [8], where P is pulsar period, I_{tot} is moment of inertia of the star, L_g is angular momentum of pinned superfluid. Such precession seems to be rapidly damped and incompatible with existence of the long-period precession [8]. In this paper we consider a model, proposed in [9]. It allows the coexistence of long-term precession and quasi-glitch events. We also take into account the influence of the small-scale magnetic field on pulsar braking.

2. Base equations

We assume that neutron star consists of 3 components, named as crust (c-component), g-component and r-component:
The crust (c-component) . We suppose that it rotates as a rigid body with angular velocity

$\vec{\Omega}_c$. It is the outer component so its rotation is observed as a pulsar rotation and hence $\vec{\Omega}_c$ can be considered as the angular velocity of pulsar $\vec{\Omega}$. We suppose that

$$\vec{M}_c = I_c\vec{\Omega}_c \text{ and } \dot{\vec{M}}_c = \vec{K}_{ext} + \vec{N}_{gc} + \vec{N}_{rc} \tag{1}$$

where $\vec{M}_c$ is the angular momentum of the crust, I_c is moment of inertia, $\vec{K}_{ext}$ is the external torque (magnetospheric origin), acting on the crust, $\vec{N}_{gc}$ and $\vec{N}_{rc}$ are torques acting on the crust due to its interaction with g and r components correspondingly.

g-**component**. It's the one of two inner componets. We assume that it consists of normal matter rotating as a rigid body with angular velocity $\vec{\Omega}_g$ and superfluid matter firmly pinned to normal matter so that superfluid vortices rotate together with normal matter with angular velocity $\vec{\Omega}_g$.

$$\vec{M}_g = I_g\vec{\Omega}_g + \vec{L}_g \text{ and } \dot{\vec{M}}_g = \vec{N}_{cg} + \vec{N}_{rg} \ , \ \dot{\vec{L}}_g = \left[\vec{\Omega}_g \times \vec{L}_g\right] \tag{2}$$

where $\vec{M}_g$ is total angular momentum of g-component, I_g is moment of inertia of its normal matter, L_g is angular momentum of pinned superfluid, $\vec{N}_{cg}$ and $\vec{N}_{rg}$ are torques acting on the crust due to its interaction with the crust and with r-component correspondingly.

r-**component** . It is inner component too. We assume that it rotates as a rigid body with angular velocity $\vec{\Omega}_r$.

$$\vec{M}_r = I_r\vec{\Omega}_r \text{ and } \dot{\vec{M}}_r = \vec{N}_{cr} + \vec{N}_{gr} \tag{3}$$

where $\vec{M}_r$ is angular momentum of r-component, I_r is its moment of inertia, $\vec{N}_{cr}$ and $\vec{N}_{gr}$ are torques acting on the crust due to its interaction with the crust and with g-component correspondingly. In the frame of references, related to the crust, the equation of rotation (1)-(3) can be rewritten as

$$\begin{aligned}
\dot{\vec{\Omega}} &= \vec{S}_{ext} + \vec{R}_{gc} + \vec{R}_{rc} && (4)\\
\dot{\vec{\mu}}_{cg} &+ [\vec{\Omega} \times \vec{\mu}_{cg}] + [\vec{\Omega} \times \vec{\omega}_g] + [\vec{\mu}_{cg} \times \vec{\omega}_g] = \vec{R}_{cg} + \vec{R}_{rg} - \vec{R}_{gc} - \vec{R}_{rc} - \vec{S}_{ext} && (5)\\
\dot{\vec{\mu}}_{cr} &+ [\vec{\Omega} \times \vec{\mu}_{cr}] = \vec{R}_{cr} + \vec{R}_{gr} - \vec{R}_{rc} - \vec{R}_{gc} - \vec{S}_{ext} && (6)\\
\dot{\vec{\omega}}_g &= [\vec{\mu}_{cg} \times \vec{\omega}_g] && (7)
\end{aligned}$$

where $\vec{\mu}_{ij} = \vec{\Omega}_j - \vec{\Omega}_i$, $\vec{N}_{ij} = -\vec{N}_{ji}$, $\vec{R}_{ij} = \vec{N}_{ij}/I_j$ $(i,j = c,g,r)$, $\vec{S}_{ext} = \vec{K}_{ext}/I_c$ and $\vec{\omega}_g = \vec{L}_g/I_g$.

In this paper we consider only small perturbation of the equilibrium state $\vec{\Omega}_c = \vec{\Omega}_g = \vec{\Omega}_r$, $\vec{L}_g = L_g\vec{e}_\Omega$, where $\vec{e}_\Omega = \vec{\Omega}/\Omega$ and we neglect any terms quadratic in $\vec{\mu}_{ji}$. For simplicity we suppose that

$$\vec{N}_{ij} = -I_j\left(\alpha_{ij}\mu_{ij}^{||}\vec{e}_\Omega + \beta_{ij}\vec{\mu}_{ij}^{\perp} + \gamma_{ij}[\vec{e}_\Omega \times \vec{\mu}_{ij}^{\perp}]\right) \tag{8}$$

where $\alpha_{ij}, \beta_{ij}, \gamma_{ij}$ are some constants and we introduce $A^{||} = \vec{A}\cdot\vec{e}_\Omega$, $\vec{A}^{\perp} = \vec{A} - A^{||}\vec{e}_\Omega$ for any value $\vec{A}$.

In order to calculate the magnetospheric torque $\vec{K}_{ext}$ acting on the crust we use the model proposed in [13]. It is assumed that neutron star is braking simultaneously by both magnitodipolar and current losses. Hence

$$\vec{K}_{ext} = -\frac{I_{tot}}{\tau_0}\left(\vec{e}_\Omega - (1-\alpha)\cos\chi\,\vec{e}_m - R_{eff}[\vec{e}_\Omega \times \vec{e}_m]\right) \tag{9}$$

where $\vec{m} = m\vec{e}_m$ is the dipolar magnetic moment of neutron star, χ is the inclination angle (angle between $\vec{e}_\Omega$ and $\vec{e}_m$, see fig.3), $\tau_0 = \frac{3}{2}\frac{c^3}{m^2\Omega^3}I_{tot}$, the coefficient R_{eff} is related to inertia of the magnetic field [14, 15]. In this paper we accept that $R_{eff} = \frac{9}{10}\frac{c}{\Omega r_{ns}} \sim 5\cdot 10^3\left(\frac{P}{1s}\right)$ [16], where

r_{ns} is neutron star radius. The coefficient α is related to the value of the current flowing through the pulsar tubes. In this paper we assume that there are only "inner gaps" with free electron emission from neutron star surface in pulsar tube. Hence the current depends on the value of surface small-scale field, see fig. 1. The value of α averaged over precession angle ϕ_Ω, see fig 3, $<\alpha>(\chi) = \frac{1}{2\pi}\int_0^{2\pi}\alpha(\chi,\phi_\Omega)d\phi_\Omega$ calculated in [13] is shown in fig 2. Here $\nu = B_{sc}/B_{dip}$, B_{sc} is induction of small-scale magnetic field, $B_{dip} = 2m/r_{ns}^3$ is induction dipolar field on magnetic pole of neutron star.

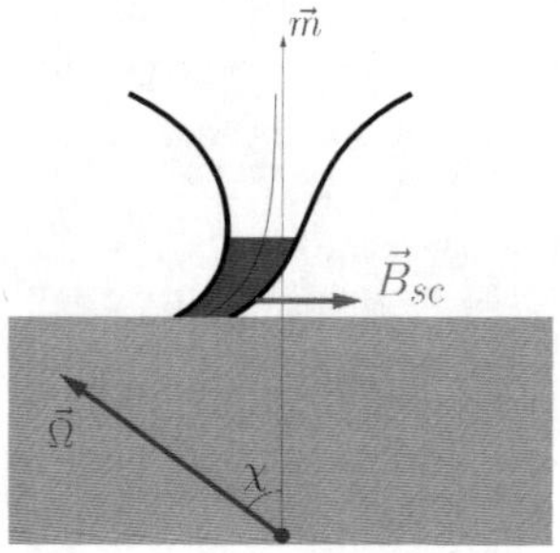

Figure 1. It is shown the direction of small scale magnetic field $\vec{B}_{sc}$ and the direction of dipolar magnetic moment $\vec{m}$ in model considered by [13]. "Inner gap" is shown by brown area, neutron star is shown by grey area.

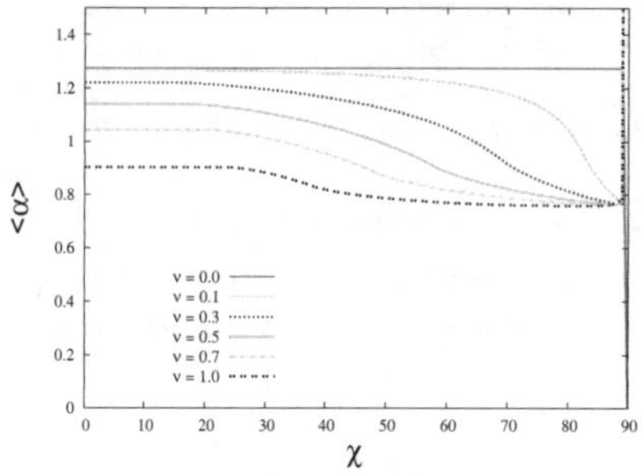

Figure 2. The dependence of the parameter $<\alpha>$ on the inclination angle χ taken from [13]

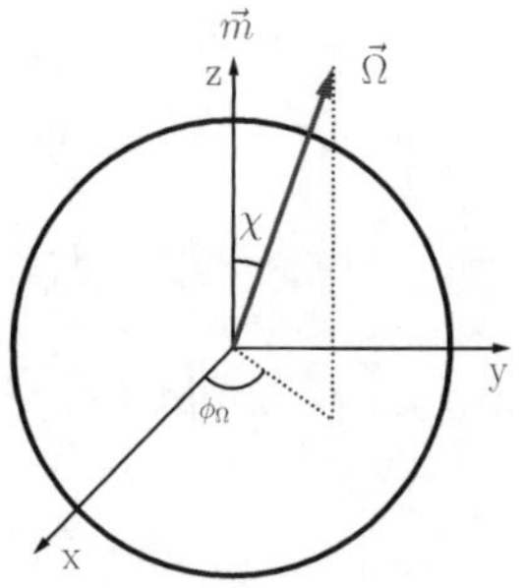

Figure 3. Definition of angles χ and ϕ_Ω.

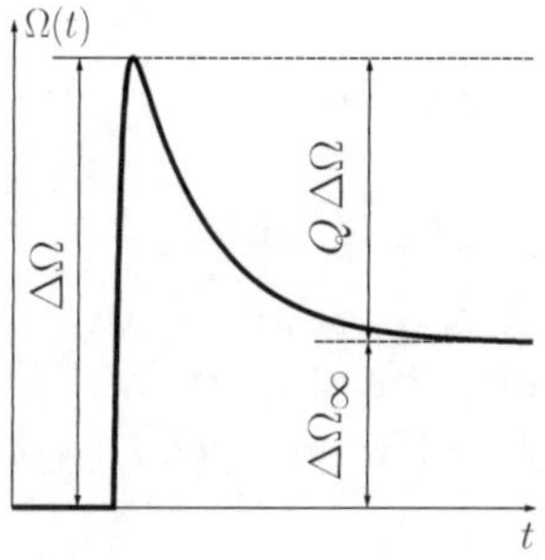

Figure 4. The evolution of angular velocity $\Omega(t)$ during quasi-glitch event. The pulsar braking is neglected.

3. Quasi-glitch events

Taking into account (5) and (6) the equation (7) can be rewritten as

$$\dot{\omega}_g^{||} = 0 \quad \text{and} \quad \frac{d}{dt}(\vec{\omega}_g^{\perp}) = -\frac{\omega_g^{||}}{\Omega}\left(\vec{S}_{ext}^{\perp} + \vec{R}_{gc}^{\perp} + \vec{R}_{rc}^{\perp}\right) - \omega_g^{||}\left[\vec{e}_\Omega \times \vec{\mu}_{cg}^{\perp}\right] \tag{10}$$

Hence, during precession the direction of $\vec{\omega}_g$ is adjusted so that pinned superfluid vortices will be directed almost along $\vec{e}_\Omega$. But its value ω_g does not change at all and due to pulsar braking the difference between ω_g and Ω will grow. So at some moment the glitch must occur in g-component. Assume that during the glitch the angular momentum $\Delta\vec{L}_g = \Delta L_g \vec{e}_\Omega$ is trasfered from superfluid part to the normal part of g-component. If we neglect pulsar braking term $\vec{S}_{ext}$ and assume that before glitch neutron star rotates as a rigid body then the solution of (4)-(7) gives us, see fig. 4

$$\Omega(t) = \Delta\Omega\left(1 - e^{-p_+ t} - Q(1 - e^{-p_- t})\right) \quad \text{where } \Delta\Omega = \frac{\Delta\Omega_\infty}{1-Q} \text{ and } \Delta\Omega_\infty = \frac{\Delta L_g}{I_{tot}} \tag{11}$$

the coefficients p_+ and p_- ($p_+ > p_-$) are roots of the equation

$$\begin{aligned} p^2 &- (\alpha_{cg} + \alpha_{rg} + \alpha_{cr} + \alpha_{gr} + \alpha_{gc} + \alpha_{rc})\, p + \\ &+ (\alpha_{gc} + \alpha_{rg} + \alpha_{cg}) \cdot (\alpha_{cr} + \alpha_{gr} + \alpha_{rc}) + (\alpha_{rc} - \alpha_{rg}) \cdot (\alpha_{gr} - \alpha_{gc}) = 0 \end{aligned} \tag{12}$$

and $Q = (I_{tot}\alpha_{cg} - I_c p_+)/(I_{tot}\alpha_{cg} - I_c p_-)$. In the case of $\alpha_{cg} \gg \left(1 + \frac{I_c}{I_r}\right)\alpha_{rc}, \left(1 + \frac{I_g}{I_r}\right)\alpha_{rg}$ we have $p_+ \approx \left(1 + \frac{I_g}{I_c}\right)\alpha_{cg}$, $p_- \approx \frac{I_{tot}}{I_c + I_g}(\alpha_{cr} + \alpha_{gr})$ and $1 - Q \approx \frac{I_c + I_g}{I_{tot}}$. If we try to relate these quasi-glitch events to observed glitches then $1/p_+$ determine the glitch growth time and $1/p_-$ is related to glitch relaxation time we receive restrictions $1/p_+ \le 1$ min [10, 2] and $1/p_- \sim 1 - 10^2$ days [1]. Unfortunately, in our model $1 - Q \sim 10^{-2} - 10^{-1}$ that may be not so bad for glitches in some pulsar like Grab $Q \ge 0.8$ [2] and J0205+6446 $Q \approx 0.77$ [11], but obviously contradicts glitches in majority of pulsars $Q \ll 1$ [11, 12]. In particular, our model does not describe glitches in Vela pulsar $Q \le 0.2$ [2].

4. Quasistatic approximation

Firstly we take into account that $\tau_{rel} \ll \tau_0$, where $\tau_{rel} \sim \max(1/\alpha_{ij}, 1/\beta_{ij}, 1/\gamma_{ij}) \sim 1 - 10^7 s$ is relaxation time [2]. Hence we can consider the rotation of neutron star under small slowly varying torque $\vec{K}_{ext}$. In this case it's possible to neglect terms $\dot{\mu}_{ij}$ in (4)-(7). And consequently we have

$$\dot{\Omega} = -\frac{\Omega}{\tau_0}\left(\sin^2\chi + \alpha\cos^2\chi\right) \tag{13}$$

$$\dot{\chi} = -\frac{I_{tot}}{I_c}\frac{1}{\tau_0}\sin\chi\,\cos\chi \cdot (B\,(1-\alpha) - \Gamma\,R_{eff}) \tag{14}$$

$$\dot{\phi}_\Omega = -\frac{I_{tot}}{I_c}\frac{1}{\tau_0}\cos\chi \cdot (B\,R_{eff} + \Gamma\,(1-\alpha)) \tag{15}$$

where coefficients B and Γ are defined as

$$B + i\Gamma = \frac{\Omega(\Omega - i(\xi_{cr} + \xi_{gr}))}{\Omega^2 - i\Omega\,(\xi_{cr} + \xi_{gr} + \xi_{rc} + \xi_{gc}) - (\xi_{gr}\xi_{gc} + \xi_{gr}\xi_{rc} + \xi_{cr}\xi_{gc})} \tag{16}$$

where $\xi_{pq} = \beta_{pq} + i\gamma_{pq}$. In the case of weak viscosity $|\xi_{pq}| \ll \Omega$ we have $B \approx 1 - \frac{\gamma_{gc} + \gamma_{rc}}{\Omega}$ and $\Gamma \approx \frac{\beta_{gc} + \beta_{rc}}{\Omega}$. It's worth to note that equation (13) has the same form as in the case of absolutely rigid star. According to (15) as long as inclination angle χ differs from 0° or 90° the star will precess with period $T_p \sim (\tau_0/R_{eff}) \cdot (I_c/I_{tot}) \sim 10 - 10^4$ years. We suppose that the period of precession $T_p \ll (I_c/I_{tot}) \cdot \tau_0 \ll \tau_0$ hence we can average (13) and (14) over precession.

$$\frac{d\chi}{dP} = -\frac{1}{P} \cdot \frac{I_{tot}}{I_c} \cdot \sin\chi\cos\chi \cdot \frac{B\,(1 - <\alpha>) - \Gamma\,R_{eff}}{\sin^2\chi + <\alpha>\cos^2\chi} \tag{17}$$

We consider only one model of components interaction, when the crust and g-component interact with r-component like a normal matter with superfluid and there are normal viscosity between the crust and g-component [9, 2]

$$\beta_{cr} = \beta_{gr} = \Omega\frac{\sigma}{1+\sigma^2}, \quad \alpha_{cr} = \alpha_{gr} = 2\beta_{gc}, \quad \gamma_{cr} = \gamma_{gr} = -\sigma\beta_{gc}, \quad \beta_{cg} = \alpha_{cg} \text{ and } \gamma_{cg} = 0 \quad (18)$$

The evolution of the inclination angle χ for different initial inclination angle χ_{start} in the case of $\sigma = 10^{-6}$, $\alpha_{cg} = 10^{-1}s^{-1}$ and $\nu = 0.5$ is shown in fig 5. We can see that in the majority of cases inclination angle became equal to its equilibrium value $B\,(1- < \alpha >) - \Gamma\, R_{eff} \approx 0$ very rapidly and all subsequent evolution is related with the slow changing of R_{eff} value due to pulsar braking. The evolution of the inclination angle χ for various values ν in the case of starting angle $\chi_{start} = 45°$ and $\sigma = 10^{-6}$, $\alpha_{cg} = 10^{-1}s^{-1}$ is shown in fig 6. Left graph corresponds to $I_c = 10^{-1}I_{tot}$, $I_g = 10^{-5}I_{tot}$ right graph to $I_c = 10^{-2}I_{tot}$, $I_g = 10^{-3}I_{tot}$. Observed values of inclination angles β_2 taken from [17] are shown by red dots ($C > 0$) and blue dots ($C < 0$). Pulsar parametres are taken from [18]. It seems that the case of small g-component is in the better agreement with observations.

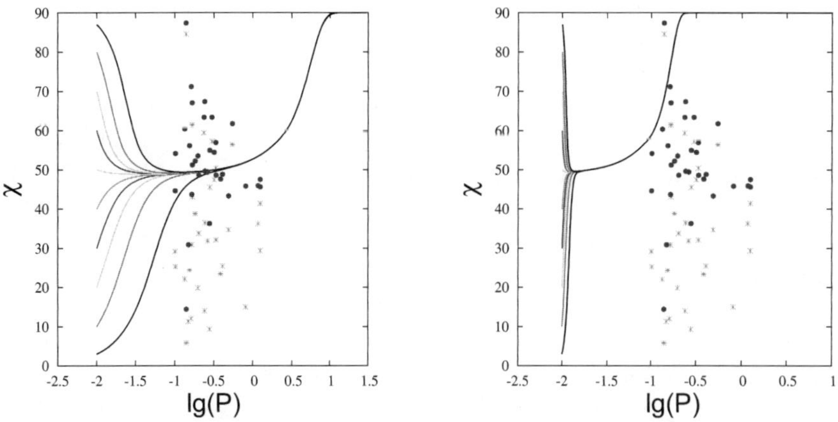

Figure 5. The evolution of inclination angle χ in case of $\sigma = 10^{-6}$, $\alpha_{cg} = 10^{-1}s^{-1}$, $\nu = 0.5$ for different initial inclination angles χ_{start}. Left graph corresponds to $I_c = 10^{-1}I_{tot}$, $I_g = 10^{-5}I_{tot}$, right corresponds to $I_c = 10^{-2}I_{tot}$, $I_g = 10^{-3}I_{tot}$. The observed values of inclination angles β_2 taken from [17] are shown by red dots ($C > 0$) and blue dots ($C < 0$). Pulsar parametres are taken from [18].

5. Conclusion

We consider a model, proposed in [9], that allows simultaneously the long-period precession and quasi-glitch events with taking into account the influence of the small-scale magnetic field on pulsar braking. For simplicity we consider only the case of axial symmetric precession and don't take into account that the presence of the small scale magnetic field makes precession triaxial [15]. The main problem of this model is the nature of g-component. It may be related with existence of rigid star core where superfluid is pinned [19] or with long living tangle of closed fluxoids flowing inside liquid core on which superfluid vortices are pinned, compare with [20].

Acknowledgments

K.Y.Kraav and D.P.Barsukov thank O.A.Goglichidze and D.M.Sedrakian for usefull discussion which leads to writting this paper, A.N.Biryukov for usefull discussion of about pulsar precession

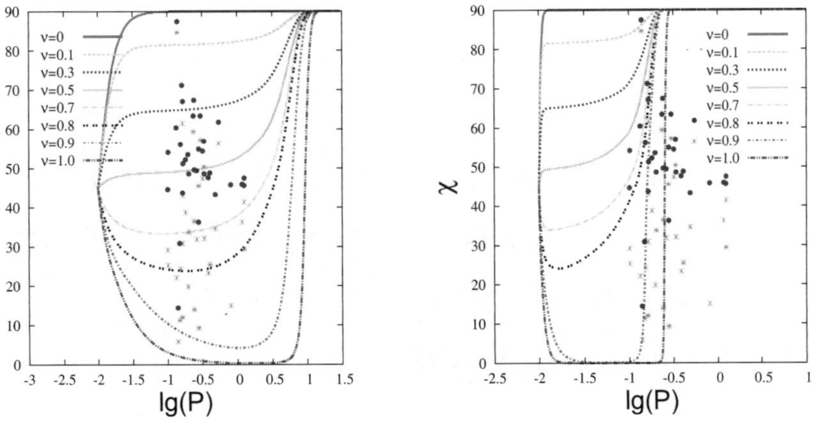

Figure 6. The same as in fig. 5 but for initial angle $\chi_{start} = 45°$ and different values for parameter ν.

and V.S.Beskin, I.F.Malov, E.B.Nikitina, A.N.Kazantsev, V.A. Urpin and A.I.Chugunov for help and useful discussions.

References

[1] F. D'Alessandro 1996 *Astrophysics and Space Science* **246** 73
[2] B.Haskell, A.Melatos 2015 *International Journal of Modern Physics* D **24** id. 1530008
[3] G.Ashton, D.I.Jones, R.Prix 2016 *MNRAS* **458** 881
[4] M.Kerr, G.Hobbs, S.Johnston, R.M.Shannon 2016 *MNRAS* **455** 1845
[5] L.Arzamasskiy, A.Philippov, A.Tchekhovskoy 2015 *MNRAS* **453** 3540
[6] V.M.Malofeev, O.I.Malov, S.V.Logvinenko, D.A.Teplykh 2012 *Odessa Astronomical Publications* **25** 194
[7] A.Biryukov, G.Beskin, S.Karpov 2012 *MNRAS* **420** 103
[8] B.Link 2006 *A& A* **458** 881
[9] A.Sedrakian, I.Wasserman, J.M. Cordes 1999 *ApJ.* **524** 341.
[10] R.G.Dodson, P.M.McCulloch, D.R.Lewis 2002 *ApJ.* **564** L85
[11] E.Gurgercinoglu 2017 *MNRAS* **469** 2313
[12] M.Yu, R.N.Manchester, G.Hobbs et. al 2013 *MNRAS* **429** 688
[13] D.P.Barsukov, P.I.Polyakova, A.I.Tsygan 2009 *Astronomy Reports* **53** 1146
[14] V.S.Beskin, A.A. Zheltoukhov 2014 *Physics Uspekhi* **57** 799
[15] O.A.Goglichidze, D.P.Barsukov, A.I.Tsygan 2015 *MNRAS* **451** 2564
[16] A.Melatos 2000 *MNRAS* **313** 217
[17] I.F.Malov, E.B.Nikitina 2011 *Astronomy Reports* **55** 19
[18] R.N.Manchester et al 2005 *Astron. J.* **129** 1993
http://www.atnf.csiro.au/research/pulsar/psrcat
[19] T.Takatsuka, R.Tamagaki 1989 *Progress of Theoretical Physics* **82** 945
[20] K.Glampedakis, P.D.Lasky 2015 *MNRAS* **450** 1638

IOP Conf. Series: Journal of Physics: Conf. Series **929** (2017) 012008 doi:10.1088/1742-6596/929/1/012008

The angular and energy distributions of neutrons, gamma- and x-rays produced in solar flares. Related changes in the isotopic composition of the photosphere and the solar wind

G I Vasiliev[1] and M E Chakchurina[1,2]

[1] Ioffe Institute, St. Petersburg, Russia
[2] Peter the Great St. Petersburg State Polytechnic University, St. Petersburg, Russia

E-mail: maria.chakchurina@yandex.ru

Abstract. Nuclear interactions of protons and helium ions accelerated in solar flares with solar atmosphere are simulated using Geant4 toolkit.The long term average spectrum of solar cosmic rays is taken from measurements of cosmogenic isotopes production in lunar soil. Solar atmospheric profiles of ^{6}Li, ^{7}Li and ^{14}C generation are calculated. These isotopes outflow into interplanetary space with coronal mass ejections and following implantation into lunar rocks are considered to explain anomaly isotopic composition of thin lunar surface layer. Also, generated energy spectra and angular distributions of neutrons, gamma- and x-rays in space are presented for different spectra of solar flares.

1. Introduction

Accelerated during solar flares, protons and helium isotopes enter the photosphere and interact with the ambient material. In nuclear reactions stable and radioactive nuclides appear and excited nuclei emit characteristic gamma lines. With a time delay, there emefge lines formed by the decay of radionuclides. Part of the gamma-rays (scattered and noninteracting) leaves the solar atmosphere. These rays and gamma- and x-rays from the electron bremsstrahlung form a continuous spectrum with characteristic lines. Intensity of these lines depends both on flare power, and on the thickness of solar atmosphere, passed by gamma quanta. The greater the thickness traversed by the particle, the more likely it is to be scattered or absorbed. In it's turn, the distance, penetrated by the registered particle, depends on the angle of observation and the depth of its generation. Investigation of neutron characteristics provides an additional tool for understanding the processes which take place during solar flares.

2. Method

Using the Geant4 software toolkit, designed for simulating the passage of particles through matter with Monte Carlo approach, we have performed modeling of interaction of protons and helium isotopes accelerated in solar flares with solar atmosphere. The initial chemical and isotopic composition of the photosphere are taken from [1]. The energy spectra of the accelerated particles are power laws in kinetic energy per nucleon with spectral index γ. The

IOP Conf. Series: Journal of Physics: Conf. Series **929** (2017) 012008 doi:10.1088/1742-6596/929/1/012008

depth of isotopes formation, the energy and angles of particles leaving the photosphere were fixed in the simulation.

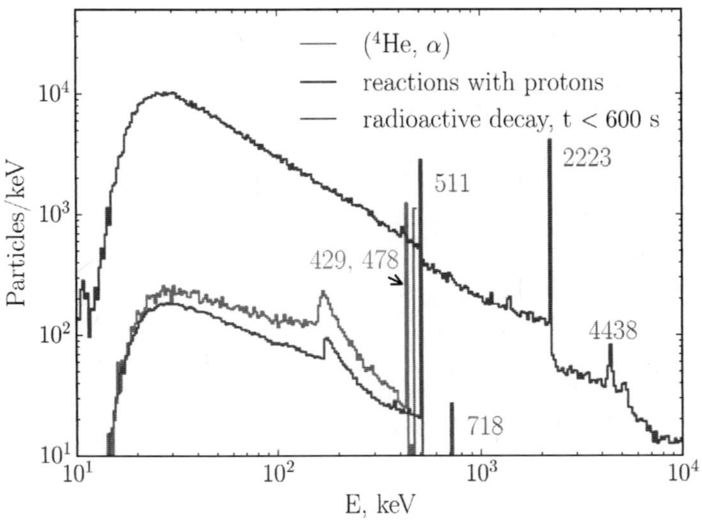

Figure 1. Gamma- and x-rays spectra.

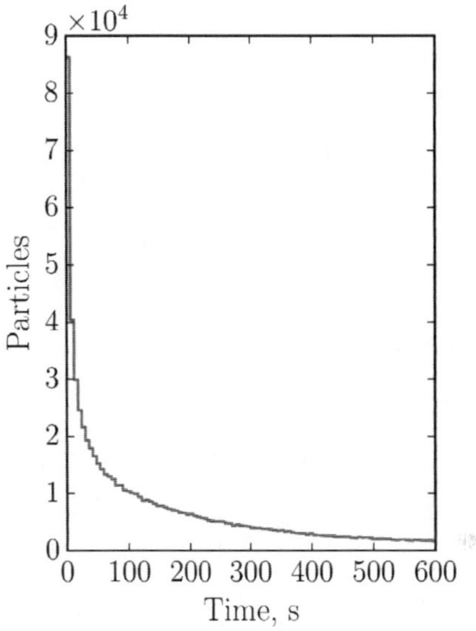

Figure 2. Temporal distribution of gamma-rays and x-rays resulting from radioactive decays.

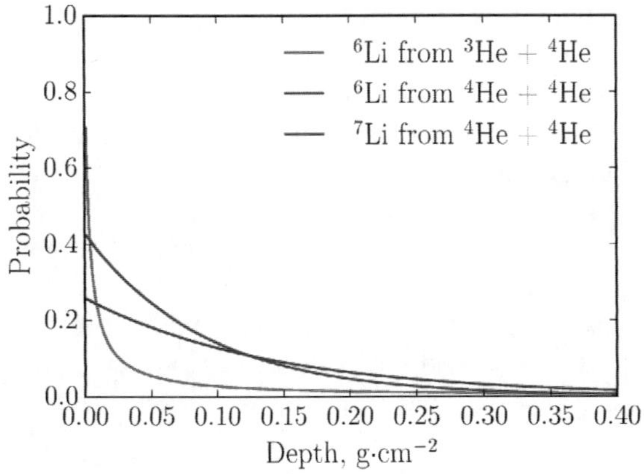

Figure 3. The depth distribution of Li yield.

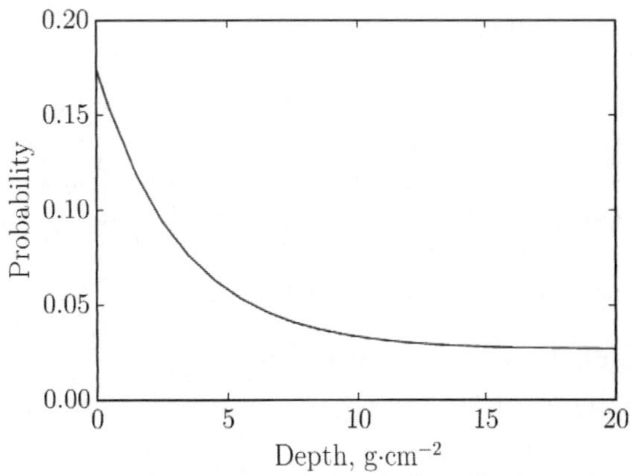

Figure 4. The depth distribution of ^{14}C yield.

3. Results

Figure 1 reports spectra of prompt gamma and x-rays, which arise in direct nuclear reactions of $6.6 \cdot 10^8$ protons ($\gamma = 3$, $E > 30$ MeV) with the photospheric substance (blue curve), and of the gamma quanta which are formed as a result of disintegration of short-lived radionuclides

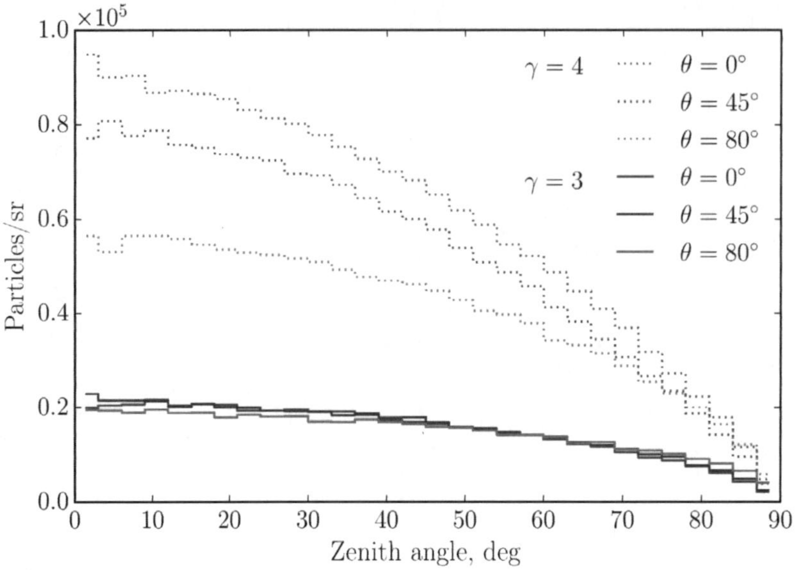

Figure 5. Angular distribution of gamma- and x-rays with energy exceeding 10 keV.

with decay times less then 10 minutes (green) and in the interaction of $1.8 \cdot 10^8$ alpha particles ($\gamma = 3$, $E > 30$ MeV/nucl) with ^{4}He which produces ^{7}Be and ^{7}Li in the excited state (red).

We have paid special attention to ^{7}Be, ^{6}Li, ^{7}Li and ^{14}C isotopes formation with their subsequent involvement into solar wind outflow. The abnormal ratios (relative to solar prevalence) of the implanted isotopes of solar wind, which were observed in samples of lunar soil at depths of several tens of nm [2,3], can be accounted for changes in isotopic composition of solar atmosphere. There are two possible exit mechanisms which draw isotopes into interplanetary space along with solar wind. In the first case isotopes produced in a thin layer (0.01–1 $\mathrm{g \cdot cm^{-2}}$) of photosphere are rapidly released with coronal mass ejections. In the second case, long-living isotopes are stored in the solar convection zone and escape later on with the solar wind. The depth distribution of ^{6}Li yield in the interaction of accelerated nuclei ^{3}He with ^{4}He at a kinetic energy of more than 7.14 MeV (threshold) and ^{7}Li in the interaction of accelerated alpha-particles with ^{4}He at a kinetic energy of more than 35 MeV for spectral index $\gamma = 3$ are shown in figure 3. Yield of isotopes in a layer with thickness of 0.1 $\mathrm{g \cdot cm^{-2}}$ per one primary particle is $0.5 \cdot 10^{-3}$, $1.2 \cdot 10^{-3}$ and $0.2 \cdot 10^{-3}$ respectively. The depth distribution of ^{14}C yield in the interactions of accelerated protons is shown in figure 4. The yield of isotopes per one primary proton is $4.6 \cdot 10^{-8}$.

The angular distribution (relative to the vertical) of gamma-, x-rays and neutrons leaving the photosphere are shown in the figures 5 and 6. Figures show that as the spectral index γ increases, the angular distribution of gamma rays leaving the photosphere depends less on proton's angles of incidence. It indicates that protons interact in a thin layer of matter, where they do not have enough time to be stopped by electromagnetic interactions. The obtained angular distributions for individual gamma lines are similar to the distributions for gamma particles of all energies.

The analysis of depth distributions of cosmogenic isotopes in the lunar soil makes it possible to determine the average spectrum of solar cosmic rays at very long time intervals [2]. After recalculation of the flux of solar cosmic rays on the Earth's orbit and assuming that only about 1% of the accelerated ions in the solar flare flow out into interplanetary space, it is possible to

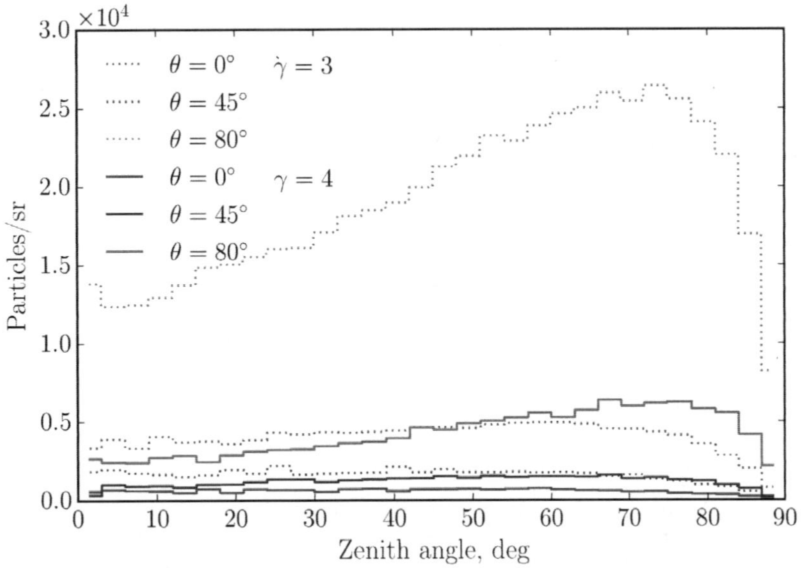

Figure 6. Angular distribution of neutrons with energies exceeding 1 MeV.

obtain an average particle flux into the photosphere. Assuming that the number of accelerated alpha-particles is about 2% of the number of protons, we can determine the lithium production rate. The resulting value shows good agreement with the data obtained in the lunar soil studies.

4. Conclusion

It may be concluded that during solar flares with comparable numbers of accelerated ions of helium and protons observations of lines from excited isotopes ^{7}Be and ^{7}Li are possible, which is confirmed by the experiment [4].

The results of our simulation show that observed enrichment of ^{14}C and ^{6}Li in lunar soil can be explained by generation of these isotopes in solar flares and their subsequent release into interplanetary space with coronal mass ejections. The accompanying gamma radiation can be used for experimental verification of validity of the model.

The involvement of experimental data on neutrons and protons produced in decay of neutrons in Earth's orbit gives an additional survey method not only for the processes at the Sun's surface, but in the solar magnetosphere as well.

References

[1] Asplund M et al. 2009 *Ann. Rev. Astron. Astrophys* **47** 481
[2] Chaussidon M, Robert 1999 *F. Nature* **402** 270
[3] Jull T et al. 1998 *Geochimica et Cosmochimica Acta* **62** 3025
[4] Kotov Yu D et al. 1996 *Astrophysical J.* **473** 514

IOP Conf. Series: Journal of Physics: Conf. Series **929** (2017) 012009 doi:10.1088/1742-6596/929/1/012009

Numerical studies of instability of generalized polytropic models of stellar disks

A. A. Smirnov [1,2], **N. Ya. Sotnikova**[1]

[1] St. Petesburg State University, Universitetskaya emb. 7-9, 199034, St. Petesburg, Russia
[2] Pulkovo observatory, Pulkovskoye chaussee 65, 196140, St. Petesburg, Russia

E-mail: zeleniikot@gmail.com

Abstract. The distribution function for generalized polytropic models has been used to construct a series of numerical models of anisotropic disks. It has been shown with the help of simulation that such systems are unstable with respect to bar formation at any degree of radial elongation of star orbits. The result is completely at variance with the conclusions of earlier works, where similar models were studied. The pattern speeds and amplitudes of the forming bar were found, and the initial distributions of orbit precession rates were calculated. Such systems have been shown to fulfill all conditions for the onset of radial orbit instability, responsible for the formation of a slow bar.

1. Introduction

About 50% of spiral galaxies are known by the observation data to have a typical elongated large-scale structure, the bar. The bar properties are dependent on the Hubble type of the host galaxy [3]. As a rule, the surface brightness profile of the bar is planar in galaxies of earlier types and exponential in those of later ones. The bars are typically highly prolate. For instance, the bar in the Galaxy has the ratio of the major to minor semi-axes of about 3:1.

Some interrelated issues are still at the focus of astronomers' attention. What are the mechanisms of bar formation? Why do they have the properties (pattern speed, size, density profile), which are observed?

As to the pattern speed, bars are divided into two types, fast and slow. The pattern speed of a slow bar is on the order of the average precession rate of star orbits in the disk. The pattern speed of a fast bar exceeds the maximal precession rate of orbits. Fast bars are intensively studied in a lot of theoretical and numerical experimental works. As a rule, the bars observed in galaxies are fast ones [11]. As to slow bars, which may form in the central regions of stellar disks, there are practically no works on the topic, either theoretical or experimental.

The mechanism of slow bar formation was suggested by Lynden-Bell [7] to be the radial orbit instability (ROI). In Lynden-Bell's approach, orbits with small precession rates are considered, i. e. those having precession rate much less than the angular frequency of star revolution, $\Omega_{\rm pr} \ll \Omega$. In this case, an orbit can be thought as a "spoke" with a small precession rate. By virtue of Poincaré's theorem on invariants, the integral to generalized moments and coordinates

$$2J_f = \frac{1}{2\pi}\int \vec{p}\cdot d\vec{q}, \qquad (1)$$

will conserve; here J_f is the value of the adiabatic invariant over a single radial oscillation of a star. Since a star completes two radial oscillations during a revolution around the galaxy center, the integral in Eq. (1) is twice the value of the adiabatic invariant. Lynden-Bell [7] demonstrated that such orbits would coalesce and form a bar in the case when adiabatic invariant J_f has a specific dependence on angular momentum L of a star:

$$\left(\frac{\partial\Omega_{\rm pr}}{\partial L}\right)_{J_f} > 0\,, \tag{2}$$

An efficient method of studiyng bar-like instabilities is simulations, allowing one to investigate various nonlinear effects, which are hard to take into account in a theory. Nonetheless, numerical simulations face several problems. The first one is the limited computing power. The number of stars in actual galaxies is $10^9 - 10^{11}$. Calculation of the evolution of such number of particles on a modern computer would take years. For this reason, the system is approximated with a smaller number of particles, $10^5 - 10^7$. The second problem growing out of the first one is the decrease of the system relaxation time with falling number of particles. After Chandrasekhar [14], the relaxation time is the time it takes a star to essentially change its characteristics (energy or trajectory). In the interval equal to the relaxation time, the system "forgets" the initial conditions. Such an effect arises due to discreteness of the matter distribution in a galaxy. Relaxation time $T_{\rm rel}$ for three-dimensional systems grows with the number of particles, N, in the system [3]:

$$T_{\rm rel} \propto \frac{N}{\log(N)}. \tag{3}$$

The estimate Eq. (3) proves to be quite incorrect for planar systems. In two-dimensional models, the relaxation time is of the order of the crossing time and is independent of the number of particles because relaxation in such systems is dominated by close encounters [10], whereas in three-dimensional systems, distant encounters contribute to particle trajectory deviations. Rybicki [10] demonstrated that the dependence of the relaxation time on the number of particles in two-dimensional systems arises if one invokes the potential smoothing procedure:

$$\Phi(r) \propto \frac{1}{r} \longmapsto \Phi(r) \propto \frac{1}{\sqrt{r^2+\varepsilon^2}} \tag{4}$$

where ε is the smoothing parameter of the system. The smoothing of the two-dimensional model potential allows to consider a two-dimensional flat disk as a three-dimensional disk of a finite small thickness ε. In this case the relaxation time does depend on the number of particles [10] [4]:

$$T_{\rm rel} \propto N. \tag{5}$$

The relaxation time in galaxies is greater than the Hubble time. In numerical simulations using the number of particles smaller by orders of magnitude than those in real galaxies, one has to find an estimate of the relaxation time and use it to set the integration time of the model.

The present work studied the equilibrium two-dimensional generalized polytropic models of stellar disks with the help of simulation. The models are described in detail in Section 2. Here we will focus on their most important property. Such models have constant degree of anisotropy, β, in the entire disk

$$\beta = \frac{\overline{V_\varphi^2}(r)}{\overline{V_r^2}(r)} - 1 \tag{6}$$

where $\overline{V_\varphi^2}(r)$ and $\overline{V_r^2}(r)$ are the average squared velocities of stars at given radius r in the azimuthal and radial directions, respectively; $\beta = 0$ corresponds to the isotropic velocity distribution, and $\beta = -1$ — to strictly radial orbits. Parameter β is chosen within minus one to zero at the initial stage of the model construction. Earlier, such models were considered by [9]. The authors found the instability boundary for $\beta \approx -0.5$. The present work revises the result completely. With models of high spatial resolution as an example, such systems are shown to be unstable at any degree of anisotropy of orbits, β. To test the ROI theory, we have evaluated the initial distributions of particle precession rates and shown that all conditions for the onset of the ROI are fulfilled in such systems.

2. Construction of equilibrium generalized polytropic model of a stellar disk

Generalized polytropic models are defined by the following phase density profile:

$$f(E,L) = A\,\Theta(L_z)L^{\beta}(E_0 - E)^{\alpha}. \tag{7}$$

Here L is the angular momentum of a star ($L = |\vec{L}| = |\vec{r} \times \vec{V}|$), $\Theta(L_z)$ is the Heaviside function, E is the total energy of a star ($E = \Phi(r) + V^2/2$), $E_0 = \Phi(R)$ is the potential energy at the system boundary, R is the system size, V is the star velocity, $r = |\vec{r}|$ is the star radius vector magnitude, α, β are dimensionless parameters, A is the proportionality factor, dependent on the total system mass.

The model construction proceeded in several stages. It started with finding gravitational potential $\Phi(r)$ to match the phase density profile Eq. (7) and involved one-parameter family of models where $\beta = -\alpha/2$ (for details see [9]). The model sizes were derived from the condition of the potential function being monotone. Then, particles were distributed in the coordinate and velocity spaces by methods of acceptance-rejection and the inverse transform. The initial coordinates and velocities of particles were chosen so that the model would remain planar in the course of simulations. Here, all particles have the same mass $m = M_{\text{tot}}/N$ (where M_{tot} is the total system mass, N is the number of particles) and rotate counterclockwise. To simplify the calculations, the system of units was used, where $G = 1, A \cdot C \cdot G = 1, \Phi(R) - \Phi(0) = 1$. Table 1 gives the resulting model sizes for different β. Fig. 1 shows the view of generalized polytropic models in the xy plane for different degrees of anisotropy, β, and $N = 10^5$. Fig. 2 shows the mass density profiles for the corresponding models, which were evaluated by two methods: velocity space integration of phase density Eq. (8) for the corresponding β [9]

$$\Sigma(r) = ACr^{\beta}(E_0 - \Phi(r))^{\beta/2+\alpha+1}\,, \tag{8}$$

and summation of particle masses over concentric layers in numerical implementations of models.

Table 1. Model sizes R at different parameters β.

β	0.0	−0.1	−0.2	−0.3	−0.4	−0.5
R	$4.38 \cdot 10^{-1}$	$3.51 \cdot 10^{-1}$	$2.63 \cdot 10^{-1}$	$1.78 \cdot 10^{-1}$	$1.03 \cdot 10^{-1}$	$4.63 \cdot 10^{-2}$

3. Solution of the N-body problem

For simulations we used the $NEMO$ package [13] and its extension gyrfalcON [6]. The $NEMO$ package allows numerical solution of particle motion equations and fast evaluation of gravitational forces, acting between particles. "Fast" in this case means that the calculation of forces in $NEMO$ employs an algorithm with computing effort less than that of direct summation. Namely, the treecode algorithm [2] has the complexity $O(NlogN)$ and its combination with fast multipole expansion [5] has the complexity $O(N)$. Smoothing parameter ε was chosen in the optimal manner [8]. The integration step was adjusted to the chosen smoothing parameter [12].

$$dt = \frac{\varepsilon_{\text{opt}}}{5v_{r_{\frac{1}{2}}}}, \tag{9}$$

where $v_{r_{\frac{1}{2}}}$ is the escape velocity from the radius of the circle, containing a half of the model mass.

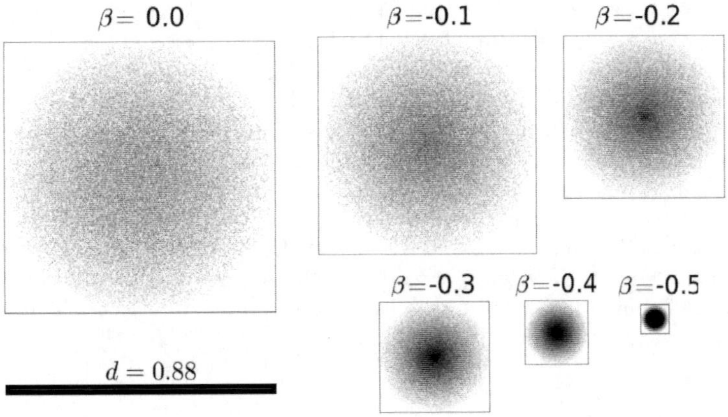

Figure 1. View of generalized polytropic models in the xy plane for different degrees of anisotropy, β, and $N = 10^5$.

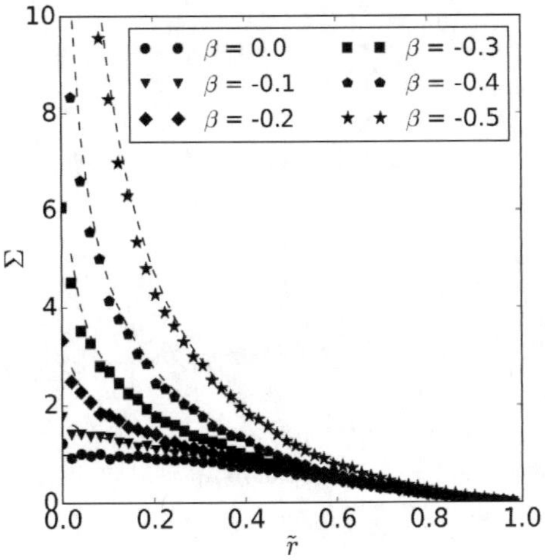

Figure 2. Disk surface density Eq. (8) for $\beta = 0.0, 0.1, 0.2, 0.3, 0.4, 0.5$ (dash lines) as a function of distance, normalized to the system size, $\tilde{r} = r/R$. Different symbols plot the density values, found by summation of particle masses over radial system layers at $N = 10^5$.

4. Bar registration

A bar was registered with the help of the Fourier transform of the density profile:

$$\bar{\Sigma}(p, m, t) = \frac{1}{N} \sum_{j=1}^{N} \exp\left\{i\left[p \ln(r_j) + m\varphi_j\right]\right\}, \tag{10}$$

where m is the azimuthal wavenumber, and p is the radial wavenumber. The bar is identified as the harmonic whith azimuthal number $m = 2$ radial number $p = 0$. Position angle $\alpha(t)$ and bar pattern speed Ω_p were found from the following relationships:

$$\begin{cases} \tan\alpha(t) = \dfrac{\operatorname{Im}(\bar{\Sigma}(2,0,t))}{\operatorname{Re}(\bar{\Sigma}(2,0,t))}, \\ \sin\alpha(t) = \dfrac{\operatorname{Im}(\bar{\Sigma}(2,0,t))}{|\bar{\Sigma}(2,0,t))|}. \end{cases}, \ \Omega_p = \frac{1}{m}\frac{d\alpha(t)}{dt}. \tag{11}$$

5. Instability of generalized polytropic models of stellar disks

Work [9] dealt with similar models for $\beta = 0.4$ and $\beta = 0.6$. Basing on the Fourier analysis of the density, the authors concluded the stability of models at $\beta > 0.5$, whereas a bar formed at $beta0.5$. However, the authors used a small number of particles $N \approx 10^3$ and disregarded the possible numerical relaxation of the models. The present work evaluated the relaxation times of generalized polytropic models at different β (see Fig. 3), basing on the analysis of deviations of test particles in a fixed potential of the numerical implementation of the model [1]. It follows from our estimates that at the number of particles, considered in [9], various effects arising due to the discrete initial distribution manifest themselves even before the bar formation. Our calculations of the evolution of generalized polytropic models with great number of particles $N = 10^5$ and $N = 5 \cdot 10^5$ for different β have shown that the models prove to be unstable at any degree of orbit elongation, β, and the bar forms (see Fig. 4).

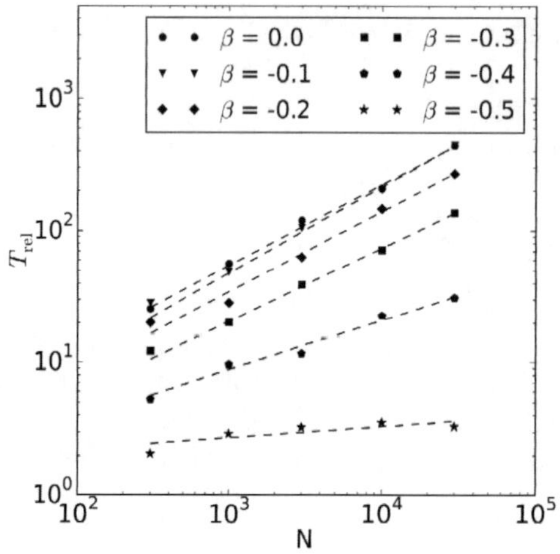

Figure 3. Relaxation time $T_{\rm rel}$ vs. the number of particles in the system, N, at different values of β, found by deviations of particles in the fixed initial potential of the numerical implementation of the model [1].

6. The initial set of orbits

The present work tested the ROI conditions for generalized polytropic models of stellar disks. The initial distributions of precession rate $\Omega_{\rm pr}$, adiabatic invariants J_f, orbit eccentricities e, and dependence $\Omega_{\rm pr}(L)$ were found from the analysis of particle motion in a fixed initial potential. Fig. 5 shows the ratio of the average particle precession and revolution rates for given eccentricities and the number of particles with given eccentricity. Even in the most isotropic case at $\beta = 0$, most orbits are highly elongated. In addition, the greater the orbit eccentricity, the better the fulfillment of condition of a small precession rate, which is necessary for the onset of the ROI. Fig. 6 plots precession rate isolines in "adiabatic invariant – angular momentum" coordinates. The direction of the precession rate growth corresponds exactly to the second ROI condition. Similar dependences are observed for all other considered models.

7. Conclusions

Numerical simulations of the evolution of generalized polytropic models of stellar disks with great number of particles reveal that the models are unstable with respect to the bar-like perturbation. At that, the initial distributions of particle precession rates fulfill all necessary conditions for the onset of the ROI, which were formulated by Lynden-Bell [7]. Namely, almost all particles prove to

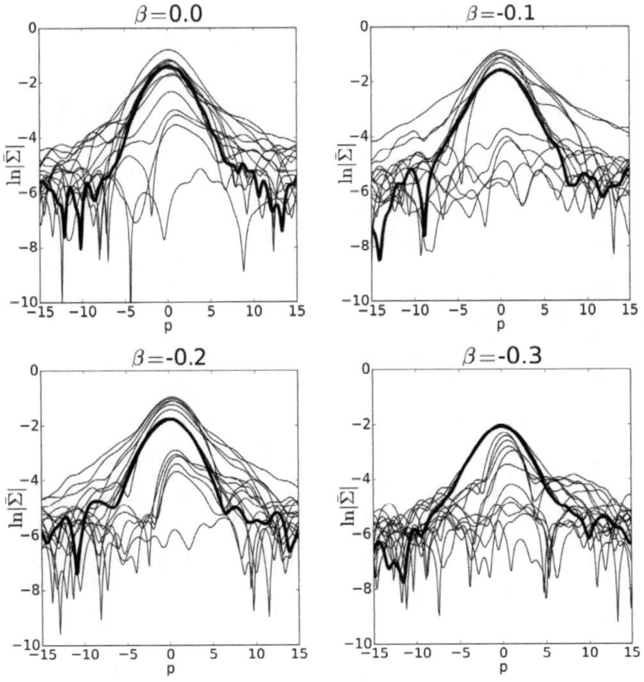

Figure 4. Evolution of the logarithm of amplitude magnitude of the Fourier transform of the disk surface density, $\ln|\bar{\Sigma}(p, m = 2, t)|$, for models with different β. Heavy lines are the profiles of the density Fourier transform, which correspond to the final time instances ($T = 250$). The maximum of the density Fourier transform shifts with time to $p = 0$, which corresponds to the bar formation.

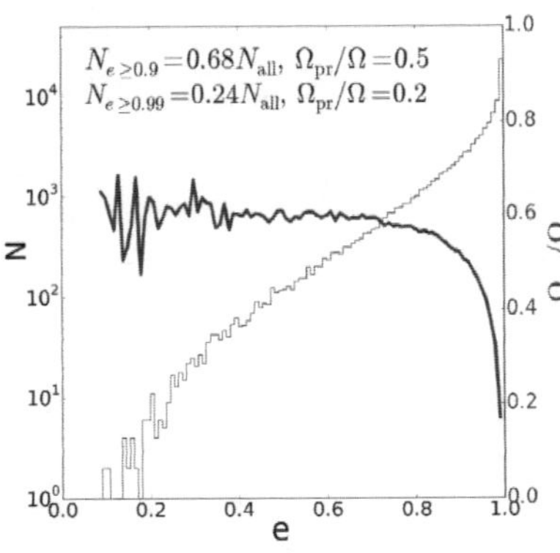

Figure 5. Average ratio of particle precession and revolution rates (heavy line) and the number of particles with given eccentricity (fine line) in each eccentricity layer with step $\Delta e = 0.01$ for the model with degree of anisotropy $\beta = 0$ and the number of particles $N = 10^5$.

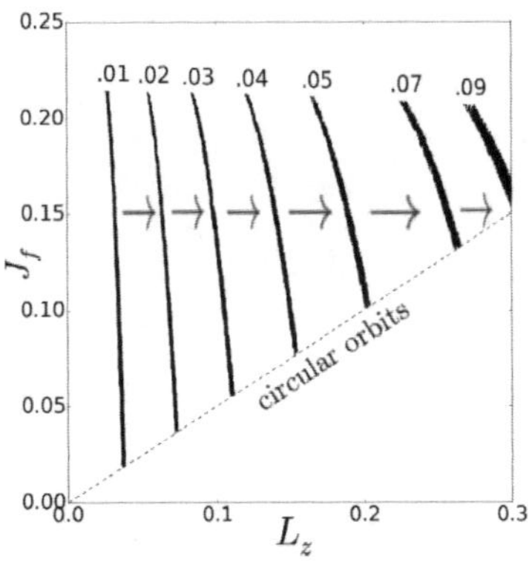

Figure 6. Precession rate isolines for particles at degree of anisotropy $\beta = 0$. The adiabatic invariant is laid off as ordinate, and angular momentum is laid off as abscissa. The dash line corresponds to circular orbits ($J_f = L/2$). Arrows mark the direction of precession rate growth at a fixed adiabatic invariant.

have small precession rates and the rates to be a monotone function of the angular momentum at a fixed adiabatic invariant. Deriving the sufficient conditions, i. e. finding the resonant particles, precessing with the bar pattern speed in the self-consistent problem, is planned to be worked on later.

References

[1] Athanassoula E, Vozikis Ch L, Lambert J C 2001 *A&A* **376**, 1135-46

[2] Barnes J, Hut P 1986 *Nature* **324**, 446-49

[3] Binney J, Tremaine S 2008 *Galactic Dynamics: Second Edition*, Princeton University Press

[4] White R 1988 *Astrophys. J.* **330**, 26-37

[5] Greengard L, Rokhlin V 1997 *Journal of Computational Physics* **135**, 280-292

[6] Dehnen W 2002 *Journal of Computational Physics* **179**, 27-42

[7] Lynden-Bell D 1979 *MNRAS* **187**, 101-107

[8] Merrit D 1996 *Astronomical J.* **111**, 2462-2464

[9] Polyachenko V L, Polyachenko E L 1994 *Astronomy Letters* **20**, 416-423

[10] Rybicki G 1972 *Proc. of IAU Colloq. 10 "Gravitational N-Body Problem"* ed. M. Lecar (Reidel Publishing Company) p 22

[11] Sellwood J 2014 Reviews of Modern Physics **86**, 1-46

[12] Rodionov S A, Sotnikova N Ya 2005 *Astronomy Letters* **49**, 470-476

[13] Teuben P 1995 *The Stellar Dynamics Toolbox NEMO* (Astronomical Data Analysis Software and Systems IV vol 77) ed. R. Shaw, H. Payne, and J. Hayes (ASP Conference Series) p 398

[14] Chandrasekhar S 1942 *Principles of stellar dynamics* (The University of Chicago press)

The impact of heat blanketing envelopes on neutron stars cooling

M V Beznogov[1], D G Yakovlev[2], M Fortin[3], P Haensel[3], J L Zdunik[3]

[1] St. Petersburg Academic University, 8/3 Khlopina Street, St. Petersburg 194021, Russia
[2] Ioffe Physical Technical Institute, 26 Politekhnicheskaya, St. Petersburg 194021, Russia
[3] Nicolaus Copernicus Astronomical Center of the Polish Academy of Sciences, Bartycka 18, Warsaw 00-716, Poland

E-mail: mikavb89@gmail.com

Abstract. The goal of this work is to investigate the effects of chemical composition of heat blanketing envelopes of neutron stars on their thermal states and thermal evolution. To this purpose, we employ newly constructed models of the envelopes composed of binary ion mixtures (H–He, He–C, C–Fe) varying the mass of lighter ions (H, He or C) in the envelope. The results are compared with those calculated using the standard "onion-like" envelope. For illustration, we apply these results to estimate the internal temperature of the Vela pulsar and to study cooling of neutron stars. We show that uncertainties in the chemical composition of the envelopes can lead up to ~ 2.5 times uncertainty of the internal temperature of the star which significantly complicates theoretical reconstruction of the internal structure of cooling neutron stars from observations of their thermal surface emission.

1. Introduction

An investigation of the properties of the matter of nuclear ($\rho_0 = 2.4 \times 10^{14}$ g cm^{-3}; see, e.g., [1]) and supranuclear densities is most important not only for astrophysics but for many other areas of physics. For example, such investigation is needed in nuclear physics, elementary particle physics and condensed matter physics. It is a challenging task for a lot of reasons and the study of neutron stars plays an important role in solving this problem because neutron stars are unique natural laboratories of superdense matter (see, e.g., [1, 2]). However, the properties of the superdense matter determine the internal structure of neutron stars which is not directly observable. Consequently, the ability to study the properties of the superdense matter is based on the ability to infer the internal structure of neutron stars from observations of their surface emission.

One of a few ways to do this is to construct theoretical models of thermal evolution of neutron stars and compare the theory with observational data (see, e.g., [2, 3]). However, here appears another problem. Such theories utilize the internal temperature of a neutron star (T_b) whereas from observations one can only extract its surface temperature (T_s). The $T_s - T_b$ relation is an important and still not fully resolved issue in the neutron star physics. It is determined by a thin but very significant heat blanketing envelope of the star (see, e.g., [4, 5]). And although the microphysics of these envelopes is more or less clear (unlike the microphysics of neutron star cores), their chemical composition is rather uncertain. This leads to uncertainties in $T_s - T_b$ relations and, in turn, in our understanding of the properties of the superdense matter in neutron star cores.

Thus, the main goal of our work is to investigate how the uncertainties of the chemical composition of heat blanketing envelopes affect the inference of the parameters of neutron stars from observations. To achieve this, we employ our recently developed models of heat blanketing envelopes composed of binary ion mixtures (H–He, He–C and C–Fe) [5]. We also use the "onion-like" model (layers of H, He, C and Fe) developed by Potekhin et al [4] as a reference for comparison with our models. That model (hereafter, the PCY97 model) is widely used in the astrophysical literature and has de-facto become a standard model of the heat blanketing envelope. Our objects of study are not too young ($t \gtrsim 100$ yr) isolated cooling neutron stars. Such stars are thermally equilibrated and isothermal inside. The main temperature gradient still persists in the heat blanketing envelopes (see, e.g., [1, 2]), so that the $T_s - T_b$ relations are determined only by these envelopes.

In the following sections we briefly discuss formation scenarios of heat blanketing envelopes, thermal states of the Vela pulsar and neutron star cooling. More details on this topic are published elsewhere [6].

2. Formation scenarios

A few decades ago it was commonly thought that the outer envelopes and the atmospheres of neutron stars consist of heavy (iron-like) elements due to the formation of the envelopes in very hot and young neutron stars where all light elements had burnt into heavier ones in thermonuclear reactions. However, an analysis of the observed spectra has demonstrated that some neutron stars are better described by hydrogen atmosphere models (see, e.g., [7]) while some other stars are better described by carbon atmosphere models (see, e.g., [8]). This gives observational evidence that heat blanketing envelopes of neutron stars can consist not only of heavy elements but of light elements as well.

Possible formation scenarios include:

- The fallback of material on the neutron star surface after the supernova explosion;
- The accretion of hydrogen and/or helium from interstellar medium or from a binary companion (if the neutron star has passed the compact binary stage, e.g., [9]);
- Accreted hydrogen can burn into helium, then helium can burn into carbon and carbon into heavier elements (e.g., [10]).

All these processes contribute to the uncertainty of chemical composition of heat blanketing envelopes making it largely unknown.

3. Thermal states of the Vela pulsar

Let us consider the Vela pulsar as an example and analyze its thermal states using the available observational data. Vela is a middle-aged pulsar, its characteristic age is $t \approx 11$ kyr. Therefore, its thermal relaxation has occurred long ago and its redshifted internal temperature $\widetilde{T}$ is constant throughout the star (excluding the heat blanketing envelope). The redshifted effective surface temperature of the Vela pulsar was estimated by Pavlov et al [11] to be $T_s^\infty = 0.68 \pm 0.03$ MK (at 68% confidence level for a magnetic hydrogen atmosphere model and "canonical" model of a neutron star with the mass $M = 1.4\,M_\odot$ and radius $R = 10$ km). Specifically, we use the value $T_s^\infty = 0.68$ MK, which corresponds to $T_s = T_s^\infty/\sqrt{1-x_g} = 0.89$ MK, with $x_g = 2GM/(Rc^2)$. Then we employ $T_s - T_b$ relations from [4, 5] to calculate T_b and the redshifted internal temperature $\widetilde{T} = T_b\sqrt{1-x_g}$. Although Vela possesses a magnetic field $B \sim 3 \times 10^{12}$ G, this field has no strong influence on the $T_s - T_b$ relations [12].

The results are presented on Fig. 1 which shows the inferred internal Vela's temperature $\widetilde{T}$ calculated with different envelope models. For a fixed envelope model, $\widetilde{T}$ is plotted as a function of ΔM, an accumulated mass of light elements in the envelope. In particular, for the H–He

IOP Conf. Series: Journal of Physics: Conf. Series **929** (2017) 012010 doi:10.1088/1742-6596/929/1/012010

envelope (the dotted curve) $\Delta M = \Delta M_{\rm H}$ is the accumulated mass of hydrogen; for the He–C envelope (the dashed curve) $\Delta M = \Delta M_{\rm He}$ is the mass of helium; for the C–Fe envelope (the dot-dashed curve) $\Delta M = \Delta M_{\rm C}$ is the mass of carbon and, finally, for the PCY97 envelope model (the solid gray curve) ΔM is the mass of light elements (H+He+C).

It is important to note that the chemical composition of any envelope has some natural limitations due to thermonuclear and pycnonuclear reactions and electron captures [4, 5]. At high temperatures and/or densities hydrogen transforms into helium, then helium into carbon, etc. Such temperatures and densities are accessible in the heat blanketing envelopes thus limiting their chemical compositions. The consequences are two-fold. First, different envelopes may have different densities at the bottom ($\rho_b = 10^8$ g cm^{-3} for the H–He envelope and $\rho_b = 10^{10}$ g cm^{-3} for the other envelopes). Second, maximum possible value of ΔM is also different for different envelopes.

As seen from Fig. 1 the uncertainties in the chemical composition of the envelope can cause up to ~ 2.5 times variation in the internal temperature for a fixed observable surface temperature. The largest variation of the internal temperature occurs in the PCY97 model because it takes into account a wider range of elements than any binary envelope model. Another notice is the anomalous behavior of the H–He curve (explained in [5]). For all other curves $\widetilde{T}$ decreases when ΔM increases because light elements increase heat transparency (as detailed in [5]).

As Vela is at the neutrino cooling stage (e.g., [2]) the main factor which determines its thermal evolution is the neutrino cooling function $\ell(\widetilde{T}) = L_\nu^\infty(\widetilde{T})/C(\widetilde{T})$, where L_ν^∞ is the neutrino luminosity of the star and C is its heat capacity (see, e.g., [13]). The neutrino cooling function is a fundamental parameter of superdense matter in a neutron star core, and it is determined by the internal temperature. A convenient unit of ℓ is $\ell_{\rm SC}$ which is the so-called "standard neutrino candle" (a non-superfluid star which cools via neutrinos generated in the modified Urca process). Using the fit expression for the neutrino cooling function from [14], which is independent of the equation of state (EOS) of nucleon matter in the stellar core, we can calculate the quantity $f_\ell = \ell(\widetilde{T})/\ell_{\rm SC}(\widetilde{T})$, the neutrino cooling function of Vela expressed in terms of standard candles.

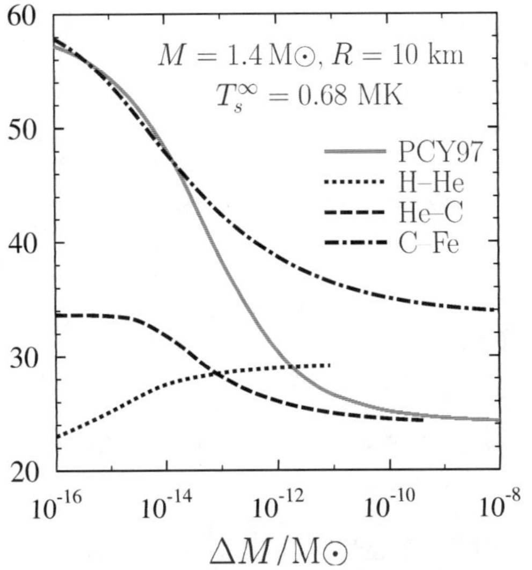

Figure 1: Redshifted internal temperature $\widetilde{T}$ of the Vela pulsar as a function of the accumulated mass ΔM for the H–He, He–C, C–Fe and PCY97 models of the heat blanketing envelopes. See text for details.

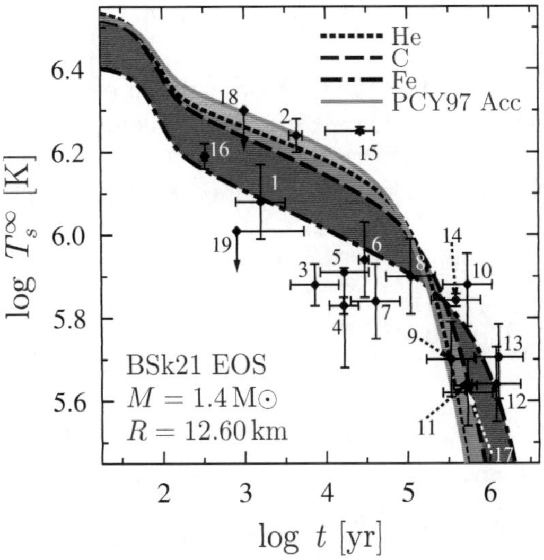

Figure 2: Cooling curves for a neutron star with $f_\ell = 1$ and different chemical chemical compositions of the heat blanketing envelope. See text for details.

We have demonstrated that a factor of ~ 2.5 uncertainty in the internal temperature translates into a factor of ~ 200 uncertainty in f_ℓ (see [6] for details). This greatly complicates an investigation of the properties of the superdense matter and emphasizes once more the importance of chemical composition of heat blanketing envelopes.

4. Neutron star cooling

Now let us consider the cooling of isolated neutron stars using a simple example. Specifically, we study the thermal evolution of the star with the mass $M = 1.4$ M⊙ and BSk21 EOS [15]. Such star has the radius $R = 12.60$ km. For illustration, we assume $f_\ell = 1$ (the standard neutrino candle) and focus on the effects of the envelopes. After fixing f_ℓ and an envelope model but varying ΔM, we obtain a sequence of cooling curves $T_s^\infty(t)$ [redshifted effective surface temperatures versus stellar age]. These curves are almost universal, i.e. independent of the EOS in the core and of the mass M [16].

Some examples of our calculations are presented in Fig. 2. The thick curves correspond to the envelopes composed of pure elements (He, C, Fe) and the fully accreted PCY97 envelope (the solid gray curve). The dark-gray area is filled with the cooling curves for the C–Fe envelopes with different ΔM; the medium-gray area is filled with the cooling curves for the He–C envelopes, and the light-gray area with the cooling curves for the PCY97 envelopes (if they are not covered by binary envelopes). The H–He envelopes are excluded for a better visual representation of the figure. We have also added the observational data on isolated neutron stars (see [14] and references therein). Neutron star labels are as follows: (1) PSR J1119–6127; (2) RX J0822–4300 (in Pup A); (3) PSR J1357–6429; (4) PSR B0833–45 (Vela); (5) PSR B1706–44; (6) PSR J0538+2817; (7) PSR B2334+61; (8) PSR B0656+14; (9) PSR B0633+1748 (Geminga); (10) PSR B1055–52 ; (11) RX J1856.4–3754; (12) PSR J2043+2740; (13) RX J0720.4–3125; (14) PSR J1741–2054; (15) XMMU J1732–3445; (16) Cas A neutron star; (17) PSR J0357+3205 (Morla); (18) PSR B0531+21 (Crab); (19) PSR J0205+6449 (in 3C 58).

As seen from Fig. 2, variations in the chemical composition of the heat blanketing envelopes significantly affect the cooling curves. A star with an iron envelope, being a standard neutrino candle ($f_\ell = 1$; the dot-dashed curve), cannot explain many observational sources. Taking into account different chemical compositions of the envelope, we can explain more sources but not all of them. To explain the remaining sources one needs to consider deviations from the standard neutrino candle. Warm sources require $f_\ell < 1$ ($f_\ell \sim 0.01$ for XMMU J1732–3445; see, e.g., [14]), while cold sources require $f_\ell > 1$ ($f_\ell \sim 10^2 - 10^4$ for the Vela pulsar depending on the exact composition of the envelope [6]). Furthermore, Fig. 2 implies that observations of all isolated neutron stars at the photon cooling stage (i.e., with the age $t \gtrsim 100$ kyr; see, e.g., [2]) are compatible with $f_\ell \lesssim 1$.

5. Conclusions

We have analyzed the effects of uncertainties in the chemical composition of the heat blanketing envelopes on thermal states and cooling of neutron stars. We have briefly outlined possible formation scenarios of the envelopes with different compositions (Sec. 2). Then, taking the Vela pulsar as an example, we have demonstrated that the uncertainties in the chemical composition of the envelopes can translate up to ~ 2.5 times uncertainties in the internal stellar temperature for a fixed observable surface temperature and up to ~ 200 times uncertainties in the neutrino cooling function which is the most important parameter of superdense matter in neutron star cores (Sec. 3). Finally, we have presented some examples of cooling curves for a neutron star with $f_\ell = 1$ (standard neutrino candle) and different heat blanketing envelopes (Sec. 4). The effects of the envelopes are clearly seen from Fig. 2. To sum up, all these results demonstrate that a proper analysis of the chemical composition of neutron star heat blanketing envelopes is essential in the order to reliably infer the properties of superdense matter in the neutron star cores.

Needless to say, our consideration is not complete. More work is required to resolve this problem. For example, one can consider multicomponent (> 2) ion mixtures either diffusively equilibrated or not. The evolution of initially not diffusively equilibrated envelopes can be also modeled. Moreover, it will be useful to improve the existing models of diffusive nuclear burning in the envelopes [10] and take into account the effects of magnetic fields. Evidently, any additional reliable information on the formation history and evolution of the envelopes would be most welcome. However, all these problems are far beyond the scope of the present paper.

Acknowledgments
The work of MB was partly supported by the Dynasty Foundation, the work of DG by the Russian Foundation for Basic Research (grant 16-29-13009-ofi-m), and the work of MF, PH, and LZ was supported by the Polish NCN research grant no. 2013/11/B/ST9/04528.

References

[1] Haensel P, Potekhin A Y and Yakovlev D G 2007 *Neutron Stars. 1. Equation of State and Structure* (*Astrophysics and Space Science Library* vol 326) (Springer, New York)
[2] Yakovlev D G and Pethick C J 2004 *Annu. Rev. Astron. Astrophys.* **42** 169–210
[3] Page D, Lattimer J M, Prakash M and Steiner A W 2009 *Astrophys. J.* **707** 1131–1140
[4] Potekhin A Y, Chabrier G and Yakovlev D G 1997 *Astron. Astrophys.* **323** 415–428
[5] Beznogov M V, Potekhin A Y and Yakovlev D G 2016 *MNRAS* **459** 1569–1579
[6] Beznogov M V, Fortin M, Haensel P, Yakovlev D G and Zdunik J L 2016 *MNRAS* **463** 1307–1313
[7] Potekhin A Y 2014 *Phys.-Usp.* **57** 735
[8] Ho W C G and Heinke C O 2009 *Nature* **462** 71–73
[9] Blaes O M, Blandford R D, Madau P and Yan L 1992 *Astrophys. J.* **399** 634–645
[10] Chang P, Bildsten L and Arras P 2010 *Astrophys. J.* **723** 719–728
[11] Pavlov G G, Zavlin V E, Sanwal D, Burwitz V and Garmire G P 2001 *Astrophys. J. Lett.* **552** L129–L133
[12] Potekhin A Y, Yakovlev D G, Chabrier G and Gnedin O Y 2003 *Astrophys. J.* **594** 404–418
[13] Yakovlev D G, Ho W C G, Shternin P S, Heinke C O and Potekhin A Y 2011 *MNRAS* **411** 1977–1988
[14] Ofengeim D D, Kaminker A D, Klochkov D, Suleimanov V and Yakovlev D G 2015 *MNRAS* **454** 2668–2676
[15] Goriely S, Chamel N and Pearson J M 2010 *Phys. Rev. C* **82** 035804
[16] Yakovlev D G, Kaminker A D, Gnedin O Y and Haensel P 2001 *Phys. Rep.* **354** 1–155

IOP Conf. Series: Journal of Physics: Conf. Series **929** (2017) 012011 doi:10.1088/1742-6596/929/1/012011

The γ-ray pulsar J1932+1916 in X-rays

A Karpova, P Shternin, D Zyuzin, A Danilenko, Yu Shibanov
Ioffe Institute, Politekhnicheskaya 26, St. Petersburg, 194021, Russia
E-mail: annakarpova1989@gmail.com

Abstract.
We report the results of a simultaneous analysis of the archival *Suzaku* and *Swift* X-ray data of the γ-ray pulsar J1932+1916 field. A point-like source and diffuse emission around it were detected at the γ-ray position of the pulsar. Their spectral properties allow one to suggest that *Suzaku* and *Swift* detected the X-ray counterpart of PSR J1932+1916. Using of the interstellar absorption–distance relations, we constrain the distance to the pulsar in the range of 2–6 kpc. We also discuss a possible association of the pulsar with the supernova remnant G54.4−0.3.

1. Introduction
Follow-up X-ray studies of newly discovered γ-ray pulsars are a natural step for detailed investigations of their properties. Here we report the analysis of the X-ray *Swift* and *Suzaku* observations of the field of the γ-ray pulsar J1932+1916 (hereafter J1932) discovered with *Fermi* [1]. The pulsar period $P = 208$ ms, the characteristic age $\tau = 35$ kyr, the spin-down luminosity $\dot{E} = 4 \times 10^{35}$ erg s^{-1}, the magnetic field $B = 4.5 \times 10^{12}$ G and coordinates R.A.=$19^{\mathrm{h}}32^{\mathrm{m}}19\overset{\mathrm{s}}{.}70(4)$ and Dec.=$+19^{\circ}16'39''(1)$[1] were obtained from timing analysis. No radio counterpart of the pulsar was detected [1].

2. Spatial analysis
Swift and *Suzaku* images of J1932 field are presented in Fig. 1. The γ-ray position of J1932 is shown by the '+' symbols. In the *Swift* X-ray Telescope (XRT) Point Source Catalogue [2], we found the pulsar X-ray counterpart candidate, a source 1SXPS J193219.4+191635, which is marked in the left panel of Fig. 1. The *Suzaku* image also reveals an X-ray emission spatially coincident with the *Fermi* position of J1932. We compared the radial profile of the *Suzaku* point spread function (PSF) with the radial brightness profile of this emission and found that its extent is significantly larger than the PSF half-power diameter of $2'$. We considered this extended emission as the pulsar wind nebula (PWN) powered by J1932. We found that the PWN spatial brightness distribution can be described by the 2D-Gaussian function with the full width at half maximum (FWHM) of $\approx 4.5'$.

3. Spectral analysis
For the spectral analysis, we used the data from all *Suzaku* X-ray imaging spectrometers (XIS). We extracted the spectra from three regions shown in the right panel of Fig. 1: the polygon region was used for the astrophysical background while the $1'$-radius circle and the ring with radii of

[1] Numbers in parentheses are 1σ uncertainties related to the last significant digits quoted.

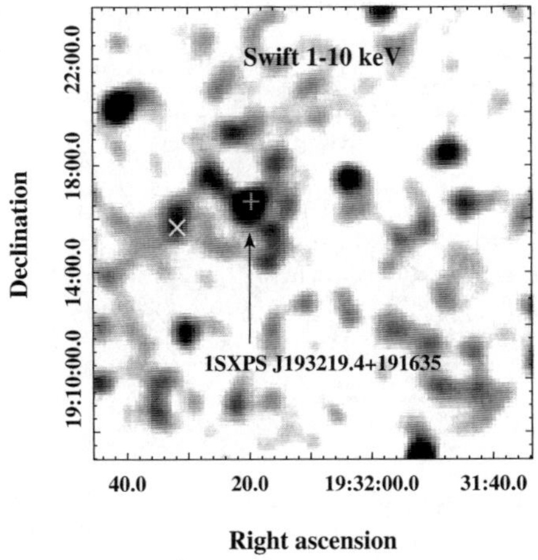

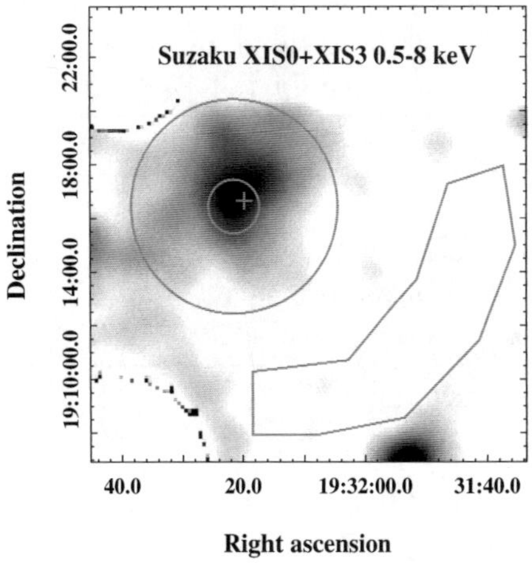

Figure 1. 1–10 keV *Swift* XRT (*left*) and 0.5–8 keV *Suzaku* combined XIS0 and XIS3 (*right*) images of the PSR J1932+1916 field. The 1SXPS J193219.4+191635 source is marked in the left image. The *Fermi* position of PSR J1932+1916 [1] is shown by the '+' symbols. In the right panel, regions used for spectral analysis are shown. The soft source 1SXPS J193231.7+191540 falling in the extraction region is marked in the left panel by the 'x' symbol.

$1'$ and $4'$ were used for the source. The soft point-like source 1SXPS J193231.7+191540 marked by the 'x' symbol in the left panel of Fig. 1 was excluded using $1'$-radius circle. We assume that both circle and ring regions contain contributions from the astrophysical background, the PWN and the pulsar. For the PWN emission, the Ancillary Response Files (ARFs) were generated basing on the broad Gaussian spatial model with FWHM=$4.5'$ (see Sect. 2). For the pulsar and astrophysical background components, we generated ARFs for the point source and for a uniformly extended source, respectively. The non-X-ray instrumental background spectra were simulated for each region and detector and subtracted.

We also extracted spectrum from the *Swift* data using the $30''$-radius aperture centered at the 1SXPS J193219.4+191635 position. For the background, we used the source-free region in the *Swift* image. Two XRT ARFs were generated: for the extended emission and for the point source to account for the PWN and pulsar contributions, respectively. The PWN contribution to the source spectrum was estimated to be of 4% of the total PWN flux basing on the spatial model.

We used the unbinned spectra and fitted them simultaneously in the 0.5–10 keV band. The astrophysical background model consists of the Cosmic X-ray Background (CXB) and absorbed three-temperature optically thin thermal plasma component (3-T APEC[2]). The CXB component was modelled by the absorbed power law (PL) with photon index Γ=1.4, the surface brightness of 5.4×10^{-15} erg s^{-1} cm^{-2} arcmin^{-2} in 2–10 keV band [3] and the hydrogen column density $N_{\rm H}$ of 2.8×10^{22} cm^{-2}. The latter value was obtained using the total selective extinction $E(B-V)$ of $\approx$4 in the J1932 direction [4] and the empirical $E(B-V)$–$N_{\rm H}$ relation, $N_{\rm H} = (0.7\pm0.1)\times E(B-V)\times10^{22}$ cm^{-2} [5]. For the pulsar and PWN components, the absorbed PL models with a common value of the column density were applied.

To obtain the spectral parameters and their uncertainties we used the Markov chain Monte-Carlo (MCMC) sampler [6] implemented as a python package `emcee` in [7]. The resulting parameters of the pulsar and PWN emission are presented in Table 1. We estimated the

[2] https://heasarc.gsfc.nasa.gov/xanadu/xspec/manual/XSmodelApec.html

Table 1. Best-fit spectral parameters for the presumed PWN and pulsar emission. All errors correspond to 90% credible intervals. Unabsorbed fluxes are given for the 0.5–8 keV band. Luminosities are given for the 0.5–8 keV band and the distance of 4 kpc.

Parameter	PWN	pulsar
Column density $N_{\rm H}$, cm^{-2}	$1.2^{+0.3}_{-0.3} \times 10^{22}$	
Photon index Γ	$1.8^{+0.4}_{-0.3}$	$1.4^{+1.0}_{-1.0}$
Unabsorbed flux f_X, erg s^{-1} cm^{-2}	$1.2^{+0.3}_{-0.2} \times 10^{-12}$	$1.3^{+0.9}_{-0.5} \times 10^{-13}$
Luminosity L_X, erg s^{-1}	$2.3^{+0.6}_{-0.4} \times 10^{33}$	$2.5^{+1.7}_{-1.0} \times 10^{32}$

goodness-of-fit using the Pearson χ^2 test statistics that is appropriate for the Poisson data. 256 spectra were simulated for the model parameters obtained from the posterior distributions. For the 68% of simulations, the test statistics value was smaller than for the observed data indicating that our model is consistent with the data.

We also tried to search for the softening of the PWN emission with increasing radius as observed for other PWNe. No softening was found likely due to low count statistics.

4. Discussion

The distance $D \sim 1.5$ kpc (with an upper limit of 6.6 kpc) to J1932 was estimated from an empirical correlation between pulsar γ-ray fluxes and distances [1]. We independently estimated the distance using the interstellar extinction maps of the Galaxy and $E(B-V)$–$N_{\rm H}$ relation [5] (see Section 3). From the spectral analysis we obtained $N_{\rm H} = (1.2 \pm 0.3) \times 10^{22}$ cm^{-2} (Table 1) which corresponds to $E(B-V) = 1.7 \pm 0.4$ (1σ errors). Thus basing on the extinction–distance dependence along the pulsar line of sight following from the maps [8, 9], the distance to J1932 can be constrained in 2–6 kpc range.

Using the obtained distance range and results of the spectral analysis we compare J1932 and its PWN with other pulsar+PWN systems presented in [10]. The PWN produced by J1932 is somewhat brighter than PWNe of other pulsars with similar $\dot{E}$. However, the X-ray efficiencies $\eta_X = L_X/\dot{E}$ for both pulsar and PWN lie within typical range of 10^{-5}–10^{-1}. The ratio $L_{\rm PWN}/L_{\rm pulsar} \approx 10$ is also typical for the pulsar+PWN systems supporting the pulsar and PWN interpretation of the detected sources.

J1932 is located near the radio shell of the supernova remnant (SNR) G54.4−0.3 which radio and X-ray images are presented in the Fig. 2 with the *Fermi* position of J1932 shown by the '+' symbols. This is likely to be a mixed-morphology SNR where X-ray emission is concentrated within the radio shell. No diffuse radio emission from the PWN was detected. Junkes [11] obtained the absorption column density of $\approx 10^{22}$ cm^{-2} in the SNR direction using the *ROSAT* observations. This value is consistent with the $N_{\rm H}$ value for J1932 (Table 1) making possible the association of these objects. If it is real, then the pulsar proper motion is $\approx$34 mas yr^{-1} and its transverse velocity is $\approx 650 D_{\rm 4kpc}$ km s^{-1} assuming the J1932 age of 35 kyr; here $D_{\rm 4kpc}$ is the distance in units of 4 kpc. Such value is typical for pulsars [12]. Determination of the proper motion direction could help to resolve the question about the association.

5. Summary

We analysed the archival X-ray *Swift* and *Suzaku* data on the γ-pulsar J1932 field. At the pulsar position, the point-like source was revealed with *Swift* and the diffuse emission was detected with *Suzaku*. These sources have spectral properties typical for the pulsar+PWN systems. Using the interstellar extinction–distance relations, we constrain the distance to the pulsar in the 2–6 kpc

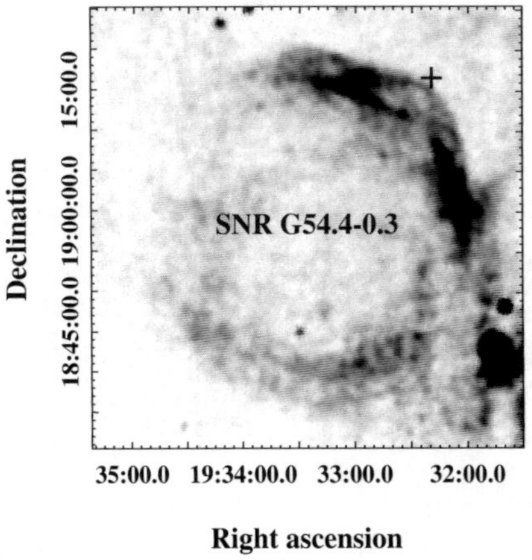

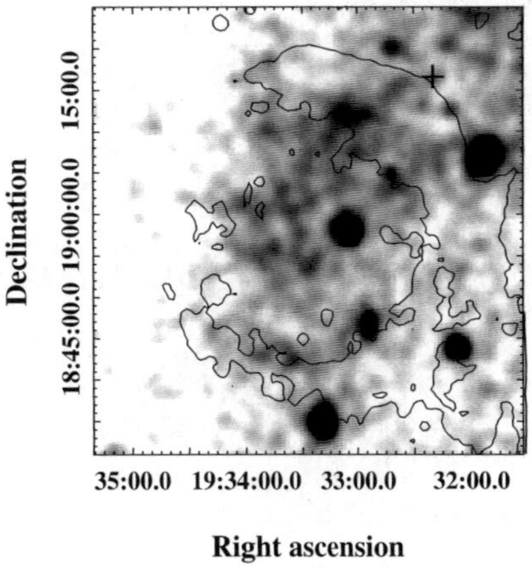

Figure 2. *Left*: 1420 MHz continuum image of the SNR G54.4−0.3 obtained from the VLA Galactic Plane Survey (VGPS; [13]). *Right*: *ROSAT* PSPC survey image of G54.4−0.3 field in the 0.1–2.4 keV band with overlaid VGPS contours. The *Fermi* position of J1932 is marked by the '+' symbols.

range. The proximity of J1932 to the SNR G54.4−0.3 and comparable values of the interstellar absorption towards both objects make possible their association. Deeper observations with better spatial resolution are needed to study the morphology and spectral properties of the X-ray source and to confirm its pulsar nature.

Acknowledgments
For the Figure 2 we used the data from the VGPS survey conducted by the National Radio Astronomy Observatory (NRAO) instruments. NRAO is a facility of the National Science Foundation operated under cooperative agreement by Associated Universities, Inc.

References
[1] Pletsch H J and et al 2013 *ApJ* **779** L11
[2] Evans P A and et al 2014 *ApJS* **210** 8
[3] Ueda Y and et al 1999 *ApJ* **518** 656–671
[4] Schlafly E F and Finkbeiner D P 2011 *ApJ* **737** 103
[5] Watson D 2011 *A&A* **533** A16
[6] Goodman J and Weare J 2010 *Comm. App. Math. Comp. Sci.* **5** 65
[7] Foreman-Mackey D, Hogg D W, Lang D and Goodman J 2013 *PASP* **125** 306–312
[8] Green G M and et al 2015 *ApJ* **810** 25
[9] Drimmel R, Cabrera-Lavers A and López-Corredoira M 2003 *A&A* **409** 205–215
[10] Kargaltsev O and Pavlov G G 2008 *40 Years of Pulsars: Millisecond Pulsars, Magnetars and More* (*American Institute of Physics Conference Series* vol 983) ed Bassa C, Wang Z, Cumming A and Kaspi V M (Melville, New-York) pp 171–185
[11] Junkes N 1996 *Unsolved Problems of the Milky Way* (*IAU Symposium* vol 169) ed Blitz L and Teuben P J (Dordrecht: Kluwer) p 627
[12] Hobbs G, Lorimer D R, Lyne A G and Kramer M 2005 *MNRAS* **360** 974–992
[13] Stil J M, Taylor A R, Dickey J M, Kavars D W, Martin P G, Rothwell T A, Boothroyd A I, Lockman F J and McClure-Griffiths N M 2006 *AJ* **132** 1158–1176

Thermal evolution of old white dwarfs

Andrew Kozhberov
Ioffe Institute, Politekhnicheskaya 26, St. Petersburg, Russia
E-mail: kozhberov@gmail.com

Abstract. This work is devoted to a description of thermodynamic properties of Coulomb crystals which are expected to form in white dwarf interiors. Effects of magnetic field, isotopic impurities, polarization of the electron background and crystal lattice type on the thermal evolution of white dwarfs are discussed. It is shown that the electron polarization could play a noticeable role in the cooling process. While other parameters in concern do not make a significant impact.

1. Introduction

White dwarfs are compact stars with masses comparable to that of the Sun (but not more than $1.38M_{\odot}$) and with radii 100 times less than the Solar radius. Thus, in their inner layers the density (ρ) may reach 10^9 g/cm^3. It is thought that matter in white dwarf interiors at densities $\gtrsim 10^4$ g/cm^3 consists of strongly degenerate electron gas and various fully ionized atomic nuclei. These ions are mostly represented by carbon ^{12}C and oxygen ^{16}O isotopes but depending on the formation history, inner layers may contain minor quantities of other elements such as neon and iron. Impurities of light elements like hydrogen and helium may occur in outer layers [1]. At early stages of the evolution this mixture of ions is in a liquid state (so-called Coulomb liquid). As star cools the ion plasma crystallizes forming a Coulomb crystal. A partially crystalized white dwarf has been recently observed [2].

The Coulomb crystal could be described as a system of point charges immersed into a uniform neutralizing background of electrons [3, 4]. It is important to study thermodynamic properties of Coulomb crystals: the heat capacity of white dwarfs determines the cooling rate and allows one to derive the age of these stars from observations of their surface temperatures [1, 3]. Typically, thermodynamic properties of a one component crystal have been considered analytically in the harmonic lattice model or by other numerical methods (e.g., [5, 6, 7, 8, 9, 10]).

For the first time the problem of Coulomb crystals in white dwarf envelopes was taken into account in 1960s (e.g., [11, 12]) and in following works was widely developed (e.g., [1, 13, 14]). Most authors were interested in the chemical composition of the interior of these stars and processes in its atmosphere. Apparently precise calculations of the heat capacity of Coulomb crystals have never been used to study the thermal evolution of white dwarfs. In this work we try to address this issue and aim to investigate how the accuracy of approximations, magnetic field, isotopic impurities, polarization of the electron background and the type of the crystal lattice affect the heat capacity of the star envelope and, consequently, the cooling rate of these stars.

2. The heat capacity of Coulomb crystals

The potential energy of a one component Coulomb crystal without polarization of the electron background is given by

$$U_{\rm exact} = \sum_{i=1}^{N}\sum_{j=1}^{i-1} \frac{Z^2 e^2}{|\mathbf{r}_i - \mathbf{r}_j|} - \sum_{i=1}^{N} \int \frac{Zen_e d\mathbf{r}}{|\mathbf{r}_i - \mathbf{r}|} + \frac{1}{2}\int\int \frac{n_e^2 d\mathbf{r} d\mathbf{r}'}{|\mathbf{r} - \mathbf{r}'|}, \qquad (1)$$

where Z is the charge number of ions, while N, $n_e \equiv Zn$ and n are the total number of ions, the electron and the ion number densities, respectively; $\mathbf{r}_i$ is the radius vector of the i-th ion. We expand the potential energy in powers of ion displacements $\mathbf{u}_i$ from their equilibrium positions $\mathbf{X}_i$ (so that $\mathbf{r}_i = \mathbf{X}_i + \mathbf{u}_i$) and retain terms up to the second order (harmonic approximation)

$$U_{\rm exact} \approx U_0 + \frac{1}{2}\sum_{i,j=1}^{N} u_i^\alpha u_j^\beta \frac{\partial^2 U_{\rm exact}}{\partial u_i^\alpha \partial u_j^\beta}\bigg|_{u_i^\alpha, u_j^\beta = 0}. \qquad (2)$$

The zero-order term in this series is the electrostatic energy, U_0. The second order term determines the phonon spectrum of the crystal $\omega_\nu(\mathbf{k})$. At any fixed wavevector $\mathbf{k}$ it could be calculated by solving the dispersion equation: $\det\{D_{pp'}^{\alpha\beta}(\mathbf{k}) - \omega_\nu^2(\mathbf{k})\delta^{\alpha\beta}\delta_{pp'}\} = 0$, where indices α, β denote Cartesian components, indices p, p' enumerate ions in the elementary cell, ν enumerates oscillation modes at given $\mathbf{k}$ ($\nu = 1, \ldots, 3N_{\rm cell}$), $N_{\rm cell}$ is the number of ions in the primitive cell. The expression for the dynamic matrix $D_{pp'}^{\alpha\beta}(\mathbf{k})$ can be found in [15].

Any phonon thermodynamic function can be calculated based on the information about the phonon spectrum. For instance, the heat capacity is $C = N\sum_{\mathbf{k}\nu} w^2 (4\sinh^2(w/2))^{-1}$, where $w \equiv \hbar\omega_\nu(\mathbf{k})/T$, summation is over all ν and $\mathbf{k}$ in the first Brillouin zone. It is well known that $C \approx 3N$ at high temperatures ($T > T_{\rm p}$, where $T_{\rm p} \equiv \hbar\omega_{\rm p}$ is the ion plasma temperature, $\omega_{\rm p} \equiv \sqrt{4\pi n Z^2 e^2/M}$ is the ion plasma frequency, M is the mass of ions). At low temperatures ($T \ll T_{\rm p}$), Debye T^3-law must take place $C \approx NH(T/T_{\rm p})^3$, where H is independent of T.

Among all lattices the body-centered cubic (bcc) lattice is known as the lattice with the lowest Helmholtz free energy and it is thought that this lattice forms in white dwarf cores, but the difference between the energies of different lattices is very small [16]. If the electron polarization is taken into account the face-centered cubic (fcc) lattice can become more energetically favorable than the bcc lattice [17]. In the present paper we study bcc, fcc and hexagonal-close packed (hcp) lattices. The last one is the lattice with the highest packing efficiency (along with fcc lattice).

In Fig. 1A the ratios of heat capacities of the bcc, fcc, and hcp lattices are plotted as functions of $t \equiv T/T_{\rm p}$. The difference between lattices at low and intermediate temperatures is noticeable. In particular, in the quantum limit $C_{\rm hcp}/C_{\rm bcc} \approx 1.190$ and $C_{\rm hcp}/C_{\rm fcc} \approx 1.089$. It is also clear that the heat capacity of the hcp lattice approaches the Debye limit at much lower temperatures than those of the bcc and fcc lattices. At intermediate temperatures all ratios have minima. For the hcp to fcc ratio the minimum is equal to 0.9 at $t \approx 0.018$ while $C_{\rm hcp}/C_{\rm bcc} \approx 0.75$ at $t \approx 0.025$ [15]. Notice that the heat capacity of more complex lattices can differ considerably from the heat capacity of the bcc lattice but it happens only at $T \ll T_{\rm p}$. In addition we compared the previous fit of the heat capacity of the Coulomb crystal $C_{\rm fit}$ from the Monte Carlo simulations [1] (it isn't connected with any particular lattice type) with our precise calculations. The ratio $C_{\rm fit}/C_{\rm bcc}$ is plotted in Fig. 1A by dashed line. It lies between 0.86 and 1.12 and at low temperatures equal 0.977.

The phonon spectrum of the magnetized Coulomb crystal can be found from the dispersion equation: $\det\{D_{pp'}^{\alpha\beta}(\mathbf{k}) - \omega_\nu^2(\mathbf{k})\delta^{\alpha\beta}\delta_{pp'} - i\omega_\nu(\mathbf{k})\omega_{\rm B}\varepsilon^{\alpha\beta\gamma}h^\gamma\delta_{pp'}\} = 0$, where $\omega_{\rm B} \equiv ZeB/(Mc)$ is the ion cyclotron frequency, $\mathbf{h}$ is the direction of the magnetic field ($\mathbf{B} = B\mathbf{h}$). The phonon

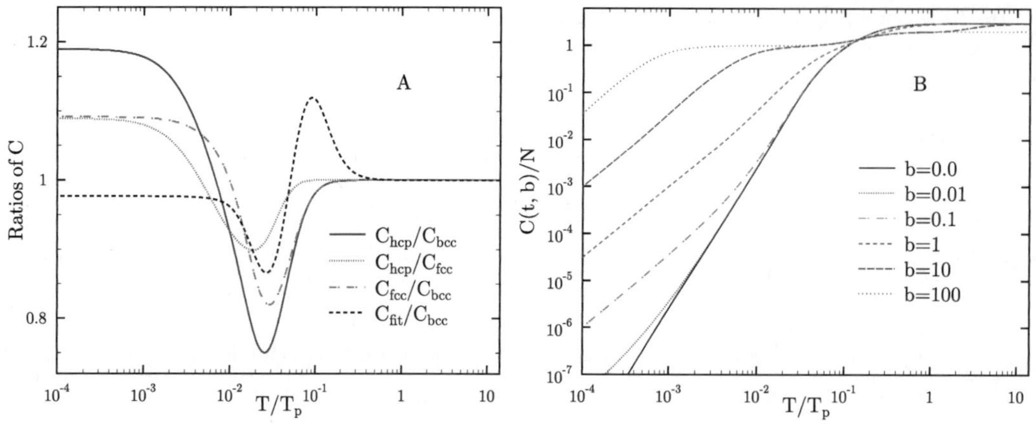

Figure 1. Panel A. Ratios of heat capacities of bcc, fcc, and hcp lattices. Panel B. The heat capacity of the bcc lattice in the magnetic field.

spectrum and thermodynamics of such systems were widely investigated in [18] and [19]. The heat capacity of the bcc lattice in the external magnetic field is plotted in Fig. 1B for different $b \equiv \omega_B/\omega_p \approx B_{15}/\sqrt{\rho_8}$, where $B_{15} = B/10^{15}$ G and $\rho_8 = \rho/10^8$ g/cm^3. One can see that magnetic fields significantly change the heat capacity only at $b \gtrsim 0.1$. Such values of b are not attained in white dwarfs. Their magnetic fields do not exceed 10^8 G. These calculations are important for magnetars, neutron stars with fields 10^{14} G and more.

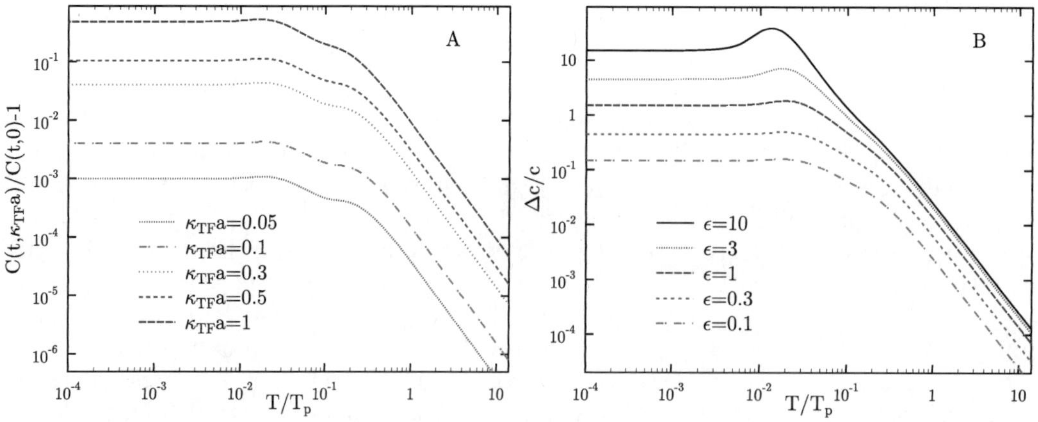

Figure 2. Panel A. Ratios of heat capacities of the bcc lattice with and without polarizable electron background. Panel B. The relative contribution of impurities to the heat capacity of the perfect bcc lattice.

Previously in this paper we thought that the electron background is uniform but it is not quite correct. The electron density changes in response to the presence of ions. The simplest way to take it into account is to use the linear response formalism and the static longitudinal dielectric function of electrons $\varepsilon(\mathbf{k}) \equiv 1 + \kappa_{\rm TF}^2/k^2$, where $\kappa_{\rm TF}^2 \equiv 4\pi e^2 \partial n_e/\partial \mu_e$, μ_e is the electron chemical potential (e.g., [5, 17] and references therein). In Fig. 2A we plot for the bcc lattice the ratio $(C(t, \kappa_{\rm TF}a) - C(t, 0))/C(t, 0)$ as function of t for different $\kappa_{\rm TF}a$, where $a = (4\pi n/3)^{-1/3}$ is the ion-sphere radius and $C(t, 0)$ is the heat capacity of the bcc lattice with the uniform background. This ratio is always greater than 0. At high temperatures it decreases as t^{-2}. At low temperatures it is independent of t and $C(t,\kappa_{\rm TF}a)/C(t,0) - 1 \propto (\kappa_{\rm TF}a)^2$ at $\kappa_{\rm TF}a \ll 1$. The linear response formalism assumes that $\kappa_{\rm TF}a < 1$ which is appropriate for the white dwarfs interior. For example, in fully ionized ^{12}C crystal at density $\rho = 10^6$ g cm^{-3} $\kappa_{\rm TF}a \approx 0.425$, while $\kappa_{\rm TF}a = 1$ at $\rho \approx 3000$ g cm^{-3}. At $\kappa_{\rm TF}a = 1$, $C(t,\kappa_{\rm TF}a)/C(t,0)$ does not exceed 1.48, at $\kappa_{\rm TF}a = 0.5$ it is below 1.105.

Another factor influencing thermodynamic properties of white dwarfs' cores is the presence of impurities and dislocations in the crystal structure. If all impurities are isotopic and substitutional, the heat capacity of the crystal can be calculated by the perturbation theory of disordered crystal spectra [20]. The ratio of an impurity contribution to the crystal heat capacity (Δc is the heat capacity change per one impurity ion) over the perfect crystal heat capacity (c is the perfect crystal heat capacity per one ion) decays as t^{-2} in the classic regime as seen in Fig. 2B [21]. At low temperatures, this ratio tends to a constant 1.5ϵ, where $\epsilon \equiv (M'-M)/M \geq -1$, M' is the impurity mass and M is the mass of "basic" ions. If $\epsilon \gg 1$ the maximum impurity effect takes place at temperatures $10^{-2}T_{\rm p} < T < 10^{-1}T_{\rm p}$. For impurities with $M' = 11M$ maximum $\Delta c/c \approx 40$ reaches at $t \approx 0.0124T_{\rm p}$. Hence, if the concentration of such impurities is 100 times less than the concentration of "basic" ions the total heat capacity will be amplified by 40%. On the other hand formation of such massive impurities in white dwarf envelopes is unlikely. Contribution of more realistic impurities is less noticeable. When the mass of impurities is not considerably different from "basic" ions the heat capacity changes don't exceed a few per cent. For instance, at low temperatures even 10% of impurities with $M' = 0.9M$ diminish the total crystal heat capacity only by 1.5%.

3. Cooling of white dwarfs

Now we can estimate how isotopic impurities, polarization of the electron background and the type of the crystal lattice affect the thermal evolution of old white dwarfs. Consider a simple model of an isothermal white dwarf with completely ionized crystallized carbon core and helium surface. All effects associated with ion mixtures and crystallization process are discarded (we assume that envelopes are fully crystalized). Crystallized white dwarfs have virtually no own sources of thermonuclear energy and shine exclusively due to cooling. Consequently, their luminosity and thermal evolution are determined by thermodynamic properties of envelopes: $-L_{\rm s} = C(T)dT/dt$, where $C(T)$ is the total heat capacity of the star.

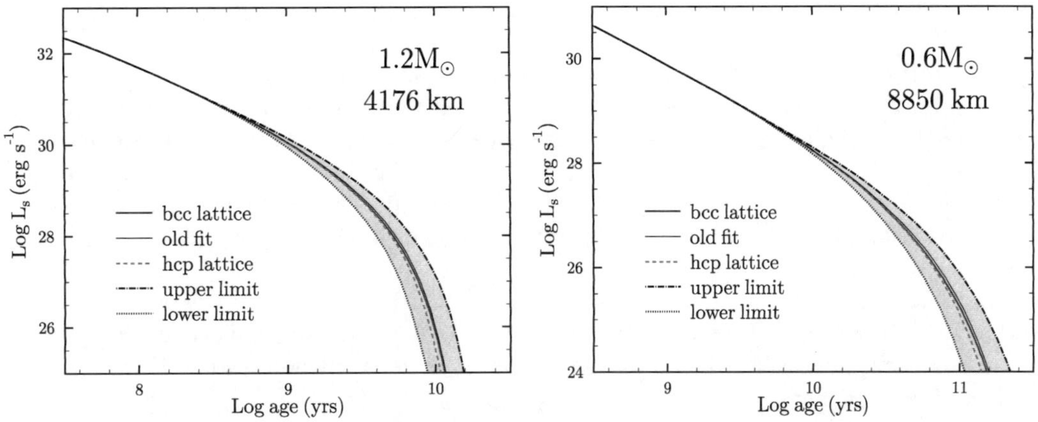

Figure 3. The thermal evolution of old white dwarfs with mass $1.2M_\odot$ (left panel) and $0.6M_\odot$ (right panel). The corresponding WD radii are given in the upper right corners.

At high temperatures heat capacities of all lattices tend to $3N$ and distinctions between them can be neglected. At low and intermediate temperatures the heat capacity of crystals can significantly differ from the heat capacity of the bcc lattice which is taken as a basis. For instance, $C_{\rm hcp}/C_{\rm bcc}$ and $C_{\rm fcc}/C_{\rm bcc}$ lay between 0.75 and 1.2. The first ratio was fitted in [15], while $C_{\rm bcc}$ in [16]. Using these equations we have calculated the thermal evolution of white dwarfs (masses $1.2M_{\odot}$ and $0.6M_{\odot}$) if their cores are modeled by crystals with the bcc lattice (thick solid line in Fig. 3) or by crystals with the hcp lattice (dashed line in Fig. 3). The massive white dwarfs ($\sim 1.2M_{\odot}$) crystalize at age $\sim 10^7$ years. At an age $\sim 10^9$ years crystals become quantum and the cooling rate of stars increases (it could be see in the left part of Fig. 3). Less massive stars crystallize at age $> 10^7$ years. The increase of the heat capacity leads to a decrease of the cooling speed and vice versa. Since $T > 0.0068T_{\rm p}$ $C_{\rm bcc} > C_{\rm hcp}$ and at $T < 0.0068T_{\rm p}$ $C_{\rm bcc} < C_{\rm hcp}$ the total difference between the bcc and hcp lattices is relatively small and does not exceed a few per cent (~ 3 % at an age of 10^{10} years for the massive white dwarf). Also we have used the old fit of the heat capacity to calculate the thermal evolution of stars [1]. Results are plotted by thin solid lines in Fig. 3. They lie even closer to new calculations for the bcc lattice than curves for the hcp lattice.

As shown previously the heat capacity can increase by about fifty present at low temperatures and densities due to the electron polarization. For simplicity we will not take into account that $\kappa_{\rm TF}a$ depends on ρ. The maximum possible variation of the heat capacity of the bcc lattice can be described like a Heaviside function with the discontinuity at $t \sim 0.5$ (it is equal 1.5 at $t < 0.5$ and 1 at $t > 0.5$). For this situation the cooling curves are plotted by dot-and-dash lines. The opposite case is described as a Heaviside function equal 1/1.5 at $t < 0.5$ and 1 at $t > 0.5$ and illustrated by dotted lines. It is not connected with any physical parameters but fully covers any possible isotopic impurities. All other cooling lines (due to the uncertainty in the lattice type or other parameters in consideration) will lie in the shaded (green) region. Hence the surface temperature of a 10^{10} years old massive white dwarf can vary less than 1.2 times. It is a very optimistic range. Realistic isotopic impurities can change the surface temperature by less than 1 percent. Uncertainties in the heat capacity values will affect the thermal evolution of less massive stars ($0.6M_{\odot}$) at an age of $\sim 10^{11}$ years which is greater than the age of the Universe.

For a more realistic analysis, one should take into account that white dwarfs may consist of a number of different chemical elements. Also we have't discussed the process of crystallization of stars and how it depends on the lattice type. We plan to address these problems in the future work.

4. Summary

We have studied heat capacity of different Coulomb crystals. The effects of the magnetic field, isotopic impurities, polarization of the electron background and the type of the crystal lattice are discussed. At high temperatures, the difference between lattices is not significant. At low temperatures, the heat capacity of some crystals may differ from the heat capacity of the bcc lattice with uniform background by up to 50%. The magnetic field does not play any role in the thermal evolution of white dwarfs while isotopic impurities and the type of the crystal lattice change the cooling rate by a few per cent only. The polarization of the electron background play more noticeable role in the cooling process of massive white dwarfs at age of $10^{9.5}$ years or more and should be taken into account. It was also shown that the difference between the previously used heat capacity model and more precise calculations does not noticeably affect the thermal evolution of white dwarfs.

Acknowledgments

The author is deeply grateful to D.A. Baiko and D.G. Yakovlev for help and discussions.

References

[1] Segretain L, Chabrier G, Hernanz M, Garcia-Berro E, Isern J and Mochkovitch R 1994 *Astrophys. J.* **434** 641–651
[2] Kanaan A and et al 2005 *A&A* **432** 219–224
[3] Shapiro S L and Teukolsky S A 1983 *Black Holes, White Dwarfs, and Neutron Stars* (New York: Wiley-Interscience)
[4] Haensel P, Potekhin A Y and Yakovlev D G 2007 *Neutron Stars 1: Equation of State and Structure* (New York: Springer)
[5] Pollock E L and Hansen J P 1973 *Phys. Rev.* A **8** 3110–3122
[6] Ichimaru S, Iyetomi H and Ogata S 1988 *Astrophys. J. Lett.* **334** L17–L20
[7] Chabrier G 1993 *Astrophys. J.* **414** 695–700
[8] DeWitt H and Slattery W 2003 *Contr. Plasma Ph.* **43** 279–281
[9] Horowitz C J and Berry D K 2009 *Phys. Rev.* C **79** 065803
[10] Hughto J, Horowitz C J, Schneider A S, Medin Z, Cumming A and Berry D K 2012 *Phys. Rev.* E **86** 066413
[11] van Horn H M 1968 *Astrophys. J.* **151** 227–238
[12] van Horn H M 1971 *IAU Symp. 42: White Dwarfs* (Dordrecht: Springer) ed Luyten W J 97–115
[13] Hansen B M S 1999 *Astrophys. J.* **520** 680-695
[14] Althaus L G, Corsico A H, Isern J and Garcia-Berro E 2010 *A&A Rev* **18** 471
[15] Kozhberov A A and Baiko D A 2015 *Ap&SS* **359** 50
[16] Baiko D A, Potekhin A Y and Yakovlev D G 2001 *Phys. Rev.* E **64** 057402
[17] Baiko D A 2002 *Phys. Rev.* E **66** 056405
[18] Baiko D A 2009 *Phys. Rev.* E **80** 046405
[19] Baiko D A and Yakovlev D G 2013 *MNRAS* **433** 2018–2027
[20] Lifshitz I M 1952 *Uspekhi Mat. Nauk* **7** 171–180
[21] Kozhberov A A and Baiko D A 2014 *Contr. Plasma Ph.* **54** 859–867

IOP Conf. Series: Journal of Physics: Conf. Series **929** (2017) 012013 doi:10.1088/1742-6596/929/1/012013

The method for celestial bodies' center of mass position relative to their figures determination on the basis of harmonic analysis of the expansion in spherical functions in order to refine the physical libration parameters

N Y Demina[1,2], A O Andreev[2], S A Demin[1,2], Y A Nefedyev[1,2]
[1] Kazan Federal University, Institute of Physics, Kazan, 420008 Russia
[2] Kazan Federal University, Engelhardt Astronomical observatory, Zelenodolski region, st. Observatory, 422526 Russia

E-mail: vnu_357@mail.ru

Abstract. In this paper the problem of the lunar center of mass relative to the center of its figure determination on the basis of space observations is considered, since the Moon is the most studied celestial object and there is a complete database on it. The future prospects for lunar laser ranging and radio interferometry require development of adequate theoretical support for modern technologies. The aim of these studies is the distances' measurement between the Moon and the Earth with an accuracy of 1 mm. Thus, determination of the lunar center of mass position, represented in this paper, and development of the selenocentric system will allow to solve the above mentioned problem more accurately and reliably. The new values of the lunar center of mass relative to its center of figure in orthogonal selenographic coordinate system $\Delta\xi$, $\Delta\eta$, $\Delta\zeta$ have been determined; they are: -1.75, -0.75, 0.11 km respectively.

1. Introduction

The modern software used for space researches apply methods of structural analysis of celestial bodies' internal structure demonstrating advantages and capabilities of an interdisciplinary approach: astrophysical observations and theoretical simulation are complemented by geophysical methods. The special role is played by physical librations parameters analysis, especially in terms of numerical theories construction. The basic problem of such theories is still planetary core detection and its main parameters determination, particularly a celestial body's center of mass position relative to its center of figure. In case of the Earth solution for this problem is usually obtained on the basis of significant amount of ground and space observations, but for other celestial bodies the capabilities are quite limited. In the present work the method for the position determination of the lunar center of mass (LCM) relative to the center of the lunar figure (LCF) is represented, since for our natural satellite there is the most complete database on space observations.

It should also be noted that the numerical approach to construct a theory of lunar physical libration (LPL) [1, 2] allows to consider many elements of the lunar body's internal stratigraphy. This gives an opportunity to get the high-precision modern data from Lunar laser ranging (LLR). Also this idea is worth developing for the analytical approach to LPL development as well for the unique opportunities to separate forced and free librations and to identify many subtle effects of lunar rotation with the physical nature of their origin. The similar approach to obtain some information on the basis of various models of LPL has promoted a series of papers in which these problems were solved [3, 4].

Today LLR, which has been implemented for more than 45 years, is one of the most efficient sources of information about the Moon. Precision of laser measurements has reached the sufficient level to define even relativistic effects in the «Earth-Moon» system. Analysis of laser data at definition of lunar rotation parameters has allowed to not only refine numerical characteristics of the dynamical lunar figure (nondimensional moments of inertia and coefficients of elasticity), but to clearly define amplitudes and phases similar to Chandler wobble (or variation of latitude) modes in free librations and at the same time to reveal the presence of strong dissipation of the rotation. Though the problem of long-term free libration existence at the presence of strong rotational dissipation is not yet solved, still at the present time a number of significant steps are made. As it was proved in the series of works [1, 3], one of the reasons for dissipation is not only tidal friction, but the possible presence of lunar liquid core as well. Dissipation processes at differential core rotation and viscoelastic mantle cause a deviation of lunar rotation from Cassini laws: as a result of the dissipation the lunar axis of rotation is out of plane, in which the axis of ecliptic and the lunar equator lie, for a constant of 0.26 arc seconds [5].

Precision of LPL description is primarily determined by the accuracy of selenopotential description. The dynamical model depends on Stokes coefficients obtained from gravimetric satellite measurements and nondimensional moments of inertia defined from LLR. The prospects of laser ranging and radio interferometry of the Moon within the international cooperation of space powers both in field of observation and processing and interpretation observational data require the development of adequate theoretical support for the modern technologies of determination the distance between Earth and Moon. Thus, the determination of the lunar center of mass, represented in the present work, and as a result – the construction of the selenocentric system, will allow to solve the above mentioned problems with higher accuracy and reliability.

2. Formulation of the problem and the method for selenocentric model construction on the basis of expansion in spherical functions

Currently all the data on selenophysics may be divided into 2 types. The first one is obtained by the lunar physical surface scanning with «board satellites», this well describes the physical surface, however does not allow to determine coordinates of reference objects situated on the Moon [6]. The other type is obtained on the basis of lunar objects direct binding to the stars; this allows to determine coordinate of reference objects but does not describe the lunar surface with a sufficient accuracy. The mentioned systems have different reference systems and axes of orientation. It is well known that data obtained as a result of all space missions refers to quasi-dynamical coordinate system in which the origin is the lunar center of mass with axes not coinciding with the lunar axes of inertia.

In the majority of modern selenodetic catalogues the quasi-dynamical coordinate systems are used as well. In these systems either an origin does not coincide with LCM or axes do not coincide with the lunar axes of inertia. It should also be noted that currently there is no dynamical selenocentric reference system constructed on the basis of satellite observations covering sufficient part of the lunar surface [7].

The problem of LCM determination relative to LCF is important for the lunar evolution and its structure investigation as well as for the construction of spin-orbital theory; it is also important for the increase in precision of the near-Moon navigational tasks solution. If LCF is determined by geometrical data (altitudes of objects) and LCM is defined on the basis of the external gravitational field (orbital etc.), then the position of the LCM may be defined relative to the center of a sphere approximating the Moon. Firstly, these results have been obtained in Englehardt Astronomical Observatory on the basis of processing of meridian observations made in Royal Observatory (Greenwich) [8] and analysis of the lunar marginal zone's maps data [9]. The recent space missions [10-11] have also given some large informational material to define relative position of LCM and LCF.

We suggest solving the presented problem by constructing selenocentric hypsometrical model based on the catalogue of 272 931 reference objects [7]. For the construction of the selenocentric model various observations and robust methods may be used [12]. In this work as a model describing the relief on the lunar surface the expansion of the altitude function in a series of spherical harmonics in the form of regression has been used [6]:

$$h(\varphi, \lambda) = \sum_{n=0}^{N} \sum_{m=0}^{n} (\bar{C}_{nm} \cos m\lambda + \bar{S}_{nm} \sin m\lambda) \cdot \bar{P}_{nm}(\cos\varphi) + \varepsilon, \quad (1)$$

where:

$h(\varphi, \lambda)$ is altitude function;

φ, λ are latitude, longitude – known parameters of the lunar objects;

$\bar{C}_{nm}$, $\bar{S}_{nm}$ are normalized harmonic amplitudes;
$\bar{P}_{nm}$ are normalized associated Legendre functions;
ε is random regression error.

With the existing software on the basis of formula (1) and harmonic analysis the Lunar Selenocentric Dynamical System (LSDS), which later was used as a work tool for the analysis of three-dimensional distribution of the center of mass position, had been constructed. At the construction of the model (1) estimates of precision and reliability of observance of the Least Square Method conditions have been carried out. In case of their violations the corresponding adaptation methods have been applied. Solution of the redefined system for various sources of hypsometrical information has been implemented within the approach of regression simulation involving in addition to ordinary stages (postulation of the model (1) and amplitudes $\bar{C}_{nm}$, $\bar{S}_{nm}$ estimation) the use of a number of quality statistics including external measures, diagnostics of observance of the Least Square Method basic conditions, adaptation in case of their violation. As calculation schemes of the Least Square Method the schemes of Gauss-Jordan and Householder have been used. Noise harmonics are removed by step by step regression.

3.The method of harmonic analysis for position of lunar center of mass determination

There are several methods to determine the position of the lunar center of mass based on observations:

1) The position of LCM gets determined relative to the center of the sphere approximating the Moon, if LCF is defined by geometrical data (objects' altitudes) and LCM – by the external gravitational field (motion of the lunar artificial satellites etc.). LCM has been fixed as a center of a system of objects' on its surface coordinates catalogue obtained from various sources, and hypsometrical data has been used to represent altitude (as a function of spherical coordinates) in the form of the expansion in spherical harmonics. After that by the amplitudes of first order the position of LCF relative to LCM is defined [13];
2) The second method is based on direct use of selenocentric catalogue of objects' being identified coordinates which are uniformly distributed on the surface of the Moon. In the present work this method is used. When conducting the present investigation, it is assumed that all the sources of hypsometrical data used in calculations are equally accurate.

In connection with the fact that initial data is unevenly distributed on far and near side of the Moon, determination of mean radius requires determination of mean radius for both sides and calculation of their arithmetic mean [13]:

$$R = R_0 + \frac{1}{2}\left(\Delta R_{near} + \Delta R_{far}\right),$$

where $R_0 = 1\,738$ km, $\Delta R = \sum_{i=1}^{n} H_i p_i \cos\varphi_i / \sum_{i=1}^{n} p_i \cos\varphi_i$. Here H_i are altitudes of points relative to mean sphere with a center in LCM; n is number of the measured points; $p_i \cos\varphi_i$ are points' statistical weight ($p_i = 1$ or 0.1). We can write:

$$\sqrt{p_i \cos\varphi_i} \sum\nolimits_{n=1}^{N} \sum\nolimits_{m=0}^{n} S_{nm}\, i = \sqrt{p_i \cos\varphi_i}\,(H_i - \Delta R),$$

where $S_{nm}\, i = (\bar{C}_{nm} \cos m\lambda + \bar{S}_{nm} \sin m\lambda) \cdot \bar{P}_{nm}(\cos\varphi_i)$.

$\bar{P}_{nm}(\cos\varphi_i)$ – normalized associated Legendre polynomials, which can be defined according to the formula:

$$\bar{P}_{nm}(\cos\varphi) = \sqrt{\frac{\varepsilon_m (2n+1)(n-m)!}{(n+m)!}} P_{nm}(\sin\varphi)\,, \varepsilon_0 = 1, \varepsilon_0 = 2, m = 1{,}2\,...$$

As the system of spherical functions is orthogonal, the coefficients of each order $N = 1 - N_k$ are defined by consistent solution of N_k sets of equations of the following type:

$$\sqrt{p_i \cos\varphi_i} \sum_{m=0}^{N} S_{nm}\, i = \sqrt{p_i \cos\varphi_i} \left(H_i - \Delta R - \sum_{n=1}^{N-1} \sum_{m=0}^{n} S_{nm}\, i \right).$$

First order coefficients define the displacement of LCF relative to LCM [13]:

$$h\Delta\xi = \sqrt{3C_{11}}, \Delta\eta = \sqrt{3S_{11}}, \ \Delta\zeta = \sqrt{3C_{10}}, \tag{2}$$

where ξ is an axis directed towards the Earth; η – equatorial axis directed orthogonally to ξ; ζ coincides with the axis of lunar rotation; $\bar{C}_{11}, \bar{S}_{11}, \bar{C}_{10}$ – normalized amplitudes of first order of expansion in spherical functions.

The stage of mathematical models' parameters estimation is one of the most important computational procedures at processing of observations [14]. The problem of mathematical processing correctness when combining requirements for the results accuracy and reliability exist at the process description along with the problem of choice between formal (approximating) and geometrical (cause-effect) models. Unfortunately, the traditional approach to parameters estimation involving a strictly fixed model and the least square method (LSM) application [15] does not satisfy the requirements of practice and capabilities of the methodology based on statistical computer simulation. Some attempts to go beyond the standard LSM [16] focus on solution of particular problems and do not involve any systematical approach to the task. The typical limits at ground observations and space data processing may be the presence of insignificant, uninformative and duplicate (dependent on each other) terms of the expansion, violation of the LSM conditions – the normal scheme of Gauss-Markov [14]. Noise leads to decrease of the parameters determination and the LSM estimates accuracy. The problem is getting more complicated due to the fact that, when estimating, there is no check for the model adequacy to the observations. Researchers in most cases are unaware of the actual state of affairs and do not apply any adaptive computational schemes, so the LSM conditions observance does not get checked.

At the present work, when estimating the selenocentric models, the methodology of regression modeling as an alternative to the traditional approach [14], involving in the tasks of estimation the regression analysis, check of the conditions, adaptation in case of their violation, and providing the presence of a special software, automating processes of computation and analysis, has been used. Regression modeling is a systematic approach at which correctness of the system's any element application (sampling, model, method for parameters estimation, method for structures estimation, quality measure, a set of conditions) may be questioned and checked with a corresponding adaptation application in case of the given conditions violation.

Comparison of the calculated amplitudes values and the recent results is a subject of considerable interest. The digital relief model for the full sphere of the Moon has been obtained from «Clementine» space mission [10], «ULCN 2005» [17], «KAGUYA» [11] and compared with LSDS.

The table 1 provides the normalized coefficients of the first order of the lunar relief's function expansion for 4 sources of hypsometrical information.The line "0, 0" contains amendment to the accepted mean radius of the Moon. The fifth and sixth table lines contain:The first horizontal cell there are n and m values responsible for order and degree of the expansion (1) respectively; the second horizontal cell of the table contains estimates of harmonic amplitudes $\bar{C}_{nm}$ and $\bar{S}_{nm}$ obtained on the basis «Clementine» space mission data ([6] provides expansion up to the 40th order); the third horizontal cell contains estimates of harmonic amplitudes $\bar{C}_{nm}$ and $\bar{S}_{nm}$ obtained on the basis of the project «ULCN» data ([17] provides expansion up to 359th order); the fourth horizontal cell contains estimates of harmonic amplitudes $\bar{C}_{nm}$ and $\bar{S}_{nm}$ obtained on the basis of «KAGUYA» space mission ([18] provides expansion up to the 180th order); the fifth horizontal cell contains estimates of harmonic amplitudes $\bar{C}_{nm}$ and $\bar{S}_{nm}$ obtained on the basis of «LSDS» model (as a result, expansion of 5th order is model). For «LSDS» model the fifth order of the expansion is sufficient, since its increase does not lead to any significant change of the mean square error.

Table 1. Normalized coefficients of the first order of the lunar relief's function expansion for four sources of hypsometrical information, km.

n, m	**CLEMENTINE [6]**		**ULCN [17]**		**KAGUYA[18]**		**LSDS**	
	$\bar{C}_{nm}$	$\bar{S}_{nm}$	$\bar{C}_{nm}$	$\bar{S}_{nm}$	$\bar{C}_{nm}$	$\bar{S}_{nm}$	$\bar{C}_{nm}$	$\bar{S}_{nm}$
1	2		3		4		5	
0, 0	-0.83		-1.07		-0.94		-0.87	
1, 0	-0.37	0	0.15	0	0.14	0	0.11	0
1, 1	-1.04	-0.43	-0.99	-0.42	-1.02	-0.45	-0.86	-0.40

Based on equation (1) and data from Table 1 the coordinates of the center of mass relative to the lunar center of figure have been determined (Table 2). In order to obtain mean square errors of displacements, the amplitudes in Table 1 should be multiplied by $\sqrt{3}$.

Table 2. Coordinates of the Moon's center of mass relative to its center of figure for four sources of hypsometrical information.

	CLEMENTINE (km)	**ULCN (km)**	**KAGUYA (km)**	**LSDS (km)**
$\Delta\xi$	-1.80	-1.71	-1.77	-1.75
$\Delta\eta$	-0.74	-0.73	-0.78	-0.75
$\Delta\zeta$	-0.64	0.26	0.24	0.11

In Table 2 values of $\Delta\xi, \Delta\eta, \Delta\zeta$ are the differences in positions of the lunar center of mass relative to its center of figure in orthogonal selenographic coordinate system. Analysis of this data shows that selenocentric model LSDS, whose system is brought to the center of mass and to main axes of inertia of the Moon, is in good agreement with the results of the recent studies. In this work the method of direct harmonic analysis of the selenocentric catalogue of the lunar objects' coordinates in order to determine the position of the lunar center of mass relative to its center of figure has been used. This approach may be applied for an analysis of other celestial objects.

4. Summary and conclusions

During this work the data on mutual positions of the lunar center of mass and its center of figure has been obtained as a basis for construction a numerical theory of the Moon. At the same time the new method of direct harmonic analysis of the investigated systems on the basis of the analysis of the known reference points' radius vectors has been developed. It should also be noted, time series of observations of celestial systems have a complex dynamics, that is why at the analysis the methods of non-equilibrium statistical physics [19–21] as well as comparative stochastic analysis [22] may be used.

In prospect, the obtained results and methods will be used at the analysis of other bodies' of the Solar system spin-orbital characteristics. In particular, at the present time a problem of Mars' and its 2 moons' Phobos and Deimos spin-orbital characteristics analysis is a subject of great interest, which is confirmed by launch of global multibillion projects in Russia as well as in ESO, NASA and other space agencies of Japan and China.

Acknowledgments

The work is performed according to the Russian Government Program of Competitive Growth of Kazan Federal University. This work was supported by the Russian Foundation for Basic Research, project nos. 15-02-01638 a, 16-02-00496 a, 16-32-60071 mol_a_dk (N.D.).

References

[1] Rambaux N and Williams J G 2011 *Celestial Mechanics and Dynamical Astronomy* **109** 85
[2] Williams J G, Boggs D H, Yoder Ch F, Ratcliff J T and Dickey J O 2001 *Journal of Geophysical Research: Planets* **106** 27933
[3] Williams J G, Turyshev S G, Boggs D H and Ratcliff J T 2006 *Advances in Space Research* **37** 67
[4] Dickey J O, Bender P L, Faller J E, Newhall X X, Ricklefs R L, Ries J G, Shelus P J, Veillet C, Whipple A L, Wiant J R, Williams J G and Yoder C F 1994 *Science* **265** 482
[5] Williams J G, Boggs D H and Folkner W M 2008 *Jet Propulsion Laboratory, California Institute of Technology, Interoffice Memorandum*
[6] Nefedyev Yu, Valeev S, Mikeev R, Varaksina N and Andreev A 2012 *Advances in Space Research* **50** 1564
[7] Andreev A O, Demina N Y, Demin S A, Nefedyev Y A and Churkin K O 2016 *Nonlinear Phenomena in Complex Systems* **19** 271
[8] Rizvanov N G, Nefed'ev Yu A and Kibardina M I 2007 *Solar System Research* **41** 140
[9] Nefedjev Yu A and Rizvanov N G 2002 *Astronomische Nachrichten* **323** 135
[10] Smith D E, Zuber M T, Neumann G A, Lemoine F G 1997 *Journal of Geophysical Research: Planets* **102** 1591
[11] Araki H, Tazawa S, Noda H, Ishihara Y, Goossens S, Sasaki S, Kawano N, Kamiya I, Otake H, Oberst J, Shum C 2009 *Science* **323** 897
[12] Nefedyev Y A, Bezmenov V M, Demin S A, Andreev A O and Demina N Y 2016 *Nonlinear Phenomena in Complex Systems* **19** 102
[13] Chuikova N 1975 *Astronomicheskii Zhurnal* **52** 1279
[14] Valeev S G 2001 *Regression modeling for data processing* (Kazan: FEN) pp 1–296
[15] Lawson C L and Hanson R J 1995 *Solving Least Squares Problems* (Philadelphia: Society for Industrial and Applied Mathematics) pp 1–336.
[16] Gubanov V S 1997 *Generalized Least Squares. Theory and Application in astrometry* (SPb.: Science) pp 1-318
[17] Archinal A, Rosiek M R, Kirk R L and Redding B L 2006 *U.S. Geological Survey: Open-File Report* **1**
[18] Araki H, Tazawa S, Noda H, Tsubokawa T, Kawano N and Sasaki S 2008 *Advances in Space Research* **42** 317
[19] Demin S A, Panischev O Y and Nefedyev Y A 2014 *Kinematics and Physics of Celestial Bodies* **30** 63
[20] Demin S A, Panischev O Y and Nefedyev Y A 2014 *Nonlinear Phenomena in Complex Systems* **17** 177
[21] Demin S A, Panischev O Y and Nefedyev Y A 2015 *Journal of Physics: Conference Series* **661** 012003
[22] Sokolova M, Kondratyeva E and Nefedyev Y 2013 *Advances in Space Research* **52** 1217

IOP Conf. Series: Journal of Physics: Conf. Series **929** (2017) 012014 doi:10.1088/1742-6596/929/1/012014

Particle-in-cell simulation of shock waves with different kinds of ions.

V I Romanky[1,2], A M Bykov[1], S M Osipov[1] and P E Gladilin[1]

[1] Ioffe Institute, 26 Politekhnicheskaya st., St. Petersburg 194021, Russia

[2] Sternberg Astronomical Institute, Moscow State University Universitetsky pr., 13, Moscow 119234, Russia

E-mail: romanskyvadim@gmail.com

Abstract. The problem of Cosmic Ray (CR) origin and acceleration is a widely discussed problem of modern astrophysics. It is considered that most likely CR sources are supernova remnant shock waves. One of the methods to study particle injection and acceleration by shock waves is a particle-in-cell computer simulation. In this paper we present results of shock wave simulations based on our self-developed implicit particle-in-cell code. First, we test the code with the well-known two stream and Weibel instabilities and second, simulate shock waves in plasma with different particle composition. Using the obtained results we study the process of charged particle acceleration at the shock front.

1. Introduction

One of the possible sources of high-energy cosmic rays are shock waves in astrophysical collisionless plasma due to the diffusive shock acceleration (DSA) mechanism, see [1], [2]. Strong magnetic fields are necessary for effective acceleration of charged particles and the observations of the synchrotron emission from supernova remnants (SNRs) show, that the magnetic field there is about 100 times stronger then the interstellar field (see for example [3],[4]). So strong fields can be produced due to instabilities in anisotropic plasma. One of the most powerful methods to explore processes in collisionless plasma is the Particle-in-Cell simulation. In this work we explore instabilities in the plasma using this method.

2. Model of implicit particle-in-cell

We developed the implicit particle-in-cell (PIC) code, based on the scheme suggested by Lapenta et al. [5] and improved for the relativistic case by Noguchi et al.[6]. This scheme is described in details in these papers and below we present only a short description. The common assumption for all particle-in-cell codes is the consideration of the plasma as a superposition of a large number of finite elements, called super-particles. Distribution function can be represented as the sum of members for all super-particles:

$$F(\mathbf{x}, \mathbf{p}, t) = \sum_{p=1}^{N_s} S(x - x_p) S(y - y_p) S(z - z_p) \delta(\mathbf{p} - \mathbf{p}_p), \tag{1}$$

where $\mathbf{x}$ and $\mathbf{p}$ are coordinates and the momentum of a particle , respectively, S is a shape-function, and δ is the Dirac's delta-function. The shape-function is chosen in form of b spline.

Also we introduce interpolation function W which represents the part of a particle which is contained in the grid cell:

$$W(x_c - x_p) = \int_{-\infty}^{+\infty} S(x - x_p)\phi(x)dx, \tag{2}$$

where c and p are indexes of the cell and the particle, respectively, and ϕ equals 1 inside the cell and 0 outside. Using this function we can determine electric and magnetic field for each particle as

$$\begin{aligned} \mathbf{E}_p &= \sum_c \mathbf{E}_c W(x_c - x_p), \\ \mathbf{B}_p &= \sum_c \mathbf{B}_c W(x_c - x_p). \end{aligned} \tag{3}$$

Then we use particle fields to solve equations of motion for particles:

$$\begin{aligned} \frac{d\mathbf{x}_p}{dt} &= \mathbf{v}_p, \\ \frac{d\mathbf{p}_p}{dt} &= q_p\left(\mathbf{E}_p + \frac{\mathbf{v}_p \times \mathbf{B}_p}{c}\right), \end{aligned} \tag{4}$$

where q_p is the particle charge and $\mathbf{p}_p$ is the momentum.

At the same time, using interpolation functions we can determine macroscopic plasma parameters, such as the electric flux and the electric density :

$$\begin{aligned} \rho_c &= \sum_p q_p W(x_c - x_p), \\ \mathbf{J}_c &= \sum_p q_p \mathbf{v}_p W(x_c - x_p). \end{aligned} \tag{5}$$

The difficulty comes from the fact that particles coordinates and fields are coupled and we can not separate the equations of motion and Maxwell equations. The first possible solution is the explicit method: equations of fields use particle velocities on the previous time step and vice versa. Explicit approach is rather simple, but it has strong restrictions on stability. The other possible method is the implicit approach which is more complicated, but at the same time more stable. The main idea of it is to predict particle velocity at the next time step, using the field at the next time step and then use velocity in the implicit scheme for the field.

In the implicit approach we should introduce intermediate values $F^{n+\theta} = \theta F^{n+1} + (1-\theta) F^n$, where F^n is the value at the n-th time step and $F^{n+\theta}$ the value at the intermediate time step. θ is the parameter of the scheme and should be greater or equal to $\frac{1}{2}$ for stability. Using this denotations we write time discretisation of the Maxwell equations as:

$$\begin{aligned} &\nabla \times \mathbf{E}^{n+\theta} + \frac{\mathbf{B}^{n+1} - \mathbf{B}^n}{c\Delta t} = 0, \\ &\nabla \times \mathbf{B}^{n+\theta} - \frac{\mathbf{E}^{n+1} - \mathbf{E}^n}{c\Delta t} = \frac{4\pi}{c}\mathbf{J}^{n+\theta}, \\ &\Delta \cdot \mathbf{E}^{n+\theta} = 4\pi\rho^{n+\theta}, \\ &\Delta \cdot \mathbf{B}^n = 0. \end{aligned} \tag{6}$$

Time discretization of equations of particle motion reads as

$$\mathbf{x}_p^{n+1} = \overline{\mathbf{v}}_p \Delta t,$$
$$\mathbf{p}_p^{n+1} = \mathbf{p}_p^n + q_p \Delta t \left(\mathbf{E}_p^{n+\theta} \left(\mathbf{x}_p^{n+\theta} \right) + \frac{\overline{\mathbf{v}}_p \times \mathbf{B}_p^n \left(\mathbf{x}_p^{n+\theta} \right)}{c} \right), \tag{7}$$

where $\overline{\mathbf{v}}_p$ is the average velocity. It should be noted that the electric field is evaluated at the moment $n+\theta$, while the magnetic field at the moment n. If $\theta = \frac{1}{2}$ the scheme has the second order accuracy in Δt. Finally we can express average velocity explicitly using $\mathbf{v}_p^n$ and $\mathbf{E}_p^{n+\theta}$:

$$\overline{\mathbf{v}}_p = \widehat{\mathbf{v}}_p + \frac{q_p \Delta t}{2 m_p \Gamma_p} \alpha_p^n \mathbf{E}_p^{n+\theta} \left(\mathbf{x}_p^{n+\theta} \right), \tag{8}$$

where we use following notations:

$$\widehat{\mathbf{v}}_p = \alpha_p^n \gamma_p^n \mathbf{v}_p^n, \tag{9}$$

$$\alpha_p^n = \frac{1}{\Gamma_p \left(1 + \left(\frac{q_p \Delta t \mathbf{B}_p^n}{2 m_p \Gamma_p} \right)^2 \right)} \left(I - \frac{q_p \Delta t}{2 m_p \Gamma_p} I \times \mathbf{B}_p^n + \left(\frac{q_p \Delta t}{2 m_p \Gamma_p} \right)^2 \mathbf{B}_p^n \mathbf{B}_p^n \right), \tag{10}$$

$$\Gamma_p = \frac{q_p \Delta t \mathbf{B}_p^n}{2 m_p} \mathbf{E}_p^n \cdot \mathbf{v}_p^n + \gamma_p^n, \tag{11}$$

where I is the identity tensor and γ_p^n is the gamma-factor of the particle at the moment n. Using expression(8) for the average velocity we can expand in series expressions for the electric flux and density (5) by term $\mathbf{x}_p^n - \mathbf{x}_p^{n+\theta}$ and substitute them into the Maxwell equations (6). After solving algebraic equations, combining all terms and excluding magnetic field we finally obtain the implicit second-order elliptic equation for $\mathbf{E}^{n+\theta}$:

$$(c\theta\Delta t)^2 \left(-\nabla^2 \mathbf{E}^{n+\theta} - \nabla\nabla \cdot \left(\mu^n \cdot \mathbf{E}^{n+\theta} \right) \right) + \epsilon^n \mathbf{E}^{n+\theta} =$$
$$\mathbf{E}^n - c\theta\Delta t \left(\frac{4\pi \widehat{\mathbf{J}}}{c} - \frac{\Delta t}{2} \nabla \cdot \widehat{\Pi} - \nabla \times \mathbf{B}^n \right) - (c\theta\Delta t)^2 4\pi \nabla \widehat{\rho}, \tag{12}$$

where we used following notations:

$$\epsilon^n = I + \mu^n, \tag{13}$$

$$\mu^n \doteq - \sum_p \frac{2\pi q_p^2 \theta \Delta t^2}{m_p} \alpha_p^n W \left(x - x_p^n \right), \tag{14}$$

$$\widehat{\Pi} = \sum_p q_p \widehat{\mathbf{v}}_p \widehat{\mathbf{v}}_p W \left(x - x_p^n \right), \tag{15}$$

$$\widehat{\mathbf{J}} = \sum_p q_p \widehat{\mathbf{v}}_p W \left(x - x_p^n \right), \tag{16}$$

$$\widehat{\rho} = \rho^n - \theta \Delta t \nabla \cdot \left(\widehat{\mathbf{J}} - \frac{\Delta t}{2} \nabla \cdot \widehat{\Pi} \right). \tag{17}$$

The spacial discretization of the equation of the electric field (12) is performed with a finite-volume scheme and then the system of linear equations is solved using general minimal residual algorithm. After that we can explicitly evaluate magnetic field using the first Maxwell equation (6) and finally, when fields computed, we can solve equations of motion (7) using iterative non-linear solver.

Our code is fully three dimensional and parallelized with MPI technology, which is adapted for distributed computing and can be executed on a wide class of computers.

3. Verification of numerical model

We chose two well-known instabilities for testing our code: the two stream instability and the Weibel instability. Electromagnetic two stream instability is caused by two fluxes of electrons moving in opposite directions. Modes with the wave vector, normal to electron velocities, grow with the increment $\frac{v}{c}\frac{\omega_p}{\sqrt{\gamma}}$, where ω_p is the electron plasma frequency and γ is the gamma-factor of both electron fluxes. We study the case with electron velocity $v = 0.99c$. As shown in Figure 1 magnetic energy growth is very similar to linear growth with analytically computed increment. When the fraction of the magnetic energy becomes close to 10 percent of the full initial energy, the instability reaches its saturation limit. The total energy deviation at the end of the simulation does not exceed 5 percent of the instability energy.

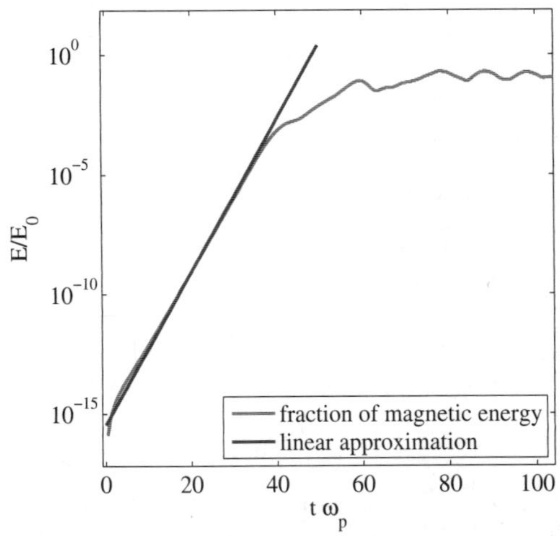

Figure 1. Time dependence of the ratio of the magnetic energy to the full initial energy in case of the two stream instability.

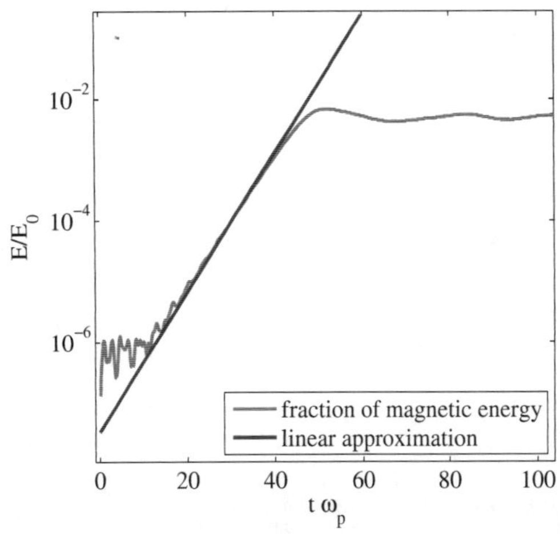

Figure 2. Time dependence of the ratio of the magnetic energy to the full initial energy in case of the Weibel instability.

The second test of the code is the Weibel instability [7], which occurs when the temperature in one spatial direction is smaller then in the other two. This instability has many different cases, we chose the one which was studied by Yoon et al. [8]. In this case electrons have delta distribution in transverse momentum and uniform one in parallel. Electron distribution function is:

$$F\left(p_\perp^2, p_\|\right) = \frac{1}{2\pi p_{0\perp}}\delta\left(p_\perp - p_{0\perp}\right)\frac{1}{2p_{0\|}}H(p_{0\|}^2 - p_\|^2), \tag{18}$$

where δ is the delta-function, H is the Heaviside function, $p_\perp, p_\|$ are momentum of particles in perpendicular and parallel directions , respectively, and $p_{0\perp}, p_{0\|}$ are constants which determine dispersion of distribution. Dispersion relation for this case can be directly solved, while the expression for the increment is rather cumbersome and will not be presented here for brevity. As shown in Figure 2 the numerical simulation with parameters $p_{0\perp} = 20m_ec$ and $p_{0\|} = 2m_ec$ is in good correspondence with the theory.

These results show that our PIC code gives consistent results in the simulation of plasma instabilities and can be applied to more complicated cases.

4. Shock wave simulation

We studied the influence of fractions of heavy ions on the shock wave, in particular, the spectrum of particles and the shape of the shock wave. We present results of two simulations with different compositions of the plasma : pure protons and electrons in the first case and with the admixture of alpha particles in the second case. Initially the homogeneous flux is moving from the right free boundary to the left. On the left side there is a reflecting super-conducting wall, which causes formation of shock wave. Simulations are one-dimensional and have following parameters: initial velocity $v = 0.2c$, number densities $n_e = 10^{-4}\text{cm}^{-3}$, $n_p = 10^{-4}\text{cm}^{-3}$ in the first simulation and $n_e = 10^{-4}\text{cm}^{-3}$, $n_p = 0.6 \cdot 10^{-4}\text{cm}^{-3}$, $n_\alpha = 0.2 \cdot 10^{-4}\text{cm}^{-3}$ in the second, temperature $5 \cdot 10^8$K, magnetic field $B_\| = 10^{-4}$G, $\text{B}_\perp = 0.7 \cdot 10^{-4}$G, the full size of the box $L = 1 \cdot 10^{12}$cm, the number of cells $N = 2 \cdot 10^4$. Electron mass is reduced to $m_e = \frac{m_p}{20}$. The full time of simulation is $T = 2000\omega_p{}^{-1}$.

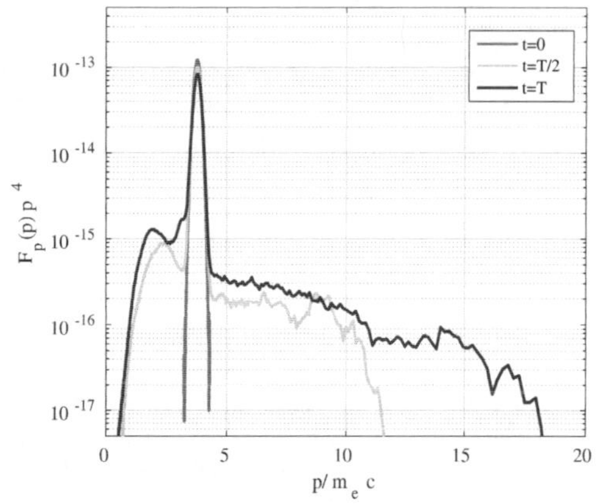

Figure 3. Distribution of protons in the simulation without alpha particles.

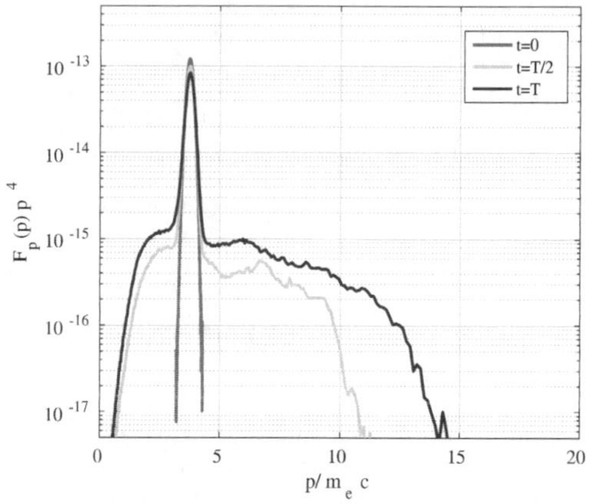

Figure 4. Distribution of protons in the simulation with alpha particles.

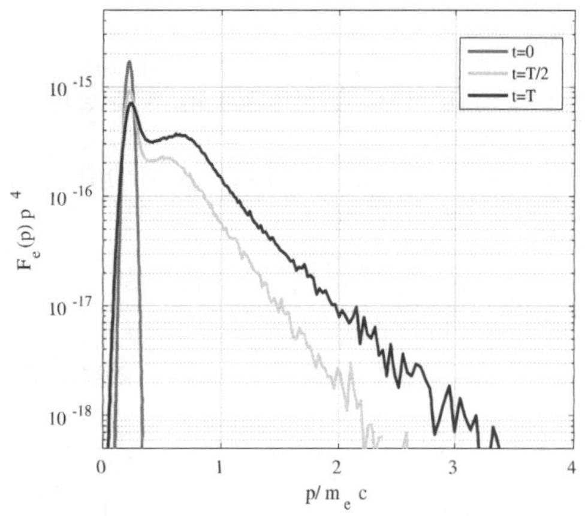

Figure 5. Distribution of electrons in the simulation without alpha particles.

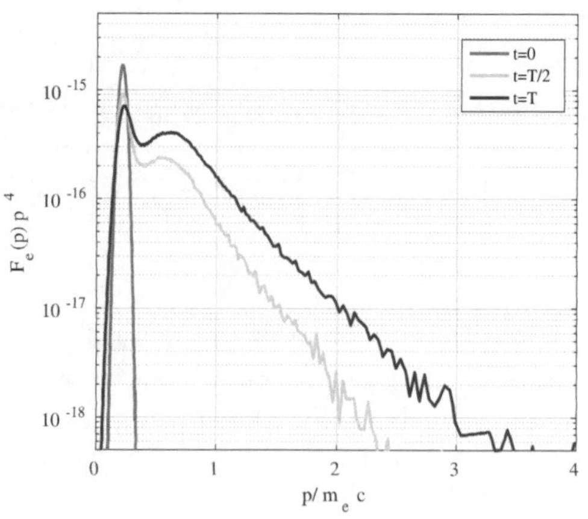

Figure 6. Distribution of electrons in the simulation with alpha particles.

The results presented in Figures 3-6 show that the values of $F(p)p^4$ for high energies are much greater for protons, than for electrons in both simulations. Also electrons need much more time to be accelerated and to form spectrum $\propto p^{-4}$. It means that protons are injected into the acceleration process more efficiently, and it is consistent with the work of Park et al. [9]. Also we have shown that minority of heavy ions increases the spectrum of protons and do not have influence on the spectrum of electrons. It can be explained by the fact, that heavy ions form the large scale turbulence and protons can efficiently scatter on this turbulence. Otherwise for electrons turbulence produced by protons is already enough large-scale and the influence of heavy ions is neglectable.

5. Conclusions

In this paper we presented the developed 3D implicit particle-in-cell code, which can be applied for various problems in the collisionless plasma. We have tested it and obtained results, that are consistent with theoretical predictions. We studied the influence of admixture of heavy ions on the evolution of the shock wave and simulations have shown that the presence of heavy ions increases the number of accelerated particles. Further research is necessary in order to study this phenomena in detail.

Acknowledgments

A M Bykov, P E Gladilin and S M Osipov have developed the implicit fields solver module, which provides conservation of the divergenceless magnetic field. V I Romansky acknowledges support from RSF grant 16-12-10519 which was used to develop the particle mover, parallelization and perform the testing of the code.

The results of the work were obtained using computational resources of Peter the Great Saint-Petersburg Polytechnic University Supercomputing Center (http://www.spbstu.ru).

References

[1] Bell A R 1978 *MNRAS* **182** 147
[2] Blandford R D and Ostriker J P 1978 *Astrophys. J.* **221** L29
[3] Berezhko E G, Ksenofontov L T and Völk H J 2003 *A&A* **412** L11
[4] Uchiyama Y, Aharonian F A, Tanaka T, Takahashi T and Maeda Y 2007 *Nature* **449** 576
[5] Lapenta G, Brackbill J U and Ricci P 2006 *Phys. Plasmas* **13** 055904
[6] Noguchi K, Tronci C, Zuccaro G and Lapenta G 2007 *Phys. Plasmas* **14** 042308
[7] Weibel E S 1959 *Phys. Rev. Lett.* **2** 83
[8] Yoon P H and Davidson R C 1987 *Phys. Rev. A* **35** 2718
[9] Park J, Caprioli D and Spitkovsky A 2015 *Phys. Rev. Lett.* **114** 085003

IOP Conf. Series: Journal of Physics: Conf. Series **929** (2017) 012015 doi:10.1088/1742-6596/929/1/012015

Modeling of relativistic pulsar wind interaction with the interstellar medium

A E Petrov, A M Bykov and S M Osipov

Ioffe Institute, 26 Politekhnicheskaya st., St. Petersburg 194021, Russia

E-mail: a.e.petrov@mail.ioffe.ru

Abstract. The consistent interpretation of the multiwavelength pulsar wind nebula observations require studying of interaction of a relativistic pulsar wind with the interstellar medium (ISM). In this paper we present the results of the Monte-Carlo modeling of the test particle propagation in a pulsar wind nebula beyond the pulsar wind termination shock. The non-thermal electron-positron spectra in the pulsar wind nebula and in the vicinity of the pulsar bow shock are derived in case of the supersonic proper motion of the pulsar through the ISM. The synchrotron emission images of the pulsar wind nebula at photon energies in the far ultraviolet and soft X-ray ranges are constructed. We discuss a possible interpretation of the recent HST and Chandra observations of the pulsar wind nebula associated with the PSR J0437-4715 in the frame of the model.

1. Introduction

Multiwavelength observations of the pulsar wind nebulae (PWNe) produced by the relativistic pulsar winds (PWs) with modern orbital and ground-based telescopes make possible investigations of efficient processes of energy conversion in these systems. These studies, of course, need a consistent interpretation of the observational data. In case of the supersonic motion of the pulsar through the ISM it requires modeling of an energy dependent transport of relativistic particles from the PW termination shock to the ISM.

Indeed, in this case a bow shock (BS) occurs, and the bubble of the shocked PW is separated from the shocked ISM matter by a contact discontinuity (CD) as observed by the *Hubble Space Telescope* and *Chandra* (see e.g. [1]). The wind flow, which is heated and decelerated down to a subsonic velocity at its termination shock (TS), collides with the flow of the shocked ISM matter driven by the pulsar motion. Thus, a complex picture of the flows occurs (see, e.g., the results of the simulations in [2, 3]). The magnetic fields in the pulsar wind plasma as well as in the heated postshock flow downstream the BS are thought to be strongly turbulent. Thus, electrons and positrons of the PW plasma injected in the PWN at the TS experience multiple scatterings at the stochastic magnetic field fluctuations of different scales frozen into the plasma flows. Therefore, the motion of the PW particles can be described via the transport equation in terms of diffusion and advection. The presence of the shocked ISM matter flow with the turbulent magnetic field enhanced at the bow shock can result into a confinement of the low energy PW particles in the PWN – BS system. Moreover, the colliding flows may provide an acceleration of the PW particles due to their multiple diffusion across the CD, which was found in [4] to be very fast and efficient. The described effects can strongly influence on the distribution of the synchrotron emission intensity from the source. Hence, the processes related to the PW particles transport through the PWN with the BS (the BSPWN) may affect the observed picture significantly.

The morphologies of the sources associated with the BSPWNe can differ in various energy ranges. An interesting example of such source is the nebula related to the pulsar PSR J0437-4715. A bow-like structure observed in $H\alpha$ and in the far ultraviolet (FUV) range (6—10 eV) is absent in the soft X-ray range (0.5—7 keV) [1].

In the present work the results of the Monte-Carlo modeling of the relativistic PW particles propagation into the ISM are presented. The test particles transport through the system with specified geometry, diffusion parameters and flows is considered. The results of the particle spectra modeling in the vicinity of the PWN and the BS are presented. Possible interpretation of the observational data for the source related to PSR J0437-4715 in the framework of the developed model is discussed.

2. The Monte-Carlo particle transport model

As was pointed in [5], the modeling of the plasma flows in the BSPWNe with an adequate account for the magnetic field structure is a complicated problem (see e.g. [2, 3]) and it is not fully resolved yet. A non-trivial behaviour of the magnetized flows near the CD imposes some complexities on the modeling of the particle transport there. This problem is alleviated for the rather high energy electrons and positrons responsible for the UV and X-ray synchrotron emission given their large mean free paths. Together with the dissipation of the magnetic inhomogeneities during their transport to the CD and the deceleration of the flows this allows a significant simplification of the problem.

The model presented here is discussed in more details in [5]. Its essential feature is taking into account some geometrical properties of the BSPWNe.

2.1. Geometry

The simulation area is presented as the composition of the axisymmetric regions and is bounded by the cylinder with radius ρ_{sys} and height h_{sys} (see the sketch shown in Figure 1). At this cylinder the free escape boundary conditions are imposed. In each region the mean free path (mfp) of a particle is energy-dependent but homogeneous in the coordinate space. The innermost region bounded by the sphere with radius r_{ts} corresponds to the cold PW. Through its surface (the TS) the particles are injected into the system. The outer boundary of the region "1" corresponding to the shocked pulsar wind (the PWN) has the following shape:

$$r(\theta) = \begin{cases} a_{\mathrm{in}}, \text{ if } \theta \leq \pi/2 \\ a_{\mathrm{in}}/\sin\theta, \text{ if } \theta > \pi/2. \end{cases} \tag{1}$$

Here the spherical coordinates (r, θ, ϕ) with the origin at the pulsar and the zenith direction aligned with the pulsar velocity vector are introduced. The scaling factor a_{in} is a parameter of the geometrical model. The flow velocity downstream of the TS is directed radially and has the value $u_w(r) = \frac{c}{3} r_{ts}^2 / r^2$. The label "3" marks the region corresponding to the postshock flow of the ISM matter. Its boundaries are described by the equation (2) which was derived in [6] for the BS shape in the thin-shell limit:

$$r(\theta, \sigma) = \sigma r_0 \csc\theta \sqrt{3(1 - \theta \cot\theta)} \tag{2}$$

The scaling factor r_0 is the standoff distance to the CD determined by the pulsar spin-down power, the proper velocity of the pulsar and the ISM mass density (see e.g. [6]). The model parameter σ determines the positions of the inner and outer boundaries of the region. Their apexes are denoted as a_{out} and a_{sys}, respectively. The region labeled with "4" corresponds to

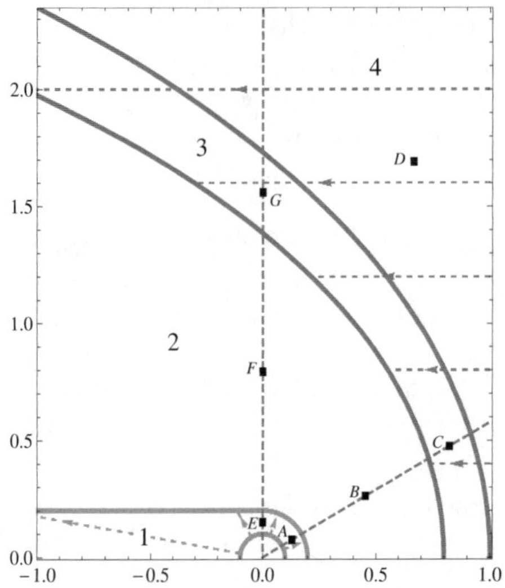

Figure 1. Sketch of the model geometry: half of the axial section of the system. Region 1 (between the orange curves) – the shocked PW, 2 — the region near the contact discontinuity, 3 (between the red curves) — the postshock flow of the ISM matter, 4 — the ISM. The black squares labeled with letters $A - G$ mark the bins chosen to show the simulation results for the particle spectra. The angles between the symmetry axis and the gray dashed lines are 30° and 90°. The chosen bins are located at the midpoints of the segments between the intersections of these lines with the surfaces separating the regions.

the unperturbed ISM. The velocity of the flow through the BS in the pulsar rest frame is chosen uniform, directed oppositely to the pulsar velocity and equal to $u_{\rm PSR} = 104$ km/sec in both third and fourth regions. In the zone near the CD (region "2") the advection is neglected.

According to the Monte-Carlo model of the relativistic particle propagation (see, e.g. [7]), the phase space of the system is divided into bins – small volumes of the phase space used to 'detect' the particle spatial and energy distribution. The model we developed allows to get the mean value of the distribution function in each bin. In the simulations discussed below the spatial binning is chosen as follows. The cylinder containing the system in the coordinate space is splitted into N_x equal pieces along the axis. Due to the axial symmetry of the system, each piece is divided into N_ρ coaxial cylindrical layers, whose outer radii are $\rho_n = n\rho_{\rm sys}/N_\rho$, where $n = 1, 2, \ldots, N_\rho$. To compensate the effect of the growth of the bin volumes with the growth of the distance from the axis on the detected values, the latter are divided by the ratio of the current bin volume to the volume of a bin with $n = 1$. In the momentum space the logarithmic binning is used.

2.2. Resonant PW particle confinement at BS

The cosmic ray protons acceleration at the BS can generate the cosmic ray driven current instabilities which provide a growth of the magnetic field fluctuations intensities in some range of the wavenumber. This can produce decreasing of the mfp for particles of the corresponding energy range. It can be shown that the acceleration of 100 GeV protons gives the sizeable effect in the energy range matching the synchrotron emission of electrons and positrons in FUV in the magnetic field of a few units of the mean induction of the turbulent ISM field $B_{\rm ISM} = 3$ μG. Low mfp near the BS causes a longtime confinement of such particles in the system. This can give rise to the enhancement of the observed structures brightness in the FUV range.

2.3. Magnetic field and mfp

A spatially uniform mean value of the magnetic field is specified in each of the regions "1"-"4". The mfp in the PWN, λ_1, is taken proportionally to the gyroradius r_g, in the ISM — in the form of (4) and in the regions "2"–"3" near the CD and close to the BS — in the form of (5)

$$\lambda_1 (B, \eta, \gamma) = \eta r_g (B, \gamma) = \eta \frac{mc^2 \gamma}{eB} \tag{3}$$

$$\lambda_{\text{ism}} (\gamma) = 3 \times 10^{18} \left(\frac{E}{1 \text{ GeV}} \right)^{1/3} \text{cm} \tag{4}$$

$$\lambda_{\text{bow}} (B, \eta, \gamma) = \begin{cases} \eta r_g (B, \gamma), \text{ if } \gamma \leq \gamma_1 \\ \eta r_g (B, \gamma) \gamma / \gamma_1, \text{ if } \gamma_1 < \gamma \leq \gamma_2 \\ \lambda_{\text{ism}} (\gamma), \text{ if } \gamma > \gamma_2 \end{cases} \tag{5}$$

Here e, m are the charge and the mass of an electron (positron), c – the speed of light, $E = mc^2\gamma$ — the particle energy, γ – the particle Lorentz-factor; $\eta \geq 1$ and γ_1 are the free parameters of the model. The value of γ_1 regulates the energy range in which the particles are confined longer in the system owing to the resonant scattering mechanism. The particles with the Lorentz-factor above γ_1 are weakly scattered by the short scale waves. The parameter γ_2 is defined by the energy at which the mfp reaches the mean ISM value, given by (4).

3. Particle spectra modeling

Here the results of the Monte-Carlo modeling of the particle distribution functions are presented. The chosen parameters of the system geometry, the mfp values and the mean magnetic field magnitudes in the regions of the system are summarized in Table 1. All lengths are given in

Table 1. The parameters of the system geometry, the magnetic fields and the scattering parameters.

r_{ts}	a_{in}	a_{out}	a_{sys}	h_{sys}	ρ_{sys}
0.1	0.2	0.8	1.0	2.0	2.35

Region number, i	$\langle B_i \rangle$, μG	mfp
1	30.0	$\lambda_1 (\langle B_1 \rangle, 2.0, \gamma)$
2	8.0	$\lambda_{\text{bow}} (\langle B_2 \rangle, 2.0, \gamma)$
3	30.0	$\lambda_{\text{bow}} (\langle B_3 \rangle, 2.0, \gamma)$
4	3.0	$\lambda_{\text{ism}} (\gamma)$

units of the distance from the pulsar to the BS apex, chosen to be equal to $r_{apex} = 2.3 \times 10^{16}$ cm, which is likely close to the corresponding value for the source related to the PSR J0437-4715. In the simulation we used $\gamma_1 = 2 \times 10^6$.

The energy distribution of the injected particles is a power-law with an exponential cutoff $f_0 (p) \propto p^{-4.2} \exp (-\gamma/\gamma_c)$ in the range $\gamma_{min} = 4 \times 10^5 \leq \gamma \leq \gamma_{max} = 1.2 \times 10^8$, where $p = mc\gamma$, $\gamma_c = 3 \times 10^7$. Here the distribution function f_0 is normalized in such a way that the mean particle number density $n_0 = \int f_0 d^3 p$.

The numbers of the spatial bins are $N_x = 200$ and $N_\rho = 100$. The range of the particle Lorentz-factors which could be detected by the system $\gamma_{min}^{\text{det}} = 4 \times 10^4 \leq \gamma \leq \gamma_{max}^{\text{det}} = 6 \times 10^8$ is wider than the injected particles energy range; it is logarithmically binned into $N_m = 400$ bins.

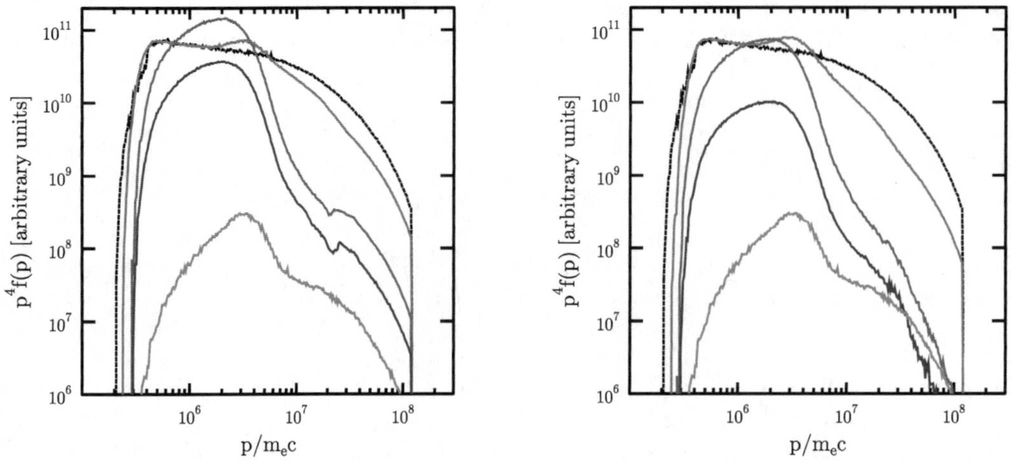

Figure 2. The particle spectra detected in the bins shown in Figure 1. **Left:** red, green and blue curves — the dependence of the detected momentum distribution function in arbitrary units, multiplied at p^4, on the momentum value p normalized at mc, for bins A, B and C, respectively. Magenta curve – the same for bin D. **Right:** the same for bins E, F, G and D. The dashed black curves show the spectrum at the termination shock, which is close to the injected one.

In Figure 2 the results of the Monte-Carlo particle spectra modeling are presented. One can see that inside of the BSPWN (regions "1" – "3") the particle spectrum deforms in comparison with the injected one. The behaviour of the particle distribution function reflects the confinement of the low energy particles in the system due to the resonant scattering diffusion coefficient at the BS and the particle re-acceleration in the colliding shock flows (see [5] for detailed discussion). In region "4" (the ISM) the spectrum becomes somewhat closer to the injected one due to a low probability for a low energy particle to pass through the region against the plasma flow.

4. Synchrotron images

To compare the results of the modeling with the observations we constructed the images of the spatial distributions of the synchrotron emission intensity in the modeled source. We used the approximations given in [8] to obtain the intensity of the synchrotron emission averaged over the orientations of the magnetic field. The results of the simulation at the photon energies $E_{\rm ph} = 8$ eV and 1 keV are presented in Figure 3. In the image for $E_{\rm ph} = 8$ eV (Figure 3, left panel) one can see a bright bow-like structure, corresponding to the inner part of region "3" in the model. Just a very faint counterpart can be found in the image for $E_{\rm ph} = 1$ keV (Figure 3, right panel). The ratio of the intensity from the vicinity of the TS to the intensity from the bow-like structure is about 1 for the FUV image and about 10 for the X-ray one.

5. Conclusions

In this paper the Monte-Carlo model of the transport of the relativistic PW particles accelerated at the termination shock through the pulsar wind nebula is presented. The model accounts for the supersonic proper motion of the pulsar through the ISM. The cosmic-ray driven instabilities at the bow shock are supposed to cause the decrease of the particle mfp in the certain energy

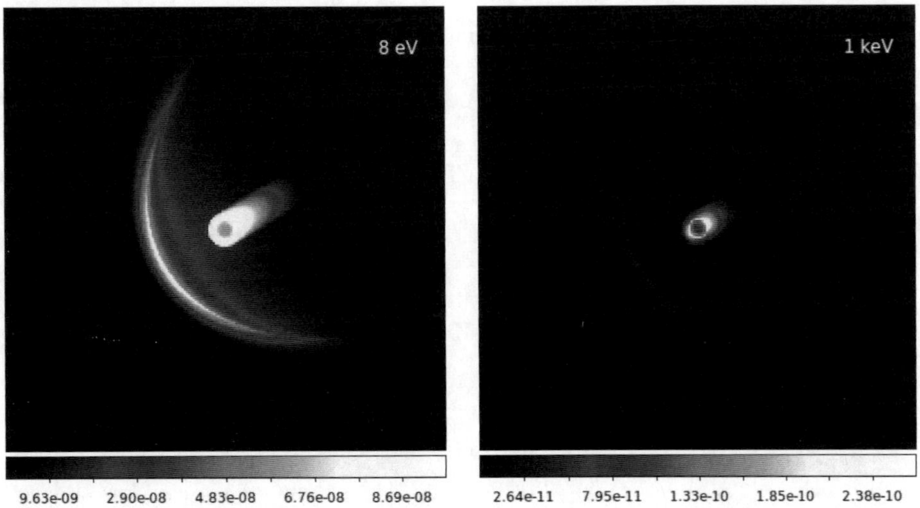

Figure 3. The simulated synchrotron images of a source similar to J0437-4715, for $E_{\rm ph} = 8$ eV (left) and 1 keV (right). The absolute intensity is given in normalized units to represent the contrast change through the image.

range, corresponding to the energies of the particles emitting the synchrotron radiation in the far ultraviolet range. The physical processes that can be responsible for the observed difference in the morphology of the source related to the PSR J0437-4715 at different frequencies are discussed. The results of the particle spectra modeling for the parameters resembling ones for the BSPWN in this source are presented.

The results show that due to mentioned decreasing of the mfp the low energy particles are confined by the system for a longer time providing the reshaping of the particle spectra. The synchrotron emission images are simulated for the modeled source at the photon energies 8 eV (FUV) and 1 keV (soft X-rays). A bright bow-like structure is apparent in the simulated synthetic FUV image, while its counterpart in the X-ray image is very faint. The ratio of the intensity from the area close to the TS to the intensity from the bow-like structure is about 1 for the FUV image and about 10 for the X-ray one.

The presented synchrotron continuum model can qualitatively reproduce most of the observed features of the frequency-dependent morphology of the source similar to one related to the PSR J0437-4715. The detailed analysis of the line spectra associated with the bow shock will be addressed elsewhere.

Acknowledgments

The authors acknowledge support from RSF grant 16-12-10225.

References

[1] Rangelov B, Pavlov G G, Kargaltsev O, Durant M, Bykov A M and Krassilchtchikov A 2016 *ApJ* **831** 129
[2] Bucciantini N 2002 *A&A* **393** 629-35
[3] Bucciantini N, Amato E and Del Zanna L 2005 *A&A* **434** 189-99
[4] Bykov A M, Gladilin P E and Osipov S M 2013 *MNRAS* **429** 2755-62
[5] Bykov A M, Amato E, Petrov A E, Krassilchtchikov A M and Levenfish K P 2017 *Space Sci. Rev.* DOI: 10.1007/s11214-017-0371-7 (*Preprint* arXiv:1705.00950 [astro-ph.HE])
[6] Wilkin F P 1996 *ApJl* **459** L31
[7] Vladimirov A 2009 PhD thesis, North Carolina State Univ.
[8] Zirakashvili V N and Aharonian F 2007 *A&A* **465** 695-702

IOP Conf. Series: Journal of Physics: Conf. Series **929** (2017) 012016 doi:10.1088/1742-6596/929/1/012016

Method of recording bioelectrical signals using a capacitive coupling

V A Simon, V A Gerasimov, D K Kostrin, L M Selivanov and A A Uhov
Department of electronic instruments and devices, Saint Petersburg Electrotechnical University "LETI", 197376, Saint Petersburg, Russia

E-mail: vsev.simon@gmail.com

Abstract. In this article a technique for the bioelectrical signals acquisition by means of the capacitive sensors is described. A feedback loop for the ultra-high impedance biasing of the input instrumentation amplifier, which provides receiving of the electrical cardiac signal (ECS) through a capacitive coupling, is proposed. The mains 50/60 Hz noise is suppressed by a narrow-band stop filter with an independent notch frequency and quality factor tuning. Filter output is attached to a ΣΔ analog-to-digital converter (ADC), which acquires the filtered signal with a 24–bit resolution. Signal processing board is connected through universal serial bus interface to a personal computer, where ECS in a digital form is recorded and processed.

1. Introduction

The commonly used contact method of ECS registration is a reliable and widely applicable technique for the diagnosis of many heart diseases. However, this method has several disadvantages. First, standard conductive gels can dry over time and thus don't provide long-term reliability of the electrical contact. Second, metal contact electrodes can cause skin irritations in case of a metal allergy. Third, the patient's mobility is often limited during the registration process.

However, the electrical contact can be replaced by a capacitive coupling, as shown in [1–8]. It represents a capacitor formed by an electrode surface, skin surface and dielectric layer (e. g. cotton textile). The coupling capacitance depends on the electrode's size and the type of a fabric. In [1], where electroencephalographic electrodes are used, the coupling capacity is in the range from 1 to 50 pF. In [2], where electrocardiographic (ECG) electrodes are used, the coupling capacity is 92 pF.

Thus, the impedance of the coupling capacitor can be as much as 10^{10} Ω in the typical ECG frequency range (from fractions to tens of Hz). In order to receive the ECS via capacitive coupling the instrumentation amplifier (IA) with ultra-low input bias current and high input impedance can be used. One of the best examples of IA is the INA116 amplifier with the typical value of the input bias current as low as $3 \cdot 10^{-15}$ A and with input impedance of more than 10^{15} Ω.

However, the ECS itself doesn't contain the constant component. Therefore, the inputs of the IA should have a direct current (DC) biasing to prevent saturation of the amplifier. The value of a biasing resistor should be at least 10^{10} Ω for the reliable reception of the low-frequency harmonics [8]. Unfortunately, the resistors with an ultra-high value are quite expensive and hard-to-buy goods.

2. Feedback loop for the input stage amplifier

In [3] an original circuit, which appears as an equivalent to the ultra-high value biasing resistance, was described. This circuit contains only inexpensive general purpose components (figure 1).

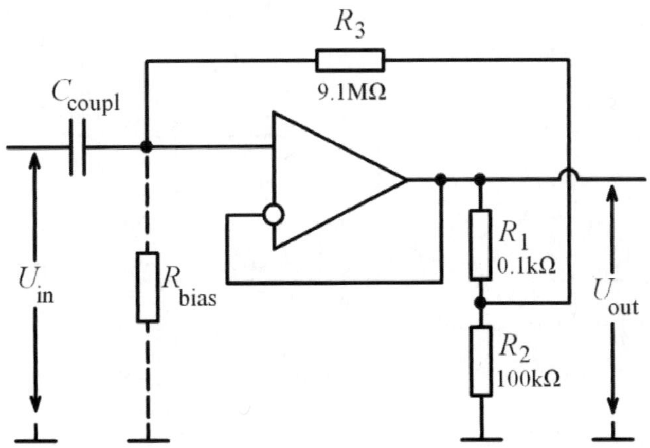

Figure 1. Circuit of the feedback loop around an operational amplifier.

A resistive divider with the transmission factor $a = R_2/(R_1 + R_2) = 0.999$ is connected to the output of the operational amplifier (op-amp). A 9.1 MΩ resistor is connected between the input of op-amp and the output of the resistive divider. The current through the resistor R_3 can be represented as the current through the virtual resistor R_{bias} with resistance value given by the formula

$$R_{\text{bias}} = \frac{R_3}{1-a}.$$

The operation of the biasing feedback loop was verified through SPICE simulations and experiments on the breadboard. Then, the biasing circuit was connected to the INA116 amplifier, as shown on figure 2.

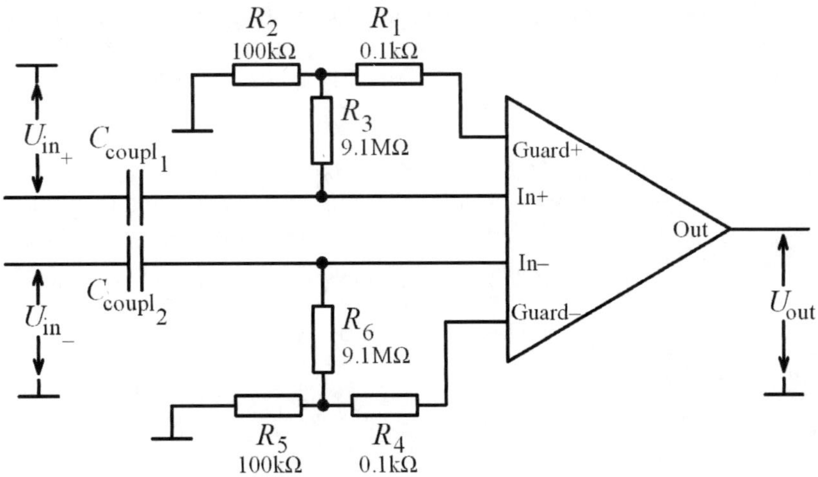

Figure 2. Feedback loops, connected between the inputs and guard pins of the IA.

3. Mains noise filter

One of the main problems with the registration of bioelectrical signals (such as ECS) is the suppression of various interferences. This problem becomes much more important when the ECS is received via capacitive coupling.

The greatest negative impact on the quality of the received ECS has the mains 50/60 Hz noise [7]. Usually a stop filter is used to remove this type of interference. The filter design should provide an independent tuning of the quality factor and notch frequency. In this work an original stop filter comprising of two cascaded all-pass filters, symmetrical voltage divider and a feedback loop with a variable depth (figure 3) was designed.

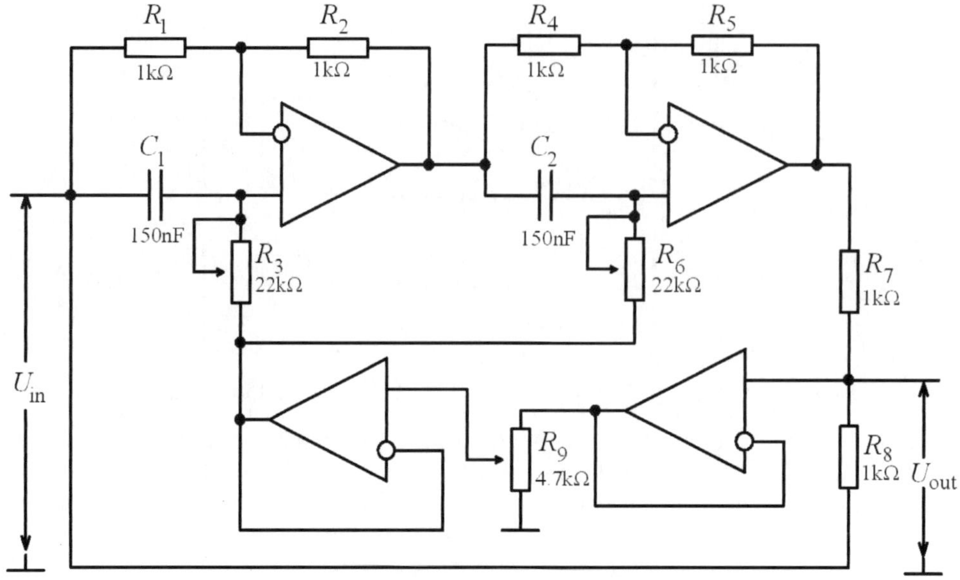

Figure 3. Circuit of the narrow-band stop filter.

After the signal had passed through the all-pass filters, it is added to its own original by a symmetric divider (R_7–R_8). Herewith the amplitudes of all harmonics of the signal are reduced according to the changes in the phase of the signal.

The phase of the harmonic with maximum suppression is changed by 180°. The frequency of the harmonic with maximum suppression is equal to the geometric mean of the two frequencies: $\omega_1 = R_3C_1 - 1$ and $\omega_2 = R_6C_2 - 1$.This is the notch frequency of the filter

$$\omega_0 = \sqrt{\omega_1\omega_2}.$$

The output signal of the filter is applied through the buffer op-amps to the potentiometers R_3 and R_6. The resistor R_9 varies the depth of the feedback d (from 0 to 100 %) and thus changes the quality factor Q, given by the equation

$$Q = \frac{0.5}{1-d}\frac{2\sqrt{\omega_1\omega_2}}{\omega_1+\omega_2}.$$

The gain-transfer characteristic of the filter is

$$H(s) = \frac{s^2+\omega_0^2}{s^2+\frac{\omega_0}{Q}s+\omega_0^2},$$

where $s = i\omega$; $i = \sqrt{-1}$.

4. Experimental verification of the proposed method

An experimental setup for the ECS acquisition includes the following (figure 4):

- cotton T-shirt with two embedded capacitive sensors;
- input stage circuit with IA and two feedback loops (as shown on figure 2);
- mains 50/60 Hz noise filter (see figure 3);
- ADC with a 24-bit resolution and ΣΔ architecture;
- personal computer with Audacity software.

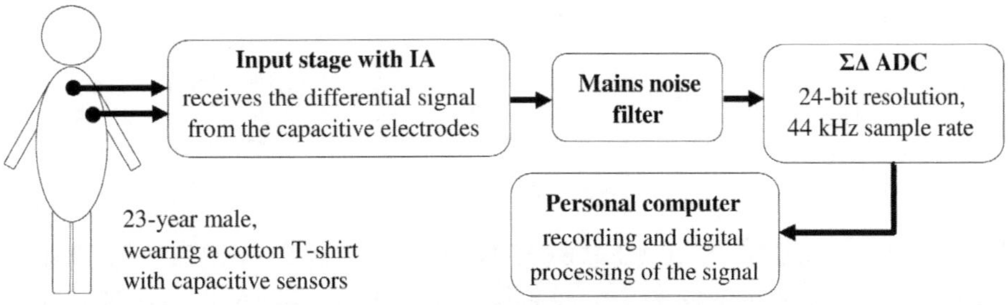

Figure 4. Layout of the experimental setup.

The output signal from the mains noise filter is acquired and digitized by the ADC. The digital signal is recorded on a personal computer using the Audacity software. Higher harmonics of the 50/60 Hz interference are removed by the digital processing. The form of the processed ECS in 1–100 Hz frequency range is presented in figure 5. The QRS complex, T–wave and P–wave [9] are all visible on the electrocardiogram record.

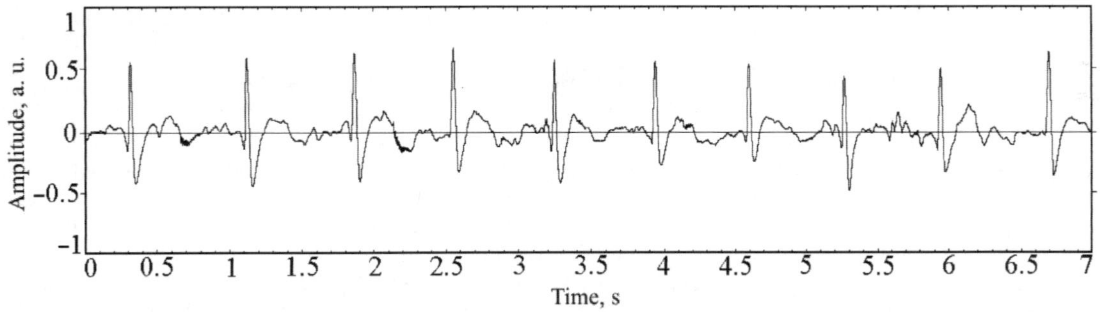

Figure 5. Electrocardiogram record obtained using a capacitive coupling.

5. Conclusion

The technique of the electrophysiological activity registration using a capacitive coupling is useful for real-time monitoring, as well as for continuous data collection. This method can be used for monitoring the health status of people working in stress conditions (rescue workers, firefighters, truckers), in sport medicine and fitness.

References

[1] Chi Y M and Cauwenberghs G 2009 *Proc. 31st Annual Int. Conf. of the IEEE Engineering in Medicine and Biology Society (Minneapolis)* pp 4218–21
[2] Aleksandrowicz A and Leonhardt S 2007 *Acta Polytechnica* **47** 68–71
[3] Prance R J, Debray A, Clark T D, Prance H, Nock M, Harland C J and Clippingdale A J 2000 *Measurement Science and Technology* **11** 291–7
[4] Harland C J, Clark T D and Prance R J 2002 *Measurement Science and Technology* **13** 163–9
[5] Sullivan T J, Deiss S R and Cauwenberghs G 2007 *Proceedings of the IEEE Biomedical Circuits and Systems Conference (Montreal)* pp 154–7
[6] Chi Y M and Cauwenberghs G 2010 *Proc. Int. Conf. on Body Sensor Networks (Singapore)* pp 297–301
[7] Gerasimov V A, Afanasjev P V, Bokhov O S, Kostrin D K, Luchinin V V, Selivanov L M and Uhov A A 2015 *Biotechnosphere* **3** 2–6
[8] Uhov A A, Gerasimov V A, Selivanov L M, Kostrin D K, Simon V A 2016 *Proc. IEEE NW Russia Young Researchers in Electrical and Electronic Engineering Conf.* pp 797–9
[9] Gacek A and Pedrycz W 2011 *ECG Signal Processing, Classification and Interpretation: A Comprehensive Framework of Computational Intelligence* (New York: Springer)

IOP Conf. Series: Journal of Physics: Conf. Series **929** (2017) 012017 doi:10.1088/1742-6596/929/1/012017

The development of the experimental setup for measuring the cell membrane electrical potential by Sucrose-Gap Technique

Yuzhakov AD[1], Nosarev AV[2], Aleinik AN[1]

[1]National Research Tomsk Polytechnic University, 634050, Tomsk, Lenin Avenue, 30 Russia

[2]Siberian State Medical University, 634050, Tomsk, Moscowski Trakt, 2, Russia

E-mail: alx-142@yandex.ru, avnosarev@yandex.ru, aleinik@tpu.ru

Abstract. This article describes the development of the experimental setup for measuring the cell membrane electrical potential by Double -Sucrose-Gap Technique. The double-gap isolation method allows the simultaneous measurement of electrical activity and tension output from contracting segments of muscle fibers. This technique has been widely used as a convenient tool for recording of the membrane activities from myelinated or unmyelinated nerves and muscle preparations. This device can be an effective way to provide undergraduate biomedical engineering students with invaluable experiences in neurophysiology. The installation design and its main characteristics are described. The advantages of the described device are the simplicity of the experiment, relatively low cost, the possibility of long-term experiment.

1. Introduction

The study of the various cells in the body is a very important aspect in the investigation of any tissue, organ or organism as a whole, because the actions of cells underlie responses of vast majority of biological systems. However, studies at the cellular level are extremely complex due to the small size and relatively small reaction time of objects. The sucrose gap method allows to measure extracellular action potential parameters fairly easy. Using the intracellular insulating sections of sucrose channels allows to restrict the extracellular shunting and to record the parameters of bioelectric potentials confidently enough.

The main advantages of the method are due to the fact that experimenter does not operate with a single cell, but with a tissue fragment, for example, muscle or nerve fiber [1]. This greatly simplifies experimentation and also reduces the cost of the equipment, because there is no need to use microelectrodes. Today, the devices of this type are not produced in Russia. Devices of this type can be custom-made, however, they are very expensive and designed to work with the nervous tissue, therefore do not allow to obtain data about the relationship between contractile and electrical activity of the investigated muscle tissue. This paper examines the basic principles of the cell membrane potential registration by double sucrose gap method and describes the installation design and its main characteristics.

2. Electrochemical processes in cells

There are a lot of electrochemical processes in living cells, but most clearly they are expressed on the cell membrane surface, which separates the inner cell volume from of the intercellular substance and other cells. All living cells have membrane electrical potential. It is an electric potential difference, existing on inner and outer sides of the membrane and caused by uneven concentrations of various ions in cells and in the extracellular space. Some cells, such as nerve and muscle are able to generate rapidly changing electrochemical impulses, which are used to transmit signals along the membranes of these cells. The membrane potential can be changed under the influence of different origin stimuli (chemical, electrical and others).

The shift in the membrane potential may occur in the negative (hyperpolarization) or positive (depolarization) side. If the cell stays in an unexcited state, the resting potential is formed on the membrane surface. For warm-blooded organisms, it ranges from -55 to -100 mV [1,2]. If resting potential is shifted to the sub-threshold value, then passive electrotonic potential arises. The action potential is a wave of excitement, which moves through the membrane of the living cell in the form of short-term changes in the membrane potential, as a result the outer surface of the membrane becomes negative with the respect to the inner surface of the membrane. An action potential is a physiological basis for a number of electrochemical processes in the cells. The main reason for the initiation of the action potential is achieving a threshold value of the membrane potential bias. The threshold value is the minimum value of the stimulus that can cause a physiological response: muscle contraction, nerve impulse or a secretion from glands. Only upon reaching the threshold value the stimulation causes a self-sustaining process of changing the biopotential value in the cell [Figure 1]. The threshold is determined by multiple conditions. For example, they include the magnitude and duration of the stimulus.

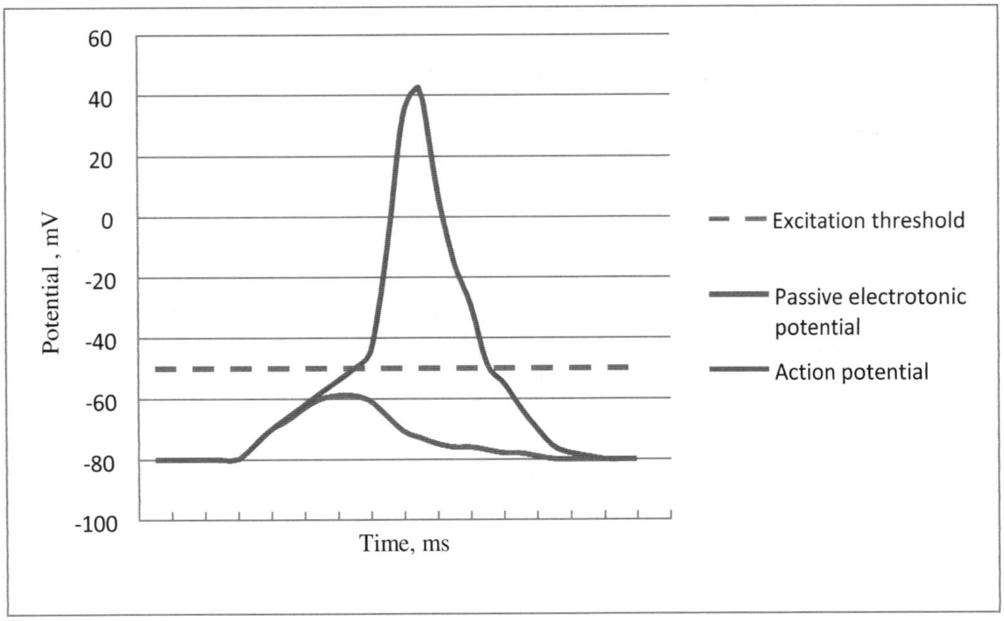

Figure 1. Changing of the potential depending on the strength of stimulation

The sucrose gap method is an extracellular way for the registration of the action potential parameters. When placing electrodes on the surface of nerve or muscle fiber, which is located in saline, the diversion will be ineffective because of the shunting effect of the solution. To eliminate the electrical bypass, fiber section, which is located between electrodes, perfused with a solution containing no ions [Figure 2]. Under one of the electrodes the saline is replaced with an isotonic solution of potassium chloride which eliminates transmembrane potential in this region which is equivalent to the electrode connection to the intracellular content of the fiber. To allow abduction of the transmembrane potential and its management, the fiber is divided into 5 sections. The middle portion is tested and perfused with saline. Through the adjacent portions on both sides sucrose solution is passed and the edge portions on both sides are perfused with an isotonic potassium chloride solution. A pair of measuring electrodes is located on both sides of one of the sucrose sections and a pair of exciting ones on both sides of the other [3]. In multicellular specimen there is an effective electrical connection between neighboring cells by which smooth muscle tissue has distinct conductive properties. The spatial constant is 1-3mm. [4]. Thus, the sucrose gap method can be used for measuring the membrane potential of smooth muscle or nerve fiber. This method has several advantages. One of them is the possibility of continuous measurement of the membrane electrical parameters changes caused by different influences in particular physiologically active substances, temperature, electric current. In addition, the sucrose gap makes it relatively easy to implement simultaneous recording of electrical and contractile activity of smooth muscle.

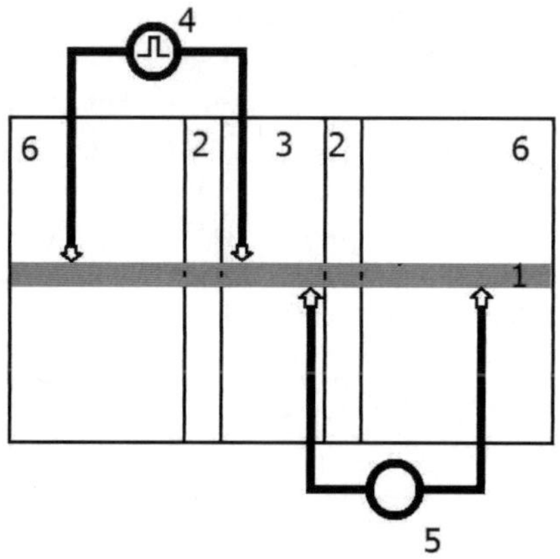

Figure 2. The scheme of the double sucrose gap device. 1 – a muscle fiber; 2 - sucrose gaps, insulating intercellular areas; 3 - the working chamber filled with Krebs solution; 4 - measuring electrodes; 5 - stimulating electrodes; 6 - working chamber filled with isotonic KCl solution.

3. Development of the device

All electronic components of the device are placed in the single housing. On the top of the housing there is a working area, where a sample of nerve or smooth muscle tissue is situated. At the bottom of the working chamber located the control and recording electrodes, which are made of platinum wire. To maintain the purity of the solutions in the chamber sucrose solutions, Krebs' solution, and KCl solution supplied to the working sections through the special channels throughout the experiment. To maintain the desired temperature the heating element and temperature sensor are placed in the working area [Figure 3]. Abduction of the solutions implemented by the internal channels. Used solutions gather in the special containers, which are removable to facilitate maintenance. The main elements of the device are made by three-dimensional printing technology.

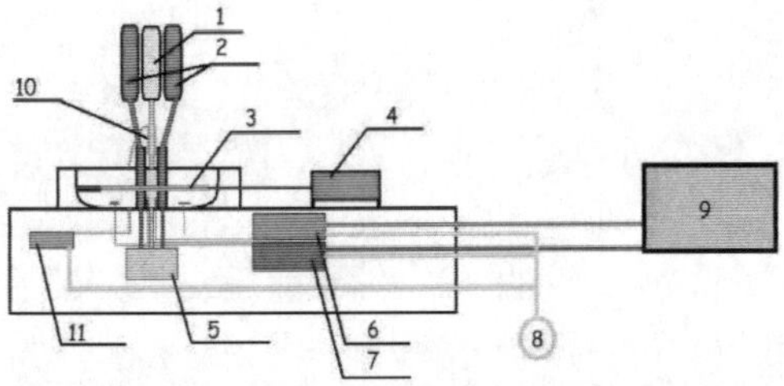

Figure 3. Schematic diagram of the device. 1- Krebs solution; 2 - sucrose; 3 - the study tissue; 4 - tension sensor; 5 - tank for waste collection; 6 – oscilloscope; 7 - pulse generator; 8 - power supply; 9 - output data to a personal computer; 10 - The heating element and temperature sensor; 11 - Temperature control board

During the development of the device, two variants of working sections were considered. In the first variant [Figure 4] each sucrose section is arranged as the single element in the form of a parallelepiped with the openings on opposite sides. On the lateral sides of the sections a latex thin film is attached. Between the sucrose sections the oblique strip with thickness of 1.5 mm is installed. In the formed gap comes the Krebs solution. The advantage of this design is the simplisity of its production. The disadvantage is the complexity in the preparation of the experiment, the low reliability of the sealing system, the need for long-term preparations for the experiment. The second option [Figure 5] suggests the production of individual walls. Each wall is hollow inside and is connected to a pneumatic system. The baffle has two round-shaped holes of 8 mm diameter on opposite sides in which two clamping rings are installed [Figure 6]. The rings fix the sealing rubber membrane of cylindrical shape. The air channels within the walls allow to regulate the pressure inside the membrane modifying the final dimensions of the transverse channel inside the baffle. The main advantages of this technical solution are the simplicity of setting up and using of the installation and its enhanced reliability. The disadvantage is the difficulty in its production. The small size of the clamping rings requires the high accuracy in the production, as well as the careful selection of the materials.

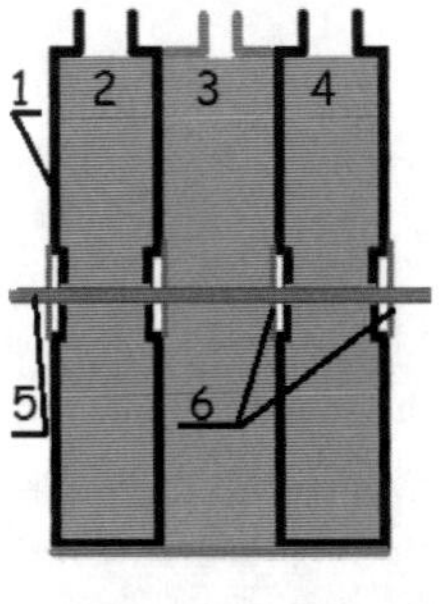

Figure 4. The section embodiment type 1.
1 - Sucrose section module; 2.- Sucrose solution;
3 - Krebs solution; 5 - Sample; 6 - Rubber seals

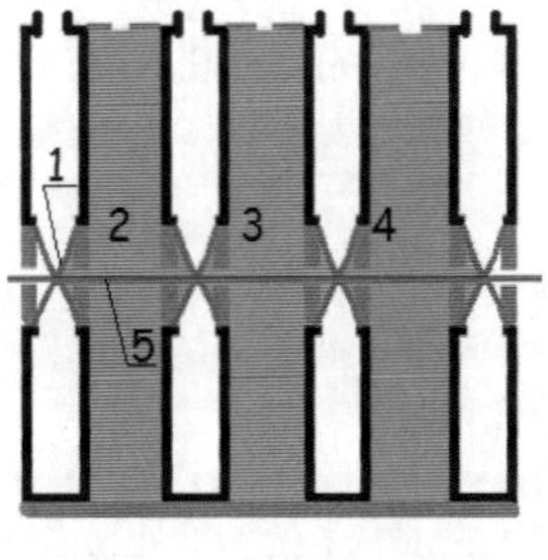

Figure 5. - The section embodiment type 2.
1 - The rubber sealing ring; 2 - sucrose solution;
3 - Krebs solution; 5 – Sample.

Figure 6. The section embodiment type 2. 1 - general view; 2 - view in section; 3- partition together with a sealing ring

4. Specifications

The design of the device provides the ease of use and high reliability. Reliable installation is due to the fact that all electronic components are placed in a single robust housing and protected against dust, moisture and strokes. Electronic components provide high accuracy and reliability of the data acquisition.

Table 1. Installation parameters.

Parameter	Value
The accuracy of the electric potential measurement	10^{-6} V
Time resolution	10^{-4} sec
Diameter of the study sample	0,1 – 4 mm
Temperature stability	0,1° C

5. Conclusions

The sucrose gap method has a number of advantages. This method makes it relatively easy to implement simultaneous recording of electrical and contractile activity of smooth muscle. The obtained data allow us to investigate the relationship between changes in the cell membrane parameters and temperature settings, physical effects, effects of damaging factors of various origins, including side effects of medications. The advantages of the described device are the simplicity of the experiment, relatively low cost, the possibility of long-term experiment. Similar devices will be produced for higher education and research institutions for the purpose of students training in medical and biological fields, as well as research.

6. Acknowledgements

This work was partly supported by Competitiveness Enhancement Program of Tomsk Polytechnic University.

IOP Conf. Series: Journal of Physics: Conf. Series **929** (2017) 012017 doi:10.1088/1742-6596/929/1/012017

References

[1] MF Shuba, NI Gokina, AV Gurkovskaya 1991 *Mechanisms of excitation and contraction of smooth muscles of cerebral vessels* (Kiev: Naukova Dumka) pp 19-37 (in Russian)
[2] L Squire, D Berg, F Bloom, Sascha du Lac, A Ghosh, N. Nicolas 2002 *Fundamental Neuroscience* (Amsterdam: Elsevier.) p 89
[3] T. Mert 2007 Sucrose-Gap Technique: Advantages And Limitations *Neurophysiology* Vol 39 No 3 pp 237-241
[4] MF Shuba Physical electron in smooth muscle *Biophisika* Vol 1 1961 (Moscow: Science) pp 14-18 (in Russian)
[5] https://www.blender.org

IOP Conf. Series: Journal of Physics: Conf. Series **929** (2017) 012018 doi:10.1088/1742-6596/929/1/012018

Specialized nanowhisker probes for high-precision investigation of native bio-objects in liquid by means of atomic force microscopy

MV Zhukov[1,2], IV Kukhtevich[3], FE Komissarenko[1,4], VI Chubinskiy-Nadezhdin[5]

[1] Department of nanotechnology and material science, ITMO University, 197101 Kronverkskiy pr., 49, Saint-Petersburg, Russian Federation
[2] Laboratory of scanning probe microscopy and spectroscopy, Institute for Analytical Instrumentation RAS, 198095 Ivan Chernykh, 31-33, Saint-Petersburg, Russian Federation
[3] Laboratory of information-measuring bio- and chemosensory microsystems, Institute for Analytical Instrumentation RAS, 198095 Ivan Chernykh, 31-33, Saint-Petersburg, Russian Federation
[4] Laboratory of renewable energy sources, St Petersburg Academic University RAS, 194021 Khlopina, 8/3, Saint-Petersburg, Russian Federation
[5] Laboratory of Ionic Mechanisms of Cell Signaling, Institute of Cytology RAS, 194064 Tikhoretsky Avenue 4, Saint-Petersburg, Russian Federation

E-mail: cloudjyk@yandex.ru

Abstract. Fabrication and investigation of specialized Pt-C single nanowhisker probes are carried out for high-precision study of objects with organic and inorganic nature in liquids by atomic force microscopy. The best modes and buffered media were revealed on calibration lattice TGQ1 choosing as reference inorganic structure. Preliminary results of the study of the structure of bacteria E.Coli in the native state were received in the PBS liquid. Fixation of the bacteria was performed on agar-agar with a mass fraction of 1.8%. The significant improvement of image contrast and resolution was found when using nanowhisker probes compare to standard Si tips.

1. Introduction

The studies devoted to investigation of inorganic and biological objects (cells, bacteria, viruses, proteins, DNA, etc.) in a liquid by atomic force microscopy (AFM) are of great interest nowadays [1, 2]. This is explained by the possibility to carry out precision «in situ» studies of biological objects in the native state with nanometer resolution. It's observable that reliable interpretation of the features of investigated native bio-structure greatly depends on the resolution ability of the method used.

The principal element in AFM is a probe (needle or tip) since the resulting image presents a convolution between geometric parameters of the probe and the surface of the sample [3]. Therefore, it is necessary to fabricate a new tool to ensure high-precision investigation of such objects with great accuracy. To improve the accuracy of AFM investigations the research works are carried out to develop the new types of probes based on nanowhisker structures [4, 5]. Such probes have a high aspect ratio, strength and stiffness with a small transverse dimension [6] that makes them valuable for probes application.

Therefore, the main aim of this article was a fabrication and approbation of the probes with nanowhiskers for high-precision research of native organic bio-objects in liquids by atomic force microscopy.

2. Experimental setup

The Pt-C nanowhisker structures (NW) were grown at the tip of the standard probes under the electron beam in a high vacuum and injection system of precursor gases [6]. Fabrication and geometries control of the NW were performed by scanning electron microscope (SEM) CrossBeam Neon 40 (Carl Zeiss, Germany) with integrated inlet gas system $C_9H_{16}Pt$.

The investigation of probes in selected liquids (PBS - phosphate-buffered saline, NaOH 0,1M/ 0,5M) was carried out in semi-contact mode. The investigation of the samples was performed by means of scanning probe microscope Ntegra Aura (NT-MDT, Russia). The studies were conducted on SNL-10 tips (Veeco, USA).

3. Results and Discussions

The NW probes with a length of about 400-600 nm and diameter of about 30-40 nm was fabricated with tip radius of curvature less than 10 nm. NW probes retain their configuration after 2-3 cycles of scanning, which is comparable with standard Si probes (Fig. 1). NW was produced by an electron beam in a vacuum in the presence of the precursor gas $C_9H_{16}Pt$ [7]. Composition of NW is 30% of platinum Pt, 70% are residues of the precursor gas and carbon particles. NW were made at an angle of inclination relative to the tip axle about ~ 22-25 degrees to reduce possible vibrations when a double-tap contact with the surface (peculiarity of the tip holder).

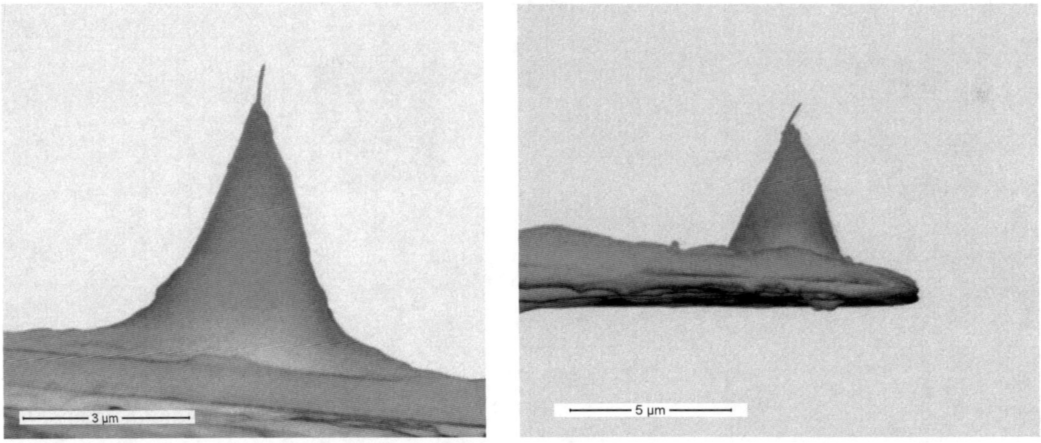

Figure 1. SEM images of the cantilevers with NW at the tips of standard Si probes obtained after scanning of the liquid medium (front and side view).

The best solution for measurement was revealed as sodium phosphate buffer (PBS). The improvement of height differences visualization in 3-4 times was revealed on calibration grating TGQ1 using NW probes compare to standard probes (from 3.9 nm to 18.7 nm on the X axis, from 37.3 to 11.8 nm in Y axis). The surface roughness using NW probes was more than such at standard probes (Ra = 0,317 nm for the standard probes, Ra = 0,887 nm for NW probes), that indicating enlargement of penetrating ability of probes. The improvement of resolution using NW probes in NaOH buffers was found as well, but alkaline features of this media can defect upper layers of biological samples during the scanning process.

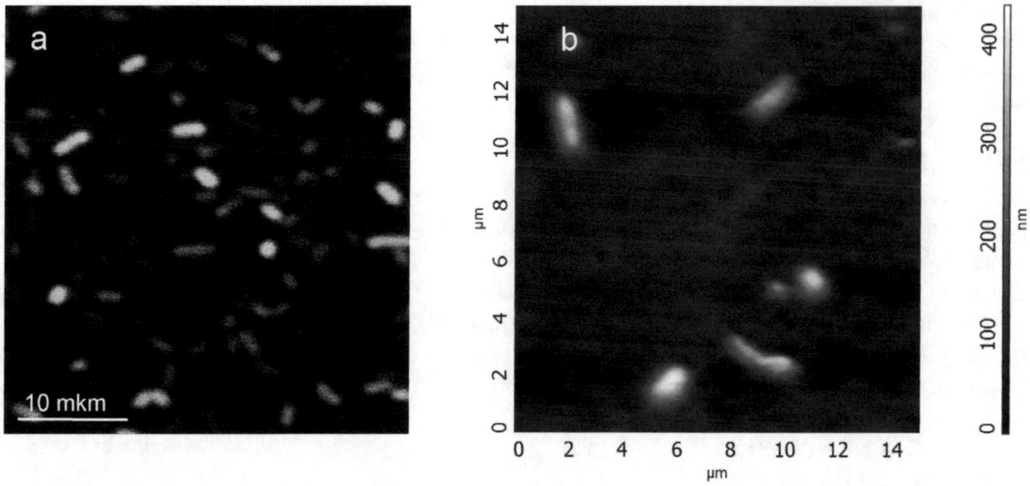

Figure 2. CLSM image (a) and AFM image (b) of E.Coli bacteria groups on a mica substrate with agar 1.8% in PBS buffer.

For immobilization of native bacteria E.Coli gelatin and agar-agar substances were used. Investigation of fixation results was carried out by confocal laser scanning microscopy (CLSM) by TCS SL (Leica, Germany). It was revealed that agar-agar with a mass fraction of 1.8% as an initial solution on a mica substrate provide best immobilization of the bacteria (Fig. 2).

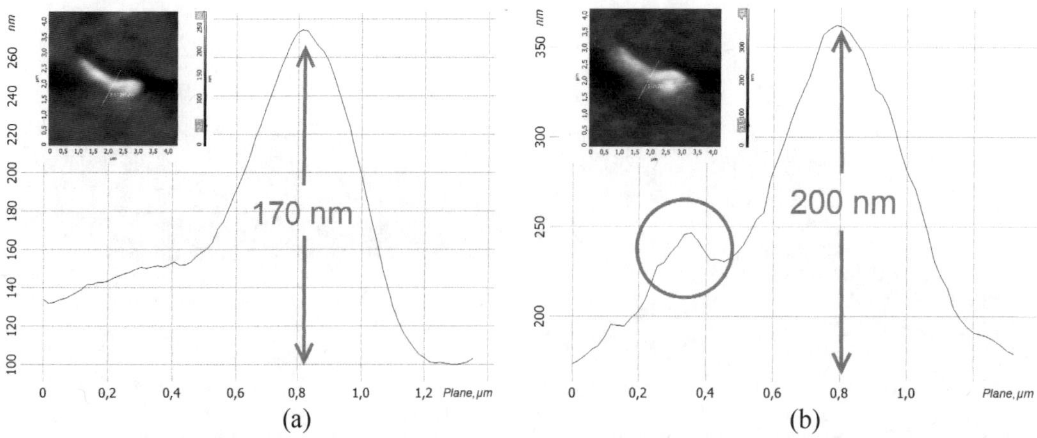

(a) (b)

Figure 3. AFM image of a single bacteria E.Coli and cross-section obtained by standard probe (a) and NW probe (b) in PBS buffer.

In the study of individual bacteria E.Coli the significant improvement of contrast and resolution was found using NW probes (Fig. 3). Thus, on the cross-sections derived from the specific single bacterium E.Coli in the same place, the increase of height differences (penetration ability) with NW probes (200 nm) and imaging of thin nanoscale structure along the bacterium (left peak on Fig. 3b) was observed. Presumably, the structure can introduce group of fibers around bacterium E.Coli. The standard probes cannot allow to reveal this particular nanoscale structure.

For detailed imaging and studying of fiber structure around E.Coli the single bacterium was visualized by NW probe (Fig. 4, a), and a longitudinal section was derived along the studied structure

(Fig. 4, b). It was found that the size of such structures comprises a value of about 400-500 nm in diameter and 20-30 nm in height. However, it should be mentioned that the measurements were carried out in the liquid and such structures can rapidly adhere together in these environment. In addition, AFM image always represents the convolution of the probe and the surface being studied (fibers), which can be smaller compared to the dimensions of the probe.

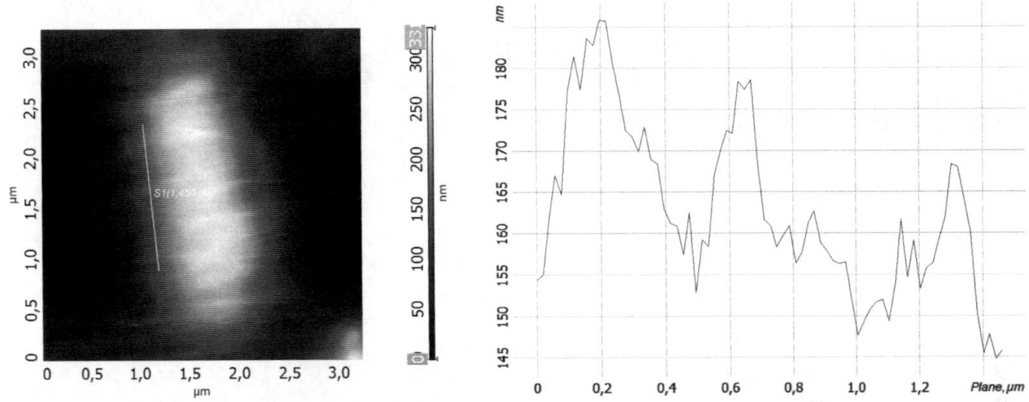

Figure 4. AFM image of single bacteria E.Coli (a) and the sectional structure along the fibers (b) obtained by NW probe in PBS buffer.

The investigation shows that the application of specialized SNW probes has better results than standard silicon probes in the study of native organic and inorganic objects in liquids by atomic force microscopy.

4. Conclusions

Thus, application of probes with NW in liquid media improves the spatial resolution and contrast of the images in the study of native biological samples, in particular, the native bacteria E.Coli. The results obtained in this study can be used for a more detailed and accurate study of native objects with complex structure in liquid by means of atomic force microscopy.

Acknowledgments

Work carried out with the support of Russian Foundation for Basic Research (16-32-00806), the leading universities of the Russian Federation (grant 074-U01).

References

[1] Ahmed Touhami, Manfred H Jericho, Terry J Beveridge 2004 *Jurnal of bacteriology* **186** 3286–3295

[2] Yuri L Lyubchenko 2011 *Micron* **42** 196–206

[3] Winzera A T, Kraftb C, Bhushanb S, Stepanenkoc V, Tessmerb I 2012 *Ultramicroscopy* **121** 8–15

[4] Beard J D and Gordeev S N 2011 *Nanotechnology* **22** 175303

[5] Levichev V V, Zhukov M V, Mukhin I S, Denisyuk A I, Golubok A O 2013 *Technical Physics* **58** 1043–1047

[6] Branner S S 1958 *Science* **128** 569–575

[7] Zhukov M V, Kukhtevich I V, Levichev V V, Mukhin I S and Golubok A O 2014 *J. Phys.: Conf. Ser.* **541** 012042

IOP Conf. Series: Journal of Physics: Conf. Series **929** (2017) 012019 doi:10.1088/1742-6596/929/1/012019

Temperature sensitivity of ligand-gated ion channels: ryanodine receptor case

B.I. Iaparov[1,2], **A.S. Moskvin**[1], **O.E.Solovyova**[2,3]

[1] Department of Mathematical and Theoretical Physics, Institute of Natural Sciences and Mathematics, Ural Federal University, Ekaterinburg, Russia

[2] Laboratory of Mathematical Modelling in Physiology and Medicine,Ural Federal University, Ekaterinburg, Russia

[3] Laboratory of Mathematical Physiology,Institute of Immunology and Physiology, Ural Branch, Russian Academy of Sciences, Ekaterinburg, Russia

E-mail: bogdan.iaparov@urfu.ru

Abstract. Temperature influences all biochemical processes, in particular, excitation-contraction coupling(ECC) in cardiac cells. In this work we propose a theoretical explanation of temperature effects on an isolated ryanodine receptor calcium release channel (RyR channel) within the electron-conformational (EC) model. We show that the EC model with an Arrhenius-like temperature dependence of the "internal" and "external" frictions and a specific thermosensitivity of the tunnelling "open $\leftrightarrow$ closed" transitions can provide both qualitative and quantitative description of the temperature effects for isolated RyR channels. Interestingly that a small change of the activation energy for the "internal" friction can make an ion channel either heat-inhibited or heat-activated while the "external" friction doesn't play a key role in temperature sensitivity: neglect of "external" friction doesn't change the channel's temperature sensitivity qualitatively.

Introduction

Temperature influences all biochemical processes and the excitation-contraction coupling (ECC) in cardiac cells is not a special case. Experiments [1] show, that hypothermia increases action potential duration, intracellular Ca^{2+} transients and muscle contraction amplitude. The key elements of ECC are ryanodine receptors (RyRs), the ligand gated ion channels which govern the Ca^{2+} release from sarcoplasmic reticulum (SR), intracellular calcium storage. Abnormal RyR gating leads to cardiac arrhythmias and heart failure [2], hence the understanding the RyR gating is mandatory in order to understand the excitation-contraction coupling.

According to single RyR channel patch clamp experiments [3], the RyR channel gating changes significantly with temperature. Low temperatures (5°C) are characterized by longer openings ($\langle t_{open} \rangle \approx 50\,ms$) with smaller ion current amplitude ($\approx 2\,pA$) while high temperatures (23°C) are characterized by shorter openings ($\langle t_{open} \rangle \approx 2\,ms$) with higher ion current amplitude($\approx 4\,pA$). From the viewpoint of the condensed matter physics it seems to be absolutely inexplicable how the ≈ 6 % change in absolute temperature can produce manyfold effect in the channel activity. How can the temperature sensitivity of the RyR channel's gating be so high?

IOP Conf. Series: Journal of Physics: Conf. Series **929** (2017) 012019 doi:10.1088/1742-6596/929/1/012019

In this work we propose a theoretical explanation of the temperature effects for an isolated ryanodine receptor calcium release channel (RyR channel) within the electron-conformational(EC) model.

Methods

Electron-conformational model

Despite intensive stuctural studies of the RyR channels [4-8], our knowledge of molecular mechanisms for the RyR channel functioning is still limited. The RyR channel is the biggest known ion channel (more than 300000 atoms, molecular weight is $\approx$ 2 MDa) that makes very hard to study its molecular structure using both experimental methods such as low-temperature electron microscopy and theoretical methods such as molecular dynamics (MD).

Recently it was demonstrated that the electron-conformational model (see, e.g., Refs. [9, 10, 11]) is able to capture important features of the individual and cooperative behaviour of the RyRs in ventricular myocytes and pacemaker cells. The EC model of RyR functioning under Ca^{2+} stimuli is based on a biophysical adaptation of the well-known theory of photo-induced structural phase transitions, which has been successfully applied to different solids [12]. Within the EC model one start reducing a large variety of the RyR degrees of freedom to only two: a fast and a slow one, conditionally termed electronic and conformational, respectively. Bearing in mind the main function of the RyR channel as an ion channel, opening under stimuli and conducting ions, we assume only two actual electronic RyR channel states: open and closed, and a single conformational degree of freedom, Q, to be described by a classical continuous variable.

The key element of the EC model is the so called energetic approach, in other words, namely the energy is supposed to be a main characteristic of the RyR channels state. We introduce a simple effective Hamiltonian for an isolated RyR channel as follows:

$$H_s = -\Delta\hat{s}_z - h\hat{s}_x - pQ + \frac{K}{2}Q^2 + aQ\hat{s}_z\,, \tag{1}$$

where s_z and s_x are Pauli matrices, and the first term describes the bare energy splitting of "up" and "down" (electronically "open" and "closed") states with an energy gap Δ, while the second term describes a quantum mixing effect. It is worth noting that given $h = 0$ we arrive at a classical approach with a dichotomic electronic variable. The third and fourth terms in (1) describe the linear and quadratic contributions to the conformational energy. Here, the linear term formally corresponds to the energy of an external conformational stress, described by an effective stress parameter p. The last term describes the EC interaction with the coupling parameter a. Hereafter we make use of the dimensionless conformational variable Q; therefore all of the model parameters $(\Delta,\ h, p, K, a)$ are assigned energy units. Two eigenvalues of our Hamiltonian

$$E_{\pm}(Q) = \frac{K}{2}Q^2 - pQ \pm \frac{1}{2}\left[(\Delta - aQ)^2 + h^2\right]^{1/2} \tag{2}$$

define two branches of the adiabatic, or conformational potential (CP), attributed to electronically closed (E_-) and electronically open (E_+) states of the RyR, respectively. Given $h = 0$ we arrive at two diabatic potentials $E_{\pm}(Q)$ for electronically closed and open states, shown in Fig. 1 at $\Delta = 0$. Two minima are separated by a intersection point at $Q = 0$.

Ion channel's dynamics in EC model

The RyR channel's dynamics in EC model includes:

- Franck-Condon electronic transitions triggered by a subspace Ca^{2+} [9];

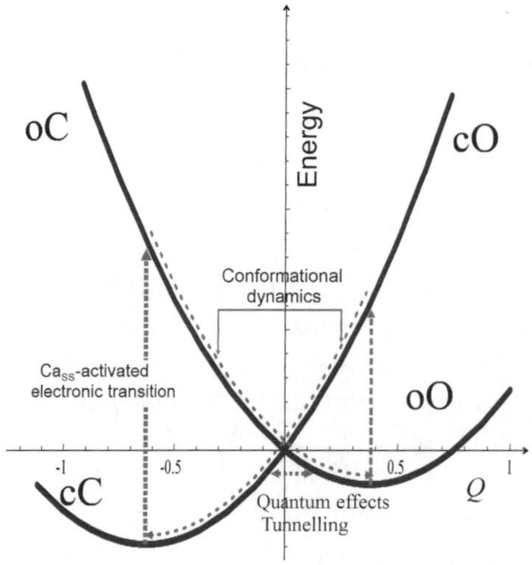

Figure 1. Conformational potential in the EC-model of the RyR channel given $h = 0$ with a global minimum for a closed state. Small letters c, o are used for electronically closed and open states, respectively, and capital letters C, O for conformationally closed and fully open states, respectively. The energy separations δE_{-}, δE_{+} are the potential barriers seen from the bottom of two diabatic potentials, respectively. Vertical arrows point to Ca^{2+}-induced Franck-Condon (FC) electronic transitions, horizontal arrow points to a non-FC tunneling transition, downhill arrows point to a conformational dynamics.

- Resonant quantum tunneling that takes place between two branches of the conformational potential with a probability obeyed the effective Gamov law as follows

$$P_{tun} = P_0 e^{-A_{tun}\Delta Q\sqrt{\Delta E}}, \tag{3}$$

where ΔQ is the width, ΔE is the height of the energy barrier, or the energy separation between the tunneling points and the intersection point (see Fig. 1), and P_0 effective tunneling attempt frequency;

- Slow conformational Langevin dynamics which does not obey the fluctuation-dissipation theorem and implies both internal or non-solvent friction(caused by interaction between RyR residues and interaction of RyR with the membrane) [13] and conventional thermal fluctuation forces:

$$M\ddot{Q} = -\frac{\partial}{\partial Q}E(Q) - \Gamma\dot{Q} + \sqrt{2\gamma k_b T}\,R(t), \tag{4}$$

where first term describes a total systematic conformational force with M being an effective RyR mass (below M let to be unity), Γ is an effective dimensionless "internal" friction damping constant (relaxation rate). The last term describes the thermal fluctuation force (Gaussian-Markovian noise), γ is a random force parameter, T is a temperature, $R(t)$ is a delta-correlated stationary Gaussian process with a zero-mean value.
Hereafter we neglect the temperature dependence of the main EC-model parameters except the friction parameter Γ and the random force parameter α, these are assumed to obey the Arrhenius type temperature dependence:

$$\Gamma(T) = \Gamma_0 e^{\frac{E_\Gamma}{k_B T}};\ \gamma(T) = \gamma_0 e^{\frac{E_\gamma}{k_B T}} \tag{5}$$

with activation energies E_Γ and E_γ, respectively.

IOP Conf. Series: Journal of Physics: Conf. Series **929** (2017) 012019 doi:10.1088/1742-6596/929/1/012019

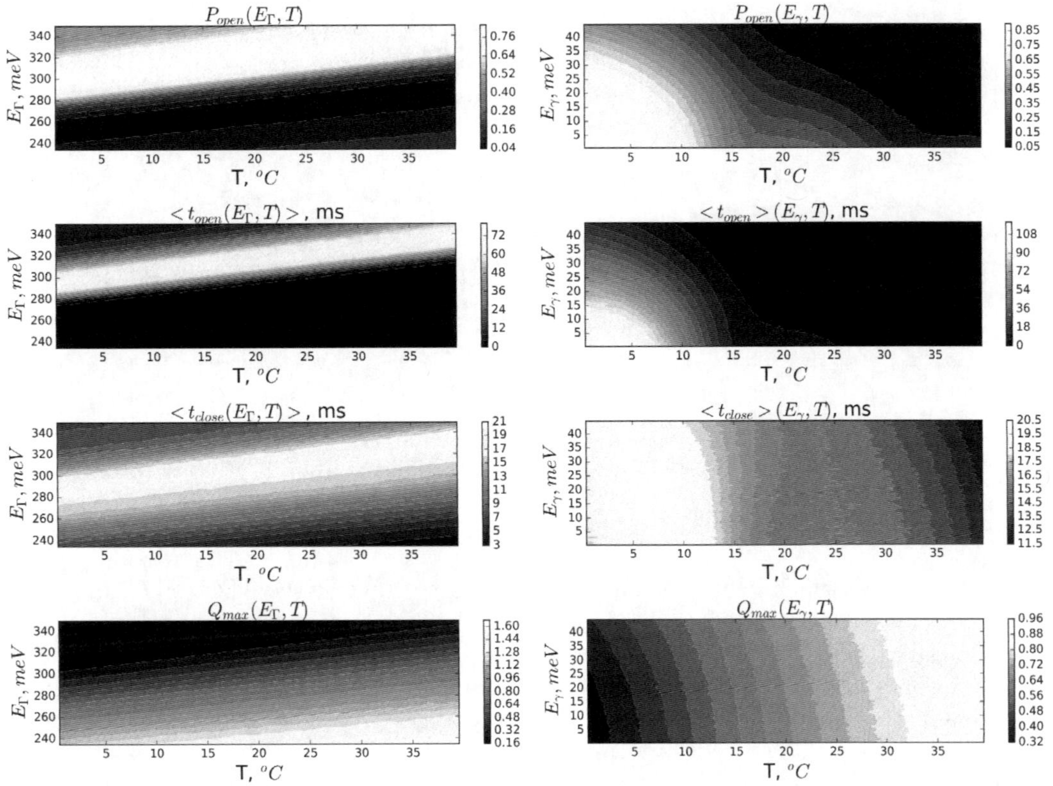

Figure 2. The temperature dependences for P_{open}, $\langle t_{open} \rangle$, $\langle t_{close} \rangle$, and maximal Q given different values of activation energies for internal E_Γ (left) and external E_γ (right) friction, respectively.

Conductance in EC model
We assume the RyR channel's conductance to obey a simple power dependence

$$D(Q) = D\,Q^\beta \tag{6}$$

when RyR channel is electronically and conformationally open, otherwise $D(Q) = 0$.

Results
In our previous works [14, 15] we have shown that the features of the tunneling transitions between CP branches together with the concept of the effective temperature can provide a qualitative and quantitative explanation of the main temperature effects for the RyR gating. Hereafter, in the paper we focus on an influence of the activation energies E_Γ and E_γ on the temperature dependence of the main kinetic characteristics of the RyR channel gating, such as open time probability, P_{open}, mean open and mean closed times, $\langle t_{open} \rangle$ and $\langle t_{close} \rangle$, respectively. We have performed a series of computer simulations of the RyR channel gating making use of a typical protocol of the EC model [9, 10, 11, 14] using the same set of parameters as in Ref. [15]. They were taken so that they provide a quantitative description of experimental data [3]. In Fig.2 we show the results of the (T, E_Γ) and (T, E_γ) parametric analysis for an isolated RyR channel. It is clearly seen that as a whole the change of E_Γ and E_γ shifts significantly the temperature dependence of the RyR channel kinetic characteristics (P_{open}, $\langle t_{open} \rangle$, $\langle t_{close} \rangle$) and

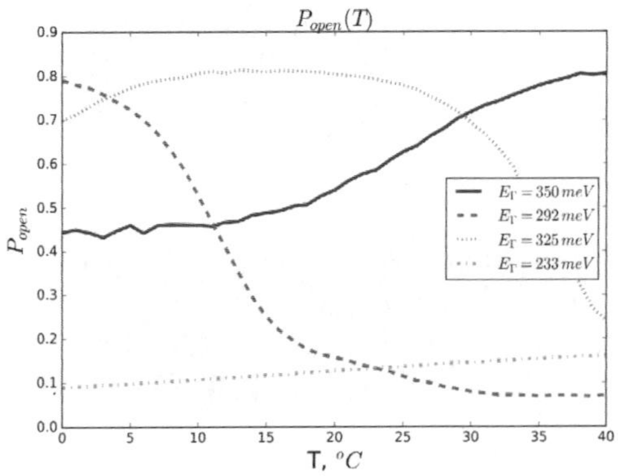

Figure 3. Temperature dependence $P_{open}(T)$ given different E_Γ.

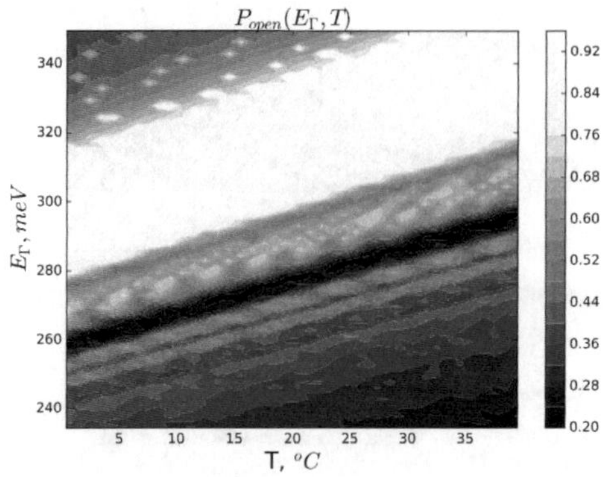

Figure 4. The temperature dependence P_{open} if the external friction is neglected ($\gamma_0 = 0$).

maximal Q ($\langle Q_{max}\rangle$) which defines channel's conductance, though the effect of E_γ on $\langle t_{close}\rangle$ and $\langle Q_{max}\rangle$ seems to be marginal. It should be noted that the temperature dependence of main kinetic characteristics may be non-monotonic. Furthermore, the change of the activation energy E_Γ, in other words, the internal friction, can turn channel from the cold-activated to the heat-activated one with a significant change of the thermosensitivity Q_{10} (Fig.3). This theoretical result is partly supported by recent experimental data [16], which show that the changes in the membrane lipid composition, which certainly do influence the internal friction, markedly change the temperature behaviour of the RyR channels. Interestingly, the internal friction due to the RyR channel's inter-residue interactions seems to play a more important role in temperature effects as compared with the external friction. Indeed, Fig.4 shows that even if the external friction be neglected ($\gamma_0 = 0$) the main kinetic characteristic P_{open} reveals a typical temperature behavior.

Conclusion

We have demonstrated that a simple electron-conformational model with an assumption of an Arrhenius-type temperature dependences for the "internal" and "external" frictions, together with a specific "thermosensitivity" of the "open" → "closed" tunneling probability [14, 15], is able to describe different types of the temperature behavior for the ion channel gating. Relatively small change of the activation energy E_Γ for the internal friction can turn channel from the cold-activated to the heat-activated one with a significant change of the thermosensitivity Q_{10}. We believe that similar mechanism can work for other ligand-gated ion channels.

Acknowledgments

The research was supported by the Ministry of Education and Science of the Russian Federation, projects nos. 1437 and 2725 (elaboration of the EC model), and the Russian Science Foundation, Project no. 14-35-00005 (application to the temperature effects in the RyR channel functioning).

References

[1] Robin H. Shutt, Susan E. Howlett.Hypothermia increases the gain of excitation-contraction coupling in guinea pig ventricular myocytes.*Am J Physiol Cell Physiol.*, 2008,**295**,C692 - 700.

[2] Sandor Györke and Cynthia Carnes.Dysregulated sarcoplasmic reticulum calcium release: Potential pharmacological target in cardiac disease.*Pharmacol Ther.*, 2008,**119**,340 - 354.

[3] R.Sitsapesan, R.A. Montgomery, K.T. MacLeod, A.J. Williams.Sheep cardiac sarcoplasmic reticulum calcium-release channels: modification of conductance and gating by temperature. *J Physiol.*, 1991,**434**,469 - 488.
[4] R. G. Efremov, A. Leitner et al. Architecture and conformational switch mechanism of the ryanodine receptor. *Nature*, 2015,**517**,39 - 43.
[5] R. Zalk, O. B. Clarke et al. Structure of a mammalian ryanodine receptor. *Nature*, 2015,**517**,44 - 49.
[6] Zhen Yan, Xiao-chen Bai et al. Structure of the rabbit ryanodine receptor RyR1 at near-atomic resolution. *Nature*, 2015,**517**,50 - 55.
[7] Risheng Wei1, Xue Wang et al. Structural insights into Ca(2+)-activated long-range allosteric channel gating of RyR1. *Cell Res.*, 2016,**26**,977 - 994.
[8] Wei Peng, Huaizong Shen et al. Structural basis for the gating mechanism of the type 2 ryanodine receptor RyR2. *Science*, 2016, Sep. 22.
[9] A.S. Moskvin, M.P. Philipiev et al. Electron-Conformational Model of Nonlinear Dynamics of the Ryanodine Channel Lattice in Cardiomyocytes, Doklady Biochemistry and Biophysics, **400**, 32 (2005); *J. Phys. Conf. Ser.* **21**, 195 (2005); Electron-conformational model of ryanodine receptor lattice dynamics,*Prog. Biophys. Mol. Biol.* **90**, 88 (2006).
[10] A.S. Moskvin, A.M. Ryvkin et al. Electron-conformational transformations in nanoscopic RyR channels govern both the hearts contraction and beating, *JETP Letters* **93**, 403 (2011).
[11] A.M. Ryvkin, A.S. Moskvin et al. Simulation of the auto-oscillatory calcium dynamics in cardiomyocytes in terms of electron conformational theory, *Doklady Biological Sciences*, **444**, 162 (2012).
[12] K. Koshino, T. Ogawa Theory of the photoinduced spin-state transitions in spin-crossover complexes *J. of Luminescence*, **87-89**, 642 (2000); N. Nagaosa and T. Ogawa Theory of photoinduced structure changes, *Phys. Rev. B* **39** 4472 (1989).
[13] Stephen J. Hagen. Solvent Viscosity and Friction in Protein Folding Dynamics. *Curr Protein Pept Sci*, 2010, **11**,385 - 395.
[14] Alexander S. Moskvin, Bogdan I. Iaparov et al. Electron-conformational transformations govern the temperature dependence of the cardiac ryanodine receptor gating. *JETP Letters*, 2015, **102**,62 - 68.
[15] Alexander S. Moskvin, Bogdan I. Iaparov et al. The Temperature Effect on Cardiac Ryanodine Receptor Gating and Conductance: Mathematical Modeling. *Biophys.*, 2016, **61**,614 - 621.
[16] S. El-Ajouz, E. Venturi, R. Sitsapesan Differential Effects of Temperature and Lipids on the Gating of RyR and SR K+ Channels. *Biophys.J.*, 2016, **110**,266a.

IOP Conf. Series: Journal of Physics: Conf. Series **929** (2017) 012020 doi:10.1088/1742-6596/929/1/012020

Interaction of indole-papaverine with DNA in solutions of various ionic strength

V I Travkina[1], E B Moroshkina[1] and D N Osinnikova[1]
[1] St. Petersburg State University, 1 Ul'ianovskaia str., Peterghof, St. Petersburg 198504, Russia

E-mail: travkinaveronika@gmail.com

Abstract. Interaction of synthetic alkaloid of isoquinoline series, which is an analogue of the biologically active compound papaverine, was studied by spectral, microcalorimetric, optical and hydrodynamic methods at different ionic strengths of medium. It was found that the investigated compound may interact with DNA in various ways depending on the ratio of ligand - DNA concentrations and ionic strength of solution (μ). When $\mu = 0.001$, indole-papaverine intercalates into the double helix of DNA. The increase of μ resulted in a decrease of the affinity of the compound to DNA and a change its binding method.

1. Introduction

The study of the formation mechanism of non-covalent reversible complexes of biologically active substances with DNA is an actual task, since the understanding of these processes plays an important role in the improvement of existing drugs, and may serve as the first step in the design of new ones.

Intercalation and groove binding are mainly caused by van der Waals interactions and hydrogen bonds that are non-covalent interactions. Forasmuch as DNA in solution at neutral pH is a polyanion with negatively charged phosphate groups of the sugar phosphate backbone, electrostatic interactions also play a significant role in the formation of such complexes [1]. At intercalation there is embedding of planar aromatic molecules of ligand between DNA base pairs [2]. At external interactions there is a fitting of ligand molecules into the grooves or a formation of aggregates on the surface [3].

This paper presents the results of a study of interaction of DNA with the indole derivatives of isoquinoline - indole-papaverine, which has a positive charge on the isoquinoline chromophore, at various ionic strengths of medium.

2. Materials and methods

There were used calf thymus DNA from the company «Sigma» (USA) with molecular weight $M = 10^7$ Da. DNA concentration was determined by the spectrophotomical method described in [4]. Indole-papaverine was synthesized at the Research Institute of Hygiene, Occupational Pathology and Human Ecology. [5] Aqueous solutions of NaCl of different ionic strength, μ, and pH 6.0 - 6.5 were used as a solvent. Complexes were prepared by direct mixing of DNA and the substance at a certain concentrations of the latter.

The stoichiometry of complexes, defined as the amount of bound ligand per pair of nitrogenous bases of DNA (r), was obtained from the spectrophotometric titration data (SFT). The absorption spectra of the solutions were recorded using «Specord UV-Vis» and «Shimadzu UV-1800" spectrophotometers. Circular dichroism spectra were recorded at dichrographs MarkV.

The enthalpy of interaction and thermodynamic parameters of the indole-papaverine binding were determined by isothermal titration calorimetry (ICT) using microcalorimeter TA Instruments Nano ITC 2G at the Resource Center of St. Petersburg State University "Thermogravimetric and calorimetric methods of investigation". [6] The calculation of the thermodynamic binding parameters in this case was carried out on the bases of a model with two independent binding sites [7].

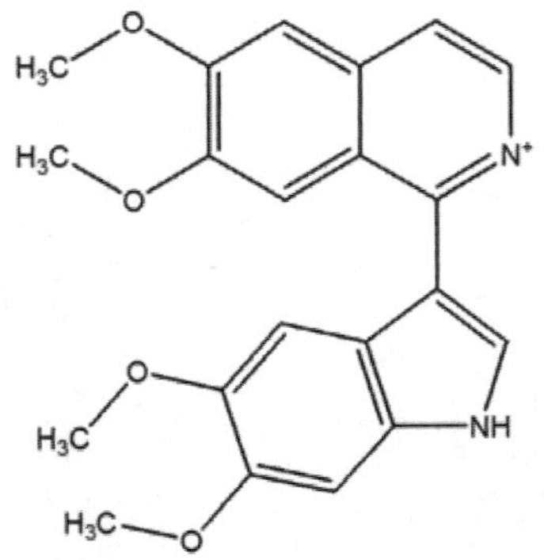

Figure 1. Structure of indole-papaverine.

The structure of complex and the way by which the compound and DNA interact during the formation of complex were established by means of viscosimetry and birefringence. Analysis of changes in the intrinsic viscosity, $[\eta]$, and the optical anisotropy of statistical segments of the macromolecule, $(\alpha_1-\alpha_2)$, was performed according to methods described previously [1,8]. We measured the dependence of the relative viscosity of solution, η_r, on DNA concentrations in the solution, C_{DNA}, using a magnetic rotational viscometer [9]. Thus, $[\eta]$ was determined as

$$[\eta] = lim\ (\eta_r\text{-}1)/C_{DNA},\ \text{at}\ g\rightarrow 0\ \text{and}\ C_{DNA}\rightarrow 0,$$

where g is the velocity gradient.

To determine the changes in the magnitude of $(\alpha_1-\alpha_2)$, resulted from the formation of the complex, Peterlin's relationship was used [10], which is proportional to the optical anisotropy of the statistical segment of the macromolecule:

$$\Delta n/(g\ (\eta-\eta_0))\sim(\alpha_1-\alpha_2),$$

where Δn - the magnitude of the birefringence, η_0 - viscosity of the solvent.

Δn was measured in the titanium dynamo-optimeter using the experimental optical setup with a mica half- shadow compensator [11].

3. Results and discussions

3.1. Spectrophotometric titration

At $\mu = 0.001$ there are two isosbestic points (Fig.2), which reflect not only two methods of binding, but also the high affinity of the ligand to DNA ($k_{bind} > 10^7\ M^{-1}$). The limiting spectrum is observed at $C_{ligand}\ /\ C_{DNA} \leq 0.1$, which confirms the possibility of a primary monomer binding. The availability of isosbestic point at a wavelength $\lambda = 377$ nm and at $C_{ligand}\ /\ C_{DNA} \leq 2.0$ indicates that there is no considerable amount of free ligand in the system. Changes in the intensity and spectrum shift occur due to the transition of monomeric binding type to dimeric one [12, 13]. Isobestic point at $\lambda = 413$ nm corresponds to the band of spectra which comprises spectrum of free ligand and complexes at $C_{ligand}\ /\ C_{DNA} \leq 2.0$, and characterizes the equilibrium between free ligand and dimeric bound.

Two isosbestic points blurred with the increase of ionic strength to $\mu = 0.003$, (Fig.3), which may indicate both the decrease of the affinity of the compound to DNA and the lack of a balance between the two types of monomer-dimer binding methods.

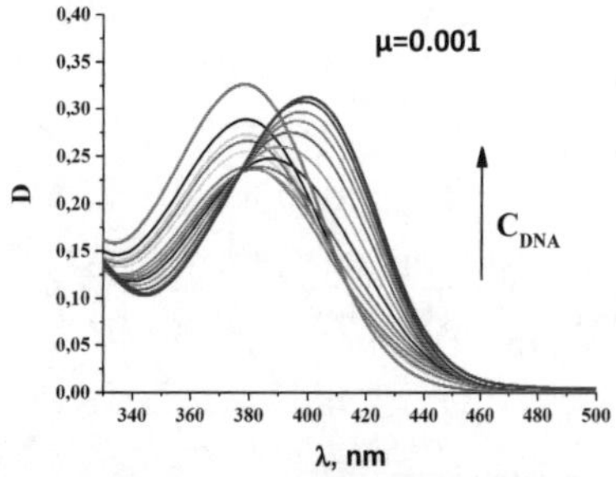

Figure 2. The absorption spectra of the compound during the titration with a constant concentration $C_{\text{ligand}} = 4.20\cdot10^{-5}$ M, $\mu = 0.001$, $0.12\cdot10^{-5}$ M $\leq C_{\text{DNA}} \leq 4.63\cdot10^{-4}$ M.

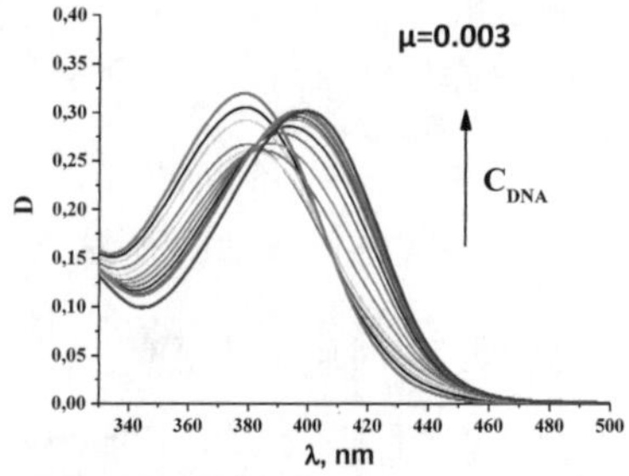

Figure 3. The absorption spectra of the compound during the titration with a constant concentration $C_{\text{ligand}} = 4.00\cdot10^{-5}$ M, $\mu = 0.003$, $0.57\cdot10^{-5}$M $\leq C_{\text{DNA}} \leq 4.70\cdot10^{-4}$ M.

3.2. Isothermal microcalorimetric titration

The results of ICT at $\mu = 0.001$ (Fig. 4) are consistent with the SFT data. The monomer binding occurs at $C_{\text{ligand}} / C_{\text{DNA}} \leq 0.2$, while dimer binding takes place at $0.2 \leq C_{\text{ligand}} / C_{\text{DNA}} \leq 1.5$. If we have $C_{\text{ligand}} / C_{\text{DNA}} \geq 1.5$, it is possible to talk about the external aggregative nonspecific binding. When the ionic strength of the medium is increased, the general form of thermal image changes significantly (Fig. 5).

The thermodynamic parameters obtained from the ICT data by the processing program of microcalorimeter assuming the model of two independent types of binding sites are shown in Table 1.

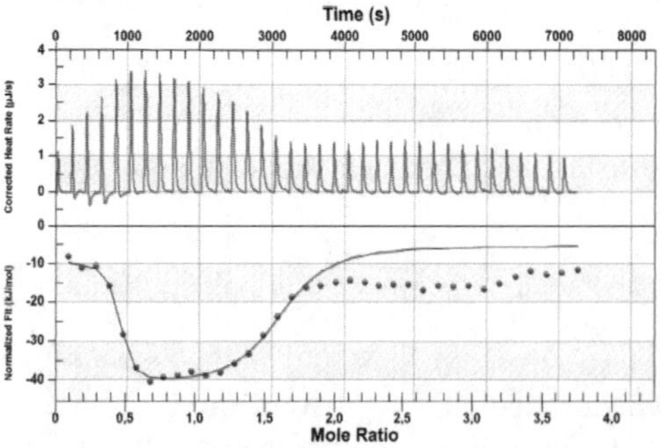

Figure 4. Thermogram of titration of DNA by ligand (upper part) and dependence of the amount of heat per one mole of the ligand from C_{ligand} / C_{DNA} at $\mu = 0.001$ (lower part). $C_{\text{DNA}} = 2 \cdot 10^{-5}$ M. The black solid line reveals the approximation by the model of two types of independent binding sites.

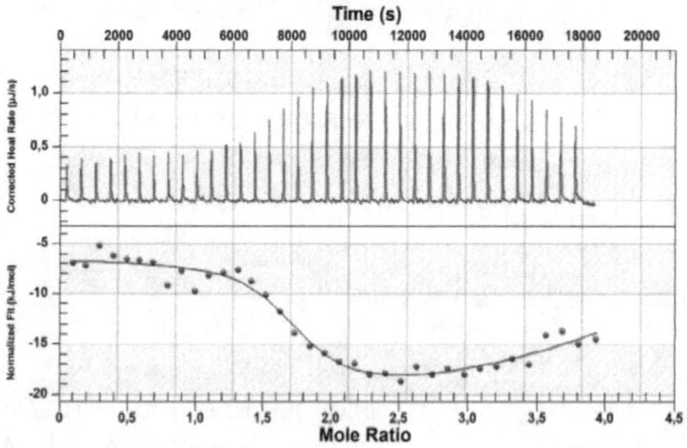

Figure 5. Thermogram of titration of DNA by ligand (upper part) and dependence of the amount of heat per one mole of the ligand from C_{ligand} / C_{DNA} at $\mu = 0.003$ (lower part). $C_{\text{DNA}} = 2 \cdot 10^{-5}$ M. The black solid line reveals the approximation by the model of two types of independent binding sites.

Table 1. Thermodynamic parameters of complexes DNA and indole-papaverine.

Complex	$\Delta H_1, \frac{kcal}{mol}$	$\Delta H_2, \frac{kcal}{mol}$	$\Delta S_1, \frac{kcal}{mol\cdot K}$	$\Delta S_2, \frac{kcal}{mol\cdot K}$	k_{bind1}, M^{-1}	k_{bind2}, M^{-1}	n_1	n_2
μ=0.001	-1.50	-9.60	35.3	4.5	$1.3\cdot10^{8}$	$1.3\cdot10^{6}$	0.4	1.0
μ=0.003	-1.56	-5.24	28.4	64.6	$2.3\cdot10^{7}$	$2.0\cdot10^{5}$	1.7	3.1

The comparison of colorimetric titration curves at ionic strengths, equal to 0.001 and 0.003, indicates that the secondary binding at $\mu = 0.003$ occurs when it is a large excess of the ligand and its thermal characteristics very similar to the non-specific aggregative binding of the ligand at μ = 0.001 and C_{ligand} / $C_{\text{DNA}} \geq 1.5$. At the same time the binding of ligand dimers observed at μ = 0.001,while at μ = 0.003 does not occur.

3.3. Circular dichroism

Parallel to SFT were made spektropolyarimetric studies at μ = 0.001 and μ = 0.003 in regeon of C_{ligand} / $C_{\text{DNA}} \leq 1.0$ (Fig. 6,7). At μ = 0.001 ICD appears when C_{ligand} / C_{DNA} > 0.4 and has the exciton form of spectrum corresponding to the interaction of the bound ligand molecules, which confirms SFT and ICT data about the emergence of dimer binding under these conditions. At μ = 0.003 ICDs exciton nature is not observed, indicating the absence of dimers in the system.

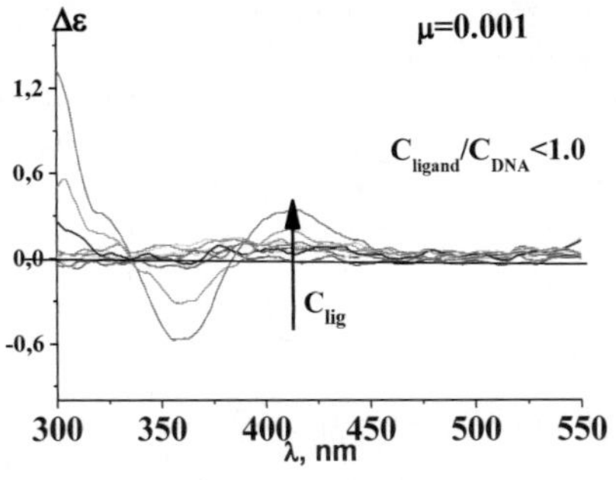

Figure 6. Circular dichroism spectra at μ = 0.001. $0.88 \cdot 10^{-6}$ M ≤ $C_{\text{ligand}} \leq 1.85 \cdot 10^{-5}$ M, $C_{\text{DNA}} = 2.00 \cdot 10^{-5}$ M.

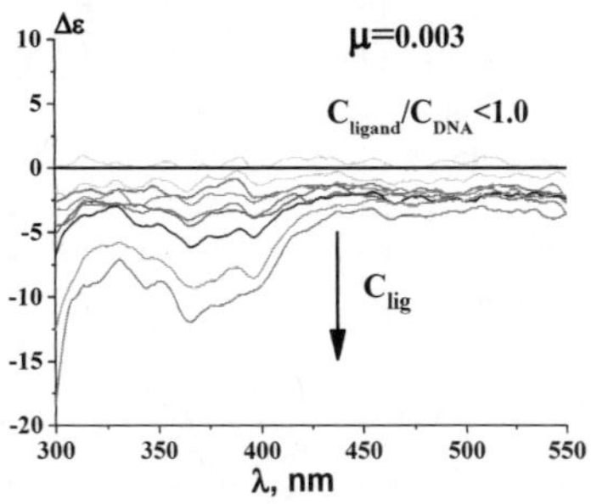

Figure 7. Circular dichroism spectra at μ = 0.003. $0.10 \cdot 10^{-6}$ M ≤ $C_{\text{ligand}} \leq 1.78 \cdot 10^{-5}$ M, $C_{\text{DNA}} = 2.00 \cdot 10^{-5}$M.

3.4. Viscometry and dynamic birefringence

To determine way of the compound binding there have been identified the characteristic viscosity and the Peterlin's ratio for free DNA and its complex with a ligand at r = 0.2 (tab.2).

At μ = 0.001, there is the increase of intrinsic viscosity during the formation of the DNA complex with indole-papaverine. In the case of high molecular weight of DNA, such changes can be caused by an increase in contour length (*L*), and an increase in the thermodynamic rigidity of the macromolecule which determined by the length of the statistical segment (*A*) [14].

Table 2. Hydrodynamic and optical parameters of DNA and its complexes with the compound.

Complex	[η], m³/kg	$\Delta n/g(\eta - \eta_{\text{DNA}})$, m·s²/kg
	μ=0.001	
DNA	15±1	26 ± 2
r=0.2	20±1	27 ± 2
	μ=0.003	
DNA	7.0±0.5	22±2
r=0.2	5.5±0.5	22±2

The value of Peterlin's ratio, which is proportional to the thermodynamic rigidity of macromolecules, remains constant within the error. Therefore, the intrinsic viscosity increase upon complexation DNA with ligand at μ = 0.001 occurs as a result of change its contour length. The calculation of this change was performed using the Flory's formula for coiled macromolecules [14]

$$[\eta] = \Phi \frac{(\overline{h_0^2})^{3/2}}{M} \gamma^3,$$

where Φ – Flory constant for a given polymer-solvent system, h_0 - the distance between the ends of the macromolecule, γ - linear swelling ratio. As a result, it was received L_{ligand} / L_{DNA} = 1.2 ± 0.1, which

IOP Conf. Series: Journal of Physics: Conf. Series **929** (2017) 012020 doi:10.1088/1742-6596/929/1/012020

corresponds to the relative change in the contour length of the macromolecule, expected at the ligand intercalation into the double helix of DNA at $r = 0.2$.

At $\mu = 0.003$ the intrinsic viscosity of the complex with the ligand content is even slightly smaller than the intrinsic viscosity of pure DNA which eliminates intercalative binding.

4. Conclusions

The investigated compound has a positive charge on the isoquinoline chromophore, so the ionic strength of medium plays a significant role in the complexation process with DNA.

On the basis of obtained experimental data we can propose the model of ligand binding to DNA depending on the ionic strength of medium and amount of ligand molecules in the complex. When $\mu = 0.001$, the interactions with DNA at $0.1 \leq r \leq 2.0$ come in two ways: monomeric and dimeric. These ways of binding are interdependent: associated monomer is a place of secondary ligand binding and, as a result, there forms dimer. Formation of the latter is accompanied by the release of heat. Free ligand and the third non-specific binding mode as external units appeared at $r \geq 1.5$. Intercalative way of binding of monomers to DNA is shown using birefringence and viscosimetry tests.

Binding way of monomers become non-intercalative with the increase of ionic strength of medium. This binding is characterized by the negative ICD, the absence of enthalpy change and the reduction of intrinsic viscosity of macromolecules during the formation of complex. For both $r \geq 1.2$ and low ionic strength of medium, there is the nonspecific aggregation binding way that result in an excess of the ligand in the system.

References

[1] Veselkov A, Moroshkina E, Soboleva O, Frisman E 1984 *J Mol. Biol.* **18** 481 – 87
[2] Lerman L 1961 *J Mol. Biol.* **3** 18–30
[3] Chen W, Qin Y, Cai Z, Chan C, Luo G and Jiang Z 2005 *J Bioorganic & medicinal chemistry* **13** 1859–66
[4] Spirin A *J Biochem. - Moscow*1958 **23**. 617 – 22
[5] Krivorotov D, Moroshkina E and Leont'eva T 2011 *Proc. of IV Moscow Int.Congress "Biotechnology: Prospects for Development"* Moscow 1 70-1
[6] Velázquez-Campoy A, Ohtaka H., Nezami A Muzammil S and Freire E 2004 *Current protocols in cell biology* 17.8.1 – 24
[7] Buurma N.J. and Haq I 2007 *Methods* **42** 2 162–72
[8] Moroshkina E, Shishov A, Krivtsova M, Zhadin N and Frisman E 1975 *J Mol.Biol.* **6** 668 – 74
[9] Frisman E, Schagina L, Vorobiev V 1965 *J Biorheology* **2** 189-94
[10] Peterlin A 1954 *J of Polymer Science* **12** 45–51
[11] Frisman E and Tsvetkov V 1952 *Zhurnal Eksperimentalnoi I Teoreticheskoi Fiziki* 26 690 – 702
[12] Moroshkina E and Saf'yannikova M 1999 *J Biofizika* **44** 425 – 29
[13] Tuite E and Norden B1994 *J of the American Chem. Society* **116** 7548–56
[14] Flory P 1953 N.-Y. *Cornell Univ. Press.*

Fluorescent spectroscopy features in the study of green leaves of plants

I S Kudryashova[1], V Yu Rud[2,3], Yu V Rud[2], V Ch Shpunt [2], A P Glinushkin[3], N N Bykova[4].

[1]Department of Quantum Electronics, Peter the Great Saint Petersburg Polytechnic University, Saint Petersburg 195251, Russia

[2]Department of Solid State Physics, Ioffe Physico-Technical Institute, Russian Academy of Sciences 26 Politekhnicheskaya, Saint Petersburg 194021, Russia

[3]Department of Ecology, Russian Research Institute for Phytopathology, Moscow Region, Odintsovo district, B.Vyazyomy, 143050, Russia

[4]Center of innovation learning technologies, Saint-Petersburg State University of Economics, 21 Sadovaya street , Saint Petersburg 191023, Russia

E-mail: rudvas.spb@gmail.com

Abstract. The article gives an overview of present results of a study of the luminescence of living plants in room temperature and also the luminescence of green leaves in a living state and taken from the parent plant. The analysis of obtained results allows to conclude that the photoluminescence spectra for green leaves in all cases represent the two closely spaced bands. Temporal intensity of photoluminescence of green leaves of plants were studied.

1. Introduction

The rapid development of biological and agricultural sciences because of the need of food security growing population of the planet dictates the importance of in-depth study of the structure and function of living systems. Knowledge of these issues allows us to improve plant breeding, sustainable agriculture, improves the life of mankind. In this regard, the need to use the research proven in other scientific questions, methods and experimental techniques increases. Recently there has been increased the use of research methods of various characteristics of the optical properties of plants.

With all these virtually unknown experiments in which green plant objects were investigated like conventional solid state semiconductor samples. The analysis results can be performed based on the established solid object representations. That is why the use of well-developed solid state physics methods of application for the investigation of photoluminescence [1] for the study of living green plants seems to us entirely innovative and promising demand [2, 3, 4, 5]. Ideas about the electrical and optical properties of green plants and the microscopic mechanism responsible for the regularities in the behavior of these biological objects when exposed to light and the influence of an electric field can be significantly expanded in the presence of such important characteristics as optical transmittance and electrical resistance. Studies of the optical transmittance spectra and photoluminescence of green leaves of various plants in the system with a live plant were conceived and carried out. But they were also derived from a single system with a live plant. Also the study of the temperature dependence of the electrical resistance of these objects was done. Also temporal patterns of the photoluminescence intensity for green leaves of plants were researched.

2. Experimental study and discussion of the results

The study of electrical resistance was carried out in a planar design pattern. Clamping electrodes made of copper plates provide ohmic contact in the investigated range of temperatures and voltages. The measured sample was placed on a quartz plate in such a way as to prevent mechanical damage to the sample. Temperature range studies were selected on the basis of the existing provisions, namely

IOP Conf. Series: Journal of Physics: Conf. Series **929** (2017) 012021 doi:10.1088/1742-6596/929/1/012021

metabolic reactions in the body is mainly enzyme reactions, in which the determining factor is the presence of enzymes, without which all the vital functions ceases [6]. It is known that the enzymes lose their activity upon heating and at temperatures 50-60°C inactivated. Inactivation of enzymes – the phenomenon is irreversible. Therefore, the study of temperature dependence was carried out in the range from 50 to 30°C, to maintain important functions of a green plant. Samples for testing consisted of green sheets with thicknesses from 70 to 150 microns. Study of the optical transmittance was carried out on the plant, which includes a monochromator SPM-2 with a quartz prism. Recombination radiation spectra were studied under excitation of an argon laser ILA-120-1 Carl Zeiss (either He-Ne laser) with excitation energies hw_{exc} = 2,1; 2,50; 2,54; 2,6; 2,71 eV. For each of the above energies, the power density was 50-100 mW/cm^2. Emission photoluminescence further went on monochromator MDR-3 grating with 600 lines per mm and was detected with photomultiplier. The spectral resolution of the experimental setup was not worse than 1 meV. All studies were performed at temperature T=300 K. In the result, it was found that green leaves under their excitation with the excitation energies listed above, show bright photoluminescence in the red.

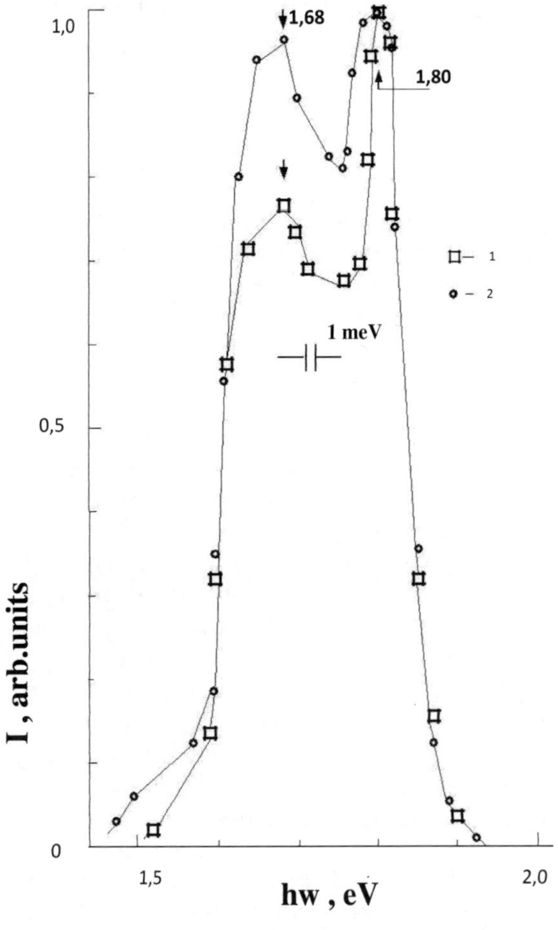

Figure 1. Spectral dependence of the photoluminescence of the leaves of Calendula officinalis L. at T=300 K (excitation energy hw_{exc}, eV : for curve 1 – 2. 705, for curve 2 – 2.410).

The photoluminescence spectra for green leaves in all cases represent the two closely spaced bands. It is important to note that the energies of both bands for different types of plants did not differ from each other (Fig. 1). The study of electrical resistance for all objects (Plantage lanecolata L (ρ= 2*10^{-6} Ohm cm at T=300 K), Taraxacum officinato Web (ρ= 1,5*10^{-6} Ohm cm at T=300 K), Arctium lappa L (ρ= 1,2*10^{-6} Ohm cm at T=300 K), Tilia platyphylla Scop (ρ= 2*10^{-6} Ohm cm at T=300 K), Cratacgus oxyacantha L. (ρ= 2*10^{-6} Ohm*cm at T=300 K), Populus nigra L. (ρ= 1,5*10^{-6} Ohm cm at T=300 K), Flagaria vesca L. (ρ= 2*10^{-6} Ohm cm at T=300 K)) showed good coincidence of the obtained values of resistivity, which is well reproduced by area, samples of leaves and shows that they have high electrical homogeneity. This may testify in favor of the conclusion about the important role of homeostasis and sustaining life. As temperature decreases, resistance in the samples studied followed a power law and is proportional to the exp($E_ρ$/kT), where $E_ρ$ is the activation energy of the resistance.

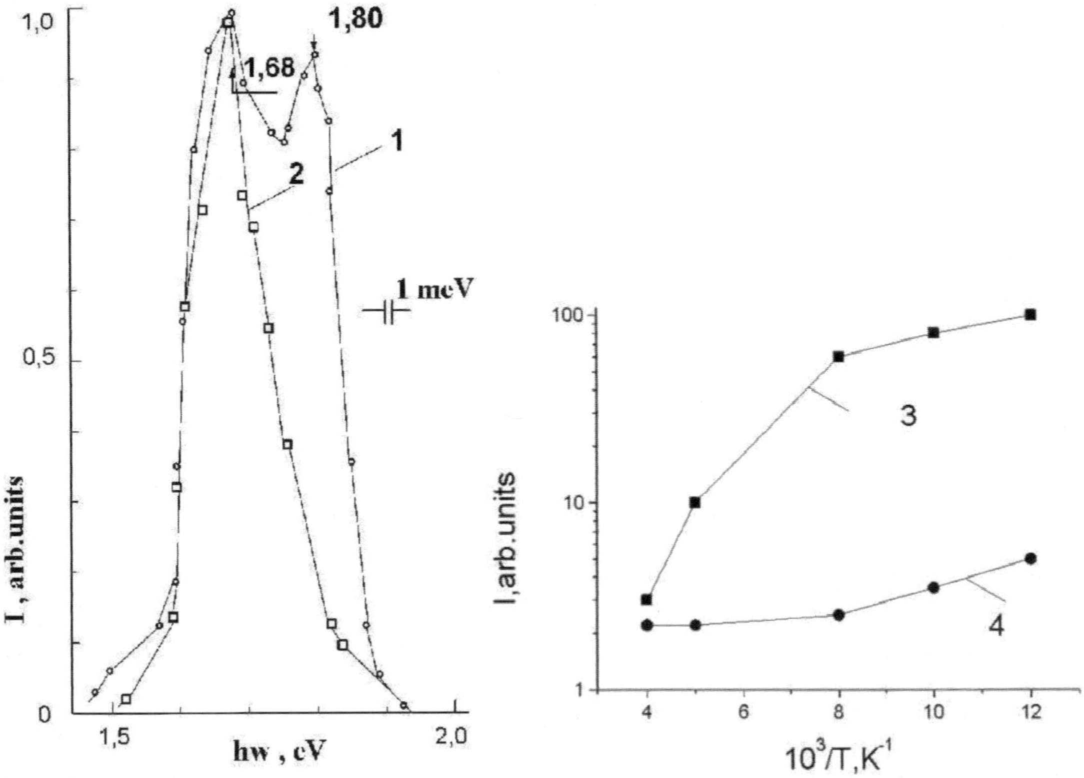

Figure 2. The effect of temperature on the spectral dependence of the luminescence (curve 1 - T= 300 K, curve 2 – T=77 K) and photoluminescence intensity of leaves of Calendula officinalis L. (curve 3- the short-wave component of the spectrum, curve 4 – wavelength component of the spectrum).

This value of activation energy of the resistance shows the value of the order of 1 eV in the temperature range from 230 to 300 K. The repeated thermal cycling led to a good coincidence of these dependences with each other in the temperature range from 77 to 300 K. This fact may indicate that at low temperatures a large part of the enzyme protein may not have lost their properties, and with the warming of metabolic reactions are restored. The process of drying the green sheet has led to an increase in resistance from 6 orders of magnitude (from 10^4 to 10^{10} Ohms). Study of optical transmittance and radiative recombination in living leaves in the composition of plants showed a similar dependence for all the studied plants. In the wavelength region of the spectral range of the optical transmittance did not exceed 20 percent. With increasing photon energy to 1.5 eV was observed a sharp decline in shortwave optical transmittance to zero at a threshold value of the energy close to 1.8 eV. The spectrum of radiative recombination region is localized near the edge of optical transmittance of green leaves. For the analysis of the photoluminescence in solid state physics especially important is the ratio of the photoluminescence peaks and the values of half-widths were also close for all types of plants. When we change the flux density of the exciting radiation, a linear law dependence of the photoluminescence intensity from the flux density was observed. Thus the spectral contour of the luminescence and energy peak positions and the ratio of their value remained unchanged (Fig. 1).

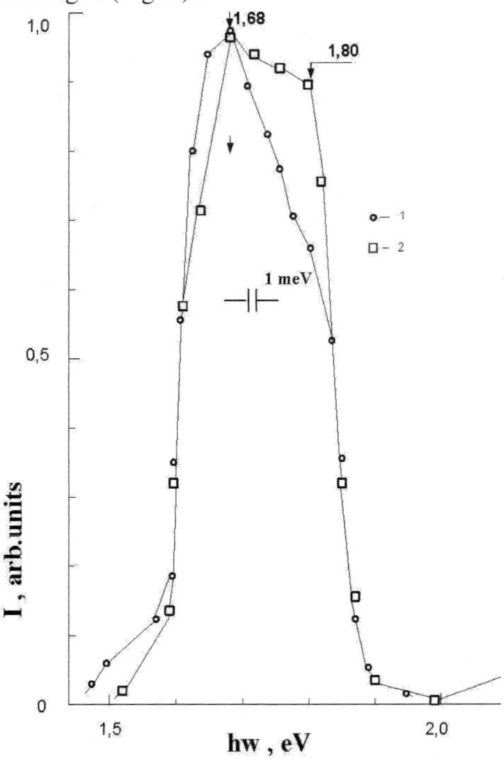

Figure 3. The spectral dependence of the photoluminescence intensity leaves : curve 1 – in green leaf system of the plant (T=300K, the excitation energy of 1.97 eV), curve 2 – the sheet is not in the system living plants in the drying process (T=300 K, the excitation energy of 2.815 eV).

These patterns in semiconductor physics are characteristic as for the radiant transitions between the free zones and for transitions from levels in the forbidden zone.

The analysis of the research results for fragments of the flower shows that the efficiency drops by 2-3 orders of magnitude compared to a green plant with a maximum significantly shifted to shorter wavelengths.

For the green leaf part of plants was discovered the dependence of the spectra of recombination radiation from a top side of the sheet is excited with the luminescence. When all studies was typical of that during the transition from the excitation of the photoluminescence from the upper to the lower surfaces of all samples is always observed the increased contribution of the short wavelength component of the spectrum. This is observed on all samples, the dependence may indicate that the recorded luminescence of the biochemical processes on the thickness of the sheet is uneven.

Studing of the photoluminescence from the temperature of the leaves was carried out in a wide temperature range, also included are not quite peculiar to living plants to low temperature (300-77 K). It was found that with decreasing temperature was observed a fire longwave components of the spectrum of radiative recombination. The intensity of the short wavelength components in the spectral range of the temperatures grew much weaker. Thus, in the temperature dependence of the intensity of the long wavelength components of the spectrum from 300 to 150 K we can see a clear exponential plot with activation energy close to 50 meV. At temperatures below 150 K the increase of the photoluminescence intensity in the region of the maximum is practically terminated. These features of the spectrum of radiative recombination when the temperature was typical for all studied samples of green leaves in live plants. It should be noted a good reproducibility of results with repeated thermal cycling of the samples in the temperature range from 300 to 77 K were stable characteristic for the research topogram the surface and from sample to sample. The characteristic behavior of the spectrum of the photoluminescence in dependence on temperature for the short-wave spectrum component different from the corresponding dependence for the long-wave component of the spectrum.

The energy position of the maximum of the long wavelength component of the photoluminescence for green plants with decreasing temperature shifted to shorter wavelengths with coefficient of $5*10^{-5}$ eV/K (Fig. 2), and the band itself is narrowed. At the same time, the temperature change has little influence on the position of the maximum of the short wavelength component of the spectrum. This may indicate differences in responsible for the long-wave and short-wave components in the mechanisms of radiative recombination of the investigated objects.

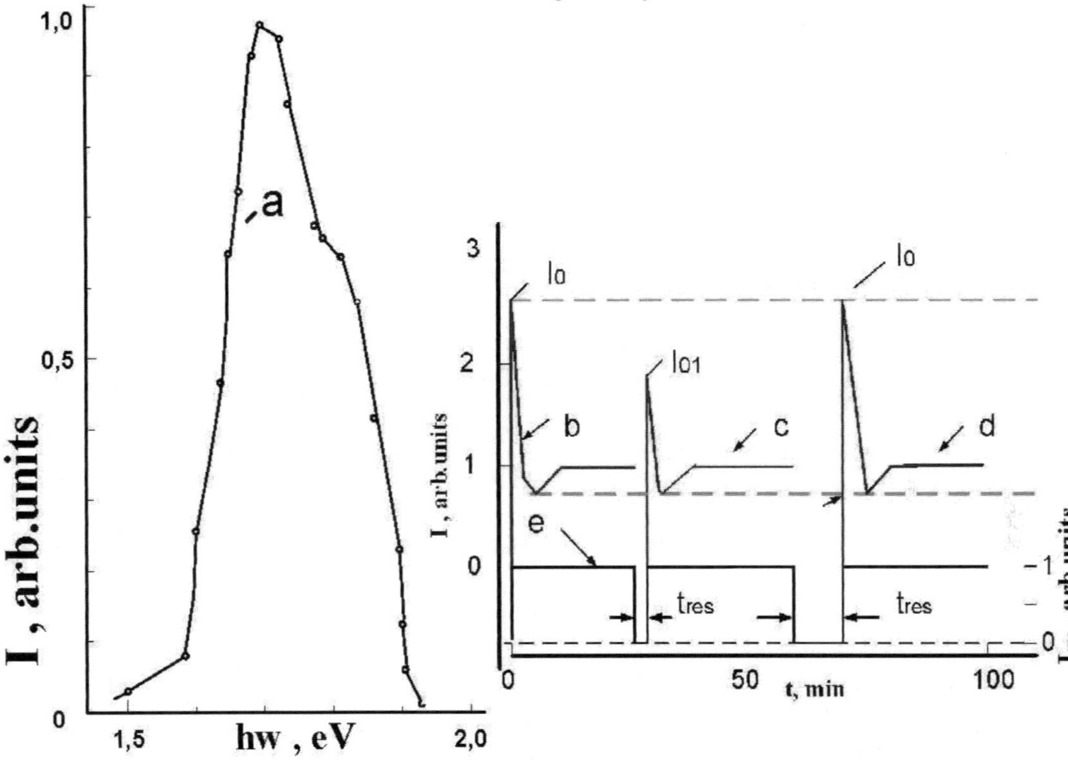

Figure 4. Spectral (a) and temporal (b, c, d) dependence of the photoluminescence intensity of the green leaves of Fragaria vesca L. in the composition of the parent plants, and the dependence of the intensity of the exciting radiation time (e). T=300 K, $\hbar\omega_{exc}$= 1.97 eV, t_{res} - the time delay of the exciting radiation.

Drying and separation of the leaves from green plants leads to quenching of luminescence and particularly its long-wave components (Fig. 3). In this case the emission spectrum is shifted to shorter wavelengths of the spectrum. In samples of leaves, taken from the green plants, prolonged process of drying the disappearance of the characteristic luminescence images, which include long-wave and short-wave peaks. Instead of this, there is the fire radiative recombination in the region of photon energies above 2 eV (Fig.3).

So the photoluminescence spectra of living leaves (which are a part of a living plant) has shown that the spectral luminescence circuit insensitive to the energy of the exciting photons. In this case the radiation intensity in all cases there were proportionately on the radiation intensity. Studied temporal patterns of photoluminescence intensity of green leaves, which are a part of the plant.

It turned out that this dependence has two stages. The first stage corresponds to a decrease in intensity for 30-40 seconds after the exciting radiation. After reaching its minimum photoluminescence begins to increase, reaching its equilibrium state.
The observed stage of recovery of luminescence indicates the kind of mobilization of vital resources plants to suppress the drop in intensity of the photoluminescence, with a consequent increase in the photoluminescence intensity under continuous light excitation laser (Fig.4). Studied laws allow in terms of mature methods of experimental physics to reveal the nature of living things that could expand the range of the experimental hardware quality control techniques and repeatability of resistance of plants, as well as increase the efficiency of agriculture.

3. Conclusion

The results show that the use of well-developed experimental techniques of semiconductor physics to the study of nature and the green leaves of various plants in the system with a live plant, but also derived from a single system with a live plant, gives new knowledge about the structure and functioning of green plants. These studies can be interesting for biologists, doctors [7], and agricultural workers, allows expanding the hardware a set of experimental techniques, quality control reproducibility of the plants at any stage of its development, connects solid state physics, biology and agricultural sciences.

References

[1] Rud' V Yu, Shpunt V Ch and Rud' Yu V 2016 The 10-th International Conference on Establishment of Cooperation between Companies and Institutions in the Nordic Countries, the Baltic Sea Region and the World. Book of Abstracts. Linneaus University, 21-23 Nov. 2016, Kalmar, Sweden 73-75
[2] Samoilov V O 2004 *Medical Biophysics* SPb. 479
[3] Pryakhin Yu Yu, Kartabaeva B B and Sidorova O P 2015 *Russian Agricultural Science Review* **6** 6-2 29
[4] Suvorov N V, Bikteev R Sh, Ragulin V S, Glinushkin A P and Duskaev G K 2015 *Biotics.* **3** 2. 15
[5] Rud' V Yu, Rud' Yu V and Shpunt V Ch 2016 *Optical Memory and Neural Networks.* 1 40
[6] Kudryashova I S, Rud V Yu, Shpunt V Ch, Rud Yu V and Glinushkin A P 2016 Journal of Physics: Conference Series volume 741 (2016) 012106– proceeding the 3st International School and Conference «Saint – Petersburg OPEN 2016» on Optoelectronics, Photonics, Engineering and Nanostructures (28–30 March 2016, St Petersburg, Russia).
[7] Rud' V Yu, Glynushkin A P, Lyapischev V A, Shpunt V Ch and Rud' Yu V 2016 The international science environmental conference «ECOBALTICA - 2016». 20 – 26 Oct. 2016. Proceedings Book. 2016. Moscow region. Russia 21

Design of microwave antenna system on planar Yagi-Uda elements and microstrip coupler

K V Petrovnin[1], R R Latypov[1]

[1]Kazan Federal University, Kazan, 420008 Russia

E-mail: petrovninkirill@gmail.com

Abstract. Paper presents results of calculation, electromagnetic modelling and measurements of manufactured antenna system on planar Yagi-Uda elements and microstrip coupler. System has summary and subtract modes. Center frequency of system is 1532 MHz with 96 MHz bandwidth. Gain of system is 8 dB in main lobe direction (in-phase mode) and 5 dB (antiphase mode).

1. Introduction

Due to progress in information technology increases the need to create specialized transceiver antenna systems for transmission/receiving in the microwave frequency range and having orientation [1]. The paper presents the description of the process and outcomes of a directional antenna system agreed planar Yagi-Uda elements and microstrip coupler with the center operating frequency $f = 1.532$ GHz and bandwidth $\Delta f = 96$ MHz.

Antenna system has in-phase and antiphase operation modes; each mode has its own indicators of the standing wave ratio (SWR) and the radiation pattern. Design process is divided into 3 phases: geometrical parameters calculation of elements [2, 3], electromagnetic simulation and optimization model using CST Studio Suite, creating a physical prototype, measurement and parameter analysis. The main features of this antenna system include its small geometric dimensions, manufacturing simplicity and planarity, two modes of operation and the possibility of increasing the number of elements to create a prototype of a phased array antenna with preservation of topology.

2. Calculation and modelling

It was originally calculated the geometric parameters of the coupler (Fig. 1) in the equivalent circuit of the ring power coupler (also named “rat-race”) for glass-reinforced epoxy laminate sheet (manufacturer name - FR-4) thickness $d = 1.5$ mm with the measured relative permittivity $\varepsilon_r = 4.67$ according to the formula (1) [2]:

$$\frac{W}{d} = \begin{cases} \dfrac{8e^{A}}{e^{2A} - 2} & \text{for } \dfrac{W}{d} < 2 \\ \dfrac{2}{\pi}\left[B - 1 - \ln(2B - 1) + \dfrac{\varepsilon_r - 1}{2\varepsilon_r}\left\{\ln(B - 1) + 0.39 - \dfrac{0.61}{\varepsilon_r}\right\}\right] & \text{for } \dfrac{W}{d} > 2 \end{cases}, \quad (1)$$

Where

$$A=\frac{Z_0}{60}\sqrt{\frac{\varepsilon_r+1}{2}}+\frac{\varepsilon_r-1}{\varepsilon_r+1}\left(0.23+\frac{0.11}{\varepsilon_r}\right) \quad \text{and} \quad B=\frac{377\pi}{2Z_0\sqrt{\varepsilon_r}}.$$

Calculated coupler offshoots have width 2.7 mm (Z_0 = 50 Ohm), circle width is 1.39 mm ($Z_1 = \sqrt{2}Z_0$ = 70.7 Ohm).

Figure 1. Coupler schematic in CST Studio

Next it was calculated the planar Yagi-Uda antenna with three directors and balancing transformer (Fig. 2).

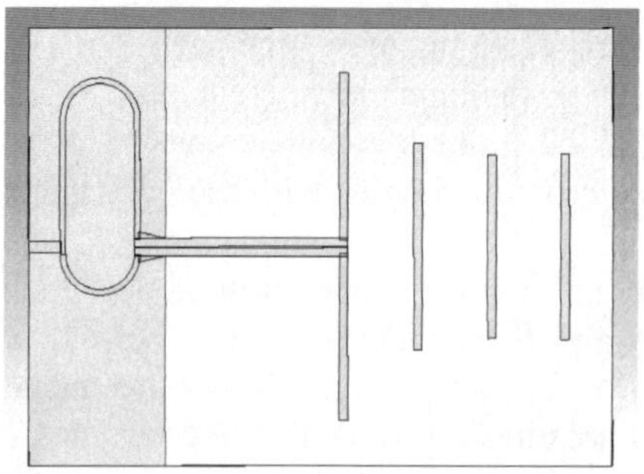

Figure 2. Antenna schematic in CST Studio

Active antenna driver has a 2 mm width and a length $l=\lambda/\sqrt{\varepsilon_{eff}}$; ε_{eff} - effective permittivity, λ – free space wavelength at 1532 MHz frequency. Directors are selected shorter than the driver. Reflector performs a ground in the unbalanced input and microstrip line of transformer; so it has the geometric dimensions: printed circuit board (PCB) width, the length is from PCB border to the distance should be approximately 0.5l. The distance from the vibrator to the directors and between subsequent directors still 0.15l-0.25l. The lengths of the directors should be less than l. Transformer is made on microstrip lines (Fig. 2, left). The input 50 Ohm line is divided into two with a resistance of 70.7 Ohms, like the coupler circuit. One of the lines longer than the other to $\lambda/2$, at the junction of microstrip lines pass into coplanar 50 Ohm stripline, gradually passing into slotline impedance of 75 Ohms. At this point, the ground-reflector is interrupted, thereby providing a balanced coordination of the driver of the antenna impedance 73 Ohm [4]. Using CST Studio environment, geometry parameters were optimized. The next step was mirroring elements, with a distance of 100 mm between them, shown in Figure 3. The result of the simulation of the two elements is a graph of the S-parameters as a function of frequency, which is most clearly interpreted voltage standing wave ratio

(VSWR) dependent on the frequency (Fig. 4), and the gain at a frequency of 1532 MHz for the two modes of operation (Fig. 5, left and right). The geometrical parameters of the microstrip lines are calculated using the formula (1), the calculation of coplanar stripline and slotline - represented in the literature [2]. General view of the system shown in Fig. 3.

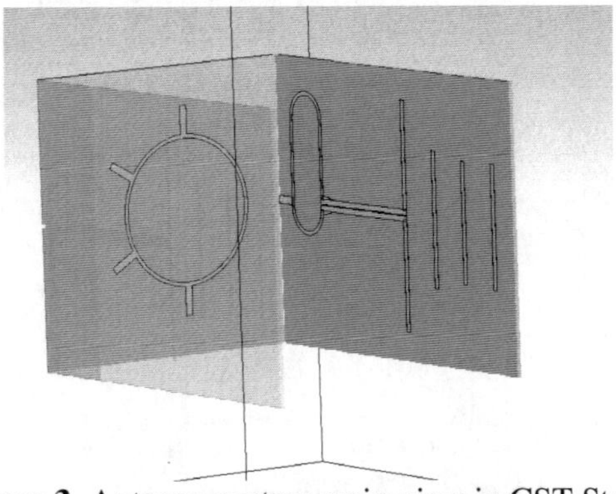

Figure 3. Antenna system main view in CST Studio

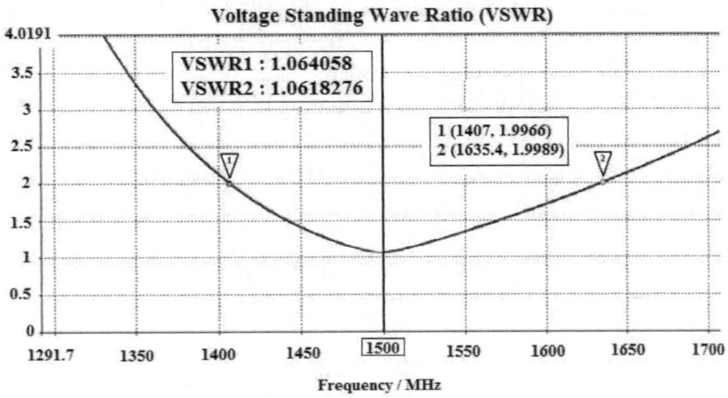

Figure 4. Results of VSWR for 2 operation modes (model)

As simulation results is bandwidth 228 MHz (assuming VSWR≤2) with center frequency 1500 MHz. The gain for the in-phase mode is 9.4 dB, width of the main lobe 60^0 (Fig 5, left), the gain for antiphase mode is 7.6 dB (Fig 5, right). Radiation pattern on antiphase mode contains two lobes, deflected to the side of the main line 43^0.

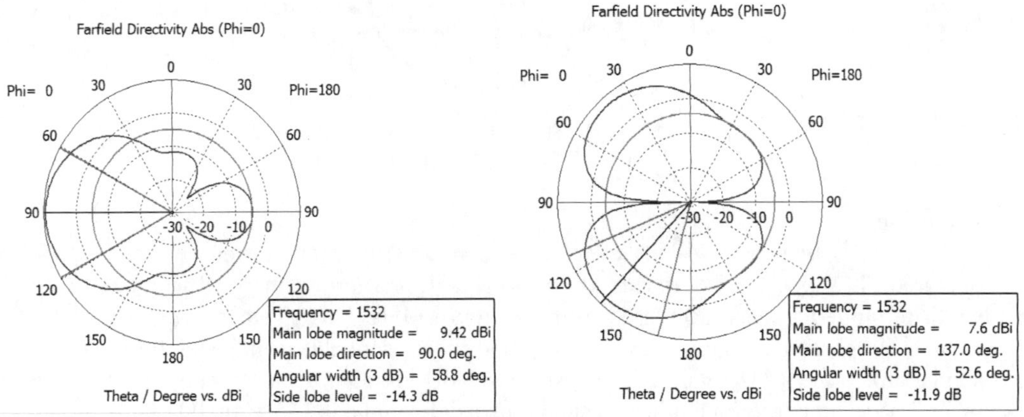

Figure 5. Results of gain for 2 operation modes (model)

IOP Conf. Series: Journal of Physics: Conf. Series **929** (2017) 012022 doi:10.1088/1742-6596/929/1/012022

3. Crafted antenna measurements

The antenna system communicated by cable segments with impedance of 50 Ohms, then the device was measured Agilent PNA-X N5247A. Fig. 6 is a plot of SWR of the system, for two operation modes.

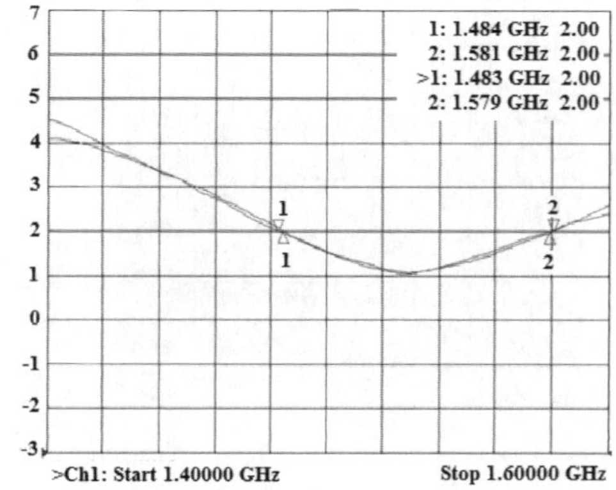

Figure 6. SWR of antenna system

SWR minimums of measured system differ from simulation results to 40 MHz. Bandwidth is 96 MHz on two modes of operation. Measurements of the gain coefficients were carried out in an open space operation with the help of a portable system for the measurement of electromagnetic fields, Rohde & Schwarz TS-EMF. Measured gain for frequency 1532 MHz is presented below (Fig.7), the solid line shows the in-phase operation mode, dashed – an antiphase mode. In-phase mode gain in the direction of the main lobe is 8 dB, the width of the lobe 60^0. Antiphase mode gain - 5 dB, width of each lobe 45^0.

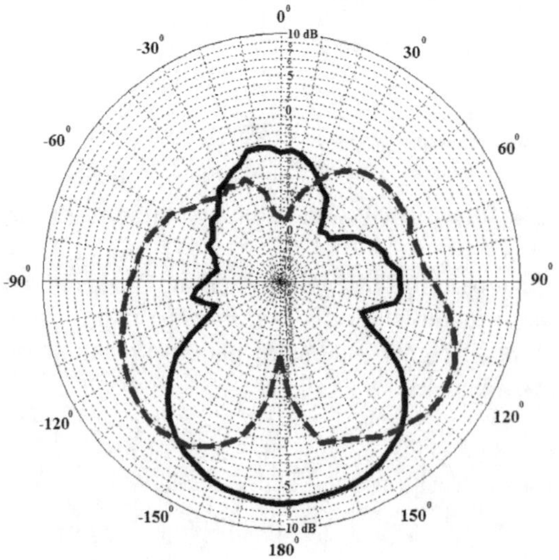

Figure 7. Gain of antenna system

4. Conclusion

Antenna system on planar Yagi-Uda elements and microstrip coupler was designed, modelled and manufactured. It supports two modes of operation. Deviation from VSWR model is 40 MHz; bandwidth of crafted antenna decreased from 228 to 96 MHz with a center frequency of 1532 MHz. Lines for the in-phase and antiphase operation modes are identical, it indicates little phase deviation characteristic of coupler. The gain of the system is 2 dB less than from model pattern of the antenna system and made similar to each other. System dimensions range is 135:100:100 mm.

The work is performed according to the Russian Government Program of Competitive Growth of Kazan Federal University.

References

[1] Bystrov R P, Potapov A A 2009 *Functional devices and electronic components of radar system* 1-2,43-58
[2] David M Pozar 2012 *Microwave Engineering 4th Edition* 148
[3] Bahl I J, Bhartia P and Stuchly S S 1982 Design of Microstrip Antennas Covered with a Dielectric Layer *IEEE Trans. Antennas Propagation* **30** 314-318
[4] Balanis 2008 C A *Modern antenna handbook* 67

IOP Conf. Series: Journal of Physics: Conf. Series **929** (2017) 012023 doi:10.1088/1742-6596/929/1/012023

Investigation of self-generation of broadband microwave chaotic and noise signals in microwave photonic ring oscillator

A.B. Ustinov, A.V. Kondrashov, B.A. Kalinikos

Department of physical electronics and technology, St. Petersburg Electrotechnical University, St. Petersburg, Russia

e-mail: ustinov_rus@yahoo.com

Abstract. Nonlinear dynamics of a microwave optoelectronic oscillator was investigated for the first time with the use of time series analysis. The detailed study of the generated microwave waveforms showed a route from stable monochromatic oscillations to noise through a series of bifurcations. The oscillator demonstrated the periodic and chaotic dynamics in the intermediate regimes of self-generation. Peculiarities of the signals and their spectra for the chaotic and noise regimes were found. The chaotic and noise dynamics were proven with the Grassberger-Procaccia method.

1. Introduction

During the past decades a special interest of researchers is devoted to the development of the microwave (MW) generators of dynamic chaos. It is caused by the possibility of using chaotic signals in telecommunications [1], radar systems [2], and in random number generators [3]. Special attention is paid to the frequency spectrum bandwidth of the generated chaotic signal. An increase of the spectrum width allows to increase speed of the communication system or to achieve better resolution of the radar.

In recent papers the microwave generators of the broadband chaotic signals based on different physical principles were described. For example, in [4] the microwave generator, which uses an interaction of semiconductor superlattices with a bulk or microstrip resonator, was investigated. In [5] the generator of chaotic oscillations based on a bipolar MW transistor was elaborated. In [6-9] the electronically tunable ring oscillators based on magnetic films were studied. In the ring oscillators the

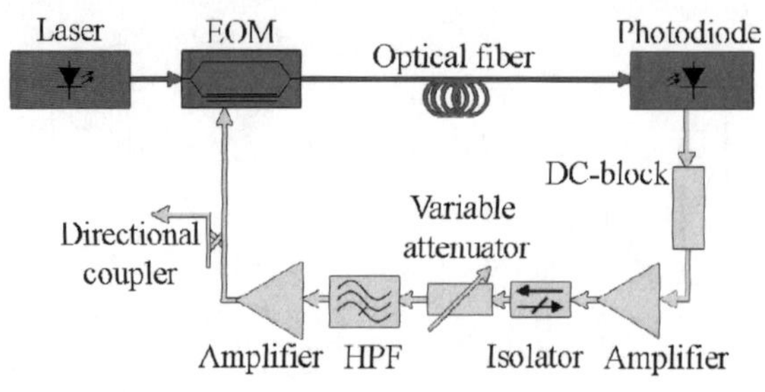

Fig. 1. Block-diagram of the experimental setup.

ferromagnetic films and ferrite-ferroelectric structures played a role of nonlinear elements. These generators allow controlling parameters of self-generated dynamic chaos by varying the magnetic or electric field applied to the nonlinear element.

Nowadays the microwave photonic devices are under extensive investigation [10-12]. Particular attention is given to the optoelectronic oscillators (OEO) for microwave signal generation [13-18]. The microwave photonics technology provides a vast advantage for microwave signal processing and generation in comparison with conventional microelectronics. One of the advantages of the optoelectronic components is their broad frequency bandwidth. Thus, it is relatively simple to create a broadband optoelectronic generator of microwave dynamic chaos [19].

The aim of this work is a detailed investigation on the nonlinear dynamics of the microwave optoelectronic oscillator. In contrast to work [19] we show, for the first time, that the optoelectronic oscillator demonstrates both chaotic and noise dynamics.

2. Experimental setup

A block-diagram of the oscillator is shown in Fig. 1. The main components of the oscillator are a laser diode, an electro-optic modulator (EOM) of Mach-Zehnder type, an optical fiber, a photodetector, a DC-block, microwave amplifiers, an isolator, a variable attenuator, a high pass filter (HPF), and a directional coupler. Coaxial cables connect components of the microwave path. Optical radiation wavelength of the laser was 1.55 μm. Its power was 10 mW. Bandwidth of the EOM was 10 GHz. Half-wave voltage of the EOM was 3.3 V. The optical fiber had a length of 100 m and a diameter of 8 μm. Upper operating frequency of the photodetector was 25 GHz. The amplifiers bandwidth was 2-8 GHz. The HPF cut-off frequency was 3.5 GHz.

A principle of operation of the oscillator is as follows. The laser emits a continuous wave optical radiation, which is used as a carrier signal. The electro-optical modulator modulates the optical radiation by a microwave signal coming from the MW path. The modulated optical signal propagates through the optical fiber and is delayed by a time of about 521 ns. The photodetector demodulates the optical signal and transmits its envelope to the MW path. The DC-block removes the direct current level from the output photocurrent. The resulting alternative current signal is amplified by the microwave amplifiers and supplied to the driving port of the electro-optic modulator, closing the ring circuit. A pair of the microwave amplifiers compensates total propagation losses of the microwave signal in the ring. The variable attenuator is used for control of these losses. A small portion of the microwave signal is extracted from the ring by the directional coupler for monitoring the generated waveforms. The monitoring was carried out by means of a fast microwave diode detector connected to a digital oscilloscope and a microwave spectrum analyzer.

It is worth mentioning that the frequency band of the generated chaotic oscillations is determined by the bandwidth of the microwave path. In the described circuit, the lower cut-off frequency of the chaotic generation is determined by the HPF characteristics. The higher cut-off frequency is determined by the microwave amplifiers. Therefore, the frequency spectrum of the chaotic signal is in the range of 3.5-8 GHz.

The electro-optic modulator was the nonlinear element of the generator. Its cosine transmission characteristic causes the nonlinearity of conversion of the microwave signal to an envelope of the optical signal.

3. Experimental results

The experimental study of the route to the noise signal self-generation was carried during a smooth increase of the ring gain coefficient G. The increase of G was caused by reducing the attenuation of the microwave signal in the variable attenuator. The value of G = 0 dB was chosen as the reference point at the threshold of the microwave signal self-generation. An increase of the ring gain led to a change of the self-generation regime. Nonlinear dynamics of the optoelectronic oscillator was caused by the nonlinear transmission characteristic of the EOM. All the other components, including microwave amplifiers, operated during the measurements in a linear regime.

Generation of a monochromatic microwave signal started at the self-generation threshold. The frequency of the generated harmonic was 3.676 GHz and its power was 0.3 mW. A number of small frequency peaks around the generated harmonic was observed. They were due to noise amplification of the frequencies corresponding to eigen-modes of the ring. These harmonics were weakened for about 40-50 dB. Since these harmonics were not phase correlated, the self-generation of signal with a constant level of microwave power was observed.

With increasing the ring gain G above 7.3 dB, the power of the microwave signal supplied to the electro-optical modulator exceeds 55 mW. A bistability occurred above this power level. The single-frequency spectrum was replaced by a frequency-comb spectrum, and the center frequency changed to 3.652 GHz. The observed spectrum consisted of harmonics with two values of detuning frequencies, approximately equal to 1.92 and 178.6 MHz.

In this regime, the fast oscillations of different duration appeared. We will call them the trains of fast oscillation. The period of the fast oscillations T_1 was approximately 5.6 ns. The duration of the trains periodically varied and repeated after the formation of a number of the trains. In the observed sequence of the trains it was easy to identify a certain period T_2 = 520.8 ns (we will call it as a period of slow oscillations), which did not depend on G. Thus, each period of the slow oscillations contained several trains of the fast oscillations. The frequencies of the generated harmonics were multiples of each other. The microwave signal had an amplitude modulation of period T_2. Therefore, the generation regime was a periodic one. The duration and profile of the trains appearing during one period were changed with an increase of the ring gain.

Further increase of the ring gain above 11.7 dB led to stochastization of the oscillations and the spectrum became broadband. On the left side of Fig. 2 the spectrum and fragment of the chaotic waveform measured at G = 15.1 dB are shown. It can be seen that the spectrum of the microwave

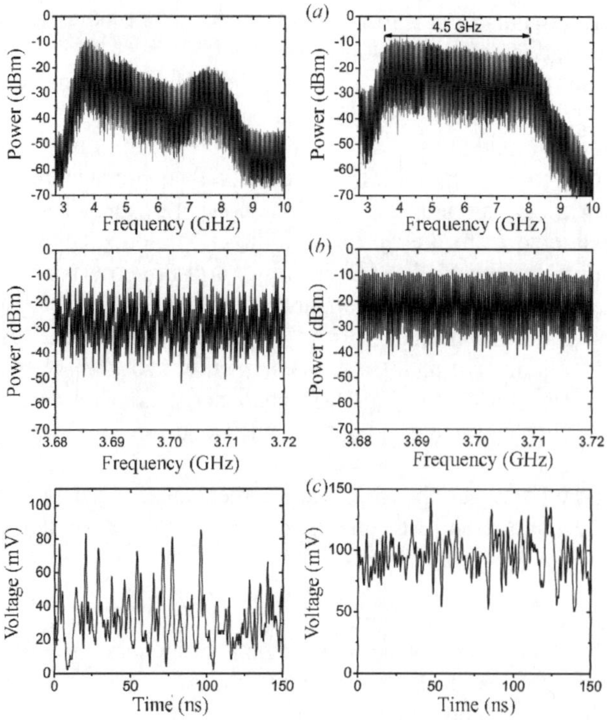

Fig. 2. Spectra (a), their fragments (b), and waveforms (c), measured in the chaotic generation regime for G = 15.1 dB (left panel) and in the noise generation regime for G = 22.9 dB (right panel).

signal consisted of the broadband noise pedestal with separate harmonics. Frequencies of these harmonics corresponded to ring eigen-modes. The envelope of the generated microwave signal had a form of a sequence of well-defined chaotic pulses.

A change of the generation regime occurred above G = 18.7 dB. The separated harmonics disappeared. The frequency spectrum of the microwave signal became broadband. The waveform in time domain took non-periodic form. The frequency spectra and oscillograms recorded for G = 22.9 dB are shown on the right side in Fig. 2.

4. Analysis of the nonlinear dynamics from time series

Analysis of the waveforms allowed us to determine, whether the generated non-periodic signal is a quasi-periodic signal, dynamic chaos or random noise [20]. The waveforms with duration of 10 μs and the resolution between points of 100 ps were recorded for analysis of the generation regime.

At the first stage phase portraits of the observed generation regimes were constructed by the time delay method. Fig. 3 (a) demonstrates the phase portrait corresponding to the periodic generation regime for G = 9.4 dB. The dynamics of the optoelectronic system was defined as evolving of the system state in time yielding two orbits. One of them was in the form of a closed loop, and another was in the form of a point. The shape of the loop was determined by the shape of the fast oscillation pulses. The point on the phase portrait corresponded to the constant power level of the waveform. For G = 9.4 dB, this point had coordinates (10 mV, 10 mV). The jumps were periodic. The phase portrait had a similar form for the other values of the ring gain, corresponded to a periodic generation regime. The time, during which system state was following the loop, was growing with an increase of the gain coefficient. In contrast, time, during which system state was staying in the point, was decreasing.

The form of the attractor was changed significantly above G = 11.7 dB due to stochastization of oscillations. Fig. 3 (b) illustrates the phase portrait obtained for G = 15.1 dB. In this regime, the chaotic pulses with the base level near 0 mV were observed. Thus the phase portrait extends to the

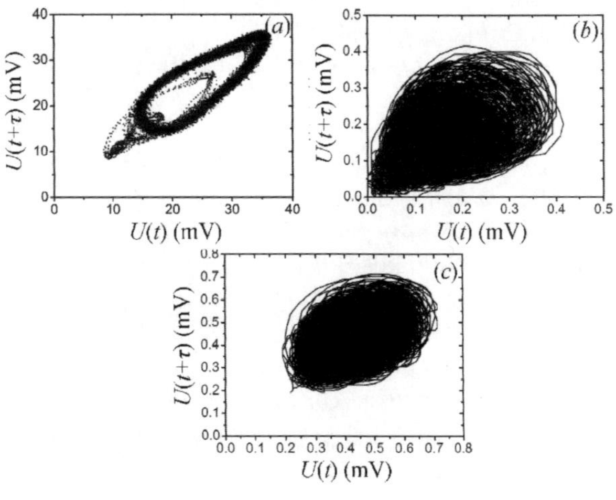

Fig. 3. 2D phase portraits of microwave signal generated at $G = 9.4$ dB(a), chaotic signal at $G =$ 15.1 dB (b) and noise signal at $G = 22.9$ dB (c).

origin of coordinates. The bifurcation occurring at G = 18.7 dB changed the waveform and led to changes of the phase portrait form. The phase trajectories start to rotate around a certain point of the phase space (see Fig. 3 (c)).

Unstructured phase portraits, shown in Fig. 3 (b,c), are typical for systems, which generate a chaotic signal with a high fractal dimension or a noise signal with infinite fractal dimension [16]. To clarify the origin of the generated signals, the values of correlation dimension, fractal dimension, minimum embedding dimension, and largest Lyapunov exponent of the obtained attractors were calculated. The results show that for $G = 15.1$ dB the minimum embedding dimension was equal to 28; the fractal dimension was equal to 11.5. The largest Lyapunov exponent was equal to 2.8 μs^{-1}. These values characterizes observed signal as a dynamical chaos.

Analysis of the phase portraits for G > 18.7 dB demonstrated that increase of phase portraits dimension up to 55 leads to increase of correlation dimension. Thus, the value of fractal dimension for these values of ring gain coefficient is infinite. This specifies observed signals as a noise. Therefore, increase of the ring gain coefficient above 18.7 dB leads to bifurcation during which chaotic generation is changing by the noise generation.

5. Conclusion

The complex nonlinear dynamics of the microwave optoelectronic oscillator was investigated. The oscillator demonstrated periodic, chaotic, and noise dynamics. A bifurcation leading from chaotic to noise dynamics was found. The chaotic and noise origins of the generated microwave signals were confirmed with the time series analysis.

The work was supported by the Russian Science Foundation (project № 16-12-10440).

References

[1] 2013 *Chaotic Signals in Digital Communications* ed M Eisencraft, R Attux, R Suyama (Boca Raton: CRC Press)
[2] 2013 *Chaotic Signal Processing* ed H Leung (Philadelphia: SIAM)
[3] Li W et al 2013 *Phys. Rev. Lett.* **111** 044102
[4] Hramov A E et al 2014 *Phys. Rev. Lett.* **112** 116603
[5] Dmitriev A S et al 2014 *Tech. Phys. Lett.* **40** 48–51
[6] Wu M Kalinikos B A, Patton C E 2005 *Phys. Rev. Lett.* **95** 237202
[7] Grishin S V et al 2014 *Tech. Phys. Lett.* **40** 853–856
[8] Ustinov A B et al 2014 *Appl. Phys. Lett.* **104** 234101
[9] Hagerstrom A et al 2011 *Phys.Rev.B.* **83** 104402
[10] Capmany J, Novak D 2007 *Nature Photon.* **1** 319–330
[11] Seeds A J and Williams K J 2006 *J. Lightw. Technol.* **24**, 4628–4641
[12] Zhang W and Yao J 2016 *IEEE Journal of Quantum Electronics* **52** 0600412
[13] Yao X S, Maleki L 1996 *Journal of the Optical Society of America B* **13** 1725-1735
[14] Matsko A B et al 2003 *Journal of modern optics* **50** 2523-2542
[15] Fortier T M 2011 et al 2011 *Nature Photonics* **5** 425-429
[16] Ustinov A B, Nikitin A A, Kalinikos B A 2015 *IEEE Magnetics Letters* **6** 3500704
[17] Belkin M E et al 2011 *Microwave and Optical Technology Letters* **53** 2474-2477
[18] Liang W et al 2015 *Nat. Communications* **6** 7957
[19] Callan K E et al 2010 *Phys. Rev. Lett.* **104** 113901
[20] Kantz H, Schreiber T 2004 *Nonlinear Time Series Analysis* (Cambridge: University press,)

IOP Conf. Series: Journal of Physics: Conf. Series **929** (2017) 012024 doi:10.1088/1742-6596/929/1/012024

A device for water content measuring of oil

A Gorbunov[1], **N Evstegneev**[1]

[1] Department of Thermodynamics and Heat Engines, Gubkin Russian State University of Oil and Gas (National Research University), Moscow, Russia

Evstegneev.n@gmail.com

Abstract. The authors of the paper describe the method of water content measuring using microwaves for oil transported by pipeline. There is a description of the experimental stand used for the determination of water content of oil by using microwaves and of the methodology of the planned experiments. There is also an algorithm for the mass concentration of water in gas-saturated water-oil emulsions determination according to the results of experimental research on the experimental stand.

1. Introduction

1.1. Relevance

Relevance of quality control of producing oil increases every passing year. Period of water-free operation for almost all of oil wells is short, and water-free production takes a small part from the lifetime of the entire production process.

Introduction of GOST R 8.615 – 2005 in the Russian Federation specified accuracy of water content measurements of oil and gas and required creating of brand new devices for flow rate and water content measurements both for specific well and the whole license block.

The solution of these objectives is associated with using new measuring devices like multiphase flowmeters, which are able to ensure control of production rate. This direction of oil and gas instrumentation has got a good development in many countries (Roxar, Micro Motion, Ultraflow, Agar, Schlumberger, Framo, Ozna, Mera, Argossi, etc.) [1,2].

At the Department of Thermodynamics and Heat Engines of Gubkin Russian State University of Oil and Gas (National Research University) research work which is dedicated to estimation of possibility of microwave using for water content determination is carried out. The work includes the following steps: theoretical justification of possibility of using microwaves for water content determination, creating of an experimental stand, conduction of experiments, development of method of water content calculation, application for a patent.

1.2. Theory of microwave heating

It is well-known that heat selectivity is one of the main features of wicrowaves [3]. Water molecules are dipoles, and when dipoles get into electromagnetic waves, they line up to the direction of magnetic inductivity vector of the waves. Frequency of microwaves is high, therefore that direction changes million time per second and molecules of water start rotating. This rotation causes friction between the molecules, and their kinetic energy transforms to heat energy. As a result, temperature of water molecules increases. This doesn't happen with oil molecules because they are not polar.

Considering that fact, it should be predicated that only water phase of oil-water mixture transporting by a pipeline will be heated by microwaves. After the microwave processing heat exchange between water and oil occurs, and their temperatures become equal. Measuring of mass flow rate and temperatures before and after the processing makes possible to calculate water content of oil.

1.3. Ultrasonic preparation

In real conditions oil transporting by a pipeline contains water inclusions of different shapes and sizes, that causes to uneven disposition of water cross section of pipeline and requires significant amount of time for temperature equalizing after the microwave processing. Authors of the paper propose to solve this problem by preparation of flow with ultrasonic waves. That transforms oil-water mixture into highly-dispersed emulsion for better heat exchange between oil and water.

Water content may be determined with high accuracy by analysis of temperatures before and after the processing, and ultrasonic preparation allows to make sector of pipeline, where microwave processing occurs, shorter. Within permanent values of microwaves capacity subtraction between temperatures depends on water and oil properties and their ratio in initial mixture.

2. Experimental stand

For realization of proposed method of water content determination experimental stand was designed and constructed (pic. 1). Stand consists in sector of a pipeline (1), through which oil-water emulsion runs. On the begging of the pipeline flowmeter (2) and ultrasonic reactor (3) are placed. On some distance after ultrasonic reactor microwave generator (5) is located which is connected with a waveguide (6). In space inside the waveguide test area (4) of the pipeline is placed. Test area (sector) made of material pervious for microwaves, in our case it is quartz. Temperature sensors (7,9) are placed before and after test area. Data from temperature sensors and flowmeter is transmitted to unit of parameters control (8).

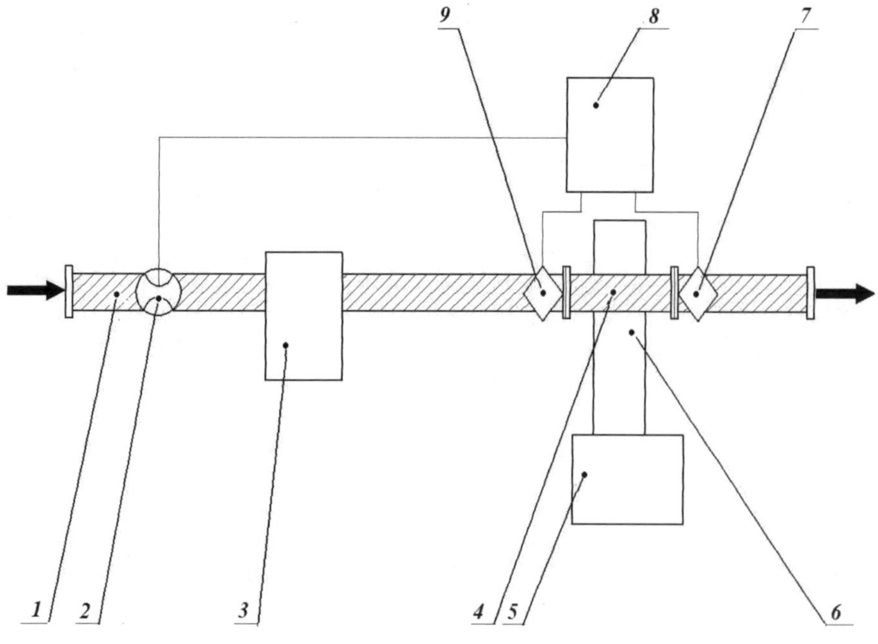

Picture 1 – The experimental stand for determination of water content of oil

Processing of the results which are gotten from experiments is conducted with considering of influence of gas factors of water and oil.

Conduction of the experiments on the stand is planned on 2017 year.

3. Water content calculation

Under microwaves processing temperature of water will increase from T_0 to T_{mw} and its internal energy will increase up to a value of ΔU_{mw}:

$$\Delta U_{mw} = G_w \cdot c_{pw} \cdot (T_{mw} - T_0) \cdot K_w, \tag{1}$$

where G_w - mass flow rate of water, kg/s; c_{pw} – isobaric heat capacity of water, $J/(kg \cdot K)$; K_w – coefficient gas factor of water, $K_w = f$ (gas factor).

Considering that there are no heat leaks to the environment, it is need to find the length of pipeline, within which temperatures of water and oil in the end of control area become equal to T_1.

Thereby, water which is heated from temperature T_0 to T_{mw} will transmit a part of obtained heat to oil Q_1 and then will cool down to T_1 in the end of the pipeline. Transmitted heat will be equal

$$Q_1 = G_w \cdot c_{pw} \cdot (T_{mw} - T_1) \cdot K_w, \tag{2}$$

Quantity of heat which will be transmitted to oil on the control area for its heating to temperature T_1 can be determined considering oil gas factor

$$Q_2 = G_o \cdot c_{po} \cdot (T_1 - T_0) \cdot K_o, \tag{3}$$

where G_o – mass flow rate of oil, kg/s; c_{po} – isobaric heat capacity of oil, $J/(kg \cdot K)$; K_o – coefficient gas factor of oil, $K_o = f$ (gas factor).

Considering that there are no heat leaks to the environment, from the law of energy conservation it should be noticed, that heat transmitted from water to oil is equal to heat obtained from water by oil

$$Q_1 = Q_2 \ . \tag{4}$$

Considering (2) and (3) equation (4) can be written as

$$G_w \cdot c_{pw} \cdot (T_{mw} - T_1) \cdot K_w = G_o \cdot c_{po} \cdot (T_1 - T_0) \cdot K_o. \tag{5}$$

Because mass flow rate of emulsion is equal to addition of mass flow rates of water and oil

$$G = G_o + G_w, \tag{6}$$

It is possible to determine mass flow rate of water from equation (5)

$$G_w = \frac{G \cdot c_{po} \cdot (T_1 - T_0) \cdot K_o}{c_{pw} \cdot (T_{mw} - T_1) \cdot K_в - c_{po} \cdot (T_1 - T_0) \cdot K_o}, \tag{7}$$

and water content of emulsion

$$m_w = \frac{G_w}{G} = \frac{c_{po} \cdot (T_1 - T_0) \cdot K_o}{c_{pw} \cdot (T_{mw} - T_1) \cdot K_в - c_{po} \cdot (T_1 - T_0) \cdot K_o}, \tag{8}$$

It is necessary to specify described theoretical relation ΔU_{mw} (1) for each specific construction and sizes of a device and determine relation of T_{mw} from different flow rate values. Coefficients K_w and K_o are features of specific oilfield, and they are determined empirically in every single case.

Obtained theoretical and experimental relations are entered to unit of parameters control. Program for water content calculation is specified after calibration of a device.

References

[1] Yermolkin O.V. Modern measuring devices and equipment for oil and gas production (interview). Gazovaya promyshlennost', 2014, no. 1 (701), pp. 56–60 (In Russian).

[2] Rekomendatsiya MI 2825-2003. Sistemy izmereniy kolichestva i pokazateley kachestva nefti. Metrologicheskiye i texnicheskiye trebovaniya k proyektirovaniyu [Recommendation MI 2825-2003. measurement system oil quantity and quality. Metrological and technical requirements for the design]. Moscow, FGUP VNIIR Publ., 2003. 52 p.

[3] Rushchits A.A., Shcherbakova Ye.I. The use of microwave heating in the food industry and public catering. Vestnik Yuzhno-ural'skogo gosudarstvennogo universiteta, 2014, vol. 2, no. 1 (In Russian)

Investigation of frequency-selective properties of microwave wideband bandpass filters

S A Khodenkov[1] and N M Boev[2]

[1] Physics Department, Reshetnev Siberian State Aerospace University, 31 Krasnoyarskiy Rabochiy pr., Krasnoyarsk 660037, Russia
[2] Laboratory of electrodynamics and microwave electronics, Kirensky Institute of Physics, 50/38 Akademgorodok, Krasnoyarsk 660036, Russia

E-mail: hsa-sibsau@mail.ru

Abstract. A new approach allowing to improve frequency-selective properties for a fixed order of *N* filter is suggested. In all the studies, conducted with the help of numerical electrodynamic analysis of 3D models of microstrip filters based on a multimode resonator, the same substrate with dielectric constant ε=2.8 and thickness *h*=2 mm (material - FLAN) was used in calculations. The central bandpass frequency of microwave structures $f_0 \approx 1.4$ GHz was registered and as well as relative bandwidth $\Delta f/f_0 \approx 80\%$. The strip conductor of central multimode resonator in six studied filters of the sixth and eighth orders has the shape of an irregular meander being electromagnetically connected with four single-mode resonators, a pair of which is located to the left of it, and the other pair is to the right. It is shown that in single-mode quarter-wave resonators building-up the number of portions of identical parallel strip conductors, connected to a screen at one end and connected with each other by a strip conductor jumper, can increase the power of suppression at low-frequency stop band by more than 15 dB, as well as near high-frequency slope of passband by more than 10 dB. Therefore, the level of maximums of return losses in the passband of wideband bandpass filter ranges within a few dB.

1. Introduction

During the process of development and study of new frequency-selective devices, including microstrip filters, developers traditionally make an attempt to increase their frequency-selective properties, to enhance the manufacturability and to reduce the size and cost of the finished product [1-5].

While constructing a broadband microstrip filter [6, 7], multimode resonator with strip conductors in the shape of an irregular meander [8] proved themselves. The use of such resonator in microwave structure as central electromagnetically associated with multiple single-mode resonators at the input and the output, two of which are connected to a screen, allows to observe among the amplitude-frequency characteristics (AFCs) plurality of attenuation poles, a pair of which is on the left, and a pair - on the right near passband [8]. As a result, frequency-selective properties of such filter are improved significantly. Besides the increase of number of periods in «irregular meander» allows to add one resonance into the passband of wideband filter based on multimode resonator. Moreover, the increase of number of single-mode resonators in microstrip construction allows to increase *N* filter order. Such complexity of the structure is accompanied by additional improvements of its frequency-selective properties. In this connection, if the research is carried out by means of numerical electrodynamic analysis of 3D models, the obtained data coincide with measurements made on constructed layouts.

However, in fact, a substantial building-up of resonances' number, forming passband of filter is accompanied by strong complexification of «manual» parametric synthesis of the latter with set characteristics, especially while using multimode resonator with the number of working oscillation modes exceeding four in number.

Therefore, in the present paper a new approach that will improve the frequency-selective properties of the wideband filter without increasing its *N* order is suggested.

2. The study of frequency-selective properties of wideband bandpass filters

For objective comparison of amplitude-frequency characteristics of all the investigated broadband filters the same substrate with dielectric constant ε=2.8 and *h*=2 mm thickness (material - FLAN) were used in calculations. Also, center bandpass frequency was recorded as f_0≈1.4 GHz and the relative bandwidth was $\Delta f/f_0$≈80%. All the data stated in the publication are obtained by electrodynamic numerical analysis of 3D model structures. In this case all strip conductors of the considered filters have axial symmetry.

Figure 1a shows the original bandpass filter of the sixth order *N*=6 [5]. In such microstrip structure central resonator with three joined portions of strip conductors *3*, *4*, as it was mentioned above, is a multimode or rather – two-mode one. It is electromagnetically connected with four single-mode resonators having extended rectangular strip conductors *1*, *2*, while conductors *2* are connected to a screen at one end.

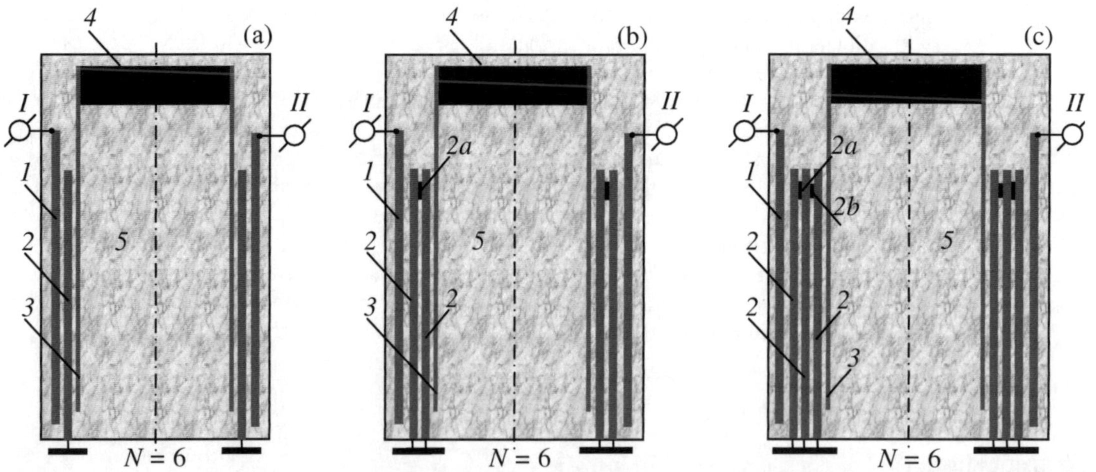

Figure 1. Wideband bandpass filters of the sixth order: original filter (a), filter with four portions of strip conductors connected to a screen (b), filter with six portions of strip conductors connected to a screen (c). Regular portions of conductors are designated by numbers (*1-4*) and located on the dielectric substrate *5*. *I*, *II* – ports of the filters.

In such filter the internal quarter-wave single-mode resonators (figure 1b and figure 1c) can be transformed, increasing the number of, connected to a screen, identical portions of the strip conductors *2*, connected with each other by strip conductor jumper *2a*. It is necessary to note that the dimensions of the strip conductor *1* and portions *3*, *4* of the strip conductor structure were not changed, while the bridge dimensions, its location and width of the conductors' portions *2* were chosen in order to enhance power suppression in stop bands of broadband filter (figure 2).

Moreover, two- or three-times increase of the number of strip conductors of such portions increases capacity suppression by 10.8 dB (11.7 dB) at frequencies of low-pass stop band and by 10.1 dB (11.6 dB) near high-pass slope of the passband.

It should be noted that the level of maximums of return loss *R(f)* in the passband structure ranges in its turn within several dB, so such change of single-mode resonators, in fact, does not make the wideband filter's setting more complicated.

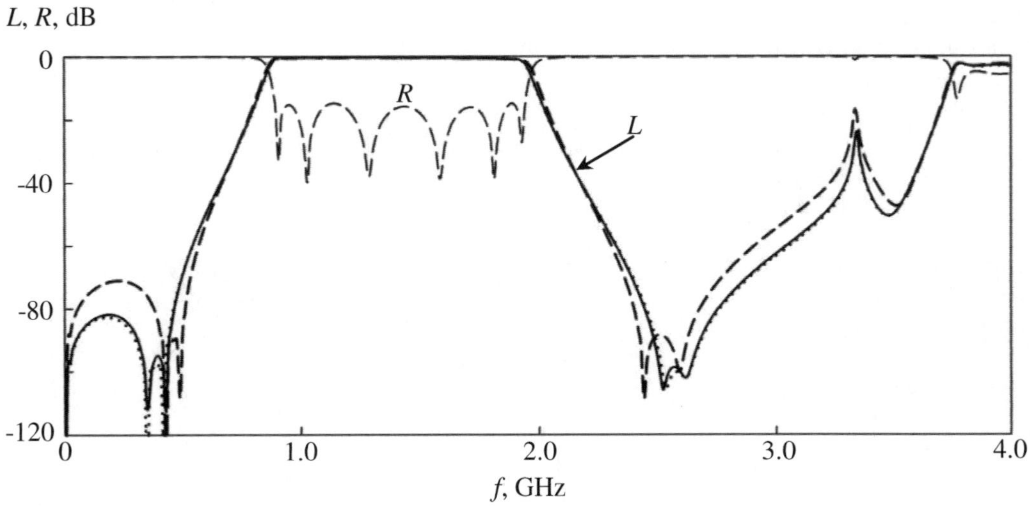

Figure 2. Amplitude-frequency characteristics of filters of the sixth order: AFCs of filter shown in figure 1a (dashed lines), AFCs of filter shown in figure 1b (line), AFCs of filter shown in figure 1c (points). *L*(*f*) – frequency dependence of direct losses, *R*(*f*) – frequency dependence of losses in reflection.

It is necessary to mention that the proposed approach aimed at improving the frequency-selective properties of wideband constructions can be applied dealing with higher order of *N* filter. It is shown by the example of bandpass filter of the eighth order that has already been implemented at four-mode resonator (figure 3).

In such resonator the portions of strip conductors *3-6* are joined in the form of irregular meander with a large number of foldings.

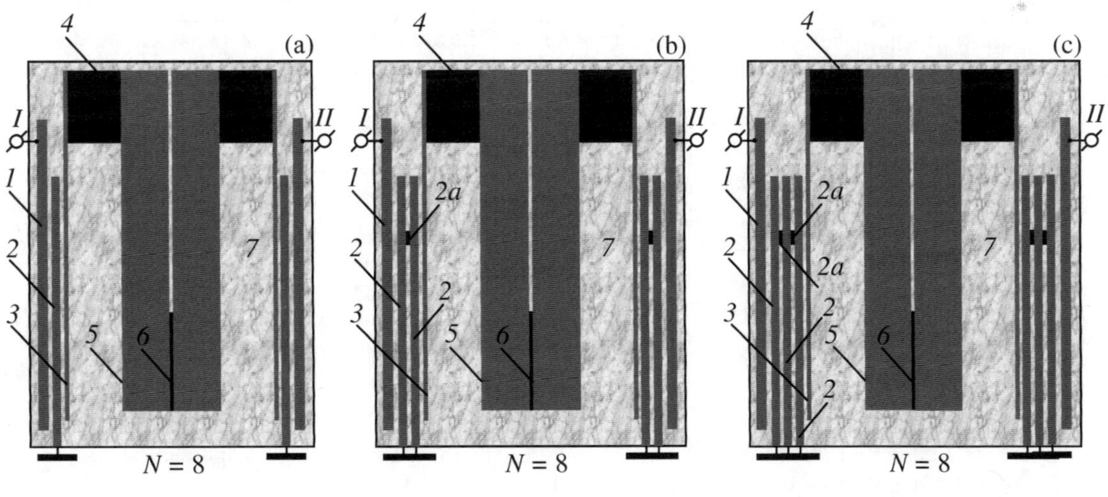

Figure 3. Wideband bandpass filters of the eighth order: original filter (a), filter with four portions of strip conductors connected to a screen (b), filter with six portions of strip conductors connected to a screen (c). Regular portions of conductors are designated by numbers (*1-6*) and located on the dielectric substrate *7*. *I*, *II* – ports of the filters.

For objective comparison of construction characteristics, the same number of additional portions of strip conductors was used. Moreover, two-times increase of the number of portions in the strip conductors *2* (figure 3) increases power attenuation by 12.3 dB (16.2 dB) at frequencies of low-pass stop band and by 5.7 dB (10.1 dB) near the high-pass slope of passband (figure 4).

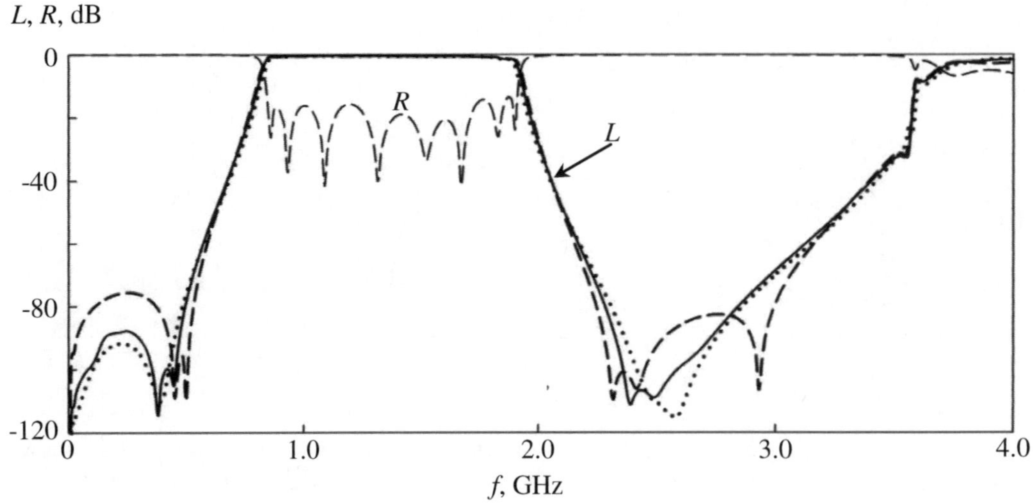

Figure 4. Amplitude-frequency characteristics of filters of the eighth order: AFCs of filter shown in figure 3a (dashed lines), AFCs of filter shown in figure 3b (line), AFCs of filter shown in figure 3c (points). *L*(*f*) – frequency dependence of direct losses, *R*(*f*) – frequency dependence of losses in reflection.

Dimensions of strip conductors in resonator as well as the gaps between them for all six studied structures are presented in table 1.

Table 1. Dimensions of filters conductors.

Figure with appropriate for filter conductors	Position of strip conductor or portions in figure	The area of strip conductor or portions (mm^2)	The gaps between strip conductors or portions (mm)
figure 1a, figure 1b, figure 1c	*1*	38.70×0.80	*1* and *2* – 0.25
figure 1a, figure 1b, figure 1c	*3*	45.70×0.70	2 and *3* – 0.35
figure 1a, figure 1b, figure 1c	*4*	19.90×5.60	
figure 1a	*2*	35.90×0.75	
figure 1b	*2*	35.90×0.60	
figure 1b	*2a*	1.30×0.05	
figure 1c	*2*	35.90×0.40	
figure 1c	*2a*	0.90×0.05	
figure 1c	*2b*	0.70×0.05	
figure 3a, figure 3b, figure 3c	*1*	42.90×1.00	*1* and *2* – 0.20
figure 3a, figure 3b, figure 3c	*3*	48.40×0.30	2 and *3* – 0.40
figure 3a, figure 3b, figure 3c	*4*	10.20×7.90	
figure 3a, figure 3b, figure 3c	*5*	47.10×6.70	
figure 3a, figure 3b, figure 3c	*6*	13.60×0.20	
figure 3a	*2*	37.40×0.70	
figure 3b	*2*	37.40×0.55	
figure 3b	*2a*	0.70×0.15	
figure 3c	*2*	37.40×0.45	
figure 3c	*2a*	0.70×0.15	

3. Conclusions

Consequently, it is shown that the improvement of frequency-selective properties of microstrip wideband bandpass filters based on resonator with strip conductors in the shape of irregular meander is observed not only while increasing the number of rollup of the strip conductor multimode resonator as well as while increasing the number of single-mode resonators in constructions, but also while building-up the number of connected to a screen portions of strip conductors in single-mode resonators.

Acknowledgments

This study was supported by the Ministry of Education and Science of the Russian Federation, grant MK-9119.2016.8 of the President of the Russian Federation for State Support of Young Russian Scientists.

References

[1] Pursiainena O L J and Baumberg J J 2005 *Appl. Phys. Lett.* **87** 101902-1
[2] Belyaev B A, Voloshin A S and Shabanov V F 2005 *Dokl. Phys.* **50** 337
[3] Tyurnev V V and Serzhantov A M 2011 *Progress in Electromagnetics Research C* **23** 151
[4] Belyaev B A, Khodenkov S A and Shabanov V F 2016 *Dokl. Phys.* **61** 155
[5] Hejazi Z M, Omar A A, Scardelletti M C, Van Keuls F W and Al-Zayed A 2008 *Progress In Electromagnetics Research* **85** 261
[6] Dovbysh I A and Tyurnev V V 2010 *Progress In Electromagnetics Research M* **12** 17
[7] Yao C, Zhang W, Han L and Ma R 2007 *Int J Infrared Milli Waves* **28** 859
[8] Belyaev B A, Khodenkov S A, Galeev R G and Shabanov V F 2015 *Dokl. Phys.* **60** 95

Microstrip filters based on 2D electromagnetic crystal

S A Khodenkov[1] **and B A Belyaev**[1,2]

[1] Physics Department, Reshetnev Siberian State Aerospace University, 31 Krasnoyarskiy Rabochiy pr., Krasnoyarsk 660037, Russia
[2] Laboratory of electrodynamics and microwave electronics, Kirensky Institute of Physics, 50/38 Akademgorodok, Krasnoyarsk 660036, Russia

E-mail: hsa-sibsau@mail.ru

Abstract. Construction of microstrip bandpass filters is suggested on the basis of 2D electromagnetic crystal consisting of *P* horizontal and *K* vertical rows of resonators. Adjacent vertical rows of resonators are connected by additional m_{K-1} resonators arranged above and below the outside 2D crystal. Parametric synthesis of six bandpass filters based on crystals with the dimension of 1×2, 2×2, 2×3, 3×1, 3×2 and 3×3 was conducted with use of electrodynamic numerical analysis of 3D models in studied constructions. While calculating substrates with relative permittivity ε=80, *h*=1 mm thick were used, in this connection, center frequency of bandwidth was registered as follows - f_0≈1.0 GHz, as well as their relative bandwidth $\Delta f/f_0$≈20%. High frequency-selective properties of the filters are connected not only with the increase of their *N* order, but also with the presence of amplitude-frequency characteristics of power attenuation poles located near the bandwidth.

1. Introduction

Electromagnetic crystals are usually periodic structures having the ability to inhibit the spread of electromagnetic radiation going through them in certain frequency bands, called forbidden zones [1, 2]. The devices based on these structures [3-6] proved themselves and are of specific interest for microwave technology developers nowadays. They are manufacturable, miniature, and besides have high-frequency selective properties [7].

In the present paper a new model of 2D arrangement of strip conductors on dielectric substrate, which allows constructing microstrip bandpass filters with high frequency-selective properties is proposed.

2. Model

Let us consider the structure which is periodic on two coordinates, consisting of half-wave microstrip resonators. The number of horizontal rows of resonators is marked as *P*, while vertical ones - as *K* (figure 1). Strip conductors of single-mode resonator have rectangular shape elongated along the *x* axis. It is well known [8] that the interaction between the resonators in one row is much stronger than the interaction between the resonators of different rows in such two-dimensional structures. Therefore, signal propagates sequentially from one resonator to another in vertical rows of resonators. Due to additional m_{K-1} microstrip resonators located below and above outside the crystal, the signal is sent to the next vertical row.

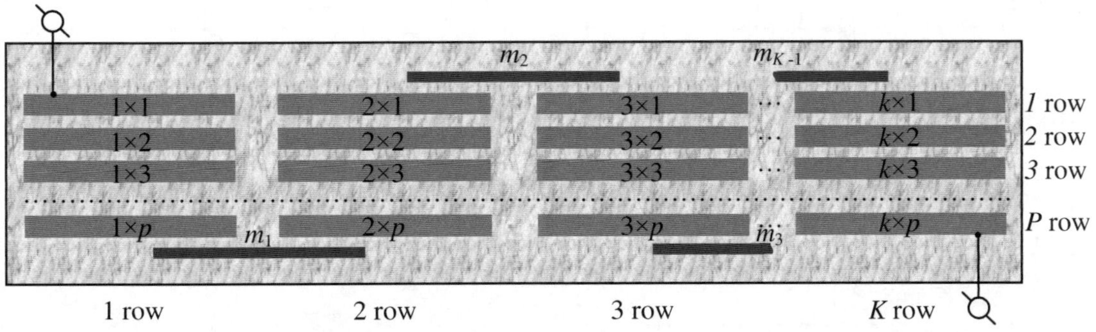

Figure 1. Model of two-dimensional location of strip conductors on dielectric substrate.

The input 50 Ω port is conductively connected with the strip conductors of the first resonator within the first row in the proposed structure. When constructing structures with even number of vertical rows of resonators $K = 2, 4, 6, ...$, the output filter port is connected with the strip conductors of the first resonator of the last vertical row. Both ports are located on the same edge of the substrate. When being placed on a substrate of an odd number of vertical rows of resonators $K = 3, 5, 7$ ... output 50 Ω port is conductively connected with the strip conductors of the last resonator within the last row. Accordingly, the ports are located relative one to another, on the opposite sides of the microstrip substrate.

Consequently, in the proposed two-dimensional microstrip crystal, the signal propagates on odd vertical rows of resonators in one direction (from top to bottom), and in even ones - in the opposite direction (from bottom to top).

Taking into consideration the fact that only one resonance is involved in the formation of microstrip filter passband, the increase of the number of horizontal P rows and vertical K rows, as well as the increase of the N filter order, is connected by the following relation

$$N = K(P+1)-1. \tag{1}$$

It is obvious in this connection that the use of the dimension $k×p$ of multimode resonators [9] instead of single-mode in such structure, allows not only to improve the frequency-selective properties of developing filter, but also to miniaturize them significatly.

3. Numerical filters' analysis

In order to test the proposed model six microstrip bandpass filters, based on electromagnetic crystal dimension 2×1, 2×2, 2×3, 3×1, 3×2 and 3×3 (figure 1), have been developed. Parametric synthesis of these structures was carried out with the help of electrodynamic numerical analysis of 3D models, which, as it is known [8, 9], is in good agreement with measurements made on the fabricated structures.

For objective comparison of their amplitude-frequency characteristics (figure 2) the same substrates with dielectric penetrability ε=80 and thickness h=1 mm (material - TBNS ceramics) were used in calculations. Also center frequency of the filter passband f_0≈1.0 GHz and their relative bandwidth $\Delta f/f_0$≈20% was registered. For this purpose, the length and width of strip conductor resonators as well as the gaps between them were selected, when needed separation between strip conductors was used.

It is necessary to note that high frequency-selective properties of the proposed microwave constructions arise not only because of the increase of filter N order, but also because of power attenuation poles observed in amplitude-frequency characteristics. In this connection, bandpass filter, based on the two-dimensional crystal with the dimension of only 3×3, has power suppression at low frequencies of around 150 dB, near bandwidth - approximately 110 dB, at minimum loss in the passband - 1.9 dB.

IOP Conf. Series: Journal of Physics: Conf. Series **929** (2017) 012026 doi:10.1088/1742-6596/929/1/012026

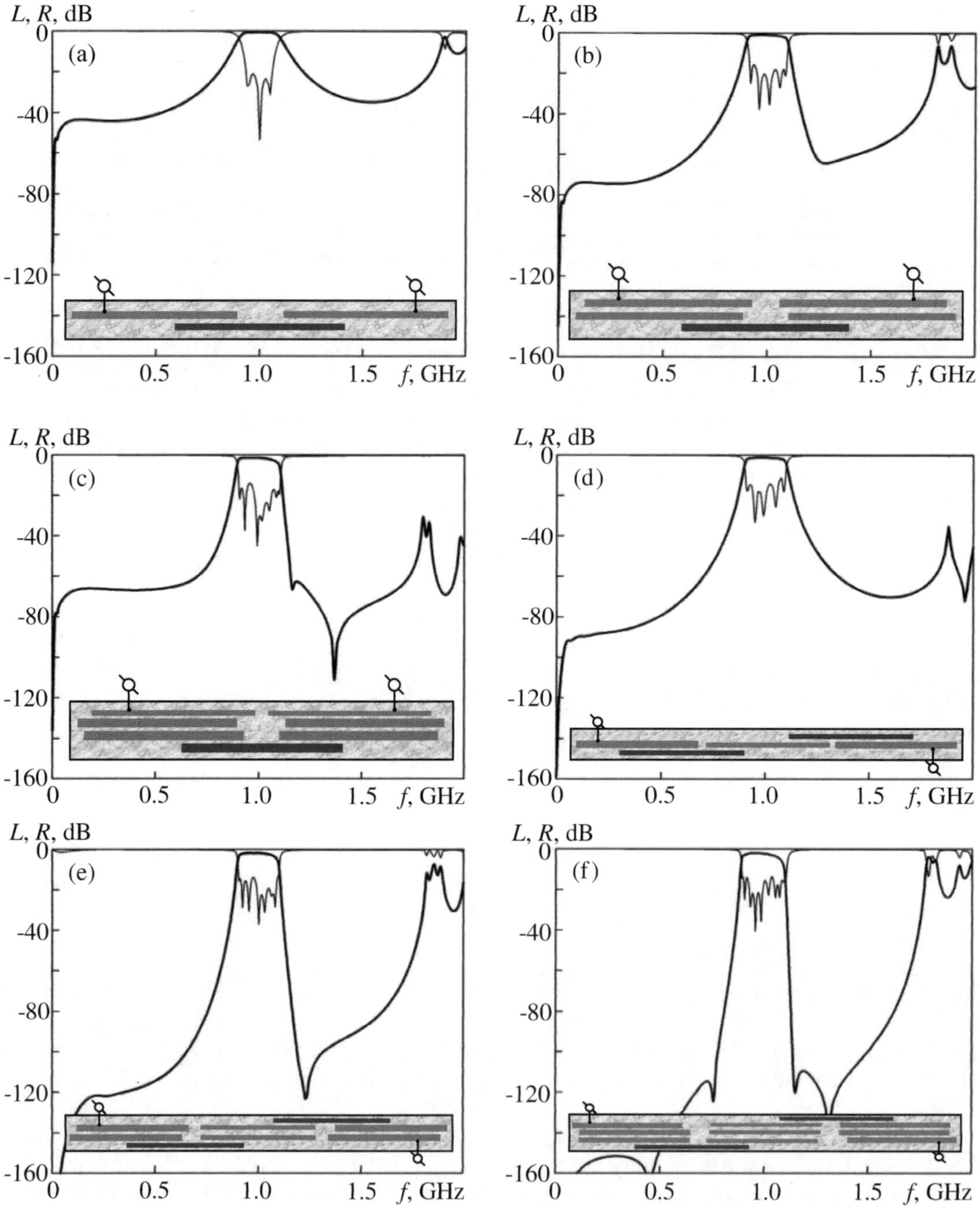

Figure 2. Comparison of amplitude-frequency characteristics of filters based on 2D electromagnetic crystal dimensions 2×1 (a), 2×2 (b), 2×3 (c), 3×1 (d), 3×2 (e) and 3×3 (f). On inserts relevant amplitude-frequency characteristics of the topology of strip conductor filters.

Dimensions of strip conductors' resonator and the gaps between them for all six investigated structures are presented in table 1.

IOP Conf. Series: Journal of Physics: Conf. Series **929** (2017) 012026 doi:10.1088/1742-6596/929/1/012026

Table 1. Dimensions of microstrip filter conductors' topology.

The dimension of the crystal on the basis of which the filter is implemented	Figure with appropriate for filter amplitude-frequency characteristics	Position of strip conductor in figure 1	The area of strip conductor (mm^2)	The gaps between strip conductors (mm)
2×1	figure 2a	1×1	20.00×1.00	1×1 and 2×1 – 5.40
2×1		2×1	20.00×1.00	1×1 and m_1 – 0.40
2×1		m_1	20.80×0.70	2×1 and m_1 – 0.40
2×2	figure 2b	1×1	20.50×0.80	1×1 and 2×1 – 3.20
2×2		2×1	20.50×0.80	1×2 and 2×2 – 5.40
2×2		1×2	20.50×1.00	1×1 and 1×2 – 0.50
2×2		2×2	20.50×1.00	2×1 and 2×2 – 0.50
2×2		m_1	20.80×0.80	1×2 and m_1 – 0.50
2×2				2×2 and m_1 – 0.50
2×3	figure 2c	1×1	20.90×0.45	1×1 and 2×1 – 1.80
2×3		2×1	20.90×0.45	1×2 and 2×2 – 6.20
2×3		1×2	20.50×0.85	1×3 and 2×3 – 4.00
2×3		2×2	20.50×0.85	1×1 and 1×2 – 0.45
2×3		1×3	20.80×0.90	2×1 and 2×2 – 0.45
2×3		2×3	20.80×0.90	1×2 and 1×3 – 0.90
2×3		m_1	21.00×0.90	2×2 and 2×3 – 0.90
2×3				1×3 and m_1 – 0.55
2×3				2×3 and m_1 – 0.55
3×1	figure 2d	1×1	20.60×0.65	1×1 and 2×1 – 2.60
3×1		2×1	20.80×0.70	2×1 and 3×1 – 2.60
3×1		3×1	20.60×0.65	1×1 and m_1 – 0.60
3×1		m_1	21.00×0.75	2×1 and m_1 – 0.50
3×1		m_2	21.00×0.75	2×1 and m_2 – 0.50
3×1				3×1 and m_2 – 0.60
3×2	figure 2e	1×1	20.40×0.85	1×1 and 2×1 – 2.20
3×2		2×1	20.80×0.70	2×1 and 3×1 – 3.70
3×2		3×1	20.60×0.95	1×2 and 2×2 – 3.70
3×2		1×2	20.60×0.95	2×2 and 3×2 – 2.20
3×2		2×2	20.80×0.70	1×1 and 1×2 – 0.50
3×2		3×2	20.40×0.85	2×1 and 2×2 – 1.00
3×2		m_1	21.20×0.60	3×1 and 3×2 – 0.50
3×2		m_2	21.20×0.60	1×2 and m_1 – 0.65
3×2				2×2 and m_1 – 0.60
3×2				2×1 and m_2 –0.60
3×2				3×1 and m_2 – 0.65
3×3	figure 2*f*	1×1	20.40×0.95	1×1 and 2×1 – 5.20
3×3		2×1	20.90×0.75	2×1 and 3×1 – 3.10
3×3		3×1	20.60×1.05	1×2 and 2×2 – 3.60
3×3		1×2	20.50×0.90	2×2 and 3×2 – 3.60
3×3		2×2	20.60×0.60	1×3 and 2×3 – 3.10
3×3		3×2	20.50×0.90	2×3 and 3×3 – 5.20
3×3		1×3	20.60×1.05	1×1 and 1×2 – 0.30
3×3		2×3	20.90×0.75	1×2 and 1×3 – 0.75
3×3		3×3	20.40×0.95	2×1 and 2×2 – 1.05
3×3		m_1	21.20×0.75	2×2 and 2×3 – 1.05
3×3		m_2	21.20×0.75	3×1 and 3×2 – 0.75
3×3				3×2 and 3×3 – 0.30
3×3				1×3 and m_1 – 0.60
3×3				2×3 and m_1 – 0.55
3×3				2×1 and m_2 – 0.55
3×3				3×1 and m_2 – 0.60

4. Conclusions

Consequently, a model of constructing microstrip bandpass filters based on 2D electromagnetic crystal with the dimension of $k \times p$ is suggested. In vertical rows of resonators, the signal transmits sequentially from one resonator to another, and due to additional m_{K-1} microstrip resonators located at the top and bottom beyond the crystal, the signal is directed towards the next vertical row. The increase of number of horizontal P rows and vertical K rows and the increase of N filter order while using single-mode resonator is connected with the following relation $N=K(P+1)-1$. Significant improvement of frequency-selective properties in the constructed filters based on 2D electromagnetic crystal filters is connected not only with the increase of their N order, but also with the appearance of power attenuation poles on the amplitude-frequency characteristic near the passband. So, bandpass filter on the basis of crystal with the dimension of 3×3 has power suppression at low frequencies of around 150 dB, near the bandwidth - approximately 110 dB.

Acknowledgments

This study was supported by the Ministry of Education and Science of the Russian Federation, grant MK-9119.2016.8 of the President of the Russian Federation for State Support of Young Russian Scientists.

References

[1] Escorcia-Garcıa J and Mora-Ramos M E 2009 *PIERS Online* **5** 36
[2] Belyaev B A, Voloshin A S and Shabanov V F 2005 *Dokl. Phys.* **50** 337
[3] Belyaev B A, Tyurnev V V and Shabanov V F 2014 *Russian Physics Journal* **56** 1378
[4] Wu X, Yamilov A, Liu X, Li S, Dravid V P, Chang R P H and Cao H 2004 *Appl. Phys. Lett.* **85** 3657
[5] Passos M G, Fernandes H C C and da F. Silva P H 2007 *PIERS Online*. **3** 695
[6] Pursiainena O L J and Baumberg J J 2005 *Appl. Phys. Lett.* **87** 101902-1
[7] Hong J S and Lancaster M J 1996 *IEEE Transactions on Microwave Theory and Techniques* **44** 2099
[8] Belyaev B A, Khodenkov S A and Shabanov V F 2016 *Dokl. Phys.* **61** 155
[9] Belyaev B A, Khodenkov S A, Galeev R G and Shabanov V F 2015 *Dokl. Phys.* **60** 95

IOP Conf. Series: Journal of Physics: Conf. Series **929** (2017) 012027 doi:10.1088/1742-6596/929/1/012027

Investigation of spin waves in magnonic crystal based on periodically metallized YIG film

V.V. Vitko[1], A.A. Nikitin[1,2], A.B. Ustinov[1], A.V. Drozdovskii[1], A.A. Semenov[1] and E. Lähderanta[2]

[1]St. Petersburg Electrotechnical University, Department of Physical Electronics and Technology, St. Petersburg, 197376 Russia
[2]Department of Mathematics and Physics, Lappeenranta University of Technology, Lappeenranta, 53850 Finland

E-mail: vitaliy.vitko@gmail.com

Abstract. A theoretical model of ferrite film magnonic crystals made with periodic metallization of the film surface was developed taking into account finite metal conductivity. Results show that finite metal conductivity affects to the insertion losses. The theoretical calculations have a good agreement with the experimental results.

In recent years, magnonic crystals made with periodical variation of magnetic waveguide properties has received significant attention in both physics and engineering. Distinguished features of the spin-wave devices are low insertion losses and magnetic field tunability in a wide frequency range. Performance characteristics of such devises are determined by the dispersion of spin waves (SWs) [1], [2]. One of the important features of the magnonic crystal (MC) is the presence of band gaps in the SW spectrum, i.e. frequency bands in which the propagation of the SW is forbidden [3]. The periodical variation of waveguide properties in this case can be implemented in a form of ferrite thickness and width modulation, external field modulation, periodical metallization etc. [4]. The MCs with periodical metallization of ferrite surfaces are perspective for observation of a variety of different linear and nonlinear SW phenomena [5]. The present work is devoted to the investigation of the SWs dispersion and transmission characteristics of the MCs based on periodically metallized YIG film taking into account conductivity and thickness of the metal electrodes.

The first step of the investigation is deriving of a dispersion relation of the surface SWs taking into account conductivity and thickness of metal electrodes. Investigated structure shown in Fig. 1 consists of the metal layer and ferrite film on the dielectric substrate stacked along the z-axis. The ferrite layer has a thickness of a_f, a magnetic permeability of μ_f, and a saturation magnetization of $4\pi M_s$. The dielectric substrate has a thickness of a_d and a relative dielectric permittivity ε_d. The metal layer has a thickness of a_m and a complex relative dielectric permittivity $\dot{\varepsilon}_m = 1 - i\sigma/(\varepsilon_0\omega)$, where ω is an angular frequency and σ is a metal conductivity.

The investigated structure is infinite in the xz plane and magnetized to the saturation along z-axis. We assume that SWs propagate along the x-axis that is perpendicular to the external magnetic field H_0 in order to execute conditions for the surface SW propagation.

IOP Conf. Series: Journal of Physics: Conf. Series **929** (2017) 012027 doi:10.1088/1742-6596/929/1/012027

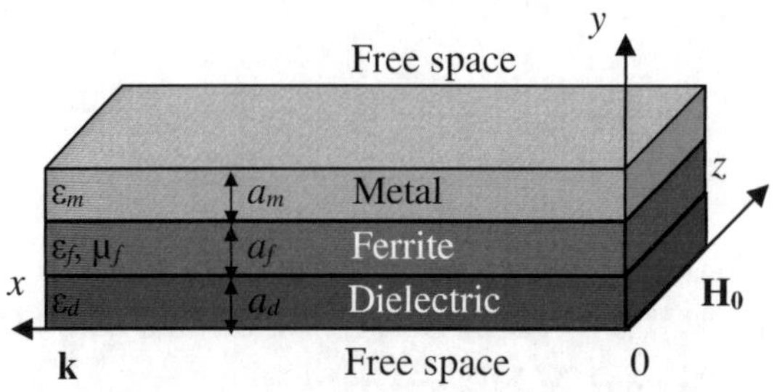

Figure 1. The dielectric-ferrite-metal structure

It is well known that Maxwell's equations can be divided on two systems of differential equations describing two fundamental sets of *TE*- and *TM*-modes with following field structures $\mathbf{E}(0,0,E_z)$, $\mathbf{H}(H_x,H_y,0)$ and $\mathbf{E}(E_x,E_y,0)$, $\mathbf{H}(0,0,H_z)$, respectively. The Helmholtz equations are derived from the Maxwell`s equations for each layer: a free space above and below structure, a ferrite, a dielectric and a metal with the complex dielectric permittivity $\dot{\varepsilon}_m$. The application of electrodynamic boundary conditions for the two tangential component of the electric and magnetic fields gives six homogeneous equations, which is equal to the number of undetermined constants. These homogeneous equations have solutions only if the determinant of the system is vanished. Therefore, the transcendental dispersion relation described spectra of the SW in the dielectric-ferrite-metal structures taking into account metal conductivity was obtained.

Numerical modelling was carried out for typical parameters for an epitaxial yttrium iron garnet (YIG) film on a gadolinium gallium garnet (GGG) substrate. The conductivity of metal electrodes was varied in a range from 50 kS/m to 700 kS/m. Following parameters were chosen for numerical calculation: a_f= 9.8 μm, ε_f= 14, $4\pi M_s$ = 143398 A/m, a_d= 500 μm, ε_d= 14, a_m= 2 μm, H_0= 82919 A/m. The result of the numerical solution of the obtained dispersion relation for real frequencies provided complex wavenumbers. Imaginary part describes Joule's losses in metal electrodes and real part defines a phase shift. Dispersion characteristics and group velocity of the surface SWs in the metallized ferrite for various metal conductivity are shown in Fig. 2(a, b).

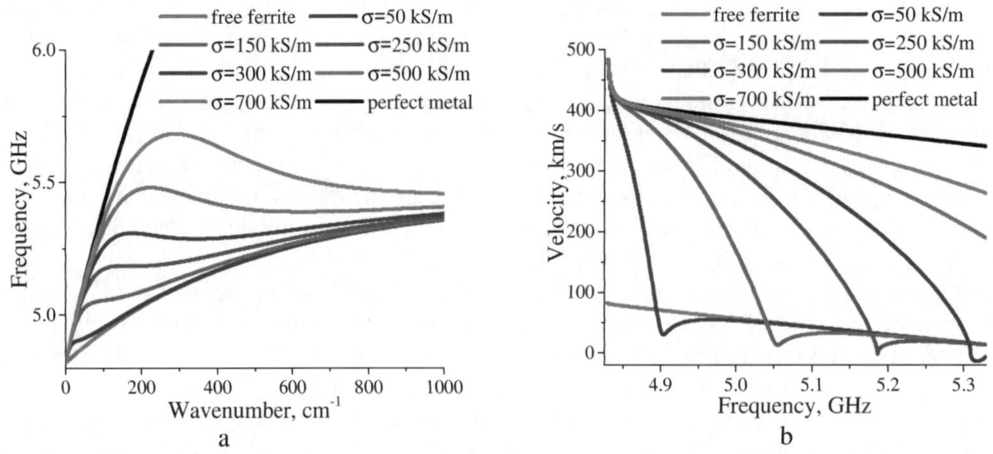

Figure 2. Dispersion characteristics of SWs (a) and dependences of group velocity on frequency (b) at different metal conductivity

One can see from Fig. 2 (a) that the increase from zero to infinity of the metal conductivity leads to change the behaviour of the dispersion characteristic from dispersion in free ferrite to perfectly metallized one. The same behaviour was also observed for the group velocity (see Fig.2 (b)). Note that the wavenumber increase provides electromagnetic field localization inside the ferrite film that leads to reduce of metal screen influence [6].

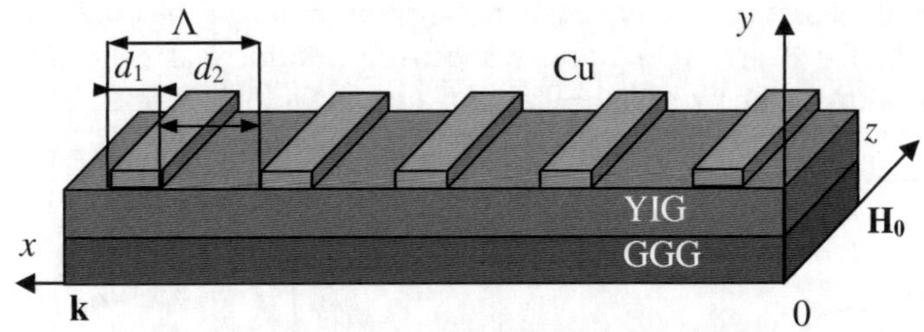

Figure 3. The MC metal-ferrite structure

The second step is the investigation of the transmission characteristics of the MC with periodical metallization that is shown in Fig. 3. The investigated structure is consisted of a metal grating on the surface of the ferrite film. The width of the metal strips is d_1, the distance between metal strips is d_2, and the period of the structure is $\Lambda = d_1 + d_2$.

The developed dispersion relation was used for a numerical calculation of transmission characteristics of the MC. According the transfer-matrix method [7] the wave propagation through the one period Λ of the investigated MC was described by the multiplication of T-matrixes, that had the following form: $T_\Lambda = T_1\, T_2\, T_3\, T_4$, where $T_1 = \begin{bmatrix} e^{-K_m(\omega)d_1} & 0 \\ 0 & e^{K_f(\omega)d_1} \end{bmatrix}$, $T_2 = \begin{bmatrix} 1/(1-\Gamma) & \Gamma/(1-\Gamma) \\ \Gamma/(1-\Gamma) & 1/(1-\Gamma) \end{bmatrix}$,

$T_3 = \begin{bmatrix} e^{-K_f(\omega)d_2} & 0 \\ 0 & e^{K_f(\omega)d_2} \end{bmatrix}$, $T_4 = \begin{bmatrix} 1/(1+\Gamma) & -\Gamma/(1+\Gamma) \\ -\Gamma/(1+\Gamma) & 1/(1+\Gamma) \end{bmatrix}$. Here $K_m(\omega) = ik_m(\omega) - \alpha_m(\omega)$, $K_f(\omega) = ik_f(\omega) - \alpha_f(\omega)$, $k_f(\omega)$ and $k_m(\omega)$ are the wavenumbers of SW in the free ferrite film and the metallized one, $\alpha_f(\omega)$ and $\alpha_m(\omega)$ are the dumping decrements in corresponding sections, Γ is a reflection coefficient between these sections. Therefore, the propagation of the SW in the MC with N periods was described by the following multiplication $T_\Sigma = \prod_{w=1}^{N} T_\Lambda = (T_\Lambda)^N$. The transmission characteristics was obtained as $S_{21} = 20\log_{10}\left(1/T_{\Sigma_{11}}\right)$. This method is suitable for the calculation of transmission characteristics of the finite-length periodic structure taking into account insertion losses.

Transmission characteristics of the investigated MCs were calculated using the described model. In order to check the adequacy of the developed theory, we carried out an experimental research for MCs that were fabricated using an epitaxial YIG film of a_f= 9.8 μm thickness on a_d= 500 μm thick GGG substrate. The ferromagnetic resonance linewidth of the YIG film was ΔH= 0.4 Oe at 5 GHz. A copper grating was formed on the YIG film surface by a thermal evaporation followed by a wet photolithography. It constituted by 8 copper stripes with a period of Λ= 600 μm and 16 copper stripes with a period of Λ=300 μm. Every strip has the width of d_1= 25 μm, the thickness of a_m=2 μm, and resistivity of 0.1 μΩ·cm. Two 50-μm-wide and 3-mm-long microstrip antennas with a separation of 7.75 mm are used for excitation and detection of SWs. Values of magnetic field H_0 and a saturation magnetization $4\pi M_s$ were the same as in the previous section.

The comparison of theoretical and typical experimental results of the transmission characteristic are shown in Fig.4 (a, b) for 8 and 16 stripes, respectively. Numerical calculations for perfect metal electrode and metal with conductivity of σ=98 kS/m are shown by blue and red lines, respectively.

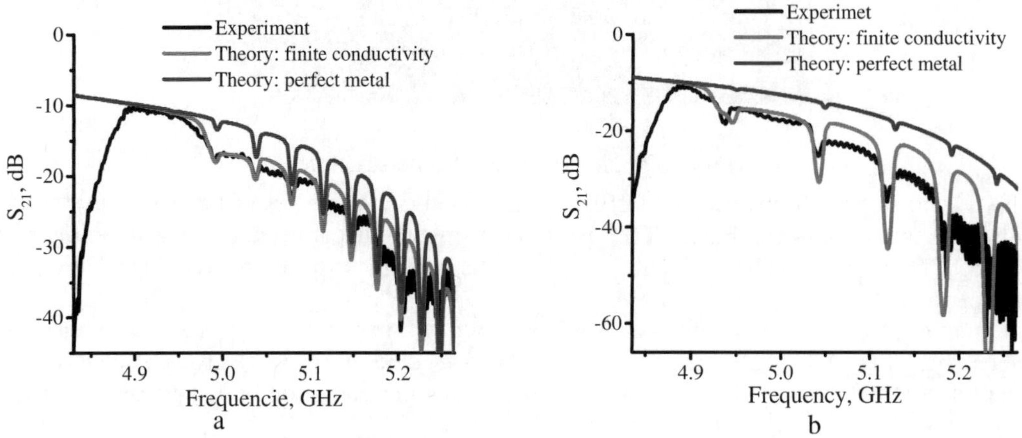

Figure 4. The comparison of experimental and theoretical results of transmission characteristics for MC with 8 periods of 600 μm (a) and with 16 periods of 300 μm (b)

One can see from Fig. 4 (a, b) well pronounced band gaps on transmission characteristics. The YIG film had unpinned surface spins that lead to standing SW resonances and appearance of additional dips in experimental transmission characteristics. Note that increasing of number of periods provides to increase insertion losses. Besides the reduction of the MCs period from 600 to 300 μm leads to the increase of the frequency range between band gaps (from 46 MHz to 106 MHz). It should be noted numerical results obtain for the metal conductivity have a good agreement with experimental ones.

In conclusion, the developed theoretical model provides possibility to investigate a wide range of problems connected with the propagation of SWs in metallized ferrite films and MCs. The dispersion characteristics of SWs in ferrite film with metallization were calculated taking into account the metal conductivity. Transmission characteristics for two types of MCs were experimentally investigated and numerically simulated. It was shown that theoretical calculations have a good agreement with experimental results.

This work was supposed in part by the project Goszadanie, the Russian Foundation of Basic Research (project № 16-32-00715 mol_a), and the Academy of Finland.

References

[1] Damon R W and Van De Vaart H Propagation of Magnetostatic Spin Waves at Microwave Frequencies in a Normally-Magnetized Disk 1965 *J. Appl. Phys.* **36** No 11 3453-3459
[2] Kalinikos B A Excitation of propogating spin waves in ferromagnetic films 1980 *IEE Proc.* **127** Pt.H. No 1 4-10
[3] Klos J W, Krawczyk M, Dadoenkova Yu S et al Photonic-magnonic crystals: Multifunctional periodic structures for magnonic and photonic applications 2014 *J. Appl. Phys.* **115** 174311
[4] Demokritov S O and Slavin A N Magnonics: From Fundamentals to Applications 2012 *Springer* **125**
[5] Krawczyk M and Grundler D Review and prospects of magnonic crystals and devices with reprogrammable band structure 2014 *Journal of Physics: Condensed Matter* **26** No 12 123202
[6] Stancil D D and Prabhakar A Spin Waves: Theory and Applications 2009 *Springer*
[7] Li Zhi-Yuan and Lin Lan-Lan Photonic band structures solved by a plane-wave-based transfer-matrix method 2003 *Phys. Rev. E* **67** 046607

Magnetically tunable high-Q spin-wave optoelectronic single-loop resonator

V V Vitko, A A Nikitin, A V Kondrashov, M I Martynov and A B Ustinov

St. Petersburg Electrotechnical University, Department of Physical Electronics and Technology, St. Petersburg, 197376 Russia

E-mail: vitaliy.vitko@gmail.com

Abstract. Resonant properties of a spin-wave optoelectronic single-loop resonator are experimentally investigated. The resonator is composed of a spin-wave delay line and a microwave photonic delay line based on a single-mode optical fiber. This circuit is equivalent to a comb filter. The microwave photonic delay line provides the main contribution to the delay time and defines mostly the free spectra range (FSR). The spin-wave delay line realizes the reconfiguration of resonator bandwidth and its frequency tuning. Influences of delay time and amplifier gain coefficient on quality-factor are analyzed.

Increase in total bandwidth of microwave devices is one of the strictest demands to modern communication systems. Microwave photonics based on combination of electric and photonic circuitries provides practical methods for achievement of subterahertz bandwidth [1, 2, 3]. The use of the photonics and the functional electronics in the same device provides enhancement of functionality and performance characteristics of the device. One of the examples of microwave photonic circuits is closed-loops with delay lines (DLs) which demonstrate wide variety of effects. Closed-loop circuit consisting of series connected active or passive microwave photonic DL and spin-wave DL realize classical configuration of the spin-wave OEO.

Closed-loop circuits with DL operate in two modes due to active or passive regime. The active regime leads to self-oscillations and the circuit operates like an oscillator. Spin-wave optoelectronic oscillators (OEOs) are demonstrative example of the enhancement. It combines advantage of low phase noise of OEOs [4] and wide magnetic tuning of spin-wave devices [5, 6]. In the case of passive regime, losses in the DL exceed the amplification that leads to saturation and the circuit operates like a resonator. Closed-loop circuit with passive regime is called an active ring resonator (ARR). The frequency response of the ARR has a comb shape with a pass bands corresponding to the resonant frequencies [7]. Position of the resonant frequencies is defined by geometry of the DL and a dispersion of the eigen-waves in it. The spin-wave DL is strongly dispersive therefore the frequency response is a function of its design parameters. Microwave photonic DL usually utilizes single-mode fibers with extremely low insertion losses that yield the delay times in the tens or even hundreds of microseconds and eventually extremely high quality factors.

The purpose of this work is the experimental investigation of the frequency response of a spin-wave optoelectronic single-loop resonator based on cascade connection of the spin-wave and microwave photonic DLs.

The block diagram of the spin-wave optoelectronic single-loop resonator experimental prototype is shown in Fig. 1. This resonator is a typical ARR.

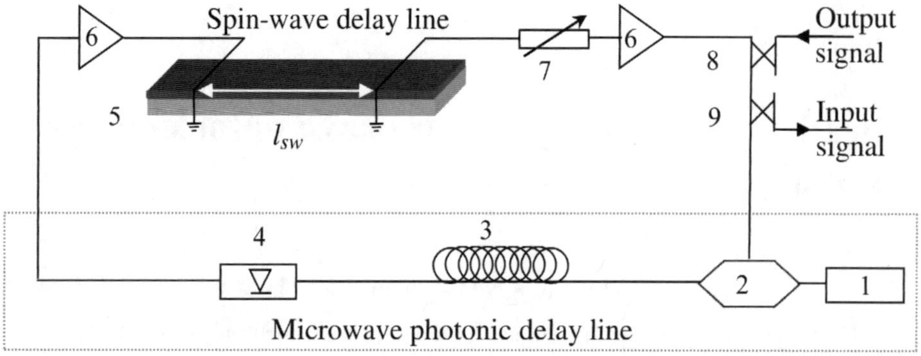

Figure 1. A block diagram of the spin-wave optoelectronic single-loop resonator

The resonator is a closed-loop circuit consisting of microwave photonic and spin-wave DLs. Microwave photonic delay line is formed by a semiconductor laser (1) continuously emitting at a wavelength of 1.55 μm, the intensity electro-optical modulator (2) of Mach-Zehnder type with a bandwidth of 20 GHz, the single-mode optical fiber (3) with various lengths l_{opt}, and a high-speed photodetector (4) with the bandwidth of 25 GHz. Spin-wave DLs (5) is based on various YIG films with variety of antenna geometries. Two identical microwave amplifiers (6) are used for circuit losses compensation. The amplifiers demonstrate 1 dB suppression at output power of 30 dBm. The gain of each amplifier is 32 dB. These values were almost constant in the nominal frequency range of 4-9 GHz. The ring gain is regulated by a variable attenuator (7). Input and output of a microwave signal is carried out by directional couplers (8) and (9) with a couple coefficient of –10 dB.

The investigated resonator operates as follows. A test microwave signal with frequency $\omega_{mw} = 2\pi f_{mw}$ is supplied through the directional coupler (9) to the microwave photonic DL. It accumulates delay time $\tau_{opt} = l_{opt}\, n / c$, where l_{opt} is a length of optical fiber, the n is its refractive index, c is the speed of light in vacuum. The phase shift of the microwave signal in microwave photonic DL is $\Phi_{opt} = k_{opt}\, l_{opt}$, where k_{opt} is a wave-number of the signal envelope, which is equal to $k_{opt} = \omega_{mw}\, n / c$. After microwave photonic DL the microwave signal is fed in the spin-wave DL accumulating the delay time $\tau_{sw} = l_{sw} / V_g$, where l_{sw} is the propagation path that is equal to the distance between the antennae, V_g is group velocity of the spin wave in YIG film. The phase shift in spin-wave DL is $\Phi_{sw} = k_{sw}\, l_{sw}$, where k_{sw} is wavenumber of spin waves at a frequency f_{mw}. The signal is amplified and supplied to the input of the microwave photonic DL. Thus the loop is closed and the microwave signal starts circulating in the circuit.

The resonance condition in the loop is determined by phase balance $\Phi_{opt} + \Phi_{sw} + \varphi_e = 2\pi m$, where φ_e is phase shift in the electronic interconnections, which is usually small as compared with Φ_{opt} and Φ_{sw}, and m is an integer number. The value of transfer coefficient is defined by values of insertion losses introduced by DLs and a value of amplifier gain coefficient. Damping decrement of the signal in microwave photonic DL is defined by losses in fiber, modulator, and photodetector. Single-mode fiber damping decrement is approximately $\alpha_{opt} = 0.2$ dB/km at 1.55 μm. However, in experiment series the power loss of the microwave signal between electro-optical modulator input and the photodetector output was about -50 dB. This value was almost constant in the frequency range of 4-9 GHz. Contrariwise, the transfer coefficient of the signal in spin-wave DL is defined by the antennae geometry and damping decrement that is $\alpha_{sw} = 2\pi |g| \Delta H / V_g$, where $|g| = 2.8\ 10^{10}\ s^{-1}\cdot T^{-1}$ is gyromagnetic ratio for electron and ΔH is half of ferromagnetic resonance linewidth.

Basic application of ARR is the comb filters. In contrast to comb filters based on single microwave photonic DL introduction of additional spin-wave DL provides an opportunity to implement reconfiguration of frequency response. The FSR is defined by the delay time in the microwave photonic DL. The filter bandwidth is determined by the bandwidth of the spin-wave DL. Changing of its design parameters allows varying of spin-wave DL bandwidth in the range from 10 MHz to 1.5 GHz. These parameters include the thickness of the YIG film, its saturation magnetization, external bias magnetic field, the geometry of the spin-wave antennae, as well as the distance between them and the distance between the antenna and the film surface.

At the first step, microwave photonic DL and spin-wave DL with close delay time values $\tau_{opt} \approx \tau_{sw}$ were chosen in order to demonstrate features of frequency response formation in the spin-wave optoelectronic single-loop resonator. The measured frequency response of the resonator in absence of the spin-wave DL for the case of fiber length l_{opt} = 20 m is shown in Fig. 2a by curve 1. One can see that the frequency response had a large number of resonant peaks corresponding to the eigen-modes of the closed-loop. Curves 2, 3, and 4 show the frequency responses of the spin-wave DLs with different design parameters. Curves 2 and 3 demonstrate frequency responses of DLs with YIG film thickness equal 16.9 μm (curve 2) and 5.4 μm (curve 3). Both these curves were measured in the case of two 50-μm-wide and 2-mm-long microstrip antennae. Curve 3 and 4 display the effect of application of two two-element antennae instead of single-element antennae. In experiment 50-μm-wide meander-type antennae with 160-μm distance between elements of each antenna were used. For all curves the distance between antennae l_{sw} was 3 mm and bias magnetic field was H_0 = 2090 Oe.

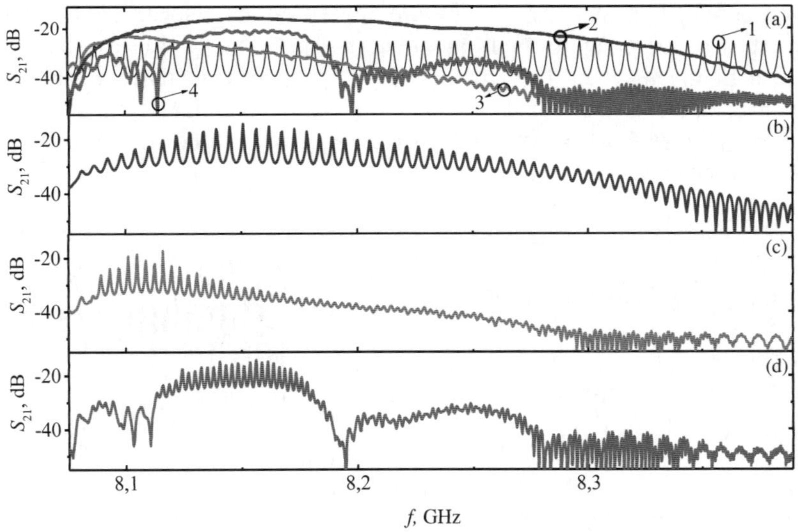

Figure 2. Experimental frequency response characteristics of: single-loop resonator based on microwave photonic DL in absence of spin-wave DL and spin-wave DL on YIG film various thicknesses and antenna geometries (a); spin-wave optoelectronic single-loop resonator with spin-wave DL on YIG film with thickness of 16.9 μm and microstrip antennae (b), spin-wave DL on YIG film with thickness of 5.4 μm and microstrip antennae (c), spin-wave DL on YIG film with thickness of 5.4 μm and meander-type antennae (d).

As shown in the Figs. 2b, 2c, and 2d the resulting frequency response of the investigated microwave photonic resonator with spin-wave DL contains resonant peaks only within the bandwidth of the spin-wave DL. The FSR decreases in comparison with curve 1 due to additional delay time τ_{sw} introduced by the spin-wave DL. Decrease of the film thickness provides higher delay time and smaller FSR as it is shown in Figs. 2c and 2d. The change of antenna geometry offers an additional feature of the frequency response reconfiguration. Fig. 3 represents the tuning of the resonant frequencies by variation of the bias magnetic field in the range from 1990 Oe to 2090 Oe. These characteristics were measured for the case of spin-wave DL made of 5.4-μm YIG film with microstrip antennae and microwave photonic delay line with length of 20 m.

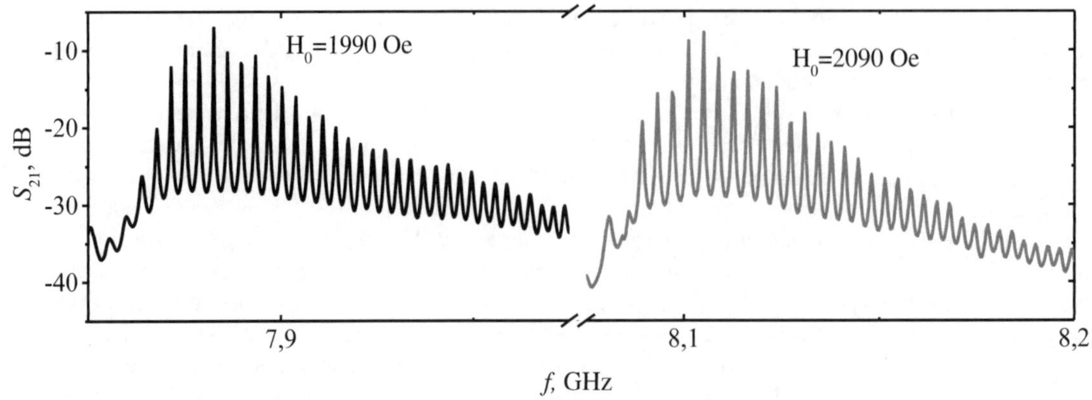

Figure 3. Magnetic tuning of experimental frequency response of the spin-wave optoelectronic single-loop resonator

Frequency response of the spin-wave optoelectronic single-loop resonator with increased distance between spin-wave antennae l_{sw} up to 4 mm is shown in Fig. 4(a). Fig. 4b demonstrates the effect of increase of optical fiber length up to 60 m. One can see that the increase of delay time provides decrease of the FSR and increase of quality factor. While the delay time in spin-wave DL is limited by an acceptable level of insertion losses -20-30 dB/μs, the delay time in microwave photonic DL achieves hundreds of microseconds.

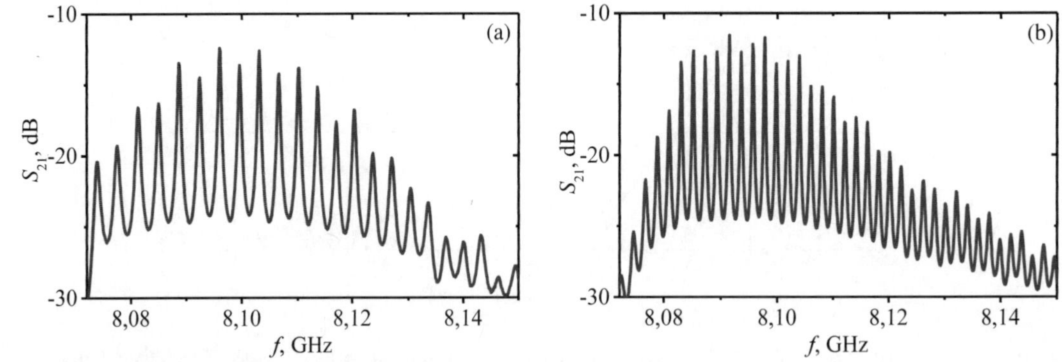

Figure 4. Experimental frequency response characteristics of spin-wave optoelectronic single-loop resonator based on the spin-wave DL with 4 mm distance between spin-wave antennae and microwave photonic DL with fiber length of 20 m (a) and 60 m (b)

As a second step, spin-wave optoelectronic single-loop resonators with delay time of the microwave photonic DL that is much higher than in the spin-wave DL $\tau_{opt} > \tau_{sw}$ were analyzed. It is clear that contribution of the delay time of the microwave photonic DL to total delay time increases with the increase of optical fiber length. For condition $\tau_{opt} \gg \tau_{sw}$ the FSR is defined by delay time in the microwave photonic DL. The decrease of contribution of spin-wave delay time in FSR due to increase of optical fiber length is shown in Fig. 5, where FSR_{Σ} is FSR of the resonator based on spin-wave and microwave photonic DLs and FSR_{opt} is FSR of resonator measured in absence of spin-wave DL. The operation frequency for Fig. 5 is 8.15 GHz. Therefore, the spectrum of resonant frequencies in the investigated resonator based on spin-wave and microwave photonic DL becomes equidistant at fiber length more than 1000 m.

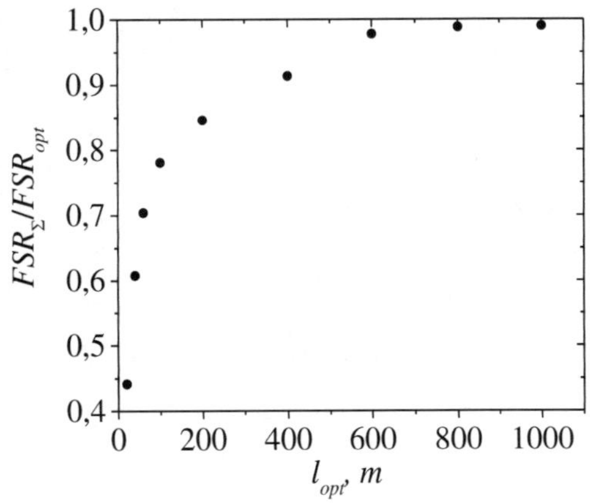

Figure 5. Dependence of relative FSR on optical fiber length

Fig. 6a shows the fragments of frequency responses measured for different optical fiber length. Fig. 6b shows the dependence (black line) of the equivalent quality factor Q of the resonant peak with minimum losses on the ring gain G in the case of fixed optical fiber length l_{opt} = 600 m and dependence (blue line) of the equivalent quality factor Q of the resonant peak on fiber length l_{opt}. It is evident that the quality factor increases with increasing l_{opt} and G and achieve the several tens of thousands. Note that $G = 0$ is self-oscillation threshold that limited the values of Q-factor in practice.

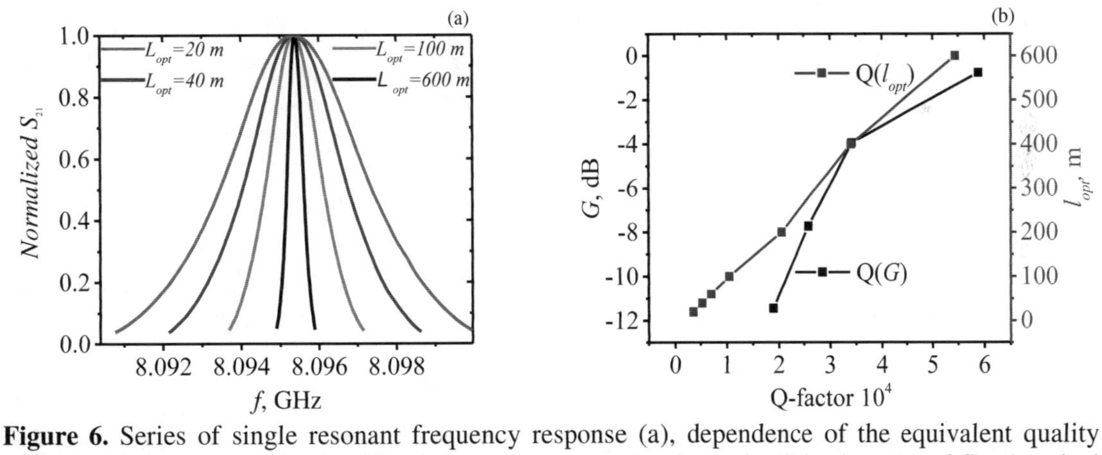

Figure 6. Series of single resonant frequency response (a), dependence of the equivalent quality factor Q of the resonant peak with minimum losses on the ring gain G in the case of fixed optical fiber length l_{opt} = 600 m and dependence of the equivalent quality factor Q of the resonant peak on fiber length l_{opt} (b)

In conclusion, the frequency response of spin-wave optoelectronic single-loop resonator is reconfigurable. The bandwidth is defined by YIG-film thickness and geometry of antennae. FSR is defined by delay time in spin-wave and microwave photonic delay lines. Increase of optical fiber length provides frequency comb spectrum in bandwidth allowed by spin-wave delay line. Change of bias magnetic field realizes frequency tuning. Quality factor increases with increase of ring gain and fiber length and archives tens of thousands. Performance of described resonator corresponds to reconfigurable tunable comb filter.

IOP Conf. Series: Journal of Physics: Conf. Series **929** (2017) 012028 doi:10.1088/1742-6596/929/1/012028

References

[1] Federici J and Moeller L *Journal of Applied Physics* 2010 **107** No 11 111101

[2] Marpaung D, Roeloffzen C, Heideman R, Leinse A, Sales S, Capmany J. *Laser Photonics Rev.* 2013 **7** 506-538

[3] Grigor'yants V V, Dvornikov A A, Il'in Yu B, Konstantinov V N, Prokof'ev V A and Utkin G M *Sov. J. Quantum Electron.* 1984 **14** 516-522

[4] Ji Y, Yao X S and Maleki L *Electronics Letters* 1999 **35** No 18 1554-1555

[5] Harris V G *IEEE Transactions on Magnetics* 2012 ***48*** No 3 1075-1104

[6] Porokhnyuk A A, Ustinov A B, Kovshikov N G, Kalinikos B A *Technical Physics Letters.* 2009 **35** No 9 843-846

[7] Stokes L F, Chodorow M, Shaw H J *Optics Letters* 1982 **7** No 6 288-290

IOP Conf. Series: Journal of Physics: Conf. Series **929** (2017) 012029 doi:10.1088/1742-6596/929/1/012029

Positron annihilation spectroscopy studies of bronze exposed to sandblasting at different pressure

S Kurdyumov[1,2*], K Siemek[3,4] and P Horodek[3,4]

[1] Saint-Petersburg Peter the Great Polytechnic University, 29 Polytechnic Str. 195251 Saint-Petersburg, Russian Federation
[2] Department of Nanophotonics and Metamaterials, ITMO University, 197101 St.Petersburg, Russia
[3]Joint Institute for Nuclear Research, 6 Joliot Curie Str., 141980 Dubna, Moscow region, Russian Federation
[4]Institute of Nuclear Physics of the Polish Academy of Sciences, 152 Radzikowskiego Str., 31-342 Kraków, Poland
*corresponding author: Tel.:(+7)9112459937 e-mail: se.kurdjumov@gmail.com

Abstract. An application of Doppler broadening of annihilation line spectroscopy to samples of beryllium bronze DIN-CuBe2 exposed to sandblasting is presented in performed studies. It is familiar that sandblasting introduces open-volume defects. Samples were sandblasted under different pressure for 1 minute using 110 μm particles of Al_2O_3. For a non-defected sample the constant value of S-parameter was detected. In the cases of sandblasted samples, S-parameter decreased when the depth enhanced. In our studies the thicknesses of defected zones were determined (it was c.a. 30 μm for a sample blasted under pressure of 1 bar and 110 μm – for 5 bar), and it was also observed that if sandblasting pressure is higher the defected zone is larger.
Key words: positron annihilation spectroscopy, sandblasting, defects, bronze

1. Introduction

Positron Annihilation Spectroscopy (PAS) is a non-destructive technique of detecting open-volume defects in solids, such as vacancies, vacancy clusters, microvoids or dislocations. [1,2] It can be used in cases where other popular methods such as scanning electron microscopy (SEM) or X-ray diffraction are not applicable. [3]

In this paper studies of beryllium bronze DIN-CuBe2 exposed to sandblasting are presented. Sandblasting is a cold-working process especially used for removing rust with a stream of fast particles of abrasive powder. The stream hits the surface generating local plastic deformations. As a result, lattice defects appear and they can be detected by PAS techniques. The presence of defects has a direct impact on the properties of surface and subsurface regions. Thus, this study is important in application because the presence of defects influences on tribological and mechanical properties of material. [4,5]

There are few papers about PAS studies [3-6] of different materials exposed to sandblasting. Pure aluminum and aluminum alloy sandblasted with 0.5 mm in diameter silicon carbide particles under 6.5 bars were investigated by Dryzek. [6] Both materials had similar defected zone thickness of about 300 μm. Divacancies were found in pure aluminum while larger vacancy clusters were recognized in the aluminum alloy. Vacancies on the edge of dislocations in stainless steel sandblasted with 110 μm Al_2O_3 particles were detected. [3] This kind of defects appeared independent on pressure and time of blasting. Additionally, reported range of defects in beryllium bronze DIN-CuBe2 created by blasting with 250 μm particles of Al_2O_3 under 6 bar for 30 sec was c.a. 250 μm. [8] In our work the impact of pressure on thickness of damaged layer and defect profile in the same kind of bronze was investigated using Doppler broadening (DB) spectroscopy.

2. Experimental technique of PAS

2.1. Samples preparation

Seven samples of beryllium bronze DIN-CuBe2 (Be – 2%, Co – 0.3%, Ni – 0.3%, Fe – 0.2% and Cu in the rest), in the shape of cylinder 6 mm high and 1 cm in diameter were studied. Firstly, they were annealed at 800°C for two hours in vacuum conditions of 10^{-5} Torr. Then they were cooled down to the room temperature in a closed furnace. This procedure allowed to remove defects and to get all samples in the same state.

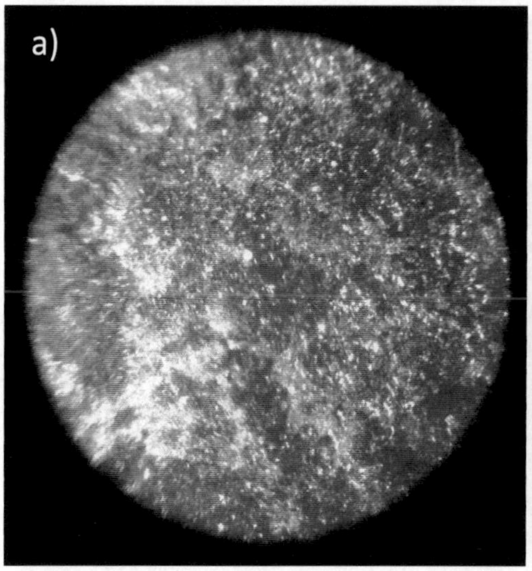

Figure 1a. An optical microscope image (magnification is equal to 180) of beryllium bronze DIN CuBe2 sample before sandblasting

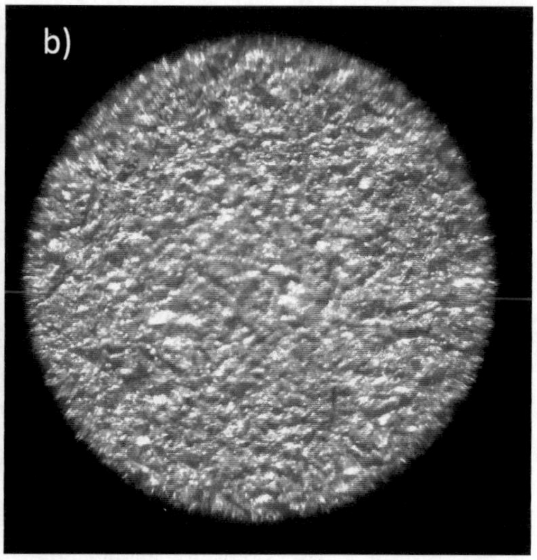

Figure 1b. An optical microscope image (magnification is equal to 180) of beryllium bronze DIN CuBe2 sample after sandblasting under 5 bar

One sample was left as a reference, non-defected specimen. Another 5 samples were sandblasted for 1 minute using 110 μm particles of Al_2O_3 at different pressure ranging from 1 to 5 bar with the step of 1 bar. In Fig.1 surfaces of sample before and after sandblasting are shown. Some differences in surface topography are visible because sandblasting increases roughness on the surface.
One of the samples was not sandblasted. It was left for studying the defect distribution from depth exposed to pressing. Thus, the thickness was reduced up to 35%.

2.2. Experiment

Encapsulated isotope of ^{22}Na with activity about 15 μCi was used in this experiment. Positrons were emitted through 4 mm in diameter and 5 μm thick titanium window. The source was placed in a special holder with the window directed towards the top. The investigated sample was put on the top of this source. One of two annihilation photons was detected in the HpGe detector. Using LYS-1 code the number of positron which annihilate in source could be estimated [7]. Around 48% of all emitted positrons annihilated in studied sample. The experimental procedure was similar to the one reported in Ref. [9].

The energy of positrons and, consequently, the implantation depth are limited. The average implantation depth can be calculated from the formula [9]

$$\bar{z} = \frac{1}{29.3Z^{0.15}\rho}, \tag{1}$$

where $\bar{z}$ is the mean positron implantation depth in [cm], Z is the atomic number of the dominating element in the alloy (for bronze, this element is copper and its atomic number is 29), and $\rho = 8.93$ g/cm^3 stands for density of bronze. The estimated average implantation depth is equal to 23 μm. About 63% positrons annihilate in the layer of such thickness from the enter surface, thus main information about positron annihilation comes only from this depth. In order to get information from deeper regions we applied the etching technique. The sample was sequentially etched in nitric acid to remove a layer of about 10 μm thick and after the measurement was performed. Etching does not damage or introduced additional defects, which could disturb the initial state. [10, 11]
The energy resolution of applied detector (FWHM) was equal to 1.20 keV at 511 keV. The spectra were analyzed by SP-16K program. [12] This program allows to obtain two characteristic parameters: S-parameter and W-parameter, which are defined as the ratio of area under the central part of the annihilation line to the whole area under the line and the ratio of area under the wing parts of the annihilation line to the whole area, respectively. According the trapping model S-parameter can be directly linked with the defect concentration, which localize positrons. Briefly, its value increases when concentration increases [8], however the dependency is not linear and can be also depend in the size the defect. The occurrence of different type of defects is reflected in the value of the W-parameter. [13] The energy windows for calculation of S parameter was $|E_\gamma - 511 \text{ keV}| < 1.38$ keV while for W parameter was determined at the energy range between 515 keV and 519 keV.

3. Results

The plot of S-parameter versus compressed depth is presented in Fig. 2. This measurement is accessory and it was done to illustrate the limits of PAS technique sensitivity to defect concentration induced by pressing. S-parameter increases while thickness reduction is less than 15%. After that a plateau appears, however, defect concentration still increases. In this region the increment of S-parameter is comparable with its error. So information about defect concentration cannot be distinguished using PAS. Thus, the limit of PAS applicability was demonstrated.

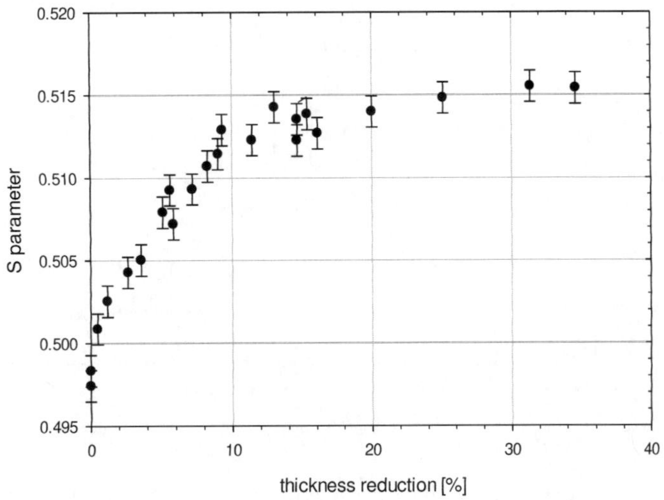

Figure 2. S-parameter in dependency on thickness reduction for a sample of beryllium bronze DIN-

CuBe2 exposed to compressing under the load of up to 15 MPa.

In Fig. 3a the dependency of the S-parameter on etched depth is depicted. For the reference sample (black squares), S-parameter values obtained in different depths are constant within the measurement accuracy (marked area). These results are expected because the reference sample contains only residual defects and etching does not introduce defects, as mentioned above. [10,11]

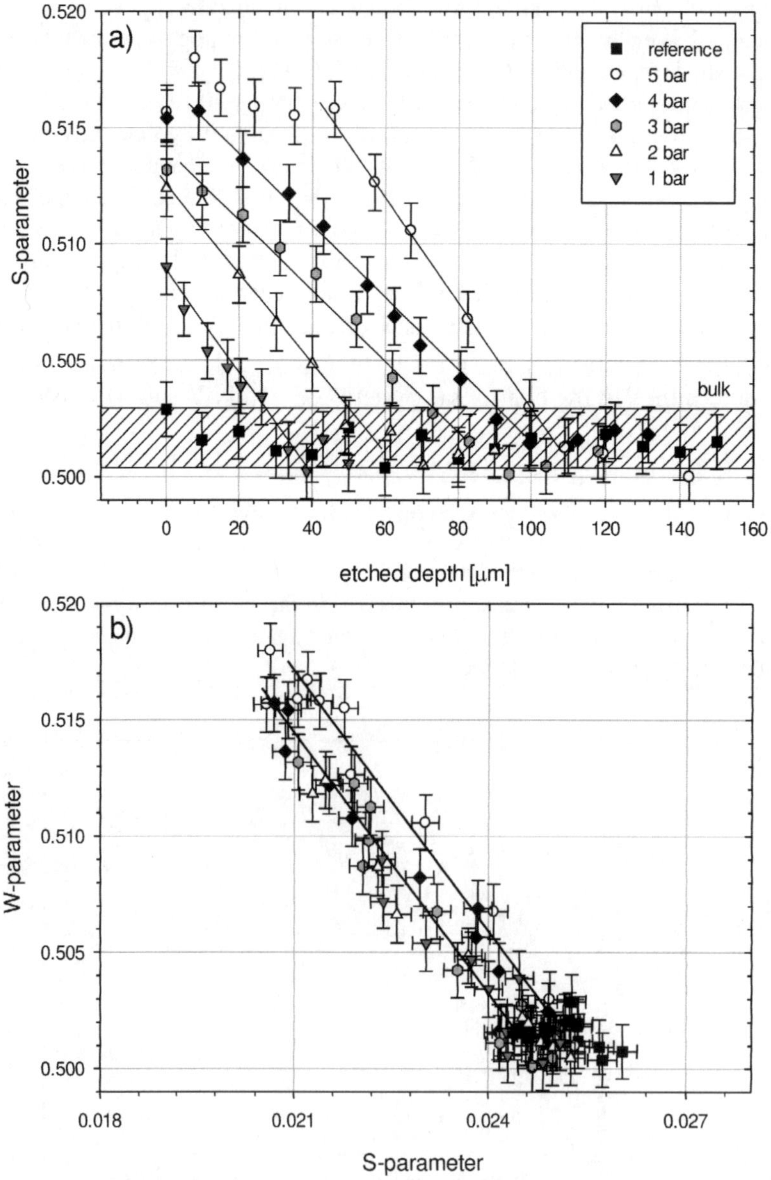

Figure 3. S-parameter in dependency on etched depth (a) and W-parameter versus S-parameter (b) for samples of beryllium bronze DIN-CuBe2 exposed to sandblasting for 1 minute with 250 μm particles of

Al_2O_3 under different pressure ranging from 1 to 5 bars with step 1 bar. The marked region is referred to the well-annealed sample in which only residual defects present (black squares).

The value of S-parameter at the surface for all sandblasted samples is larger than for the reference sample because there are induced by defects introduced by this process. Then, for samples sandblasted with pressure from 1 to 4 bar, S-parameter decreases linearly with the growth of the depth. This means that the defect concentration decreases with the increasing depth.

There is depth, special for each sample, in which the value of S-parameter reaches its value for reference sample. This depth is the total thickness of damage zone in the sandblasting process. It depends on the pressure applied during sandblasting. For sample, sandblasted with pressure of 1 bar, this depth is equal to 25 μm, for 2 bars – 50 μm, for 3 bars – 85 μm and for 4 bars – 100 μm.

Results differ for the sample sandblasted under pressure of 5 bars. At the beginning, the value of S-parameter is almost the same as for 4 bars and it is similar to value of maximum S-parameter for the compressed sample. This is the value of S-parameter which reflects to the limiting concentration of defects from which PAS technique does not give information about defect concentration. It is approximately a constant line until the etched depth is less than 50 μm. This means that at the depth of up to 50 μm there are so many defects that all positrons are trapped in the defects and S-parameter saturates. Only after 50 μm the decreasing of the S-parameter is observed. In this sample the estimated thickness of defected zone is 110 μm.

In Fig. 3b S-parameter in dependency on W-parameter is shown for all studied samples. Points representing pressures in the range 1-4 bar are situated along a straight line and it means that in cases the same type of defects is observed. The second slope between surface and bulk appears for sample blasted at 5 bar. It can point out the creation of another kind of defect. Unfortunately, type of defect cannot be defined using Doppler spectroscopy. Using positron lifetime spectroscopy can help to identify them.

4. Conclusion

The beryllium bronze DIN-CUBe2 exposed to sandblasting was investigated using PAS. The defect concentration decreased with the growth of depth, which was reflected in the plot of S-parameter versus depth. Furthermore, it was shown that increasing of defected area was caused by higher pressure. Also, thickness of damage zone in each sample was determined. It was c.a. 30 μm for a sample blasted under pressure of 1 bar and 110 μm – for the sample blasted under 5 bars.

For the compressed sample the borders of PAS method's applicability were demonstrated. It was shown that defect concentration can be so high that PAS technique is unable to determine it precisely.

5.Acknowledgements

This work was done in the frames of Students' Summer Program 2016 in JINR, Dubna.

References

[1] I. Makkonen, E. Korhonen, V. Prozheeva, F. Tuomisto 2016 J. Phys. Condens. Matter **28** 1-7

[2] I. Prochazka 2001 Materials Structure **8** 55-60
[3] P. Horodek, M. Eseev, A. Kobets 2015 Nukleonika **60** 721-724
[4] X.S. Guan, Z.F. Dong, D.Y. Li 2005 Nanotechnology **16** 2963-2971
[5] X.Y. Wang, D.Y. Li 2003 Wear **255** 836-845
[6] E. Dryzek 2003 J. Mater. Sci. **38** 3755–3763
[7] K.Siemek, J.Dryzek, Acta Phys. Polon. A, **125** (2014) 833-836
[8] P. Horodek, J. Dryzek 2016 Acta Phys. Pol. B Proc. Supp. **9** 1001-1007
[9] P. Horodek, J. Dryzek 2010 Nukleonika **55** 17-19
[10] J. Dryzek, Tribol. Lett. 2014 **54** 229-236
[11] J. Dryzek, Tribol. Lett. 2011 **42** 9-15
[12] http://www.ifj.edu.pl/~mdryzek/page_r18.html
[13] P.V. Kuznetsov, Yu.P. Mironov, et.al. 2014 AIP Conference Proceeding **1623** 327-330

X-ray diffraction investigations of the thermal expansion of iron borate $FeBO_3$ crystals

A Zamkovskaya, E Maksimova, I Nauhatsky, M Shapoval
Crimean Federal V.I. Vernadsky University, Simferopol, Crimea, Russian Federation

E-mail: trabem.z@gmail.com

Abstract. The lattice parameters a, c of iron borate $FeBO_3$ polycrystals have been measured in the temperature range 25–600°C by the X-ray diffraction method. The thermal expansion coefficients α are calculated from the measured values of the parameters. A substantial anisotropy of the thermal expansion is found. It is shown that the thermal expansion coefficient α_c along the c axis is greater than the thermal expansion coefficient α_a in a plane perpendicular to this axis. By X-ray peak profile analysisthe crystallite size was estimated from Scherrer's formula. We have also investigated the temperature dependence of crystallite size in $FeBO_3$.

1. Introduction

Iron borate $FeBO_3$ is a model object for research in solid state physics as a result of the unique combination of a number of physical properties, such as transparency in the visible region and magnetic ordering et al. [1, 2]. The performance characteristics of iron borate are highly dependent on temperature. Under the influence of the temperature there are changes in the phase composition of the samples, in the structural characteristics of the crystal lattice, in crystallite size, etc. For example, in $FeBO_3$, heated to temperatures of about 700-800°C there are new phases: hematite Fe_2O_3 and iron oxoborate Fe_3BO_6, [3].

The aim of this work: a study of structural and disperse characteristics of iron borate $FeBO_3$ on temperature.

2.Object and method of research

Iron borate has a calcite structure with the point group symmetry $\bar{3}m\,(D_{3d})$ and the space group $R\bar{3}C\ (D_{3d}^6)$. The space group has an R-type Bravais lattice; thus, the unit cell parameters can be specified in both hexagonal and rhombohedral settings. The structural rhombohedral parameters: a_r=5.52 Å, α_r=49.54°, V_r= 89.532Å^3. The structural hexagonal parameters: a_h=4.626(1) Å, c_h=14.496(6) Å, V_h=268.596 Å^3, [4]. The rhombohedral and hexagonal $FeBO_3$ unit cells contain two and six formula units, respectively. Oxygen atoms surround each iron atom forming a slightly distorted $[FeO_6]$ octahedron whose O atoms belong to different $[BO_3]$ unites [5]. The boron atoms are located in the center of plane of equilateral oxygen triangles. The borate groups are distributed in layers in such a way that the triangular $[BO_3]$ groups present reversed orientations in alternating layers; while $[FeO_6]$ octahedron share corners with other six octahedra: three from the upper layer and three from the lower layer, figure 1.

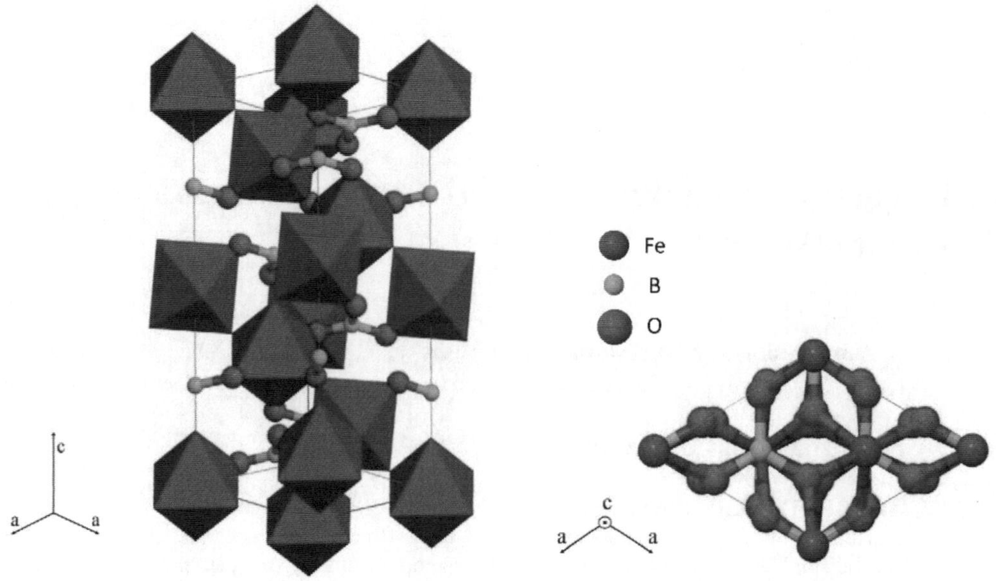

Figure1. The hexagonal unit cell of the $FeBO_3$.An $[FeO_6]$ octahedron is represented in green. Fe, B, and Oatoms are represented are green, yellow and red spheres, respectively.

For structural studies, samples of polycrystalline iron borate were ground into powder by standard technology. The temperature behavior of $FeBO_3$ have been determined in situ by high-temperature powder X-rays diffraction method (SmartLabRigaku, CuK_α) in the temperature range from 25°C to 600°C. The XRD investigation of $FeBO_3$ was performed in the range 2θ = 20° - 100° in steps h(2θ)=0.02°, figure 2.

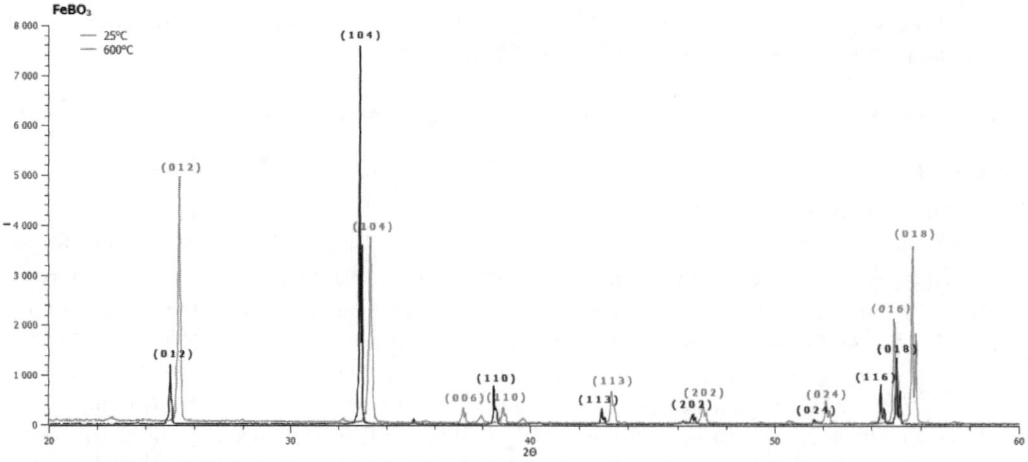

Figure 2. Observed X-ray powder diffraction patterns for $FeBO_3$ at 25 °C and 600 °C.

3. Analysis and discussion

The structure parameters of $FeBO_3$ were determined according to Bragg's law, [6]:

$$n\lambda = 2d \sin\theta \tag{1}$$

where n - order of diffraction,

λ - X-ray wavelength,
d - interplane distance,
θ - Bragg's angle.

The plane spacing d is related to the lattice constants *a*, *c* of hexagonal structure by the following relation, [5]:

$$\frac{1}{d_{hkl}^2} = \frac{4}{3}\frac{h^2+hk+k^2}{a^2} + \frac{l^2}{c^2} \tag{2}$$

were *h*, *k* and *l* - Miller indices.

The unite cell parameters calculated on the program [7] as a function of temperature are shown at table 1.

Table 1. Temperature dependence of hexagonal lattice constants *a*, *c* and volume of cell *V* of $FeBO_3$.

t, °C	a, Å	c, Å	V, Å^3
25	4,6285±0,0003	14,4878±0,0011	268,790±0,03
400	4,6620±0,0008	14,6313±0,0011	275,411±0,03
500	4,6642±0,0003	14,6445±0,0011	275,910±0,03
600	4,6691±0,0003	14,6544±0,0014	276,681±0,03

The dependence of the lattice constants and volume of cell on temperature has been expressed by polynoms of second order, the respective coefficients being found by the least-squares method, figure 3.

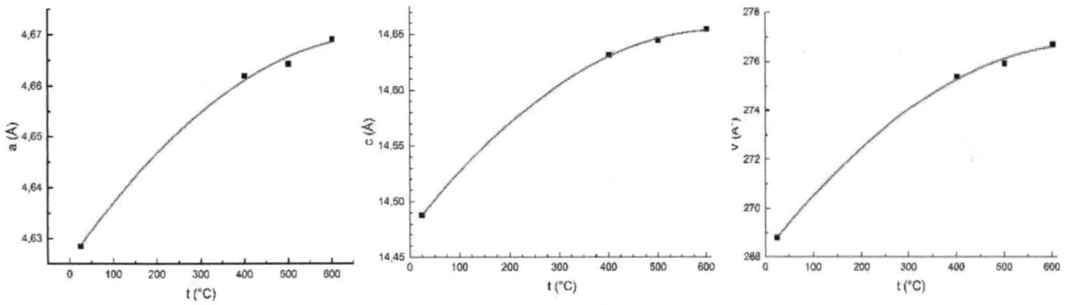

Figure 3. $FeBO_3$ cell parameters and volume between 25 and 600 °C. Experimental values are shown as points (error bars are smaller than the symbols used).

The full fit of the data to a second-order approximation:

$$a(t) = a_0 + a_1 \cdot t + a_2 \cdot t^2 \tag{3}$$

were a_0- lattice constants at 0 °C,
a_1, a_2 - characteristic constants,
t - temperature in °C.

The values of characteristic constants determined by least-squares method are shown in table 2.

Table 2. Dependence of the hexagonal lattice constants a, c and volume V of $FeBO_3$ on temperature.

Parameter	a_0	a_1	a_2
a, Å	4,6255	1,2290$\cdot 10^{-4}$	-8,5442$\cdot 10^{-8}$
c, Å	14,4740	5,6931$\cdot 10^{-4}$	-4,4978$\cdot 10^{-7}$
V, Å^3	268,189	0,0250	-1,835$\cdot 10^{-5}$

With respect to the definition of the coefficient of the thermal expansion, [6]:

$$\alpha = \frac{1}{a_0} \cdot \frac{da}{dt}, \tag{4}$$

were da – difference of lattice parameters a for the interval dt,

dt - difference of temperature

we obtain from (3) the dependence of thermal expansion coefficient on temperature:

$$\alpha(t) = a_1 + a_2 \cdot t \tag{5}$$

The results of similar fits to obtain the linear thermal expansion coefficients of the two axes of the unite cell together with the resulting of volume thermal expansion, a_1 and a_2 are shown in table 3.

Table 3. Linear and volumetric thermal expansion parameters for $FeBO_3$.

Coefficients α, C^{-1}	a_1	a_2
α_a	$2{,}6571 \cdot 10^{-5}$	$-3{,}6943 \cdot 10^{-8}$
α_c	$3{,}9333 \cdot 10^{-5}$	$-6{,}2151 \cdot 10^{-8}$
α_V	$9{,}3255 \cdot 10^{-5}$	$-1{,}3683 \cdot 10^{-7}$

It can be seen that the thermal expansion is anisotropic, being greatest for the c-axis (i.e. for the direction along the axis of the 3-order symmetry) and smallest for the direction perpendicular to the c-axis, e.g. $\alpha_a = 8{,}0995 \cdot 10^{-6}$ C^{-1} and $\alpha_c = 8{,}2575 \cdot 10^{-6}$ C^{-1} for Δt=200 °C (400 °C – 600 °C).

The three-dimensional thermal expansion diagrams of $FeBO_3$ drawing according to [8] is shown in figure 4.

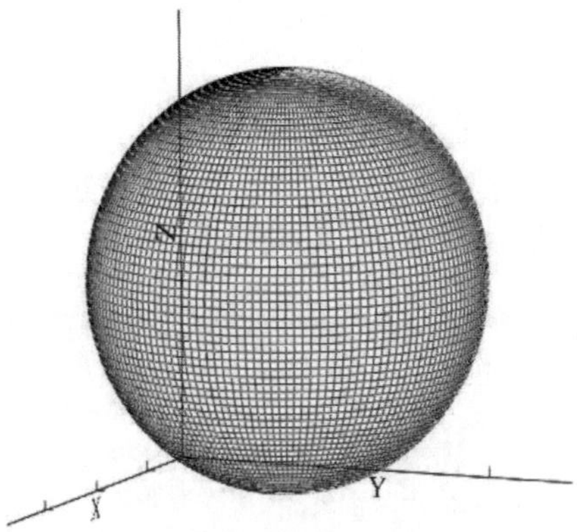

Figure 4. Indicatory surface of thermal expansion of $FeBO_3$.

We have investigated the parameters of the micro-structural quantities in $FeBO_3$ by X-ray peak profile analysis. The approximation method was used to study the individual contributions of crystallite sizes and lattice strain on the peak broadening of $FeBO_3$. The results showed that the diffraction peaks broadening inversely proportional to the cosine of the reflection angle caused by size of particles of crystals, [9].

We have estimated the coherent crystallite size (D) on the Debye–Scherrer equation, [10]:

$$D = K\lambda / B_{hkl} \cos\theta \tag{6}$$

were K - constant equal to 0.90,

λ - wave length of the incident X-ray,

B - the width of the diffraction peak, in radians, at a height half-way between background and the peak maximum,

θ - Bragg's angle.

The temperature dependence of average crystallite size was calculated from XRD peak width of (102) are shown in figure 5.

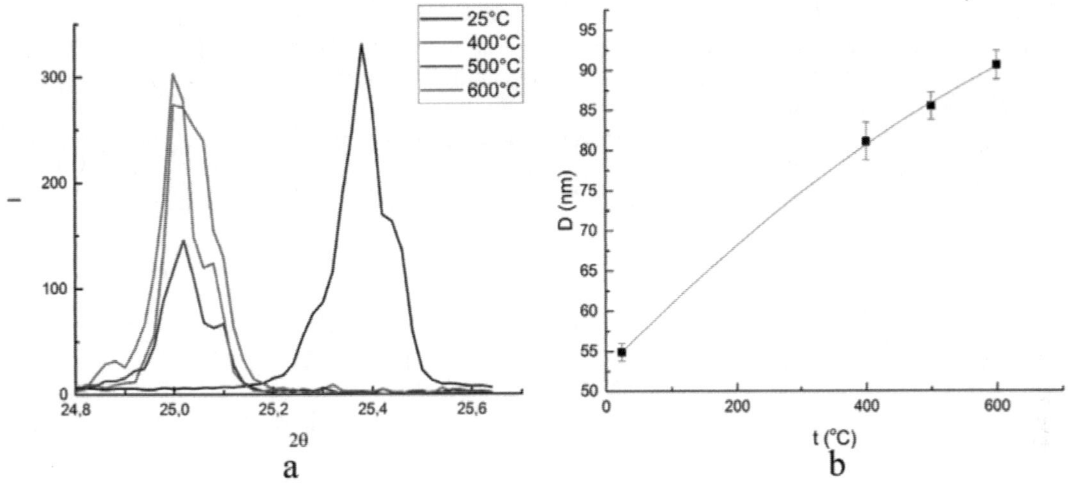

Figure 5.
a) Peak shape of (102) in $FeBO_3$ at 25, 400, 500 and 600 °C.
b) Temperature dependence of crystallite size of $FeBO_3$.

4. Conclusion

In this paper we report the experimental study of the thermal behavior of iron borate $FeBO_3$ by X-ray diffraction measurements and calculations. Correct evaluations of the thermal expansion coefficients could be done in the temperature range from 400 to 600°C, since below Neel point T_N = 75°C, $FeBO_3$ is an easy plane weak ferromagnet with the anomalous temperature variation of the thermal expansion near the phase transition temperature [1]. Our results show that the thermal expansion of $FeBO_3$ along the *c* axis is more than along the *a* axis at temperatures above T_N.

It was found that the sample temperature increase leads to a monotonic and quite a substantial increase in the value of coherent crystallite size, indicating that the ordering of crystal structure in iron borate with heat processing.

References

[1] Kalashnikova A M,Markovin P A, Pisarev R V, RasingTHM 2007 Optical Study of the Electronic Structure and Magnetic Ordering in a weak Ferromagnet $FeBO_3$ *Journal of Experimental and Theoretical Physics Letters* **86** pp 712-717

[2] Maksimova E M, Nauhatsky I A, Strugatsky M B, Zubov M E 2010 Surface magnetism of real iron borate monocrystals *Journal of Magnetism and Magnetic Materials* **322** pp 383-480

[3] Joubert J C, Shirk T, White W B, Rustum Roy 1968 Stability, infrared spectrum and magnetic properties of $FeBO_3$ *Mat. Res. Bull.* **3** pp 671-676

[4] Diehl R 1975 Crystal Structure Refinement of Ferrite Borate, FeBO3. *Sol. Stat. Com.***17**

[5] Pérez S, Hilario D G, Sans O, Ortiz J A, Vegas H M, Errandonea A, Ruiz-Fuertes D, Martínez-García J et al 2014 Compressibility systematics of calcite-type borates : An experimental and theoretical structural study on ABO3 (A = Al, Sc, Fe and In) *Journal of Physical Chemistry*

[6] Gerasimov V N, Dolivo-Dobrovol'skaya E M, Kamentsev I E, Kordat'eva V V, Kosoi A L, Lesyuk G I, Rozhdestvenskaya I V, Stroganov E V, Filatov S K, Frank-Kamenetskaya O V et

IOP Conf. Series: Journal of Physics: Conf. Series **929** (2017) 012030 doi:10.1088/1742-6596/929/1/012030

al. 1975 *A Manual on X-Ray Diffraction Investigations of Minerals* ed V A Frank-Kamenetskii (Leningrad: Nedra)
[7] Holland T J B and Redfern S A T 1997 Unit cell refinement from powder diffraction data: the use of regression diagnostics *Mineralogical Magazine*
[8] Zamkovskaya A and Maksimova E 2015 Anisotropic thermal expansion in crystals of different categories *Journal of Physics: Conference Series* **661** (2015)012067, pp. 1-6.
[9] Shapoval M A and Zamkovskaya A I 2016 Determination of the parameters of the micro-structural quantities in iron borate by the approximation method *Proc. of the 2th Sci. Conf. "Days of science at the Vernadsky CFU"* pp 78-80
[10] R. Jenkins & R.L. Snyder 1996 *Introduction to X-ray Powder Diffractometry* (New York: John Wiley & Sons) pp 89-91

IOP Conf. Series: Journal of Physics: Conf. Series **929** (2017) 012031 doi:10.1088/1742-6596/929/1/012031

Boson peak in two-dimensional random matrix models

D A Conyuh[1], Y M Beltukov[2], and D A Parshin[1]

[1]Peter the Great St.Petersburg Polytechnic University, 195251 Saint-Petersburg, Russia
[2]Ioffe Institute, 194021 Saint-Petersburg, Russia

E-mail: conyuh.dmitrij@yandex.ru

Abstract. We present a stable random matrix approach to the old standing problem of the boson peak in amorphous dielectrics (glasses). In this paper, we consider a two-dimensional case. We show that in the 2d case the boson peak in the density of vibrational states appears in a natural way in different random matrix models similar to the 3d case. Changing the parameters of random matrices one can shift the boson peak to higher or to lower frequencies depending on the strength of disorder in the model. In all investigated cases the position of the boson peak is correlated with the Young modulus of the lattice.

1. Introduction

It is well known that all glasses independently of their chemical composition have universal properties in a wide range of frequencies (or temperatures) [1]. One of these properties is a well-known boson peak in the reduced density of vibrational states $g(\omega)/\omega^{d-1}$, where d is a space dimension [2]. It characterizes an additional to Debye vibrational density at low frequencies far below the Debye frequency.

This peak is well seen in the measurements of the heat capacity, in the light and X-ray scattering, and in the inelastic neutron scattering. The boson peak frequency ω_b is correlated with Ioffe-Regel crossover frequency $\omega_{\rm IR}$ when the phonon mean free path becomes of the order of the phonon wavelength [3, 4, 5, 6, 7]. Despite a big number of articles about the boson peak, its physical origin is still unclear. In the present paper, we apply a new approach to the boson peak problem based on random matrices. We show that in many random matrix models the boson peak appears in a natural way.

2. Random matrix model

The symmetric dynamical matrix M determines the eigenfrequencies squared. It is natural to suppose that in amorphous solids due to local disorder the dynamical matrix has some disorder as well. It also has some general regular properties. First of all, it has *translational symmetry* [8]. In a scalar model with equal atomic masses $m_i = 1$, the sum of elements in each row and each column of the matrix M should be zero

$$\sum_{i=1}^{N} M_{ij} = \sum_{j=1}^{N} M_{ij} = 0, \tag{1}$$

where N is the number of atoms. Secondly, the translation invariance conditions (1) are *necessary* but not sufficient for existing low-frequency acoustic modes (acoustic phonons) in amorphous solids. For existing of these modes, one needs nonzero macroscopic rigidity (Young modulus). Finally, the matrix M should be *mechanically stable*. It means that all eigenfrequencies squared should not be negative. Then in such a system in the framework of standard elasticity theory the low-frequency sound wave can propagate. We consider below three different simplest two-dimensional scalar models (but sufficiently general) where all these conditions are satisfied and we have a boson peak in the reduced density of states $g(\omega)/\omega$.

The mechanical stability of the random symmetric matrix M is guaranteed if we take it in the form [9]

$$M = AA^T. \tag{2}$$

Here A is some real and not necessary symmetric matrix. In the simplest case, the matrix A is a square $N \times N$ random matrix. For simplicity let us build this matrix A on a quadratic lattice (with lattice constant $a_0 = 1$) with identical atoms in the nods. Let us take into account the interaction between nearest neighbors only [9]. Non-zero non-diagonal matrix elements A_{ij} for nearest neighbors i and j are random Gaussian values with unit variance and zero mean. The diagonal elements are defined as

$$A_{kk} = -\sum_{i \neq k} A_{ik} \tag{3}$$

to satisfy the condition of the translational invariance (1).

As a result, we have satisfied the most important properties of the dynamical matrix M. Nevertheless, the acoustic phonons can not propagate in this lattice and there is no boson peak in the density of states. The reason is that the Young modulus of the lattice is zero and the system has zero rigidity in the thermodynamic limit. The same takes place in the 3d case [10].

3. Bond counting

According to the Maxwell counting rule [11], the above-mentioned system is isostatic [12]. As a result, it is critically soft and has zero macroscopic rigidity. It follows from the point that the number of degrees of freedom N is exactly equal to the number of bonds K. Indeed, using the dynamical matrix M we can write the potential energy of the system U as a quadratic form of atomic displacements from the equilibrium, u_i ($i = 1...N$)

$$U = \frac{1}{2}\sum_{i=1}^{N}\sum_{j=1}^{N} M_{ij}u_i u_j. \tag{4}$$

Since the dynamical matrix M can be presented as a product (2), the potential energy U can be written as a sum over k of N quadratic terms

$$U = \frac{1}{2}\sum_{k=1}^{N}\left(\sum_{i=1}^{N} A_{ik}u_i\right)^2. \tag{5}$$

It proves that the quadratic form (4) is positive definite. Summation over *rows* of the matrix A in Eq. (5) is taken over i from 1 to N, which is the number of atoms or the number of degrees of freedom (in the scalar model). Summation over *columns* of the matrix A in Eq. (5) is taken over k from 1 to N. Therefore, the index k enumerates different quadratic terms.

Each quadratic term in the potential energy U is determined by some column of the matrix A and can be considered as a *mechanical bond* (in a general sense) in the system [13]. The number of these bonds K is the number of columns in the matrix A. In our scalar model, it coincides with the number of atoms N, i.e. with the number of degrees of freedom since the matrix A is a square matrix. Therefore, the system $M = AA^T$ with the random square matrix A is mechanically stable but has zero macroscopic rigidity.

To make our system macroscopically rigid one must add additional bonds. Since each bond corresponds to some column in the matrix A we must increase the number of columns in the matrix A and make $K > N$. As a result, the square $N \times N$ matrix A becomes a rectangular $N \times K$ matrix $\mathcal{A}$. In order to increase the number of bonds, one can add to the matrix AA^T some other square matrix $N \times N$ which is translational invariant. The sum of these two matrices will have finite rigidity. It can be done in many ways. We consider below three of them.

4. Three random matrix models with non-zero rigidity

First method. Let us take the dynamical matrix M in the form

$$M = \mathcal{A}\mathcal{A}^T = AA^T + \mu M_0. \tag{6}$$

Here A is the same square random matrix $N \times N$, built on a quadratic lattice and having translation invariance. The matrix M_0 is a regular crystalline dynamical matrix $N \times N$, built on the same lattice with unit springs between nearest neighbors and having non-zero rigidity. The parameter $\mu \geqslant 0$ is a rigidity of these springs. If $\mu = 0$ we go back to the previous case (2) of the system with zero rigidity.

It is easy to check without writing of the exact form of the rectangular matrix $\mathcal{A}$, that its number of rows is equal to N (i.e. to the number of degrees of freedom). The number of columns (i.e. the number of bonds) is equal to $3N$: N columns are columns of the matrix A, and additional $2N$ columns (bonds) are determined by the crystalline matrix M_0. As a result, the matrix $\mathcal{A}$ is $N \times 3N$ matrix. We have shown that the Young modulus E in this system for $\mu > 0$ is different from zero and $E \propto \sqrt{\mu}$ (for $\mu \ll 1$). Therefore, phonons can propagate in this lattice.

Normalized to unity the density of states $g(\omega)$ for the model (6) is shown in Fig. 1(a) for $N = 1000^2$ and different values of the parameter μ. We used the Kernel Polynomial Method (KPM) to calculate the density of states [14]. The straight lines show the Debye contribution

$$g_D(\omega) = \frac{1}{2\pi}\frac{\omega}{E}, \tag{7}$$

calculated using the Young modulus $E(\mu)$ for non-zero values of μ.

In the case of $\mu = 0$ we obtain the density of states of non-rigid $M = AA^T$ system, $g_0(\omega)$. It does not satisfy the Debye law (7) with any value of the Young modulus. One can show that $g_0(\omega)$ has a logarithmic singularity at $\omega = 0$.

The reduced density of states $g(\omega)/g_D(\omega)$ shows the boson peak for nonzero values of the parameter μ (Fig. 1(d)).

Second method. Let us write the matrix M as

$$M = \mathcal{A}\mathcal{A}^T = AA^T + M_0(p). \tag{8}$$

It is different from the previous case. Now $\mu = 1$ and all springs in the crystalline matrix M_0 have unit rigidity. To decrease the number of bonds we randomly cut out springs from the matrix M_0. The cutting out of one spring is equivalent to removing of one bond, i.e. one column of the matrix $\mathcal{A}$. The part of remaining springs in the system is denoted by p. The matrix M_0 with cut out springs is denoted by $M_0(p)$. It is easy to see that the matrix $\mathcal{A}$ has N rows and $N(1 + 2p)$ columns. We have shown that the Young modulus $E(p)$ in this system for $p > 0$ is different from zero and phonons can propagate in the system. We have shown that the Young modulus $E \propto p$ for $p \ll 1$.

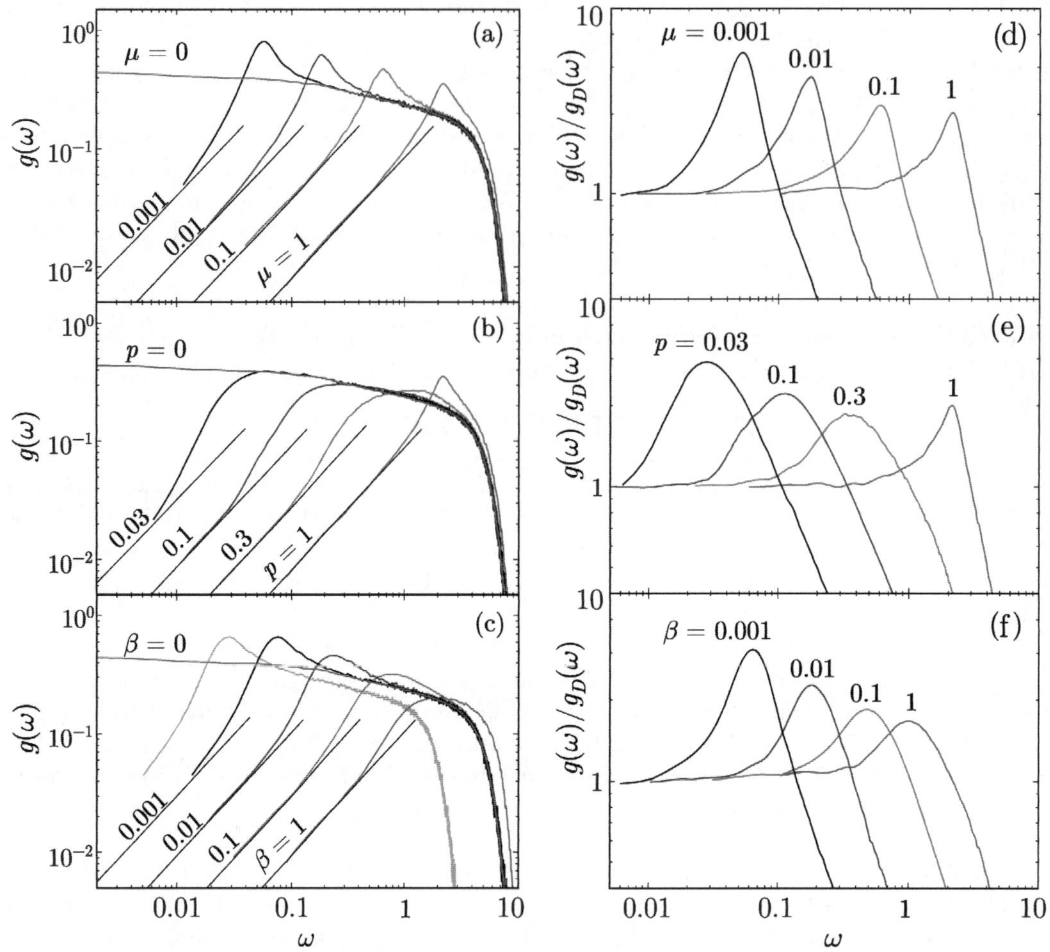

Figure 1. Left column: (a), (b), and (c) show the normalized to unity density of states $g(\omega)$ for three random matrix models (6), (8) and (9) respectively and different values of parameters μ, p and β. The random matrix A was built on the quadratic lattice with periodic boundary conditions and $N = 1000^2$. The straight lines show the Debye density of states (7) with independently calculated values of Young modulus for each of three random matrix models and different values of the above-mentioned parameters. Right column: (d), (e), and (f) show the boson peaks in the reduced density of states $g(\omega)/g_D(\omega)$ calculated from the density of states $g(\omega)$, shown in (a), (b), and (c) respectively.

Normalized to unity the density of states $g(\omega)$ in this model is shown in Fig. 1(b) for different values of the parameter p. The straight lines show the Debye contribution (7) calculated from the Young modulus $E(p)$. The case $p = 0$ corresponds to the density of states of the non-rigid system, $g_0(\omega)$. The boson peak in the reduced density of states $g(\omega)/g_D(\omega)$ is shown in Fig. 1(e) for non-zero values of p.

Third method. Let us take the matrix M in the form

$$M = \mathcal{A}\mathcal{A}^T = AA^T + \beta BB^T. \tag{9}$$

Here A is again the same random $N \times N$ matrix built on a quadratic lattice and having translational invariance. The matrix B is independent from A random $N \times N$ matrix made by the same rules as the matrix A. The parameter $\beta \geqslant 0$ can vary. For $\beta = 0$ we come back to the case of non-rigid system (2).

It is easy to see that the number of rows in the matrix $\mathcal{A}$ is equal to the number of degrees of freedom N as before, and the number of columns (i.e. the number of bonds) is equal to $2N$. It is clear that N bonds (columns) come from the matrix A, and N bonds come from the matrix B. We have shown that there are phonons in such a system since Young modulus for $\beta > 0$ is different from zero and $E \propto \sqrt{\beta}$ (for $\beta \ll 1$). It is interesting to note that each term in Eq. (9) taken separately has zero rigidity but together the Young modulus is different from zero.

Normalized to unity the density of states $g(\omega)$ in this model is shown in Fig.1(c) for different values of the parameter β. The straight lines show the Debye contribution (7) which was calculated from the Young modulus $E(\beta)$ for non-zero values of β. The case $\beta = 0$ corresponds to the density of states of the non-rigid system, $g_0(\omega)$. The boson peak in the reduced density of states $g(\omega)/g_D(\omega)$ is shown in Fig.1(f) for non-zero values of β.

5. Boson peak frequency

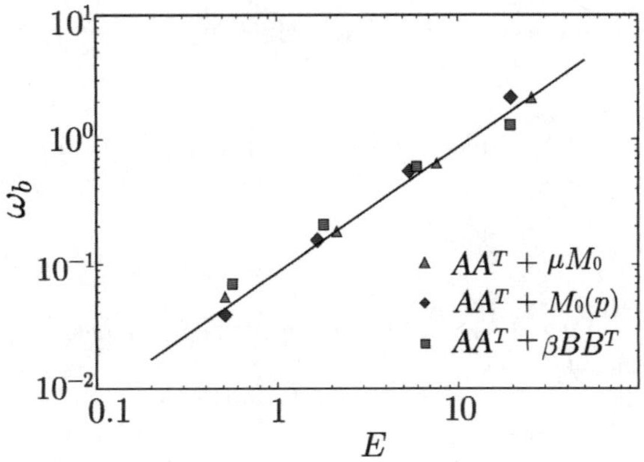

Figure 2. The correlation of the boson peak frequency ω_b and Young modulus E in three considered random matrix models for four different values of parameters μ, p and β shown in Fig.1.

As follows from the analysis of the data, the boson peak frequency ω_b is correlated with the Young modulus of the lattice E. Figure 2 shows positions of the boson peak ω_b in three considered matrix models (6), (8) and (9) together with the corresponding Young modulus value E for four non-zero values of parameters μ, p and β. As follows from this figure, the boson peak frequency ω_b is proportional to the corresponding Young modulus value E in all considered cases.

$$E \sim \omega_b. \tag{10}$$

The same behavior was obtained in the 3d case of the same random matrix models [13]. A similar correlation between the boson peak ω_b and the shear modulus G was observed in the granular medium near the jamming transition point [15]. At this point the shear modulus goes to zero. The correlation between boson peak frequency and the shear modulus was also observed experimentally in modified borate glasses [16]. This correlation makes it possible to explain qualitatively in a unique way the physical origin of the boson peak in our random matrix models.

On the one hand, at low frequencies $\omega < \omega_b \ll 1$ the density of states $g(\omega)$ is given by the Debye contribution $g_D(\omega_b)$. Taking into account that $\omega_b \sim E$, we get $g_D(\omega_b) \sim 1$. On the other hand, at high frequencies $\omega > \omega_b$ the density of states $g(\omega) \approx g_0(\omega)$ as follows from Fig.1.

However, these two asymptotics have different values at $\omega = \omega_b$ since $g_0(\omega)$ logarithmically diverge at $\omega = 0$. For small enough values of ω_b, the value of $g_0(\omega_b)$ should overcome the constant value of $g_D(\omega_b)$. Such behavior is responsible for the peak in the reduced density of states, which height is determined by the ratio $g(\omega_b)/g_D(\omega_b) \sim g_0(\omega_b)$. The height of the peak logarithmically grows with decreasing of ω_b. Such behavior differs from the 3d case [13], where the height of the peak has the power-law behavior.

Summarizing, we have shown that the boson peak appears naturally in different models of stable random matrices with translational invariance under rather general conditions applied to the system. In all investigated cases the boson peak frequency is proportional to the Young modulus of the system. According to the 3d model investigated in more details in [10], the boson peak in the 3d model appears at frequencies around Ioffe-Regel crossover frequency. The phonons experience strong scattering on quasilocal vibrations and, as a result, transforms to diffusons. In other words, near the boson peak the character of propagation of vibrational modes is changing drastically [10]. In the 2d model, the density of states is similar to the 3d model: the boson peak frequency is the crossover between the Debye law $g_D(\omega)$ and the high-frequency density of states $g_0(\omega)$. Therefore, there is reason to believe that the character of vibrations in all our 2d model is similar to the 3d model. The detailed calculations of the vibrational properties will be presented in a subsequent paper.

One of the authors (Y.M.B.) thanks the Dynasty Foundation and the Council for grants of the President of the Russian Federation (grant SP-3299.2016.1) for the financial support.

References

[1] D. A. Parshin, Phys. Solid State **36**, 991 (1994).
[2] V.L. Gurevich, D.A. Parshin, H.R. Schober, Phys.Rev. B, **67**, 094203, (2003).
[3] V.L. Gurevich, D.A. Parshin, J. Pelous and H.R. Schober, Phys.Rev. B **48**, 16318 (1993).
[4] D.A. Parshin and C. Laermans, Phys.Rev. B, **63**, 132203 (2001).
[5] B. Rufflé, G. Guimbretière, E. Courtens, R. Vacher and G. Monaco, Phys.Rev.Lett. **96**, 045502 (2006).
[6] B. Rufflé, D.A. Parshin, E. Courtens and R. Vacher, Phys.Rev.Lett. **100**, 015501 (2008).
[7] H. Shintani and H. Tanaka, Nature Materials **7**, 870 (2008).
[8] A.A. Maradudin, E.W. Montroll, G.H. Weiss, and I.P. Ipatova, Theory of Lattice Dynamics in the Harmonic Approximation (Academic Press, New York, 1971).
[9] Y. M. Beltukov and D. A. Parshin, JETP Lett. **93**, 598 (2011)
[10] Y.M. Beltukov, V.I. Kozub, D.A. Parshin, Phys.Rev.B, **87**, 134203 (2013).
[11] J.C. Maxwell, Philosophical Magazine **27**, 294 (1865).
[12] C.S. O'Hern, L.E. Silbert, A.J. Liu, S.R. Nagel, Phys. Rev.E **68**, 011306 (2003).
[13] Y. M. Beltukov, D. A. Parshin. JETP Letters **104**, 552 (2016).
[14] Y.M. Beltukov, C. Fusco, D. A. Parshin, A. Tanguy, Phys.Rev.E, **93**, 023006, (2016).
[15] V. Vitelli, N. Xu, M. Wyart, A. J. Liu, S. R. Nagel, Phys.Rev.E **81**, 021301 (2010).
[16] S. Kojima, M. Kodama, Physica B: Cond. Mat. **263**, 336 (1999).

Stability investigation of implicit parametrical schemes for the systems of kinetic equations

Elizaveta A Prokhorova, Gerasim V Krivovichev

Faculty of Applied Mathematics and Control Processes Saint Petersburg State University, 7/9 Universitetskaya nab., Saint Petersburg, 199034, Russian Federation

E-mail: proxliza@mail.ru, g.krivovichev@spbu.ru

Abstract. Systems of parametrical lattice Boltzmann equations (LBE's) are considered. The formulae for the apparent viscosity for the general representation of these systems is obtained by Chapman — Enskog asymptotic expansion on Knudsen number. Obtained expression represents viscosity as a function of the relaxation parameter and parameter of the LBE's. Necessary stability conditions in form of inequalities are derived from the non-negativity condition of the apparent viscosity. The validity of the stability conditions are demonstrated by the solution of lid-driven cavity flow problem.

1. Introduction

In last years lattice Boltzmann method (LBM) established itself as a powerful tool for the numerical solution of different problems of mechanics and physics of gases and fluids [1, 2, 3]. The main feature of the method is an application of *kinetic* equations, such as Boltzmann and Bhatnaghar – Gross – Krook (BGK) equations to the modelling of flows, instead of the equations of continuum mechanics (such as Navier – Stokes equations), as it is realized in traditional computational fluid mechanics. Great advantage of LBM consists with high parallelism of its computational algorithms (e.g. see [4, 5]).

One of the problems of the theory of LBM is a conditional stability of its main computational difference scheme named lattice Boltzmann equation (LBE), which has an explicit nature. This problem restrict the values of the relaxation parameter of the modelled system, so the range of the considered physical processes is restricted. The problem may be solved by the construction of implicit schemes.

Implicit schemes for LBM constructed by finite-difference and finite-element discretisation procedures are considered in [3, 6, 7, 8]. In [10] implicit scheme constructed by θ-method is applied to the solution of steady-flow problems. P. Asinari in [9] construct implicit schemes for multi-relaxation-time LBM with various orders of approximation. In [11] implicit LBE is constructed and applied to the solution of phase-change problem with Stefan condition. In [12] the second order LBE for multiphase systems is obtained. The equation is constructed by application of trapezoidal quadrature rule to the integral form of the system of kinetic equations.

In the presented paper the system of parametrical LBE's derived in [13] is investigated. The expression for the apparent viscosity is obtained by the method of Chapman – Enskog asymptotic expansion. Necessary stability conditions in form of inequalities on relaxation and scheme parameters are introduced. The validity of the proposed conditions is demonstrated by the numerical solution of lid-driven cavity flow problem.

The structure of the paper is follows. In section 2 the system of parametrical LBE's is presented. In section 3 the expression for the apparent viscosity is derived and stability conditions are formulated. In section 4 lid-driven cavity problem is considered. Some conclusion remarks are made in section 5.

2. Parametrical lattice Boltzmann equations

The system of BGK kinetic equations for the ensemble of particles with velocities $\mathbf{V}_i$ is written as:

$$\frac{\partial f_i}{\partial t} + \mathbf{V}_i \nabla f_i = -\frac{1}{\lambda}(f_i - f_i^{(eq)}), \tag{1}$$

where $f_i = f_i(t, \mathbf{r})$ are the distribution functions of particles with velocities $\mathbf{V}_i = V\mathbf{e}_i$, where $V = l/\delta t$, l is a mean free path, δt is a mean free time, $\mathbf{e}_i$ are the dimensionless vectors, t is a time, $\mathbf{r}$ is a vector of space variables, λ is a relaxation time, $f_i^{(eq)}$ are the equilibrium distribution functions. In two-dimensional case the following set of vectors named $D2Q9$ pattern may be used [1]: $\mathbf{e}_1 = (0,0)$, $\mathbf{e}_2 = (1,0)$, $\mathbf{e}_3 = (0,1)$, $\mathbf{e}_4 = (-1,0)$, $\mathbf{e}_5 = (0,-1)$, $\mathbf{e}_6 = (1,1)$, $\mathbf{e}_7 = (-1,1)$, $\mathbf{e}_8 = (-1,-1)$, $\mathbf{e}_9 = (1,-1)$.

System (1) may be rewritten in integral form on the time interval $[t, t+\delta t]$ [14]:

$$f_i(t+\delta t, \mathbf{r}+\mathbf{V}_i\delta t) - f_i(t, \mathbf{r}) = -\frac{1}{\lambda}\int\limits_0^{\delta t}(f_i(t+\xi, \mathbf{r}+\mathbf{V}_i\xi) - f_i^{(eq)}(t+\xi, \mathbf{r}+\mathbf{V}_i\xi))d\xi. \tag{2}$$

System of integral equations (2) in [13] is rewritten in equivalent form with parametrical coefficients:

$$f_i(t+\delta t, \mathbf{r}+\mathbf{V}_i\delta t) - f_i(t, \mathbf{r}) = -\frac{1-\sigma}{\lambda}\int\limits_0^{\delta t}(f_i(t+\xi, \mathbf{r}+\mathbf{V}_i\xi) - f_i^{(eq)}(t+\xi, \mathbf{r}+\mathbf{V}_i\xi))d\xi -$$

$$-\frac{\sigma}{\lambda}\int\limits_0^{\delta t}(f_i(t+\xi, \mathbf{r}+\mathbf{V}_i\xi) - f_i^{(eq)}(t+\xi, \mathbf{r}+\mathbf{V}_i\xi))d\xi, \tag{3}$$

where $\sigma \in [0,1]$ is a dimensionless parameter.

Parametrical LBE's may be obtained from system (3) by discretization of (3) on the time and space grids and by application of quadrature formulas to the computation of integrals in right part of (3). In [13] six systems of parametrical LBE's are obtained by application of simple quadrature formulas of low algebraic accuracy. These difference equations may be presented in following form:

$$f_i(t_j+\delta t, \mathbf{r}_{kl}+\mathbf{V}_i\delta t) - f_i(t_j, \mathbf{r}_{kl}) = A(f_i(t_j, \mathbf{r}_{kl}) - f_i^{(eq)}(t_j, \mathbf{r}_{kl})) +$$

$$+B(f_i(t_j+\delta t, \mathbf{r}_{kl}+\mathbf{V}_i\delta t) - f_i^{(eq)}(t_j+\delta t, \mathbf{r}_{kl}+\mathbf{V}_i\delta t)), \tag{4}$$

where $A = A(\sigma, \tau)$, $B = B(\sigma, \tau)$, $\tau = \lambda/\delta t$ is a dimensionless relaxation time, t_j is a node of time grid constructed with step δt, $\mathbf{r}_{kl}$ is a node of space grid constructed with step l. System (4) represent systems of implicit difference equations which are defined by following dependencies of A and B on σ:

1) system 1: $A = -(1-\sigma)/\tau$, $B = -\sigma/\tau$.
2) system 2: $A = -\sigma/\tau$, $B = -(1-\sigma)/\tau$.

3) system 3: $A = -(1-\sigma)/2\tau, \quad B = -(1+\sigma)/2\tau.$
4) system 4: $A = -(1+\sigma)/2\tau, \quad B = -(1-\sigma)/2\tau.$
5) system 5: $A = -\sigma/2\tau, \quad B = (\sigma-2)/2\tau.$
6) system 6: $A = (\sigma-2)/2\tau, \quad B = -\sigma/2\tau.$

3. Apparent viscosity of parametrical LBE's

The expression for apparent viscosity plays an important role in practical computations as a tool for the stability analysis and as a value of kinematic viscosity, used in computations of real flows, different from the real value of the viscosity, which in case of $D2Q9$ pattern is presented as [15]:

$$\nu = \frac{\lambda V^2}{3}. \tag{5}$$

The formulae for the apparent viscosity may be derived from differential approximation of (4) by the method of Chapman — Enskog asymptotic expansion [15].

Differential approximation of (4) is obtained by application of Taylor formulae:

$$f_i(t+\delta t, \mathbf{r}+\mathbf{V}_i\delta t) = f_i(t,\mathbf{r}) + \frac{\partial f_i(t,\mathbf{r})}{\partial t}\delta t + \frac{\partial f_i(t,\mathbf{r})}{\partial x_\alpha}V_{i\alpha}\delta t + \frac{(\delta t)^2}{2}\frac{\partial^2 f_i(t,\mathbf{r})}{\partial t^2} +$$

$$+\frac{(\delta t)^2}{2}\frac{\partial^2 f_i(t,\mathbf{r})}{\partial x_\alpha \partial x_\beta}V_{i\alpha}V_{i\beta} + (\delta t)^2\frac{\partial^2 f_i(t,\mathbf{r})}{\partial t \partial x_\alpha}V_{i\alpha} + o(\delta t^2). \tag{6}$$

where the Einstein summation rule on Greek indices α and β is realised and $x_1 = x$, $x_2 = y$, $V_{i1} = V_{ix}$, $V_{i2} = V_{iy}$.

The method of Chapman — Enskog expansion is based on the following expression on Knudsen parameter $\varepsilon = l/L$, where L is a typical length of the flow domain:

$$f_i \approx f_i^{(0)} + \varepsilon f_i^{(1)} + \varepsilon^2 f_i^{(2)}, \tag{7}$$

where $f_i^{(0)} = f_i^{(eq)}$, $f_i^{(1)}$ and $f_i^{(2)}$ satisfy the following properties:

$$\sum_{i=1}^{n} f_i^{(1)} = \sum_{i=1}^{n} f_i^{(2)} = 0, \quad \sum_{i=1}^{n} f_i^{(1)}\mathbf{V}_i = \sum_{i=1}^{n} f_i^{(2)}\mathbf{V}_i = \mathbf{0}. \tag{8}$$

Derivatives on independent variables are represented by following multiscale expansions [15, 16]:

$$\frac{\partial}{\partial t} = \varepsilon\frac{\partial}{\partial t_1} + \varepsilon^2\frac{\partial}{\partial t_2}, \quad \frac{\partial}{\partial x_\alpha} = \varepsilon\frac{\partial}{\partial x_{1\alpha}}, \tag{9}$$

where t_1, t_2, x_{11}, x_{12} are the new variables.

Macrovariables, such as density $\rho(t,\mathbf{r})$ and velocity $\mathbf{u}(t,\mathbf{r})$ are represented by following formulas [16]:

$$\rho(t,\mathbf{r}) = \sum_{i=1}^{n} f_i(t,\mathbf{r}), \quad \rho(t,\mathbf{r})\mathbf{u}(t,\mathbf{r}) = \sum_{i=1}^{n} \mathbf{V}_i f_i(t,\mathbf{r}). \tag{10}$$

By substitution of (6),(7),(9) into (4) and by application of (8) and (10) the following equations for macrovariables are obtained:

$$\frac{\partial \rho}{\partial t} + \frac{\partial}{\partial x_\alpha}(\rho u_\alpha) = 0, \tag{11}$$

$$\frac{\partial(\rho u_\alpha)}{\partial t} + \frac{\partial(\rho u_\alpha u_\beta)}{\partial x_\beta} = -\frac{\partial p}{\partial x_\alpha} + \nu \frac{\partial}{\partial x_\beta}\left(\rho\left(\frac{\partial u_\beta}{\partial x_\alpha} + \frac{\partial u_\alpha}{\partial x_\beta}\right)\right), \tag{12}$$

where ν is an apparent viscosity represented in case of $D2Q9$ pattern by following formulae:

$$\nu = \left(1 + \frac{A-B}{2}\right)\frac{\tau}{3}\frac{l^2}{\delta t}. \tag{13}$$

System (11)–(12) may be characterized as *quasihydrodynamical* system for modelling of semi-compressible flows. In case of incompressible regime $\rho = const$ and system (11)–(12) is rewritten in form of Navier — Stokes system:

$$\frac{\partial u_\alpha}{\partial x_\alpha} = 0, \qquad \frac{\partial u_\alpha}{\partial t} + u_\beta \frac{\partial u_\alpha}{\partial x_\beta} = -\frac{1}{\rho}\frac{\partial p}{\partial x_\alpha} + \nu \Delta u_\alpha.$$

As it can be seen, expression (13) is different from the expression for real viscosity (5) due to the presence in (13) of the fictitiuos term $\tau l^2(A-B)/(6\delta t)$, which is named numerical viscosity. This fact must be considered in computations of real flows.

Necessary stability conditions may be obtained from the condition $\nu \geq 0$ and are presented by following theorem:

Theorem. Let system (4) is stable on initial conditions. So the following conditions on coefficients σ and τ are realized:

1) For system 1:

$$\sigma \geq \frac{1}{2} - \tau,$$

is valid $\forall\sigma \in [1/2, 1]$, $\forall\tau > 0$.

2) For system 2:

$$\sigma \leq \tau + \frac{1}{2},$$

is valid $\forall\sigma \in [0, 1/2]$, $\forall\tau > 0$.

3) For system 3:

$$\sigma \geq -2\tau,$$

is valid $\forall\sigma \in [0, 1]$, $\forall\tau > 0$.

4) For system 4:

$$\sigma \leq 2\tau,$$

is valid $\forall\sigma \in [0, 1]$ when $\tau > 1/2$ $\forall\sigma \in [0, 2\tau]$ when $\tau < 1/2$.

5) For system 5:

$$\sigma \leq 1 + 2\tau,$$

is valid $\forall\sigma \in [0, 1]$ when $\forall\tau > 0$.

6) For system 6:

$$\sigma \geq 1 - 2\tau,$$

is valid $\forall\sigma \in [0, 1]$ when $\tau > 1/2$ and $\forall\sigma \in [1-2\tau, 1]$ when $0 < \tau < 1/2$, when $\sigma = 1$ is valid $\forall\tau > 0$.

IOP Conf. Series: Journal of Physics: Conf. Series **929** (2017) 012032 doi:10.1088/1742-6596/929/1/012032

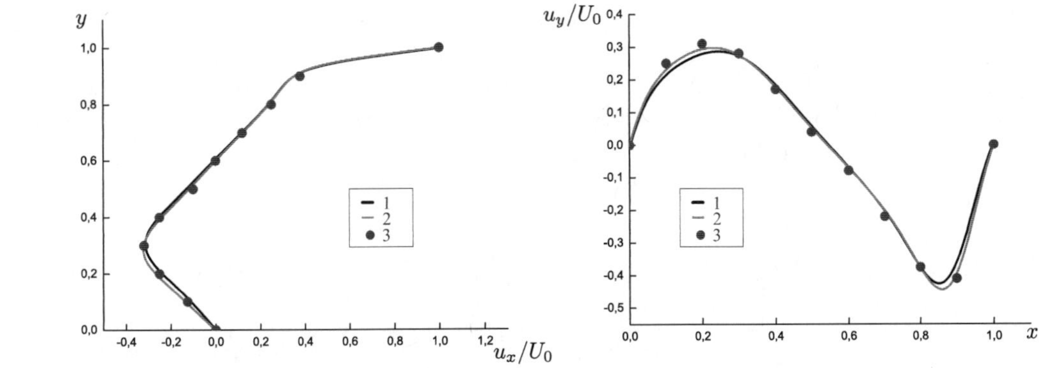

Figure 1. The results of computation at $Re = 400$. Black line — the case of $\sigma = 1$, red line — case of $\sigma = 1/2$, results from [17] are represented by blue dots

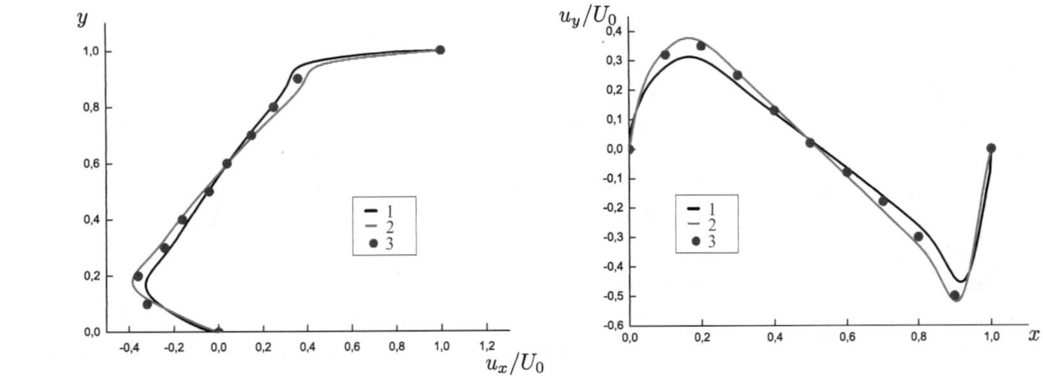

Figure 2. The results of computation at $Re = 1000$. Black line — the case of $\sigma = 1$, red line — case of $\sigma = 1/2$, results from [17] are represented by blue dots

4. Numerical solution of lid-driven cavity flow problem

The problem is stated in rectangular domain $\{(x, y)|x \in [0, P], y \in [0, P], P > 0\}$. Boundary conditions are written as [17]:

$$u_x(t, x, 0) = u_y(t, x, 0) = 0, \quad u_x(t, x, P) = U_0, \quad u_y(t, x, P) = 0, \quad x \in [0, P],$$

$$u_x(t, 0, y) = u_y(t, 0, y) = u_x(t, P, y) = u_y(t, P, y) = 0, \quad y \in [0, P),$$

where $U_0 = const \neq 0$. Boundary conditions for distribution functions f_i reproduced presented boundary conditions on velocity are realized by the approach described in [18].

Results of computations on the space grid of 100×100 nodes are compared with the results from [17] at cases of different values of Reynolds number Re. The scheme based on system 1 is applied at cases of $\sigma = 1$ (first order) and $\sigma = 1/2$ (second order). This results are presented at fig. 1–2. The systems of nonlinear equations at every grid node are solved by Newton method. Two schemes obtained from system 1 at cases of $\sigma = 1$ and $1/2$ are considered. At fig. 1 case of Reynolds number $Re = 400$ is considered at case of 6000 time nodes, when at $\sigma = 1$ $\tau = 0.1126$ and at $\sigma = 1/2$ $\tau = 0.6126$, at fig. 2 — the case of $Re = 1000$ with 1000 time nodes, when at $\sigma = 1$ $\tau = 0.2354$ and at $\sigma = 1/2$ $\tau = 0.7354$. The plots of u_x/U_0 are presented at the line

$\{x = 0.5P, y \in [0, P]\}$, plots of u_y/U_0 — at the points of line $\{x \in [0, P], y = 0.5P\}$. As it can be seen, the obtained results are demonstrated the validity of presented stability conditions. At table 1 the values of the norm of the vector of mean square errors for u_x and u_y are presented. As it can be seen, the results obtained by scheme of second order are closest to the null at all values of *Re* and may be considered as a numerical solution obtained with higher order of accuracy.

Table 1. The values of the norm of the vector of mean square errors of the velocity vector components at different values of *Re*.

σ	$Re = 50$	$Re = 100$	$Re = 400$	$Re = 1000$
1	$5.0514 \cdot 10^{-4}$	$1.1237 \cdot 10^{-4}$	$5.1427 \cdot 10^{-4}$	$2.3744 \cdot 10^{-3}$
1/2	$5.8641 \cdot 10^{-5}$	$3.0376 \cdot 10^{-5}$	$3.4718 \cdot 10^{-5}$	$2.2343 \cdot 10^{-4}$

5. Conclusion

In the presented paper parametrical LBE's are considered. The formulae for the apparent viscosity is obtained by Chapman — Enskog asymptotic expansion. Obtained expression represents viscosity as a function of the relaxation parameter and parameter of the LBE's. Necessary stability conditions in form of inequalities are derived from the non-negativity condition of the apparent viscosity. The validity of the stability conditions are demonstrated by the solution of lid-driven cavity flow problem.

The proposed idea may be realized in analytical investigation of parametrical LBE's with many time steps, obtained by application of quadrature formulas of different algebraic orders.

Acknowledgements

The reported study was funded by Russian Foundation of Basic Research according to research project No. 16-31-00021 mol_a.

References

[1] Chen S and Doolen J D 1998 *Ann. Rev. Fluid Mech.* **30** 329
[2] Krivovichev G V and Panchenko M S 2016 *Appl. Math. Sci.* **10** 947
[3] Krivovichev G V 2014 *Appl. Math. Sci.* **8** 1605
[4] Hong P-Y, Huang L-M, Lin L-S and Lin C A 2015 *Computers and Fluids* **110** 1
[5] Ye Y, Li K, Wang Y and Deng T 2015 *Computers and Fluids* **110** 114
[6] Cao N, Chen S, Jin S and Martinez D 1997 *Phys. Rev. E* **55(1)** 55R21
[7] Li Y, LeBoeuf E J and Basu P K 2004 *Phys. Rev. E* **69** 065701
[8] Seta S and Takahashi R 2002 *J. Stat. Phys.* **7** 557
[9] Asinari P 2006 *Phys. Rev. E* **73** 056705
[10] Bernaschi M, Succi S and Chen H 2001 *J. Sci. Comp.* **16** 135
[11] Eshlaghi M and Felicelli S D 2012 *Int J Heat Mass Transf.* **55** 2420
[12] Sankaranarayanan B, Shan X, Kevrekidis I G and Sundaresan S 2002 *J. Fluid Mech.* **452** 61
[13] Krivovichev G V 2014 *Appl. Math. Sci.* **8** 5003
[14] He X and Luo L-S 1997 *Phys. Rev. E* **55(6)** R6333
[15] Sofonea V 2003 *J. Comp. Phys.* **184**, 422
[16] Wolf-Gladrow D A 2000 *Lattice-Gas Cellular Automata and Lattice Boltzmann Models* (Berlin: Springer)
[17] Ghia U, Ghia K N and Shin C T 1982 *J. Comp. Phys.* **48** 387
[18] Le Coupanec E and Verschaeve J C G 2011 *Math. Comp. Simul* **81** 2632.

IOP Conf. Series: Journal of Physics: Conf. Series **929** (2017) 012033 doi:10.1088/1742-6596/929/1/012033

Analysis of parametric finite-difference schemes for the system of linear advection equations

Elena S Marnopolskaya, Gerasim V Krivovichev

Faculty of Applied Mathematics and Control Processes Saint Petersburg State University, 7/9 Universitetskaya nab., Saint Petersburg, 199034, Russian Federation

E-mail: marnlena@yandex.ru, g.krivovichev@spbu.ru

Abstract. Explicit finite-difference schemes for the linear advection equations are considered. The schemes may be used in the advective step of the splitting method for the kinetic equations. The time derivative is approximated with first order by special asymmetric approximation. Schemes from first to fourth orders of approximation on space are constructed. Stability analysis is realized by von Neumann method. Stability criteria in the form of inequalities on Courant parameter are obtained. As it is demonstrated, obtained schemes demonstrate better stability properties in cases of high order asymmetric approximations in comparison with well-known explicit schemes.

1. Introduction

Mathematical models, based on Boltzmann or Bhatnaghar — Gross — Krook (BGK) kinetic equations are considered for the description of fluid and gas flows at systems with different space scales. These models are considered in many practical problems, e. g. in the construction of Knudsen pump [1] or in porous systems, where the size of pores varies from nanometer to millimeter, such as coal [2]. In last decades lattice Boltzmann method which is based on the finite-difference (FD) schemes for BGK equations has established itself as a powerful tool for numerical modelling of fluid and gas flows on multiprocessor and multicore systems [3, 4].

Many of numerical algorithms for kinetic equations are based on the splitting on physical processes. At every time step this method is realized by two steps named advection and collision. During the advection step linear advection equation is solved. During the collision step equation with nonlinear collision term is solved. The main advantage of the splitting method is based on the opportunities for the stability improvement by the application of various FD schemes on its steps. As it is demonstrated by S. V. Bogomolov [5] and T. Ohwada [6], the splitting procedure is realized only with first order on time step.

In the presented paper FD schemes for the advective step of splitting procedure are investigated. In section 2 modified splitting method for the kinetic equations with discrete velocities is considered. In section 3 FD schemes for linear advection equations are constructed. In section 3 stability analysis of parametric schemes is performed and necessary stability conditions are formulated. Some concluding remarks are made in section 4.

2. Splitting method

Let us consider the kinetic equation without body force term:

$$\frac{\partial f}{\partial t} + \mathbf{v}\nabla f = I(f), \tag{1}$$

where $f = f(t, \mathbf{r}, \mathbf{v})$ is a particle distribution function, t is a time, $\mathbf{r}$ is a vector of space variables, $\mathbf{v}$ is a velocity, $I(f)$ is a collision term. After the discretization of (1) by discrete velocity method [7] on velocity grid $\mathbf{v}_i$, $i = \overline{1, n}$, the following system of equations for $f_i(t, \mathbf{r}) = f(t, \mathbf{r}, \mathbf{v}_i)$ is obtained:

$$\frac{\partial f_i}{\partial t} + \mathbf{v}_i\nabla f_i = I_i(\mathbf{f}), \tag{2}$$

where $\mathbf{f} = (f_1, \ldots, f_n)$.

The splitting method for solution of (2) is realized on time interval $[t_j, t_{j+1}]$, when $\Delta t = t_{j+1} - t_j < \delta t$, where δt is a mean free time. The method consists of two steps:

1. Advection. The following system of linear advection equations is solved:

$$\frac{\partial \tilde{f}_i}{\partial t} + \mathbf{v}_i\nabla \tilde{f}_i = 0, \tag{3}$$

with the initial condition: $\tilde{f}_i(t_j, \mathbf{r}) = f_i(t_j, \mathbf{r})$.

2. Collision. The system without space derivatives is considered:

$$\frac{\partial \widehat{f}_i}{\partial t} = I_i(\widehat{\mathbf{f}}), \tag{4}$$

with the initial condition: $\widehat{f}_i(t_j, \mathbf{r}) = \tilde{f}_i(t_{j+1}, \mathbf{r})$. As it is demonstrated in [5, 6], this computational procedure provide to obtain results only of *first* order on time step Δt. Results on space steps may be obtained with higher order by means of special approximations constructed by finite-difference, finite-element or finite-volume techniques [8, 9, 10, 11]. Advantage of splitting method consist with the opportunities for improvement of its stability by application of various discretization techniques.

Solution of the problem for (4) may be realized by implicit Euler method, which is A-stable:

$$\widehat{f}_i(t_{j+1}, \mathbf{r}) = \widehat{f}_i(t_j, \mathbf{r}) + \Delta t I_i(\widehat{\mathbf{f}}(t_{j+1}, \mathbf{r})). \tag{5}$$

This scheme discretise (4) only with first order on time step. But the accuracy order of splitting method is equal to unity, so the application of schemes of higher order for (4) does not make sense.

Nonlinear system (5) may be solved by following iterative method:

$$\widehat{f}_i^{(s+1)}(t_{j+1}, \mathbf{r}) = \widehat{f}_i(t_j, \mathbf{r}) + \Delta t I_i(\widehat{\mathbf{f}}^{(s)}(t_{j+1}, \mathbf{r})), \tag{6}$$

where $s = 0, 1, 2, \ldots$ is the iteration number.

One of the main difficulties (except the convergence) of the realization of (6) is a choice of initial approximation $\widehat{f}_i^{(0)}(t_{j+1}, \mathbf{r})$ at every j. To avoid this problem, the following modification of the splitting method is proposed — at every time step instead of the solution of the Cauchy problem for (4) the following *sequence* of the problems is solved:

$$\frac{\partial \widehat{f}_i^{(q)}}{\partial t} = I_i(\widehat{\mathbf{f}}^{(q)}), \quad t \in (t_j, t_{j+1}], \quad q = 1, \ldots, \quad \widehat{f}_i^{(q+1)}(t_j, \mathbf{r}) = \widehat{f}_i^{(q)}(t_{j+1}, \mathbf{r}), \quad q = 0, \ldots, \tag{7}$$

where $\widehat{f}_i^{(0)}(t_{j+1},\mathbf{r}) = \tilde{f}_i(t_{j+1},\mathbf{r})$. The solution of these problems is realized by implicit Euler method (5)–(6), but with *only one* iteration step with initial approximation $\widehat{f}_i^{(q+1,0)}(t_{j+1},\mathbf{r}) = \widehat{f}_i^{(q+1)}(t_j,\mathbf{r}) = \widehat{f}_i^{(q)}(t_{j+1},\mathbf{r})$:

$$\widehat{f}_i^{(q+1)}(t_{j+1},\mathbf{r}) = \widehat{f}_i^{(q+1,0)}(t_{j+1},\mathbf{r})+\Delta t I_i(\widehat{\mathbf{f}}^{(q+1,0)}(t_{j+1},\mathbf{r})) = \widehat{f}_i^{(q)}(t_{j+1},\mathbf{r})+\Delta t I_i(\widehat{\mathbf{f}}^{(q)}(t_{j+1},\mathbf{r})). \quad (8)$$

The cycle on q for (7)–(8) may be stopped, when the values of the density ρ at moment t_{j+1} stop changed. These values are computed by following formulae [12]:

$$\rho^{(q)}(t_{j+1},\mathbf{r}) = \int f(t_{j+1},\mathbf{r},\mathbf{v})d\mathbf{v} \approx \sum_{i=1}^{n} f_i^{(q)}(t_{j+1},\mathbf{r}).$$

So the convergence condition of the procedure (7)–(9) can be formulated by following inequality:

$$\frac{||\rho^{(q+1)}(t_{j+1},\mathbf{r}) - \rho^{(q)}(t_{j+1},\mathbf{r})||}{||\rho^{(q)}(t_{j+1},\mathbf{r})||} \le Tol,$$

where value of Tol is set in advance. It must be noted, that other approaches for the completion of the iteration procedure may be considered.

3. Finite-difference schemes for the linear advection equation

As it can be seen, equations of the system (3) are independent of each other. So difference schemes for scalar linear advection equation can be considered. This equation in the one-dimensional case is written as:

$$\frac{\partial u}{\partial t} + c\frac{\partial u}{\partial x} = 0, \quad (9)$$

where $u = u(t,x)$, $c = const$ is a velocity of the advection along the characteristic. In all further formulas the case of $c > 0$ is considered.

The following approximation of time derivative is proposed:

$$\frac{\partial u}{\partial t}(t_j,x_n) \approx \frac{u_n^{j+1} - \frac{1}{2}\left(u_n^j + u_n^{j-2}\right)}{2\Delta t}, \quad (10)$$

where $u_n^j \approx u(t_j,x_n)$, t_j is a node of time grid constructed with step Δt, x_n is a node of grid on space variable constructed with step h. Scheme with (10) approximate eq. (9) only with the *first* order on time step. As it is demonstrated in [13, 14] stability properties of schemes for nonlinear BGK equations with approximation (10) are better, than for the schemes with second order central approximation.

Let us consider FD schemes for (9) obtained after approximation on space variable x in following form:

$$u_n^{j+1} = \frac{1}{2}\left(u_n^j + u_n^{j-2}\right) - 2\Delta t\mathcal{L}_h(u_n^j), \quad (11)$$

where $\mathcal{L}_h$ is a discrete operator, which approximate term $\partial u/\partial x$ at node (t_j,x_n). The following representations of this term are considered:

1) First order backward difference:

$$\mathcal{L}_h(u_n^j) = \frac{c}{h}\left(u_n^j - u_{n-1}^j\right). \quad (12.1)$$

2) Second order differences:

$$\mathcal{L}_h(u_n^j) = \frac{c}{2h}\left(u_{n+1}^j - u_{n-1}^j\right), \tag{12.2}$$

$$\mathcal{L}_h(u_n^j) = \frac{c}{2h}\left(3u_n^j - 4u_{n-1}^j + u_{n-2}^j\right). \tag{12.3}$$

3) Third order difference:

$$\mathcal{L}_h(u_n^j) = \frac{c}{6h}\left(u_{n-2}^j - 6u_{n-1}^j + 3u_n^j + 2u_{n+1}^j\right). \tag{12.4}$$

4) Fourth order differences:

$$\mathcal{L}_h(u_n^j) = \frac{c}{12h}\left(u_{n-2}^j - 8u_{n-1}^j + 8u_{n+1}^j - u_{n+2}^j\right), \tag{12.5}$$

$$\mathcal{L}_h(u_n^j) = \frac{c}{12h}\left(3u_{n+1}^j + 10u_n^j - 18u_{n-1}^j + 6u_{n-2}^j - u_{n-3}^j\right). \tag{12.6}$$

According to the approach proposed in [15], let us approximate term $c\partial u/\partial x$ in following form:

$$c\frac{\partial u}{\partial x}(t_j, x_n) \approx \mathcal{L}_h(u_n^j) = \varepsilon F_1 + (1-\varepsilon)F_2,$$

where F_1 and F_2 are the FD approximations of this term and $\varepsilon \in [0,1]$ is a dimensionless parameter. So schemes with parametric coefficients represented by following formulae are constructed:

$$u_n^{j+1} = \frac{1}{2}\left(u_n^j + u_n^{j-2}\right) - 2\Delta t\left(\varepsilon F_1 + (1-\varepsilon)F_2\right), \tag{13}$$

where F_1 and F_2 are presented by formulas (12.1)–(12.6). The existence of the parameter ε in the right part of (13) provide possibilities to regulate different properties of the scheme, such as stability, numerical dispersion and numerical dissipation [16].

4. Stability analysis

Stability analysis is realized by application of von Neumann method, which is proposing to obtain *necessary* stability conditions of multistep schemes represented by (11) and (13). The method is based on the following representation of the solution:

$$u_n^j = \lambda^j(\varphi)\exp(in\varphi), \tag{14}$$

where $i^2 = -1$, $\varphi \in [0, 2\pi)$ and λ is a spectral function. Stability condition is represented by following inequality:

$$|\lambda(\varphi)| \le 1, \quad \forall\varphi \in [0, 2\pi).$$

After the substitution of (14) into (11) or (13), the following third-order equation on λ is obtained:

$$\lambda^3 + a\lambda - \frac{1}{2} = 0, \tag{15}$$

where $\gamma = c\Delta t/h$ is a Courant number, $a = a(\gamma, \varphi)$ for schemes from (11) and $a = a(\gamma, \varphi, \varepsilon)$ for schemes represented by (13). Roots of eq. (15) are obtained by Cardano's formulas.

IOP Conf. Series: Journal of Physics: Conf. Series **929** (2017) 012033 doi:10.1088/1742-6596/929/1/012033

4.1. Stability conditions

The following stability conditions are obtained for the schemes represented by eq. (11): $\gamma \leq 0.25$ for the case of approximation (12.1), $\gamma \leq 0.5$ for (12.2), $\gamma \leq 0.12$ for (12.3), $\gamma \leq 0.25$ for (12.4), $\gamma \leq 0.36$ for (12.5) and $\gamma \leq 0.15$ for (12.6).

Stability criteria for the schemes presented by (13) are obtained in following form: $\gamma \leq \tilde{\gamma}(\varepsilon)$, where $\tilde{\gamma}(\varepsilon)$ represent the function, which approximate the boundary of the stability domain in the space of parameters ε and γ. The expressions for $\tilde{\gamma}(\varepsilon)$ are obtained as a polynomials of low degree by the application of Matlab's procedure polyfit() and are written as:

$$\tilde{\gamma}(\varepsilon) = -0.11\varepsilon^3 + 0.32\varepsilon^2 - 0.46\varepsilon + 0.49,$$

for the case of the scheme, where F_1 is approximated by (12.1), F_2 — by (12.2),

$$\tilde{\gamma}(\varepsilon) = -0.47\varepsilon^3 + 1.15\varepsilon^2 - 1.05\varepsilon + 0.49,$$

for the case, where F_1 is approximated by (12.3) and F_2 by (12.2),

$$\tilde{\gamma}(\varepsilon) = -0.11\varepsilon + 0.36,$$

for the case, where F_1 is approximated by (12.4) and F_2 by (12.5),

$$\tilde{\gamma}(\varepsilon) = 0.1\varepsilon^2 - 0.31\varepsilon + 0.35,$$

for the case, where F_1 is approximated by (12.5) and F_2 by (12.6).

4.2. Comparison with other schemes

Despite the fact that an accuracy order of the schemes (11) and (13) on time step is equal to unity, these schemes provide more opportunities to practical computations. It corresponds to their stability for the cases of high order asymmetric approximations on space variable. Let us demonstrate this fact in comparison with well known schemes.

For the explicit schemes of the first order on time step with following representation of time derivative:

$$\frac{\partial u}{\partial t}(t_j, x_n) \approx \frac{u_n^{j+1} - u_n^j}{\Delta t},$$

only the scheme with approximation (12.1) is stable with the stability condition $\gamma \leq 1$. Schemes with higher orders of approximation on spatial steps are unstable [17], while the proposed schemes are conditionally stable.

In the cases of the schemes with the second order of approximation on time step, when:

$$\frac{\partial u}{\partial t}(t_j, x_n) \approx \frac{u_n^{j+1} - u_n^{j-1}}{2\Delta t},$$

only the schemes with *symmetric* approximations (12.2) and (12.5) (with stability conditions $\gamma \leq 1$ and $\gamma \leq 0.75$ respectively) are stable. Schemes with other approximations are unstable. So, the presented schemes (11) and (13) with approximations (12.1)–(12.6) provided opportunities to realize calculations with high orders on spatial steps in case of *asymmetric* spatial approximations. It may be useful in the realization of splitting method for kinetic equations.

5. Conclusion

Explicit finite-difference schemes for the linear advection equations are considered. The schemes may be used in the advective step of the splitting method for the kinetic equations. The time derivative is approximated with first order by special asymmetric approximation. Schemes from first to fourth orders of approximation on space are constructed. Stability analysis is realized by von Neumann method. Stability criteria in the form of inequalities on Courant parameter are obtained. As it is demonstrated, obtained schemes demonstrate better stability properties in cases of high order asymmetric approximations in comparison with well-known explicit schemes.

Acknowledgements

The reported study was funded by Russian Foundation of Basic Research according to research project No. 16-31-00021 mol_a.

References

[1] Dodulad O I, Kloss Yu Yu, Savichkin D O and Tcheremissine F G 2014 *Vacuum* **109** 360
[2] Rodrigues C F and de Sousa M J L 2002 *Int. J. Coal Geol.* **48** 245
[3] Krivovichev G V and Panchenko M S 2016 *Appl. Math. Sci.* **10** 947
[4] Hong P-Y, Huang L-M, Lin L-S and Lin C A 2015 *Computers and Fluids* **110** 1
[5] Bogomolov S V 1988 *USSR Comp. Math. Math. Phys.* **28(1)**, 79
[6] Ohwada T 1998 *J. Comp. Phys.* **139** 1
[7] Broadwell J E 1964 *J. Fluid Mech.* **19** 401
[8] Sofonea V 2003 *J. Comp. Phys.* **184** 422
[9] Krivovichev G V 2014 *Appl. Math. Sci.* **8** 1605
[10] Krivovichev G V 2014 *Proceedings of Int. Conf. on Comp. Tech. in Phys. and Eng. Appl, ICCTPEA 2014*, 83
[11] Patel S and Lee T 2016 *J. Comp. Phys.* **327** 799
[12] Wolf-Gladrow D A 2000 *Lattice-Gas Cellular Automata and Lattice Boltzmann Models* (Berlin: Springer)
[13] Mikheev S A and Krivovichev G V 2014 *Proceedings of Int. Conf. on Comp. Tech. in Phys. and Eng. Appl, ICCTPEA 2014*, 118
[14] Krivovichev G V and Mikheev S A 2015 *Proceedings of Int. Conf. "Stabil. Contr. Proc." in Memory of V.I. Zubov (SCP)*, 399
[15] Guo Z and Zhao T S 2005 *Phys. Rev. E* **71** 026701
[16] Krivovichev G V and Marnopolskaya E S 2016 *Appl. Math. Sci.* **10** 2381
[17] Kalitkin N N 1978 *Numerical Methods* (Moscow: Nauka) (in Russian)

IOP Conf. Series: Journal of Physics: Conf. Series **929** (2017) 012034 doi:10.1088/1742-6596/929/1/012034

Functional differential equations of neutral type with integrable weak singularity: hydrogen thermal desorption model

Yury V Zaika and Ekaterina K Kostikova

Institute of Applied Mathematical Research, Karelian Research Centre, Russian Academy of Science, 11 Pushkinskaya street, Petrozavodsk, 185910, Russia

E-mail: zaika@krc.karelia.ru

Abstract. One of the technological challenges for hydrogen materials science (including the ITER project) is the currently active search for structural materials with various potential applications that will have predetermined limits of hydrogen permeability. One of the experimental methods is thermal desorption spectrometry (TDS). A hydrogen-saturated sample is degassed under vacuum and monotone heating. The desorption flux is measured by mass spectrometer to determine the character of interactions of hydrogen isotopes with the solid. We are interested in such transfer parameters as the coefficients of diffusion, dissolution, desorption. The paper presents a thermal desorption functional differential equations of neutral type with integrable weak singularity and a numerical method for TDS spectrum simulation, where only integration of a nonlinear system of low order ordinary differential equations (ODE) is required. This work is supported by the Russian Foundation for Basic Research (project 15-01-00744).

1. Introduction

Interaction of hydrogen and its isotopes with solids has many applications [1–5]. It is sufficient to mention power engineering, protection of metals from hydrogen corrosion, design of chemical reactors, rocket and missile engineering. Hydrides help to retain substantial amounts of hydrogen: environmental-friendly energy source. Hence high expectations are attached to hydrogen batteries and motors that avoid using high pressures or low temperatures. Some special topics are considered in [6, 7]. Enthusiasts speak not only of hydrogen energy but also of hydrogen economy [4]. Transfer parameters depend on the process characteristics of producing the material batch, and one needs effective algorithms for processing measured curve instead of focusing on 'tabular data'. We focus on the thermal desorption method and take into account only the limiting factors and the information capacities of the TDS experiment explicated below.

2. Mathematical model of TDS-experiment

We consider hydrogen transfer through a test metal or alloy sample (plate of thickness ℓ). Suppose heating is relatively slow and uniform so that the diffusion flow can be assumed to be proportional to the concentration gradient. The material is sufficiently homogeneous to neglect interaction of H with traps (micro-defects). Let us assume a standard model for diffusion:

$$c_t(t,x) = D(T)c_{xx}(t,x), \quad (t,x) \in Q_{t_*}, \tag{1}$$

where t is time, $Q_{t_*} = (0,t_*) \times (0,\ell)$; $c(t,x)$ is concentration of dissolved atomic hydrogen. Diffusion coefficient D is an Arrhenius function of temperature $T(t)$: $D = D_0 \exp\{-E_D/[RT(t)]\}$, D_0 and E_D are the preexponential factor and activation energy.

Suppose the plate contacts with gas H_2 and surface is potential barrier (see [2, pp. 177–206]). Taking into account (de)sorption processes, we obtain the following boundary conditions:

$$c(0,x) = \bar{c}(x), \;\; x \in [0,\ell], \;\; t \in [0,t_*], \tag{2}$$

$$c_0(t) = g(T)q_0(t), \;\; c_\ell(t) = g(T)q_\ell(t), \tag{3}$$

$$\dot{q}_{0,\ell}(t) = \mu s(T)p_{0,\ell}(t) - b(T)q^2_{0,\ell}(t) \pm Dc_x|_{x=0,\ell}\,, \tag{4}$$

$b(T) = b_0 \exp\{-E_b/[RT]\}$, $s(T) = s_0 \exp\{-E_s/[RT]\}$. Here $c_0(t) \equiv c(t,0)$, $c_\ell(t) \equiv c(t,\ell)$ are boundary bulk concentrations of diffusing atomic hydrogen; $q_0(t)$, $q_\ell(t)$ are concentrations on the surfaces ($x = 0,\ell$); $g(T)$ is the coefficient of local equilibrium between concentrations on the surface and near-surface bulk; μ is the gas kinetic coefficient; $s(T)$ is the parameter indicating that only a small part of incident hydrogen find its way on the surface (for brevity, we call s the adhesion coefficient and keep in mind that this coefficient describes the result of concurrent physisorption, dissociation, and chemisorption of molecular gas on the surface); $p_0(t)$, $p_\ell(t)$ are pressures of molecular gas. Although hydrogen can be both in atomic and molecular forms, we use atoms as the unit of amount: $[c] = 1/\text{cm}^3$, $[q] = 1/\text{cm}^2$, $[Dc_x]=[J]=1/\text{cm}^2\text{s}$ $\big(J=bq^2\big)$. The quick dissolution model (3) is obtained from the more universal formulas

$$k^-(T)\big[1 - c_{0,\ell}(t)c^{-1}_{\max}\big]q_{0,\ell}(t) - k^+(T)\big[1 - q_{0,\ell}(t)q^{-1}_{\max}\big]c_{0,\ell}(t) = \mp D(T)c_x\big|_{0,\ell}\,.$$

Coefficients k^-, k^+ describe the intensity of the dissolution processes in the bulk and transfer to surface. If these processes are significantly faster than diffusion (for considered temperature range) then it is assumed that $Dc_x \approx 0$. Hence (if $c,q \ll \max$) we obtain (3), where $g = k^-/k^+$. If the surface is isotropic ($E_{k^-} \approx E_{k^+}$) then the coefficient $g(T)$ has weak temperature dependence. We can formally use the Arrhenius equation $g(T) = g_0 \exp\{-E_g/[RT\}]$, $E_g = E_{k^-} - E_{k^+}$. Here activation energy E_g is not necessarily positive. The value $\theta(t) = q(t)/q_{\max}$ means the degree of the surface covering. Flux density of atom H adsorption (dissociative hydrogen chemisorption on the surface) can be simulated by formula $\mu s(T)p(t)[1-\theta(t)]^2$ in the balance equations (4). But we have $\theta \ll 1$ for low level of hydrogen. In the context of current TDS-experiment it is convenient to take $[p]$ = Torr and so $\mu(T) = (2\pi mkT)^{-1/2} \approx 2.484 \cdot 10^{22}/\sqrt{T}$. Here we have $[\mu] = 1_{H_2}/(\text{Torr}\,\text{cm}^2\text{s})$, $[T] = \text{K}$, k is the Boltzmann constant, m is mass of a hydrogen molecule.

Saturation means that pressure $\bar{p}$ of H_2 on both sides of the plate is constant at $T = \text{const}$. Then (after a time) saturation concentration ($\bar{c}$) of dissolved atomic diffusivity hydrogen is established. Assuming zero derivatives in (4), we obtain $\bar{c} = \Gamma\sqrt{\bar{p}}$, $\Gamma \equiv g\sqrt{\mu s/b}$ ($\bar{c} = g\bar{q}$). Consequently the model is according to the Sieverts law adequacy range ($\bar{c} \propto \sqrt{\bar{p}}$).

Let us clarify the TDS-experiment. Plate of the material is placed into a chamber. Hydrogen gas is pumped into the chamber under relatively high pressure. The sample is heated to hasten sorption until the sample absorbs enough hydrogen (equilibrium gas saturation). Then heating current is shut off and the material is cooled. In such a case, the physico-chemical processes slow down sharply and significant amount of hydrogen remains inside the sample. The sample is monotonically and rather slowly reheated in vacuum. The heating rate is low and constant. Molecular hydrogen pressure in the chamber is measured by a mass-spectrometer. The desorption flux from the surface is defined by pressure. The desorption flux density is denoted by $J(t) = b(t)q^2(t)$ (usually $J(t) \propto p(t)$). From now on the contracted notation will be used for simplification: $b(t) \equiv b(T(t))$, $D(t) \equiv D(T(t))$, $s(t) \equiv s(T(t))$. Parameter g of rapid dissolution (local surface-bulk equilibrium) usually is taken to be constant ($E_{k^-} = E_{k^+}$) near the TDS-peak. Symmetry conditions are fulfilled: $p = p_{0,\ell}$, $q = q_{0,\ell}$, $c_0 = c_\ell$, $\bar{c}(x) = \bar{c} = \text{const}$.

IOP Conf. Series: Journal of Physics: Conf. Series **929** (2017) 012034 doi:10.1088/1742-6596/929/1/012034

3. Functional differential TDS-equation

Suppose the TDS-degassing model: $r(t) \equiv \mu s(t)p(t)$, $J(t) \equiv b(t)q^2(t)$,

$$c_t = D(t)c_{xx}\,,\ c(0,x) = \bar{c},\ c_{0,\ell}(t) = gq(t),\ \dot{q}(t) = r(t) - b(t)q^2(t) + D(t)c_x(t,0).$$

Vacuum system is assumed to be powerful enough. Hence resorption is neglected ($r(t) = 0$) for degassing. Values of $J(t)$ are input data for the inverse problem of parametric identification.

In the paper we only present numerical simulation of TDS-spectrum $J = J(T)$. Heating is usually linear ($T(t) = T_0 + vt$). Heating rate v is less than 1 K/s. Heating is stopped (although degassing is continued) if upper temperature limit $T(t) = T_{\max}$ is reached.

Let us formulate the mathematical problem. Difference schemes for the boundary-value problem (with different types of defects) solution was presented in [8]. But a curve of TDS-spectrum is determined by surface concentration ($J = bq^2$). Hence, we try to avoid the iterative solution of the boundary-value problem for the current approximations of the model parameters D_0, E_D, b_0, E_b, s_0, E_s, g. For this aim we perform the problem transformation. Finally, the modeling requires only integration of a system of low order ordinary differential equations (ODE).

We replace time by $t' = \int_0^t D ds$ (keep previous notation t):

$$c_t(t,x) = c_{xx}(t,x),\ c(0,x) = \bar{c},\ c_{0,\ell} = gq(t), \tag{5}$$

$$c_x|_0 = -c_x|_\ell = \dot{q}(t) + [J(t) - r(t)]D^{-1}(t). \tag{6}$$

Let $q(t)$ be functional parameter, formula (6) is additional relation to the linear problem (5). Let us change variables so that boundary conditions in (5) become homogeneous:

$$\hat{c} = c(t,x) - gq(t),\ \hat{c}_t(t,x) = \hat{c}_{xx}(t,x) + f(t),\ f(t) = -g\dot{q}(t),\ \hat{c}(0,x) = \hat{\varphi}(x) = 0,\ \hat{c}|_{0,\ell} = 0.$$

Write the solution using function of instantaneous source (Green's function):

$$\hat{c}(t,x) = \int_0^t \int_0^\ell G_1(x,\xi,t-\tau)f(\tau)\,d\xi\,d\tau,\ G_1(x,\xi,t) = \frac{2}{\ell}\sum_{n=1}^{\infty}\exp\Big\{-\frac{n^2\pi^2}{\ell^2}t\Big\}\sin\frac{n\pi x}{\ell}\sin\frac{n\pi\xi}{\ell}.$$

Dynamical boundary conditions contain $\hat{c}_x(t,0)$:

$$\hat{c}_x|_0 = -\frac{4g}{\ell}\int_0^t \dot{q}(\tau)\sum{}'\exp\Big\{\frac{n^2\pi^2}{\ell^2}(\tau - t)\Big\}\,d\tau,\quad \sum{}' \equiv \sum_{n=1,3,5\ldots}.$$

The series is divergent for $\tau = t$ so that term-by-term integration is meant. For original time t we obtain $c_x(t,0) = \hat{c}_x(t,0) = c_x(t,\ell) = \hat{c}_x(t,\ell)$,

$$\dot{q}(t) = r(t) - b(t)q^2(t) - \frac{4gD}{\ell}\sum{}'\int_0^t \dot{q}(\tau)\exp\Big\{-\frac{n^2\pi^2}{\ell^2}\int_\tau^t D(s)\,ds\Big\}\,d\tau. \tag{7}$$

The equation and the original boundary-value problem are equivalent. This means that the solution $q(t)$ uniquely determines the solution $c(t,x)$. An analogy to functional-differential equations of neutral type is due to impossibility to eliminate derivative $\dot{q}$ in the right side of the equation due to divergent series. We are interested in the time interval $[t_1,t_2] \subset (0,t_*)$ near activity peak of thermal desorption (measurements for t near to $0, t_*$ are uninformative). Resorption is neglected ($r(t) = 0$) for powerful vacuum system, thus the equation is simplified.

Dimensionless parameters are more convenient for numerical modeling. We use replaces: $t' = \int_0^t D(s)ds/\ell^2$, $x' = x/\ell$, $v = q/\bar{q}$ ($\bar{c} = g\bar{q}$). But notation of t is not changed. We obtain:

$$\dot{v}(t) = -\tilde{b}(t)v^2(t) - 4g\ell\sum{}'\int_0^t \dot{v}(\tau)\exp\{-n^2\pi^2[t-\tau]\}d\tau,\ v(0) = 1,\ \dot{v} \equiv \frac{dv}{dt},\ \tilde{b} \equiv \frac{b(t)\ell^2\bar{q}}{D(t)}. \tag{8}$$

Initial saturation is conducted at high temperature $T = \overline{T} = \text{const}$ and pressure $\bar{p} = \text{const}$ (to stimulate sorption). After establishing of equilibrium concentration we obtain

$$\mu(\overline{T})s(\overline{T})\bar{p} = b(\overline{T})\bar{q}^2, \ \ \bar{c} = g\bar{q} \ (g = \text{const}) \ \Rightarrow \bar{c} = g b_0^{-1/2}\sqrt{\mu s\bar{p}}\,|_{\overline{T}} \exp\{E_b[2R\overline{T}]^{-1}\},$$
$$\tilde{b}(t) = \bar{q}\, b(t)\ell^2 D^{-1} = \bar{c}\, b(t)\ell^2[gD(t)]^{-1} = \tilde{b}_0 \exp\{-E_{\tilde{b}}[RT(t)]^{-1}\},$$
$$E_{\tilde{b}} \equiv E_b - E_D, \ \tilde{b}_0 \equiv \ell^2 D_0^{-1}\sqrt{b_0\mu s\bar{p}}\,|_{\overline{T}} \exp\{E_b[2R\overline{T}]^{-1}\}.$$

Formally let $\tilde{b}(t) \equiv \tilde{b}(T(t))$ be Arrhenius parameter, here $E_b - E_D$ is activation energy. Usually $E_b > E_D$ and surface processes are faster speeded up on heating. ($\dot{b} = \ldots E_b\dot{T}$, $\dot{D} = \ldots E_D\dot{T}$), thus, $E_{\tilde{b}} > 0$. The pre-exponential factor $\tilde{b}_0$ depends on all parameters D_0, b_0, E_b, s_0, E_s (except E_D). Hence the coefficient $\tilde{b}$ changed monotonically and limited $0 < \tilde{b}^- \le \tilde{b}(t) \le \tilde{b}^+$.

Hereinafter let us suggest that the limit of surface concentration is at $q_{\max} \sim 10^{15} - 10^{16}$ (monolayer in the context of geometrical statics). Where at the stage of initial saturation for selected model parameters we have $\bar{q} > 10^{14}$, for numerical simulations we must take into account the surface filling degree. In such a case, the formula $\mu s\bar{p} = b\bar{q}^2$ is substituted with the ratio $\mu s\bar{p}\,[1 - \bar{q}q_{\max}^{-1}]^2 = b\bar{q}^2$ to estimate the initial value of $\bar{q}$.

For extracting singularity let us clarify why the quite accurate approximation of the thermal desorption functional differential equation (8) still requires a large number of terms in $\sum'$. The function

$$\Theta(s) - 4{\sum}' \exp\{-n^2\pi^2 s\}, \quad {\sum}' \equiv \sum_{n=1,3,5\ldots},$$

has finite values for $s > 0$. The series is rapidly converging for large s. Formally, substituting $s = 0$ (the integration variable reaches the upper limit of t) we obtain a divergent series. This can be fixed by term-by-term integration. We have $\max|z_i(t)| = O(n^{-2})$ ($n = 2i - 1$, $i \ge 1$), wherefore convergence is slow. Let us run some transformations using the theory of Jacobi theta functions. More specifically, we will focus on

$$\theta_3(t, x) = 1 + 2\sum_{n=1}^{\infty} \exp\{-n^2\pi^2 t\}\cos(2n\pi x), \ t > 0.$$

We have an alternative presentation [10] for $x = 0$: $\theta_3(t, 0) = \sum_{-\infty}^{+\infty} \exp\{-n^2/t\}/\sqrt{\pi t}$.

The series on the left is rapidly converging for large t. But the series on the right is rapidly converging for small t. Here we are only concerned with small t. Defining $\theta(t) = \sum \exp\{-\pi n^2 t\}$ ($n \in \mathbb{Z}$, $t > 0$) we obtain the formula $\theta(1/t) = \sqrt{t}\theta(t)$ or $\sqrt{t}\sum\exp\{-\pi n^2 t\} = \sum\exp\{-\pi n^2/t\}$ ($n \in \mathbb{Z}$). This relation is known as functional equation for theta function [11].

Let us run auxiliary transformations. The series is split into odd and even terms:

$$[\pi t]^{-1/2}\sum_{-\infty}^{+\infty} \exp\{-n^2/t\} = 1 + 2\sum_{n=1}^{\infty}\exp\{-n^2\pi^2 t\}$$
$$= 1 + 2{\sum}' + 2\sum_{k=1}^{\infty}\exp\{-k^2\pi^2 4t\} = 2{\sum}' + [4\pi t]^{-1/2}\sum_{-\infty}^{+\infty}\exp\{-n^2/4t\}.$$

With this in mind, the last series is subtracted from the first series and the result is doubled. The following expansion is obtained for $s > 0$:

$$\Theta(s) \equiv 4{\sum}'\exp\{-n^2\pi^2 s\} = \frac{1 - Q}{\sqrt{\pi s}}, \quad Q(s) \equiv -2\sum_{n=1}^{\infty} q^{n^2}, \ q \equiv -\exp\{-(4s)^{-1}\}.$$

The series Q is rapidly converging for small s. As $s \to +0$ we have $Q \to 0$ and the integrable singularity $\Theta \approx 1/\sqrt{\pi s}$. Since the series is alternating, we have the inequalities $Q > 0$ (more exactly $Q \in (0, 1)$) and $Q \le 2|q|$. The graph for function $Q(s)$ is an S-shaped saturation curve and $Q(1) \approx 0.999$. Using the above-described presentation of $\Theta(s)$ we obtain the TDS functional differential equation of neytral type (terminology from [9]):

$$\dot{v}(t) = -\tilde{b}(t)v^2(t) - \varkappa \int_0^t \frac{1 - Q(t-\tau)}{\sqrt{\pi[t-\tau]}} \dot{v}(\tau)\, d\tau \ \ \big(t \geq 0,\ v(0) = 1\big). \tag{9}$$

Numerical method and computer simulation

To be specific in the paper, we use experimental data for nickel and steel 12Cr18Ni10Ti [3]. The assumed values of model parameters are the following: (steel) $\ell = 0.1\,\mathrm{cm}$, $T_0 = 300\,\mathrm{K}$, $b_0 = 1.28 \times 10^{-9}\,\mathrm{cm^2/s}$, $E_b = 90 \times 10^3\,\mathrm{J/mol}$, $D_0 = 3.09 \times 10^{-4}\,\mathrm{cm^2/s}$, $E_D = 31 \times 10^3$, $g = 50\,\mathrm{cm^{-1}}$, $s_0 = 0.6$, $E_s = 110 \times 10^3$, $\bar{c} = 6 \times 10^{17}\,\mathrm{at.H/cm^3}$, $\overline{T} = 1170\,\mathrm{K}$, $\dot{T} = 0.5\,\mathrm{K}$; (nickel) $\ell = 0.1$, $T_0 = 300$, $b_0 = 1.53 \times 10^{-14}$, $E_b = 43.2 \times 10^3$, $D_0 = 7.5 \cdot \times 10^{-3}$, $E_D = 40 \times 10^3$, $g = 100$, $s_0 = 1.8 \times 10^{-2}$, $E_s = 61.4 \times 10^3$, $\bar{p} = 37.4\,\mathrm{Torr}$, $\overline{T} = 770$, $\dot{T} = 0.5$.

The main role in degassing dynamics belongs to quadratic desorption. So let us approximate the integral term in equation (9). The aftereffect horizon here is $h < 1$ ($\tau \in [t, t-h]$ for dimensionless time, $Q(1) \approx 0.999$). Let us fix $h \sim 0.3-0.4$ due to continuously differentiable function $Q(s)/\sqrt{s}$ graph and trapezoid rule for numerical integration. Then, TDS-equation (9) can be approximated by ODE system of low order one-by-one on time segments of length h.

The greatest contribution to the integral is made by the value $\dot{v}(\tau)$ ($\tau \approx t$) due to non-limited (but integrable) singularity. Thus, the quadratic approximation $\dot{v}(\tau) \approx \dot{v}(t) + A[t-\tau] + B[t-\tau]^2$ is used due to function $\dot{v}$ concavity. Let us consider the current period of time $t \in [kh, (k+1)h]$, $k \geq 0$. The conditions $\dot{v}(\tau)|_{kh} = \dot{v}(kh)$ and $\int_{kh}^t \dot{v}(\tau)\, d\tau = v(t) - v(kh)$ (here suppose $v(t) \approx v(kh) + \dot{v}(t)[t - kh]$) determine the sought values of $A(t), B(t)$ (constant on τ):

$$\dot{v}(\tau) \approx \dot{v}(t) + 2\frac{\dot{v}(t) - \dot{v}(kh)}{t - kh}[t-\tau] - 3\frac{\dot{v}(t) - \dot{v}(kh)}{[t-kh]^2}[t-\tau]^2, \quad \tau \in [kh, t]. \tag{10}$$

Represent the integral as a sum $\tau \in [0, kh]$, $\tau \in [kh, t]$. For the second additive, the singularity $\dot{v}(\tau)/\sqrt{t-\tau}$ is explicitly integrated by substituting (10). The trapezoid rule is used for the integral without singularity ($Q(+0)/\sqrt{+0} = 0$). For approximation of the first integral

$$\int_0^{kh} \frac{1 - Q(t-\tau)}{\sqrt{\pi[t-\tau]}} \dot{v}(\tau)\, d\tau = 4{\sum}' \int_0^{kh} \dot{v}(\tau) \exp\left\{-n^2\pi^2[t-\tau]\right\} d\tau$$

we use only several series terms (for definiteness $n = 1, 3$) on the right-hand side due to the exponent n^2. Thereby, the negative additives ($\dot{v} < 0$) are dropped. Let us compensate for that by replacing kh with t, and introduce in addition variables $w_{1,2}$ (see below). As the result we obtain the ODE system:

$$\dot{v}(t) = -\hat{b}(t)v^2(t) + \frac{\hat{\varkappa}}{\sqrt{\pi}}\left[\frac{2}{15} + \frac{Q(t-kh)}{2}\right]\sqrt{t-kh}\ \dot{v}(kh) \tag{11}$$

$$- 4\hat{\varkappa} \sum_{n=1,3} \underbrace{\int_0^t \exp\{-n^2\pi^2[t-\tau]\}\dot{v}(\tau)\, d\tau}_{w_{1,2}(t),\quad w_i(0)=0}, \quad \hat{b} \equiv \frac{\tilde{b}}{1 + \frac{32\varkappa\sqrt{t-kh}}{15\sqrt{\pi}}}, \ \hat{\varkappa} \equiv \frac{\varkappa}{1 + \frac{32\varkappa\sqrt{t-kh}}{15\sqrt{\pi}}},$$

$$\dot{w}_1(t) = \ldots - [4\hat{\varkappa} + \pi^2]w_1(t) - 4\hat{\varkappa}\, w_2(t), \quad \dot{w}_2(t) = \ldots - 4\hat{\varkappa}\, w_1(t) - [4\hat{\varkappa} + 9\pi^2]w_2(t).$$

The ellipsis stands for the right-hand side of formula (11). The sought function $v(t) = q(t)/\bar{q}$ ($t = t'$, $t' \in [kh, (k+1)h]$, $k \geq 0$) is computed by the standard Runge-Kutta 4th order method for integration of the ODE system (the authors used Scilab software). The numerically obtained

TDS-spectra nearly agree with experimental curves. If a TDS-spectrum has two peaks, then $w_{3,4}$ should be additionally used to improve modeling accuracy. For a comparison, lines with marks in Fig. 1,2 illustrate TDS-spectra simulated by numerical solution (based on difference scheme) of the original distributed boundary-value problem with dynamical boundary conditions (the algorithm, including the model with reversible trapping in the bulk, was presented in Ref. [8]).

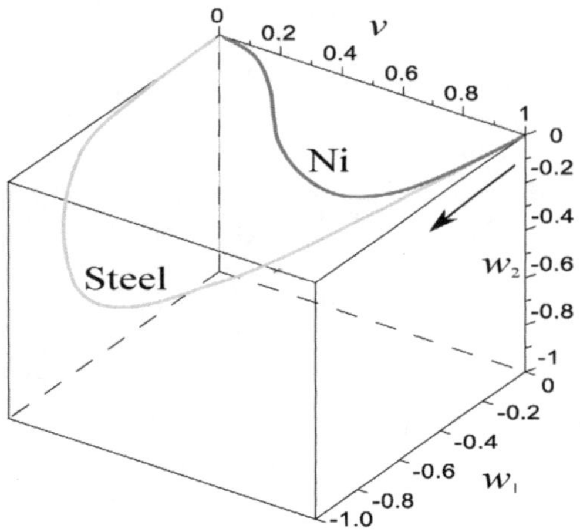

Figure 1. Phase trajectories

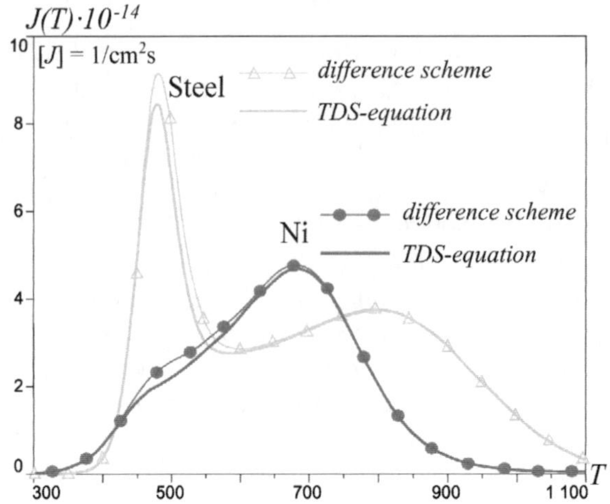

Figure 2. TDS-spectrum, numerical simulation

Conclusions

The paper presents a thermal desorption functional differential equation of neutral type with integrable weak singularity for modeling the degassing TDS-spectrum of hydrogen-presaturated structural materials. The proposed numerical method allows using a standard procedure for low order ODE system solving as an alternative to solving the original non-linear boundary-value problem with dynamical boundary conditions. One can thus quickly model TDS-spectra for a wide range of parameters, reveal the limiting factors and explore the material also under extreme conditions (where a natural experiment would be problematic).

The main physical sense of the results is the following. One cannot a priori rule out fluctuating thermal desorption, not even for defect-free materials with minor trapping of the diffusing hydrogen (the diffusion model in this paper is free of additives associated with traps). The problem for future research is to differentiate the peaks arising from "surface-volume" interaction dynamics from those induced by hydrogen release from traps with different binding energies.

References

[1] Alefeld G and Völkl J Eds. 1978 *Hydrogen in metals* (Berlin: SpringerVerlag)
[2] Zakharov A P Ed. 1987 *Interactions of hydrogen with metals* (Moscow: Nauka)
[3] Yukhimchuk A A Ed. 2009 *Hydrogen isotopes. Basic and applied research* (Sarov: RFNC VNIIEF)
[4] Ball M and Wietschel M Eds. 2009 *The hydrogen economy* (Cambridge Univ. Press)
[5] Hirscher M Ed. 2010 *Handbook of hydrogen storage: new materials for future energy storage* (Wiley–VCH)
[6] Evard E A, Gabis I E and Yartys V A 2010 Kinetics of hydrogen evolution from MgH2: experimental studies, mechanism and modelling *Int. J. Hydrogen Energy* **35** pp 9060–9069
[7] Zaika Yu V and Rodchenkova N I 2013 Hydrogen-solid boundary-value problems with dynamical conditions on surface *Math. Modelling* (New York: Nova Sci. Publishers) pp 269–302
[8] Zaika Yu V, Kostikova E K 2014 Computer simulation of hydrogen thermodesorption *Advances in Materials Sci. and Applications* **3** 3 pp 120–129
[9] Hale J. Theory of Functional Differential Equations. Springer-Verlag; 1977.
[10] Whittaker ET, Watson GN. A Course of Modern Analysis. Cambridge University Press; 1996.
[11] Leng S. Elliptic functions. Addison-Wesley publishing; 1973.

IOP Conf. Series: Journal of Physics: Conf. Series **929** (2017) 012035 doi:10.1088/1742-6596/929/1/012035

The arrowhead decomposition method for a block-tridiagonal system of linear equations

P A Belov[1], **E R Nugumanov**[1,2] **and S L Yakovlev**[1]

[1] Department of Computational Physics, St. Petersburg State University, Ulyanovskaya 1, 198504 St. Petersburg, Russia

E-mail: pavelbelov@gmail.com

Abstract. The arrowhead decomposition method (ADM) for the parallel solution of a block-tridiagonal system of linear equations is presented. The method consists in rearranging the initial linear system into an equivalent one with the "arrowhead" structure of the matrix. It is shown that such a structure provides a good opportunity for parallel solving. The computational speedup of ADM with respect to the sequential matrix Thomas algorithm is analytically estimated based on the number of elementary multiplicative operations for the parallel and serial parts of the methods. A number of parallel processors required to reach the maximum computational speedup is found. A good agreement of the analytical estimations of the computational speedup and practically obtained results is observed.

1. Introduction

Many boundary value problems arising in physics and applied mathematics after discretization lead to a linear system with a block-tridiagonal matrix. As a simple example, one can consider a finite-difference approximation of the two-dimensional Poisson equation on the equidistant grids over each variable. In this case, each block of the matrix becomes large and sparse, but the block-tridiagonal structure is kept. A combination of the finite-difference method over one coordinate and the basis set method [1] over another coordinate generally produces block-tridiagonal matrix with dense blocks. The intermediate case of the basis set with a local support, namely splines [2], results in the band blocks [3, 4] with the bandwidth comparable with that one from the finite-difference method.

The direct method for solving the linear systems is the Gauss elimination. An adaptation of the Gauss elimination to the block-tridiagonal systems is known as the matrix Thomas algorithm [5] or as the matrix sweeping algorithm [6]. In this algorithm, the idea of Gauss elimination is applied to the blocks themselves. The algorithm is well defined and robust for matrices with the diagonal dominance, but its sequential nature makes it difficult to be applied for parallel calculations.

The present paper is focused on the developed arrowhead decomposition method (ADM) for efficient parallel solving the block-tridiagonal linear systems. In this method, the initial matrix is logically reduced to the "arrowhead" form, namely some new independent on-diagonal blocks, sparse off-diagonal blocks and a coupling matrix of much smaller size. The method originates from the domain decomposition [7-11] and nested dissection [12, 13], where the idea

[2] Present address: BashNIPIneft LLC, 86/1, Lenin Street, 450006 Ufa, Russia

to divide a large problem into small ones which can, to some extent, be solved independently was introduced. In Ref. [14] this idea is illustrated by the tridiagonal matrix obtained from the finite-difference discretization of the one-dimensional Laplace operator. It is shown that this method is faster than the Wang algorithm [15, 16] and following ones. In this paper, the computational speedup [17] of ADM with respect to the sequential matrix Thomas algorithm is analytically estimated based on the number of elementary multiplicative operations for the parallel and serial parts of the methods. A number of parallel processors required to achieve the maximum computational speedup is obtained. The analytical results are compared with the results of practical calculations. For the test linear systems, we use the discretized boundary value problems for the integro-differential Faddeev equations [18-20, 4, 21, 22].

2. Arrowhead decomposition method

The block-tridiagonal linear system under consideration is described by the equation

$$A_i X_{i-1} + C_i X_i + B_i X_{i+1} = F_i, \quad A_1 = B_N = 0, \tag{1}$$

where A_i, B_i, C_i, $i = 1, \dots, N$, are the blocks of the matrix and F_i are the blocks of the right-hand side (RHS) supervector $\mathbf{F}$. The unknown supervector $\mathbf{X}$ is composed of blocks X_i. The size of each block of the matrix is $n \times n$, whereas the size of each block of the supervectors is $n \times l$, where $l \geq 1$ is the number of columns.

The idea of the ADM is presented in Figure 1. The initial block-tridiagonal linear system (1) is rearranged into the equivalent "arrowhead" form which allows the parallel solving. The rearrangement is performed by interchanging the block-rows and block-columns. The interchange of block-columns also leads to the change of the elements of the unknown vector and the RHS vector. The new structure of the matrix can be represented by the 2×2 block-matrix

$$\begin{pmatrix} \mathbf{S} & \mathbf{W}_R \\ \mathbf{W}_L & \mathbf{H} \end{pmatrix} \begin{pmatrix} s \\ h \end{pmatrix} = \begin{pmatrix} F_s \\ F_h \end{pmatrix}. \tag{2}$$

Here, the unknown solution h corresponds to the moved part of the full solution. The notation is shown in Figure 1. The square superblock $\mathbf{S}$ consists of the new independent blocks S^k, $k = 1, \dots, M$, on the diagonal. The matrix element $\mathbf{H}$ is the bottom right coupling superblock. Other lateral superblocks $\mathbf{W}_R$, $\mathbf{W}_L$ present additional blocks of the matrix. The solution of the system (2) is given by the relations

$$\left\{ \begin{array}{lcl} s & = & \mathbf{S}^{-1} F_s - \mathbf{S}^{-1} \mathbf{W}_R h \\ h & = & \left(\mathbf{H} - \mathbf{W}_L \mathbf{S}^{-1} \mathbf{W}_R\right)^{-1} \left(F_h - \mathbf{W}_L \mathbf{S}^{-1} F_s\right) \end{array} \right. . \tag{3}$$

These relations contain matrix products and inverses which can, to some extent, be done in parallel. The independence of blocks S^k allows the parallel calculation of the products $\mathbf{S}^{-1} F_s$ and $\mathbf{S}^{-1} \mathbf{W}_R$. Actually, instead of inverses we solve in parallel over k the linear systems for S^k. The sparse structure of blocks $\mathbf{W}_L$ and $\mathbf{W}_R$ drastically reduces the number of matrix operations in Eq. (3). Once a part $h = (h^1, \dots, h^{M-1})^T$ of the total solution $\mathbf{X} = (s, h)^T$ is obtained from the second relation of Eq. (3), the remaining part is calculated in parallel over k by the formula

$$s^k = z^k - Z^k h^{k-1} - Z^k h^k,$$

where formally $z^k = (S^k)^{-1} F_s$, $Z^k = (S^k)^{-1} W_R^k$, and $h^0 = h^M = 0$.

To solve the independent linear systems with S^k and the equation for h, one can apply any appropriate technique. Although the ADM can be used recursively (nestedly) for solving these linear systems [23], in the present paper, we employ the matrix Thomas algorithm.

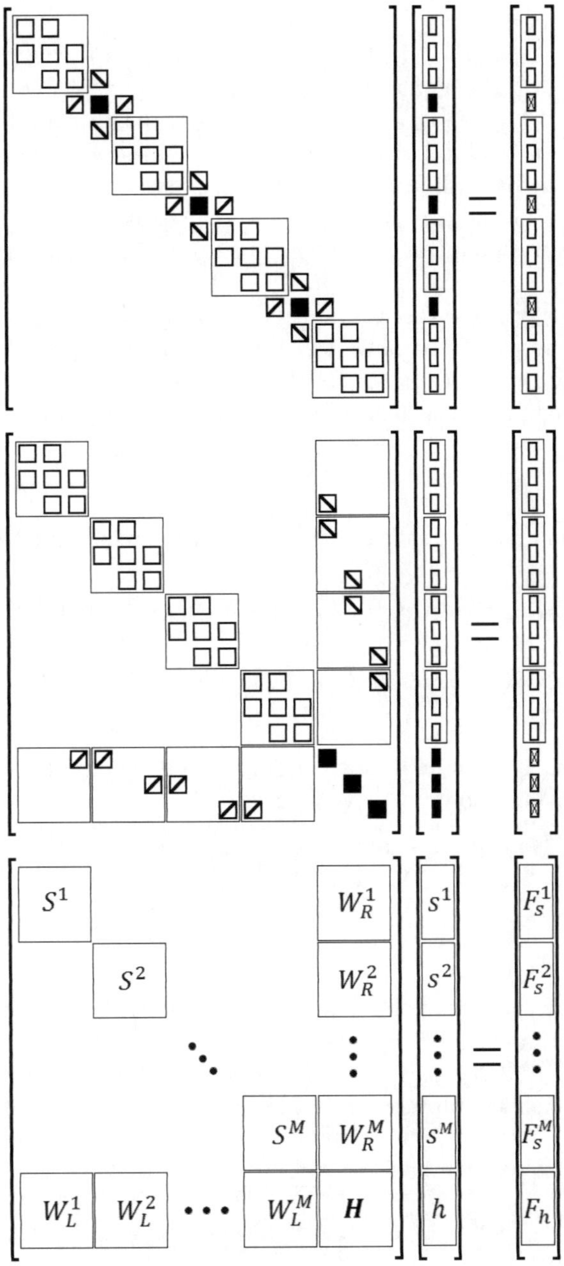

Figure 1. A graphical scheme for rearrangement of the initial block-tridiagonal linear system into the equivalent form. The nonzero blocks and vectors are denoted by thick lines. New independent superblocks and corresponding vectors at each panel are denoted by thin lines. *Top panel:* the initial linear system with the marked interchanged blocks. *Central panel:* the obtained rearranged system with the "arrowhead" matrix. *Bottom panel:* the notation of the matrix elements for the rearranged system.

3. Computational speedup

The computational speedup of the ADM with respect to the sequential matrix Thomas algorithm can be estimated as the ratio of the computation time by the matrix Thomas algorithm to that one of the ADM [17]. The computation time is directly related to the number of serial operations, namely products and divisions, of each algorithm. We calculated a number of serial multiplicative operations for both algorithms and, as a consequence, analytically estimated the computational speedup. The additive operations were not taken into account assuming that they require much less computational time. The time for the memory management is also considered to be negligible.

According to Ref. [24], we hold that for multiplication and inverse of a block of size n one needs exactly n^3 multiplicative operations. These estimations are in agreement with the conventional realizations of these algorithms, namely LINPACK and LAPACK [25, 26]. Then, the number of multiplicative operations for the sequential matrix Thomas algorithm is [27]

$$\mathrm{O}_{\mathrm{TA}} = (3N-2)(n^3+n^2 l).$$

For the ADM, the total number of multiplicative operations was found to be given by the formula

$$\begin{aligned}&\Big((4N_1-1)n^3+(4N_1-1)n^2 l\Big)+\Big((6N_M-3)n^3+(4N_M-1)n^2 l\Big)+\\&+\sum_{k=2}^{M-1}\Big(7N_k n^3+5N_k n^2 l\Big)+(3M-5)(n^3+n^2 l),\end{aligned} \tag{4}$$

where N_k, $k=1,\ldots,M$ is the number of blocks on the diagonal of S^k. In general, N_k may be different for different k, but should satisfy the relation:

$$\sum_{k=1}^{M} N_k = N-(M-1).$$

The maximum performance is achieved if the computation time for solving the linear system with the block S^k is equal for each k. In this case, the quantities N_k are equal for $k=2,\ldots,M-1$, and N_1, N_M are larger than the others. The profit in the performance of such choice of N_k with respect to the case when all N_k, $k=1,\ldots,M$ are equal is relatively small. Therefore, for the sake of simplicity, we consider the case when all N_k are the same. In such a case, if the number of parallel processors equals to the number of blocks M on the diagonal, then the number of serial operations of the ADM will be

$$O_{\mathrm{ADM}} = \left(\frac{(N-M+1)}{M}\right)(7n^3+5n^2 l)+(3M-5)(n^3+n^2 l).$$

The computational speedup, as the ratio $S=O_{\mathrm{TA}}/O_{\mathrm{ADM}}$, is easily obtained:

$$S=\frac{3N-2}{3M-5+\left(5+\frac{2}{1+l/n}\right)\left(\frac{N-M+1}{M}\right)}. \tag{5}$$

One can estimate that the dependence of the computational speedup on the ratio l/n is weak for $l \ll n$. Figure 2 shows the computational speedup (5) for $N=3071$, $n=100,400$, $l=1$. As a function of number of parallel processors, M, the speedup is linear $S=3M/7$ for $M \ll N$. As M increases further, the speedup flattens out, reaches its maximum and, then, decreases. Based on the obtained speedup (5), the parallel efficiency of the ADM can be defined as $E = S/M$ [28]. A number of parallel processors required to achieve the maximum speedup is given as

$$M^* = \sqrt{\frac{(N+1)}{3}\left(5 + \frac{2}{1+l/n}\right)}. \tag{6}$$

As a result, for using the ADM one can choose all independent blocks S^k of the same size. The optimal number of parallel processors M^* for the fixed number of blocks N on the diagonal is given by Eq. (6).

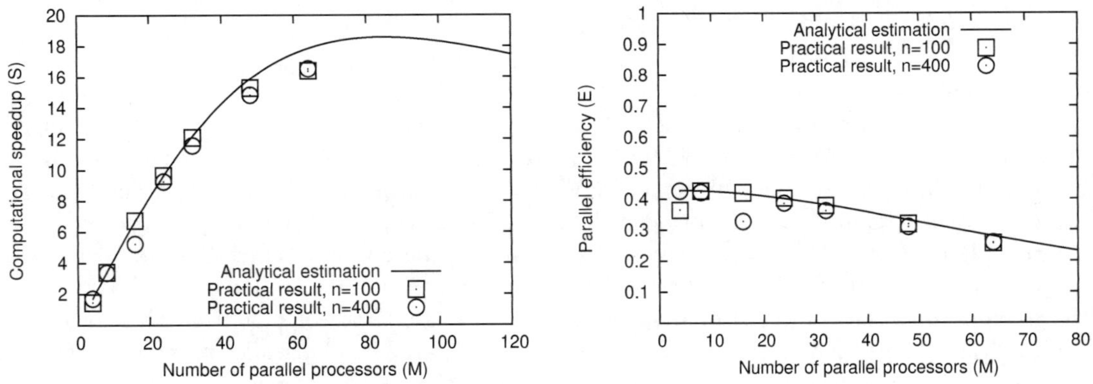

Figure 2. *Left panel:* the analytically estimated computational speedup (solid curve) of the ADM with respect to the matrix Thomas algorithm as well as the speedup obtained in practical calculations (empty squares and circles) as a function of the number of parallel processors, $M \geq 4$. *Right panel:* the parallel efficiency of the ADM for the same series of numerical experiments. The size of each block of the block-tridiagonal matrix is $n = 100, 400$, a number of blocks in the diagonal is $N = 3071$, only one RHS vector is used.

4. Practical results

The analytical estimations, given in the previous section, were validated in practical calculations. We used the Symmetric Multiprocessor (SMP) system with 64 processors (Intel Xeon CPU X7560 2.27GHz) and 2TB of shared operative memory. The ADM was implemented as an independent program written in C/C++. For the linear algebra operations, the corresponding LAPACK 3.5.0 subroutines [26] were called. The parallelization was done using the OpenMP 3.0 [29].

In validation studies of the computational speedup (5) only the solution time was taken into account. This time includes time needed for computation itself as well as time required for possible memory management during the solution process. The time for generation of the initial blocks was ignored.

The initial block-tridiagonal matrix was obtained from the well-known discretization of the two-dimensional Laplace operator of the Faddeev equations in polar coordinates [18, 19, 4]. The block-tridiagonal structure arose from an expansion of the unknown solution of these equations on a basis of the Hermite splines [3, 30] over the θ-grid and the finite-difference approximation of the second derivative over the equidistant ρ-grid [31]. Additionally, the zero elements of the blocks were filled by relatively small random nonzero values to guarantee the integrity of our studies.

In Figure 2, we show the computational speedup as well as the parallel efficiency of the ADM measured in a series of experiments with various number M of parallel processors and sizes of blocks taken to be $n = 100, 400$. The general trend of the analytical result (5) is clearly confirmed by the practical calculations. Since for a small number of processors, $4 \leq M \leq 24$, the computational speedup is linear with respect to the number of parallel processors $S = 3M/7$, the efficiency is bounded by the coefficient $3/7$. For larger M, the linear behaviour is distorted. The practical results are a little bit smaller than the analytical ones. This difference may be attributed to the nonideality of the memory management of the computational system or to additional system processes running simultaneously and affecting the performance of the studied task. These issues are especially noticeable for the case when the number of parallel processors is $M = 48$ and $M = 64$. For these cases, the practically achieved computational speedup is apparently smaller than the analytical one.

5. Conclusions

In summary, we presented the arrowhead decomposition method for the efficient parallel solution of the block-tridiagonal systems of linear equations. The method consists in rearranging the initial linear system into the equivalent one with the "arrowhead" structure of the matrix. The computational speedup of the arrowhead decomposition method with respect to the sequential matrix Thomas algorithm was analytically estimated based on the number of elementary multiplicative operations for the parallel and serial parts of the methods. We showed that the maximum of the computational speedup for a given number of blocks on the diagonal is achieved for a finite number of parallel processors. A number of parallel processors required to reach the maximum computational speedup was obtained. A good agreement of analytical estimations of the computational speedup and practically measured results was observed.

Acknowledgments

Financial support from SPbU (grant No. 11.38.241.2015) and RFBR (grant No. 16-32-00047) is acknowledged. The calculations were carried out using the facilities of the SPbU Resource Center "Computational Center of SPbU".

References

[1] Press W H, Teukolsky S A, Vetterling W T and Flannery B P 2007 *Numerical recipes: the art of scientific computing* (New York: Cambridge University Press)
[2] Ahlberg J H, Nilson E N and Walsh J L 1967 *Theory of splines and their applications* (New York: Academic Press)
[3] Prenter P M 1975 *Splines and variational methods* (New York: Wiley)
[4] Belov P A and Yakovlev S L 2013 *Phys. Atom. Nucl.* **76** 126
[5] Thomas L H 1949 Elliptic problems in linear difference equations over a network *Watson Sci. Comput. Lab. Rept.* (New York: Columbia University)
[6] Samarskii A A and Nikolaev E S 1978 *Methods for solving finite-difference equations* (Moscow: Nauka)
[7] Bjørstad P E and Widlund O B 1986 *SIAM J. Numer. Anal.* **23** 1097
[8] Il'in V P, Litvinenko S A and Sveshnikov V M 2003 Parallelization of alternating direction implicit methods for three-dimensional domains *Lect. Not. in Comp. Sci.* **2763** 89
[9] Korneev V G 2013 *Comp. Meth. in Appl. Math.* **13** 161
[10] Korneev V G and Langer U 2015 *Dirichlet-dirichlet domain decomposition methods for elliptic problems* (Singapore: World Scientific)
[11] Gurieva Y L, Il'in V P and Perevozkin D V 2016 *Vychisl. Metody Programm.* **17** 132
[12] Duff I S, Erisman A M and Reid J K 1986 *Direct methods for sparse matrices* (Oxford: Clarendon Press)
[13] Solovyev S A 2014 *Vychisl. Metody Programm.* **15** 441
[14] Ortega J M 1988 *Introduction to parallel and vector solution of linear systems* (New York: Springer)
[15] Wang H H 1981 *ACM Trans. Math. Soft.* **7** 170
[16] Volokhova A V, Zemlyanaya E V and Rikhvitskiy V S 2015 *Vychisl. Metody Programm.* **16** 281
[17] Gallopoulos E, Philippe B and Sameh A H 2016 *Parallelism in matrix computations* (Springer)

IOP Conf. Series: Journal of Physics: Conf. Series **929** (2017) 012035 doi:10.1088/1742-6596/929/1/012035

[18] Merkuriev S P, Gignoux C and Laverne A 1976 *Ann. Phys.* **99** 30
[19] Faddeev L D and Merkuriev S P 1993 *Quantum scattering theory for several particle systems* (Dordrecht: Kluwer)
[20] Belov P A and Yakovlev S L 2012 *Bull. Russ. Acad. Sci. Phys.* **76** 913
[21] Belov P A and Yakovlev S L 2014 *Phys. Atom. Nucl.* **77** 344
[22] Belov P A and Yakovlev S L 2017 *Few-Body Syst.* **58** 24
[23] Belov P A, Nugumanov E R and Yakovlev S L 2015 Decomposition method for block-tridiagonal matrix systems *E-print* arXiv:1505.06864
[24] Samarskii A A and Gulin A V 1989 *Numerical methods* (Moscow: Nauka)
[25] Dongarra J J *et al* 1979 *LINPACK Users' guide* (Philadelphia: SIAM)
[26] Anderson E *et al* 1999 *LAPACK Users' guide* (Philadelphia: SIAM)
[27] Isaacson E and Keller H B 1966 *Analysis of numerical methods* (New York: Wiley)
[28] Douglas C C, Haase D and Langer U 2003 *A tutorial on elliptic PDE solvers and their parallelization* (Philadelphia: SIAM)
[29] Adeniyi-Jones C *et al* 2008 OpenMP Application Program Interface. Http://www.openmp.org/mp-documents/spec30.pdf
[30] Pupyshev V V 2004 *Phys. Part. Nucl.* **35** 257
[31] Khramtsov E S, Belov P A, Grigoryev P S *et al.* 2016 *J. Appl. Phys.* **119** 184301

IOP Conf. Series: Journal of Physics: Conf. Series **929** (2017) 012036 doi:10.1088/1742-6596/929/1/012036

The coefficient smoothing method application to the problem of gas pipeline glaciation

S A Mikheev[1], N N Ermolaeva and G V Krivovichev

Faculty of Applied Mathematics and Control Processes Saint Petersburg State University, 7/9 Universitetskaya nab., Saint Petersburg, 199034, Russian Federation

E-mail: [1] helps2@rambler.ru, n.ermolaeva@spbu.ru, g.krivovichev@spbu.ru

Abstract. In northern seas the gas temperature in the pipeline may be lower than the water-ice phase transition temperature, so the glaciation process must be considered. The coefficient smoothing method for the problem of glaciation of the cylinder immersed in seawater is considered. The model of glaciation process is presented as a problem for the linear heat equation in domain with unknown moving boundary. The method of solution is based on the transition to the Dirichlet problem for the nonlinear two-dimensional heat equation in domain with fixed boundaries. The splitting method is applied to the solution of the problem for two-dimensional heat equation. Two approximations for the Dirac delta function are proposed. Calculation of uniform glaciation process was carried out with the using of nonlinear implicit finite-difference schemes. Time moments of the glaciation for different spatial layers are obtained. Results of calculations are compared with the results obtained by the front-tracking method.

1. Introduction

In northern seas during the process of gas transportation through the pipelines, gas temperature may be lower than the water-ice transition temperature [1, 2], so the pipeline glaciation becomes possible, when the gas flow cools down below a certain temperature. The problem of the heat transfer in the pipe and surrounding media is considered [3, 4, 5]. In [3] a model for the steady state heat transfer between the gas and pipeline surroundings is considered. The proposed model does not take into account time dependent heat accumulation. In[4] the model of unsteady heat transfer is proposed as an initial boundary value problem for the one-dimensional heat equation without phase changes. In [5] these models are compared to the cases of onshore and offshore pipelines. The results of the onshore buried pipeline demonstrate that the steady heat transfer model over predicts the amplitude of temperature changes. For the case of offshore pipeline some results were obtained only for the case of unsteady model. But it must be noted, that for the correct modelling of the pipeline glaciation, phase transitions, such as seawater-ice transition, must be taken into account. So in computational modelling of this process initial boundary value problem with moving unknown boundary and the Stefan condition must be considered [1].

There are many works dedicated to the numerical solution of Stefan problem, based on two main approaches. First is the front-tracking method (FTM), when the position of the phase boundary is continuously tracked. In this method, finite-difference or finite-element schemes with deformed or adaptive grids are employed and the explicit tracking of phase front is provided. There are many various realizations of this approach. In [6] the method with variable time step, when the phase boundary move from one space node to another is realized. Murray and Ladis in

[7] provide an approach based on space grid with variable step, but with fixed number of space intervals between fixed and moving boundary. Crank in [8] discussed various finite-difference methods with variable space and time steps and a change of space variable to fix the moving boundary. Ermolaeva and Kurbatova in [1, 2] proposed FTM to the gas pipeline glaciation problem. As it is mentioned in [9], these methods are effective only in the one-dimensional case, an extension to the two- and three-dimensional problems leads to various difficulties.

Another approach to numerical solution is based on the fixed domain formulation. An example of this approach is a coefficient smoothing method (CSM) proposed in [10, 11, 12], which transform problem for the linear heat equation with Stefan condition on moving boundary to the problem for nonlinear heat equation with fixed boundary and smoothed nonlinear coefficients. In this formulation the Stefan condition on the phase boundary is automatically satisfied and the phase boundary is defined by value of temperature equal to the temperature of phase transition.

Presented paper devoted to the application of CSM to the problem of gas pipeline glaciation. In section 2 the problem is considered. In section 3 the method is described. In section 4 the splitting method is discussed. In section 5 the results of the solution of the uniform glaciation problem are presented and discussed. Some concluding remarks are made in section 6.

2. Mathematical model of gas pipeline glaciation

Mathematical model of pipeline glaciation is based on the following assumptions [1, 2]: glaciation process only in cross-section of the pipeline is considered and normal component of heat flux vector from the seawater to the front line of glaciation is considered as constant.

The two-dimensional heat equation for the temperature distribution in an ice layer in pipeline cross-section is written in polar coordinates as:

$$S\frac{\partial T}{\partial t} = \frac{1}{r}\frac{\partial}{\partial r}\left(rk\frac{\partial T}{\partial r}\right) + \frac{1}{r^2}\frac{\partial}{\partial \varphi}\left(k\frac{\partial T}{\partial \varphi}\right), \tag{1}$$

where $T = T(t, r, \varphi)$ is the temperature, t is the time, r is a radial coordinate, φ is a polar angle, $S = c\rho$, c is a specific heat (per volume unit), ρ is the density of sea ice, k is the coefficient of thermal conductivity. Eq. (1) is considered on time interval $(0, T_{end}]$ and in space domain $\Omega = \{(r, \varphi)|\varphi \in [0, 2\pi), r \in (R, \tilde{R}(t, \varphi))\}$, where R is a radius of the pipeline and the function $\tilde{R}(t, \varphi)$ characterize the phase change boundary $G = \{(r, \varphi)|\varphi \in [0, 2\pi), r = \tilde{R}(t, \varphi)\}$ (fig. 1).

Boundary and initial conditions for eq. (1) are written as:

$$T(t, R, \varphi) = T_0(t), \quad T(t, \tilde{R}, \varphi) = T_*, \quad \tilde{R}(0, \varphi) = R, \tag{2}$$

where T_* is the phase transition temperature, $T_0(t)$ is a known temperature of gas in pipeline estimated after gas flow calculations [2]:

$$T_0(t) = \frac{m_1}{m_2 + t} + m_3,$$

where $m_1 = 15120$, $m_2 = 5040$, $m_3 = 268$.

On the boundary of phase transition G the following Stefan condition [8, 10] is stated:

$$k\frac{\partial T}{\partial \mathbf{n}}\bigg|_{G^+} - q_2 = Q\rho\frac{d}{dt}F(t, \tilde{R}, \varphi), \tag{3}$$

where Q is the latent heat of fusion of ice, q_2 is a constant normal component of heat flux from seawater to the boundary G, $dF(t, \tilde{R}, \varphi)/dt$ is a normal component of the velocity of the boundary G. In addition to conditions (2),(3) the periodic condition for T is stated:

$$T(t, r, \varphi + 2\pi) = T(t, r, \varphi). \tag{4}$$

As it can be seen, presented problem (1)–(4) is a problem for the *linear* eq. (1) in domain Ω with *unknown moving boundary* G, defined by conditions (2),(3).

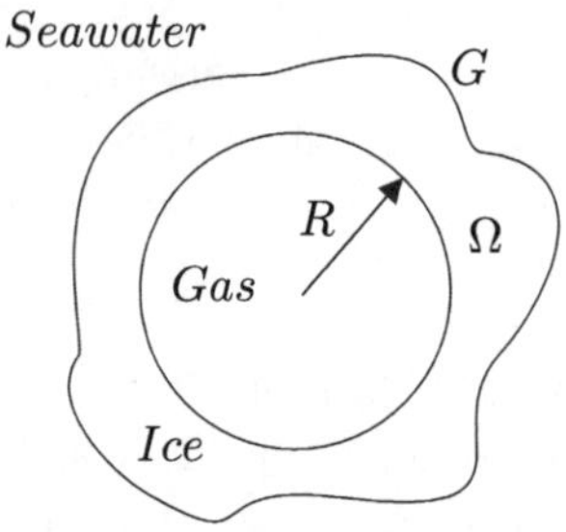

Figure 1. Schematic representation of the domain for the problem of ice growth modelling.

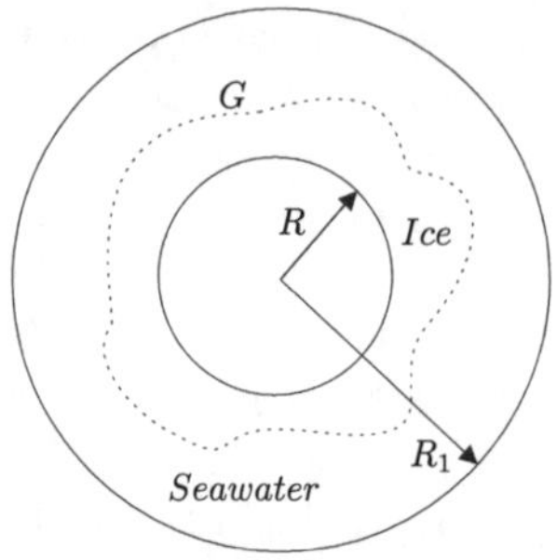

Figure 2. Space domain for the computations by the coefficient smoothing method.

3. The coefficient smoothing method

According to the approach, presented by A. A. Samarskii [10] condition (3) is implicitly included in eq. (1) and *nonlinear* equation is considered in domain with *known steady* boundaries.

Let us consider new domain $\Psi = \{(r,\varphi)|\varphi \in [0,2\pi), r \in (R,R_1)\}$, where constant R_1 characterize radial fictitious boundary situated in seawater. It is assumed that in Ψ the phase change boundary G is situated and two phases — ice and seawater are considered. The phases are defined by constant densities ρ_1 and ρ_2, specific heat coefficients c_1 and c_2 and coefficients of thermal conductivity k_1 and k_2 respectively (fig. 2).

According to [10], the new function is introduced: $\widetilde{S}(T) = S + \rho Q\delta(T - T_*)$, where the second term characterize the heat capacity of phase change, when $T = T_*$ and $\delta(T - T_*)$ is a Dirac delta function, this function is approximated (smoothed) by function $\tilde{\delta}(T - T_*, \Delta) \geq 0$, where $\Delta > 0$ characterise interval $(T_* - \Delta, T_* + \Delta)$, where $\tilde{\delta} \neq 0$. According to this approach, S and k are replaced by following nonlinear functions:

$$\widetilde{S}(T) = \begin{cases} S_1, & T < T_* - \Delta, \\ \frac{S_1+S_2}{2} + \rho Q\tilde{\delta}(T - T_*, \Delta), & |T - T_*| < \Delta, \\ S_2, & T > T_* + \Delta, \end{cases} \quad \widetilde{k}(T) = \begin{cases} k_1, & T < T_* - \Delta, \\ \frac{k_1+k_2}{2}, & |T - T_*| < \Delta, \\ k_2, & T > T_* + \Delta. \end{cases}$$

where $S_i = c_i\rho_i$, $i = 1, 2$.

In this way eq. (1) is replaced by following *nonlinear* equation:

$$\widetilde{S}(T)\frac{\partial T}{\partial t} = \frac{1}{r}\frac{\partial}{\partial r}\left(r\widetilde{k}(T)\frac{\partial T}{\partial r}\right) + \frac{1}{r^2}\frac{\partial}{\partial \varphi}\left(\widetilde{k}(T)\frac{\partial T}{\partial \varphi}\right), \tag{5}$$

in domain Ψ, with the following Dirichlet boundary conditions:

$$T(t, R, \varphi) = T_0(t), \quad T(t, R_1, \varphi) = T_w, \tag{6}$$

where T_w is a temperature of seawater.

The initial condition for eq. (5) in Ψ is stated as:

$$T(0, r, \varphi) = f(r, \varphi). \tag{7}$$

The phase change boundary G, situated in Ψ is defined in the process of calculation by following condition:

$$T(t, \tilde{R}(t, \varphi), \varphi) = T_*.$$

4. The splitting method

According to [13], the differential operator in right part of eq. (5) is represented as the sum of two operators, defined by formulas:

$$A_1(\dots) = \frac{1}{r}\frac{\partial}{\partial r}\left(r\widetilde{k}(\dots)\frac{\partial(\dots)}{\partial r}\right), \quad A_2(\dots) = \frac{1}{r^2}\frac{\partial}{\partial \varphi}\left(\widetilde{k}(\dots)\frac{\partial(\dots)}{\partial \varphi}\right).$$

Let us construct a computational grid consist of nodes (t_n, r_k, φ_l), where $r_m = mh_r$, $\varphi_p = ph_\varphi$, $t_n = n\tau$, $m = 0, \dots, M$, $p = 0, \dots, P$, $n = 0, \dots, N$, where M, P, N are positive integers, h_r, h_φ, τ are the steps on the spatial coordinates and time.

During the numerical solution on time interval $[t_n, t_{n+1}]$ two *one-dimensional* problems are considered. The first problem is considered for the function $T_1 = T_1(t, r, \varphi)$, where φ is considered as a parameter. The problem for T_1 is written as:

$$\widetilde{S}(T_1)\frac{\partial T_1}{\partial t} = A_1(T_1), \quad T_1(t_n, r, \varphi) = T(t_n, r, \varphi), \quad T_1(t, \varphi, R) = T_0(t), \quad T_1(t, R_1, \varphi) = T_w. \quad (8)$$

The second problem is considered with operator A_2 and its solution is noted as $T_2 = T_2(t, r, \varphi)$, where r considered as a parameter. The problem is stated as:

$$\widetilde{S}(T_2)\frac{\partial T_2}{\partial t} = A_2(T_2), \quad T_2(t_n, r, \varphi) = T_1(t_{n+1}, r, \varphi), \quad T_2(t, r, 0) = T_2(t, r, 2\pi). \quad (9)$$

As it can be seen, the problem (8) at fixed φ and problem (9) at fixed r are considered as one-dimensional. These problems are solved on interval (t_n, t_{n+1}) sequentially. On the first step problem (8) is solved and function T_1 is obtained at moment t_{n+1}. This function is considered as initial condition for problem (9). After the solution of (9) the value of T at moment t_{n+1} is obtained as: $T(t_{n+1}, r, \varphi) \approx T_2(t_{n+1}, r, \varphi)$.

Heat equation from problem (8) is discretized by implicit scheme, proposed in [14]:

$$\widetilde{S}(T_{1,m}^n)\frac{T_{1,m}^{n+1} - T_{1,m}^n}{\tau} = \frac{1}{h_r r_m}\left(a_{1,m+1}\frac{T_{1,m+1}^{n+1} - T_{1,m}^{n+1}}{h_r} - a_{1,m}\frac{T_{1,m}^{n+1} - T_{1,m-1}^{n+1}}{h_r}\right), \quad (10)$$

where $T_{1,m}^n \approx T_1(t_n, r_m, \varphi)$, $m = 2, \dots, M-1$, values of φ are considered on grid constructed in $[0, 2\pi)$ with step h_φ, $a_{1,m} = 0{,}5(r_m\widetilde{k}(T_{1,m}^n) + r_{1,m-1}\widetilde{k}(T_{1,m-1}^n))$. The scheme is realized with boundary conditions obtained from (8): $T_{1,1}^{n+1} = T_0(t_{n+1})$, $T_{1,M}^{n+1} = T_w$ by solution of system of linear algebraic equations by the tridiagonal matrix algorithm.

Heat equation from problem (9) is approximated by the same scheme:

$$r_m^2\widetilde{S}(T_{2,p}^n)\frac{T_{2,p}^{n+1} - T_{2,p}^n}{\tau} = \frac{1}{h_\varphi}\left(a_{2,p+1}\frac{T_{2,p+1}^{n+1} - T_{2,p}^{n+1}}{h_\varphi} - a_{2,p}\frac{T_{2,p}^{n+1} - T_{2,p-1}^{n+1}}{h_\varphi}\right), \quad (11)$$

where $T_{2,p}^n \approx T_2(t_n, r, \varphi_p)$, $p = 2, \dots, P-1$, values of r are considered on previously defined grid, $a_{2,p} = 0{,}5(\widetilde{k}(T_{2,p}^n) + \widetilde{k}(T_{2,p-1}^n))$. Scheme (11) is solved with periodic conditions: $T_{2,1}^{n+1} = T_{2,P}^{n+1}$. System of linear equations (11) for $T_{2,p}^{n+1}$ is solved by circular tridiagonal matrix algorithm.

Delta function is approximated on $(T_* - \Delta, T_* + \Delta)$ by constant and quadratic functions:

$$\widetilde{\delta}(T - T_*, \Delta) = \begin{cases} \frac{1}{2\Delta}, & |T - T_*| < \Delta, \\ 0, & |T - T_*| > \Delta, \end{cases} \quad (12)$$

$$\widetilde{\delta}(T - T_*, \Delta) = \begin{cases} \frac{3}{4\Delta^3}(\Delta^2 - (T - T_*)^2), & |T - T_*| < \Delta, \\ 0, & |T - T_*| > \Delta, \end{cases} \quad (13)$$

the value of Δ was equal to 0.01 during all calculations.

Table 1. The time of glaciation of spatial layers $[R, R+l]$ obtained by FTM (min.) [2].

Layer (l)	1 mm	1 cm	2 cm	3 cm	4 cm	5 cm	6 cm
Time	8.6	113.5	298.9	571.0	944.6	1435.9	2063.6

Table 2. The time of glaciation of spatial layers of length $R+l$ obtained by the CSM at various R_2 and delta-functions approximations (12)–(13) (min.).

R_2	1 cm		1.25 cm		1.5 cm	
Layer (l)	(12)	(13)	(12)	(13)	(12)	(13)
1 mm	8.7450	8.8110	**8.6460**	8.7120	8.7780	8.8330
1 cm	114.96	115.61	**113.55**	114.09	112.82	113.35
2 cm	302.54	304.16	**299.03**	300.19	295.92	296.88
3 cm	578.25	582.66	567.38	**569.92**	558.61	560.43
4 cm	**953.77**	962.15	931.09	936.60	912.25	915.90
5 cm	**1438.3**	1454.4	1400.4	1411.6	1367.4	1375.1
6 cm	**2061.2**	2088.2	1978.0	1998.4	1939.5	1954.5

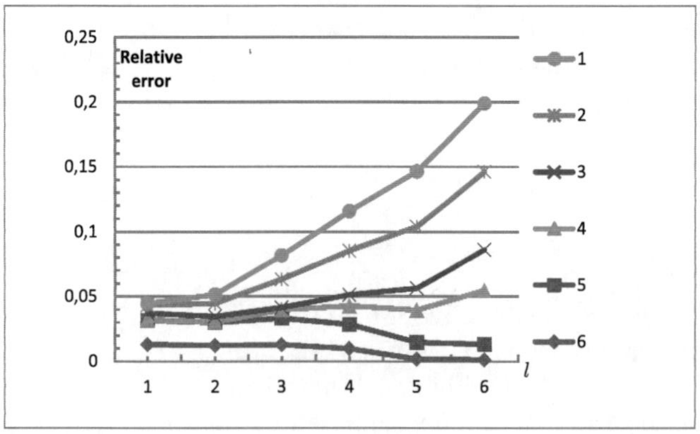

Figure 3. The relative errors of the time of glaciation of layers $[R, R+l]$. 1 and 2 — case of $R_2 = 1.5$ cm, 3, 4 — $R_2 = 1.25$ cm, 5, 6 — $R_2 = 1$ cm. 1, 3, 5 corresponds to quadratic approximation of Dirac delta function, 2, 4, 6 — to constant approximation.

5. Solution of the problem of uniform glaciation

The problem of uniform glaciation is solved in domain Ψ with the initial condition, which is independent on φ: $T(0, r, \varphi) = f(r)$, where $f(r)$ is considered as a solution of the symmetrical stationary problem for the heat equation in Ψ, when $T = T(r)$:

$$\frac{1}{r}\frac{d}{dr}\left(r\frac{dT}{dr}\right) = 0, \quad T(R) = T_0(0), \quad T(R_1) = T_w. \tag{14}$$

Solution of (14) is written as: $T(r) = A\ln(r) + B$, where constants A and B are defined from boundary conditions. The value of R_1 is defined as $R_1 = R + l + R_2$, where interval $[R, R+l]$,

$l > 0$ characterize the thickness of ice layer (e.g., 1 cm) and value of R_2 is varied.

Calculations are realized with the following values of the parameters of ice-water system, presented in SI system [2]: $R = 0.67$, $c_1 = 2100$, $\rho_1 = 928$, $Q = 335000$, $k_1 = 2.3$, $c_2 = 4200$, $\rho_2 = 1025$, $k_2 = 0.56$, $T_* = 271$.

In table 1 results of the solution of one-dimensional problem for uniform glaciation obtained by FTM in [2] are presented.

Calculations by CSM are performed on various time and spatial grids. In order to compare with the results presented in table 1, the time interval of 2200 minutes is considered. At table 2 the results of the calculation on spatial grid with 500×500 nodes and time grid with $2 \cdot 10^5$ nodes are presented. The cases of various values of R_2 and approximations of delta function (12) and (13) are considered. The results closed to results from table 1 are presented by bold font. For the comparison of the results obtained by two methods for every ice layer we compute the values of the following ratio: $\varepsilon = |t_{ft} - t_{cs}|/t_{ft}$, where t_{ft} is a glaciation time obtained by FTM, t_{cs} is a time obtained by CSM. Subroutine with realization of the algorithm of FTM is tested on different Stefan problems with analytical solutions and demonstrate a high accuracy, so the values of ε may be considered as relative errors for the results obtained by CSM. In the fig. 3 the plots of these relative errors for different values of R_2 and approximations (12) and (13) are presented. As it can be seen, the using of quadratic approximation (13) is better only for the case of $l = 3$ cm and $R_2 = 1.25$ cm. In other cases values approximately equal to the presented in table 1 are obtained in the case of (12) and $R_2 = 1$ and 1.25 cm respectively.

6. Conclusion

Presented paper is devoted to the numerical solution of the glaciation problem for the gas pipeline. CSM proposed by A. A. Samarskii is considered. The method is based on the transition to the Dirichlet problem for the nonlinear two-dimensional heat equation in domain with fixed boundaries. The splitting method is applied to the solution of the problem for two-dimensional heat equation. Two approximations of the Dirac delta function are proposed. Calculation of uniform glaciation process was carried out with the using of nonlinear implicit finite-difference schemes. Time moments of the glaciation for different spatial layers are obtained. Results of calculation are compared with the results obtained by the FTM.

References

[1] Ermolaeva N N and Kurbatova G I 2015 *Proc. 2015 Int. Conf. on Mechanics — Seventh Polyakhov's Reading* p 7106725
[2] Ermolaeva N N and Kurbatova G I 2015 *Proc. 2015 Int. Conf. on "Stability and Control Processes" in Memory of V.I. Zubov, SCP 2015* p 375
[3] Ramsen J et al 2009 it Proceedings of 40th Annual Meeting in Galvestone Texas p 0901
[4] Chaczykowski M 2009 *Applied Mathematical Modelling* **34** 1051
[5] Helgaker J and Oosterkamp A and Ytrehus T 2013 *MekIT13: Seventh national conference on Computational Mechanics* p 113
[6] Douglas J and Gallie T M 1955 *Duke Math. J.* **22** 557
[7] Murray W D and Lardis F 1959 *J. Heat Transf.* **81** 106
[8] Crank J. 1984 *Free and Moving Boundary Problems* (Oxford: Clarendon Press)
[9] Salvatori L and Tosi N 2009 *Algorithms* **2** 1177
[10] Samarskii A A and Moiseenko B D 1965 *USSR Comp. Math. and Math. Phys.* **5** 43
[11] Budak A A, Sobol'eva E N and Uspenskii A B 1965 *USSR Comp. Math. and Math. Phys.* **5** 59
[12] Fedorenko R P 1975 *USSR Comp. Math. and Math. Phys.* **15** 246
[13] Yanenko N N 1971 *The method of fractional steps (The solution of problems of mathematical physics in several variables)* (Berlin: Springer)
[14] Samarskii A A 2001 *The theory of difference schemes* (New York: Marcel Dekker)

IOP Conf. Series: Journal of Physics: Conf. Series **929** (2017) 012037 doi:10.1088/1742-6596/929/1/012037

Numerical determination of the CFT central charge in the site-diluted Ising model

P A Belov[1], A A Nazarov[1] and A O Sorokin[2]

[1]Department of Physics, St. Petersburg State University, Ulyanovskaya 1, 198504 St. Petersburg, Russia

[2]Petersburg Nuclear Physics Institute NRC Kurchatov Institute, Orlova roscha 1, Gatchina, 188300 Leningrad region, Russia

E-mail: antonnaz@gmail.com

Abstract. We propose a new numerical method to determine the central charge of the conformal field theory models corresponding to the 2D lattice models. In this method, the free energy of the lattice model on the torus is calculated by the Wang-Landau algorithm and then the central charge is obtained from a free energy scaling with respect to the torus radii. The method is applied for determination of the central charge in the site-diluted Ising model.

1. Introduction

Spin models play an important role in statistical physics. They describe real systems, namely magnets, their non-trivial properties and phenomena, especially at a critical point. These properties can be studied by various analytical and numerical approaches, so the spin models are the proper frameworks for their development and verification.

It is well known [1] that at a critical point a spin system has the conformal symmetry, so the conformal field theory (CFT) is of special importance in this case. In two dimensions (2D), CFT is especially powerful since it possess a rich symmetry represented by the infinite-dimensional Virasoro algebra [2]:

$$[L_n, L_m] = (n-m)L_{n+m} + \frac{c}{12}(n^3 - n)\delta_{n+m,0},$$

where $L_{n,m}$ are the generators of the Lie algebra with a central extension, $\delta_{n+m,0}$ is the Kronecker delta. The appearance of the central charge, c, is a result of the conformal anomaly. It arises in quantum theory in the two-point correlation function of the stress-energy tensor [3]. Models of CFT have a small number of parameters, namely the central charge and conformal weights of the primary fields, h_i, which are the eigenvalues of the operator L_0. These models allow one to obtain analytically or numerically a lot of observables, such as multi-point correlation functions, to study different boundary conditions, behavior in certain geometries and various perturbations of CFT models [3, 4, 5, 6].

For most famous lattice models the CFT counterparts are analytically known. For example, the Ising model corresponds to the CFT characterised by $c = 1/2$ [2]. There are some models like the q-state Potts model [7] and the tricritical Ising model [8, 9] whose lattice counterparts and field parameters are also known.

However, there are models whose counterparts as well as the CFT parameters are unknown. A remarkable example is the site-diluted Ising model [10] which implies presence of magnetic clusters in the whole system. This model is usually characterized by two parameters: temperature T (or exchange energy J) and the probability p of a lattice site to be magnetic. The attempts to describe the critical behavior of this system have a long history and the agreement has not been achieved so far. Here, we can mention papers of Dotsenko and Shalaev [11, 12] who theoretically obtained the critical behavior of correlation length, magnetic suspectibility and heat capacity. There are also many numerical studies of the critical behavior [13, 14, 15], some of them contradict with the theorerical results. For example, the recent paper by Najafi [15] studies the 2D lattice site-diluted Ising model at the critical point. In this study, the simulated domain interfaces were explained by the Schramm-Loewner evolution [16]. As a result, a non-linear behavior of the central charge $1/2 \leq c \leq 1$ for $p = 0.8 - 1.0$ was obtained from its relation to the diffusion parameter [17, 18]. So, Ref. [15] claims that the site-diluted Ising model is different from the pure Ising model.

In this report, we present a new method to obtain the central charge from Monte Carlo (MC) simulations. Although the classical MC techniques [19] do not allow one to extract the free energy, we apply the recently developed Wang-Landau algorithm [20] to calculate the free energy of a lattice model on a torus. The central charge is obtained from the free energy scaling with respect to the torus radii. Our method is universal and can be applied to different lattice models. We illustrate the applicability of the method by the site-diluted Ising model. For this model we calculate the central charge, c, as a function of probability, p, for $p = 0.8 - 1.0$. Our simulations show that for this model $c \approx 1/2$. This indicates that the Ising-like behavior is preserved in the site-diluted model and the central charge is universal in contrast to results of Ref. [15].

2. Determination of the central charge

We study the critical behavior of the site-diluted Ising model. Similar to the pure 2D Ising model, it is formulated on a lattice with magnetic sites at the lattice vertices. We performed simulations on a triangular lattice, though the critical behavior is independent of the lattice type. Each site has a spin $s = \pm 1$ or can be non-magnetic ($s = 0$). Sites are magnetic with the probability p, so the case of $p = 1$ corresponds to the Ising model. The Hamiltonian of the site-diluted model is the same as in the Ising model:

$$H = -J \sum_{\langle i,j \rangle} s_i s_j,\ s_i = \pm 1, \text{or } 0 \tag{1}$$

where $\langle i, j\rangle$ denotes the sum over neighbouring sites of the lattice.

The site-diluted Ising model has a critical point for any given value of p. The critical temperature changes continuously with respect to p. For example, in Figure 1 we show the phase diagram and critical indices obtained in our additional MC simulations on a square lattice by the Wolff cluster algorithm [21, 19].

In our method, we use the Wang-Landau algorithm [20] to simulate the energy distribution $g(E)$ in the partition function

$$Z = \sum_{[s]} e^{-\frac{E[s]}{kT}} = \sum_{E} g(E) e^{-\frac{E}{kT}}.$$

On each step of the algorithm, a new configuration is chosen randomly with the weight given by the accumulated value of $g(E)$. If the new configuration is accepted, the value of $g(E)$ is increased.

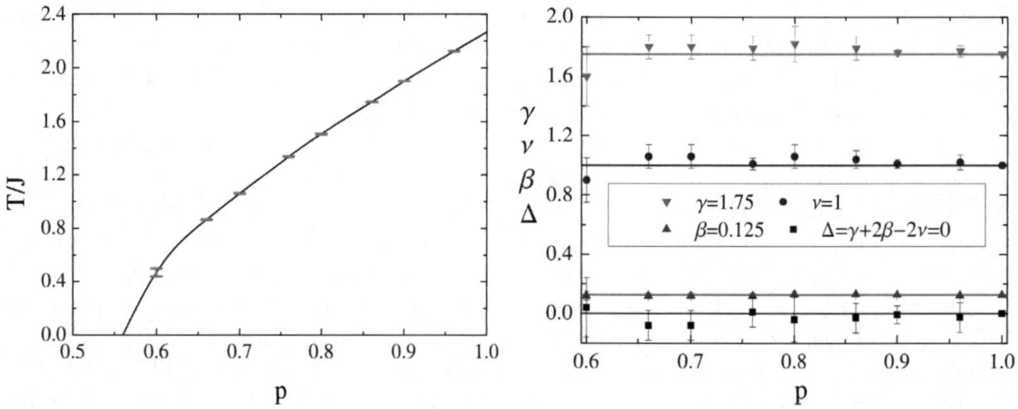

Figure 1. (*Left*) The critical line of the site-diluted Ising model obtained by the Wolff cluster algorithm; the ordered phase lies below the critical line. (*Right*) A comparison of the critical indices for the site-diluted and pure Ising models.

We perform the simulations on $M \times N$ lattice with the periodic boundary conditions. In this way, we obtain the partition function and free energy

$$f(M,N) = -\frac{T_c}{MN} \log Z(M,N) \tag{2}$$

on the torus of circumferences M and N.

The Hilbert space of CFT is a direct sum of the Virasoro algebra representations $V(c,h_i)$: $\mathcal{H} = \bigoplus_i V(c,h_i) \otimes V(c,\bar{h}_i)$. So, the partition function on the torus is given by the combination of the Virasoro characters [3]

$$\frac{Z(q)}{Z_0} = \sum_{i,\bar{i}} \mathcal{M}_{i,\bar{i}} \chi_i(q) \bar{\chi}_{\bar{i}}(\bar{q}),$$

where $q = \exp(2\pi i \tau)$ and $\tau = iM/N$ is the modular parameter of the torus. If $M \gg N$, we can expand the free energy (2) holding only leading contributions with q^{h_i}

$$f(N,M) = f_0 - \frac{T_c \pi c}{6N^2} - \frac{T_c}{MN} \left[\sum_{i \neq 0, \bar{i} \neq 0} \mathcal{M}_{i,\bar{i}} q^{h_i + \bar{h}_{\bar{i}}} \right], \tag{3}$$

since the parameter $q = \exp(-2\pi M/N)$ is very small. We performed simulations for different values of M, N and obtained the dependencies of the free energy expansion (3) on these quantities. The extrapolation to $\frac{M}{N} \to \infty$ allowed us to extract the central charge c from the second term of Eq. (3) because the third term becomes negligible.

3. Simulation results

3.1. Ising model ($p = 1$)

For the Ising model, $p = 1$, we compared the free energy simulations with the exact Onsager solution [22]. It can be seen in Figure 2, where the dependence of the free energy on the temperature is plotted. We show this dependence for different lattice sizes N together with the exact solution.

To determine the central charge, we first need to calculate the limit as $M \to \infty$ of the free energy $f(M,N)$ in Eq. (3). We obtained it by extrapolating the simulation results for

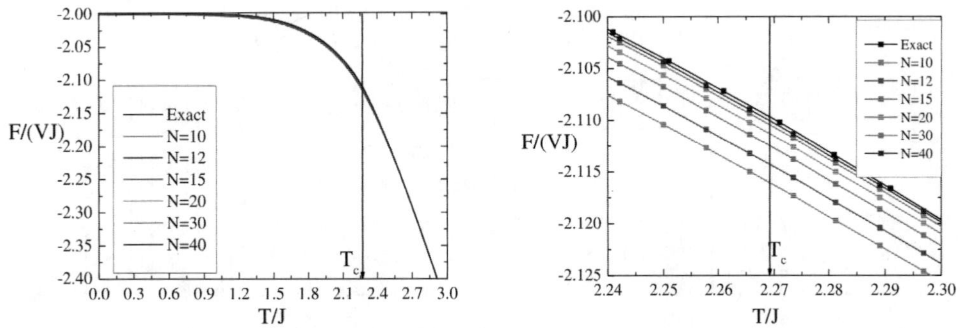

Figure 2. A thermal dependence of the free energy for different values of N in a wide range of temperatures (*left*) and near the critical point (*right*).

different M and N to the limit $M \to \infty$. For example, the dependence of the free energy on $q^h = \exp(2\pi hM/N)$ for the different values of N is shown in Figure 3. We extrapolated the results to $q \to 0$ as $M \to \infty$ and determined $f(\infty, N)$. Then we fitted the dependence (3) and obtained the value of central charge $c = 0.50 \pm 0.02$, which is in a very good agreement with the theoretical result of $c = \frac{1}{2}$.

3.2. Site-diluted Ising model ($p < 1$)

For the site-diluted Ising model, $p < 1$, we studied probabilities $p = 0.8,\ 0.9,\ 0.95$ and performed extensive simulations. The difficulty in this case lies in the dependence of the ground state energy (and resulting free energy) on the initial distribution of non-magnetic sites. So, for a correctness of results, one should repeat a simulation for a large number of initial configurations and then average the obtained values of the free energy. That is not very practical because it is too time consuming. To accurately estimate the average free energy one needs to increase a number of generated configurations for smaller p because of the distribution width broadening (see Figure 4). It makes this way to be impractical for $p < 0.9$.

It is possible to overcome this complication. If we take the initial configuration with the ground state energy value to be close enough to the mean ground state energy for a given value of p, the free energy at the critical point on this lattice will be close to the mean value of free energies. As a result, we generated many lattices with distributed non-magnetic sites, calculated the ground state energy on these lattices and ran the Wang-Landau algorithm on the

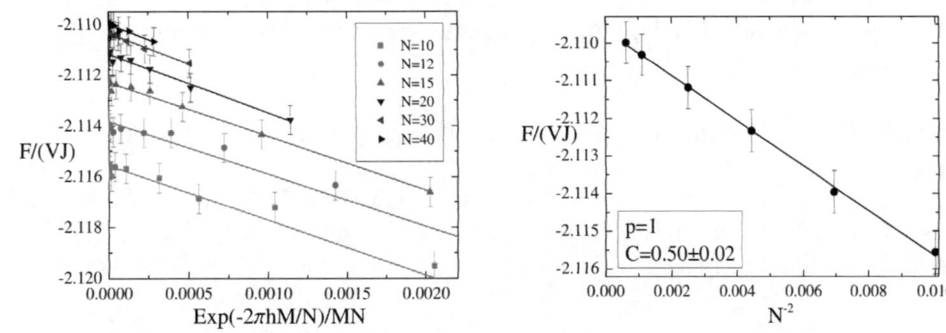

Figure 3. A finite-size scaling of the free energy $f(M, N)$ as a function of $q^h = e^{-2\pi hM/N}$ (*left*) and $f(\infty, N)$ as a function of N^{-2} (*right*).

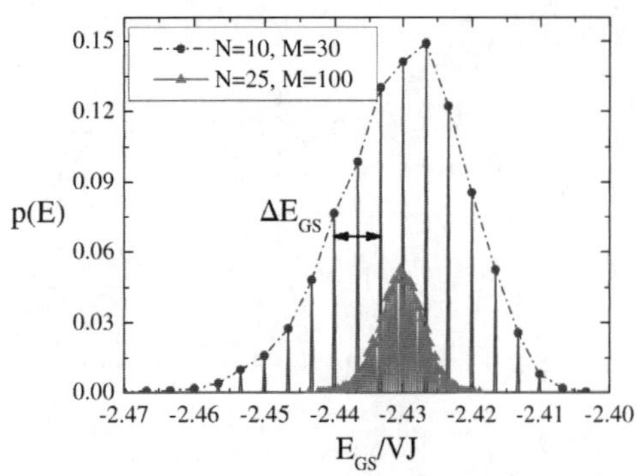

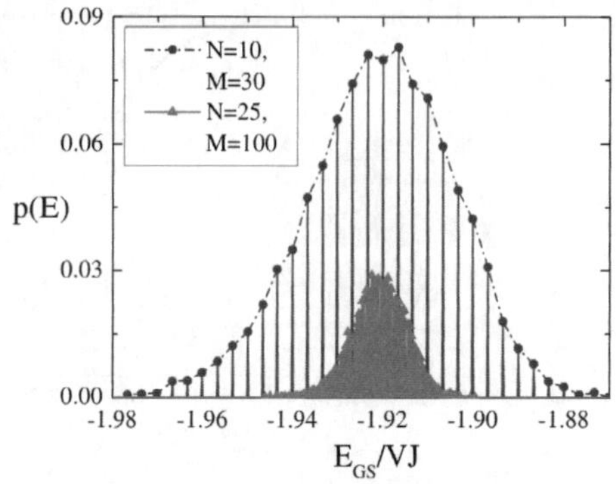

Figure 4. Additional inaccuracy in an energy estimation for the site-diluted Ising model with $p = 0.9$ (*left*) and $p = 0.8$ (*right*).

lattice with the energy closest to the mean ground state energy. Such a way gives an additional inaccuracy ΔE_{GS} for a free energy value as a difference between the ground state energy of the chosen configuration and the mean one. This difference becomes irrelevant for large lattices, but noticeable for lattices used in practice (see Figure 4). We compared these two ways for the case $p = 0.9$ and found that they give very similar central values of the central charge. However, the uncertainty obtained in the second way in our particular simulation was found to be larger by 30%.

The results for the central charge for $p \leq 1$ obtained in the second way are shown in Figure 5 and in Table 1. In contrast to Ref. [15], we see that c remains close to $\frac{1}{2}$ as p decreases, although the inaccuracy becomes larger.

p	T_c	c
1	$3.64095\ldots$	0.50 ± 0.02
0.95	3.368 ± 0.002	0.47 ± 0.06
0.9	3.084 ± 0.003	0.48 ± 0.12
0.8	2.499 ± 0.005	0.54 ± 0.19

Table 1. Summary of the results for the critical temperature and the central charge.

This result is confirmed indirectly by results of the Wolff cluster algorithm [21, 19]. The right plot of Figure 1 shows that the values of the critical indices agree with those for the pure Ising model. So, one can expect that for $0.65 < p < 1$ the site-diluted Ising model has the Ising-like critical behavior with $c = 1/2$.

4. Conclusion and discussion

In this paper, we presented a new method to determine the central charge of a conformal field theory. We employed our method to the pure two-dimensional Ising lattice model and to the site-diluted Ising model, but it can be applied to other models as well. We showed that the site-diluted Ising model has the central charge $c = \frac{1}{2}$ and demonstrates Ising-like behavior for $p < 1$.

The contradiction with the results of Ref. [15] may stem from the different approaches. In our work, we used the global property of the model – a dependence of the free energy on the torus modular parameter. The approach of Ref. [15] is based on the geometry of domain walls, which may be altered by the appearance of the non-magnetic sites. We suppose that the different

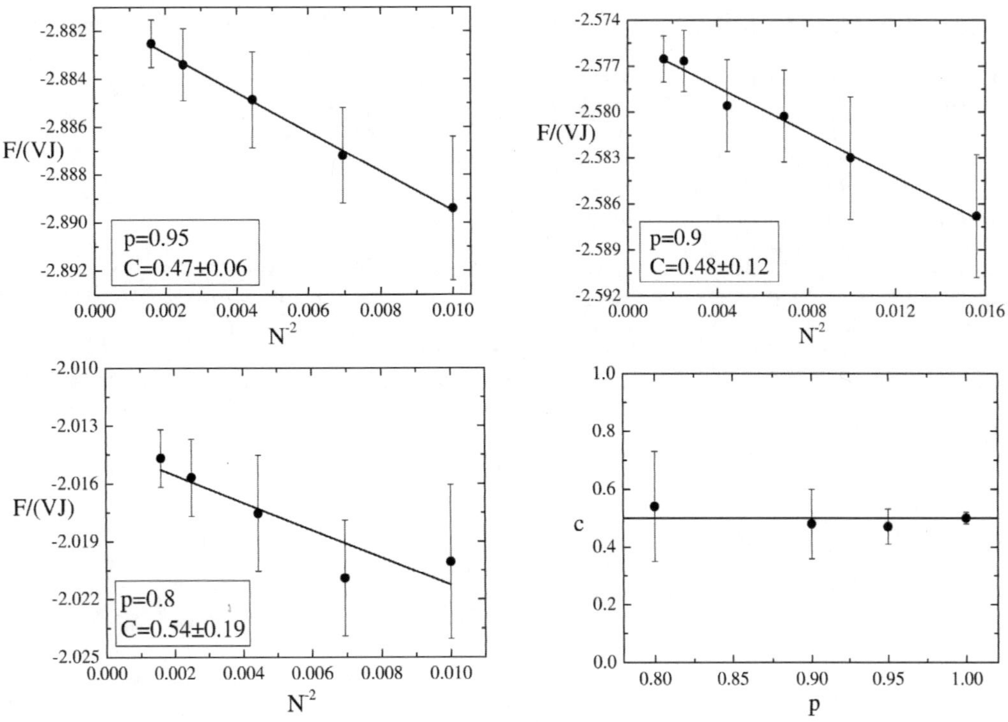

Figure 5. Estimation of the central charge as a plot slope for the site-diluted Ising model with $p = 0.95$ (*top left*), $p = 0.9$ (*top right*) and $p = 0.8$ (*bottom left*); summary of the result for the central charge values (*bottom right*).

boundary condition changing operator should be used at the tip of the growing domain wall to obtain a correct CFT description. The relation between the Schramm-Loewner evolution parameter κ and the central charge c, used in Ref. [15], is applicable, however, only if the boundary condition changing operator is unaltered [17, 18, 23]. So, for the studied model, the numerical values of κ can not be used to determine the central charge.

Acknowledgments

We are thankful to the organizers of the conference Physica.SPb/2016. Anton Nazarov acknowledges the St. Petersburg State University for a support under the Research Grant No. 11.38.223.2015. Alexander Sorokin is supported by the RFBR grant No. 16-32-60143. The calculations were partially carried out using the facilities of the SPbU Resource Center "Computational Center of SPbU".

References

[1] Polyakov A M 1970 *JETP Lett.* **12** 381–383
[2] Belavin A, Polyakov A and Zamolodchikov A 1984 *Nuclear Physics* **241** 333–380
[3] Di Francesco P, Mathieu P and Senechal D 1997 *Conformal field theory* (Springer)
[4] Cardy J 1986 *Nuclear Physics B* **275** 200–218
[5] Cardy J 1984 *Nuclear Physics B* **240** 514–532
[6] Fateev V and Zamolodchikov A 1990 *Physics and mathematics of strings: memorial volume for Vadim Knizhnik* 245
[7] Wu F Y 1982 *Reviews of modern physics* **54** 235
[8] Blume M 1966 *Physical Review* **141** 517
[9] Capel H 1966 *Physica* **32** 966–988

[10] Ballesteros H, Fernández L, Martin-Mayor V, Sudupe A M, Parisi G and Ruiz-Lorenzo J 1997 *Journal of Physics A: Mathematical and General* **30** 8379
[11] Dotsenko V S and Dotsenko V S 1983 *Advances in Physics* **32** 129–172
[12] Shalaev B 1984 *Fizika Tverdogo Tela* **26** 3002–3005
[13] Andreichenko V, Dotsenko V S, Selke W and Wang J S 1990 *Nuclear Physics B* **344** 531–556
[14] Kim J K and Patrascioiu A 1994 *Physical review letters* **72** 2785
[15] Najafi M 2016 *Physics Letters A* **380** 370–376
[16] Schramm O 2000 *Israel Journal of Mathematics* **118** 221–288
[17] Bauer M and Bernard D 2002 *Physics Letters B* **543** 135–138
[18] Bauer M and Bernard D 2003 *Communications in mathematical physics* **239** 493–521
[19] Landau D P and Binder K 2014 *A guide to Monte Carlo simulations in statistical physics* (Cambridge university press)
[20] Wang F and Landau D 2001 *Physical review letters* **86** 2050
[21] Wolff U 1989 *Physical Review Letters* **62** 361 URL `http://journals.aps.org/prl/abstract/10.1103/PhysRevLett.62.361`
[22] Onsager L 1944 *Physical Review* **65** 117
[23] Bauer M and Bernard D 2003 *Physics Letters B* **557** 309–316

Reduction of the dimension of neural network models in problems of pattern recognition and forecasting

A D Nasertdinova[1], V V Bochkarev[1]

[1]Kazan Federal University, Kremlin Street 18, Kazan, Tatarstan, 420008, Russia

E-mail: voinova.anastasija@gmail.com

Abstract. Deep neural networks with a large number of parameters are a powerful tool for solving problems of pattern recognition, prediction and classification. Nevertheless, overfitting remains a serious problem in the use of such networks. A method of solving the problem of overfitting is proposed in this article. This method is based on reducing the number of independent parameters of a neural network model using the principal component analysis, and can be implemented using existing libraries of neural computing. The algorithm was tested on the problem of recognition of handwritten symbols from the MNIST database, as well as on the task of predicting time series (rows of the average monthly number of sunspots and series of the Lorentz system were used). It is shown that the application of the principal component analysis enables reducing the number of parameters of the neural network model when the results are good. The average error rate for the recognition of handwritten figures from the MNIST database was 1.12% (which is comparable to the results obtained using the "Deep training" methods), while the number of parameters of the neural network can be reduced to 130 times.

1. Introduction

Deep neural networks are widely used for solving the problems of image recognition, forecasting and classification. To describe complex dependencies, the network should include many neurons and connections. On the other hand, the amount of empirical data is always limited and using neural networks with too many parameters results in overfitting [1]. In the process of overfitting, the neural network gives good results on the examples used for training, and unsatisfactory results on new data. In recent years, a number of approaches to the problem of overfitting was proposed. The main idea is to reduce the number of parameters of a neural network. One of the most effective methods is Dropout [2]. The main idea is to randomly remove single neurons and their connections from the network during training. This prevents excessive network adaptation. Excluded neurons do not contribute to the training process at any stage of the back propagation algorithm. Therefore, dropping out at least one of the neurons is equal to training of a new neural network. This method allowed us to reduce the recognition error of handwritten numbers from the MNIST database to 0.79%. One of the drawbacks of the dropout method is that it increases training time. Dropout network training usually takes 2-3 times more time than the standard training of neural network of the same architecture.

In [3], a method is proposed to reduce the dimension of two-dimensional data for classification. This method is implemented using the so-called generalized autoencoder (GAE), which fixes the structure of the data space by minimizing the weighted distances between the restored vectors and the original ones, for example, the method of k-nearest neighbours. It differs from the traditional auto-encoder in two aspects:

- GAE examines the compressed representation y_i for the vector x_i and builds a connection between the other data $\{x_j, x_k, ...\}$ by using y_i for reconstruction of each item in the set, not only x_i itself;
- GAE imposes a relational weight s_{ij} on the reconstruction error $\|x_j - x_i'\|$.

Therefore, the GAE fixes the structure of the data space by minimizing the weighted reconstruction error. The application of this method improved performance on the MNIST data set to 3.9%.

In this article, we propose a method of controlling overfitting by reducing the number of independent parameters of neural networks using the principal component analysis. The algorithm was tested on three sets of data: a database of images of MNIST handwritten symbols, time series of the average monthly number of sunspots and the series of the Lorentz system.

2. Use of the principal component analysis for neural networks training

The principal component analysis (PCA) is one of the main techniques for reducing the dimensionality of data and extraction of the necessary information from these data. It is assumed that the data compose a matrix (for example, they are series of measurements of several values). The PCA is based on he idea of transition to a new coordinate system in the original subspace of features where various components (columns of the matrix) will be uncorrelated. Mathematically, this is realized by means of the singular decomposition of the data matrix X, which can be described by the following formula [4]:

$$X = \sum_{r=1}^{p} \sigma_r u_r v_r^H \text{ (or } x_{ij} = \sum_{r=1}^{p} \sigma_r u_{ri} v_{rj}) \quad (1)$$

where u_r and v_r — orthonormal singular vectors, $\sigma_r \geq 0$ – are singular values (in fact, these are the root-mean-square deviations for new variables), and p is the matrix rank. The most informative components in such a coordinate system are those with the largest variance. Accordingly, the data matrix can be approximated in the best way (that is, with minimal loss of information) if we leave M terms in expression (1) corresponding to the largest singular values.

$$\tilde{X} = \sum_{r=1}^{M} \sigma_r u_r v_r^H \quad (2)$$

From now on, we assume that the components are ordered in descending singular values.

The main idea of the proposed method is to apply PCA to the weight matrices of the layers of the neural network. Using PCA, we obtain a network in which the weight matrices of the layers have a relatively small rank. Thus, we achieve regularization of the training of the neural network by significantly reducing the number of independent parameters of the neural network model.

Let's consider the matrix of network weights connecting the two layers consisting of N neurons, and assume that it can be shown as (2), where *M* is significantly smaller than *N*. Due to the fact that singular orthogonality and normalization conditions are imposed on the singular vectors u_r and v_r, the number of independent values will be considerably smaller than N^2. It is not difficult to calculate that the number of independent parameters for the weight matrix of the layer will be equal to:

$$2NM - M^2 \quad (3)$$

To train a neural network with weighted matrices of reduced rank, it is required to correct the scheme of the back propagation algorithm. Accordingly, new software libraries are needed to train the neural networks of this architecture. In this paper, two approaches are considered that allow us to achieve the desired structure of the weight matrices of network layers using the existing libraries of neural computing.

This problem can be solved using two methods:

- by adding a linear layer of smaller dimension between the layers of the network with nonlinear transfer functions;

- by transformation of matrixes of layer weight coefficients of already trained neural network using PCA, by removing components corresponding to small singular numbers. First of all, PCA is applied to the initial layers, after which it is required to train subsequent layers. Then, this procedure is performed for the next layer etc.

Let us consider the first method in more detail (Figure 1). There are two layers with nonlinear transfer functions in the source network (on the left). The layers are connected with a weight matrix ω. We insert an additional linear layer between the layers of the source network. In this case, the vector of output values of the first layer is successively multiplied by the matrix of weights of the linear layer $\omega^{(1)}$, and then - by the matrix of the weights of the second layer $\omega^{(1)}$. The matrix $\omega = \omega^{(2)}\omega^{(1)}$ obtained in such way will have a rank less than M. When using the criterion of a mean square error for the network training, the result will be completely analogous to the application of PCA.

The proposed methods were tested for the problem of recognition of handwritten numbers, as well as for the task of forecasting time series.

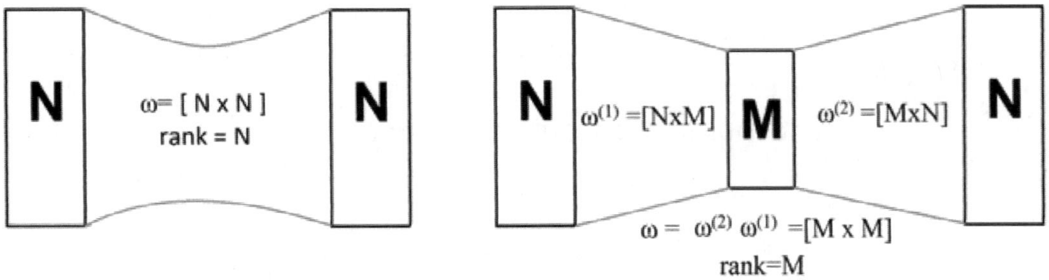

Figure 1. Structure of a neural network without an additional layer (on the left). Structure of a neural network with an additional linear layer with M neurons (on the right).

3. Recognition of handwritten number from the MNIST database

The MNIST database (available at http://yann.lecun.com/exdb/mnist) contains 60 000 black-and-white images of handwritten numbers [5]. The size of each image is 28 x 28 pixels. This database is widely used for comparative testing of training technologies and methods of pattern recognition. The data set does not require primary processing and formatting.

We trained a set of recognizers. Each recognizer checked whether a given number is shown on the successive image. The target value of the network output is 1 if the image has a specified number and is 0 if there is no specified number. The decision threshold was chosen in such a way that the probabilities of error of the first kind (it is decided that there is a number in the image when there is none) and errors of the second kind (it is decided that there is no number in the case when it exists) were equal. In this case, the quality of recognition can be quantified by a single number - the percentage of errors.

An unidirectional three-layer neural network of the form 784-800-800-1 is formed for the given task, which corresponds to 1268000 connections. We use the sigmoidal transfer function, the mean square error as the target function, the training algorithm R-prop (based on the gradient descent method) [6]. This algorithm does not require the use of one-dimensional search procedures and imposes insignificant memory requirements, works rather fast and is recommended for solving large-dimensional problems.

Let us consider the results of comparative testing taking the example of the recognizer for the number "2". The training lasted 500 epochs. The recognition error on the control set was 10.4%. This value is 90.4 times higher than the error on the training set, which indicates the presence of overfitting.

Similar calculations were performed for a neural network with an additional linear layer. To perform comparative testing, we create a neural network of the 784-800-M-800-1 form, where M is the number of neurons in the additional layer. The graph of the recognition error versus the number of neurons in the additional layer (solid lines) is shown in Figure 2. The graph shows the dashed straight lines which represent the forecast of the previous three-layer model for comparing the results. Good results are obtained at M = 50. In this case, the probability of error on the test set is 1.13%. The number of connections between the hidden layers in accordance with the formula (3) decreases by 8.3 times.

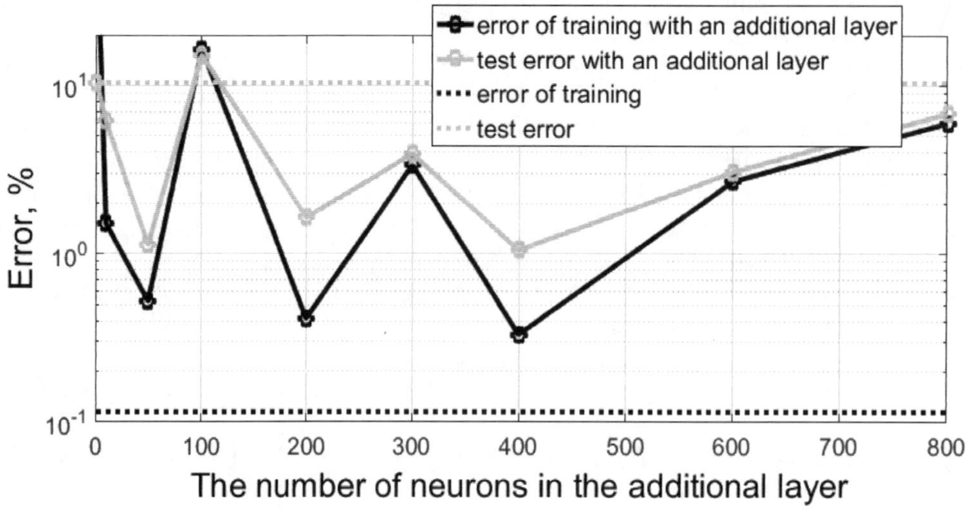

Figure 2. Probability of recognition errors of the number "2" using a neural network with an additional linear layer.

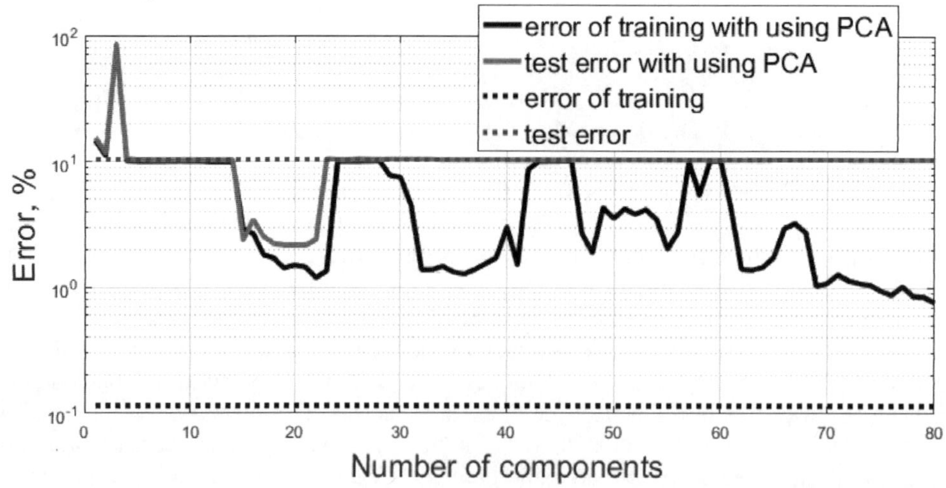

Figure 3. The probability of recognition errors of the number "2" by a three-layer neural network using the principal component analysis for adjusting the weight matrix of the hidden layer.

Also, the second method of reducing the number of independent parameters of the neural network was considered using the same example. The PCA was applied to the weighted matrix of the hidden layer (dimension - 800x800) trained by the three-layer neural network. The error probability values on the training and test set for the network with the matrix transformed by means of the PCA are shown in Fig. 3. It is interesting that by simply removing the redundant information from the network without conducting any additional training, the network results on the test set can be improved in some cases. The best result is obtained at M = 21. In this case, the probability of recognition error on the test set is 2.16%. The number of independent parameters in the weight matrix of the hidden layer can be reduced by 19.3 times.

Better results are obtained after additional training of subsequent layers. As for the given example, it is sufficient to optimize the weights of the output layer. After optimizing the weights, the frequency of recognition errors was 1.33%.

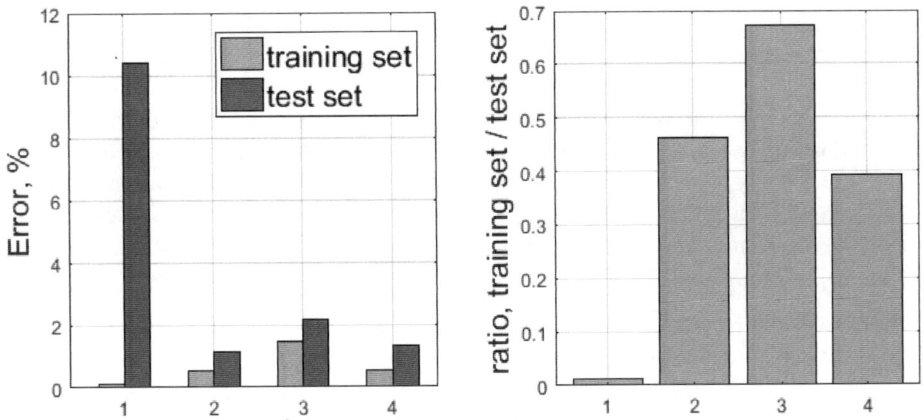

Figure 4. Comparison of recognition errors probabilities of the number "2" using different methods on the training and test set (on the left) The ratio of training error to test error (on the right). The applied method: 1 - three-layer neural network, 2 - network with an additional linear layer, 3 - correction of the weight matrix of the hidden layer using the PCA, 4 - correction of the weight matrix of the hidden layer using the PCA with further optimization of the weights of the output layer.

Comparison of the results of recognition of the number "2" by different methods is shown in Figure 2. It is necessary to pay special attention that the ratio of the error probabilities on the training and test sets in all cases of using the PCA is characterized by reasonable values in contrast to a simple three-layer network. Thus, the use of PCA allowed us to avoid overfitting.

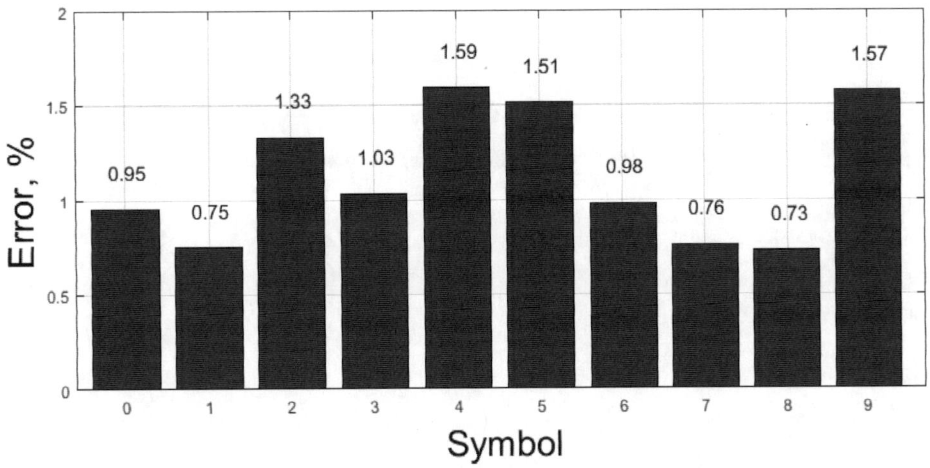

Figure 5. Percentage of recognition errors for different symbols.

The procedures described above were realised for each number from 0 to 9. The best results for each of the figures are shown in the figure 5. The best result was obtained for the figure 8, it was 0.73%. The worst result was obtained for the number 4 which was 1.59%. The average error value was 1.12%. The greatest decrease in the number of independent parameters of the neural network is also observed for the figure 4 (by 133 times).

4. Conclusion

The article suggests a method that allows reduction of the number of independent parameters of the neural network model. Our approach is based on the use of PCA and can be implemented using existing software libraries of neural computing.

The proposed method was tested for solving the problem of recognition of handwritten numbers from the MNIST database. The average percentage of recognition errors was reduced to 1.12%, which is comparable to the results obtained using more powerful methods of in-depth training. At the same time, the number of independent parameters was reduced by 130 times compared to a conventional multilayer neural network. Good results were also obtained for the problem of time series forecasting. It should be noted that the proposed approach can be used together with the methods of in-depth training.

The research of the second author was supported by the Russian Government Program of Competitive Growth of Kazan Federal University.

References

[1] Haykin S 2008 *Neural Networks and Learning Machines* 3rd ed (New Jersey: Upper Saddle River - Pearson Education)
[2] Srivastava N, Hinton G, Krizhevsky A, Sutskever I and Salakhutdinov R 2014 Dropout: A Simple Way to Prevent Neural Networks from Overfitting *J. of Machine Learning Research* **15** 1929-58
[3] Wang W, Huang Y, Wang Y and Wang L 2014 Generalized Autoencoder: A Neural Network Framework for Dimensionality Reduction *IEEE Conf. on CVPRW* **79** 496-503
[4] Gorban A and Kegl B 2007 *Principal Manifolds for Data Visualisation and Dimension Reduction,* ed Wunsch D and Zinovyev A (Springer, Berlin – Heidelberg – New York)
[5] LeCun Y, Bottou L, Bengio Y and Haffner P 1998 Gradient-based learning applied to document recognition *Proceedings of the IEEE* **86(11)** 2278-2324
[6] Riedmiller M and Braun H 1993 *IEEE Int. Conf. on NN* vol 1 586-591

IOP Conf. Series: Journal of Physics: Conf. Series **929** (2017) 012039 doi:10.1088/1742-6596/929/1/012039

Modelling of hydrogen permeability of membranes for high-purity hydrogen production

Yury V Zaika and Natalia I Rodchenkova

Institute of Applied Mathematical Research, Karelian Research Centre, Russian Academy of Sciences, Pushkinskaya Street 11, Petrozavodsk, Russia

E-mail: `zaika@krc.karelia.ru`, `nirodchenkova@krc.karelia.ru`

Abstract. High-purity hydrogen is required for clean energy and a variety of chemical technology processes. Different alloys, which may be well-suited for use in gas-separation plants, were investigated by measuring specific hydrogen permeability. One had to estimate the parameters of diffusion and sorption to numerically model the different scenarios and experimental conditions of the material usage (including extreme ones), and identify the limiting factors. This paper presents a nonlinear mathematical model taking into account the dynamics of sorption-desorption processes and reversible capture of diffusing hydrogen by inhomogeneity of the material's structure, and also modification of the model when the transport rate is high. The results of numerical modelling allow to obtain information about output data sensitivity with respect to variations of the material's hydrogen permeability parameters. Furthermore, it is possible to analyze the dynamics of concentrations and fluxes that cannot be measured directly. Experimental data for $Ta_{77}Nb_{23}$ and $V_{85}Ni_{15}$ alloys were used to test the model. This work is supported by the Russian Foundation for Basic Research (Project No. 15-01-00744).

1. Introduction

The interest in the interaction of hydrogen isotopes with structural materials is multifarious [1–3]. It is sufficient to mention problems in power production, protection of metals from hydrogen corrosion, chemical reactor design, rocket production. Hydrides are able to retain large quantities of hydrogen. Hence, high expectations are attached to hydrogen batteries that avoid using ultra-high pressures or low temperatures. Plastification and thermo-hydrogen treatment of titanium alloys are based on reversible alloying of metals by hydrogen. Some particular problems of the hydrogen materials science connected with this article are investigated in [4,5]. Enthusiasts have been speaking about hydrogen economy along with hydrogen energy [3].

Experimental results show that physical-chemical phenomena at the surface are limiting factors [1,2] along with the diffusion processes inside the metal. Transport parameters also depend on the specific technology of the material manufacturing. Therefore, it is not reasonable to target at 'tabular data'. Instead, effective algorithms for processing experimental curves are necessary. The proposed mathematical model matches the experimental method of hydrogen permeability and takes into account only the main limiting factors and informative capabilities of the experiment. This mathematical research is based on the data on the hydrogen permeability of some promising alloys [6–8]. The experimental technique from [8] was employed.

2. Mathematical model

At first, we briefly describe the experiment. The material sample is a vacuum vessel barrier, pre-heated to a fixed temperature. Degassing of the sample was performed in advance. At the initial time instant the pressure of molecular hydrogen is built up by frequency jumping at the inlet. The declining pressure in the input chamber and increasing hydrogen pressure in the output chamber are measured. The informative capabilities of the experiment are limited, therefore, in the model we shall take into account only the main factors for the filtering problem.

Consider hydrogen transfer through the sample (ℓ is the plate thickness and S is its area). The sample temperature T is constant within a single experiment. The concentration of dissolved hydrogen (in the monatomic state) is sufficiently low and the diffusion flux can be considered proportional to the concentration gradient. Some of H atoms interact with traps. The traps are micro-defects of different nature including, for instance, micro-cavities, which can capture hydrogen. The equations taken as the diffusion model with limited capture within the plate are

$$\frac{\partial c}{\partial t} = D(T)\frac{\partial^2 c}{\partial x^2} - f(T,z,c), \quad \frac{\partial z}{\partial t} = f(T,z,c), \tag{1}$$
$$f := a(T)\Big[1 - \frac{z(t,x)}{z_{\max}}\Big]c(t,x) - a_{\text{out}}(T)z(t,x),$$

where $c(t,x)$ is the concentration of diffusing (atomic) hydrogen; $z(t,x)$ is the concentration of the captured diffusant; D is the diffusion coefficient ($[D] = \text{cm}^2/\text{s}$); $a := a_{\text{in}}$ and a_{out} are the coefficients of absorption and release of H atoms by traps. Sign $:=$ is used in the sense of equality by definition. The magnitude $z_{\max} = \max z$ is considered to be sufficiently small (a metal alloy with high hydrogen permeability is investigated), so the capture is at the correction level and more detailed modelling of the capture process is not necessary. The summand $a_{\text{out}}z$ is negligible within the operating temperature range $T \in [400, 900]$ K. On the scale of the establishment of the stationary permeability mode for relatively thin membranes the traps quickly become saturated and have practically no effect on the penetrating flux. Therefore, hereinafter we suppose that $a_{\text{out}} = 0$. This coefficient is significant in the 'saturation-degassing' experiment at high temperatures, and setting of the value at $a_{\text{out}} > 0$ technically does not complicate the numerical solution of the initial boundary-value problem. Coefficients D, a depend on the sample temperature T in an Arrhenius way with pre-exponential factors D_0, a_0 and activation energies E_D, E_a (R – absolute gas constant): $D = D_0 \exp\{-E_D/[RT(t)]\}$, $a = a_0 \exp\{-E_a/[RT(t)]\}$. Initial data are determined by the fact that the sample had been preliminarily degassed $c(0,x) = 0$, $z(0,x) = 0$.

Nonlinear boundary conditions are derived from the material flux balance:

$$-\frac{dQ_{\text{in}}}{dt} = \big[\mu(T)s(T)p_0(t) - b(T)c_0^2(t)\big]S = -SD\frac{\partial c}{\partial x}\Big|_{x=0}, \tag{2}$$
$$-\frac{dQ_{\text{out}}}{dt} = \big[\mu(T)s(T)p_\ell(t) - b(T)c_\ell^2(t)\big]S = SD\frac{\partial c}{\partial x}\Big|_{x=\ell}. \tag{3}$$

Here, $Q_{\text{in}}(t)$, $Q_{\text{out}}(t)$ are the amounts of hydrogen atoms in the input chamber of volume V_{in} and output chamber of volume V_{out}, $c_0(t) := c(t,0)$, $c_\ell(t) := c(t,\ell)$. The gaseous hydrogen is in molecular form, but for consistency (atomic hydrogen diffuses through the metal) we use atoms as the unit of amount. According to the gas kinetic theory, the incident particle flux density J_p is connected with the pressure p by the Hertz–Knudsen formula: $J_p = p/\sqrt{2\pi mkT}$ (k is the Boltzmann constant, m is the mass of a hydrogen molecule). In the context of the experiment it is convenient to select the following units $[\ell] = \text{cm}$, $[p] = \text{Torr}$. Then we numerically obtain the dependence $J_p = \mu p$, $\mu(T) \approx 2.474 \cdot 10^{22}/\sqrt{T}$ ($[\mu] = 1_{H_2}/(\text{Torr}\,\text{cm}^2\,\text{s})$, $[T] = \text{K}$, the numerical value of T is under the root sign). Only a small part of 'incident' H atoms will however be absorbed into the membrane volume. This fact is reflected by the multiplier s. Thus, μsp is the resulting flux of atoms through the surface into the bulk without differentiation into more elementary stages (adsorption, chemosorption, dissociation, dissolution). One can write $2s$ instead of s and interpret s as the fraction of absorbed H atoms.

Hereinafter, $J_{0,\ell} = bc_{0,\ell}^2$ are the densities of the desorption flux from the sample (deviation from the square desorption law is significant only at extreme temperatures), b is the desorption coefficient ($[b] = \mathrm{cm}^4/\mathrm{s}$). We will usually omit the word *density* assuming that the surface has unit area. We suppose that coefficients s and b also depend on the temperature in an Arrhenius way. Formally, one can write the Arrhenius dependence for s, but in the exponent the 'activation energy' E_s may also be negative as a linear combination of the activation energies and heats of surface processes on the way 'from gas to the solution'. If the saturation pressure $\bar{p}$ on both sides of the membrane is constant at $T = \mathrm{const}$, then the equilibrium concentration $\bar{c}$ of dissolved atomic diffusive hydrogen is eventually established. From the model (2), (3), when all derivatives are zero, we have $\bar{c} \propto \sqrt{\bar{p}}$: $\bar{c} = \Gamma\sqrt{\bar{p}}$, $\Gamma := \sqrt{\mu s/b}$. Therefore, the model is consistent with the Sieverts' law adequacy range ($\bar{c} \propto \sqrt{\bar{p}}$).

The range $[p_{\min}, p_{\max}]$ is narrow, so we restrict ourselves to the relation $z_{\max} = \sigma\bar{c}$, $\sigma \leqslant 0.1$. This restriction will not lead to failure of the Sieverts' law $\left(\bar{c} + z_{\max} \propto \sqrt{\bar{p}}\right)$ while $\bar{c} + z_{\max} \approx \bar{c} = \Gamma\sqrt{\bar{p}}$ is within the experimental accuracy range. It now remains to find the magnitudes of Q_{in}, Q_{out}. Within the transfer time the gas is in the thermodynamical quasi-equilibrium with the surface, therefore we use the formula $N = pV/(kT)$. Here, N is the number of gas particles occupying the volume V at the temperature $T = T_V$ and the pressure p (in the SI system $[p] = \mathrm{Pa}$, $[V] = \mathrm{m}^3$, $[k] = \mathrm{J/K}$). Taking into account the relationships Torr $= 133.322\,\mathrm{Pa}$, $\mathrm{Pa} = \mathrm{J/m}^3$ (formally), we get the following expressions for the corresponding pressures and volumes in the boundary conditions (2), (3) $Q = 2N = \alpha pV/T$, $\alpha \approx 1.931 \cdot 10^{19}$. Here, p, V, T are the numerical values of variables in the selected units (Torr, cm^3, K).

Formally, zero initial conditions and the boundary condition ($x = 0$) of the flux balance are inconsistent at $t \to +0$. In fact, 'instant' hydrogen input lasts for some, even if negligibly short, time. This fact has been taken into account in the solution algorithm of the initial boundary-value problem, which is omitted here. Mathematically, without reference to the computational algorithm, we should talk about the problem in terms of the theory of generalized solutions.

3. Results of numerical modelling

The input data are $T = 400^\circ$ C, $\ell = 0.014\,\mathrm{cm}$, $S = 0.785\,\mathrm{cm}^2$, $V_{\text{in}} = 3000\,\mathrm{cm}^3$, $V_{\text{out}} = 1750\,\mathrm{cm}^3$. The result of approximation of experimental data for the $\mathrm{Ta}_{77}\mathrm{Nb}_{23}$ alloy [8, Fig. 1] by the model curves (inlet and outlet pressures of molecular hydrogen) is shown in Fig. 1. We comment on Fig. 2. According to the accepted model, the equilibrium concentration $\bar{c}$ under saturation conditions $p = \mathrm{const}$, $T = \mathrm{const}$ is determined (when all derivatives are zero) as $\bar{c} = \Gamma\sqrt{p}$, where $\Gamma = \sqrt{\mu s/b}$ is the solubility coefficient. But we shall keep in mind that only dissolved atomic diffusive hydrogen is considered. Degassing is thought to be complete at the concentration $\bar{c} + z_{\max} = \Gamma_{\max}\sqrt{p}$. In the accepted model $\Gamma_{\max} = \Gamma[1 + \sigma]$, and the Sieverts' law holds true $\bar{c} + z_{\max} \propto \sqrt{p}$. In Fig. 2 with a hypothetical value $z_{\max} = 10\,\bar{c}$ total solubility is fixed at an order of magnitude greater than in Fig. 1. If a steady-state flux of hydrogen permeability $J = -D\partial_x c$ is registered in a breakthrough experiment where $p_0(t) = p = \mathrm{const}$ and vacuum is created at the outlet, we have $J = \Phi\sqrt{p}/\ell$, $\Phi := D\Gamma = D\sqrt{\mu s/b}$ on the assumptions that $c_0 = \bar{c}_0 = \Gamma\sqrt{p}$, $c_\ell = 0$. If within a set of experiments ($T = \mathrm{const}$, p is varied) the proportionality $J \propto \sqrt{p}$ is registered with acceptable accuracy, then so-called hydrogen permeability coefficient Φ is determined using the values of p, J (using only the formula $J = \Phi\sqrt{p}/\ell$, and disregarding information about the accepted model and the values of D, Γ). The permeability is determined by the diffusing hydrogen (in the model $\Phi = D\Gamma$). If the permeability Φ *is calculated* from available independent values of the diffusion and solubility coefficients (derived from different experiments and with

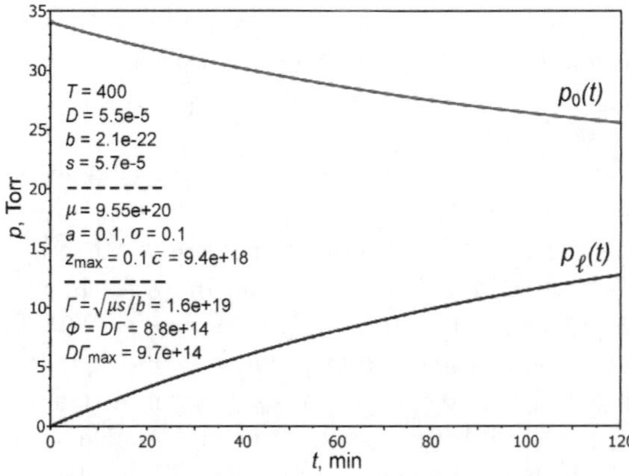

Figure 1. Approximation of pressures, 400°C.

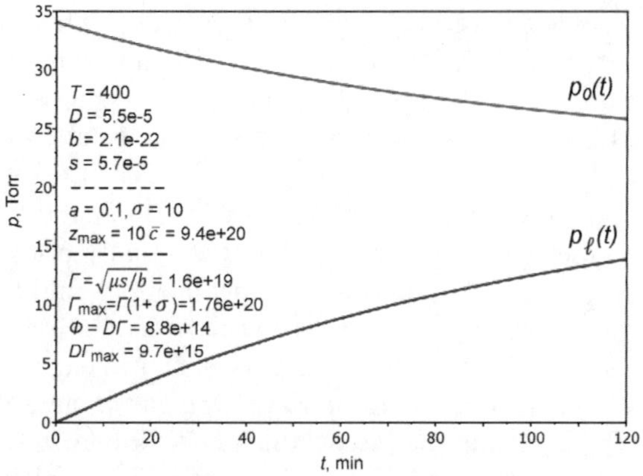

Figure 2. Pressures in the case of $z_{\max} = 10\,\bar{c}$.

total solubility including captured hydrogen), then we can obtain $D\Gamma_{\max} \gg D\Gamma$. Evidently, it is one of the main factors for the scatter in the data on the solubility and permeability coefficients. The problem is the following: for thin membranes in the permeability mode it is difficult to detect the loss of hydrogen captured by traps even if the traps' capacity is high and it essentially influences the total solubility inside the material.

Let us proceed to the analysis of the bulk concentration dynamics during the 'communicating vessels' experiment. We begin with near-to-surface concentrations ($x = 0, \ell$) – Figs. 3, 4. The

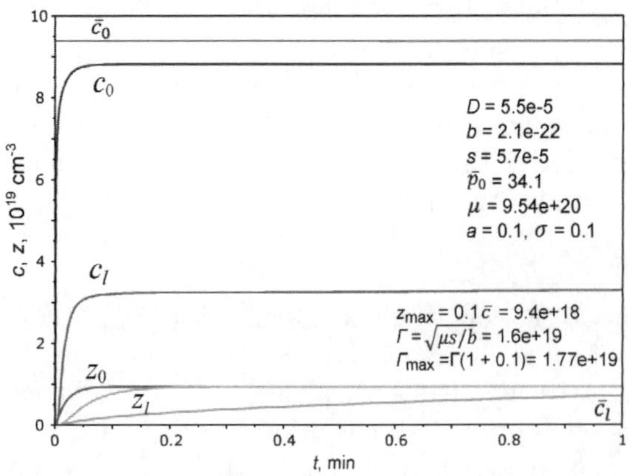

Figure 3. Near-to-surface concentrations, 1 min.

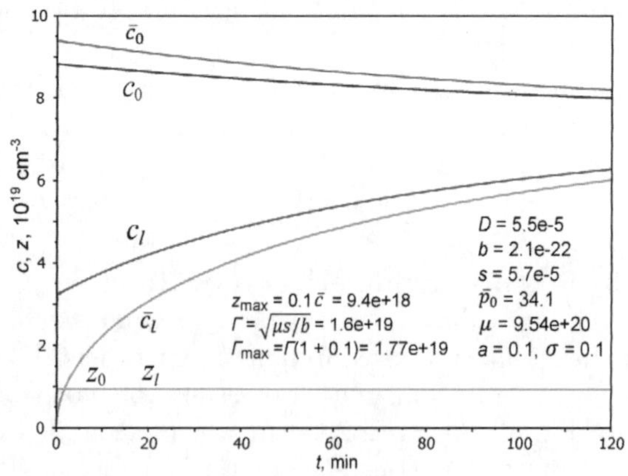

Figure 4. Near-to-surface concentrations, 120 min.

traps became saturated during one minute: $z_0 \approx z_\ell \approx z_{\max}$. The concentrations $c_0(t) = c(t, 0)$, $c_\ell(t) = c(t, \ell)$ quickly get stabilized, but one can see this curve's behavior is local on two-hour scale: the transient regime of splash is changed to a slow trend line. A comparable stabilization time c_ℓ (with respect to c_0) is due to the fast hydrogen permeability of the alloy and $\ell \ll 1$. The high velocity of the transient process is predictable, since the characteristic diffusion time ℓ^2/D ($D \sim 10^{-5}$) is within ten seconds. The total concentration is determined by the sum $c + z$. Quasi-equilibrium (Sieverts') concentrations of diffusing hydrogen $\bar{c}_{0,\ell} \propto \sqrt{p_{0,\ell}}$ are determined using the molecular hydrogen pressure by the formulas $\mu s p_{0,\ell} = b\bar{c}^2_{0,\ell}$: $\bar{c}_{0,\ell}(t) = \Gamma\sqrt{p_{0,\ell}(t)}$. Considering hydrogen capture by traps we should use the sums $\bar{c}_{0,\ell}(t) + z_{\max}$ and $c_{0,\ell}(t) + z_{0,\ell}(t)$.

IOP Conf. Series: Journal of Physics: Conf. Series **929** (2017) 012039 doi:10.1088/1742-6596/929/1/012039

Fig. 4 allows to estimate how much higher the difference of concentrations $c_\ell - \bar{c}_\ell$ at the outlet is compared to that at the inlet $\bar{c}_0 - c_0$. The difference of inlet/outlet concentration range will decline only asymptotically. In the context of the gas separation problem we are firstly interested in the penetrating flux. But the approximation of the gradient $\partial_x c$ by the difference quotient $[c_\ell(t) - c_0(t)]/\ell$ suffers when we use quasi-equilibrium estimates at both the inlet ($\bar{c}_0 > c_0$) and the outlet ($\bar{c}_\ell < c_\ell$). The results for the $V_{85}Ni_{15}$ alloy are qualitatively similar.

4. Fast hydrogen permeability model

The Richardson approximation in a quasi stationary state (when the concentration distribution of diffusing hydrogen in the volume is linear)

$$J(t) = -D\partial_x c = D\ell^{-1}\big[c_0(t) - c_\ell(t)\big] \approx J_R(t) = D\Gamma\ell^{-1}\big[\sqrt{p_0(t)} - \sqrt{p_\ell(t)}\big]$$

is usually used to analyze hydrogen permeability (see, for instance, [6, 8]). It is shown above that substitution of equilibrium concentrations leads to a significant error where the membrane is very thin and the alloy is chosen to achieve fast hydrogen permeability. We formulate the problem of modelling the concentrations $c_{0,\ell}$ using the pressures $p_{0,\ell}$ (the problem is also of interest *per se*) without quasi-equilibrium simplification $c(t) = \Gamma\sqrt{p(t)}$.

The quasi-stationary state is achieved within a time t_0, which is short compared to total experiment time (the traps become saturated and $\partial_x c = -[c_0(t) - c_\ell(t)]/\ell$):

$$\dot{p}_{0,\ell}(t) = \mp\beta_{0,\ell}\big[c_0(t) - c_\ell(t)\big], \quad \beta_{0,\ell} \equiv SD\left[\alpha V_{\text{in,out}}\ell\right]^{-1}T_V, \tag{4}$$

$$\mu s p_{0,\ell}(t) - bc^2_{0,\ell}(t) = \pm D\ell^{-1}\big[c_0(t) - c_\ell(t)\big], \quad t \geqslant t_0 > 0. \tag{5}$$

Since $\dot{p}_\ell(t) = -V_{\text{in}}V_{\text{out}}^{-1}\,\dot{p}_0(t) \Rightarrow p_\ell(t) = p_\ell(t_0) + V_{\text{in}}V_{\text{out}}^{-1}\big[p_0(t_0) - p_0(t)\big]$, it is sufficient to express $c_{0,\ell}(t) = c_{0,\ell}(p_0(t))$ from the boundary conditions (5) and substitute them into the first equation of (4) (the sign depends on the order of the sequence $0, \ell$).

For numerical modelling it is convenient to introduce dimensionless variables

$$X_{0,\ell}(t) = 1 + 2\ell c_{0,\ell}(t)bD^{-1}, \quad a_{0,\ell}(t) = 4\ell^2\Gamma^2 p_{0,\ell}(t)b^2D^{-2} - 1. \tag{6}$$

In addition, the system of equations (5) is compactly written in the symmetric form $a_0 + 2X_\ell = X_0^2$, $a_\ell + 2X_0 = X_\ell^2$. For the variable $X \equiv X_\ell$ we get an equation $\big[X^2 - a_\ell\big]^2 = 4\big[2X + a_0\big]$, which can be solved in radicals (we are interested in the positive root). However, the explicit expression is cumbersome and we will anyway have to numerically integrate the first equation of (4) in the form $\dot{p}_0 = f(p_0)$. Therefore, we shall aim to derive differential equations for $X_{0,\ell}$, since information about the dynamics of boundary concentrations $c_{0,\ell}$ is also of interest *per se*.

Differentiate the equations (5) with respect to time and substitute the pressure derivatives from (4). In the variables $X_{0,\ell}$ we obtain the system ($\mu \propto 1/\sqrt{T}$):

$$\dot{X}_0(t) = -sM_0\big[X_0 - X_\ell\big] \cdot \frac{X_\ell - V_{\text{in}}V_{\text{out}}^{-1}}{X_0X_\ell - 1}, \quad M_0 \equiv \frac{\mu ST_V}{\alpha V_{\text{in}}}, \tag{7}$$

$$\dot{X}_\ell(t) = sM_\ell\big[X_0 - X_\ell\big] \cdot \frac{X_0 - V_{\text{in}}^{-1}V_{\text{out}}}{X_0X_\ell - 1}, \quad M_\ell \equiv \frac{\mu ST_V}{\alpha V_{\text{out}}}. \tag{8}$$

Let us formulate step-by-step the numerical modelling algorithm for the current values of D, b, s (the authors used the freeware Scilab-6). We target at the 'normal' experimental conditions [6–8], including the parameters p, T, ℓ, V, S.

1. We fix $t = t_0$: fast transient processes are omitted (in the example above the duration of transient processes is about tens of seconds on an hours-long experimental time scale). For

the variable $X \equiv X_\ell$ we numerically solve the incomplete biquadratic equation $\left[X^2 - a_\ell(t_0)\right]^2 = 4\left[2X + a_0(t_0)\right]$. The mathematical package will naturally provide us with four roots, from which we choose the positive one for which $X^2 - a_\ell > 2X$ because $\bar{c}_\ell < c_\ell$. From the equations system $a_0 + 2X_\ell = X_0^2$, $a_\ell + 2X_0 = X_\ell^2$ $(t = t_0)$ we find the missing value $X_0(t_0)$. Formally, one equation is enough, but we consider averaging procedures including determination of the values of $p_{0,\ell}(t_0)$.

2. We numerically integrate the system (7), (8) $(t \geqslant t_0)$ with the obtained initial data. The precaution is the following: the data should be redefined as real, otherwise the 'imaginary' component of the concentration begins to accumulate.

3. The change of variables in (6) determines the concentrations $c_{0,\ell}(t)$ from which the model pressures $p_{0,\ell}(t)$, $t \geqslant t_0$, are calculated using the equations (5).

Computational experiments show that the model curves almost coincide $(t \geqslant t_0)$ with those generated by the original model, viz nonlinear distributed initial boundary-value problem. All the varying parameters D, b, s of the original model that influence permeability are essential when implementing the algorithm described above. Thus, the simplified fast hydrogen permeability model does not lose in informativeness concerning transport parameters.

Conclusions

The hydrogen permeability model is focused on the task of material selection for the membrane technology of high-purity hydrogen production. The physico-technical nature of the problem involves the assessment of the main integrated transport indices keeping in mind the limited experiment informativeness. In particular, the parameter of the maximum trap capacity $z_{\max}$ does not consider the variety of traps.

Iterative solution of the distributed nonlinear initial boundary-value problem for approximation of the experimental data (in order to estimate the transport parameters) requires application of specialized software. On the other hand, for thin membranes with fast hydrogen permeability the Richardson approximation (when the boundary concentrations of the dissolved hydrogen are considered as quasi-equilibrium and corresponding to the Sieverts' law) only allows to evaluate the order of the permeability coefficient.

The paper presents a fast hydrogen permeability model in the form of a system of two ordinary differential equations, which is equivalent to the original initial boundary-value problem (excluding the initial transient process). It allows to quickly 'scan' the material within a wide range of parameters: after parametric identification based on limited experimental information the material's operating conditions can be manipulated to include extreme ones. Of course, we are speaking here about mathematical modelling aimed to estimate the transport parameters.

References

[1] Alefeld G and Völkl J (Eds) 1978 *Hydrogen in Metals* (Berlin; New York: Springer) vol I Basic Properties p 427, vol II Application-oriented Properties p 387
[2] Zakharov A P (Ed) 1987 *Interactions of Hydrogen with Metals* (Moscow: Nauka) p 296
[3] Ball M and Wietschel M (Eds) 2009 *The hydrogen economy* (Cambridge University Press) p 646
[4] Zaika Yu V and Bormatova E P 2011 Parametric identification of a hydrogen permeability model by delay times and conjugate equations *Int. J. Hydrogen Energy* **36** 1295–1305
[5] Zaika Yu V and Rodchenkova N I 2013 Hydrogen-solid boundary-value problems with dynamical conditions on surface *Math. Modelling* (New York: Nova Sci. Publishers) 269–302
[6] Dolan M D 2010 Non-Pd BCC alloy membranes for hydrogen separation *J. Membrane Science* **362** 12–28
[7] Song G, Dolan M D, Kellam M E, Liang D and Zambelli S 2011 V-Ni-Ti multi-phase alloy membranes for hydrogen purification *J. Alloys and Compounds* **509** 9322–28
[8] Kozhakhmetov S, Sidorov N, Piven V, Sipatov I, Gabis I and Arinov B 2015 Alloys based on Group 5 metals for hydrogen purification membranes *J. Alloys and Compounds* **645** S36–S40

Equation of state for computer simulation of metal ablation by femtosecond laser pulses

R V Davydov, V I Antonov

Peter the Great St.Petersburg Polytechnic University, Saint Petersburg, Russia

E-mail: davydovroman@outlook.com

Abstract. In this paper a wide-range equation of state for computer simulation of metal ablation by femtosecond laser pulses is proposed. The simulation results are compared with experimental data for several metals (Al, Au, Cu, Ni). A good agreement for numerical results with experimental data shows that this equation of state together with two-temperature hydrodynamic model can be employed for choosing laser parameters to receive better accuracy in simulation of metal ablation.

1. Introduction

The rapid development of ultra-short lasers over the past decades make them very useful tool for laser materials processing. A great deal of progress in nanofabrication of materials has relied on the use of them. Production of nanoparticles can be done in several ways, one of them is laser ablation [1].

Despite extensive research work, accurate prediction of the ablation process is still lacking, because it significantly depending on laser parameters, surrounding medium and target material characteristics. Furthermore, to analyze the physical processes at high energy densities, when laser is used, an adequate description of the thermodynamic properties of matter over a broad region of states including the normal conditions and plasma at high pressures and temperatures is required.

Nowadays a two-temperature model has been widely employed for solving ultra-short laser processing of metals [2-5]. This continuous model describes the energy transfer inside a metal with two coupled generalized heat conduction equations for the temperatures of the electrons and the lattice. To describe the material removal processes two-temperature model is often inserted into a hydrodynamic code. But the choice of equations of state for materials, required for solving hydrodynamic equations, can significantly affect the results.

2. Equation of state

To solve the system of hydrodynamic equations in the two-temperature model we construct wide-range semi-empirical two-temperature equation of state. In this equation a metal is expected to consist of the same electrically neutral atomic cells with atomic weight A and charge Z. To describe thermodynamic properties (such as pressure, temperature and internal energy) we use the Helmholtz free energy. For the single atomic cell it is proposed as the sum of three terms, describing the electronic and ionic components, and the interaction between them:

$$F = F_e + F_i + F_{ei} \tag{1}$$

As the main thermodynamic parameters we use atomic cell volume V and two temperatures: T_e, T_i. The number of free electrons in one atomic cell we call the degree of ionization and denote by the letter y. The pressure of the electrons is assumed to be the pressure of an ideal Fermi gas of electrons with density y/V and temperature T_e, where y is the solution of the modified equation of ionization:

$$\mu_F\left(\frac{y}{V},T_e\right)+I(y)-B(V,T_e)=0 \tag{2}$$

Here μ_F - the chemical potential of an ideal Fermi gas, $I(y)$ - ionization potential, which is a smooth function of y and is constructed as a smoothing spline on the known experimental values of the successive stages of ionization $I(1)$, $I(2)$ etc.

$B(V,T_e)$ - negative correction, which purpose is the description of cold ionization of highly compressed matter:

$$B(V,T_e)=b\left(\frac{z}{V}\right)^{\beta}(1+\mu T_e V^{\sigma})^{-1} \tag{3}$$

where b, β, σ, μ - parameters, determined from shock compression data of metals.

For the free energy of an ideal Fermi gas we use approximation:

$$F_e=yT_e[\frac{3}{5}\varphi-\frac{3}{2}\ln\left(1+\frac{5}{2\varphi}\right)] \tag{4}$$

where

$$\varphi=\frac{1}{2}(3\pi^2)^{\frac{2}{3}}(\frac{y}{V})^{\frac{2}{3}}/T_e \tag{5}$$

From (4) we get formulas for the chemical potential, pressure and internal energy of the free electrons:

$$\mu_F=\frac{1}{2}(3\pi^2)^{\frac{2}{3}}(\frac{y}{V})^{\frac{2}{3}}+T_e[\left(1+\frac{2}{5}\varphi\right)^{-1}-\frac{3}{2}\ln(1+\frac{5}{2\varphi})] \tag{6}$$

$$P_e=\frac{1}{5}(3\pi^2)^{\frac{2}{3}}(\frac{y}{V})^{\frac{5}{3}}+T_e(\frac{y}{V})\left(1+\frac{2}{5}\varphi\right)^{-1} \tag{7}$$

$$\varepsilon_e=\frac{3}{2}VP \tag{8}$$

F_i is describing the transition from the crystal state at low temperatures to the state of an ideal ion gas at high temperatures:

$$F_i=\frac{3}{2}T_i ln\frac{1+\lambda G}{V^{\frac{2}{3}}T} \tag{9}$$

where

$$G=(\frac{4\pi}{3})^{\frac{1}{3}}\frac{y_c{}^2(V)}{V^{\frac{1}{3}}T_i} \tag{10}$$

and $y_c(V)$ assumed to be a known function of the atomic volume V, defined by the equation of ionization, λ - parameter of interpolation.

From (9) we get:

$$P_i=\frac{T_i}{V}\frac{1+3\lambda(\frac{1}{2}-\frac{dln y_c(V)}{dlnV})G}{1+\lambda G} \tag{11}$$

$$\varepsilon_i=\frac{3}{2}T_i\frac{1+2\lambda G}{1+\lambda G} \tag{12}$$

And finally, for P_{ei} we take into account that it should not affect the equation of ionization, full pressure should be equal zero at normal density and zero temperature, component of the internal energy must take into account the Coulomb attraction between the free electrons, at low densities P_{ei} must decrease rapidly, electron-ion interaction at a fixed density should decrease as the temperature increases:

$$P_{ei}=-\frac{1}{5}(3\pi^2)^{\frac{2}{3}}(\frac{y_0}{V_0})^{\frac{5}{3}}(\frac{V_0}{V})^{\frac{4}{3}}\frac{1+\delta}{1+[\delta+(1+\delta)T_e/T^*](\frac{V}{V_0})^{\gamma}} \tag{13}$$

Here V_0, γ, δ, T^* are the parameters, determined from the known values of density, sublimation energy, isothermal compressibility and thermal expansion coefficient.

3. Two-temperature model

In this work we describe the evolution of material parameters using the conservation of mass, momentum and energy of electron and ion subsystems in a two-temperature single-fluid 1D Lagrangian form:

$$\frac{\partial}{\partial t}\left(\frac{1}{\rho}\right)+\frac{\partial v}{\partial m}=0 \tag{14}$$

$$\frac{\partial v}{\partial t}+\frac{\partial P}{\partial m}=0, v=\frac{\partial x}{\partial t} \tag{15}$$

$$\frac{\partial \varepsilon_i}{\partial t}+P_i\frac{\partial v}{\partial m}=\frac{\alpha_{ei}}{\rho}(T_e-T_i) \tag{16}$$

$$\frac{\partial \varepsilon_e}{\partial t}+P_e\frac{\partial v}{\partial m}=\frac{\partial}{\partial m}\left(k\rho\frac{\partial T_e}{\partial m}\right)-\frac{\alpha_{ei}}{\rho}(T_e-T_i)+J_L \tag{17}$$

where m is the mass coordinate, $dm=\rho dx, m=\int_{x_0}^{x}\rho dx$, direction of x-axis is chosen perpendicular to the irradiated surface of the metal, v is the velocity, t is the time, ρ is the density, ρ_0 is the initial density, P_e and P_i are the pressures of electrons and ions, T_e and T_i are the temperatures of electrons and ions, ε_e and ε_i are the internal energies of electrons and ions, $P=P_e+P_i$ and $\varepsilon=\varepsilon_e+\varepsilon_i$ are the full pressure and the internal energy.

For the coefficient of electron-ion relaxation α_{ei} we use formula from work [6] and energy of the absorbed laser radiation J_L, described by:

$$J_L=\frac{F_{abs}}{\tau_L\delta\sqrt{\pi}\rho}\exp\left(-\frac{t^2}{\tau_L^2}\right)\exp\left(-\frac{x(m,t)-x(m_0,t)}{\delta}\right) \tag{18}$$

where τ_L is the laser pulse duration, δ is the skin depth of metal, F_{abs} is the laser radiation energy.

The electron wide-range thermal conductivity coefficient k is constructed as an interpolation between limiting metal and hot plasma cases [7]. The viscosity and thermal conductivity of the ions are neglected in this model. To calculate the ablation depth, we perform an integration of the mass flux through the surface using the expression:

$$d=\frac{1}{\rho_0}\int_0^t(\rho v)|_{x=x_0}dt' \tag{19}$$

4. Results

We compare simulation results (ablation depth) after a single laser shot with experimental data for aluminium, gold, copper and nickel. Ultrashort laser pulses are generated by an amplified all-solid-state Ti:Saphire laser chain [8]. Low energy pulses are extracted from a mode-locked oscillator. The pulses are then injected into an amplifying chain including: an optical pulse stretcher, a regenerative amplifier associated with a two-pass amplifier using a 20 W Nd:YLF laser as pumping source, and a pulse compressor. Linearly polarized pulses with wavelength centered on 800 nm, an energy of 1.5 mJ at 1 kHz repetition rate and typical duration of 100 fs were delivered.

The samples are mounted on a three-motorized-axes system with 0.5 μm accuracy. Experiments are performed in the image plane of an aperture placed before the objective. Focusing objectives of 25 mm or 10 mm focal lengths to obtain fluences in a range of 0.1 to 10 J/cm^2 with the same beam size, 16 μm in diameter, in the image plane.

For ablation depth measurements, grooves are machined by moving the sample, which is adapted such that each point on the groove axis undergoes 8 consecutive irradiations at each target pass. The number of times the sample passes in front of the fixed beam can be adjusted. Figure 1 shows a scanning electron microscopy (SEM) picture of the machined grooves on copper for 2, 4, 6, 8 and 10 passes.

IOP Conf. Series: Journal of Physics: Conf. Series **929** (2017) 012040 doi:10.1088/1742-6596/929/1/012040

The groove depth is measured by an optical profilometer with a 10 nm depth resolution. The ablation rates for each groove are deduced taking into account the sample speed, the repetition rate of the laser, the beam size and the number of passes. For each energy density an ablation depth averaged over a few tens of laser shots.

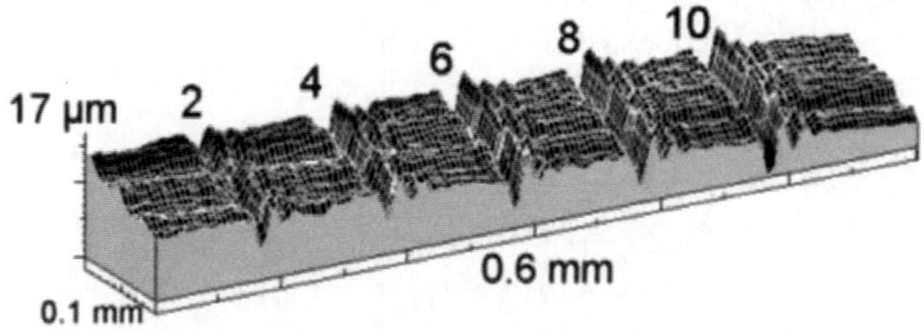

Figure 1. SEM picture of machined-groove profiles, from 2 to 10 passes, on a copper sample.

As we can see in figure 2 results of the simulation are close to experimental data for all metals in the range of 0.1 to 10 J/cm^2.

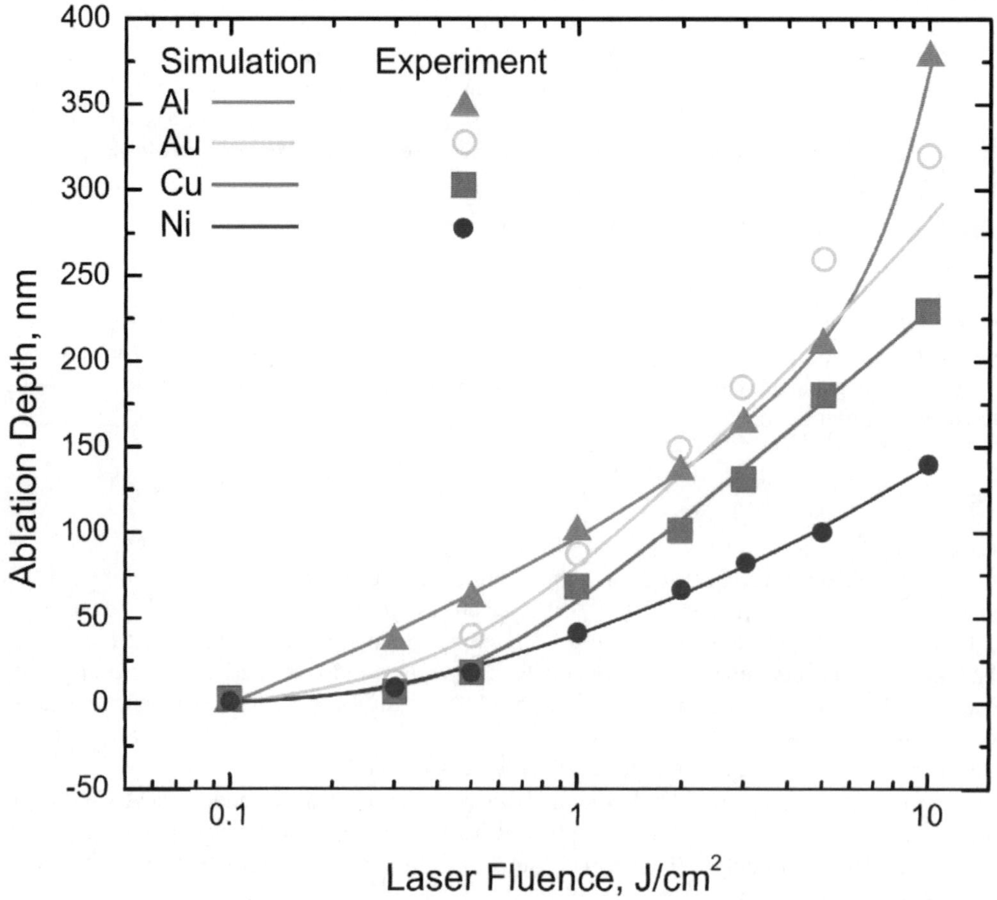

Figure 2. Experimental and numerical ablation depth on a target of different metals as a function of the laser fluence obtained with a 100 fs laser pulse.

Some disagreement is observed for gold and further investigation is required. The possible explanation of this problem is ignoring thermal excitation of d band electrons in Au [9]. These electrons can significantly contribute into heat capacity and electron-ion iteraction and provide some underestimation of ablated depth.

5. Conclusion

A wide-range equation of state for computer simulation of metal ablation by femtosecond laser pulses is constructed. The simulation results for ablation depth using this equation together with two-temperature hydrodynamic model are in good agreement with experimental data for aluminum, gold, copper and nickel. It shows that this equation of state together with two-temperature hydrodynamic model can be employed for choosing laser parameters to receive better accuracy in simulation of metal ablation.

References

[1] Makarov G N 2013 *Phys. Usp.* **56** 643–682
[2] Li Q, Lao H, Lin J, Chen Y and Chen X 2011 *Appl. Phys. A.* **105** 125-129
[3] Ren Y, Chen J and Zhang Y 2012 *J. Heat Mass Transfer* **55** 1620-1627
[4] Mishra S and Yadava V 2013 *Opt. Laser Technol.* **48** 461-474
[5] Zhang J, Chen Y, Hu M and Chen X 2015 *Appl. Phys.* **117** 063104
[6] Petrov Yu V 2005 *J. Laser and Particle Beams* **23** 283-289
[7] Povarnitsyn M E, Andreev N E, Levashov P R, Khishchenko K V and Rosmej O N 2012 *Phys. Plasmas* **19** 023110
[8] Colombier J P, Combis P, Bonneau F, Harzic R Le and Audouard E 2005 *Phys. Rev. B* **71** 165406
[9] Lin Z, Leveugle E, Bringa E and Zhigilei L 2010 *J. Phys. Chem.* **114** 5686-5699

Algorithm for detection the QRS complexes based on support vector machine

Van G.V.*, Podmasteryev K.V.

Scientific-Educational Center of "Biomedical engineering", Orel State University named after I.S.Turgenev, 29 Naugorskoe sh., Orel, 302020, Russia

*gennady.van@mail.ru

Abstract. The efficiency of computer ECG analysis depends on the accurate detection of QRS-complexes. This paper presents an algorithm for QRS complex detection based of support vector machine (SVM). The proposed algorithm is evaluated on annotated standard databases such as MIT-BIH Arrhythmia database. The QRS detector obtained a sensitivity Se = 98.32% and specificity Sp = 95.46% for MIT-BIH Arrhythmia database. This algorithm can be used as the basis for the software to diagnose electrical activity of the heart.

1. Introduction

QRS complex is a dominant of electrocardiographic signal, Figure 1. Quantitative and qualitative analysis of the QRS - complex as well as the time sequence of its appearance is one of the main tools in heart disease diagnosis. The deviation of the QRS-complex form, the change of its amplitude and time characteristics and physiological rhythms disturbance are caused by various heart diseases. The problem of exact finding of a QRS complex position on the recording and defining the diagnostics of significant characteristics becomes obvious. However, the solution of this task is complicated by personal variability, wide ranges of the norm, and also by the presence of noises and artefacts of different origin in ECG signal, Figure 2.

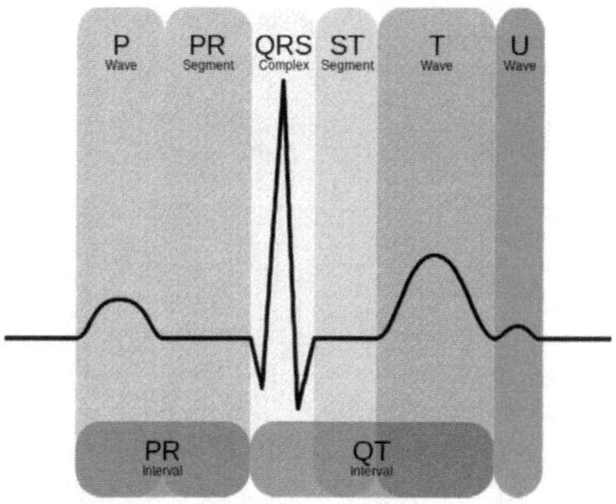

Figure 1 – Electrocardiogram of a single heartbeat. Modified from source picture (url:http://en.wikipedia.org/wiki/Electrocardiography#/media/File:EKG_Complex_en.svg)

Digital signal processing development led to significant improvement of QRS-complexes detection algorithms. These algorithms can be classified by the mathematical apparatus taken as their bases:

algorithms based on time and frequency domain transformations; syntactic algorithms; combined algorithms; algorithms based on machine learning methods [1].

However, despite the consistent development of new algorithms, the problem of fast and accurate QRS detection is still urgent. One of the most promising approaches is the use of artificial intelligence methods and machine learning. These methods are relatively easy to use, allow real-time implementation and have high sensitivity and specificity.

The aim of this study was to develop an algorithm of QRS-complexes detection. This algorithm is based on a support vector machine method because this method possesses high efficiency and high-speed performance that allows it to be used in online time [2, 3].

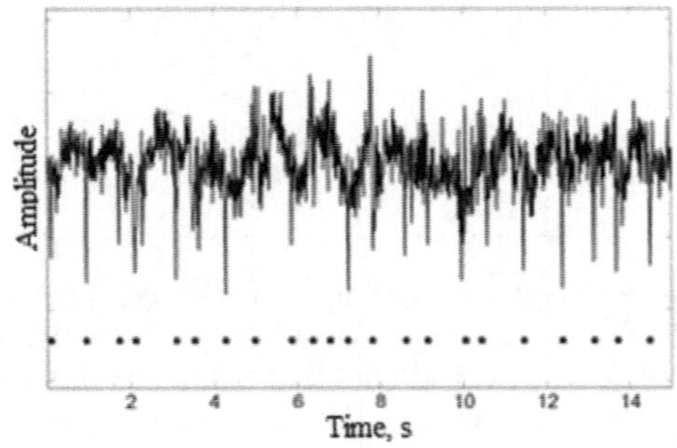

Figure 2 – Noisy ECG signal. Adapted figure from [4].

2. Methodology

In this section, we describe the proposed algorithm for the detection of QRS-complex in MIT-BIH arrhythmia database using SVM classifier. The algorithm's steps are considered in Figures 3 and 4. The implementation of the algorithm and the results of the research were received in Jupiter for the Python 3 programming language with the use of the machine learning library scikit-learn.

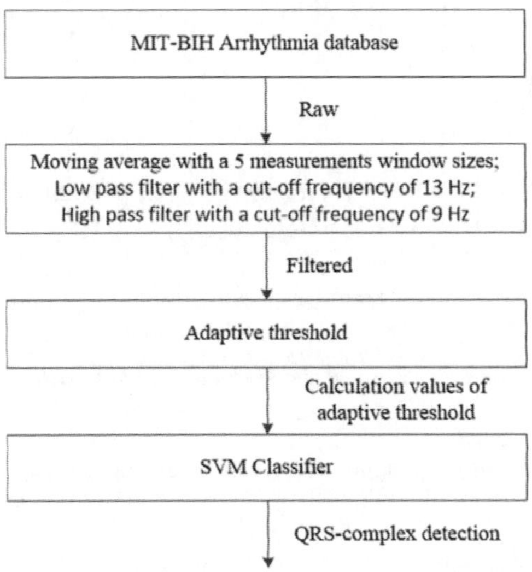

Figure 3 – Schematic representation of intermediate steps for SVM algorithm implementation.

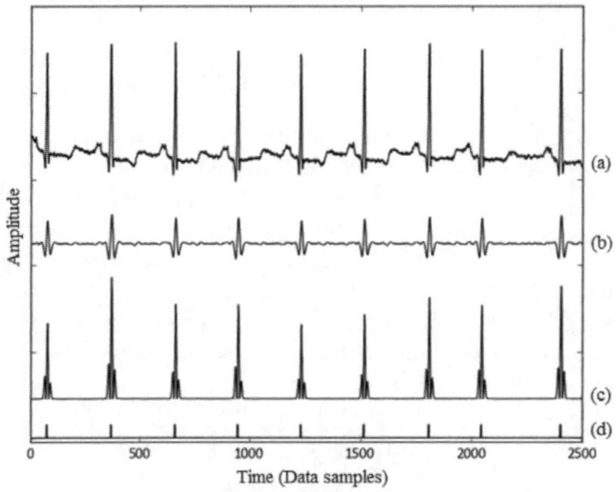

Figure 4 – Results obtained at each step of the algorithm: (a) raw ECG, (b) filtered ECG, (c) filtered signal is squared and (d) QRS locations.

2.1. ECG databases

For the training and testing the signals from the MIT-BIH Arrhythmia database are used. This database consists of 48 half-hour recordings for a total of 24 h of ECG data. Each one has a duration of 30 min and include two leads – the modified limb lead-II and one of the modified leads V1, V2, V4 or V5 [5], sampled at 360 Hz with resolution of 5 μV/bit. Two cardiologists have annotated all beats. This 24 h MIT-BIH database contains more than 109,000 beats. A typical signal from MIT-BIH Arrhythmia database is shown in Figure 5 [6, 7].

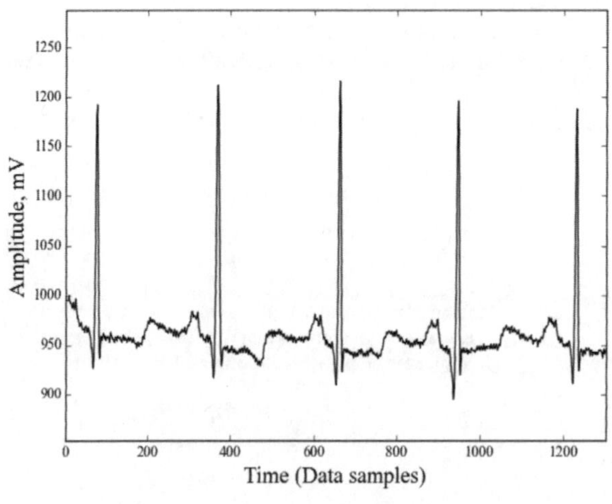

Figure 5 – Typical signal from MIT-BIH Arrhythmia database.

2.2. ECG signal pre-processing

The received raw digital ECG signal of a record is shown in Figure 4a. For the increase of efficiency of an algorithm the procedure of preliminary processing of a signal is used as the signal contains different noise and artifacts.

For the conversion a low pass filter with a cut-off frequency of 13 Hz and high pass filter with a cut-off frequency of 9 Hz are used, as well as the function of the moving average with a 5 measurements window sizes. The filtered ECG signal is shown in Figure 4b. In addition to such informative sign as rise speed of a signal, the filtered signal is squared, Figure 4c.

2.3. Informative features

In this stage, the adaptive threshold values by means of the informative features are created. For this algorithm the combination of such informative features as rise speed of a signal (as QRS complex has the greatest climb rate), and the correlation of QRS complexes forms were selected (as QRS complex has a specific characteristic form) are used. [8, 9].

2.3.1 Rise speed of signal

Informative feature as rise speed of a signal is chosen, because QRS complex possesses the greatest climb rate. This informative feature is implemented as follows: an array of the values of the slope of the tangent to each point of the filtered and squared ECG signal. After that SVM classification function is applied to the received selection.

2.3.2 Correlation of QRS complexes forms

Informative feature as correlation forms of QRS-complexes are chosen, because QRS-complex possesses a specific form. This informative feature is implemented as follows: Firstly, test pulse is created based on 1000 QRS-complexes with the R-peak in the middle lasting 51 counting (141.67 ms), which occurs after their averaging, for this procedure the signal №100 is used. Secondly, an array of correlation coefficients forms of QRS-complexes in a moving window is created. After that SVM classification function is applied to the received selection.

2.4. Training phase

For train recording signals №100, №104, №214, №200, №205 are used. The training set consists of 30,000 values for each of the signs. The function of classification was calculated for each informative feature separately. Using data that indicate the R-peak without pathology, as well as data that correspond to 120 counts (333.34 ms) before and after the R-peak without pathology.

2.5. Classification phase

The main condition for testing algorithm was the use of records which do not participate in training. This was the reason that, if testing is done with the pattern used in learning or training, the accuracy will be artificially high. Signals from MIT-BIH database with 30 min of duration are used for testing. After testing, a train of 1's is obtained at the output of SVM classifier. Then this train of 1's is picked and by using their duration, average pulse duration of 1's is evaluated. Those trains of 1's, whose duration turns out to be more than the average pulse duration are detected as QRS-complex and the others are discarded. The locations of the QRS-complexes, as detected by SVM, are shown is Figure 4d.

3. Testing and results

For the calculating of algorithm efficiency such indicators were used:

Algorithm efficiency can be estimated by calculating specificity *Sp* – a probable assessment of absence of false positive result and sensitivity *Se* – a probable assessment of correctness of a QRS complex detection:

$$Sp = \frac{TP}{TP + FP} \quad (1)$$

$$Se = \frac{TP}{TP + FN} \quad (2)$$

where TP is a correct detection of a QRS-complex; FN is when the algorithm fails to identify a QRS-complex; FP – if algorithm detects a non-QRS-complex as a QRS-complex [7]. The efficiency indicators of this algorithm are: sensitivity – 98,32%; specificity – 95,46%.

4. Conclusion

The efficiency indicators of this algorithm are: sensitivity – 98,32%; specificity – 95,46%. Besides, this algorithm possesses a high-speed performance that allows it to be used in real time.

Further increase of efficiency of this algorithm is possible due to the use of bigger quantity of informative signs; changes of training procedure and teaching selection; changes of machine training method which has been the basis for the algorithm; changes of preliminary signal processing procedure; changes of output data interpretation procedure.

References

[1] Elgendi M. et al., "Revisiting QRS detection methodologies for portable, wearable, battery-operated, and wireless ECG systems," *PLoS One*, 2014. **vol. 9**, 2014.

[2] Van G.V., Podmasteryev K.V. Review algorithm for detection the QRS-complex based on machine learning // *Symposium Proceedings of 2nd International Scientific Symposium "Sense. Enable. SPITSE." 2015*. (St.Petersburg, Russia, 22 June - 03 July 2015). - P. 61-63.

[3] Hastie T., Tibshirani R., Friedman J. The Elements of Statistical Learning: Data Mining, Inference, and Prediction. — Springer-Verlag, 2009.

[4] S. J. Redmond, Y. Xie, D. Chang, J. Basilakis, and N. H. Lovell, "Electrocardiogram signal quality measures for unsupervised telehealth environments.," *Physiol. Meas*., **vol. 33**, no. 9, pp. 1517–33, Sep. 2012.

[5] Massachusetts Institute of Technology. MIT-BIH ECG database. <http://www.physionet.org/cgi-bin/atm/ATM>.

[6] Moody G.B, Mark R.G. The impact of the MIT-BIH Arrhythmia Database // *IEEE Eng in Med and Biol*. 2001. P.45-50. **Vol 20** №3.

[7] Saini I. et al. QRS detection using K-Nearest Neighbor algorithm (KNN) and evaluation on standard ECG databases // *Journal of Advanced Research*. 2013. **Vol. 4**, 331–344, №4.

[8] Van G.V., Podmasteryev K.V. The analysis of informative features for R-peak detection // *XII Internantional Scientific Conferense «Physics and Radioelectronics in Medicine and Ecology» with Scientific Youth session PhREME'2016 (Vladimir - Suzdal, Russia, 5-7 July, 2016), Proceedings, Book II*. - P. 126-129.

[9] Hampton J.R. The ECG made easy. London: Churchill Livingstone, 2003.

IOP Conf. Series: Journal of Physics: Conf. Series **929** (2017) 012042 doi:10.1088/1742-6596/929/1/012042

Electrical and optical properties of iodide nanoparticles embedded into zeolite porous matrices

A V Cvetkov[1], S E Gango[1], M S Ivanova[1], S D Khanin[2], A E Lukin[1], V G Solovyev[1], S V Trifonov[1], V L Veisman[1]

[1]Pskov State University, Lenin square, 2, 180000 Pskov, Russia
[2]Herzen State Pedagogical University of Russia, Moyka-river embankment, 48, 191186 St. Petersburg, Russia

E-mail: aleksandr23031994@gmail.com

Abstract. Experimental study of the iodine nanoparticles effect on the optical and electrical properties of zeolite - based matrix nanocomposite materials has been carried out. Phase transitions in the nanocomposites under study have been found.

1. Introduction

Zeolites are well-known regular porous crystalline solids, which are usually referred to as molecular sieves in chemistry and chemical physics [1, 2]. On the other hand, one can use these dielectrics as host matrices to design wide variety of unique nanocomposite materials by infiltrating their voids with guest substances [3, 4]. The guest nanoparticle systems in host zeolite matrices are characterized by low dimensions (~ 1 nm), regular shapes and arrangements of guest species and their high concentration ($\sim 5 \times 10^{20}$ см$^{-3}$). These advantages are due to the crystalline structure of zeolites containing identical cages and channels of molecular dimensions regularly located in their framework.

In the last decades different porous dielectric host matrices filled with iodine guest species were studied by optic and electric methods [5-7].

In this work, we have studied some optical, electrical and thermoelectrical properties of nanocomposite materials prepared by infiltration of cages and channels of synthetic type A zeolite and zeolite-like AFI ($AlPO_4 - 5$) porous host matrices with iodine guest substance from vapour phase.

2. Structure of zeolites and zeolite-like crystals under study

Zeolite frameworks are constructed by the linking TO_4 tetrahedra (usually $T = Si$, Al) with shared oxygen atoms [1, 2]. Cages and channels of their open structures are regularly arranged and contain exchangeable alkali or alkaline-earth cations, compensating the negative electrical charge of alumina tetrahedra $[AlO_4]^-$ in their framework, as well as water molecules. Both cations and water molecules move freely inside the zeolite. Water can be reversibly removed when the temperature increases during dehydration process, while zeolite porous structure itself usually remains stable up to high temperatures. The general chemical formula of these crystalline aluminosilicates is given by [1, 2]

$$Me_{x/n}\left[Al_x Si_y O_{2(x+y)}\right] \cdot zH_2O \ ,$$

where Me^{n+} is a cation. The *Si*-to-*Al* ratio $R = y/x$ depends upon the type of zeolite and the method of its synthesis.

The ideal framework of zeolite NaA $Na_{12}[Al_{12}Si_{12}O_{48}]\cdot 27H_2O$ contains supercages and sodalite cages with inner diameters of 1.14 nm and 0.66 nm, respectively (figure 1).

Ideal unit cell framework formula of $AlPO_4-5$ aluminophosphate is $Al_{12}P_{12}O_{48}$.

This zeolite-like matrix is very convenient for infiltration of its nanoporous structure with guest substances because the concentration of aluminium atoms in its lattice is equal to that of phosphorus atoms, and therefore its one-dimensional channels are free from exchangeable cations. These parallel channels with the inner diameter 0.73 nm are oriented along the crystal ***c*** - axis (figure 2).

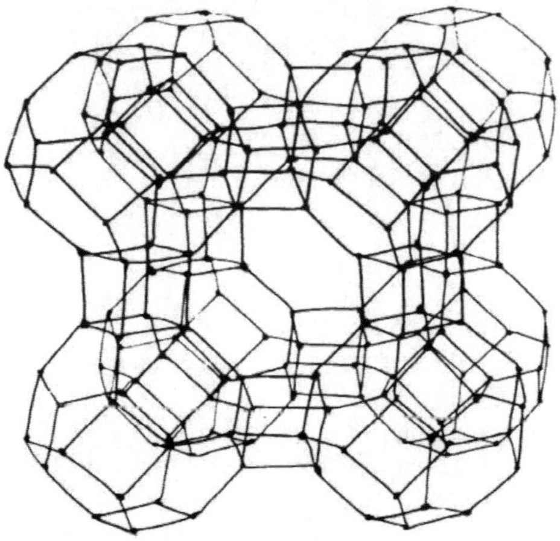

Figure 1. Framework of zeolite NaA [4].

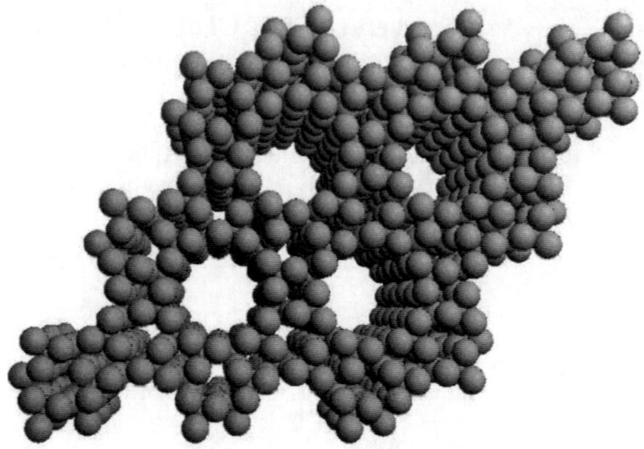

Figure 2. Structural model of zeolite-like aluminophosphate type AFI [4]. Large spheres represent oxygen atoms.

3. Experimental procedure

To prepare I/A or I/AFI nanocomposites, incorporation of iodine species into voids and channels of type A zeolite or AFI zeolite-like matrices during iodine vapour phase adsorption at 580-770 K for 6-12 h has been used.

The diffuse reflectance spectra $R_d = f(\hbar\omega)$ of the powdered samples were determined by the standard method in the photon energy $E = \hbar\omega$ spectral range from 1.05 to 5.65 eV. Magnesium oxide served as a reference sample. The measured reflectance spectra $R_d(\hbar\omega)$ were converted into the values of the Gurevič – Kubelka – Munk function, which describes the optical absorption of the sample in relative units [8, 9]:

$$F(R_d) = \frac{(1-R_d)^2}{2R_d}.$$

Alternating-current (AC) conductivity of pressed pellets prepared from zeolite powder and KBr at the pressure of 100 MPa was measured at the frequency of 1 kHz with graphite electrodes. However, this method gives no possibility to avoid disturbing effects of grain boundaries and interfaces, arising from the sample polycrystallinity. Therefore, in order to study electrical properties of individual microscopic zeolite single crystals and zeolite-based nanocomposite samples (~20-100 microns in size), different types of direct-current (DC) measuring cells have been proposed by Markov et al. [4, 10, 11]. Using these cells in our work, we have fixed the microscopic I/A or I/AFI single crystals on quartz substrates between two push indium contacts. Thermal electromotive force of small samples was measured by pulse method proposed in [11-13].

4. Results and discussion

In contrast with the pronounced optical anisotropy of I/AFI nanocomposites [5, 12, 14] in transmitted light with two perpendicular directions of incident polarization, we did not find any light absorption anisotropy of I/NaA single crystals under polarized optical microscope. This difference is obviously due to formation of quasi one-dimensional iodine chains observed in [5, 14] inside AFI parallel channels and isotropic distribution of iodine species inside three-dimensional system of cages and channels of zeolite type A nanoporous host matrix.

The calculated optical absorption spectra of the NaA initial zeolite, bulk iodine, and I/NaA nanocomposite are depicted in figure 3. It can be seen from curve 1 that the zeolite host matrix is transparent in the studied spectral range. The absorption edge of the massive iodine is observed at 1.26 eV (curve 2). This agrees with the electronic band gap of the bulk crystalline iodine (1.25 eV). However, the absorption spectrum of the I/NaA nanocomposite (curve 3) is shifted toward the ultraviolet range and demonstrates three absorption bands near 2.70, 3.44 and 4.13 eV. These maxima correlate with three lines in the absorption spectrum of molecular iodine, as well as with the absorption spectra of iodine species in I/AFI sample measured in [14].

The current-voltage characteristics $I = f(U)$ of the samples under study are linear at low voltages ($U < 10$V). At high voltages the power law $I = CU^m$ takes place, where $m > 1$. This behaviour is typical for zeolites and zeolite-based nanocomposites [4, 11].

Temperature dependences of the direct current flowing through I/AFI composite along and across the ***c*** - axis of the single crystal are plotted in figure 4.

The most striking feature of the Arrhenius plots is the pronounced peculiarity at $T \approx 343$K (vertical arrow in figure 4, left panel), which could be observed when the temperature increases slowly (~2 K/min) and the current flows through I/AFI composite along the iodine chains. One can suppose that this peculiarity is due to the iodine phase transition from the chain structures to the molecular species, which was found at this temperature by means of optical characterization and phase dynamics of the iodine species in I/AFI composite system [5, 14].

The possible mechanism of the charge transport in I/AFI composite at $T > 343$ K is the tunnelling of electrons between iodine molecular species. Obviously, the melting of iodine chain (aligned with the ***c*** - axis of the AFI single crystal) to molecular iodine has no noticeable effect on the direct current which flows (probably, also due to the tunnelling process) across the ***c*** - axis (figure 4, right panel).

Moreover, the phase transition under consideration does not occur during cooling of the "hot" I/AFI composite, filled with disordered iodine molecular vapour phase, resulting in the enhanced value of the activation energy (figure 4). After few days the ordered iodine molecular wires in I/AFI composite become partially reconstructed, and the phase transition at $T \approx 343$K during heating could be observed again.

Our measurements of thermoelectric current flowing through I/AFI nanocomposite agree with electronic mechanism of charge transport at $T < 343$ K, because Seebeck coefficient S of this material at 300 K is negative: $S_{I/AFI} \approx -1\frac{\mu V}{K}$ in contrast to that of bulk iodine, where $S_I \approx +1\frac{mV}{K}$ (typical value for p-type semiconductor).

One can expect more effective electronic charge transport in three-dimensional interconnected channel system filled with iodine species forming I/NaA nanocomposite. Indeed, Seebeck coefficient of I/NaA composite is still negative, but its absolute value is about 10 times larger, than that of I/AFI material.

At the same time, the electronic currents in I/NaA and I/AFI nanocomposites at room temperature are about 10-100 times higher than the ionic currents in initial hydrated NaA and AFI hosts (without iodine guest substance) under the same applied voltage.

On the other hand, the melting of iodine chains to molecular iodine has smaller effect on the DC or AC electric current flowing through I/NaA nanocomposite material with isotropic distribution of iodine species in comparison with the pronounced decrease of electronic conductivity due to the phase

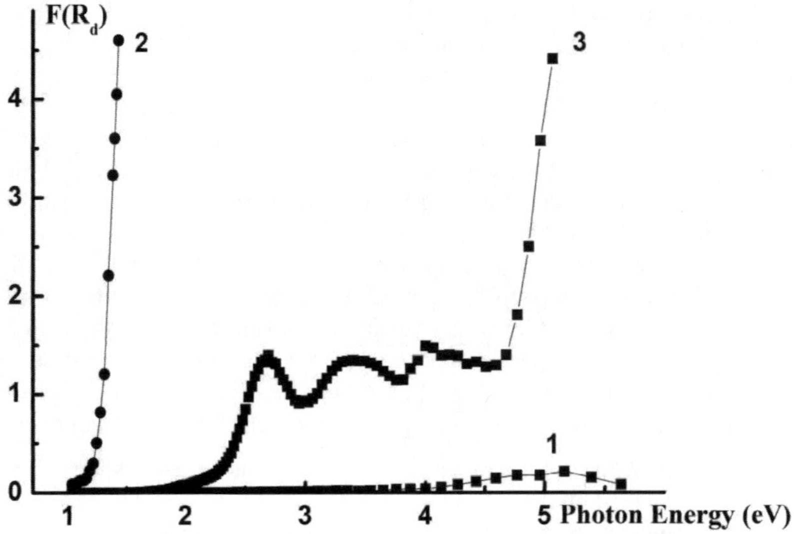

Figure 3. Graphs of Gurevič – Kubelka – Munk functions extracted from the diffuse reflectance spectra for the (1) NaA zeolite, (2) bulk iodine, and (3) I/NaA nanocomposite.

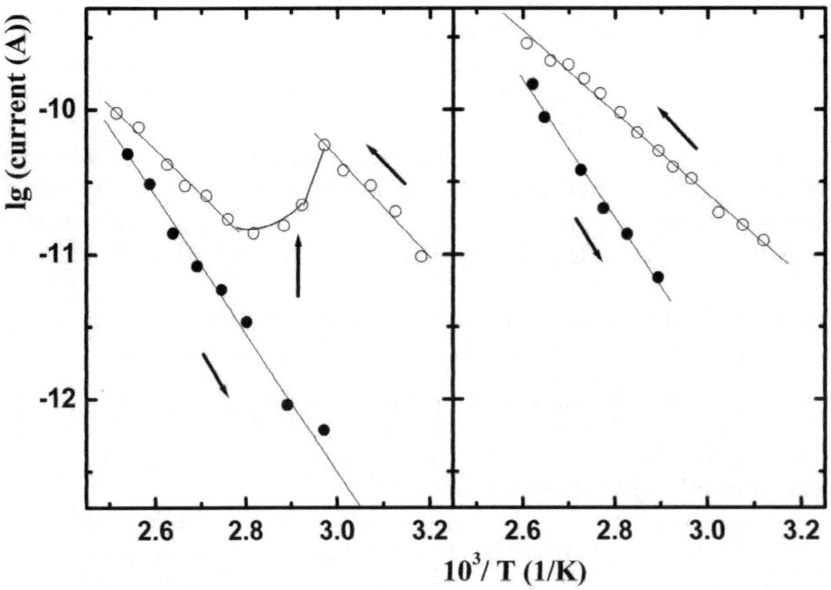

Figure 4. Arrhenius temperature dependences of the direct current flowing under constant applied voltage through I/AFI composite along (left panel) and across (right panel) parallel channels of the single crystal. Inclined arrows indicate heating and cooling of the sample. Vertical arrow indicates the temperature of iodine chains to molecular iodine phase transition.

transition in iodine chains located in the quasi-one-dimensional channels of AFI matrix. That can be clearly seen from figure 5.

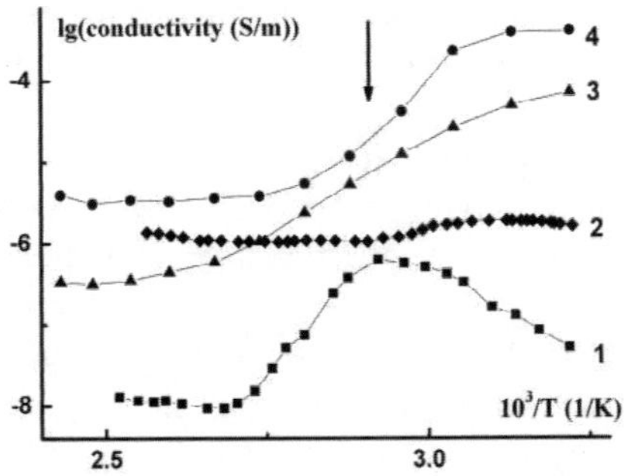

Figure 5. Temperature dependences of (1) DC conductivity of I/AFI single crystal, (2) DC conductivity of I/NaA single crystal, (3) DC conductivity of I/NaA+KBr pressed pellet, (4) AC conductivity of I/NaA+KBr pressed pellet. Vertical arrow indicates the temperature of iodine chains to molecular iodine phase transition.

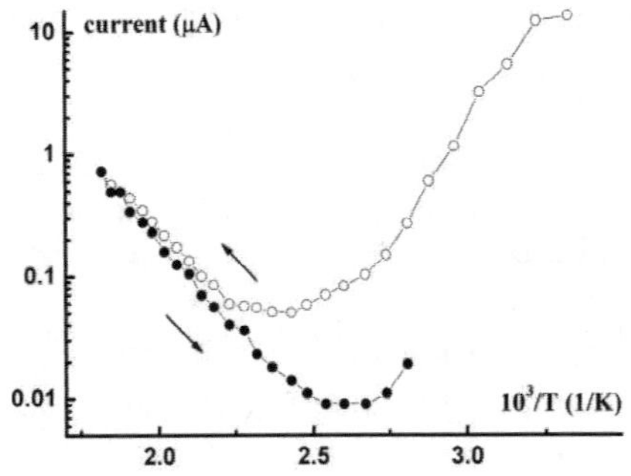

Figure 6. Temperature dependences of direct current flowing under constant applied voltage $U = 50$ V through I/NaA+KBr pressed pellet. Inclined arrows indicate heating and cooling of the sample.

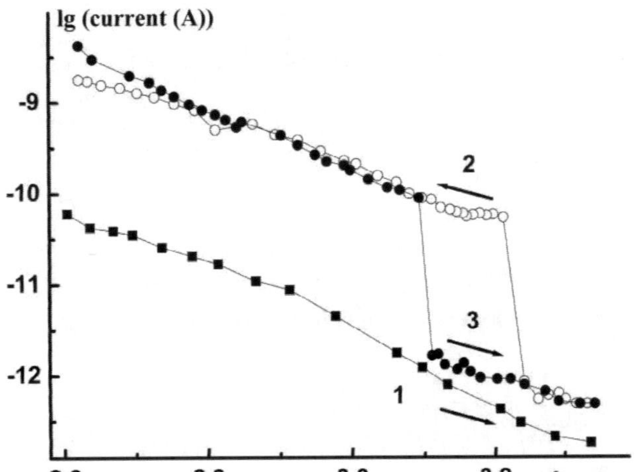

Figure 7. Temperature dependences of direct current flowing under constant applied voltage $U = 15$ V through (1) LiA single crystal; (2, 3) I/LiA single crystal. Inclined arrows indicate (2) heating and (1, 3) cooling of the samples.

In contrast with the iodine phase transition in I/AFI quasi-one-dimensional system which takes place only during heating of the sample (figure 4), one can observe iodine chains reconstruction in I/NaA nanocomposite during its cooling (figure 6) – probably, due to rather fast diffusion of iodine molecules through three-dimensional interconnected channel system of zeolite type A matrix.

We have prepared also I/KA and I/LiA nanocomposites by infiltration of three-dimensional systems of cages and channels of KA and LiA zeolite matrices, respectively, with iodine. According to our experiments, electric properties of I/KA nanocomposites are quite similar to those of I/NaA material, demonstrating phase transition in iodine subsystem at $T \approx 343\text{K}$. We suppose that after the break of iodine chains at $T > 343$ K the charge transport in I/NaA and I/KA composites becomes predominantly ionic.

The conductivity of I/LiA composite containing small Li^+ cations with high drift mobility under applied electric field is probably ionic in the entire temperature region under study: Arrhenius temperature dependence with almost the same activation energy is observed for LiA and I/LiA samples (figure 7).

Here, one cannot see any phase transition in iodine subsystem at $T \approx 343\text{K}$; instead of it, considerable conductivity enhancement occurs at $T \approx 313\text{K}$ during heating; some hysteresis takes place during cooling process (figure 7). It should be noted that the conductivity of bulk crystalline LiI is not very high due to rather low concentration of Schottky defects. On the other hand, one can increase this conductivity when infiltrates lithium iodide guest substance into porous alumina host dielectric matrix [15]. We suppose that the dramatic change of I/LiA electric conductivity near $T \approx 313\text{K}$ (figure 7) may be due to "superionic" phase transition, resulting in remarkable change of unoccupied cation positions number in this complex "host-guest" system.

5. Acknowledgments

The authors are grateful to G Cema, K Gusev and M Prokhorenko for help in sample preparation and characterization. This work was supported by the Ministry of Education and Science of Russian Federation according to the program "Development of Scientific Potential of Higher Educational Institutions".

References

[1] Breck D W 1974 *Zeolite Molecular Sieves* (New York: Wiley)
[2] *Zeolite Chemistry and Catalysis* 1976, ed J A Rabo (Washington: American Chemical Society)
[3] Kumzerov Y and Vakhrushev S 2004 *Encyclopedia of Nanoscience and Nanotechnology* vol 7 ed H S Nalwa (American Scientific Publishers) p 811
[4] Solovyev V G, Ivanova M S, Pan'kova S V, Trifonov S V, Veisman V L 2009 *Handbook of Zeolites: Structure, Properties and Applications* ed T W Wong (New York: Nova Science Publishers) p 77
[5] Ye J T, Tang Z K, Siu G G 2006 *Applied Physics Letters* **88** 073114
[6] Khanin S, Solovyev V, Trifonov S, Veisman V 2013 *Smart Nanocomposites* **4** 29
[7] Alexeeva N, Cema G, Podorozhkin D, Solovyev V, Trifonov S, Veisman V 2015 *Journal of Self-Assembly and Molecular Electronics* **2** 27
[8] Gurevič M 1930 *Physikalische Zeitschrift* **31** 753
[9] Kubelka P, Munk F 1931 *Zeitschrift für technische Physik* **12** 593
[10] Markov V N, Solov'ev V G 1990 *Instruments and Experimental Techniques* **33** 1230
[11] Solovyev V, Kumzerov Y and Khanin S 2011 *Physics of Regular Matrix Composites (Electrical and Optical Phenomena in Nanocomposite Materials Based on Porous Dielectric Matrices)* (Saarbrücken: Lambert Academic Publishing), in Russian
[12] Trifonov S V, Vanin A I, Veisman V L, Gango S E, Kondratyeva M N, Solovyev V G 2011 *Nanotechnics* **2 (26)** 78, in Russian
[13] Gango S E, Markov V N, Solov'ev V G 1998 *Instruments and Experimental Techniques,* **41**, 858
[14] Ye J T 2006 *Fabrication and Optical Characterization of Nanostructures Formed inside Zeolite Single Crystals* Ph.D. Thesis (Hong Kong: University of Science and Technology)
[15] Maekawa H, Tanaka R, Sato T, Fujimaki Y, Yamamura T 2004 *Solid State Ionics* **175** 281

IOP Conf. Series: Journal of Physics: Conf. Series **929** (2017) 012043 doi:10.1088/1742-6596/929/1/012043

Effect of tungsten implantation on the switching parameters in V_2O_5 films

S.V. Burdyukh, O.Ya. Berezina, A.L. Pergament

Petrozavodsk State University, Petrozavodsk, 185910 Russia

E-mail: burduch@gmail.com

Abstract. The paper examines the effect of doping with tungsten on switching in hydrated vanadium pentoxide films. The switching effect is associated with the metal-insulator transition in a vanadium dioxide channel that forms in the initial film due to the process of electrical forming (EF). Doping is carried out by the plasma immersion ion implantation method. It is shown that implanting small tungsten doses improves the switching parameters after EF. When implanting large doses, switching is observed without EF, and if EF is applied, the switching effect, on the contrary, disappears.

1. Introduction

In vanadium dioxide at a temperature $T_t = 341$ K, a metal-insulator phase transition is observed, which makes this material hugely promising in the search for new solutions in the field of development of oxide electronics functional devices [1]. For some practical applications, it is important to vary the metal-insulator transition (MIT) parameters, such as the transition temperature, the resistance and reflection coefficient values jump, the shape of the hysteresis loop etc. Such a variation could be made by means of doping with transition-metal elements, e.g., tungsten [2, 3]. In the structures based on V_2O_5 films, the effect of reversible switching from a high-resistance state to a low-resistance state is observed due to the formation of a channel consisting of VO_2, and accordingly the MIT therein, during electroforming (EF) [3]. In this work, we report on the study of the effect of tungsten implantation into thin films of vanadium pentoxide, by plasma-immersion ion implantation (PIII), on the parameters of the switching structures based on these films.

2. Experimental details

The main idea of PIII is that the sample is placed in the processing chamber in contact with plasma. The ion implantation from plasma is performed when applying a negative high-voltage pulse on the sample. To generate plasma in the processing camber, a hot-cathode plasma source is used. When implanting tungsten, the discharge is ignited in argon. A crucible with tungsten hexacarbonyl $W(CO)_6$ is placed in the discharge chamber and heated by a special nichrome heater up to 200°C. The melting point of tungsten hexacarbonyl is 169°C, and the boiling point is 175°C. $W(CO)_6$ sublimates in vacuum. Thus, $W(CO)_6$ vapor appears in the discharge chamber. Molecules $W(CO)_6$ are ionized in the discharge and implanted into the sample. Schematic diagram of the implantation system is shown in Figure 1.

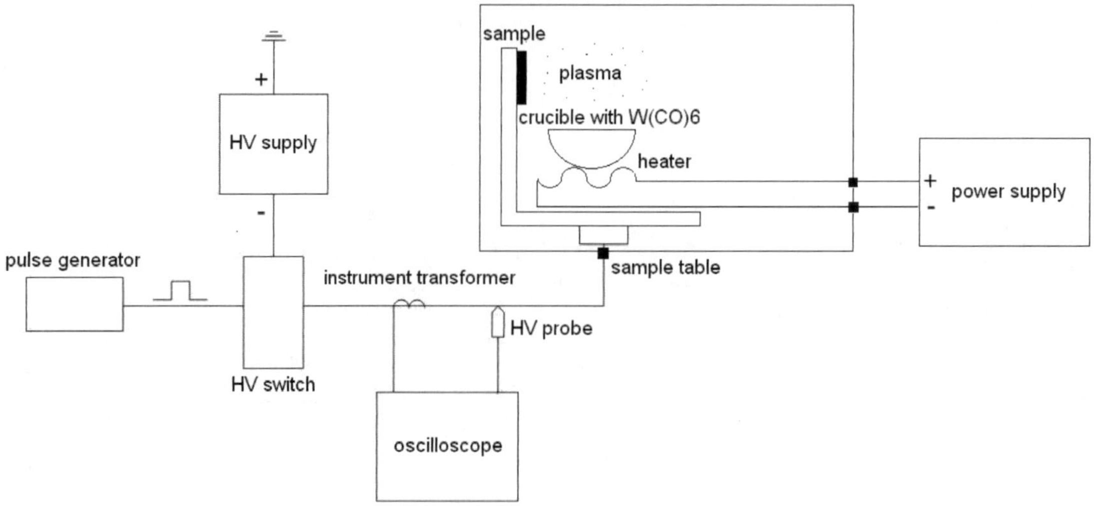

Figure 1. PIII system schematic diagram.

The gas composition during the experiment was monitored with a quadrupole mass spectrometer Thermo Scientific DSQ II. From the mass spectra (Figure 2), we determined the partial pressures of the gas mixture components.

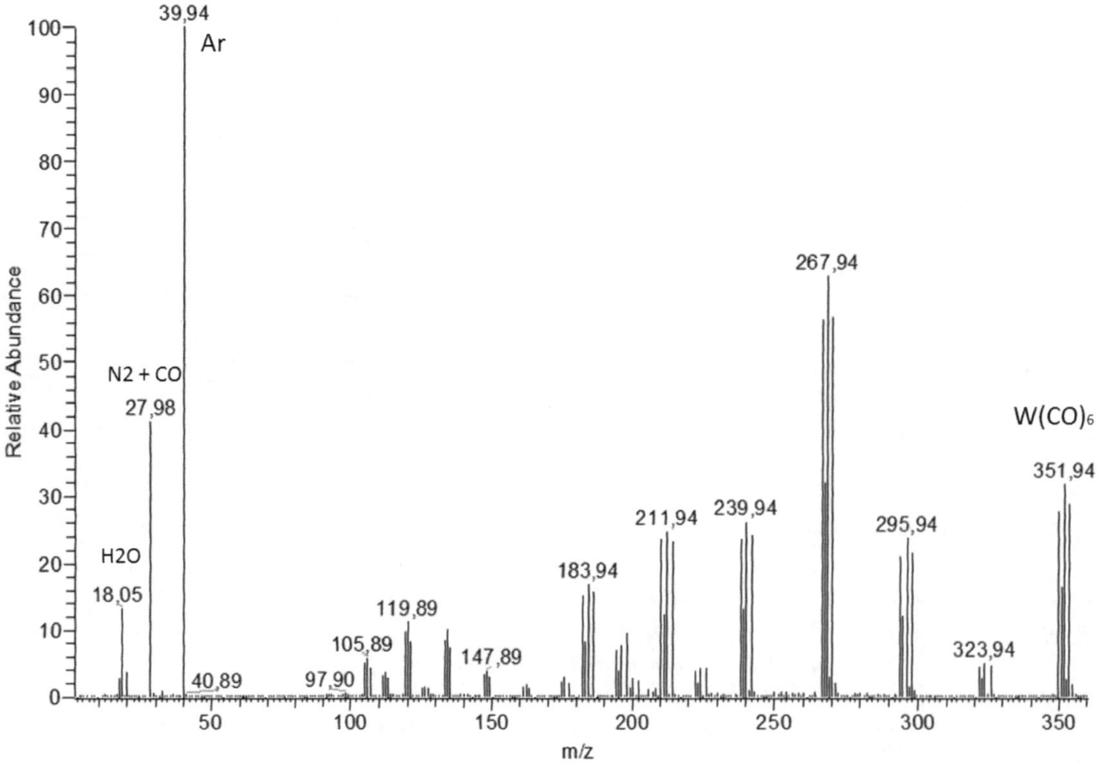

Figure 2. Mass-spectrum of gas in processing chamber during experiment.

The implantation experiment parameters were as follows: the discharge current - 10 A, discharge voltage - 55 V, cathode filament current - 60 A, argon flow through the chamber - 0.0018 m^3Pa/s. The amplitude of the negative pulse voltage applied to the sample was 2 kV, pulse current I_{imp} - 50 mA; pulse time Δt - 5 s, repetition frequency - 1.7 kHz, and the total treatment time ranged from 1 to 5 minutes. The Ar partial pressure was 1.2 Pa, the partial pressure of $W(CO)_6$ - 3.8 Pa. The switching effect was studied in a sandwich-type structures Au-$V_2O_5 \cdot nH_2O$-Au on the basis of $V_2O_5 \cdot nH_2O$ xerogel films obtained by the sol-gel method [3]. Forming of the switching structures was carried out by a Keithley 2400 source meter in the current limiting mode; the EF voltage was 30 V.

3. Results and discussion

After EF, in the film of pure hydrated vanadium pentoxide, a vanadium dioxide channel forms where the MIT can occur. The current-voltage characteristic of the structure becomes S-shaped. Thus, a stable switching structure is formed with a sufficiently large resistance jump (of up to two orders of magnitude) in the I-V curve (Figure 3).

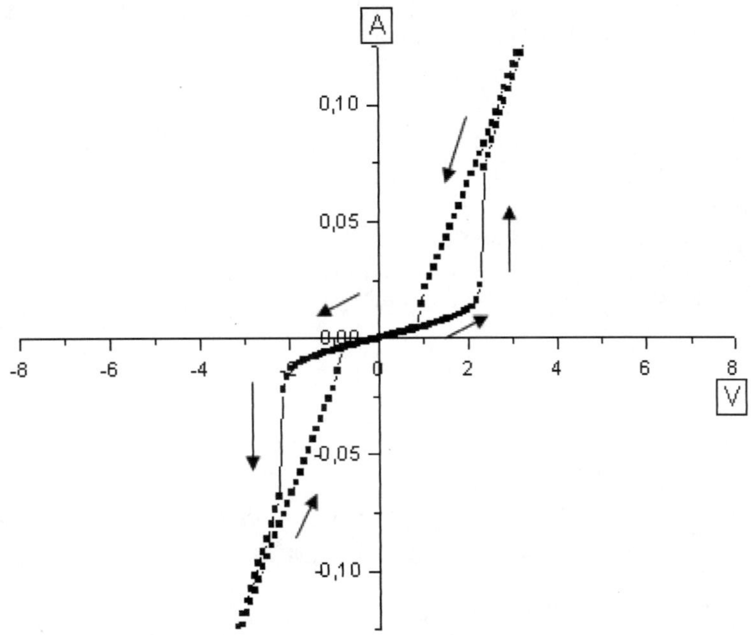

Figure 3. I-V characteristic of $V_2O_5 \cdot 1.8H_2O$ film after electrical forming.

Implantation of tungsten greatly influenced on the I-V curves of the switching structures. For the $V_2O_5 \cdot 1.8H_2O$ film after implantation of $W(CO)_6$ for 1 minute (W content in the film is 0.43 at.%), the conductivity jump when switching from a high- to low-resistance state is higher than that in the original film (Figure 4). Moreover, the switching parameters scatter is significantly lower than that in the initial films without implantation. This might indicate that the admixture of tungsten reduces the scatter due to a decrease of the MIT temperature [3].

For $V_2O_5 \cdot 1.8H_2O$ film after implantation of $W(CO)_6$ for 5 minutes (the W concentration is 7.07 at.%), switching in sandwich structures without a preliminary EF is observed. At the same film after forming, the conduction jump disappears completely, and the film in the channel after forming transits into the metallic state (Figure 5). Note that the doping of vanadium dioxide films with large doses of hydrogen by PIII also results in the film metallization [4].

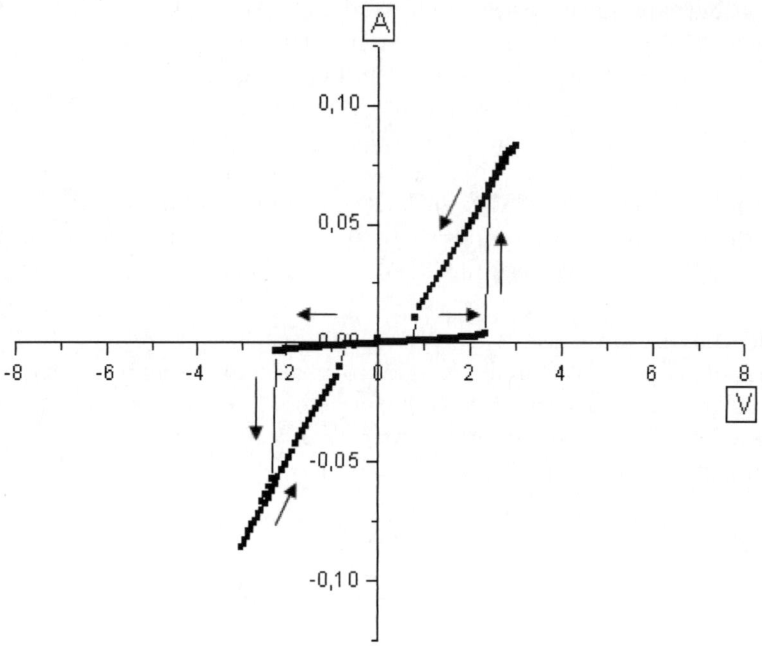

Figure 4. I-V characteristic of $V_2O_5{\cdot}1.8H_2O$ film after $W(CO)_6$ implantation for 1 min. and after subsequent EF.

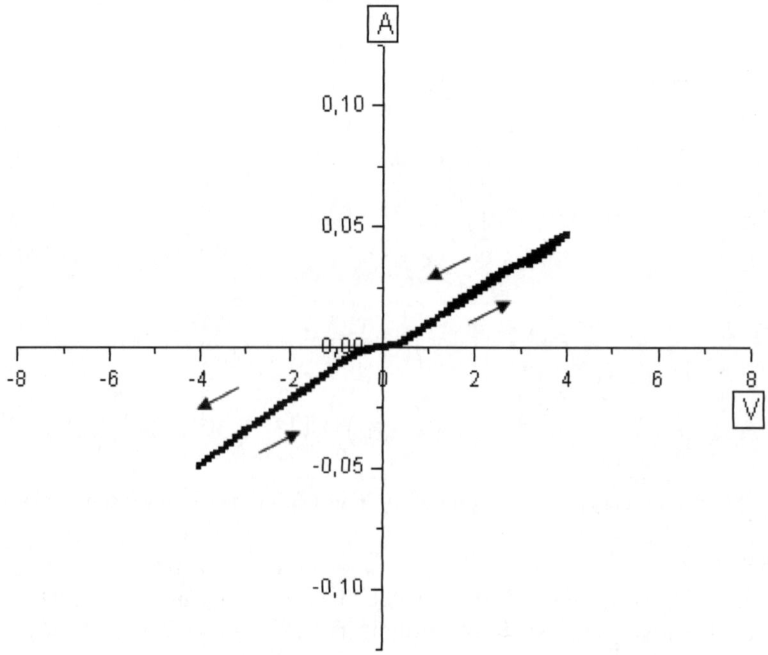

Figure 5. I-V characteristic of $V_2O_5{\cdot}1.8H_2O$ after implantation of $W(CO)_6$ for 5 min. and after subsequent EF.

Thus, the implantation of small doses of tungsten into vanadium pentoxide films significantly improves theirs switching parameters. When implanting relatively large doses, the switching effect is observed without a preliminary forming, but in case of EF, switching vanishes. This is apparently due to the fact that in this case a conventional electrical breakdown occurs instead of electrical forming [5].

Acknowledgments
This work was supported by the RF Ministry of Education and Science as a base part of state program no. 16.5857.2017/BP in the scientic field, and the Strategic Development Program of Petrozavodsk State University (2017-2021).

References

[1] Pergament A.L., Stefanovich G. B., Velichko A.A. 2013 *Journal on Selected Topics in Nano Electronics and Computing* **1** 24
[2] He X, Zeng Y, Xu X, Gu C, Chen F, Wu B, Wang C, Xing H, Chen X, Chu J. 2015 *Phys Chem Chem Phys* **17** 11638
[3] Berezina O. Ya., Velichko A. A., Lugovskaya L. A., Pergament A. L., Stefanovich G. B., Artyukhin D. V., Strelkov A. N. 2007 *Technical Physics Letters* **33** 552
[4] Burdyukh S. V., Stefanovich G. B., Pergament A. L., Berezina O. Ya., Avdeev N. A., Cheremisin A. B. 2016 *Technical Physics Letters* **42** 32
[5] Chudnovskii F.A., Odynets L.L., Pergament A.L., Stefanovich G.B. 1996 *Journal of Solid State Chemistry* **122** 95

IOP Conf. Series: Journal of Physics: Conf. Series **929** (2017) 012044 doi:10.1088/1742-6596/929/1/012044

Effect of temperature on the value of the surface potential and the phase state of 4'-n-octyl-4-p-cyanobiphenyl liquid crystal

A S Chumakov[1,2], Ammar J K Al-Alwani[1,3], I A Gorbachev[1,2], E G Glukhovskoy[1,2], N V Usoltseva[4]

[1] National Research Saratov State University named for N.G. Chernyshevsky, Astrakhanskay st. 83, 410012, Saratov, Russia

[2]Educational-Research Institute of nanostructure and biosystems

[3]Babylon University, Babylon, Iraq

[4]Nanomaterials Research Institute, Ivanovo State University, Ivanovo, Russia

E-mail: lehahares@rambler.ru

Abstract. Liquid crystals are one of the most interesting materials for various applications. The article considers the influence of temperature to the phase change and the value of surface potential of 4'-n-octyl-4-p-cyanobiphenyl liquid crystal Langmuir monolayers. These parameters have a critical role in process of transfer monolayers to the solid substrate and for further using of them. It is shown that with increasing temperature of subphase leads to the monolayer of liquid crystal becomes isotropic and exhibits a gradual decrease in the surface potential in process of compression.

1. Introduction

Liquid crystals (LC) and devices based on them have long been widely used [1, 2]. On the basis of LC are built many types of displays [3, 4], height sensitivity, low power temperature sensors [5] and many other devices [6]. Such popularity of liquid crystals gained through a unique combination of properties. In the normal state in the same time they have the fluid properties and solid state properties and LC exhibit a strong anisotropy of the properties, particularly in optic [7]. It is also well aware of the dependence of the phase state of liquid crystals on the temperature. Possibility of changes phase by the small temperature changes can be crucial in areas where thin films of liquid crystals are used. For example, when we form them using Langmuir-Blodgett technology.

In our work we obtained monolayers of 4'-n-octyl-4-p-cyanobiphenyl (8CB) liquid crystal on the surface of the aqueous subphase by the Langmuir-Blodgett technology. The choice of this material is due by the fact that 8CB molecules are composed of polar and nonpolar part (figure 1). And thus, exhibit amphiphilic properties. In addition, the polar part of the highly long and has a large dipole moment. Thus, if we will be able to control film structure in the process their formation and transfer, we will be able to control the potential relief of the film on the solid substrate.

2. Material and method

The 8CB liquid crystal was purchased from "Sigma-Aldrich and dissolved in chemically pure chloroform (produced by "Vecton", Russia). The concentration of liquid crystal solution was a

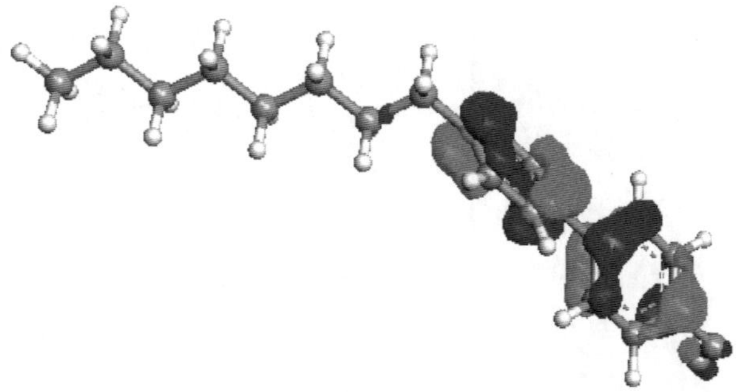

Figure 1. Schematic picture of 8CB molecule.

constant in all experiments 10^{-3}M. The changing in the surface pressure isotherms and surface potential at air-water interface of 8CB monolayer were investigated by using Langmuir-Blodgett trough (KSV Nima LB Trough Medium KN 1003, Finland) and Surface Potential Sensor (KSV NIMA, Finland) which has sensitivityof ± 1mV. In all experiments we used like a subphase deionized water with resistance of about18,2MΩ×cm. The 30μl of 10^{-3}M 8CB solution were spread on the water surface and after 5 min (it is time for chloroform evaporation) the barriers were start compression of the liquid crystal monolayer. The speed of barriers was 10 mm/min. The isotherms were recorded in few series of experiments by various values temperature of subphase - 16°C, 21°C, 33°C and 41°C. Temperature control was realised by using LOIP thermostat (Russia) connected with working trough of Langmuir-Blodgett device.

3. Results and discussion

The molecules of 8CB have polar (CN) group that demonstrate strongly interaction with the water (hydrophilic group). Previous studies demonstrated that the 8CB molecules can form monolayers at air-water interface by using LB-method [8, 9]. The investigation of surface potential ΔV of liquid crystal monolayer reported previously [10] this parameter can provide us information about molecular organization during gas phase of liquid crystal [11], and the average dipole moment for the monolayer can be presented by Helmholtz equation:

$$\Delta V = \frac{\mu_{\perp}}{A\varepsilon_1\varepsilon_0} \quad (1)$$

Where$\mu_{\perp}=\mu cos(\theta)$ (θ - the angle between the surface and the dipole moment axis), $\mu_{\perp}$ is called the effective dipole moment [12] and μ is dipole moment, A is the occupied area by molecules on the subphase of water, ε_1 and ε_0 are the dielectric constant of monolayer and the electric permittivity of vacuum, respectively.

Figure 2a shown the surface pressure isotherm (solid curve) and surface potential dependence (dotted curve) of 8CB monolayer at low temperature of subphase (16°C), obviously we can observe there three regions. The region (I) related to area larger than 168,17cm^2 coexistence of 8CB monolayer gas phase – surface pressure is almost zero and the ΔV increased gradually to 0,557V at 150,07cm^2. In the region (II) between 168,17cm^2 and 150,12cm^2 the surface pressure increased sharply to the collapse monolayer at 150,12cm^2 with ΔV of 0,642V. At region (III) between 150,12cm^2 and 45,07cm^2 the formation of liquid crystal multilayers begins [13, 14]. The surface pressure increased slightly up to 6,26mN/m at 45,07cm^2 with ΔV of 0,616V.

The effective dipole moment can be calculated from equation (1), the ε_1 =1 and the $\mu_{\perp}$= 1,3D, 1,34D, 0,385D for region (I),(II) and (III),respectively. There is a little change in values of surface pressure and ΔV isotherm between regions (II) and (III). That indicated that the orientation of liquid crystal molecules stay in the same (or near to same) direction.

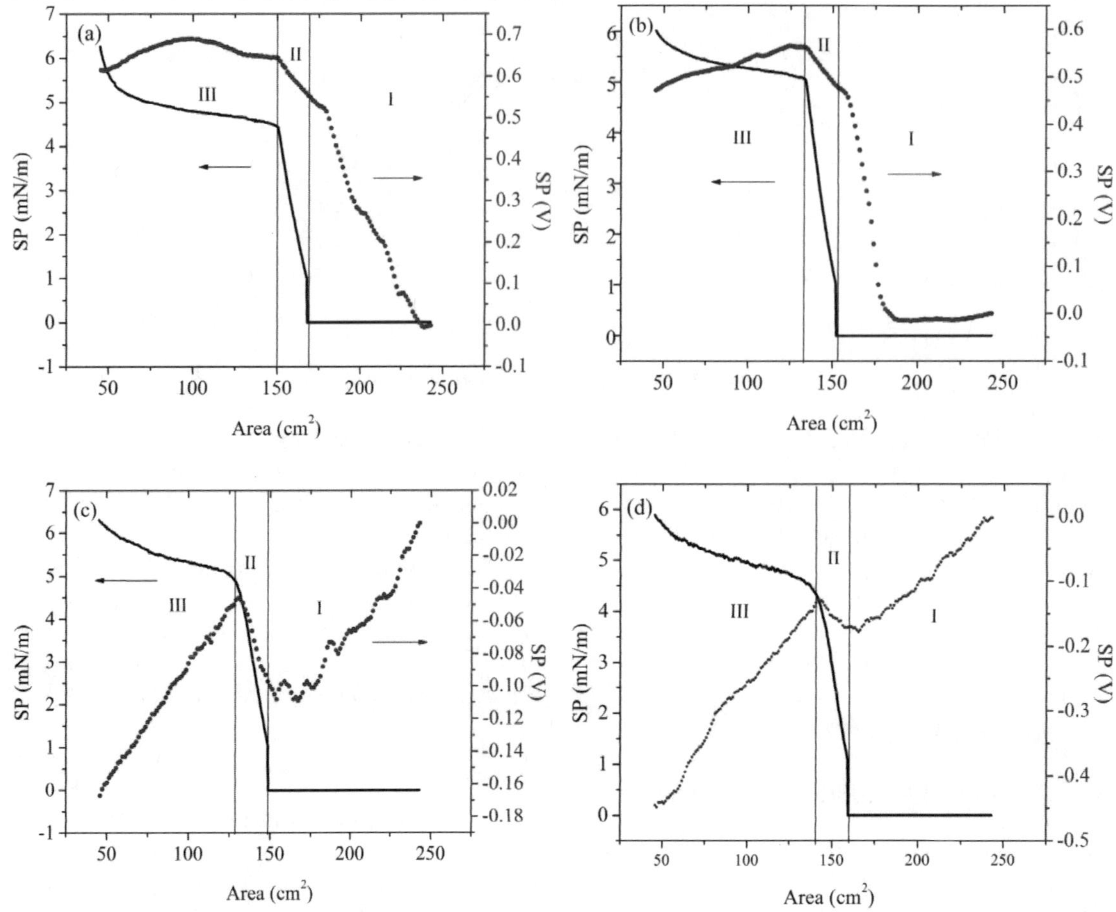

Figure 2. Surface pressure isotherms (solid curve) and surface potential (ΔV) (dotted curve) of liquid crystal (8CB) at subphase temperature of 16°C (a), 21°C (b), 33°C (c), 41°C (d).

Figure 2b shown surface pressure isotherm (solid curve) and ΔV dependence (dotted curve) for 8CB monolayer. Temperature of subphase was 21°C. The region (I) up to 152,64cm^2 shown gas phase with gradually increased surface potential to 0,481V. In this case this region of 8CB monolayer has less molecules concentration than by gas phase at temperature of subphase 16°C (figure 2a). The surface pressure value at region (II) sharply raised from about 0mN/m at 152,64cm^2 till collapse about 5mN/m at 133,89cm^2 (transition region for next phase) and ΔV was grown to 0,562V. The region (III) presents plateau at area between 133,89cm^2 and 45,28cm^2 and value of surface potential $\Delta V = 0{,}472$V. We can explain this by process of forming of regular bilayer on the top of monolayer [15]. The effective dipole moment were estimated $\mu_{\perp}$= 1,02D, 1,04D and 0,298D for region (I), (II) and (III), respectively.

In figure 2a and b the ΔV value increase slowly with reducing the occupied areain the region (III). It can be assumed that the dipole moment of first 8CB monolayer directed down towards subphase of water [15, 11] at temperature of 16°C and 21°C.

Figure 2c shown typical surface pressure isotherm (solid curve) and ΔV dependence (dotted curve) for 8CB monolayer on the 33°C water subphase. There are three distinct regions: region (I) presented the gas phase with constant surface pressure up to 148,77cm^2 with decreasing in the ΔV value to -0,098V. In the next region (II), the surface pressure isotherm sharply rise from 148,77cm^2 to the collapse point at 128,65cm^2, the ΔV value slow grown to -0,05V. In the third region (III) from collapse point to area of 45,55cm^2 the ΔV value has been minimized to – 0,17V. The effective dipole moment shown values $\mu_{\perp}$= - 0,2D, - 0,088D, - 0,1D for regions (I), (II) and (III), respectively. The results of surface potential isotherm and effective dipole moment indicated that the molecular orientations changed by increasing the temperature of subphase.Also reason for this is the changing to nematic phase in the 8CB [16, 17].

The surface pressure and ΔV isotherms at temperature of subphase 41°C presented on figure 2d. The region (I) is a gas phase up to area of 159,41cm^2. In this region the curve of ΔV clearly shows a significant decrease surface potential to -0,171V. The ΔV value increased slightly to -0,126V in collapse region at 140,7cm^2. In the region III value of surface potential sharply dropped down to -0,448V at area value of 45,19cm^2.

The 8CB molecules were arranged almost horizontally on the subphase surface and this fact can be confirmed by the results of the surface pressure and ΔV isotherms [18]. But after barriers start compression, the molecules begin to interact and they lift up from the subphase surface and moleculesare arranged parallel to one another [18].That mean the molecules orientations depends on the surface pressure, this assumption is agreement with the increasing value of ΔV isotherm and effective dipole moment.

Figure 2c and d show the surface potential value decreasing to the negative value, we can assume that is attributed to the random ordering of the chains of liquid crystal in monolayer. Most probably reason for thisis changing in phase by effect of temperature [18]. If the surface potential have negative value the liquid crystal molecules is directed upwards to the air.

Figure 2 shows that the rising temperature and decreasing the occupied area lead to more smooth transition to liquid expanded phase in the region II [17]. The effective dipole moment for 8CB molecule that calculated from equation (1) has higher value at 16°C, with raising temperature of water subphase the effective dipole moment decreasing gradually (figure 2). The changes in the effective dipole moment and ΔV values are indicating the changing in the direction and orientation angle of 8CB molecules.

4. Conclusion

The formation of 8CB liquid crystal monolayer by Langmuir-Blodgett technique was studied. The surface potential and surface pressure of 8CB monolayers were investigated at various temperature of subphase. The results revealed that the surface potential of 8CB monolayer decrease with the increase the temperature of water subphase. The first monolayer formed at air-water interface and molecules aligned down towards water (from the air to the water) at temperature of subphase 16°C and 21°C. The result revealed formation liquid-expanded phasein region II with increase temperature of subphase (figure 2). The sharp decreases in the value of surface potential of 8CB monolayer at 33°C and 41°C indicating changing the orientation of molecules in monolayers.

Acknowledgments

The work is supported by grant № 16-37-50050 of RFBR and Nanomaterials Research Institute by Ivanovo State University.

References

[1] Heilmeier G H, Zanoni L A, Barton L A 1968 *Proceedings of the IEEE* **56,** 71162–71
[2] Kelker H and Scheurle B 1969 *Angewandte Chemie* **81** 903–04
[3] Jianlong Liu, Hongmei Ma, Yubao Sun 2016 *Liquid Crystals* **43,12**
[4] Liang Zhao-yan, Fang Kun, Xuan Li and Huang Xi-min 2000 *Chin. Phys. Soc* **9,11**
[5] Algorri J F, Urruchi V, Bennis N, Sánchez-Pena J M 2014 *Sensors* **14,4** 6571-83
[6] Jong-Hyun Kim, Makoto Y and Hiroshi Y 2002 *Nature* **420** 159–162

IOP Conf. Series: Journal of Physics: Conf. Series **929** (2017) 012044 doi:10.1088/1742-6596/929/1/012044

[7] Shin Tson Wu and Uzi Efron 1986 *Appl. Phys. Lett.* **48** 624
[8] Anna M, Danuta B 2011 *Int. J. Mol. Sci.* **12** 4923–45
[9] Martynski T, Hertmanowski R, Bauman D 2001 *Liquid Crystals* **28,3** 445-50
[10] Modlinska A, Inglot K, Martynski T, Dabrowski R et. all 2009 *Liquid Crystals* **36, 2** 197–208
[11] Inglot K, Martynski T, Bauman D 2009 *Dyes and Pigments* **80** 106–114
[12] Vogel V, Möbius D1988 *J. of Coll. and Int. Sci.* **126, 2** 408–20
[13] Marc N G de Mul, J. Adin Mann Jr 1994 *Langmuir* **10, 7** 2311–16
[14] Matthew C F, Gerald G F, Curtis W F, Channing R R 1994 *Langmuir* **10, 4** 1251–56
[15] Martynski T, Biadasz A, Bauman D 2003 *Z. Naturforsch* **58a,** 97–102
[16] Jiang M, Zhong F 1997 *J. Chem. Phys.* **106,14**
[17] Suresh K A, Bhattacharyya A 1997 *Langmuir* **13** 1377-80
[18] Enderle Th, Meixner A J, Zschokke-Granacher I 1994 *The J. of Chem. Phys.* **101** 4365

IOP Conf. Series: Journal of Physics: Conf. Series **929** (2017) 012045 doi:10.1088/1742-6596/929/1/012045

Synchronization in the system of coupled oscillators based on VO_2 switches

V V Perminov, V V Putrolaynen, M A Belyaev and A A Velichko

Petrozavodsk State University, Petrozavodsk 185000, Karelian Republic, Russia

Email: teh-val@yandex.ru

Abstract. VO_2 switches with threshold voltages less than 2 V were fabricated. Synchronization between two oscillators based on VO_2 switches coupled through resistance and capacitance was investigated. For resistive coupling range of resistances in which synchronization appeared was defined and phase instability was observed. For capacitive coupling two synchronization modes were found. Simulation of changing of the relative phase for two oscillators in the case of capacitive coupling for both synchronization modes has been carried out. The relative phase variation range more than 310° was obtained.

1. Introduction

Metal-insulator transition in vanadium dioxide has been known since 70-th of the 20-th century [1]. In practice, this effect can be used in electronic oscillators, temperature and chemical sensors with frequency output. Nowadays, oscillators based on VO_2 switches arouse interest as elements of oscillatory neural networks (ONN) [2]. The advantages of these ONN over other solutions [3, 4] are compatibility with CMOS technology, scalability, and possibility of working in megahertz frequency range.

In this paper, with the aim of application such oscillators in the ONN we investigate synchronization effects between two oscillators based on VO_2 switches in case of resistive and capacitive coupling. The resistive coupling is attractive because of its compactness and possibility of tune coupling strength using MOSFET or memristor as a coupling element. On other hand, capacitive coupling is characterized by galvanic isolation between oscillators, that would provide more reliable oscillations.

As the ability to control the phase difference between oscillators is important in phase-shift keying ONN, simulation of changing of the relative phase for two oscillators has been carried out.

2. Sample preparation and characterization

The planar switches structures were formed on r-cut sapphire substrate by magnetron reactive sputtering of vanadium in Ar and O_2 atmosphere in ratio Ar/O_2=7/1 at room temperature, total pressure was 5 mTorr, DC power was set to 200 W. Annealing of obtained films was carried out in O_2 atmosphere at temperature of 480°C and pressure of 10 mTorr for 40-60 minutes. V/Au pads were formed by optical lithography and magnetron sputtering (figure 1). All switches were located on the same substrate and VO_2 thin film.

The thickness of obtained films was 200 nm. Investigation of surface morphology by atomic-force microscopy showed that annealed films had grain structure. XRD analysis indicates the presence of vanadium dioxide and vanadium pentoxide phases. The resistivity temperature dependence (figure 2) showed that increasing of annealing time leads to increase in hysteresis' loop width, transition temperature, and film resistivity both in metal and insulator state. The films which were annealed for 40 minutes were used for making VO_2 switches because they provided lower threshold voltages.

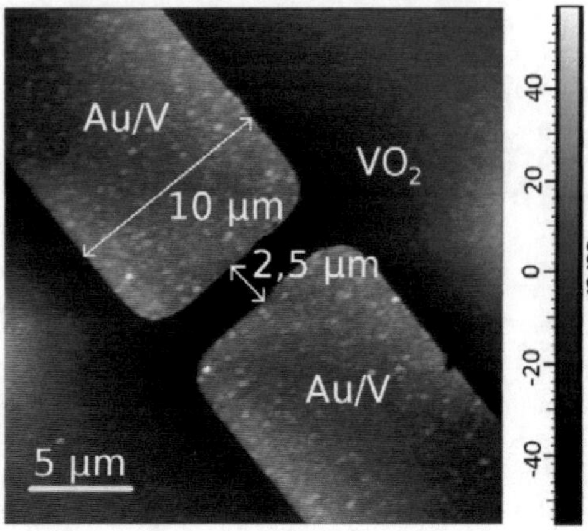

Figure 1. AFM image of the swiching structure.

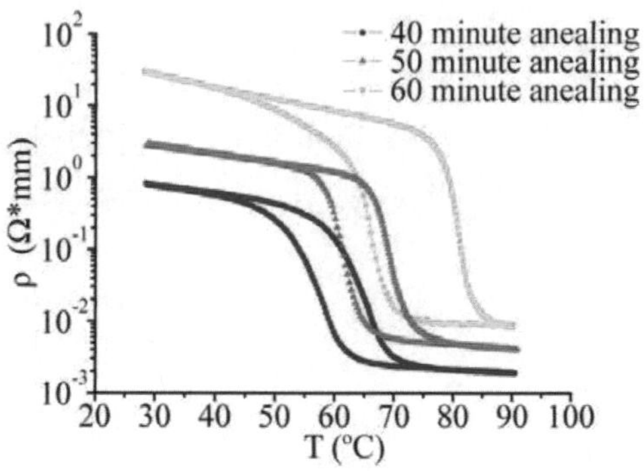

Figure 2. Resistivity versus temperature for different annealing time.

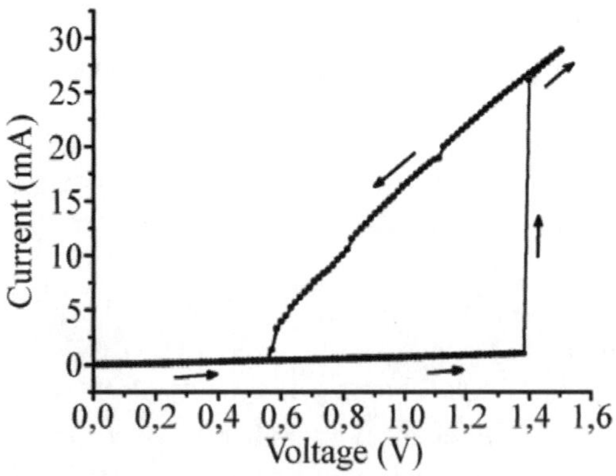

Figure 3. Static I-V characteristic of obtained switching structure.

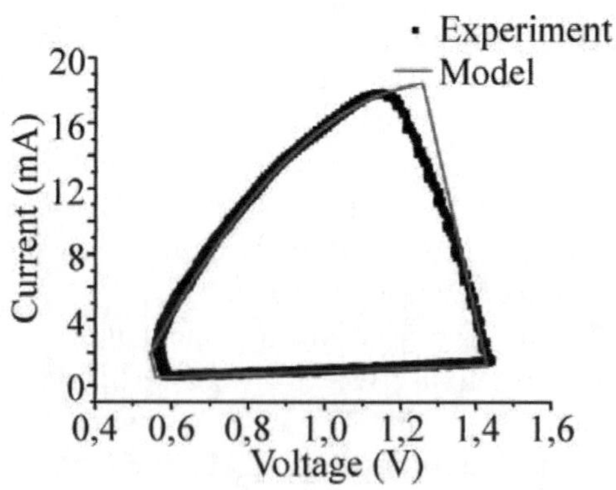

Figure 4. Dynamic I-V characteristic of obtained switching structure and developed SPICE-model.

Threshold voltages of obtained switches were less than 2 V, differential resistance was about 40 Ω in the metallic state and 1,1 kΩ in the semiconductor state (figure 3). The main parameters of the obtained switches differed in less than 10%. For circuit simulation based on the measured dynamical I-V characteristics and taking into account non-linear resistance in metallic state (figure 4) the model of VO_2 switches for Ngspice simulator was developed.

3. Experimental setup

Oscillator consisted of the voltage supply (V_{DD} = 82 V), serial resistor R_S = 50 kΩ, VO_2 switch and current resistor R_i = 10 Ω which were connected in series. The current resistor was used to measure a current through VO_2 switch. Capacitor C = 100 nF was connected in parallel to VO_2 switch and current resistor. Coupling capacitor or resistor was connected between two oscillators as shown in figure 5.

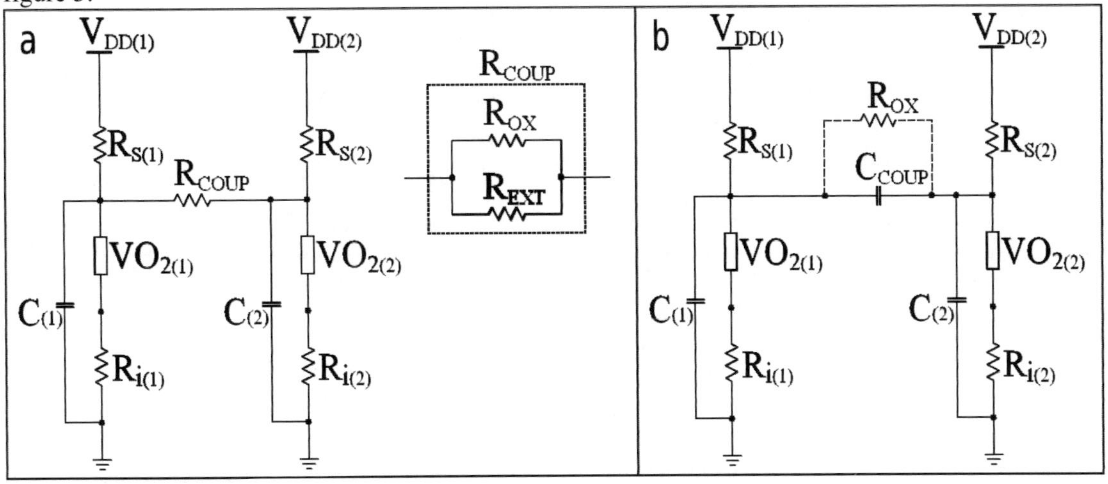

Figure 5. Equivalent circuit of coupled oscillator: a – resistive coupling; b – capacitive coupling.

It should be noted, that besides external coupling capacitor C_{COUP} or resistor R_{EXT}, there was coupling resistance R_{OX} = 10 kΩ due to the placement of switches on the common film near each other, so coupling resistance R_{COUP} consisted of R_{OX} and R_{EXT} as shown in figure 5a. But, as will be shown in the next section, R_{OX} did not affect synchronization of oscillators.

4. Results and discussion

We revealed that in the case of resistive coupling when R_{COUP} was greater than 3 kΩ, synchronization did not appear, but there was a change of signal's form and a decrease of oscillation frequency while R_{COUP} was decreasing (figure 6). Synchronization took place when R_{COUP} was in the range of 2-3 kΩ. However, synchronization was unstable, random phase failure caused premature triggering of one of the oscillators was occur. After failure, the phase recovered for several periods of oscillations. Decreasing R_{COUP} from 2,4 kΩ to 1,9 kΩ lead to an increase of relative phase from 130° to 180° (figure 7) and a decrease of oscillations frequency from 6 to 5,2 kHz (figure 6). If R_{COUP} was less than 2 kΩ then generation was unstable and after several oscillations system proceeded to state when one switch was in the metallic state and another in insulator state.

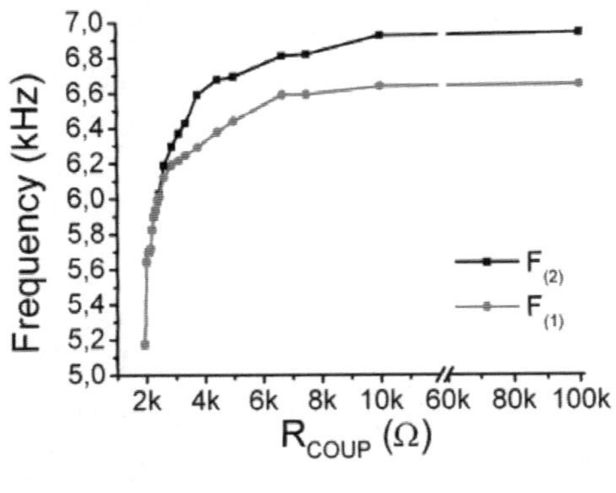

Figure 6. Oscillation frequency versus coupling resistance value (Simulated).

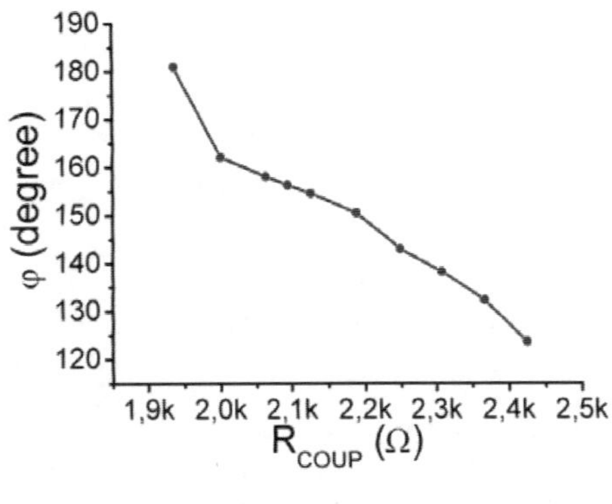

Figure 7. Relative phase φ versus coupling resistance value (Simulated).

IOP Conf. Series: Journal of Physics: Conf. Series **929** (2017) 012045 doi:10.1088/1742-6596/929/1/012045

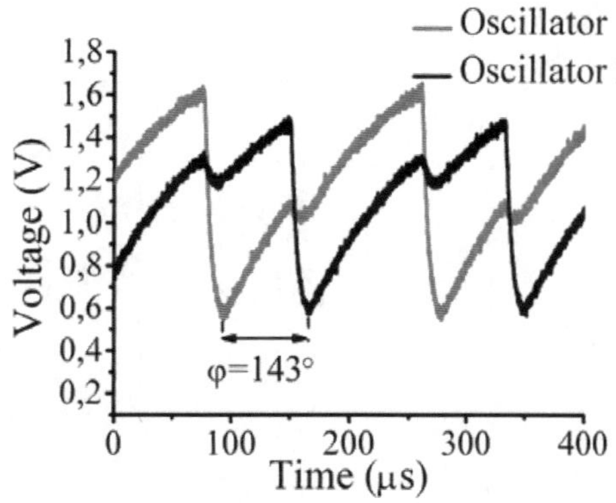

Figure 8. Voltage waveform of capacitively coupled oscillators (C_{COUP}=22 nF).

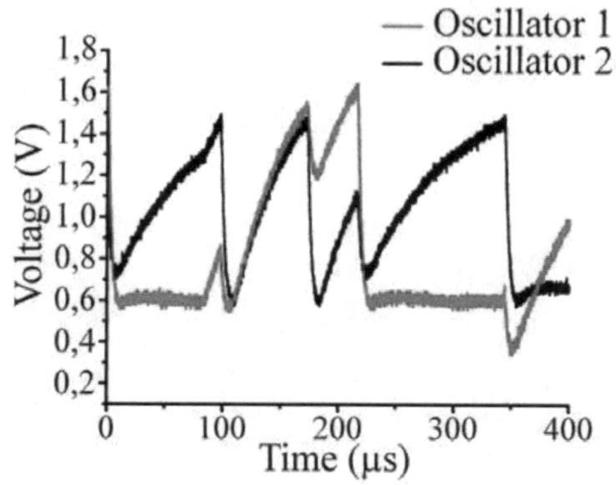

Figure 9. Voltage waveform of capacitively coupled oscillators (C_{COUP}=100 nF).

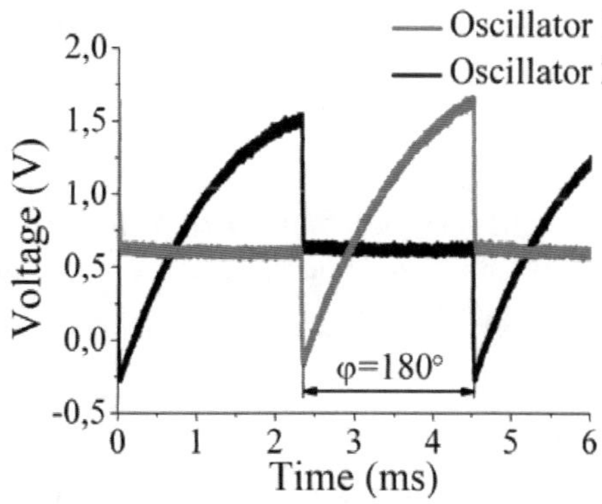

Figure 10. Voltage waveform of capacitively coupled oscillators (C_{COUP}=1 μF).

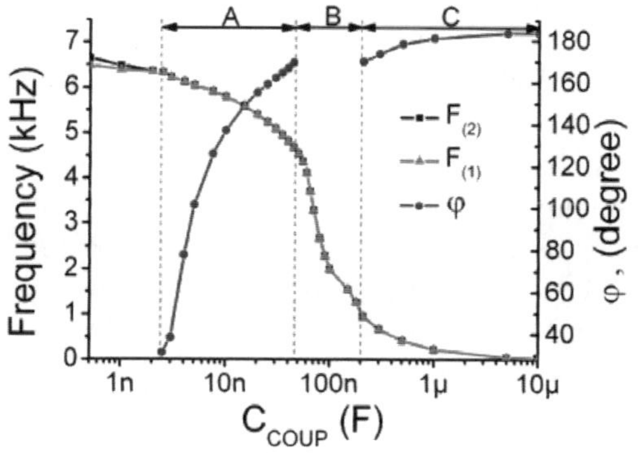

Figure 11. Oscillation frequency and relative phase φ versus coupling capacity value (Simulated).

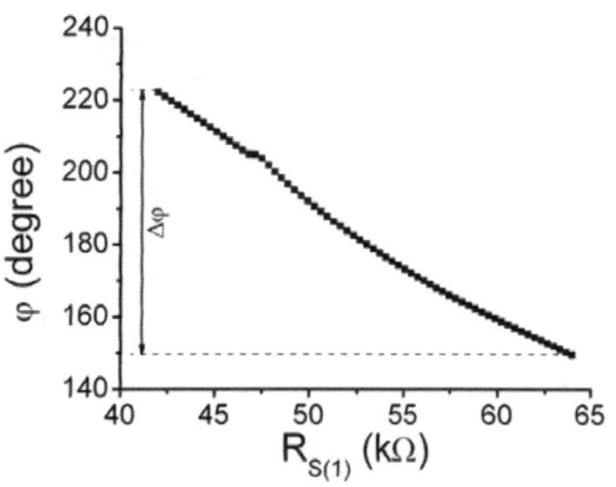

Figure 12. Relative phase φ versus serial resistor $R_{S(1)}$ value (C_{COUP}=1 μF, $R_{S(2)}$ = 53 kΩ).

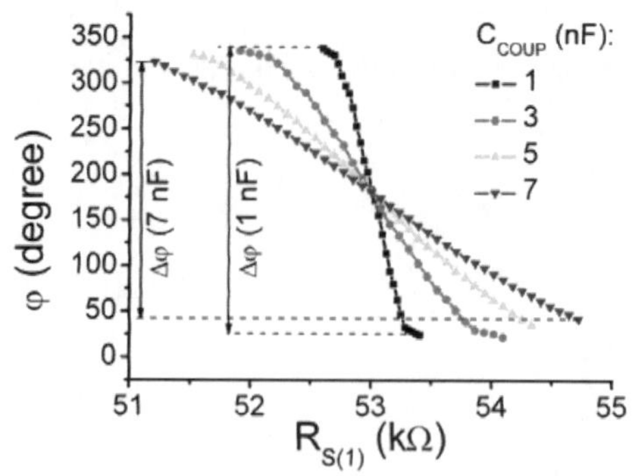

Figure 13. Relative phase φ versus serial resistor $R_{S(1)}$ value (C_{COUP} from 1 to 7 nF, $R_{S(2)}$ = 53 kΩ).

Investigation of the capacitive coupling showed that synchronization was stable and took place at the relatively small value of C_{COUP} – from 3 nF. Increasing C_{COUP} from 3 to 50 nF lead to increase of mutual influence between oscillators (figure 8) and to increase of relative phase from 30° to 170° (figure 11, zone A). In the C_{COUP}' range from 50 to 200 nF (zone B on figure 11), there was not stable relative phase (figure 9). When C_{COUP} was greater than 200 nF (zone C on figure 11) oscillators synchronized in antiphase (figure 10) and period of oscillation was determined generally by C_{COUP}. This mode can be explained by larger C_{COUP}' recharging current, that hold switch of one oscillator in ON-state while voltage on another oscillator increase. Similar oscillation mode observed in work [5]. During numerical simulation (SPICE simulation) it was revealed that oscillation dynamic in this mode was caused by the non-linear resistance of VO_2 switches in the metallic state.

Changing of the relative phase for two oscillators was implemented by the method, proposed in work [5]. The circuit similar as for capacitive coupling (figure 5b) but without R_{OX} was used. Serial resistor $R_{S(1)}$ value was changing in all range, where oscillation observed (from 42 to 64 kΩ), $R_{S(2)}$ value was fixed at 53 kΩ. The circuit and switches parameters were chosen equals: $V_{DD} = 80$ V, $R_i = 10$ Ω, C = 100 nF, $V_{th} = 0{,}99$ V, $V_h = 0{,}44$ V, $R_{off} = 1{,}17$ kΩ. Simulation with large $C_{COUP} = 1$ μF was carried out. Synchronization observed in all range of $R_{S(1)}$ and relative phase variation range Δφ was about 70° (figure 12). The simulation with small C_{COUP} from 1 to 7 nF showed that synchronization occurs in the certain range of $R_{S(1)}$ (figure 13), which increase with increasing C_{COUP} value. Relative phase changes linearly on the most of the synchronization range. On the other hand, increasing C_{COUP} value leads to decreasing relative phase variation range Δφ. The maximum of relative phase variation range Δφ appear at $C_{COUP} = 1$ nF and its value is more than 310°.

5. Conclusion

Thus, capacitive coupling is more suitable for synchronization of oscillators based on VO_2 switches, because it provides stable synchronization and galvanic isolation between oscillators. This type of coupling can be used in systems that consist of a larger number of oscillators fully coupled to each other or coupled in a star configuration.

The simulation of the relative phase changing for two capacitively coupled oscillators by tune serial resistor R_S shows that small C_{COUP} provides the larger range of available relative phase. The large C_{COUP} can be useful when synchronization of oscillators in all range of serial resistor value needed.

Acknowledgements
This work was supported by Russian Science Foundation, grant no. 16-19-00135.

References

[1] Chudnovskii F A 1976 Metal-Semiconductor Phase Transition in Vanadium Oxides and Technical Applications Sov. Phys. Tech. Phys. **20** pp 999-1012

[2] Datta S, Shukla N, Cotter M, Parihar A and Raychowdhury A 2014 Neuro Inspired Computing with Coupled Relaxation Oscillators 51st ACM/EDAC/IEEE DAC **1** pp 1-6

[3] Hoppensteadt F C and Izhikevich E M 1999 Oscillatory Neurocomputers with Dynamic Connectivity Phys. Rev. Lett. **82** pp 2983-2986

[4] Nikonov D E, Csaba G, Porod W, Shibata T, Voils D, Hammerstrom D, Young I A, Bourianoff G I 2015 Coupled-Oscillator Associative Memory Array Operation for Pattern Recognition IEEE J. Explor. Solid-State Computat. Devices Circuits **1** pp 85-93

[5] Sharma A A, Bain J A, Weldon J A 2015 Phase coupling and control of oxide-based oscillators for neuromorphic computing IEEE J. Explor. Solid-State Computat. Devices Circuits **1** pp 58-66.

IOP Conf. Series: Journal of Physics: Conf. Series **929** (2017) 012046 doi:10.1088/1742-6596/929/1/012046

Dependence of the photoconductivity of nanofibers based on indium-zinc oxide on the concentration of components

N.P. Markova, O.Ya. Berezina, A.L. Pergament

Petrozavodsk State University, Petrozavodsk, 185910 Russia

E-mail: khomlyk@mail.ru

Abstract. The paper presents results of a study of indium-zinc oxide nanofibers sensitivity to ultraviolet radiation in the range 230-290 nm. Nanofibers are synthesized by electrospinning. It is shown that the largest increase in photoconductivity is observed at the indium to zinc concentration ratio of about 1:1.

1. Introduction

UV sensors are widely used for detecting radiation in many different applications from chemical industry and medical apparatus to sunlight UV meters and various scientific equipment tools. Recently, a company Rohm has developed a UV sensor using a thin film of zinc oxide [1]. Sensors based on nanowires should be in general more sensitive because of a higher free surface area. Zinc oxide is wide gap semiconductor possessing unique electrical and optical properties; particularly, it belongs to the class of transparent conductive oxides. The low conductivity of undoped zinc oxide makes it difficult to use this material for producing transparent electrode, which leads to necessity to decrease the resistance of the functional material. Increased conductivity is achieved by introducing of donor impurities, such as group 3 elements, into zinc oxide. Embedding M^{3+} cations in the ZnO structure increases the free charge carrier density [2]. Of all the group 3 elements, Ga^{3+} and In^{3+} have the closest values of the effective ion radius to the Zn^{2+} radius, and these elements are mainly used as dopants for zinc oxide [2]. In the present paper, we report on the study of indium-zinc oxide (IZO) nanofiber photoconductivity in the UV spectral range with the aim to determine an optimum indium concentration providing a maximum photoresponse.

2. Experimental details

Fabrication of IZO fibers was performed by the electrospinning method. Formation of nanowires occurred in electrostatic field in the jet of polymer solution or molten polymer [3]. Fibers were formed from a mixture of solutions of zinc acetate dihydrate $Zn(CH_3COO)_2{\cdot}2H_2O$ and hydrated indium nitrate $In(NO_3)_3{\cdot}4.5H_2O$ in water and solution of high-molecular polyvinylpyrrolidone PVP (Mr = $1.3{\cdot}10^6$ g/mol) in ethanol. Solutions with various atomic content of In and Zn were prepared. The diameter of the obtained nanofibers ranged from 240 to 300 nm (Figure 1).

After the process of electrospinning, to remove polymer, annealing of the obtained fibers was carried out in a high-temperature vacuum furnace OTF-1200X in air at 600°C. After

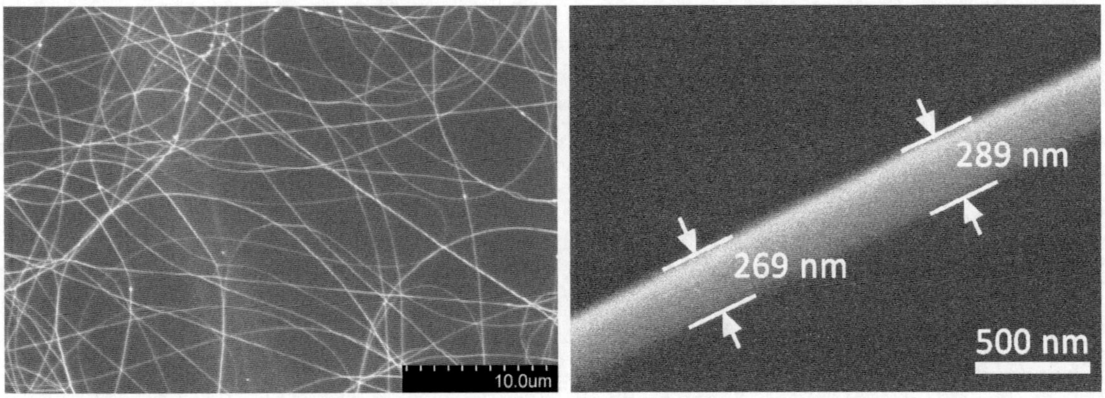

Figure 1. SEM images of nanofibers before annealing.

annealing, the nanofibers' sizes reduced down to 50-100 nm. To evaluate the PVP removal, elemental composition was investigated using a SEM equipped with an EDX system. Also, thermogravimetric analysis showed a loss of 80% weight of the fibers during annealing. Both techniques confirmed the absence of PVP in the samples after first annealing. To measure the current-voltage characteristics, nanofibers were deposited on SiO_2 substrate. Gold contacts with a diameter of 1 mm and with a distance between them of 1 - 1.5 mm were then deposited by thermal vacuum evaporation through a mask. I-V characteristics of the samples were measured using a Keithley 2410 source-meter. A device "Photon" served as the source of weak ultraviolet radiation in the wavelength range 230-290 nm. It was placed directly over the sample at a distance of 5 cm. In this case, the sample surface irradiance was 45 W/cm^2.

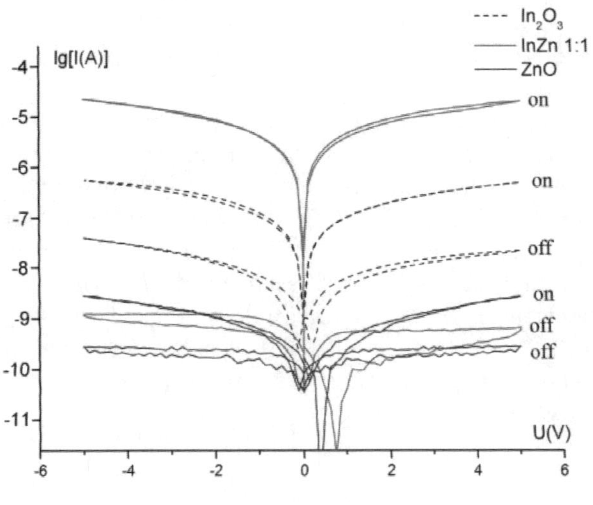

Figure 2. I-V characteristics of IZO nanofibers with different concentration of Zn without UV irradiation (UV-off) and under irradiation (UV-on).

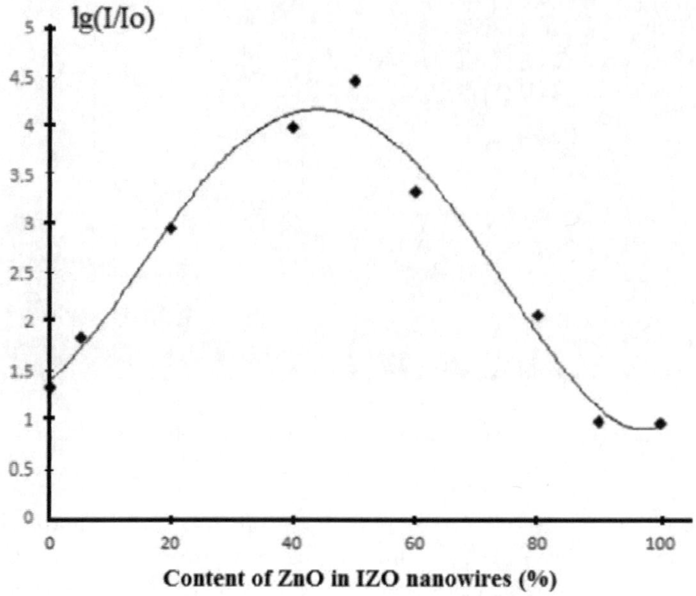

Figure 3. Sensitivity of IZO nanofibers to UV radiation IZO (I - photocurrent at UV irradiation, Io - initial current).

3. Results and discussions

I-V characteristics of the samples with different atomic ratios of In and Zn, under UV irradiation and with natural light, are presented in Figure 2. It is found that the maximum conductivity increase, almost five orders of magnitude, is achieved when a ratio of In:Zn = 1:1 (i.e., ~50 at.% of In and, accordingly, 50 at.% of Zn) - see Figure 3.

The times of current rise and decline are about 100 and 500 sec., respectively (Figure 4). The slow increase and decline in the conductivity might be associated with photochemical processes of oxygen desorption and adsorption on the fibers' surface. Oxygen molecules from the environment are easily absorbed on the surface of nanowires by capturing free electrons from the conduction band, and the molecules become negatively charged ions. When exposed to UV radiation, electron-hole pairs are generated in the IZO nanowires. Then, the photogenerated holes recombine with adsorbed oxygen ions to form oxygen molecules that desorb from the nanofiber surface [4]. Simultaneously, there is an increase in the number of electrons in the conduction band. Under the influence of the applied bias voltage, free electrons move to the anode, and the photocurrent thus appears. The larger the surface area of IZO nanofibers, the more oxygen molecules can adsorb and desorb from their surface at UV irradiation and therefore the higher is the photocurrent.

4. Conclusion

To summarize, IZO is a promising material for UV radiation sensing [2, 5]. In this work, by means of the electrospinning method, nanofibers of mixed indium (0-100%) and zinc oxide (IZO) are obtained. It is shown that the conductivity of the IZO nanofibers increases under irradiation with light in the range of λ = 230-290 nm. The maximum conductivity increase, almost five orders of magnitude, is achieved when the ratio of In:Zn is about 1:1 This effect can be used for development of efficient UV radiation sensors.

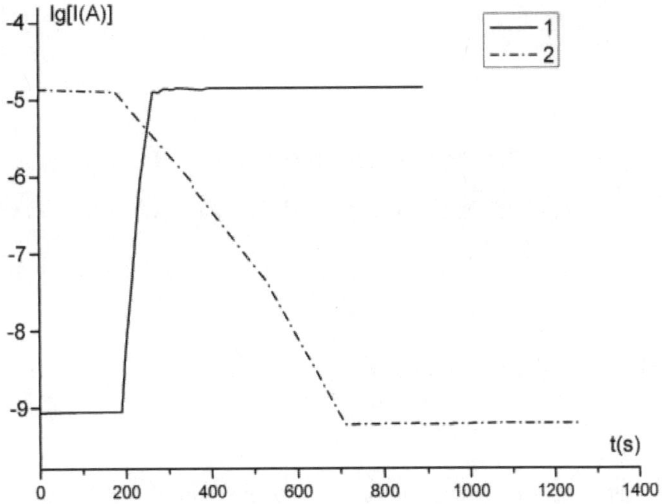

Figure 4. Photocurrent as a function of time at UV irradiation switching-on (1) and switching-off (2). Sample - IZO nanofibers with indium content of 50%.

Acknowledgments
This work was supported by the RF Ministry of Education and Science as a base part of state program no. 2014/154 in the scientific field, project no. 1426, and the Strategic Development Program of Petrozavodsk State University.

References

[1] Ultra-Slim High Precision ZnO Ultraviolet Sensor 2009 *ThomasNet Industry News* ULR: http://news.thomasnet.com/fullstory/precision-uv-sensor-is-composed-of-zinc-oxide-570113
[2] Vorobyeva N.A., Rumyantseva M.N., Vasiliev R.B., Kozlovskiy V.F., Soshnikova Yu.M., Filatova D.G., Zaytsev V.B., Zaytseva A.V., Gaskov A.M. 2015 *Vacuum* **114** 198
[3] Berezina O. Ya., Kirienko D. A., Markova N. P., Pergament A. L. 2015 *Tech. Phys.* **60** 1361
[4] Mamat M. H., Khusaimi Z., Zahidi M.M., Mahmood M. R. 2011 *Jpn. J. Appl. Phys.* **50** 241
[5] Shyju G. J., Dawn Dharma Roy S., Sanjeeviraja C. 2011 *Materials Science Forum* **671** 21

The dependence of the wavelength on MBE growth parameters of GaAs quantum dot in AlGaAs NWs on Si (111) substrate

R R Reznik[1,2,4,5], I V Shtrom[1], Yu B Samsonenko[1,4], A I Khrebtov[1], I P Soshnikov[1,3,4,6] and G E Cirlin[1,2,4,5]

[1]St-Petersburg Academic University – Nanotechnology Research and Education Centre RAS, Khlopina 8/3, 194021, St-Petersburg, Russia
[2]Peter the Great St.Petersburg Polytechnic University, Polytechnicheskaya 29, 195251, St-Petersburg, Russia
[3]Ioffe Physical Technical Institute RAS, Politekhnicheskaya 26, 194021, St-Petersburg, Russia
[4]Institute for Analytical Instrumentation RAS, Rizhsky 26, 190103, St-Petersburg, Russia
[5] ITMO University, Kronverkskiy pr. 49, 197101 St-Petersburg, Russia

E-mail: moment92@mail.ru

Abstract. The data on the growth peculiarities and physical properties of GaAs insertions embedded in AlGaAs nanowires grown on Si (111) substrates by Au-assisted molecular beam epitaxy are presented. It is shown that by varying of the growth parameters it is possible to form structures like quantum dots emitting in a wide wavelengths range for both active and barrier parts. The technology proposed opens new possibilities for the integration of direct-band $A^{III}B^{V}$ materials on silicon platform.

1. Introduction

The relevance of investigations nanowires (NWs) is explained by the necessity to solve important problem - the creation of new non-planar semiconductor nanomaterials and nanosystems with controlled properties on silicon. We have previously shown the fundamental possibility of high-quality GaN and InN NWs formation on silicon with thin SiC buffer layer [1-3].These and $A^{III}B^{V}$ materials can be, and are already used for the development of devices for microelectronics, optoelectronics, analytical biomedicine, emission cathodes, probes for scanning tunneling microscopy, highefficiency solar energy converters, etc. [4]. Of particular interest are NWs with a combined dimension, e.g., of the "quantum dot (QD) in NW" type. First, this enables the formation of size-ordered QDs in systems of III–V materials. Second, new electronic and optical properties appear in nanostructures of this kind, and, therefore, these hybrid materials are objects of intensive studies in modern physics. Previously, we showed that, in principle, nanoscale InAsP inserts can be formed within InP NWs [5, 6], and GaAs inserts, within AlGaAs NWs [7, 8]. To further develop this area, we present in this report growth data and results obtained in studies of the properties of nanostructures of the "GaAs QD in AlGaAs NW" type, grown on the Si (111) surface by molecular-beam epitaxy (MBE), with gold used as a catalyst.

2. Experiments

MBE growth was performed using a Riber 21 Compact installation equipped, in addition to the growth chamber, with a combined vacuum chamber for the deposition of Au (metallization chamber). As substrates served polished semi-insulating Si (111) wafers. Growth was performed in two stages. After the oxide layer was removed from the surface of a wafer preliminarily treated in a 10:1 aqueous solution of HF at a temperature of 850°C, the sample temperature was lowered to 550°C and a gold film with a thickness of ~0.1–0.2 nm was deposited in the metallization chamber. Upon being kept for a minute for even Au drops to be formed, the sample was cooled to room temperature and transferred to the growth chamber without the violation of ultrahigh-vacuum conditions. The temperature in the chamber was increased to growth temperature (was varied from 510°C to 560°C) , then Al, Ga, and As shutters were opened, and AlGaAs NWs were grown within 20 min under As-stabilized growth conditions. The reflection highenergy electron diffraction (RHEED) patterns indicated that a purely wurtzite NW phase was formed already upon the deposition of 50 nm of AlGaAs, which remained unchanged during the whole growth process (a typical RHEED pattern is shown in Figure. 1). After that the Al shutter was closed for several seconds (5–15), which resulted in the formation of a GaAs nanoisland (QD) at the apex of the AlGaAs NW. Further, the Al shutter was again opened for 5 min to form a cap layer. The nominal Al content x in the solid solution, measured from the ratio of the fluxes used to grow a layer on the surface of the GaAs (100) substrate, was varied within the range x = 0.3-0.6. The AlGaAs growth rate was maintained constant at 1 monolayer per second (ML/s). Thus, the growth rates of GaAs and AlAs were varied in accordance with the required molar fraction x in the $Al_xGa_{1-x}As$ solid solution (0.3/0.7, 0.4/0.6, 0.5/0.5, 0.6/0.4 ML/s).

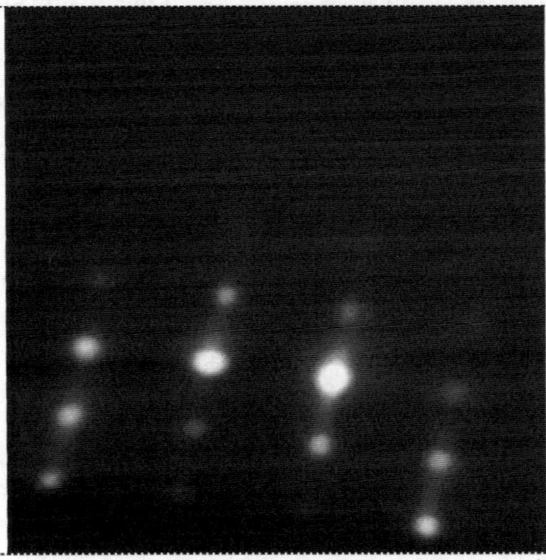

Figure 1. Reflection high-energy electron diffraction (RHEED) patterns of AlGaAs NWs on Si(111) substrate after 5 minutes of growth.

The surface morphology was studied by scanning electron microscopy (SEM). The emission properties of NW arrays were examined on a special optical test bench that included a dispersive spectrometer assembled on the basis of an MDR-204-2 monochromator (LOMO Fotonika OOO). The photoluminescence (PL) was excited by a continuous-wave neodymium laser (wavelength 532 nm) at an excitation power density of about 10 W/cm2. A Hamamatsu R928 photomultiplier served as a photodetector. The PL signal was excited and detected in the normal configuration. Measurements were made at a temperature of 10 K with nanowire-containing samples placed in a closedcycle helium cryostat.

3. Results

Figure 2 (a)-(c) shows the SEM images of NWs arrays grown at different growth temperatures. It is seen, that increasing of growth temperature leads to a deterioration of NWs morphology. Experiments have shown that the optimum growth temperature is 510°C.

At figure 2 (d)-(f) we present the SEM images of NWs arrays grown during different times. It can be seen from images, that 20 minutes of growth is enough for 2 μm nanowires.

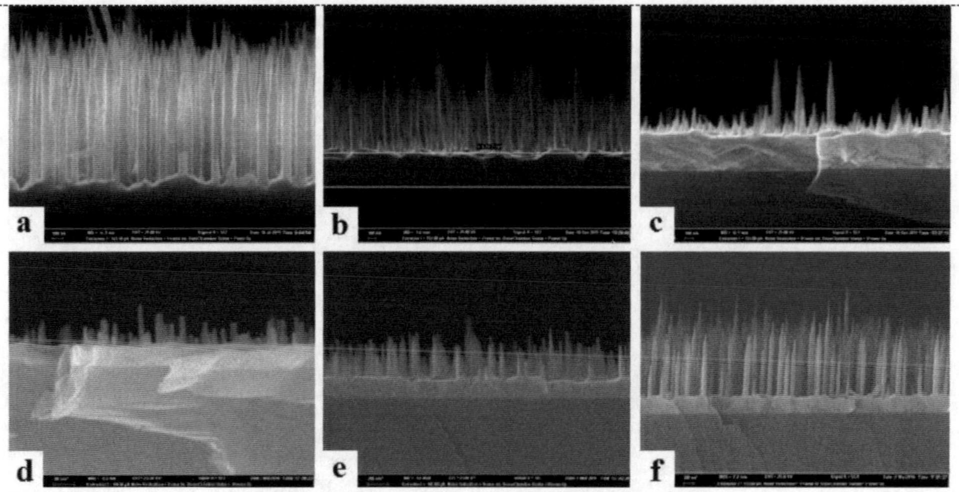

Figure 2. SEM images of hybrid AlGaAs/GaAs/AlGaAs NWs with different growth temperatures and growth time: (a) 510°C (20 min), (b) 540°C (20 min), (c) 560°C (20 min), (d) 5 min (510°C), (d) 10 min (510°C), (d) 20 min (510°C).

It follows from the images of the samples (figure. 3) that both the shape and height of the NWs change with increasing molar fraction of Al. The shape changes from "pencil-like" to "coniform," with the height of the peaked part increasing and the shape becoming almost conical at $x = 0.6$. The height also substantially decreases from 3 ($x = 0.3$) to 0.8 ($x = 0.6$) μm. These changes are characteristic of NWs whose growth is limited by various factors, such as, e.g., growth temperature [9]. In our case, this occurs due to the significant difference in the binding energies (and, consequently, in the coefficients of migration over the surface) of Al and Ga adatoms. Being more strongly bound to the surface, Al adatoms block the migration of Ga atoms, and this becomes more pronounced as the Al/Ga ratio increases. It was found that the formation of a GaAs insert within a NW does not change its configuration, with the configuration being only dependent on the molar fraction of Al.

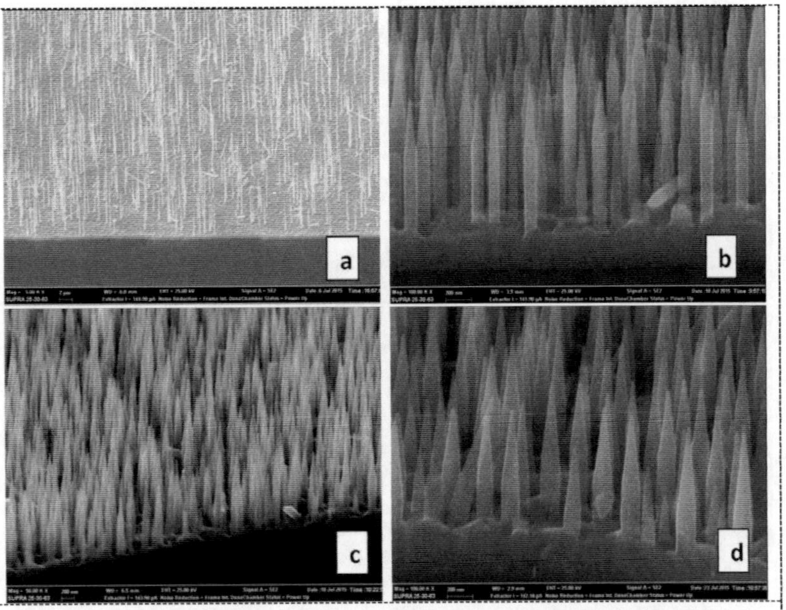

Figure 3. SEM images of hybrid AlGaAs/GaAs/AlGaAs NWs with different Al contents in the solid solution (nominal values): (a) $Al_{0.5}Ga_{0.5}As$, (b) $Al_{0.4}Ga_{0.6}As$, (c) $Al_{0.5}Ga_{0.5}As$, (d) $Al_{0.6}Ga_{0.4}As$.

A number of interesting facts follow from the PL measurements. Examples of the PL spectra are shown in figure 4, including those for different compositions of AlGaAs NWs (which is manifested in the different emission wavelengths in the short-wavelength part of the spectra) and for GaAs inserts (in the long-wavelength part). First, it was shown that, as AlGaAs NWs grow, there occurs the self-organized growth of a rather complex structure constituted by a NW core with some content of Al (always lower than that for Eg1 in figure 4) and a shell with a higher content (for Eg2 in figure. 4). Both these compositions are 2–3 times lower than the nominal composition. Using the formula Eg(AlGaAs)(x) = 1.512 + 1.455x (eV) [10], we estimated the compositions of the AlGaAs solid solution in both the core and shell of a NW, depending on the nominal composition (i.e., that measured on a GaAs (100) calibration sample before the growth of NWs); the data obtained are listed in the table. It should be noted that the similar formation of a self-organized structure of the "core–shell" type was mentioned previously in the cases of both MBE [11] and vapor-phase epitaxy [12]. However, these results contradict each other. For example, the molar composition x of the shell exceeded that in the core in the first case (similarly to the present study), whereas the opposite situation was observed in [12]. A detailed analysis of the reasons for a phenomenon of this kind was made in a separate publication [13].

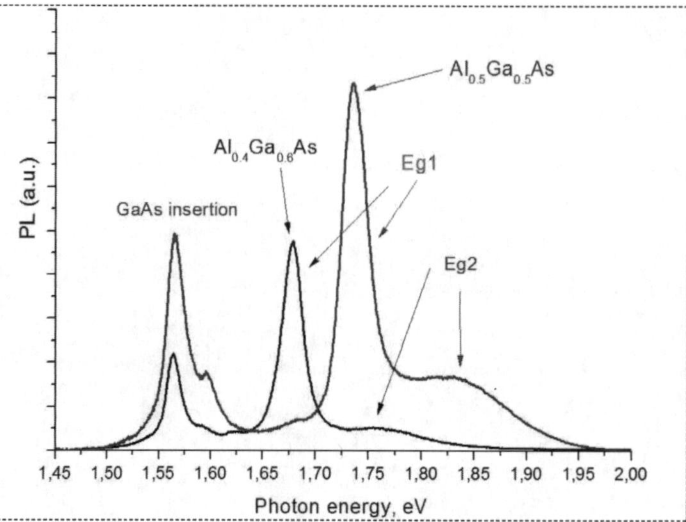

Figure 4. PL spectra measured at T = 10 K for hybrid AlGaAs/GaAs/AlGaAs NWs with different Al contents of the solid solution and different growth durations of GaAs QDs: x = 0.4, 12 s; x = 0.5, 15 s.

The PL lines associated with emission from GaAs QDs are present in the long-wavelength part of the spectrum. An important circumstance is that their position may vary with the size of the insert (its formation duration). In this case, the QD wavelength is always shifted relative to bulk GaAs to shorter wavelengths, which indicates that a quantized structure is formed. For example, the emission wavelength for the $Al_{0.3}Ga_{0.7}As$ composition lies within the range 750–780 nm (not shown in the figure), depending on the QD-growth duration (5–15 s). It also becomes possible to obtain emission from GaAs QDs at the same wavelength in the case of different AlGaAs shell compositions (Figue. 4). However, it becomes necessary to increase the GaAs growth duration in the case of a higher AlGaAs composition (15 and 12 s of GaAs growth for the nominal compositions x of 0.5 and 0.4, respectively).

4. Conclusions

In summary, AlGaAs/GaAs/AlGaAs NWs have been grown by MBE on SI(111) substrates using Au as a catalyst for a first time. Thus, it is possible to controllably obtain a hybrid system based on AlGaAs/GaAs/AlGaAs NWs with a given band diagram by varying the growth duration of the GaAs insert and the composition of AlGaAs NWs. In addition, the suggested technology enables the integration of III–V direct-gap compounds and silicon.

Acknowledgments

The nanowire samples were grown under the support of Russian Science Foundation (Project No 14-12-00393). This work was partially supported by RFBR (grant 16-29-03113 ofi_m) and Skolkovo Foundation (6 dd. 30.12.2015).

References

[1] Reznik R R, Kotlyar K P, Ilkiv I V, Soshnikov I P, Kukushkin S A, Osipov A V, Nikitina E V and Cirlin G E MBE growth and optical properties of GaN nanowires on SiC/Si(111) hybrid substrate *AIP Conference Proceedings* **1748** 040003 (2016)

[2] Reznik R R, Kotlyar K P, Ilkiv I V, Soshnikov I P, Kukushkin S A, Osipov A V, Nikitina E V and Cirlin G E The use of SiC/Si(111) hybrid substrate for MBE growth of GaN nanowires

Journal of Physics: Conference Series **741** 012027 (2016)

[3] Reznik R R, Kotlyar K P, Ilkiv I V, Soshnikov I P, Kukushkin S A, Osipov A V, Nikitina E V and Cirlin G E Growth and Optical Properties of Filamentary GaN Nanocrystals Grown on a Hybrid SiC/Si(111) Substrate by Molecular Beam Epitaxy *Physics of the Solid State* **58** no10 1952–1955 (2016)

[4] Dubrovskii V G, Cirlin G E and Ustinov V M Semiconductor nanowhiskers: Synthesis, properties, and applications *Semiconductors* **43** no 12 1539-1584 (2009)

[5] Tchernycheva M Cirlin G E, Patriarche G Travers L Zwiller L Perinetti U and Harmand J C Growth and characterization of InP nanowires with InAsP insertions *Nano Lett.* **7** no 6 1500–1504 (2007)

[6] Cirlin G E, Tchernycheva M, Patriarche G and Harmand J-C Effect of postgrowth heat treatment on the structural and optical properties of InP/InAsP/InP nanowires *Semiconductors* **46** no 2 175- 178 (2012)

[7] Kats V N, Kochereshko V P, Platonov A V, Chizhova T V, Cirlin G E, Bouravleuv A D, Samsonenko Yu B, Soshnikov I P, Ubyivovk E V, BleuseJ and Mariette H Optical study of GaAs quantum dots embedded into AlGaAs nanowires *Semicond. Sci. Technol.* **27** no 1 015009 (2012)

[8] Barettin D, Platonov A V, Pecchia A, Kats V N, Cirlin G E, Soshnikov I P, Bouravleuv A D, Besombes L, Mariette H, der Maur M A and Carlo A D Model of a GaAs Quantum Dot Embedded in a Polymorph AlGaAs Nanowire *IEEE J. Sel. Top. Quantum Electron.* **19** no 5 1901209 (2013)

[9] Dubrovskii V G, Sibirev N V, Cirlin G E, Tchernycheva M, Harmand J C and Ustinov V M Shape modification of III-V nanowires: The role of nucleation on sidewalls *Rhys. Rev. E* **77** 031606 (2008)

[10] Kuech T F, Wolford D J, Potemski R, Bradley J A, Kelleher K H, Yan D, Farrell J P, Lesser P M S and Pollak F H Dependence of the $Al_xGa_{1-x}As$ band edge on alloy composition based on the absolute measurement of x *Appl. Phys. Lett.* **51** no 7 505-507 (1987)

[11] Chen , Shehata S, Fradin C, LaPierre R, Couteau C and Weihs G Self-Directed Growth of AlGaAs Core–Shell Nanowires for Visible Light Applications *Nano Lett.* **7** no 9 2584-2589 (2007)

[12] Lim S K, Tambe M J, Brewster M M and Gradečak S Controlled Growth of Ternary Alloy Nanowires Using Metalorganic Chemical Vapor Deposition *Nano Lett.* **8** no 5 1386- 1392 (2008)

[13] Dubrovskii V G, Shtrom I V, Reznik R R, Samsonenko Yu B, Khrebtov A I, Soshnikov I P, Rouvimov S and Cirlin G E submitted to *CrystalGrowth and Design*

IOP Conf. Series: Journal of Physics: Conf. Series **929** (2017) 012048 doi:10.1088/1742-6596/929/1/012048

Thin-film fractal nanostructures formed by electrical breakdown

P O Tadtaev[1], A A Bobkov[1], V F Borodzyulya[2], I A Lamkin[1], I I Mihailov[1], V A Moshnikov[1,2], N V Permyakov[1], A V Solomonov[1], N T Sudar[2] and S A Tarasov[1]

[1] Saint-Petersburg State Electrotechnical University "LETI", Saint-Petersburg, Russia
[2] Saint-Petersburg State Polytechnical University, Saint-Petersburg, Russia

Email: tadtayevpavel@gmail.com

Abstract. This is a study of the fractal micro- and nanostructures formation caused by the electrical breakdown of the indium-tin oxide (ITO) covered with various organic coatings. The samples were created by covering a glass substrate with a 1 to 10um-thick layer of indium-tin oxide. Some of the samples were then coated with organic layers of polycarbonate, poly(methyl methacrylate) and others. In order to create high local electrical field densities a special setup based on a eutectic GaIn liquid needle was created: it allowed for the contact area of 60um in diameter and application of the step voltage swept from 20 to 300 volts. The setup also contained a spectrometer for measuring the spectra of the breakdown optical effects. The results showed that the destruction of ITO led to the formation of the spiral fractal nanostructures, parameters of which depended on the thickness of the layer and the presence of the organic cover. In case of the latter, polymer coating was shown to visualize and zoom the topography of the nanostructures which might be used as a method of "polymer photography" for such fractal formations. The analysis of the spectra showed their dependence on the parameters of the structures which proves the possibility of conducting optical diagnostics of the created structures.

1. Introduction

Formation of the fractal micro- and nanostructures is one of the relevant objectives of the state-of-the-art material science. It is mostly due to the fact that fractal nanostructures offer highly-developed surface and exert unique adsorption and catalytic properties, which may be advantageous in designing new types of optoelectronic devices, next generation gas sensors, fractal p-n junctions for photovoltaics and even structures for biomedical application. The production methods are supposed to ultimately form a brand new technique of the not-too-distant future – fractal nanolithography. The application example of this technique could be a method of local formation of the fractal nanostructures via hydrophilic modification by means of atomic-force microscopy and self-assembly of colloidal nanoparticles during evolution of the sol [1,2]. The gist of the method comprises the initial tracing of the areas of high hydrophilicity and the interaction between the substrate and the sol-gel nanoparticles along with the subsequent formation of the fractal nanostructures in those local areas in the "bottom-up" process [3,4]. This paper studies the formation of the fractal micro- and nanostructures by means of the electrical breakdown. The investigation of the breakdown process occurring in conventional conducting film materials is of big technical interest for optoelectronic applications, especially for the flexible electronics.

2. Samples and setup

The typical cross-section of the organic light-emitting structure is shown in the Figure 1(a). Generally, such structures are fabricated on the transparent glass or quartz substrates. Thus, the emission from the structure's active layers can pass through the substrate, which also imposes restrictions on the transparency of the layers between the substrate and the active medium. For the visible band light-emitting devices one of the most popular materials of the active layer is a small-molecule compound Alq3 (Tris-(8-hydroxyquinoline)aluminium); the hole-transporting layer is generally TPD (N,N′-diphenyl-bis(4-methylphenyl)-biphenyl-4,4′-diamine) – a wide-bandgap organic material, also transparent in the visible range of the spectrum. Nevertheless, the behaviour and the quality of the structure is mostly influenced by the cathode layer. In organic light-emitting structures the most commonly used cathode materials are oxide compositions of indium-tin (ITO), tin oxides doped with fluoride (FTO) and some other materials like ZnO, etc. This paper studies the processes occurring in the transparent cathode layer under high current densities, offers a method of analysis and describes the correlation between such processes and properties of the layer. What is more, the details of the fractal nanostructures formation are described and the method of polymer-based visualisation is offered.

For the purposes of the study of the breakdown processes occurring in the transparent electrode layers and fractal nanostructures formation several structures were created, as shown in the Figure 1(b).

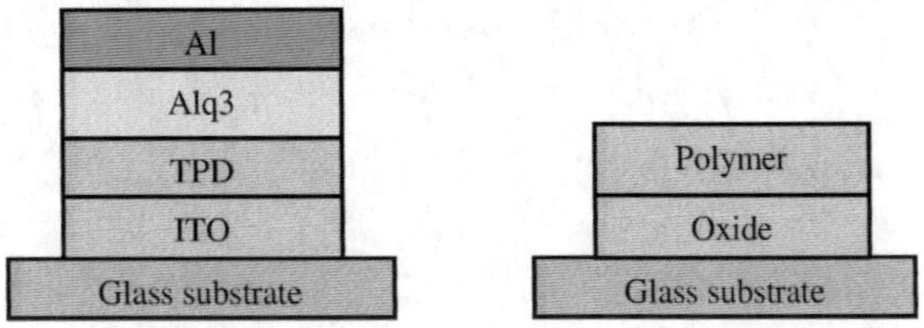

(a) (b)

Figure 1. Typical structure of the organic light-emitting diode (a) and the structures created for the study (b).

The samples were created by covering a glass substrate (0.5-3 mm thick) with a 1-10 um layer of indium-tin oxide (ITO). Some of the samples were then coated with various organic layers. Apart from the abovementioned ones, polycarbonate (PC) and poly(methyl methacrylate) (PMMA) were used. Organic layers were ether spin-coated or formed by means of vacuum thermal evaporation under 10-6 torr. In order to create high local electrical field densities a special setup based on a eutectic GaIn liquid needle was created. It allowed for the contact area of 60um in diameter. Electrode was formed by extracting the needle out of the Ga/In eutectic drop. The setup provided the application of the step voltage swept from 20 to 300 volts. ITO layer acted as the second electrode in the setup. The current-limiting resistor could be varied from 1 to 20 kOhm. 100 Ohm resistor was connected serially to the sample in order to measure the current flowing through the structure. Voltage on this resistor was also measured by a digital oscilloscope. The visual control system allowed for the digital camera footage with a resolution of approximately 5 um. The setup also included the spectrometer which provided spectral measurements of the breakdown emission with a high wavelength and timing resolution. Additionally, there was a magnetic field application capability. Fractal micro- and nanostructures created during the experiments were examined by means of high-resolution optical and atomic-force microscopy and metallographic analysis.

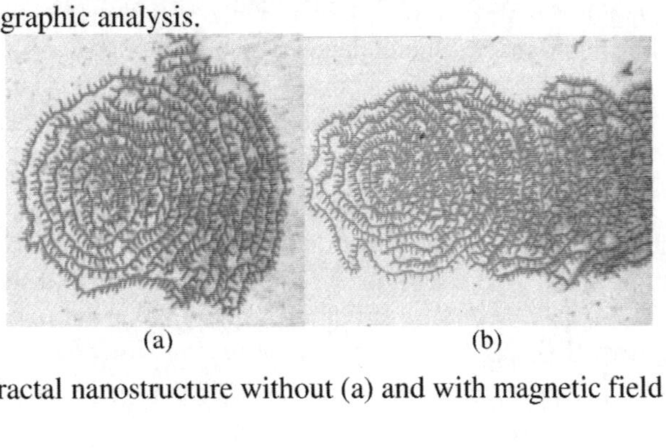

(a) (b)

Figure 2. Fractal nanostructure without (a) and with magnetic field applied (b).

IOP Conf. Series: Journal of Physics: Conf. Series **929** (2017) 012048 doi:10.1088/1742-6596/929/1/012048

3. Results

The study showed that ITO disruption was localized in the current channels (tracks) with their shape depending on the ITO layer thickness. Tracks in the thin ITO films (100-300 nm) were almost linear and 200-400 um long. As for the thicker films (more than 300 nm), tracks were shaped as a multiloop spiral with a diameter of approximately 1 mm. Figure 2 represents the view of the 500-nm thick ITO layer after a local breakdown. As shown, the breakdown track is spiral with multiple branches, mostly oriented outwards. Thus, extensive fractal nanostructures are present on the surface of the ITO after a breakdown. The shape of the structure depends on the properties of the oxide layer, its resistance and crystal structure in particular. The shape is generally symmetrical and formed as an extending spiral, as shown in the Figure 2(a). In presence of the magnetic field the structure is oriented along with its H-vector and is elongated, as shown in the Figure 2(b).

Figure 3 represents more detailed pictures of fractal microstructures formed by the breakdown of the oxide layers. Results were obtained by the method of metallographic analysis. Spiral branched structure is clearly arranged.

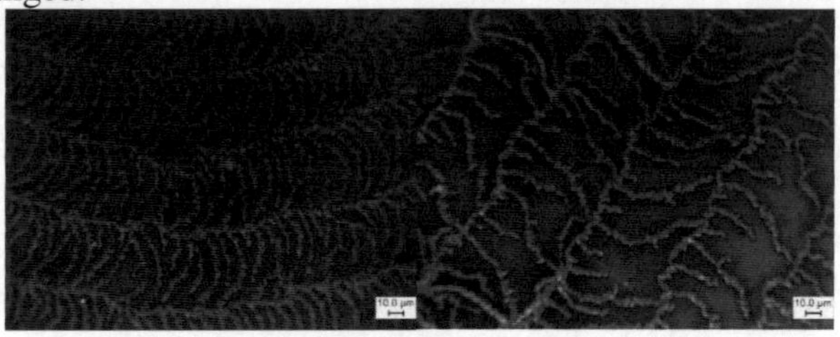

Figure 3. Images obtained by means of metallographic analysis.

In order to obtain even more detailed information about structures the samples were examined by means of atomic-force microscopy (AFM). As seen from the Figure 4, track width was about 2-5 um with a depth of 500 nm. It is important to notice the bulges on the edges of the channel about 300 nm high.

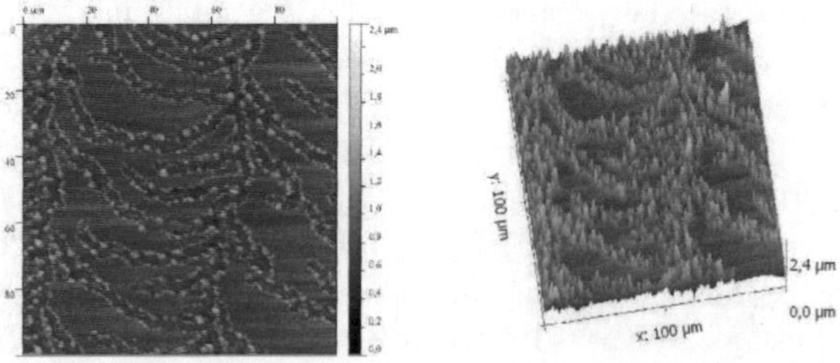

Figure 4. AFM pictures of the breakdown tracks.

The explanation of the observed results of optical and atomic-force microscopy might be as follows. In case of the breakdown of the indium-tin oxide layers placed on the glass substrate a dynamic system emerges coming amid formation of the filament path on the surface which defines the shape of the fractal microstructure. Such shape would depend on the timing and current parameters of the breakdown. Having said this, the breakdown-generated heat consumption processes actually come to the fore, therefore increasing the influence of the thermal conductivity of the film. Thermal emission causes the formation of the additional surface – the filament path – and the development of the fractal nanostructure.

Thus, the discontinuity of the oxide material takes place, followed by its partial evaporation (sublimation) and amorphisation of the circumjacent to the filament area of the layer. This leads to the formation of the "bulge – channel – bulge" structure – the shift of the material to the edges of the structure caused by the thermal shock. Depending on the crystal structure of the layer, spiral or radial tracks are obtained.

Figures 5 to 7 represent results of the optical spectroscopy. Figure 5 contains the spectrum of the breakdown emission of the ITO layer 3 um thick and with resistance of 100 Ohm per square. A linear spectrum with a dominant peak at 583 nm is obtained, secondary peaks of which depend on the properties of the oxide layer. In this case a spiral structure is created, as shown in the Figure 2(a). In case of the breakdown of the 20 Ohm per square structure, peaks at 410 and 450 nm exert higher intensity, as shown in the Figure 6, and the radial fractal structure is formed. Thus, oxide layers can be characterised by the optical spectroscopy methods.

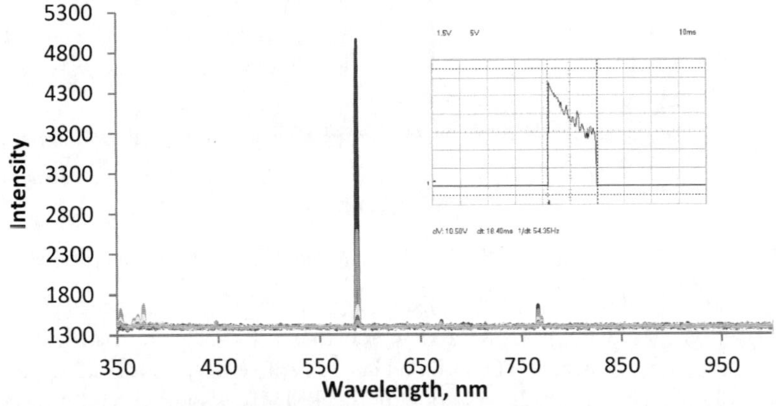

Figure 5. Spectrum of the breakdown emission (ITO, 100 Ohm/square). Inset: oscilloscope pattern of the breakdown process.

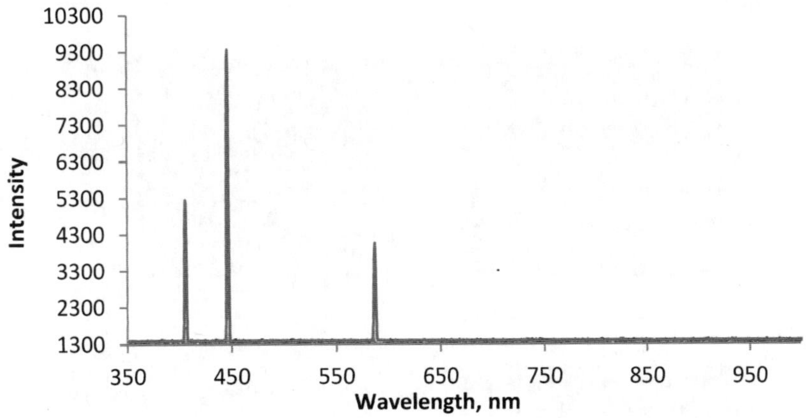

Figure 6. Spectrum of the breakdown emission (ITO, 20 Ohm/square).

Time-domain spectral analysis of the fractal nanostructures breakdown formation in the ITO layers allows for acquiring data on relevant luminescent processes. Figure 7 shows the development of the spectrum with time step of 4 milliseconds. Apart from the primary line, a considerable number of secondary peaks is present, each of which corresponds with some radiative transition during the processes of thermal shock, material sublimation and fractal nanostructure formation.

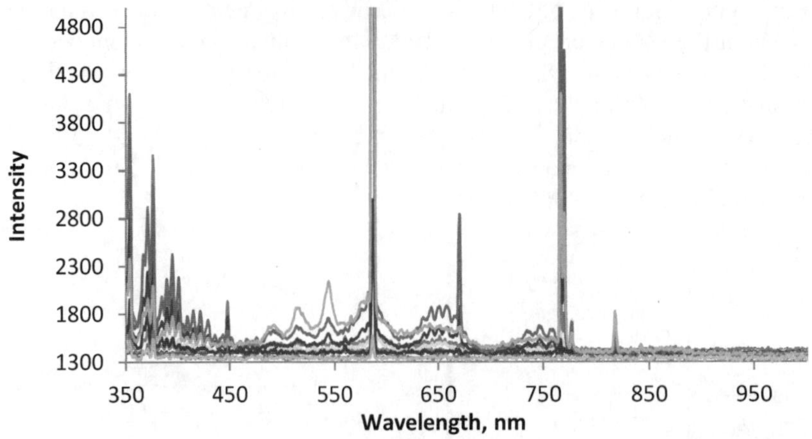

Figure 7. Development of the breakdown spectrum with a 4ms time step.

Disruption of the ITO layer is best visualisable when studying the samples coated with polymer layer (e.g., poly(methyl methacrylate), PMMA) which increases the contrast of the image obtained. Besides, an electrical contact between ITO and the Ga/In needle emerges only after the breakdown and disruption of the polymer coating itself. As soon as the PMMA was disrupted at approximately 100 volts, this voltage was applied directly to ITO which enables much higher current density flowing through the thin (around 10 um) channel in the ITO, heating processes inside which lead to the destruction and evaporation of the polymer in the track and, eventually, disruption of the ITO in the vicinity of the channel.

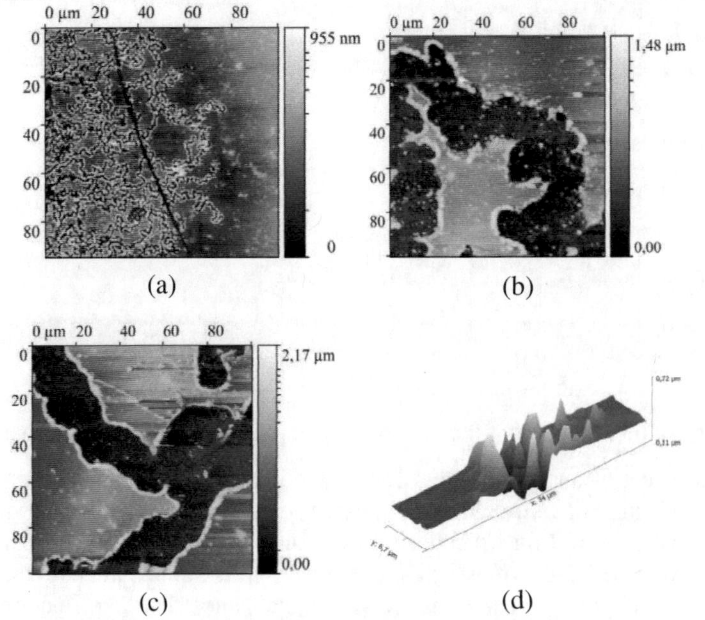

Figure 8. AFM images of disrupted ITO layer (a), breakdown tracks of the ITO layer coated with PC (b) and PMMA (c); the cross-section of the track (d).

Therefore, coating transparent oxide layers with polymers allows for visualisation of the breakdown processes in the structure. The "polymer photography" method may be used to estimate the quality of the oxide layer. Figure 9 contains several AFM images of the ITO based structures. Top left image is a typical radial fractal structure of the disrupted ITO. Top right and bottom left images represent the filaments paths in the structures coated with PC and PMMA respectively. Due to the lower activation energy of the sublimation processes, polymers' evaporation is much more intense, which leads to the considerable extension of the track. The cross section of the track is shown on the bottom right image. It is easy to notice the bulges at the edges of the channels, which are due to the shift of the polymer material as a consequence of the thermal shock.

4. Conclusion

The results of the study showed that the electrical destruction of ITO led to the formation of the spiral and fractal nanostructures, parameters of which depended on the thickness of the layer and the presence of the organic cover. In case of the latter, polymer coating was shown to visualize and zoom the topography of the nanostructures which might be used as a method of "polymer photography" for such fractal formations. The analysis of the spectra showed their dependence on the parameters of the structures which proves the possibility of conducting optical diagnostics of the created structures.

5. References

[1] Averin I A, Moshnikov V A, Maksimov A I, Pronin I A, Karmanov A A and Igoshina S E 2013 Patent RUS 2532428

[2] Moshnikov V A, Mksimov A I, Aleksandrova O A, Pronin I A, Karmanov A A, Terukov E I, Yakushova N D, Averin I A, Bobkov A A and Permyakov N V 2016 *Technical Physics Letters* **42** 18 82-87

[3] Kuzmin S V, Saha P, Sudar N T, Zakrevskii V A, Sapurina I, Solosin S, Trchova M and Stejskal J 2008 *Thin Solid Films* **516** 8 2181-87

[4] Vettegren V I, Zakrevskii V A, Smirnov A N and Sudar N T 2010 *Physics of the Solid State* **52** 8 1650-55

IOP Conf. Series: Journal of Physics: Conf. Series **929** (2017) 012049 doi:10.1088/1742-6596/929/1/012049

Study of inverse magnetostrictive effect in metallic glasses $Fe_{80-x}Co_xP_{14}B_6$

V S Severikov[1], A M Grishin[1,2], V S Ignahin[1]

[1] Petrozavodsk State University, 185910 Petrozavodsk, Karelian Republic, Russia
[2] KTH Royal Institute of Technology, SE-164 40 Stockholm-Kista, Sweden

E-mail: severva3@gmail.com

Abstract. The paper presents the possibility to build a tension gauge capable to discriminate different kinds of deformations: compression and twisting (induced by torsion strain) based on the magnetoelastic effect in new metallic glasses $Fe_{80-x}Co_xP_{14}B_6$. Applied loads increase coercive field H_c, saturation induction B_s and rectangularity of magnetic hysteresis loop. For example, hysteresis loop traced for 1 mm narrow, 50 cm long and 30 μm thick $Fe_{40}Co_{40}P_{14}B_6$ straight ribbon subjected to longitudinal stress of 346 MPa shown increased B_s from 1.24 to 1.7 T and squareness from 0.55 to 0.88 compared to unloaded specimen. For twisting, on the contrary, both squareness and coercive field vary whereas the value of B_s remains unchanged.

1. Introduction

Magnetic properties of rapid quenched $Fe_{80-x}Co_xP_{14}B_6$ metallic glasses surpass characteristics of $Fe_{40}Ni_{40}P_{14}B_6$ magnetosoft material (Metglas 2826) widespread in the commercial market. Really, the maximum relative differential permeability of as quenched $Fe_{80-x}Co_xP_{14}B_6$ ribbons is about 110000, the saturation induction B_s = 1.45 T, quasistatic coercive field as low as 4 A/m, Curie temperature above 700 K, significantly higher thermal stability due to increased by 60 K crystallization temperatures, low hysteresis loss of about 0.26 W/kg in the saturation mode at frequency of 100 Hz [1-3].

Magnetostriction and inverse magnetostrictive (magnetoelastic) effect have been thoroughly investigated only in some polycrystalline magnetic materials. Due to versatile functional properties amorphous magnetic materials promise many practical applications. Besides high magnetoelastic coupling, they exhibit large elastic thresholds, which permit the application of much larger stresses not accompanied by plastic deformations.

The aim of this article is to report experimental results on magneto-mechanical effect in amorphous rapidly quenched ribbons with different $Fe_{80-x}Co_xP_{14}B_6$ compositions.

2. Experimental

Series of $Fe_{80-x}Co_xP_{14}B_6$ (x = 25, 32, 35 and 40 at.%) ribbons were meltspun onto the massive copper wheel from the RF-melted superheated master ingots. X-ray diffraction reveals amorphous structure in as-cast specimens containing the superposition of *bcc* α-FeCo and *bct* $(Fe,Co)_3(P,B)$ nuclei with a characteristic size as small as 1.6 nm. Both *isothermal* and *isochronal* annealing of ribbons in protective atmosphere lead to the predominant growth of *bcc* α-FeCo 20-30 nm-sized crystals within the amorphous metallic matrix [4].

Characteristics that quantify magnetoelastic phenomenon were obtained from the series of hysteresis *B-H* loops recorded for $Fe_{80-x}Co_xP_{14}B_6$ ribbons under applied external stresses. Due to very high magnetic permeability, special precautions were undertaken to choose a proper orientation of ribbons in 30 cm long primary and 5 cm long two secondary (pick-up and reference) solenoidal coils of the magnetometer. It is determined by actual magnetic declination of 12.764° **E** and relatively high magnetic inclination of 74.797° in Petrozavodsk city (Karelian Republic, Russia – Latitude: 61°46'59"; Longitude: 34°19'59").

Schematic of experimental setup to record hysteresis loops under applied stresses is shown in Fig. 1. Magnetometer contains the external coil which generates the magnetic field and internal receiving coils. The specimen under test is placed in one of two receiving coils connected in series and electrically in opposition. Signal from the receiver coils is integrated with the use of an active integrator. The upper end of the tape is fixed on the stand, and the bottom one is loaded by hanging a certain weight.

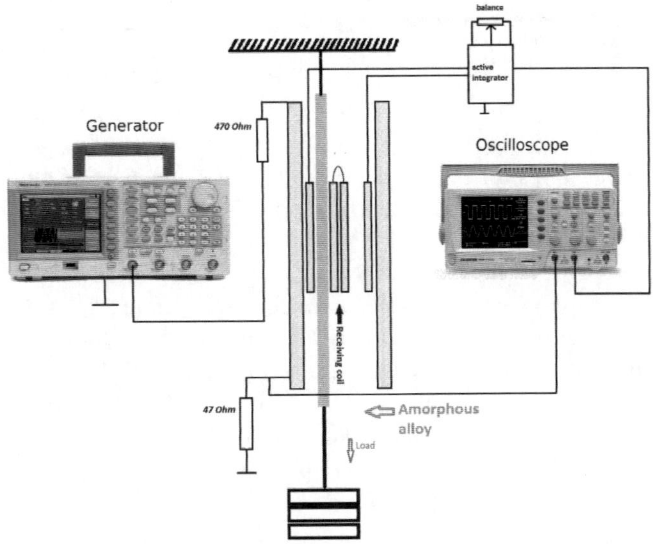

Figure 1. Installation for the measurement setup.

3. Results and discussion

1 mm wide and 30 μm thick $Fe_{80-x}Co_xP_{14}B_6$ ribbons were subjected to mechanical stresses. We neglected the change of ribbon's cross-section under load and calculated mechanical stress σ dividing applied force by cross-sectional area of the ribbon.

Series of hysteresis loops under different linear mechanical loads applied along ribbons is shown in Figs. 2 **a**,**b**. They illustrate experimental data for glass composition $Fe_{45}Co_{35}P_{14}B_6$ at magnetizing field up to H_{max} = 450 A/m and for $Fe_{55}Co_{25}P_{14}B_6$ metallic glass at H_{max} = 2200 A/m. Two insets for graphs show that saturation induction B_s increases with a load growth in both cases. It is notable that at low loads the dependences $B_s(\sigma)$ have a linear region to approximately σ = 100 MPa and then saturate with B_s remaining almost constant. At the same time, coercive field H_c slightly increases under strain deformation for both samples.

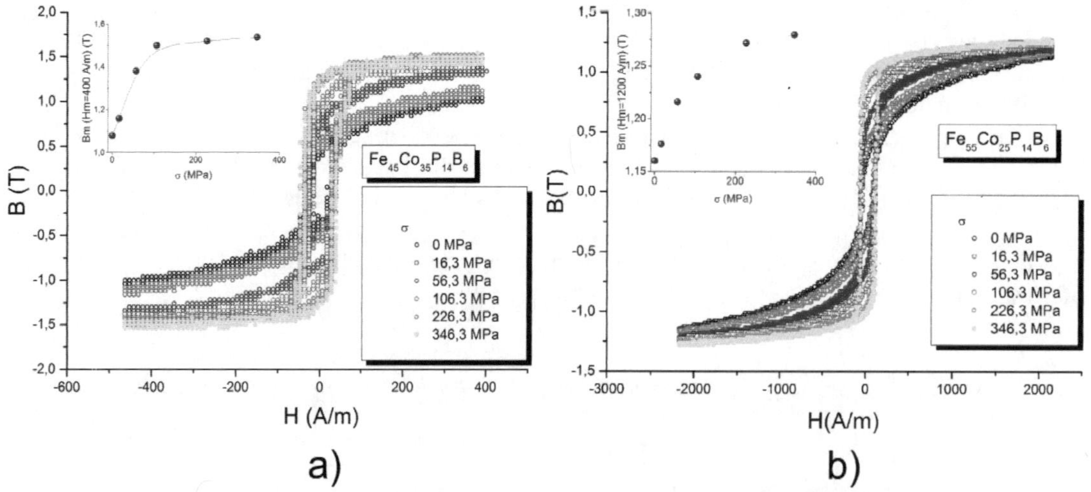

Figure 2. **a** - Series of hysteresis loops for $Fe_{45}Co_{35}P_{14}B_6$ ribbon at magnetizing field up to $H_{max} = 450$ A/m for different linear mechanical loads. **b** - Hysteresis loops for $Fe_{55}Co_{25}P_{14}B_6$ ribbon at $H_{max} = 2200$ A/m.

Besides the increase of saturation magnetization B_s, growth of mechanical load results in changing the shape of hysteresis loop. Loop squareness B_s/B_{rem} is a useful parameter to quantify the change of magnetic properties. Here B_{rem} is a remnant induction remaining at $H = 0$. Fig. 3 depicts a series of hysteresis loops for $Fe_{48}Co_{32}P_{14}B_6$ metallic glass. Inset to Fig. 3 presents loop's squareness and saturation induction B_s as functions of mechanical load. In contrast to $B_s(\sigma)$, there is a small initial region where the squareness seems to be independent on the load.

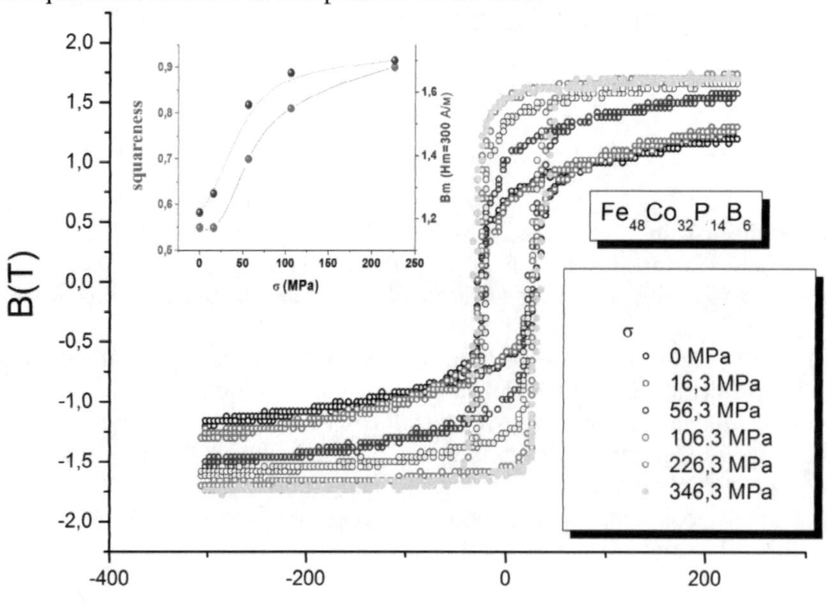

Figure 3. Series of hysteresis loops for $Fe_{48}Co_{32}P_{14}B_6$ metallic glass at magnetizing field up to $H_{max} =$ 300 A/m with different mechanical loads applied along the ribbon. Inset shows the dependences of hysteresis loop's squareness and saturation induction B_s on mechanical stress σ.

Under applied linear stress of 346.3 MPa, the saturation induction B_s increases from 1.24 to 1.7 T, loop's squareness also increases from 0.55 to 0.9, whereas the coercive field remains practically unchanged. 3D plot in Fig. 4 summarizes both dependencies of magnetic induction B in $Fe_{48}Co_{32}P_{14}B_6$ glass on magnetic field H and applied linear stress σ.

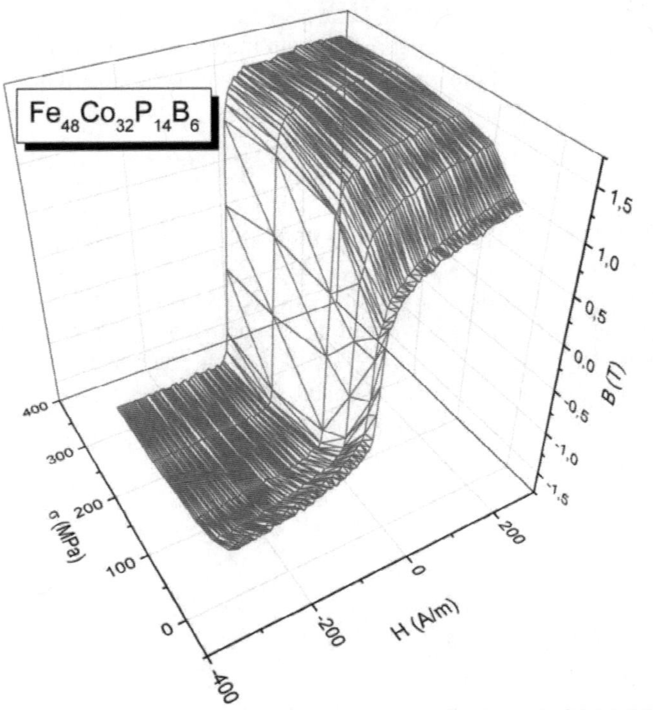

Figure 4. 3D plot for the field and stress dependent magnetic induction $B(H,\sigma)$.

We found that magnetic response in magnetosoft rapidly quenched $Fe_{80-x}Co_xP_{14}B_6$ ribbons depends also on the type of applied mechanical stress. Twisting deformations result in dramatic increase of coercive field. Fig. 5 demonstrates this effect as a series of hysteresis loops for $Fe_{48}Co_{32}P_{14}B_6$ metallic glass ribbons subjected to a torsional stress.

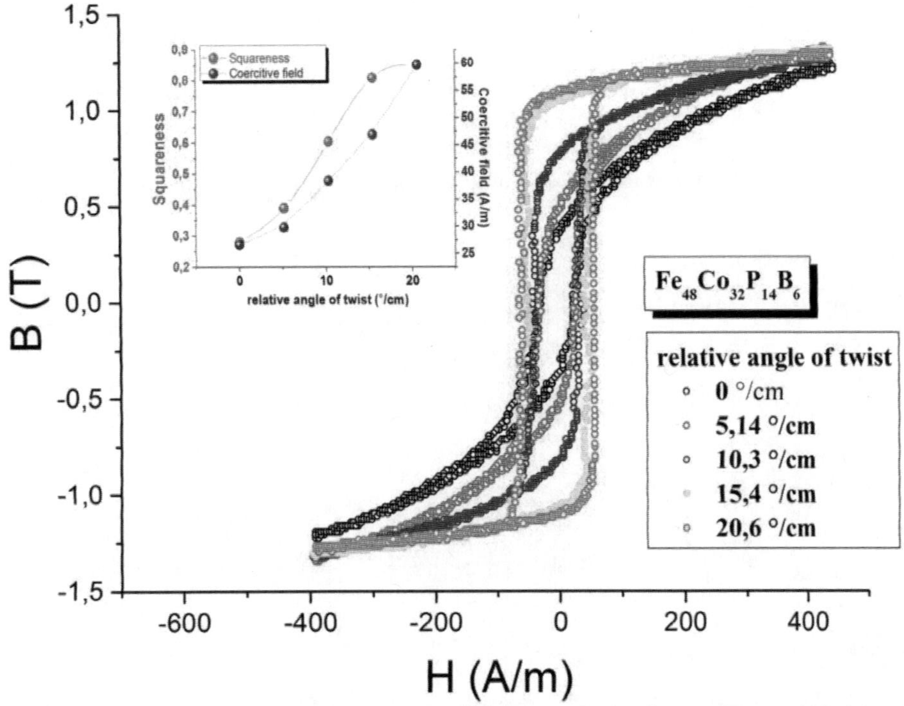

Figure 5. Series of hysteresis loops in $Fe_{48}Co_{32}P_{14}B_6$ ribbon magnetized up to H_{max} = 400 A/m and twisted at different angles.

As clearly seen in Fig. 5, opposite to longitudinal strain the torsion deformations as high as 20.6 °/cm result in huge growth of coercive field H_c from 27.5 A/m up to 60 A/m and significant increase of squareness from 0.29 to 0.85, whereas the saturation induction B_s being practically unchanged.

4. Conclusions

Presented results show that different mechanical stresses have significant influence on the shape of hysteresis loops in metallic glasses $Fe_{80-x}Co_xP_{14}B_6$. Magnetoelastic effect manifests itself by growth of coercive field H_c, saturation induction B_s and loop's squareness under applied longitudinal mechanical load. Under twisting, on the contrary, saturation magnetization remains constant whereas both coercive field H_c and loop's squareness experience a rapid increase. These observations testify possibility to discriminate different types of mechanical deformations and to build a multicomponent strain gauge based on magnetosoft $Fe_{80-x}Co_xP_{14}B_6$ metallic glasses.

References

[1] Hollmark M, Tkatch VI, Grishin AM and Khartsev SI, 2001 Processing and Properties of Soft Magnetic $Fe_{40}Co_{40}P_{14}B_6$ Amorphous Alloy *IEE Transactions on Magnetics* **37**, 2278 – 2280.
[2] Tkatch VI, Grishin AM and Khartsev SI, 2002 Delayed nucleation in $Fe_{40}Co_{40}P_{14}B_6$ metallic glass *Materials Science and Engineering* **A337**, 187 – 193.
[3] Prahova DA, Grishin AM, Ignahin VS, Lugovskaya LA, Osaulenko RN, 2016 Melt-spun Fe-Co-P-B metglasses: structure, crystallization kinetics, magnetic properties *Journal of Physics: Conference Series*, **769**, 012034.
[4] Lugovskaya LA, Osaulenko RN, Grishin AM, Ignahin VS, 2015 X-ray study of soft magnetic metallic glasses FeCoPB *Proceedings of the Kola Science Centre, Chemistry and Materials, Special Issue,* **5** (31), 395-399.

Study of functional properties of gas-sensitive cobalt-containing polyacrylonitrile films

T.V. Semenistaya, A.A. Voronova

Southern Federal University, Engineering-technological Academy, Institute of Nanotechnologies, Electronics, and Equipment Engineering, 2, Shevchenko St., Taganrog, 347922, Russia

semenistayatv@sfedu.ru

Abstract. The design of the sensor materials with challenging gas-sensitivity can be solved by materials selection and their compatibility with the manufacturing technologies that allows to operate the process of formation of nanocomposite structure and to receive the required material. The polyacrylonitrile (PAN) as the conducting polymer with a highly π-conjugated polymeric chain due to flexibility for tailoring the structure of the final products by the pyrolysis method under the influence of incoherent IR-radiation is chosen. The aim of the work was to study the peculiarities of formation procedure of cobalt-containing PAN films. The gas-sensing Co-containing PAN films have been fabricated. The different temperature and time have been used to form the films. Depending on intensity and exposure time of IR-radiation the thermostructured PAN films with resistance values of $\cdot 10^8\ \Omega$ to $10^{10}\ \Omega$ have been fabricated. It is shown that the heat-treated PAN is the p-type semiconductor. Irrespective of the level of the modifying additive in film-forming solution and the time-temperature modes little change of film resistance has been found. It has been found that the Co-containing PAN films are gas-sensing films and have high selectivity to Cl_2 and NO_2. A stationary state gas distribution method was used for testing gas-sensing properties. Obtained the Co-containing PAN films are perspective for low-temperature applications as Cl_2 and NO_2 sensors.

1. Introduction

In recent years, research in the field of nanoscience and nanotechnology has emerged due to the applications of nanocomposite and nanoscale materials.

Conducting polymers with conjugated π-bonds chains and polymer-nanocomposites are developing field of modern nanotechnology. They have become worldwide research interest for the structure-property relationships. The latter is particularly interesting classes of nanostructures for their advantages of obtaining the desired properties by incorporation of the nanoscale materials into polymer matrix [1,2]. For the discovery of conducting polymers, Alan J. Heeger, Alan G. MacDiarmid and Hideki Shirakawa were awarded the Nobel Prize in Chemistry in 2000 [3]. Typical conducting polymers include polyacetylene (PA), polyaniline (PANI), polypyrrole (PPy), polythiophene (PTh), poly(para-phenylene) (PPP), poly(phenylenevinylene) (PPV), polyfuran (PF), and their derivatives. These conjugated polymers can be electrical insulators, semiconductors or conductors, depending on the level of doping and nature of the dopants. Insulators can be converted into semi-conductive or conductive states upon treating with dopants and/or subjecting to chemical or electrochemical redox reactions. The electrical conductivity of these conjugated polymers can increase by several orders of magnitude. To extend the functions or to improve the performances of these polymers, they are frequently doped with other functional materials to form composites [3,4].

Polyacrylonitrile (PAN) and copolymers of PAN have been widely studied for almost a century for commercial and technological exploitations. PAN is still a promising object of research to create nanofibers and porous nanotubes and for a variety of applications: medicine, energy, agriculture, filtration, technical textile, protective materials, etc. [5]. Technologically PAN interesting for its ability to form stable structures of different conductivity during heat treatment forming a ladder structure through processes of cyclization of nitrile groups –C≡N and dehydrogenation. The crosslinking of the chain molecules in the form of –C=N–C=N– is a process that prevents melting during subsequent carbonization when using temperatures higher than 200 °C [5-7]. Heat-treated PAN unlike other conductive polymers, is a material with semiconductor properties without need to embed dopants.

The problem of obtaining functional materials with the best characteristics using the specified process parameters for their formation is relevant for modern micro- and nanoelectronics. Ability to change the electrical conductivity from the dielectric to metallic while chemical structure is turning from the linear form to polyconjugated upon heat treatment has driven impressive research efforts in development of optimal technologies for the organic conjugated polymer materials.

Gas-sensing properties of the PAN-based films strongly depend on microstructure, components and structure of film material. There are certain correlations between surface morphology of the PAN-based films and the PAN-polymer molecule structure that caused by processes occurring during the formation of the sensing material. The problem is to study the correlation using methods of empirical research and that is the reason for its insufficient study.

The goal of this paper is to examine the features of cobalt-containing PAN film surface and to study gas-sensing properties that are based on the selective absorption of gas molecules inheriting the adsorption-resistive effect [1-17].

Embedding transition metal of low concentration in organic matrix [18-20] or controlling molecular structure of organic polymer through using different methods of polymer material fabrication are effective techniques to improve functional characteristics of composite materials including adsorption-resistive properties of PAN-based films [21-23]. Varying the time-temperature parameters has been used to fabricate PAN-based film with various molecular structure of polymer that caused change in conductivity of thermostructured PAN polymer [24,25].

2. Experiment

To fabricate films the following components were used: polyacrylonitrile (PAN) (Aldrich 181315) as polymer organic matrix, cobalt chloride (II) $CoCl_2$ as modifying additive for increasing the selectivity and adsorption activity of PAN, dimethylformamide (DMFA) as a solvent. Film-forming solutions were deposited (centrifuged) on polycor substrates and dried in oven at T = 160 °C, t = 30 minutes. The Co-containing PAN films were formed under the influence of incoherent IR-radiation under low vacuum conditions [19,20,23,26]. IR-annealing was conducted in infrared camera, halogen lamps KG-220 with high radiation in the area of 0.8-1.2 μm were used as the source of radiation. Time-temperature modes of IR-annealing were selected experimentally, since the intensity and duration of infrared radiation provide an opportunity to control the properties of the films by changing the molecular structure of the polymer. The radiation intensity at the first stage of IR-annealing corresponds to temperatures varying from 250 to 350 °C during a 5-20-minute time period, and the intensity of radiation at the second stage corresponds to temperatures of 300 – 500 °C during a 2-10-minute period.

The film surface morphology was observed by atomic force microscopy (AFM) using scanning probe microscope Solver P47 Pro (NT-MDT). The 5 × 5 μm^2 area AFM-images of the films are processed with Image Analysis program.

Resistance measurements were carried out using E6-13A Terraommetr. The gas sensitivity (S) is defined as the ratio $S = (Ro - Rg)/Ro$, when $Ro > Rg$, where Ro is the resistance in air, Rg is the resistance in the atmosphere of analyte.

3. Results and discussion

It is known that there are control parameters to create PAN-based material with desired properties: formation conditions of the organic polymer structure and distribution of the modifying additive in the "organic matrix" composite material [2-4, 17]. Creation of gas sensing material is connected with a search for the optimal microstructures, operating temperature, response time and quantitative content of modifying agent that suit the analyte molecules.

It is known that when pyrolyzed over the temperature of 300 °C, the linear chains of dielectric PAN undergo chemical conversion of polymer with a cyclic structure which contains conjugated double bonds. As a result of the heat treatment of the films PAN acquires properties of a semiconductor structure due to chemical changes at a molecular level. The main processes that determine the unique electrical properties of PAN are intermolecular and intramolecular cyclization, intermolecular cross-linking that enable to form polymer with semiconducting properties [27].

The resistance values of fabricated films are in the range $10^8 \div 10^{10}$ Ohm. Such resistance values are explained by deferent structure of heat-treated PAN. It is found that the increase of the IR-radiation intensity and concentration of a modifying additive decreases the resistance values of fabricated films (Figure 1).

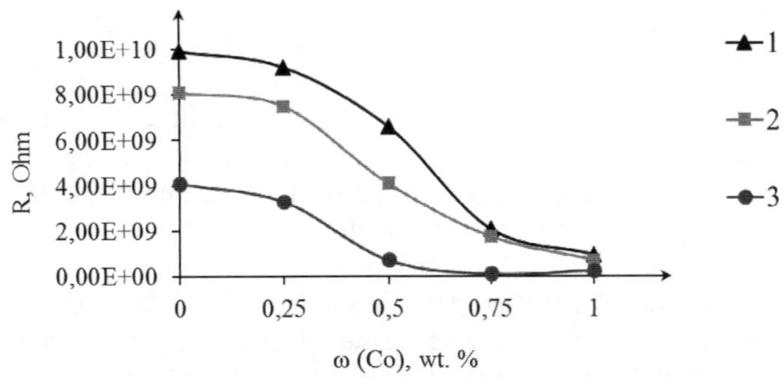

Figure 1. The dependence of the Co-containing PAN film resistance on the weight concentration of modifying additive: 1 – T_{IR2} = 400 °C; 2 – T_{IR2} = 450 °C; 3 – T_{IR2} = 500 °C

As observed from figure 2 the resistivity of the Co-containing PAN films decreases with increase in temperatures with in the experimental range that is characteristic of semiconductor materials.

Figure 3 shows the Arrhenius plots of temperature dependence of resistivity of Co-containing PAN films. It has been observed that the dependence represented by equation $R = R_0 \exp(-\Delta E/(2kT))$, where ΔE is the activation energy, k is Boltzmann's constant and the factor 2 accounts for the thermal excitation of both positive and negative charges [28]. As shown from Figure 3 the resistance of the films decreases linearly with increasing temperature in coordinates ln R – 1/T. Such dependence indicates the semiconducting behavior of the obtained films.

Regardless of the weight concentration of modifying additive in the initial solution, and temperature-time conditions of film formation the slight increase in temperature decreases resistance due to the relatively weak temperature dependence of the carrier mobility, which is typical for organic semiconductors [2].

According to Hall effect measurements revealed that the PAN films have p-type conductivity [2,19]. It is known that for the semiconductor gas sensor with inorganic material of p-type as a sensitive layer, reducing gas atmosphere causes to a decrease in its conductivity, owing to a decrease in electron concentration in the surface layer and an increase of the height of intergranular barriers [29].

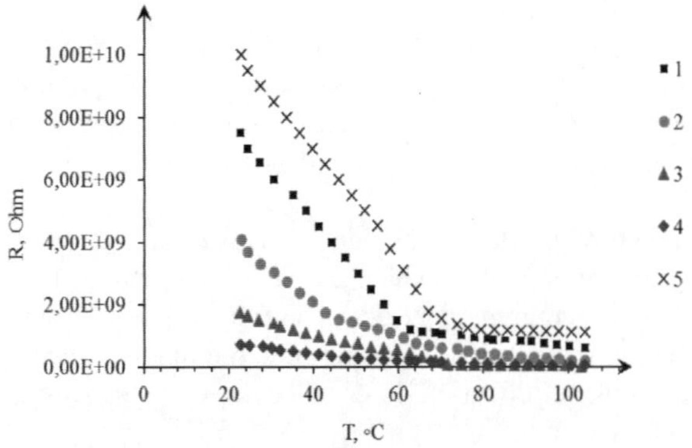

Figure 2. Temperature dependence of the Co-containing PAN film resistance on the weight concentration of modifying additive: 1 – ω (Co) = 0,25 wt.%; 2 – ω (Co) = 0,5 wt.%; 3 – ω (Co) = 0,75 wt.%; 4 – ω (Co) = 1 wt.%; 5 – ω (Co) = 0 wt.%

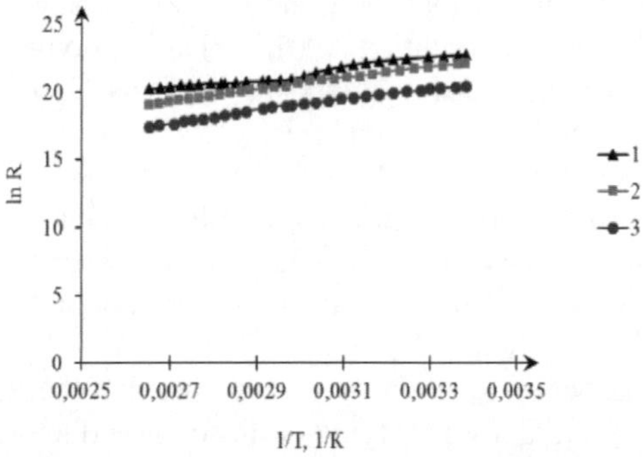

Figure 3. Arrhenius plots of temperature dependence of the resistance of the Co-containing PAN films on the weight concentration of modifying additive: 1 – ω (Co) = 0,25 wt.%; 2 – ω (Co) = 0,5 wt.%; 3 – ω (Co) = 1 wt.%

Figure 4 depicts the dependences of the Co-containing PAN film gas sensitivity on different temperature in atmosphere of chlorine c(Cl_2) = 36 ppm and nitrogen dioxide c(NO_2) = 21 ppm. Temperature-dependent adsorption studies revealed that operating temperature of the sensing film is close to room temperature, it gives the possibility to create energy-efficient sensors.

Gas sensitivity of the Co-containing PAN films was found to decrease with increasing weight concentration of a modifying additive that was shown in Figure 5. The results of the sensing tests of the films revealed the best sensitivity was observed for the sample with weight concentration of modifying additive ω (Co) = 0,5 wt.% toward nitrogen dioxide NO_2; and ω (Co) = 0,25 wt. % toward chlorine Cl_2.

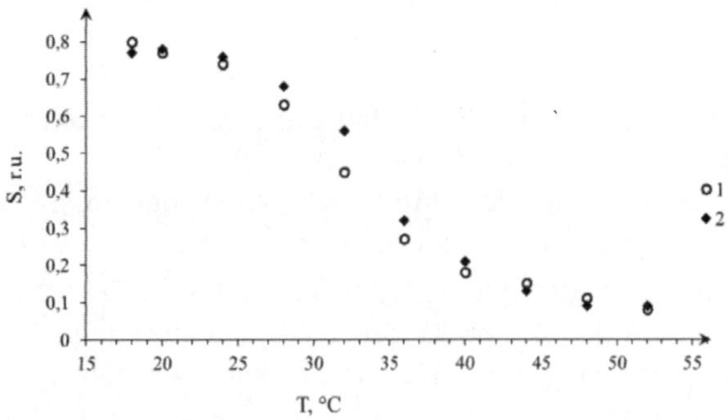

Figure 4. Sensitivity dependence on temperature of the Co-containing PAN films, the weight concentration of modifying additive and detected gas: 1 – ω (Co) = 0,5 wt.%, NO_2; 2 – ω (Co) = 0,25 wt. %, Cl_2

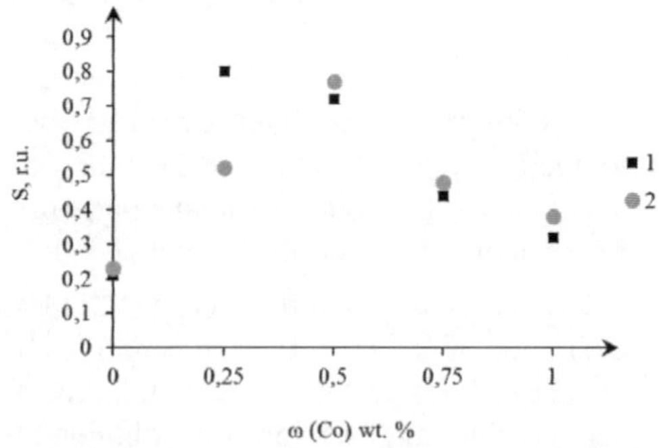

Figure 5. Sensitivity dependence on the weight concentration of modifying additive of the Co-containing PAN films: 1 – to Cl_2; 2 – to NO_2

Thus, potential applications of the Co-containing PAN films include environmental monitoring of gas emissions as non-heated semiconductor sensor. Samples which showed the best indicators of the submitted parameters are the films with 0.25 wt. % and 0.5 wt. % weight concentration of modifying additive.

4. Conclusion

The design of environmentally-sensitive products involves selecting appropriate materials, technology and manufacturing equipment. From the experimental data, it was established that the obtained Co-containing PAN films are sensitive to gas oxidizers such as NO_2 and Cl_2 at room temperature, and the sensitivity depends on their resistance values. Advantage of PAN-based material is the possibility of obtaining films with functional properties, and there is no need to use complex technological equipment [13-16, 18-26].

Acknowledgments

This work was performed with the financial support of Southern Federal University within project No. VnGr-07/2017-21. The equipment of the Multi-Access Center of Microsystem Engineering and Integral Sensor Technology of Southern Federal University was used for this study.

References

[1] Kumar A.P., Depan D., Tomer N.S., Singh R.P. Nanoscale particles for polymer degradation and stabilization – Trends and future perspectives // Progress in Polymer Science. – 2009. – Vol. 34. – P. 479-515.
[2] Oprea A., Bârsan N., Weimar U. Work function changes in gas sensitive materials: Fundamentals and applications // Sensors and Actuators B. – 2009. – Vol. 142. – P. 470-493.
[3] Lu X., Zhang W., Wang C., Wen T.-C., Wei Y. One-dimensional conducting polymer nanocomposites: Synthesis, properties and applications // Progress in Polymer Science. – 2011. – Vol. 36. – P. 671-712.
[4] Li X., Wang Y., Yang X., Chen J., Fu H., Cheng T. Conducting polymers in environmental analysis // Trends in Analytical Chemistry. – 2012. – Vol. 39. – P. 163-179
[5] Nataraj S.K., Yang K.S., Aminabhavi T.M. Polyacrylonitrile-based nanofibers – A state-of-the-art review // Progress in Polymer Science. – 2012. – Vol. 37. – P. 487-513.
[6] Zemcov L.M., Karpacheva G.P. Himicheskie prevrashcheniya poliakrilonitrila pod dejstviem nekogerentnogo infrakrasnogo izlucheniya. // Vysokomolekulyarnye soedineniya. – 1994. – V. 36. – № 6. – P. 919-924.
[7] Al'-Hadrami I.S., Korolev A.N., Zemcov L.M., Karpacheva G.P, Semenistaya T.V. Issledovanie ehlektroprovodnosti IK-pirolizovannogo med'soderzhashchego poliakrilonitrila. // Izvestiya vuzov. Materialy ehlektronnoj tekhniki. – 2008. – №1. – P. 14-17.
[8] Vihrov S.P., Bodyagin N.V., Larina T.G., Mursalov S.M. Processy rosta neuporyadochennyh poluprovodnikov s pozicij teorii samoorganizacii // Fiz. i tekhn. poluprovodnikov. – 2005. – V. 39. – №. 8. – P. 953-959.
[9] Elyuhina O.V., Sokolovskij G.S., Kuchinskij V.I. Samoorganizaciya izoehlektronnyh primesej Mg i O v ZnSe // Fiz. i tekhn. poluprovodnikov. – 2007. – V. 41. – №. 2. – P. 129-133.
[10] Avacheva T.G., Bodyagin N.V., Vihrov S.P., Mursalov S.M. Issledovanie samoorganizacii neuporyadochennyh materialov s primeneniem teorii informacii // Fiz. i tekhn. poluprovodnikov. – 2008. – V. 42. – №. 5. – P. 513-518.
[11] Yablokov M.YU., Zav'yalov S.A., Obolonkova E.S. Samoorganizaciya nanochastic palladiya pri formirovanii metall-polimernyh pokrytij // ZHurnal fizicheskoj himii. – 1999. – T. 73. – № 2. – P. 219-223.
[12] Gusev E.Y. Ageev O.A., Gamaleev V.A., Mikhno A.S., Mironenko O.O., Pronin E.A. Effect of annealing on conductivity type of nanocrystalline ZnO films fabricated by RF magnetron sputtering // Advanced Materials Research. – 2014. – Vol. 863. – P. 539-542.
[13] Semenistaya T.V. Polyacrylonitrile-based materials: Properties, Methods and Applications // Springer Proceedings in Physics. – 2016. – Vol. 175. – P. 61-77.
[14] Lu P., Ivanets V. A., Semenistaya T. V., Plugotarenko N. K. Investigation of the Influence of Structure of Films Based on Silver-Contayning Pan on their Gas-Sensitivity with Application of the Theory of Self-Organization, Information Theory and Atomic-Force Microscopy // Journal of Nano and Microsystem Technique. – 2012. – №5. – P. 21 – 28.
[15] T.V. Semenistaya, V.V. Petrov, Kh.Kh. Kalazhokov, Z.Kh. Kalazhokov, B.S. Karamurzov, Kh.V. Kushkhov, S.P. Konovalenko. Study of the properties of nanocomposite cobalt-containing

IR-pyrolyzed polyacrylonitrile films // Surface Engineering and Applied Electrochemistry. – 2015. – Vol. 51. – No. 1. – P. 9-17.

[16] Bednaya T.A., Konovalenko S.P., Semenistaya T.V., Petrov V.V., Korolev A.N. Izgotovlenie gazochuvstvitel'nyh ehlementov sensora dioksid azota i hlora na osnove kobal'tsoderzhashchego poliakrilonitrila. // Izvestiya vysshih uchebnyh zavedenij. EHlektronika. – 2012. – Vol.96. – № 4. – P. 66-71.

[17] Kozhitov L.V., Krapuhin V.V., Karpacheva G.P., Kozlov V.V. Struktura i fiziko-himicheskie svojstva organicheskogo poluprovodnika na osnove poliakrilonitrila i ego kompozita s nanochasticami medi. // Izvestiya vysshih uchebnyh zavedenij. Materialy ehlektronnoj tekhniki. – 2004. – №4. – P. 7-10.

[18] Al'-Hadrami I.S., Korolev A.N., Semenistaya T.V., Nazarova T.N., Petrov V.V. Issledovanie gazochuvstvitel'nyh svojstv med'soderzhashchego poliakrilonitrila. // Izvestiya vuzov. EHlektronika. – 2008. – №1. – P. 20 – 25.

[19] Lu Ping, Gorbatenko Yu. A., Semenistaya T. V., Vorobev E. V., Korolev A. N. Fabrication of Gas Sensors Sensing Elements Based on Polyacrylonitrile Films and Argentiferous Polyacrylonitrile Films Determining their Characteristics // Journal of Nano and Microsystem Technique. – 2011. – № 9. – P. 5-12.

[20] Semenistaya T.V., Petrov V.V., Lu P. Nanocomposite of Ag-polyacrylonitrile as a selective chlorine sensor // Advanced Materials Research. – 2013. – Vol. 804. – P. 135-140.

[21] T.V. Semenistaya, V.V. Petrov, S.P. Konovalenko. Modelirovanie ehlektrofizicheskih svojstv i koehfficienta gazochuvstvitel'nosti plenok med'soderzhashchego poliakrilonitrila // Izvestiya vuzov. Materialy ehlektronnoj tekhniki. – 2014. – Vol. 66. – № 2. – P. 116-121.

[22] Konovalenko S.P., Bednaya T.A., Semenistaya T.V., Korolev A.N. Prognozirovanie vliyaniya tekhnologicheskih parametrov formirovaniya gazochuvstvitel'nyh materialov na osnove poliakrilonitrila na ehlektrosoprotivlenie // Izvestiya vuzov. Materialy ehlektronnoj tekhniki. – 2013. – № 1. – P. 48-52.

[23] Tatiana Semenistaya, Nina Plugotarenko and Viktor Petrov. A Monte-Carlo method for simulating drying temperature of the gas-sensing material based on polyacrylonitrile // Applied Mechanics and Materials. – 2015. – Vols. 727-728. – P 145-149.

[24] Konovalenko S.P., Semenistaya T.V. Issledovanie svojstv plenok kobal'tsoderzhashchego PAN s primeneniem QSPR-metodologii // Izvestiya YUFU. Tekhnicheskie nauki. – 2013. – Vol. 138. – № 1. – P. 178-183.

[25] Nina Plugotarenko, Tatiana Semenistaya and Viktor Petrov. Multifractal detrended fluctuation analysis in examining surface properties of the gas-sensing metal-containing polyacrylonitrile // Modern Applied Science. 2015. – Vol. 9. – No. 11. – P. 213-221.

[26] T.V. Semenistaya, V.V. Petrov, T.A. Bednaya, O.A. Zaruba. Gasoline vapor sensor based on Cr-containing polyacrylonitrile nanocomposite films through artificial neural networks application // Materials Today: Proceedings. – 2015. – Vol. 2. – №. 1. – P. 77-84.

[27] Jing M., Wang C., Wang Q., Bai Y., Zhu B. Chemical structure evolution and mechanism during pre-carbonization of PAN-based stabilized fiber in the temperature range of 350 – 600 °C. // Polymer Degradation and Stability. – 2007. – Vol. 92. – P. 1737-1742.

[28] Gupta H., Mahajan A., Bedi R.K. Silver phthalocyanine films for photovoltaic applications // Atti della «Fondazione Giorgio Ronchi». – 2009. – V. 64.- № 4. – P. 597-604.

[29] Gas'kov A.M., Rumyanceva M.N. Vybor materialov dlya tverdotel'nyh gazovyh sensorov. // Neorganicheskie materialy. – 2000. – Vol. 36. – № 3. – P. 369-378.

Investigation of porous silicon obtained under different conditions by the contact angle method

A O Belorus, Y V Bukina, A I Pastukhov, D S Stebko, Yu M Spivak and V A Moshnikov

St. Petersburg State Electrotechnical University «LETI», Saint Petersburg, Russian Federation

stebel2015@yandex.ru

Abstract. This paper investigates a hydrophobicity/hydrophilicity of porous silicon by the contact angle method. Porous silicon series were obtained by electrochemical anodic etching of n-Si (100) and (111) under the current anodization density range of 5-120 mA/cm^2. For this purpose the original laboratory installation and the software «Measurement of contact angle» were developed. It is shown that, the contact angle can vary significantly (up to 80 degrees for (100)) depending on the current anodization Discussion of the results is carried out taking in account the composition of the functional groups and of surface morphology of the porous silicon. These results are important for developing porous silicon particles as nanocontainers in the targeted drug delivery.

1. Introduction

One of the modern trends in the development of medicine is targeted drug delivery. Porous materials are extremely important for this direction. The key characteristics of such materials are porosity, the type of porous texture, the morphology of the surface, the specified composition of specific functional groups on the surface, the hydrophilicity or hydrophobicity of the material [1-3]. Porous silicon (por-Si) is a promising material for use as nanocontainers for targeted drug delivery. One of the significant advantages of pore-Si is the possibility of obtaining this material in a wide range of porosity, pore diameters, surface composition, etc. by controlling the synthesis conditions.

For targeted drug delivery considerable attention must be paid to the nature of the adsorption of medicinal substances by the porous media. The character of the adsorption must obviously be different for a por-Si with different structure parameters and a functional composition of the surface. Also, the dose of the accumulated drug substance, the possibility of simultaneous functionalization by molecules of different sizes, the simultaneous attachment of hydrophilic and hydrophobic substances, and the like will depend greatly on the characteristics of the por-Si.

The study of such a complex heterophase porous brittle object as porous silicon requires a special approach to the choice of research methods. There are methods for determining the characteristics of porous silicon, which are important for its use in targeted drug delivery [4-6]. In addition to determining the geometric parameters of the porous texture, it is important to study the phase composition of the por-Si surface, its energy characteristics and the character of the adsorption centers on it [1, 3, 7, 8].

The indicator method provides a qualitative determination of the functional groups on the surface of insoluble solids [9, 10, 1]. This method is extremely sensitive and allows differing between functional groups of the same chemical composition, differing only in the electron density distribution. The detection of the surface functional groups of por-Si, prepared at certain conditions, needs much different kind of samples. It is advisable to investigate functional groups not over the entire range of pK_a (negative decimal logarithm of the acid dissociation constant), but only at determined values. The presence of certain surface functional groups has a significant influence to possibility of functionalization of the por-Si surface by biomarkers. Kind of adsorption centers on the surface of material (acidic, basic or almost neutral hydroxyls centers) is directly connected with the level of hydrophilic or hydrophobic properties. Thus, after the determination of the level, it is possible to choose the most appropriate range of pK_a for research by using method of contact angle. It allows to reduce number of samples that expedites the measurement process.

Thus, this article is devoted to the study of the energy characteristics of porous silicon, depending on the technological conditions of synthesis.

2. Experimental

Por-Si samples were obtained in the single-chamber electrochemical cell in an aqueous solution of HF and isopropanol mixture. The technological conditions of electrochemical anodization process are shown in table 1. The variable parameter was the anodization current density. It is expected that porosity, surface development, pore diameter will increase with increasing anodization current density, and the composition of functional groups on the PS surface will change (the portion of hydroxyl groups decreases, and the fraction of disordered silicon atoms increases) [1]. As a raw material, the single crystal *n*-Si (111) and (100) with a resistivity of 4.5 Ω·cm was used.

Table 1. Technological conditions of electrochemical anodization

Series	Si parameters	Si specific resistivity, Ω·cm	Anodizing time, min	Current destiny J_A, mA/cm^2
I	n-Si (100), n-Si (111)	4,5	10	5
II				15
III				30
IV				50
V				80
VI				120

Further, the samples were investigated by the contact angle method (the static sessile drop method). We used a homemade laboratory installation. The scheme of the installation is shown on Figure 1. It includes data processing device 1; digital microscope 2 with horizontal optical axis, which is at the same plane with the sample of por-Si; switch image button 3, which is also turn on/off the backlight; mount system of digital microscope 4; the lever 5 to provide changing the focal length of microscope; optical axis 6 of microscope; drop 7; substrate 8 (polymer film that planted on the glass); adjustable table 9; movable mechanism 10. The drop of water was dripped to the layers of por-Si by means of mechanical feeder; then the drop was captured by digital microscope.

To calculate the contact angles (θ), the special program «Measurement of contact angle» was written in the visual programming language on LabView environment. The main feature of «Measurement of contact angle» is that program takes data from digital image of drop, which includes values of diameter and height, and calculates contact angle. Angle calculation accuracy is estimated at 1 – 5 degrees. This virtual instrument is fully automated and, in this way, it expedites the entire process of processing the data.

3. Results and discussion

The dependencies of the contact angle on the anodization current density are shown in Fig. 2. As one can see from Fig. 2, the dependence patterns for porous silicon of (111) and (100) differ significantly. A strong nonlinear character of the dependence with a minimum is observed for por-Si (100): the angle varies from 43° to 124°. The minimum corresponds to the current density of 30 mA/cm^2. At this anodization current density, the most hydrophilic por-Si properties in this series are obtained. At high and low current densities, porous silicon exhibits hydrophobic properties (contact angles are 108° and 121° relatively).

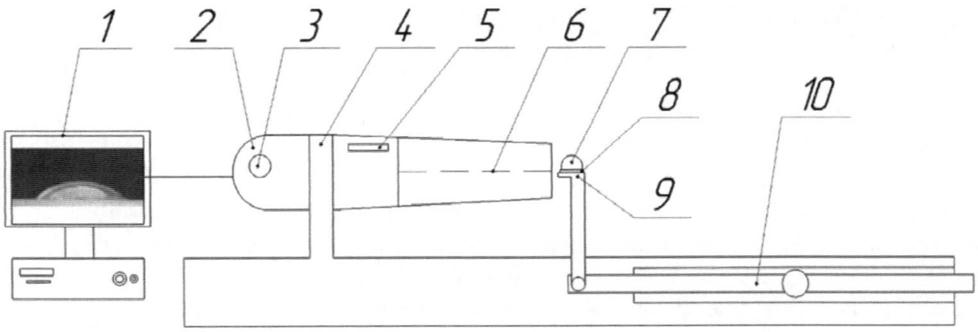

Figure 1. The scheme of the installation of the contact angle measuring

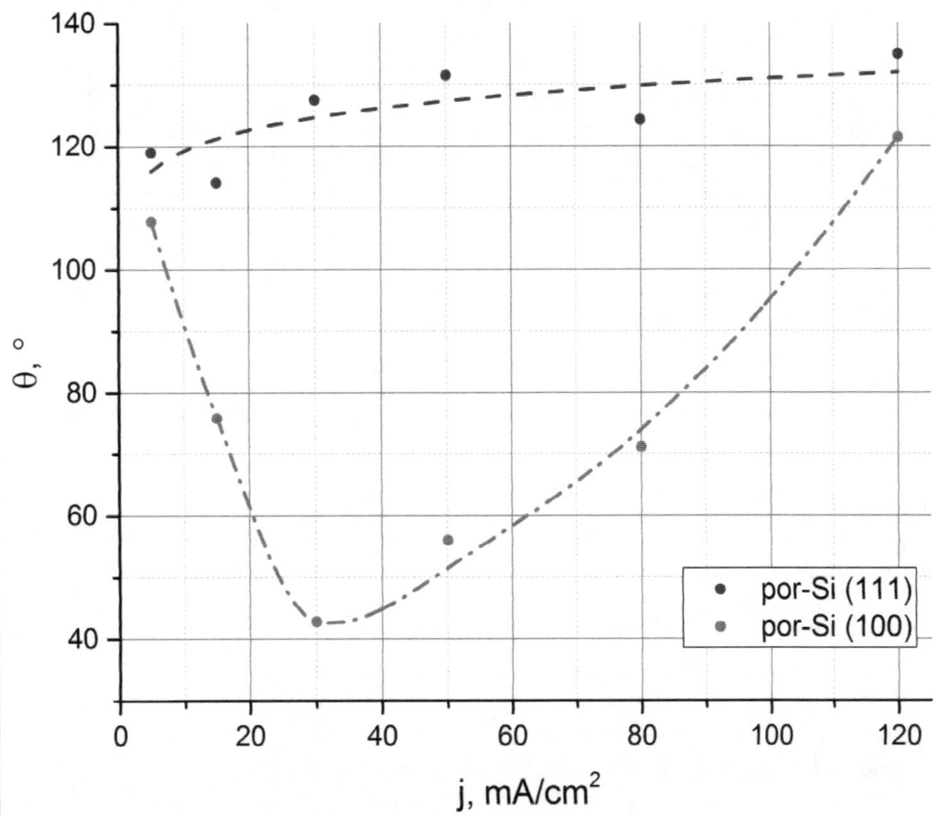

Figure 2. Contact angle, depending on the anodization current density for two types of crystallographic orientation of por-Si substrates

This kind of dependence, apparently, can be explained as follows. At low anodic current densities, the morphology of the por-Si surface, as a rule, is of the order of units of nanometers, which are orders of magnitude smaller in comparison with por-Si obtained at 20-30 mA/cm^2. Apparently, in this section of the dependence (1-30 mA/cm^2), the predominant factor is the change in the composition and proportion of the functional groups on the por-Si surface during etching with different current densities. As the anodization current density increases, por-Si surface development, porosity and pore diameter increase, which leads to an increase in the contact angle of wetting.

The dependence $\theta = f(J_A)$ for por-Si (111) samples has an slowly increasing monotonic character and shows the hydrophobic properties of the surface. It can be assumed that in the formation of porous silicon in n-Si (111), the prevailing factor determining the hydrophobic character of the surface, in this case, is the morphology of the surface.

4. Summary

In this paper, porous silicon series were obtained at various anodization current densities in the range 5-120 mA/cm^2 in n-Si with crystallographic orientation of (111) and (100). By the method of the contact angle, it has been shown that as prepared n-por-Si (111) has hydrophobic properties that increase slightly as the anodization current density increases. This is apparently associated with an increase in the porosity and development of surface morphology. For n-por-Si (111), the dependence of the contact angle on the anodization current density is stronger, non-monotonic, characterized by a minimum at $J_A = 30$ mA/cm^2. We assume that this dependence is due to the predominant influence of the functional groups composition on the por-Si surface at low anodization current densities. At high anodization current densities, evidently, that increasing porosity and morphology roughness leads to increasing the contact angle.

Thus, by varying the conditions for obtaining porous silicon and the type of substrate, it is possible to control the wettability properties of the surface over a wide range of values. Also, the results of the work will allow us to select the more precisely the required range of technological conditions for further investigation of porous silicon by the indicator method.

This is important in the development of nanocontainers based on porous silicon in the targeted delivery of drugs. The obtained dependences will help to purposefully change the properties of por-Si surface for functionalization with medical substances, thereby controlling the character of the interaction of the specified substance with the porous matrix. Obtained information of the surface wettability of por-Si should improve the encapsulation efficiency of medical substances.

Acknowledgement

The work was supported by the Task of The Ministry of Education and Science RF № 3.6288.2017/8.9 BC (basic part).

References

[1] Spivak Yu M, Myakin S V, Moshnikov V A, Panov M F, Belorus A O and Bobkov A A 2016 Surface functionality features of porous silicon prepared and treated in different conditions *J. of Nanomater.* **2016** 2629582

[2] Belorus A O, Bespalova K and Spivak Yu M 2016 Morphology and internal structure of porous silicon powders in dependence on the conditions of post-processing *Proc. IEEE NW Russia Young Researchers in Electrical and Electronic Engineering Conf. (St. Petersburg)* (St. Petersburg, Russia: Saint Petersburg Electrotechnical University "LETI") p 22-27

[3] Koshevoi V L, Pshchelko N S, Belorus A O and Levitskiy V S 2016 The study of the phase composition of polymorphous silicon film by Raman spectroscopy *Proc. IEEE NW Russia Young Researchers in Electrical and Electronic Engineering Conf. (St. Petersburg)* (St. Petersburg, Russia: Saint Petersburg Electrotechnical University "LETI") p 62-64

[4] Belorus A O, Maraeva E V, Spivak Yu M and Moshnikov V A 2015 The study of porous

IOP Conf. Series: Journal of Physics: Conf. Series **929** (2017) 012051 doi:10.1088/1742-6596/929/1/012051

silicon powders by capillary condensation *J. of Phys.: Conf. Series* **586** 12-17
[5] Spivak Yu M, Belorus A O, Somov P A, Tulenin S S, Bespalova K and Moshnikov V A 2015 Porous silicon nanoparticles for target drag delivery: structure and *J. of Phys.: Conf. Series* **643** 012022
[6] Kovalevskii A A 1999 Structure and morfology of Si films grown on Porous Silicon by reduction of Dichlorosilane *Inorganic materials* **35** 152-156
[7] Lenshin A S, Kashkarov V M, Spivak Yu M and Moshnikov V A 2012 Study of electronic structure and phase composition of porous silicon *Glass Phys. and Chem.* **38** 315-321
[8] Lenshin A S, Kashkarov V M, Spivak Yu M and Moshnikov V A 2012 Investigations of nanoreactors on the basis of p-type porous silicon: electron structure and phase composition *Materials Chem. and Phys.* **135** 293-297
[9] Vasiljeva I V, Mjakin S V, Rylova E V and Korsakov V G 2002 Electron beam modification of the surface of oxide materials (SiO_2 and $BaTiO_3$) *Russian J. Phys. Chem.* **76** 71-76
[10] Mjakin S V, Sychov M M and Vasilieva I V 2006 *Electronno-luchevoe modifitsirovanie funktsionalnykh materialov (Electron beam modification of functional materials)* (Saint-Petersburg: Petersburg State Transport University)

IOP Conf. Series: Journal of Physics: Conf. Series **929** (2017) 012052 doi:10.1088/1742-6596/929/1/012052

Investigation of helium charged beams passing through the porous alumina membranes

E N Muratova[1], **A A Shemukhin**[2]

[1]Saint Petersburg Electrotechnical University "LETI", ul. Prof. Popova, 5, Saint Petersburg, 197376, Russia
[2]Skobeltsyn Institute of Nuclear Physics, Moscow State University, 1(2), Leninskie gory, Moscow, 119991, Russia

E-mail: *sokolovaeknik@yandex.ru*, *shemuhin@gmail.com*

Abstract. Membranes of nanoporous aluminum oxide (alumina) have been obtained using the electrochemical etching technique by varying technological regimes. They were investigated by the Rutherford backscattering technique. The obtained spectra revealed distribution of elements over the entire thickness of the layer. It has been shown that almost all ions, which fly in the beam, fall into the channel - por Al_2O_3. This leads to a focusing of the charged particles beam.

1. Introduction

Over the past years, scientific interest in the creation and study of nanostructures has increased. Nanostructures are of practical and scientific interest to understanding the fundamental electronic, magnetic, optical, thermal and mechanical properties of materials with nanometer dimensions characteristic elements, and also to creating devices with previously unattainable parameters. Beams of charged particles, which are finding wide application in both fundamental studies and solving applied problems (e.g., in technology and in medicine), are attracting also great interest at present [1, 2]. A series of recent works [3–5] consider the possibility of transporting beams of accelerated charged particles via dielectric channels without energy loss and loss of the initial charge state.

To date, a number of effective methods of creating nanostructures is developed, and among them methods based on the principle of self-formation are extremely important. Nanoporous materials, of course, can be classified as nanostructured objects derived by the self-organization.

One of the most studied of structured porous materials obtained by electrochemical anodization is a porous aluminum oxide (por-Al_2O_3). The uniqueness of aluminum is in the process of electrochemical etching under certain process conditions may be obtained from self-assembled layer of Al_2O_3 pore structure of the "honeycomb". Membranes based on nanoporous alumina are required in nanotechnology, microbiology, and nuclear physics, because they have unique properties, such as mechanical strength, thermal stability, and chemical resistance.

The structure membranes based on nanoporous alumina, which has been intensely studied in recent decades [6, 7], is an ideal dielectric matrix of nanocapillaries [4]. A peculiarity of por-Al2O3 is that, during electrochemical etching, it shows the ability to self organization and forms a structure with a given geometry when technological conditions are varied [8, 9]. Studies of using por-Al2O3 membranes as a dielectric focusing matrix, which can provide transport of beams of accelerated charged particles through dielectric capillaries, are thus of interest [10,11]. It will make it possible to perform highly local analysis of the structure and composition of samples without setting up high vacuum conditions [12, 13] and exerting topologically ordered effects on a nanosize level.

Thus, formation of mechanically strong membranes based on porous alumina with ordered capillary structure on micro- and nano-scale in a thin (~ 10 μm) aluminum foil is the actual problem.

The aim of this work was the formation of free masks—membranes with the through nanocapillaries and study possibilities of their application as nanoscale templates for high-energy ion beams.

2. Experiment

We used a well-known mechanism to formation of ordered cellular structure por-Al_2O_3 [7, 9]. To obtain por-Al_2O_3 unicameral cell to electrochemical etching with the ability to control the temperature of the system has been designed.

In order to obtain a porous anodic alumina membrane [9] the electrochemical anodization of aluminum foil (40 μm) was applied in a potentiostatic mode in electrolytes based on aqueous solutions of phosphoric (H_3PO_4) and sulfuric (H_2SO_4) acids for 10–60 min at temperatures from 0 to 15°C. When using electrolytes with different acidities, the speeds of an anions introduction into the porous layer are different. This effect determines the voltage of the pore formation process and consequently the pore diameter. The anodizing voltage was within the range of 15-30 V (for H_2SO_4) or 130-160 V (for H_3PO_4) depending on the electrolyte type.

The basic structural parameters, such as the pore diameters *a* (see figure 1), the distances between pores *b*, the thickness of the porous layer *c* and the barrier layer *d* of the obtained samples were studied using the scanning electron microscopy (SEM) at low accelerating voltages (10–20 kV) under high vacuum, that allowed us to study the insulating samples without a prior metal deposition.

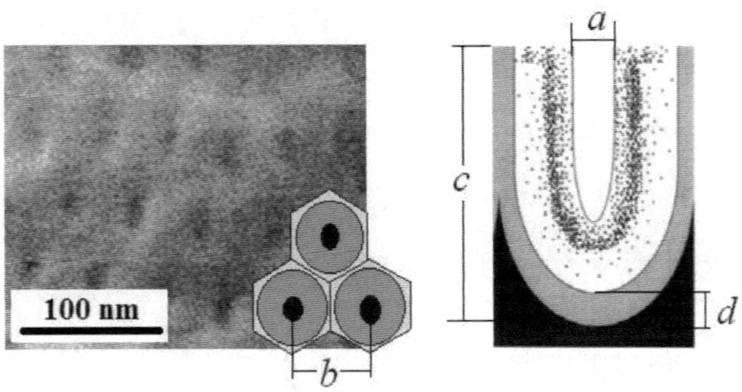

a b

Figure 1. SEM (a) and schematic images (b) of the por-Al_2O_3 membranes.

Through translucent membrane auto-fixed on aluminum substrate were obtained on aluminum foil. Membranes based on por-Al_2O_3 were obtained using the conditions of galvanostatic modes. U-t curves have a typical shape characteristic of the processes of electrochemical oxidation is illustrated in Fig. 2 The sharp rise in voltage on the curve is due to complete oxidation of the aluminum foil and through membranes formation based on por-Al_2O_3. It should also be noted that the anodizing current increasing in twice result in the anodizing time reducing approximately doubled.

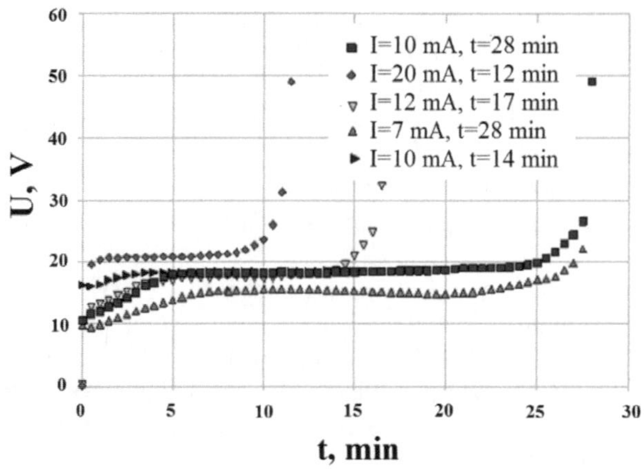

Figure 2. Galvanostatic curves of system Al / electrolyte.

During the anodizing it is necessary to maintain the process temperature (depending on the electrolyte used) to avoid the dissolution of oxide formed by acidic electrolyte. It is possible to control the thickness of the porous layer and alter its mechanical properties by controlling the temperature during the anodizing process. In this work the optimal temperature of the process for different electrolytes supported in according to Table 1.

Table 1. Technological parameters of the anodizing process.

Electrolyte	H_2SO_4	H_3PO_4
The voltage of generation, V	5–25	60–120
Process temperature, ^{0}C	20	0–2

To conduct research using the Rutherford backscattering technique (RBS) it was established a special experimental chamber (Fig.3), which includes a collimating aperture, a system of detection of backscattered ions and using a goniometer system. The diameter of the ion beam after passing through the system aperture is 1 mm. Residual pressure in the chamber did not exceed $5{\cdot}10^{-4}$ Pa. The current density on the target was maintained constant and amounted to 4 nA.

The sample is installed in the chamber on the goniometer system. The goniometer provides the rotation of the sample around all three axes and translational movement in a vertical direction, allowing to measure different portions of the surface of the sample without changing orientation relative to the beam. The angular displacement can be made using a stepper motor, and angular displacement is made using stepper motors, the rotation is carried out with a step of 0.02°. The inaccuracy of the execution of the turns at large angles does not exceed 1%. The angular position of the detector relative to the target can be changed in the scattering plane on a circle which center coincides with the center of rotation of the goniometer.

To study the passage of a beam of ions He^+ through the oxide matrix were used the RBS technique in combination with channeling. For this purpose, the beam diameter of 1 mm was directed onto the sample surface. With the manipulator in the horizontal plane α the angle of incidence of the probing beam ions was installed α = 7 and the spectrum of RBR sample was recorded in the absence of channeling. Then the angle x speed was decreased, and at each step the energy spectrum of RBS was recorded.

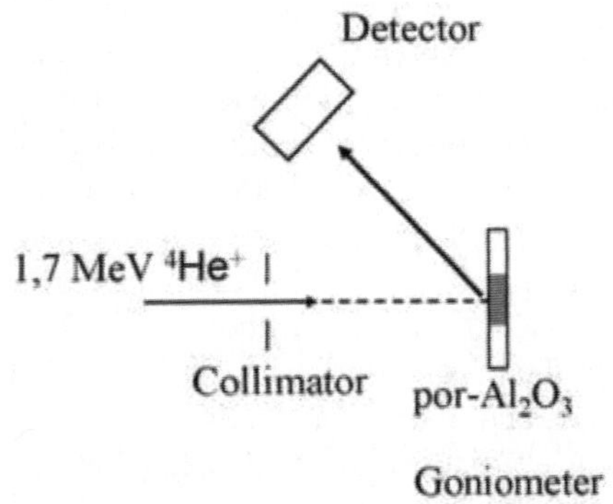

Figure 3. Schematic representation of the experimental process.

When the RBS spectra of membrane were detected, a detectable scattering angle was 165°. The semiconductor detector filmed how the signal passed through the membrane of ions of He^+ ions and backscattered ions.

3. Results and discussion

Obtained membranes were studied using the RBS. The RBS technique was used for in situ determination of the optimal geometry of the experiment. First, the position of a semiconductor detector at which the yield of the backscattered ions is uniform and minimal was found. This case corresponds to the situation when the pores in the target are through. It is worth noting that the above-mentioned ion-beam method of controlling the quality of nanoporous membranes structures is nondestructive.

The RBS spectra (figure 4) show the dependence of the number of passed particles on the slope angle of the target for the membrane containing pores of about 20 nm in diameter.

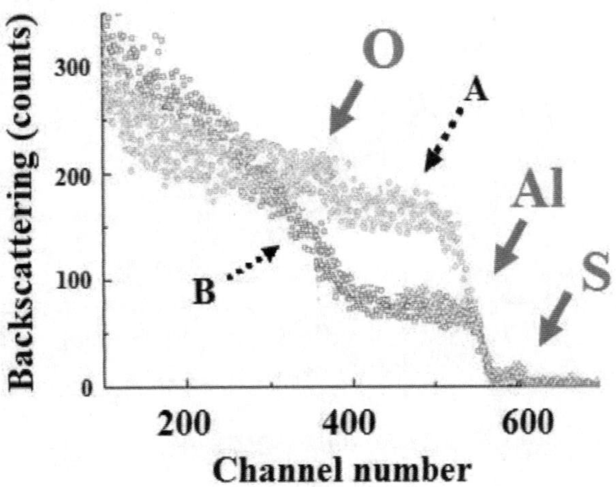

Figure 4. The energy spectrum of backscattered He+ ions with the energy of 1.7 MeV for the scattering angle of 120°. A – the structure of aluminum without etching, B – after electrochemical etching [14].

In the virgin structure close to the 400 channel (curve A), the peak of oxygen, which corresponds to the natural oxide, is observed. The thickness of the layer with disordered atoms of aluminium allows to estimate it at about 5–7 nm. The RBS spectrum of the porous membrane consists of two of flat portion, indicating the formation of a uniform oxide layer on all the analysis depth. However, it appears that the por-Al_2O_3 structure contains not only the basic elements (oxygen and aluminum), but also sulfur. After etching, an extra peak in the high energy part of the spectrum (curve B) occurred which corresponds to backscattered He^+ ions from sulfur. It is distributed rather evenly throughout the depth of the membrane, with some maximum concentration near the surface. The concentration by volume may achieve 16% close to the surface. The appearance of this impurity may be due to the foil quality, the results of washing samples after anodic oxidation, and the possibility of acid anions incorporated into the formed oxide. The increase in the yield of backscattered ions with decreasing energy is due to the dependence of the backscattering cross-section from energy.

Membranes, in which there are additional ions, are not suitable for beam focusing effect. For example, the presence of sulfur ions promotes charge leakage, instead of forming a focusing channel. For guading - effect the dielectric matrix should have a rather high resistance, since such beams transportation is provided by charging of the inner walls. However, the data of the membrane suitable for transporting bundles of high-energy charged particles. In this case, the transportation may be provided by the mirror reflection ions emitted at grazing angles, that is, due to small-angle scattering.

It should be noted that Rutherford backscattering technique in combination with channeling can be used for analyzing the transmission coefficient of oxide matrixes. Figure 5 shows RBS spectra of backscattered ions at different position of the sample. The RBS spectra were recorded in two directions: when the beam was parallel to the pores in membranes (curve B), and in directions didn't contain open channels. In the case when the direction of incidence of the ion beam coincides with the direction of the pores, there is a minimum yield of scattered ions as a result of the passage of the ions into the sample. Fig. 5 show that when orientation of the alumina membrane is perpendicular to the analyzing beam to fall (curve B), then output backscattered ions decreases sharply in 14 times, compared to the direction that didn't contain open channels (curve A). This demonstrates that most of the ions have flown through the membrane. The output of the backscattered ions from sulfur decreased in more than 20 times in the direction when the beam was parallel to the pores. It shows that the sulfur concentration decreases as the distance from the surface.

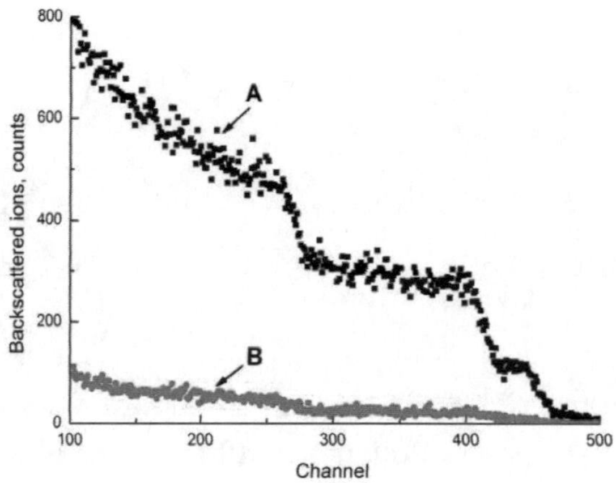

Figure 5. The energy spectrum of backscattered He^+ ions for the s cattering angle of 120.°

The work [14] has shown that the passage of ions through such dielectric matrix is 60–70%. However, according to figure 1, the area ratio of the pore square to the sample square is 0.2–0.3. This can be explained as follows: probably the charged particles when passing through the membrane create a potential on the dielectric walls of the pores (channels). Thus, the fraction of the particles goes immediately into the channel, and the part deviates from the initial direction and into the channel potential dielectric matrix. This situation leads to the good focusing of a beam of charged particles, which can be used in nanolithography. Thus, using the RBS technique for investigating nanoporous alumina membrane allows to "in situ" control of change structure by the action of ion irradiation. The beam current is controlled by an independent monitoring system. So when the ions backscattered exit in towards channeling, it leads to deterioration pores in the membrane.

4. Conclusion

Thus, the technology of producing self-forming porous membranes has been proposed. The porous membrane is a system of ordered nanoscale (pore diameter of 20 nm or more) alumina capillaries with an aspect ratio of 500.

The RBS results have shown that using of sulfuric acid as electrolyte during the anodizing process results in the incorporation of sulfur anions in the membrane structure throughout its thickness. These data are confirmed by X-ray microanalysis. The presence of sulfur makes the membrane not suitable for the transport of low energy charged particle beams. It is noted that using of phosphoric acid as the electrolyte is not conducive to the incorporation of impurity anions in the depth of the membrane. Therefore, such structures are attractive at creating of mask for ion lithography.

We propose to monitor the change of the membrane structure in the mode of "in situ" by the action of the ion irradiation using the RBS technique combined with channeling. It was found that the increase of the output of backscattered ions in the direction of channeling leads to degradation of the membrane pores.

It is shown that the nanoporous membranes based on alumina containing a nanoscale capillaries system provide the transport of high-energy flow of helium ions with energy of 1.5–2 MeV.

Acknowledgments

This work was supported by RFBR research project №16-38-60110 mol_a_dk and partly 15-07-99656 A.

REFERENCES

[1] Vokhmyanina K A, Zhilyakov L A, A. Konstanovsky V, et al., Transportation and focusing of accelerated proton beams by means of dielectric channels 2006 *Phys. A: Math. Gen.* **39**, 4775-4779

[2] Pokhil G P, and Vokhmyanina K A, Drift model of ion beam guiding using capillaries 2008 *J.Surf. Investig.: XRay, Synchrotron, Neutron Tech.* **2 (2)**, 237-240

[3] Stolterfoht N, Hellhammer R,Sobocinski P, et al., 2005 *Nucl. Instrum. Methods Phys. Res.,* B **235**, 460

[4] Komarov F F, Kamyshin A S and Grishin P A, Micro- and Nanocapillary Structures Based on Dielectric Materials to Focus the Ion Beams 2013 *J. Nano-Elektron. Fiz.* **5 (1),** 01015

[5] Vokhmyanina K A, Zhukova P N, Irribarra E F, et.al., Investigation of contactless electron transmission through dielectric channels 2013 *J. Surf. Investig.: XRay, Synchrotron, Neutron Tech.* **7 (2),** 271-275

[6] Masuda H, Abe A, Nakao M, et al. 2003 *J. Adv. Mater.,* Ordered mosaic nanocomposites in anodic porous alumina 2003. **15(2)** 161-164

[7] Eftekhari A. 2008 *Nanostructured Materials in Electrochemistry*, ed by A Eftekhari (Wiley-VCH Verlag GmbH & Co. KGaA, Weinheim) p.463

[8] Zimina T M, Muratova E N, Spivak Yu M, et al., Tehnologii formirovaniya i primenenie nanosloev i nanoporistyh kompozitsy Al2O3 dlya mikro- i nanotehniki 2012 *Nano-Mikrosist. Tekh.*, **12**, 15-24 (in rus.)

[9] Muratova E N, Spivak Yu M, Moshnikov V A, et al. Influence of technological parameters of nanoporous Al2O3 layers' preparation on their structural characteristics 2013 *Glass Phys. Chem.* **39 (3),** 320-328

[10] Luchinin V V, Muratova E N and Shemukhin A A Matritsy iz poristogo oksida alyuminiya kak kapillyarnye matritsy-shablony dlya lokalizatsii vozdeystviya ionov vysokih energy 2013 *Nano-Mikrosist. Tekh.*, **12**, 39-41 (in rus.)

[11] Luchinin V V, Moshnikov V A, Muratova E N and Samigullin R Sh Formation of ordered nanoscale capillary membranes based on anodic alumina 2015 *Journal of Physics: Conference Series* **586** 012008

[12] Shemukhin A A, Balakshin Yu V, Chernykh P N and Chernysh V S, Medium energy ion scattering spectroscopy: Study of germanium amorphization under ion irradiation 2013 *J. Surf. Investig.: XRay, Synchrotron, Neutron Tech.* **7 (2)**, 318-321

[13] Shemukhin A A, Balakshin Yu V, Chershysh V S, et.al., Fabrication of ultrafine silicon layers on sapphire 2012 *Tech. Phys. Lett.* **38 (10)**, 907-909

[14] Shemukhin A A and Muratova E N Investigation of transmission of 1.7-MeV He^+ beams through membranes of porous Alumina, 2014 *JETP Letters* **40** 219-221

Laser deposition of resonant silicon nanoparticles on perovskite for photoluminescence enhancement

E Y Tiguntseva[1], A S Zalogina[1], V A Milichko[1], D A Zuev[1], M M Omelyanovich[1,2], A Ishteev[3], A Cerdan Pasaran[4], R Haroldson[4], S V Makarov[1], A A Zakhidov[1,3,4]

[1]ITMO University, 49 Kronverksky Pr., St. Petersburg, 197101, Russia
[2]Aalto University, PO Box 13000 FI-00076, Finland
[3]National University of Science and Technology, NUST "MISIS", 2 Leninsky Pr., Moscow, 119049, Russia
[4]The University of Texas at Dallas, 800 W Campbell Rd, Richardson, TX 75080, USA

E-mail: tiga.kate@gmail.com

Abstract. Hybrid lead halide perovskite based optoelectronics is a promising area of modern technologies yielding excellent characteristics of light emitting diodes and lasers as well as high efficiencies of photovoltaic devices. However, the efficiency of perovskite based devices hold a potential of further improvement. Here we demonstrate high photoluminescence efficiency of perovskites thin films via deposition of resonant silicon nanoparticles on their surface. The deposited nanoparticles have a number of advances over their plasmonic counterparts, which were applied in previous studies. We show experimentally the increase of photoluminescence of perovskite film with the silicon nanoparticles by 150 % as compared to the film without the nanoparticles. The results are supported by numerical calculations. Our results pave the way to high throughput implementation of low loss resonant nanoparticles in order to create highly effective perovskite based optoelectronic devices.

1. Introduction

Interest in the organic-inorganic perovskites of methylammonium lead trihalides ($MAPbI_3$) family increased when their conversion efficiency of photovoltaic (PV) devices raised rapidly from 6.5% to 9.7% and promptly to 19% in 2012-2015 [1]. These impressive performance allowed perovskites to compete with the leading solar materials of the third generation. Advances in film formation and the optimized perovskite PV architectures have led to further conversion efficiency increase to 22.1% [2]. Recently, new applications of this hybrid materials have been investigated, including light emitting diodes (LED) and semiconductor optical amplifiers and lasers [3].

The trend toward increasing plasmonic amplification of mesoporous solar cells (MSCs) based on gold or silver can be traced in recent years. Increasing the efficiency of PV by such nanostructures can be conceptually obtained by plasmonic electrodes incorporated into MSCs architecture [4]. MSCs have attractive cost of composite materials, as well as the creation. However, for the visible spectral region in these plasmonic structures the losses are increased. An alternative to plasmonics is a strong magnetic response that can be obtained by the use nanoparticles of dielectric materials. In contrast to plasmon nanoparticles the first resonance of the dielectric nanoparticles is a magnetic dipole that was proved experimentally [5]. High-index dielectric nanoparticles may have an induced magnetic dipole moment and none dissipative losses because of lack of free charges in the medium compare with plasmon nanoparticles. The resonant frequency can be controlled by changing the size and shape of the nanoparticles as well as the ambient conditions [6]. Additionally, the electric and magnetic dipole resonances can be overlapped in spectral range bringing a number of unique optical properties. Namely due to these advantages the dielectric nanostructures can be a good alternative to plasmonic nanostructures. This concept paves the way to all-dielectric oligomer sensors and nanoantennas [7], dielectric waveguides [8], nonlinear optics [9], all-dielectric Huygens' metasurfaces [10] and metamaterials [11].

The aim of this work is to apply for the first time resonant silicon nanoparticles to improve optical properties of the hybrid perovskite films. Actually, according to the Mie theory, high-refractive index nanoparticles with low losses support strong local electric field enhancement within perovskite films, while gold plasmonic nanoparticles [12] would additionally strongly absorb useful signal [13]. Moreover, we use cheap and high throughput laser printing method for nanoparticles fabrication and transfer [14] to the surface of the perovskite films.

2. Results

2.1. Perovskite films preparation and characterisation

Thin films of $MAPbI_3$ perovskites can be obtained by various methods and consist essentially of two basic components: the organic methylammonium and inorganic PbI_3 lead trihalide. There are there common used of perovskite film fabrication: by vacuum deposition, solution processing method and hybrid vapor-solution process [15].

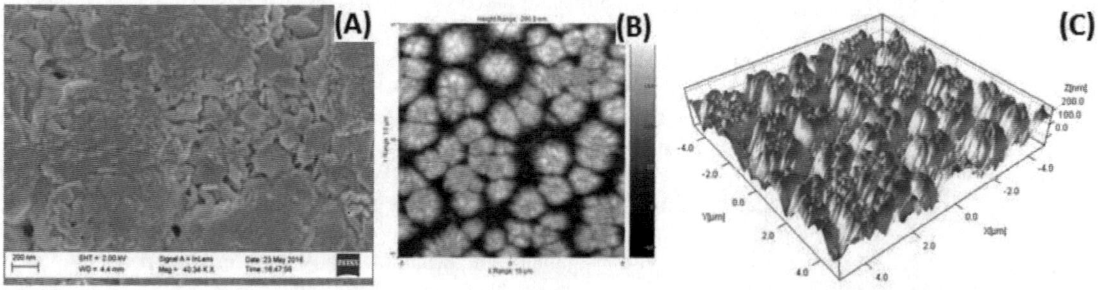

Figure 1. (a) SEM image of perovskite surface without nanoparticles. AFM images of perovskite: (b) 2D and (c) 3D view.

Perovskite films prepared by solution engineering method were investigated by scanning electron microscopy (SEM). The SEM images reveal that the surface consists of non-uniform perovskite grains with an average size from 200 to 250 nm (figure 1a). Also, the film surface was characterized using atomic force microscopy (AFM). AFM method has allowed us to obtain 2D and 3D images of the perovskite surface (figure 1b, c). The AFM images show that in the field of scanning of the perovskite film surface is rather rough and has significant unevenness of the grain surfaces.

2.2. Nanoparticles deposition

To study the possibility of increasing the efficiency of devices based on perovskite ($MAPbI_3$), the series of experiments have been held on deposition of resonant silicon nanoparticles on the surface of the $MAPbI_3$ films by the unique technique of laser printing (figure 2a) [14].

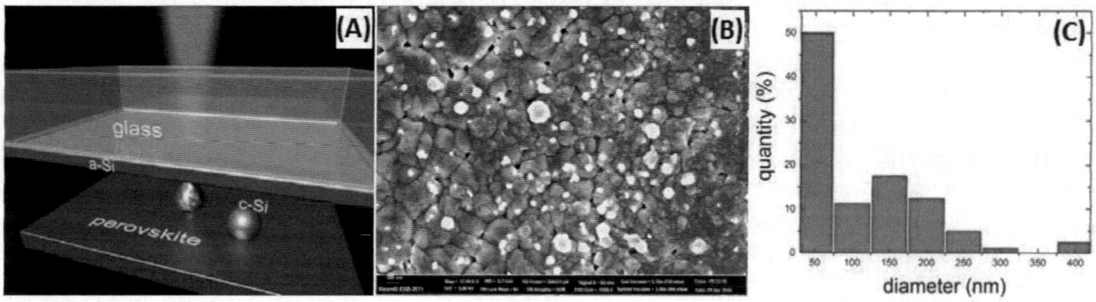

Figure 2. (a) Schematics of laser printing of silicon nanoparticles on perovskite surface. (b) SEM

image of a perovskite with silicon nanoparticles. (c) Size distribution of the silicon nanoparticle on the SEM image is shown as a histogram.

In the laser printing method [14] the nanoparticles are fabricated from a smooth surface (in a single-shot regime) in the forward-transfer geometry (figure 2a), when the receiving substrate is placed under the film with a spacing of 50 μm. This geometry has an advantage over the back-transfer geometry owing the possibility of nanoparticle transfer onto a wide variety of substrates, including opaque and structured samples. The silicon nanoparticles are fabricated at laser energies E < 2 nJ, providing laser fluence (F) range F < 550 mJ/cm^2. The obtained nanoparticles are almost of a spherical shape and their diameters are usually in the range of 50-200 nm, depending on the laser pulse fluence.

The distribution and sizes of the nanoparticles is studied by SEM (figure 2b). Based on the SEM images, it is seen that the surface perovskite layer is covered by silicon nanoparticles quite arbitrarily. Nanoparticles average size varies from 50 nm to 400 nm. Size distribution of silicon nanoparticles on the surface of the sample present at the histogram (figure 2c). The maximum size of the nanoparticles reach 400 nm in diameter, but it is a small amount as a percentage of the total number.

2.3. Photoluminescence enhancement

Characterization of the photolumiescence (PL) properties of the $MAPbI_3$ sample was performed using a multifunctional experimental setup (figure 3a). The deposited silicon nanoparticles were observed using a video camera placed in the image plane of our optical layout. The volume of the measured signal is determined by the numerical aperture of collection objective (NA = 0.9) and the confocal pinhole diameter (50 μm) [14].

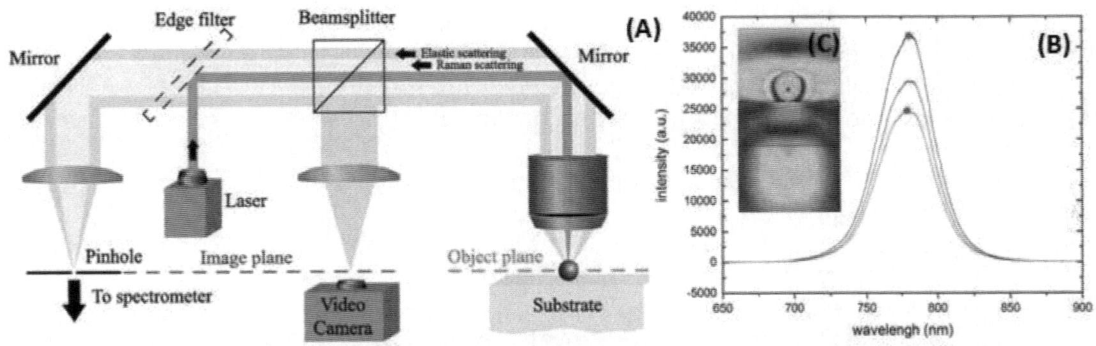

Figure 3. (a) A schematic illustration of multifunctional experimental setup for Raman for PL experiments. (b) PL enhancement from perovskite with silicon nanoparticles: maximum signal from perovskite with Si nanoparticles (black line); averaged signal from plain perovskite (red line); averaged signal from perovskite with Si nanoparticles (blue line). (c) Result of numerical calculation of electric fields near a silicon nanoparticle (D = 200 nm). The excitation wavelength is 633 nm.

After depositing the silicon nanoparticles on the surface of the perovskite we studied the PL resulting from such complex nsnostructure (Si-$MAPbI_3$). PL measurements (figure 3b) show that the PL signal in the area with silicon nanoparticles is stronger than in the area of the sample without nanoparticles. The obtained maximum enhancement in the area with nanoparticles is 150% as compared with the area without them.

Our preliminary calculations by means of CST Microwave Studio show that the silicon nanoparticle allows for near field enhancement in the perovskite film (figure 3C). In particular, we see strong (up to 10 times) increase of E-field around the resonance nanoparticles (D = 200 nm) on the surface of the perovskite. Indeed, this size of the nanoparticle supports excitation of a magnetic dipole resonance at given wavelength providing effective light localization in a subwavelength volume. Such magnetic type Mie mode is more preferable for field enhancement rather than electric one [16]. Since an electric dipole resonance can be excited for larger nanoparticle, we believe that the contribution of the magnetic one in PL enhancement is dominant according to our size distribution (figure 2c).

3. Conclusion

The obtained experimental results have demonstrated that the proposed concept of resonant silicon nanoparticles deposition on top of a perovskite thin film is valid and the increased photoluminescence quantum efficiency due to near-filed enhancement around nanoparticles in agreement with our calculations. Therefore, this concept is one of the promising directions in the development of highly efficient perovskite optoelectronic devices. In future studies, we plan to incorporate silicon nanoparticles in the bulk of perovskite films in a regular way on the patterned nanostructures such as groves and cylindrical holes created by nanoimprinting, as recently demonstrated in [17, 18].

Acknowledgments

The work has partial financial support from the Ministry of Education and Science of the Russian Federation (Grant № 14.Y26.31.0010 for optical measurements) and in the framework of Increase Competitiveness Program of NUST "MISiS" (No. K2-2015-014 for samples preparation). We also appreciate the partial support of Welch Foundation grant AT 16-17 and CONACYT for academic opportunities and academic support (visiting research of A. Cerdan).
Laser experiments were supported by Russian Science Foundation (Grant 15-19-00172).

References

[1] Kim H, Lee C, Im J, Lee K, Moehl T, Marchioro A, Moon S, Humphry-Baker R, Yum J, Moser J E, Grätzel M and Park N 2012 *Scientific Reports* **2** 591
[2] Saliba M, Matsui T, Seo J, Domanski K, Correa-Baena J, Nazeeruddin M K, Zakeeruddin S M, Tress W, Abate A, Hagfeldt A and Grätzel M 2016 *Energy and Environmental Science* **9** 6
[3] Sutherland B R and Sargent E H 2016 *Nature Photonics* **10** 295
[4] Erwin W R, Zarick H F, Talbert E M and Bardhan R 2016 *The Royal Society of Chemistry Energy Environ* **9** 1577
[5] Evlyukhin A B, Novikov S M, Zywietz U, Eriksen R L, Reinhardt C, Bozhevolnyi S I and Chichkov B N 2012 *Nano Lett.* **12** 3749
[6] Evlyukhin A B, Eriksen R L, Cheng W, Beermann J, Reinhardt C, Petrov A, Prorok S, Eich M, Chichkov B N and Bozhevolnyi S I 2014 *Scientific Reports* **4** 4126
[7] Miroshnichenko A E,and Kivshar Y S 2012 *Nano letters* **12** 12
[8] Savelev R S, Miroshnichenko A E, Sukhorukov A A and Kivshar Y S 2014 *JETP letters* **7** 100
[9] Shcherbakov M R, Neshev D N, Hopkins B, Shorokhov A S, Staude I, Melik-Gaykazyan E V, Decker M, Ezhov A A, Miroshnichenko A E, Brener I, Fedyanin A A and Kivshar Y S 2014 *Nano Letters* **14** 6488
[10] Decker M, Staude I, Falkner M, Dominguez J, Neshev D N, Brener I, Pertsch T and Kivshar Y S 2015 *Advanced Optical Materials* **10** 1002
[11] Moitra, Slovick B A, Yu Z G, Krishnamurthy S and Valentine J 2014 *Applied Physics Letters* **104** 171102
[12] Omelyanovich M, Makarov S, Milichko V, Simovski C 2016 arXiv 1601.06778
[13] Omelyanovich M, Ra'dil Y and Simovski C 2015 *J. Opt.* **17** 125901
[14] Dmitriev P A, Makarov S V, Milichko V A, Mukhin I S, Gudovskikh A S, Sitnikova A A, Samusev A K, Krasnok A E and Belov P A 2016 *Nanoscale* **8** 5043
[15] Chena Q, Marcoa N D, Yanga Y, Songa T B, Chena C C, Zhaoa H, Honga Z, Zhoua H and Yanga Y 2015 *Nano Today* **04** 009
[16] Dmitriev P A, Makarov S V, Milichko V A, Mukhin I S, Gudovskikh A S, Sitnikova A A and Belov P A 2016 *Nanoscale* **8** 9
[17] Wang H, Haroldson R, Balachandran B, Zakhidov A, Sohal S, Chan J Y, Zakhidov A and Hu W 2016 *ACS Nano* DOI: 10.1021/acsnano.6b05535

[18] Makarov S, Milichko V, Ushakova E, Omelyanovich M, Cerdan-Pasaran A, Haroldson R, Balachandran B, Wang H, Hu W, Kivshar Y and Zakhidov A 2016 *ACS Photonic* (in press)

IOP Conf. Series: Journal of Physics: Conf. Series **929** (2017) 012054 doi:10.1088/1742-6596/929/1/012054

The electron tunneling from fullerene molecule in a weak electric field

P A Golovinski[1,2], **A A Drobyshev**[2], **E A Mikhin**[2], **A V Tuchin**[2]

[1] Moscow Institute of Physics and Technology, Moscow 141700, Russia
[2] Voronezh State Technical University, Voronezh 394006, Russia

E-mail: golovinski@bk.ru

Abstract. We derive equations describing the electron tunneling ionization of the molecular orbitals without symmetry. The modified method of smooth matching the asymptotic wave functions with molecule's wave functions is suggested, to use the results of numerical calculations of the molecules in the GAUSSIAN. The spatial distribution of the current is considered for tunneling ionization the fullerene C_{60}. We obtained the ionization probability of individual orbitals and spatial distribution of the electron currents at different values of the field, a distance to the screen and orientation of the fullerene.

1. Introduction

Tunneling ionization of atoms and molecules is determined by the behavior of the valence electron at large distances from the atomic nucleus. To simplify the problem, one often used one-dimensional models. The most common variant of this model is a well known the Fowler-Nordheim formula for metals with the Coulomb correction on the image-potential [1]. The heterogeneity of the studied structure is taken into account in such a model by semi-empirical correction factors [2, 3]. A simplified approach to the description of tunneling currents from individual atoms is based on an idea of the point source [4, 5]. It allows one to calculate the spatial distribution of the electron current for a simple [6] and complex emitters [7]. This approach is useful in describing the tunneling electron in negative ions [8] and electron photodetachment in the presence a constant electric field [9-11], but is not suitable for neutral atoms and molecules with the long-range Coulomb interaction. The consistent solution to the problem is based on the fact that at large distances from the nucleus the electron wave function satisfies the Schrödinger equation with the Coulomb potential and the potential of a uniform electric field. This asymptotic behavior of the potential allowed us to obtain an analytic solution to the problems of tunneling ionization of the hydrogen atom and hydrogen-like ions [12, 13].

In this paper we consider the spatial distribution of current in the tunneling ionization of fullerene C_{60} in a uniform electric field. Note that a strong external electric field leads to additional changes in the electronic structure of the fullerene [14, 15]. In any case, the current spatial nonuniformity arises from the distribution of the local electron density over different molecular orbitals.

2. The asymptotic model of the tunnel ionization in a weak electric field

We follow the general approach to the ionization of many-electron systems [16, 17] and use a number of approximations. The first is the approach of the frozen cores. The second is the idea that the dynamics of each individual electron j can be represented as a motion in a nonlocal potential $\hat{V}_j(\mathbf{r}_j)$, in which the self-interaction is eliminated. This means that we considered atomic and molecular systems well described in the Hartree-Fock approximation, and the complete wave function can be represented as a product of antisymmetrised one-electron wave functions of the individual electrons. This means that the Hamiltonian H of the whole electronic system can be written as a sum of single-particle Hamiltonians (in atomic units):

$$H = \sum_{j=1}^{N} h_j \, , \; h_j = -\frac{1}{2}\Delta_j + \hat{V}_j(\mathbf{r}_j) \, , \tag{1}$$

where N is number of electrons in the system. After projecting onto the space of the bound states of $N-1$ electron, for the tunnel electron we obtain the equation

$$\left(-\frac{1}{2}\Delta + \hat{V}(\mathbf{r})\right)\psi = E\psi \, . \tag{2}$$

At a large distance from the center of orbitals we have

$$\hat{V}(\mathbf{r})\Big|_{r\to\infty} = -\frac{1}{r} + Fz \, . \tag{3}$$

In this region the asymptotic solution of the Schrodinger equation is factorized in the parabolic coordinate system [18].

In the inner atomic-molecular space, the quantum states are calculated numerically based on the method of molecular atom-like orbitals. The main difficulty of tunneling problem is to coordinate solutions within the molecule and asymptotically remote region. Implementation of this approach presupposes the existence an area in which both solutions are correct. However, the sample functions used in a well-known the GAUSSIAN program of the quantum chemistry simulation do not have the correct asymptotic behavior in the under-barrier region. The initial basis functions contain the usual separation of radial and angular coordinates. For a molecular orbital, the wave function in the coordinate system with the center in atom is written as

$$\psi_0(\mathbf{r}) = \sum_{LM} C_{LM} R_L(r) Y_{LM}(\theta', \varphi') \, , \tag{4}$$

where $Y_{LM}(\theta,\varphi)$ is spherical harmonic, L and M are the quantum numbers of the angular momentum and its projection to the selected axis, r,θ,φ are spherical coordinates, and the radial part at large r decreases as the Gaussian distribution, that is, the asymptotic behavior is false. Therefore, convenient functions that meet the compatibility conditions are absent. One way to overcome this difficulty is the analytic continuation of numerical solution out of the area of their formal applicability [19] with the subsequent fitting the asymptotic in sub-barrier region. A special feature of our procedure is to coordinate the amplitudes of numerical and analytical asymptotic wave functions at the point of smooth matching solutions before entering the under-barrier region by continuation of the semiclassical asymptotic of the outgoing wave to this area. The advantage of this approach is the uniqueness of the asymptotic fitting and option to account for electric field action close to the inner turning point.

3. Separation of variables in the Schrödinger equation for hydrogen atom in parabolic coordinates

The Schrödinger equation for an electron with energy E in the Coulomb attraction field and a uniform electric field of strength F in parabolic coordinates takes the form

$$\frac{4}{\xi+\eta}\left[\frac{\partial}{\partial\xi}\left(\xi\frac{\partial\psi}{\partial\xi}\right)+\frac{\partial}{\partial\eta}\left(\eta\frac{\partial\psi}{\partial\eta}\right)\right]+\frac{1}{\xi\eta}\frac{\partial^2\psi}{\partial\varphi^2}+2\left(E+\frac{2}{\xi+\eta}-\frac{F}{2}(\xi-\eta)\right)\psi=0, \quad (5)$$

where $x=\sqrt{\xi\eta}\cos\varphi$, $y=\sqrt{\xi\eta}\sin\varphi$, $z=2^{-1}(\xi-\eta)$ are parabolic coordinates [18].

We are looking for a solution of equation (5) as a product of functions

$$\psi=\Phi(\xi)\mathrm{X}(\eta)e^{im\varphi}(2\pi)^{-1/2}. \quad (6)$$

Introducing variables $\Phi(\xi)=\phi(\xi)\xi^{-1/2}$ and $\mathrm{X}(\eta)=f(\eta)\eta^{-1/2}$ we obtain two differential equations

$$-\frac{1}{2}\frac{d^2\phi}{d\xi^2}+U_1(\xi)\phi=\frac{E}{4}\phi, \quad (7)$$
$$-\frac{1}{2}\frac{d^2f}{d\eta^2}+U_2(\eta)f=\frac{E}{4}f.$$

Effective potentials that are contained in the equation (7) are equal

$$U_1(\xi)=-\frac{\beta_1}{2\xi}+\frac{m^2-1}{8\xi^2}+\frac{F\xi}{8}, \quad (8)$$
$$U_2(\eta)=-\frac{\beta_2}{2\eta}+\frac{m^2-1}{8\eta^2}-\frac{F\eta}{8},$$

with separation constants $\beta_1+\beta_2=1$. Ionization occurs in the direction $z\to\infty$, which corresponds to the coordinate $\eta\to\infty$. External weak field has no significant effect on the function ϕ. Therefore, as an approximate solution of the equation (7) we can take the well-known exact solution for $F=0$ [16], which is expressed in terms of the generalized Laguerre polynomials:

$$\phi_{nm}(\xi)=\kappa^{1/2}(\kappa\xi)^{|m|/2}e^{-\kappa\xi/2}\left(\frac{n_\xi!}{(n_\xi+|m|)!}\right)^{1/2}L_{n_\xi}^{(|m|)}(\kappa\xi), \quad (9)$$
$$n_\xi=-(|m|+1)/2+n\beta_1,\ \kappa=\sqrt{-2E}, E=-(2n^2)^{-1}.$$

For the analysis of the second equation (7) we introduce variables $\rho=\kappa\eta$, $\widetilde{F}=F/(4\kappa^{3/2})$, $\beta=\beta_2/\kappa$, and obtain

$$\left(\frac{d^2}{d\rho^2}+k^2(\rho)\right)f(\rho)=0, \quad (10)$$
$$k^2(\rho)=-\frac{1}{4}-\frac{m^2-1}{4\rho^2}+\widetilde{F}\rho+\frac{\beta}{\rho}.$$

The asymptotic solution of the equation (10) is considered in [20]. Classical turning points are $\rho_1 \sim 2\beta + \left(4\beta^2 - m^2 + 1\right)^{1/2}$, $\rho_2 \sim \left(4\tilde{F}\right)^{-1}$. In the interval between these points, the potential gradually transforms from the Coulomb potential into linear one. In the vicinity of each of the turning points the solution can be written as

$$f_{1,2}(\rho) = \frac{1}{\sqrt{v'_{1,2}(\rho)}}\left\{a_{1,2}\mathrm{Ai}\left[v_{1,2}(\rho)\right] + b_{1,2}\mathrm{Bi}\left[v_{1,2}(\rho)\right]\right\}, \tag{11}$$

where

$$v_1(\rho) = \left(\frac{3}{2}\int_{\rho_1}^{\rho} |\,k(\rho)\,|\,d\rho\right)^{2/3}, \ \rho > \rho_1, \tag{12}$$

$$v_1(\rho) = -\left(\frac{3}{2}\int_{\rho}^{\rho_1} |\,k(\rho)\,|\,d\rho\right)^{2/3}, \ \rho < \rho_1,$$

and

$$v_2(\rho) = -\left(\frac{3}{2}\int_{\rho_2}^{\rho} |\,k(\rho)\,|\,d\rho\right)^{2/3}, \ \rho > \rho_2, \tag{13}$$

$$v_2(\rho) = \left(\frac{3}{2}\int_{\rho}^{\rho_2} |\,k(\rho)\,|\,d\rho\right)^{2/3}, \ \rho < \rho_2.$$

To the left of the point ρ_1 and the right of the point ρ_2 the solution is oscillatory. In the interval $\rho_1 < \rho < \rho_2$ the solution is the sum of exponentially decreasing and increasing functions. To coordinate asymptotic we use the semiclassical approximation [18].

4. Ionization probability

The motion of the electron in the initial molecular orbitals splits into channels corresponding to certain states of the hydrogen atom [19]. The solution of the problem of electron tunneling at distance $\rho >> \rho_2$ can be written as the sum over the channels with certain parabolic quantum numbers:

$$f(\eta) = \eta^{-1/2}\sum_{\nu} A_\nu f_\nu(\eta)\Phi_\nu(\xi)e^{im\varphi}, \nu = n_\xi, m. \tag{14}$$

In the weak-field approximation coefficients A_ν are expressed in terms of coefficients of matching asymptotic of the exact and numerical solutions. The initial state $\psi_0(\mathbf{r})$, projected onto the full set of functions $\Phi_{n|m|}(\xi)e^{im\varphi}$, gives the result

$$G_{n_\xi m}(\vartheta,\eta) = \int_0^\infty\int_0^{2\pi} \Phi_{n|m|}(\xi)\frac{e^{-im\varphi}}{\sqrt{2\pi}}\psi_0(\mathbf{r})\,d\xi d\varphi, \tag{15}$$

where ϑ is the angle between the direction of the electric field and the axis of quantization of the orbital angular momentum of an electron in the molecular orbital.

The known asymptotic behavior of the wave function of a tunnel electron (6) allows us to calculate the probability of ionization, using the current density j_z in the direction of the axis z through the orthogonal surface S, in the form of

$$\Gamma = \int j_z \, dS \, . \tag{16}$$

Here, the current density for a large z can be written as

$$j_z = \frac{i}{2}\left(\psi \frac{\partial \psi^*}{\partial z} - \psi^* \frac{\partial \psi}{\partial z}\right) \approx \frac{i\,|\phi(\xi)|^2}{2\pi\xi\eta}\left(f^* \frac{\partial f}{\partial z} - f \frac{\partial f^*}{\partial z}\right). \tag{17}$$

The probability of tunneling ionization depends on the barrier permeability *P*. Calculating the integral between the turning points in the classically forbidden region of motion, we obtain [18]

$$P = \exp\left(- \int_{\rho_1}^{\rho_2} |k|\, d\rho \right) = \left(4\beta\widetilde{F}\right)^{-\beta} \exp\left(- \frac{1}{3 \cdot 4\widetilde{F}} \right). \tag{18}$$

The matching of asymptotic behavior $f_\nu(\eta)$ in the internal region of variable $\eta < \eta_1$ can be done by equating the functions $f_\nu(\eta)$ and $G_\nu(\vartheta, \eta)$, and their derivatives [21].

The total probability of the electron ionization rate in the weak field limit is given by [19, 22]

$$\Gamma(\vartheta) = \sum_{n_\xi}^{\infty} \sum_{m=-\infty}^{\infty} \Gamma_{n_\xi m}(\vartheta) \, , \tag{19}$$

where $\Gamma_{n_\xi m}(\vartheta)$ is partial ionization probability in the channel with parabolic quantum numbers n_ξ и m. The partial current density at a large distance from the source is of form

$$j_{z,n_\xi m} = \frac{\kappa}{\pi\xi\eta} \left| \frac{\phi_{n_\xi m}(\xi) G_{n_\xi m}(\vartheta, \mathrm{P}_{n_\xi m}) \sqrt{v_1'(\kappa \mathrm{P}_{n_\xi m})}}{\mathrm{Ai}(v_1(\kappa \mathrm{P}_{n_\xi m}))\sqrt{\pi}} P_{n_\xi m} \right|^2 . \tag{20}$$

It follows the expression for the partial probability

$$\Gamma_{n_\xi m}(\vartheta) = \frac{\kappa}{\pi} \left| \frac{v_1'(\kappa \mathrm{P}_{n_\xi m}) G_{n_\xi m}(\vartheta, \mathrm{P}_{n_\xi m})}{\mathrm{Ai}(v_1(\kappa \mathrm{P}_{n_\xi m}))} P_{n_\xi m} \right|^2 . \tag{21}$$

5. Conclusions

We have examined the spatial effects in the tunneling ionization of multicenter systems depending on their structure. It is important to note that the position of the atomic centers and the shape of the molecular orbitals are manifested in the electron current distribution. At a great distance from the C_{60} fullerene molecule the distribution pattern of the tunnel current density tends to image a point source. To clarify the role of the many-electron structure of complex molecules and charge delocalization in the formation of interference effects in the emission current, more research is needed.

This work is supported by the Russian Foundation for Basic Research (grant 16-32-00253).

References

[1] Nordheim L W 1928 *Roy. Soc. Proc. A* **121**(788) pp 626-639
[2] Gröning P, Ruffieux P, Schlapbach L, and Gröning O 2003 *Advanced Engineering Materials* **5** pp 541-550

IOP Conf. Series: Journal of Physics: Conf. Series **929** (2017) 012054 doi:10.1088/1742-6596/929/1/012054

[3] Wang X Q, Wang M, Li Z H, Xu Y B, and He P M 2005 *Ultramicroscopy* **102** pp 181-187
[4] Gottlieb B, Kleber M, and Krause J 1991 *Z. Phys. A* **339** pp 201-206
[5] Golovinski P A, Drobyshev A A 2014 *JETP* **118** (6), pp 863–868
[6] Donnere B and Kleber M 2005 *Am. J. Phys.* **73**(8) pp 690-700
[7] Golovinski P A, Drobyshev A A 2014 *Journal of Electromagnetic Analysis and Applications* **6** pp 8-14
[8] Demkov Yu N and Drukarev G F 1965 *Sov. Phys. JETP* **20** pp 614-620
[9] Golovinskii P A 1997 *J. Exp. Theor. Phys.* **85**(5) pp 857-861
[10] Golovinski P A, Drobyshev A A 2012 *Tech. Phys. Lett.* **38** (3) pp 272–275
[11] Golovinskii P A, Drobyshev A A 2013 *Tech. Phys.* **58** (2) pp 158–164
[12] Batishev P A, Tolstikhin O I, and Morishita T 2010 *Phys. Rev. A* **82** 023416(14)
[13] Jiang W-Ch, Tolstikhin O I, Peng L Y, and Gong Q 2012 *Phys. Rev. A* **85** 023404(15)
[14] Rappoport D and Furche F 2010 *J. Chem. Phys.* **133** 134105(13)
[15] Tuchin A V, Bityutskaya L A, and Bormontov E N 2014 *Phys. Solid State* **56**(8) pp 1685-1688
[16] Tolstikhin O I, Morishita T, and Madsen L B 2011 *Phys. Rev. A* **84** 053423(17)
[17] Tolstikhin O I, Madsen L B, and Morishita T 2014 *Phys. Rev. A* **89** 013421 (15)
[18] Landau L D and Lifshitz E M *Quantum Mechanics: Non–Relativistic Theory* (Butterworth–Heinemann, Oxford, 2005; Fizmatlit, Moscow, 2008)
[19] Madsen L B, Jensen F, Tolstikhin O I, and Morishita T 2013 *Phys. Rev. A* **87** 013406(10)
[20] Yamabe T and Tachibana A 1977 *Phys. Rev. A* **16**(3) pp 877-890
[21] Baz A I, Zel'dovich Ya B, Perelomov A M 1969 *Scattering, Reactions and Decay in Nonrelativistic Quantum Mechanics* (Jerusalem: Israel Program for Scientific Translations)
[22] Madsen L B, Tolstikhin O I, and Morishita T 2012 *Phys. Rev. A* **85** 053404(12)

IOP Conf. Series: Journal of Physics: Conf. Series **929** (2017) 012055 doi:10.1088/1742-6596/929/1/012055

The formation of multi-axis micromechanical gyroscopes and accelerometers using surface micromachining

E Yu Gusev[1], J Y Jityaeva[2], Al V Bykov[3]
Southern Federal University, Institute of Nanotechnology, Electronics and Electronic Equipment Engineering, Taganrog 347922, Russia

[1]eyugusev@sfedu.ru, [2]julia.yityaeva@gmail.com, [3]alvbykov@nt.ru

Abstract. The process flow of unified integral multi-axis micromechanical gyroscopes and accelerometers fabrication by surface micromachining was developed. The process flow for typical n-type silicon wafer Si (100) contains over 20 primary operations, including 5 lithography steps using 5 masks with minimum features size of 1 μm. Special operations were discussed. Experimental results of sacrificial SiO_2 etching for poly-Si-on-SiO_2 structures using buffered solution of hydrofluoric acid and ammonium fluoride (1:4) were obtained. The undercut etching rate was approximately of about 20 nm/sec and the minimum time required for sacrificial SiO_2 removing under beams of 0.5-2.5 width was in in the range of 11 to 62.5 seconds. Finally two and three-axis micromechanical gyroscopes and accelerometers were fabricated.

1. Introduction

The integration of micromechanical components with microelectronics has led to the emergence of a new technology - technology microelectromechanical systems (MEMS) and microtechnology. Using typical processes if the IC technology a series of mechanical elements were fabricated, like membranes, motors, pumps and valves [1-4]. Design and development of bulk and surface micromachining, high aspect ratio technology of nanotructures (nanomechanics) allowed to get micro- and nanomechanical sensors (gyroscopes and accelerometers), as well as to expand their applications from household appliances to high-tech aerospace applications [3-7]. One of the most commonly used transducers are inertial sensors like accelerometers and gyroscopes. The current requirements for these devices are – multi-axis sensitivity, small size, light weight and low power consumption.

Well-known micromechanical sensors allow to measure the characteristics of, e.g. a linear acceleration along one or two axes of sensitivity. The requirement of the multi-axis sensitivity makes it necessary to microassembly, followed by precision alignment, or an increase in the overall dimensions when several sensors with different sensitive axis directions placed on a substrate or chip. Moreover they are not always made by industrial technologies. Locality of these methods (e.g., focused ion beam and scanning probe nanolithography [8-12]) limited their usage on the level of laboratory research.

Decision of the designated problem was the development of integrated multi-axis sensors and their manufacturing technologies by surface micromachining [13,14]. Thus the basic material is polysilicon due to its structural and electrical properties and easy handling [2-4].

Previously the design two-axis micromechanical gyroscope and three-axis micromechanical accelerometer based on two layers of polycrystalline silicon with capacitive registration system) was performed [14]. The goal of this research is to develop and realize a process flow of multi-axis MEMS gyroscopes and accelerometers manufacturing based on polycrystalline silicon by surface micromachining.

2. Experimental details

The unified process flow was proposed [14] for developed designs the MEMS gyroscope and accelerometer [10]. Method of experimental fabrication for micromechanical structures is given in [10]. The photolithography was conducted using five masking coatings: aluminum, chrome, photoresists FP-051KI, MICROPOSIT SP25-10 and Ultra i-0.8 by two modes – soft and hard contact.

In order to determine the optimal modes of removing the sacrificial layer (time and composition of the etching solution) have been designed and manufactured test elements based on the structure of poly-Si/SiO_2/Si. The topologic pattern of test elements was fabricated by focused ion beam (FIB) lithography (Nova NanoLab 600). The current ion beam and accelerating voltage were 3 nA and 30 keV, respectively. The removal of the silicon dioxide sacrificial layer was carried out under continuous stirring and room temperature in buffered solution of hydrofluoric acid and ammonium fluoride (40%HF:40%NH_4F=1:4) in the range time from 10 to 60 seconds. The etch depth was monitored by scanning electron microscopy. Finally, party of experimental sample multi-axis MEMS gyroscopes and accelerometers were fabricated.

3. Results and discussion

The results of the primary science research of materials and technologies allowed to choose required modes. It allow to achieve the desired values of thickness layers, of lateral dimensions of the structural elements at lithography and to provide high quality of deposition and removal photoresist and In particular, the polycrystalline silicon layer was produced by plasma deposition and doping using diffusion of phosphorus. Values of concentration, mobility of charge carriers and surface resistivity amounted to $\sim 2\cdot 10^{20}$ cm^{-3}, ~ 30 $cm^2/(V\cdot sec)$ и 2-9 Ω/sq, , at that the microhardness and Young module of doped films were 13.7–23.6 GPa and 224.3–380.4 GPa, respectively [15-17].

Test structures were fabricated for preliminary studies of photolithography and removing the sacrificial layer.

High-quality photolithography is achieved by improving the adhesion of the photoresist, changing the time of exposure and emergence as well as gap the photomask-substrate. Photoresists that specially designed for reactive ion etching of (FP-051KI, MICROPOSIT SP25-10) were used as masks. It has been established that components of photoresist are evaporating in the resulted of etching poly-Si of 2μm and SiO_2 of 1.5 μm. The mask of aluminum and chromium thickness of 50-100 nm was formed for the etching such layers. There is a problem to remove the plasma-processed photoresist, even for oxygen plasma plasma it was difficult. Mask less exposure to dose to clear allowed to remove photoresist residues. For this reason, for structural layer etching the mask of chromium or aluminum was used.

Experimental investigation of etching of test elements in hydrofluoric acid solution allowed to determine the characteristics necessary for removing the sacrificial layer and for releasing the inertial mass. The calculated average value of etch rate SiO_2 was 20 nm/sec at room temperature. Analysis of the received SEM images (Fig. 1a, b) gave the opportunity to determine the undercut of the sacrificial layer in the horizontal direction (Fig. 1c). The dependence is linear. The presence of open windows at the structural layer significantly reduces the amount of time required to remove the sacrificial layer, and prevents sticking.

Analysis of the experimental results allowed to specify design and technological solution of multiaxial microgyroscopes and accelerometer [14], and to develop a photomasks and the manufacturing flow with a minimum element size of 1 mm (Figure 2).

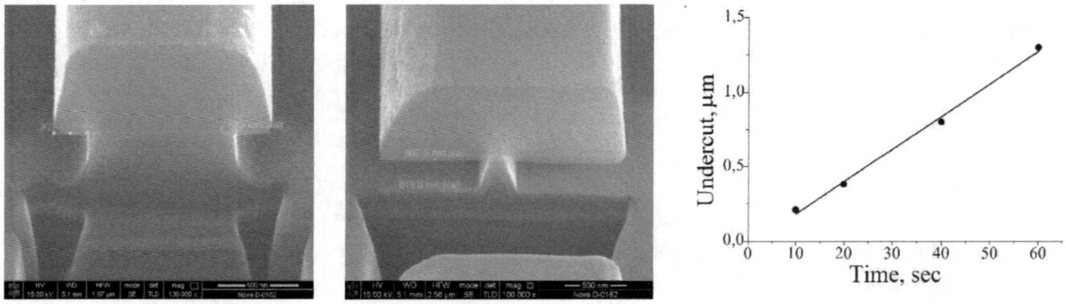

(a) **(b)** **(c)**

Figure 1(a-c). SEM images of the poly-Si/SiO_2 structure at 10 **(a)** and 40 **(b)** seconds of sacrificial etching; undercut vs etching time curve **(c)**

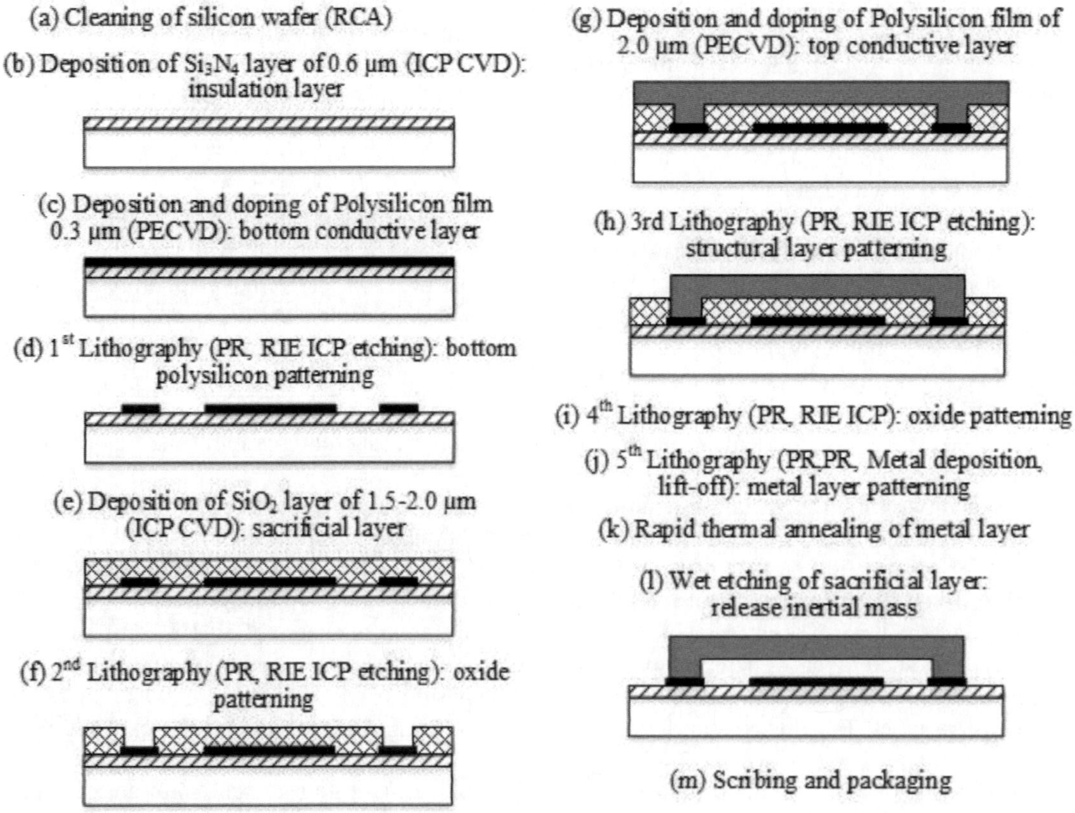

Figure 2. Process flow of the micromechanical gyroscope and accelerometer fabrication

The following it should be noted. The mask #3 contains a matrix of transparent square fields, which transfer to a photoresist layer and finally to underlying layer – square windows or openings (h), to facilitate the release of the structural layer on the sacrificial etching step (l). To decrease release stiction effect during removing sacrificial layer step the guarding areas were used (f-g). The lift-off lithography is easier to handle for low conformity coating (electron beam evaporation than sputtering).

Series of multi-axis MEMS gyroscopes and accelerometers was fabricated in accordance with the developed route (Fig. 3) and sent to the testing.

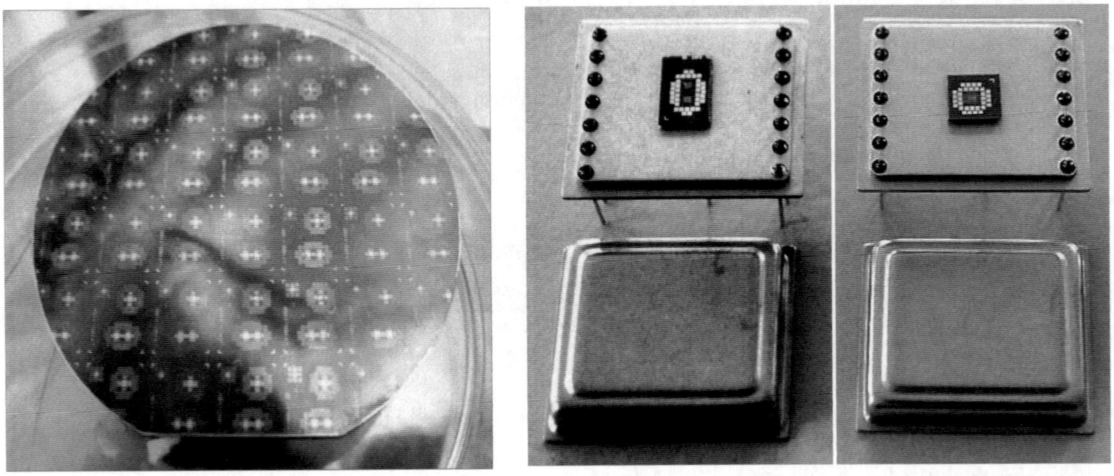

(a) **(b)** **(c)**

Figure 3(a-c). Optical image of the processed wafer **(a)**, microgyroscope **(b)** and microaccelerometer **(c)** chip on the package

It is important to emphasize that the workability of the proposed route allows to fabricate multiple types of devices (multi-axis accelerometers and gyroscopes) in a single technological cycle. Nanomechanical tunneling accelerometers can be create by the inclusion of additional operation local modification of the structure using methods of the focused ion beams [11].

4. Conclusion

The process flow of integral multiaxis micromechanical gyroscopes and accelerometers was developed applied to equipment of research and educational center "Nanotechnologies" and center of collective use of equipment of Southern Federal University. The process flow contains over 20 primary operations (substrate cleaning, plasma deposition of silicon nitride, polycrystalline silicon and silicon oxide, its plasma etching, wet etching of silicon oxide, polysilicon doping, deposition of contacts and its thermal annealing, as well as photolithography), including 5 lithography steps using 5 masks with minimum features size of 1 micron. Sacrificial etching for releasing inertial masses was discussed as well as other specific operations. Questions of inertial structures design, masks development and packaging are not affected. According to the developed flow a series of two and three-axis polysilicon micromechanical gyroscopes and accelerometers was fabricated by surface micromachining and sent for testing. The results of the work can be used for process flow developments of micromechanical gyroscopes and accelerometers and MEMS&NEMS devices.

Acknowledgment

The equipment of the Centre of Collective Use of Equipment "Nanotechnologies" of Southern Federal University was used. The study was supported by The Ministry of Education and Science of Russian Federation within the contract/project 14.575.21.0045 (unique identifier RFMEFI57514X0045).

References

[1] Sniegowski J J, Boer M P 2000 *Annual Review of Materials Research* **30** 299-333
[2] Maboudian *R* 1998 *Surface Science Reports* **30** 207-269
[3] Berman D, Krim J 2013 *Progress in Surface Science* **88** 171-211
[4] French P J 2002 Sensors and actuators A Physical **99** 3
[5] Yun W 2012 Microsystems and Nanotechnology (Springer-Verlag Berlin Heidelberg) 653

[6] Yasaitis J, Judy M, Brosnihan T, Garone P, Pokrovskiy N, Sniderman D, Limb S, Howe R, Boser B, Palaniapan M, Jiang X, Bhave S 2003 *Proceedings of SPIE - The International Society for Optical Engineering* **4979** 145
[7] Core T A, Tsang W K, Sherman S J 1993 *Solid State Technology* **36** 39
[8] Ageev O A, Smirnov V A, Solodovnik M S, Rukomoikin A V, Avilov V I 2012 *Semiconductors* **46** (13) 1616
[9] Avilov V I, Ageev O A, Kolomiitsev A S, Konoplev B G, Smirnov V A, Tsukanova O G 2014 *Semiconductors* **48** (13) 1757
[10] Ageev O A, Balakirev S V, Bykov Al V, Gusev E Yu, Fedotov A A, Jityaeva J Y, Il'in O I, Il'ina M V, Kolomiytsev A S, Konoplev B G, Krasnoborodko S U, Polyakov V V, Smirnov V A, Solodovnik M S, Zamburg E G 2016 *Springer Proceedings in Physics* (New York: Springer International Publishing) **175** 563
[11] Ageev O A, Gusev E Yu, Kolomiytsev A S, Lisitsyn S A, Bykov Al V 2016 *Journal of Physics: Conference Series* **741** 012177
[12] Vopilkin E A, Klimov A Yu, Rogov V V, Pryakhin D A, Gusev S A, Skorohodov E V, Shuleshova I Y, Shashkin V I 2014 *IEEE Sensors Journal* **14** (6) 1831
[13] Senturia S D 2002 Microsystem design (Kluwer academic publishers: New York: Boston: Dordrecht: London: Moscow)
[14] Konoplev B G, Ryndin E A, Lysenko I E, Denisenko M A, Isaeva A S 2016 Proceeding of the International Conference "Micro- and Nanoelectronics – 2016" (ICMNE-2016, Moscow-Zvenigorod, Russia) 158
[15] Gusev E Yu, Jityaeva J Y, Bykov Al V, Bespoludin V V 2015 *Izvestiya SFedU. Engineering Sciences* **9** 126
[16] Ageev O A, Gusev E Yu, Jityaeva J Yu, Kolomiytsev A S, Bykov Al V 2016 *CRC Press/Balkema* (London: Taylor & Francis Group) 13
[17] Ageev O A, Gusev E Yu, Jityaeva J Y, Ilina M V, Bykov Al V 2016 *Journal of Physics: Conference Series* **741** 012001

IOP Conf. Series: Journal of Physics: Conf. Series **929** (2017) 012056 doi:10.1088/1742-6596/929/1/012056

Molecular dynamics simulation of metal nanoislands growth

N V Kapralov[1], E S Babich[2] and A V Redkov[1, 2]

[1] Department of Physics and Technology of Nanoheterostructures, St. Petersburg Academic University RAS, St. Petersburg 194021, Russia
[2] Institute of Physics, Nanotechnology and Telecommunications, Department of Physics and Technology of Nanostructures, Peter the Great St. Petersburg Polytechnic University, St. Petersburg 195251, Russia

e-mail: nikolay.kapralov@gmail.com

Abstract. We present the atomistic model and the simulation of a self-assembled growth of a silver nanoisland film and small groups of nanoislands on a glass substrate after thermal poling of the glass with a profiled electrode. The calculations were performed in molecular dynamics simulator LAMMPS taking into account the diffusion of the metal atoms towards and along the glass surface and their clustering. Lennard-Jones potential was used to describe metal-metal and metal-glass interaction. The potential parameters were determined to provide qualitative coincidence of the simulated configurations of the metal nanostructures and the experimental ones, such as an isolated nanoisland, a pair and a set of three nanoislands and a "plasmonic molecule".

1. Introduction

Metal nanostructures such as a nanoisland film and small groups of metal nanoislands are in particular of interest for optics due to wide applications of surface plasmon resonance (SPR) phenomenon. The SPR excitation at the resonant wavelength results in an effective absorption of the corresponding electromagnetic wave and high enhancement of local electric field nearby metal surface [1]. These optical properties of metal films have been employed in surface-enhanced Raman spectroscopy [2] providing 8-9 orders of magnitude enhancement, biosensing [3], photovoltaics [4], and photocatalysis [5].

One of the perspective ways of nanoisland film fabrication is an out-diffusion of reduced silver and self-assembled growth of silver nanoislands on the surface of a silver-to-sodium ion-exchanged glass [6]. Moreover, modulation of silver ions distribution within the glass via an additional thermal poling treatment with a profiled anodic electrode allows fabricating isolated nanostructures [7]. Optical properties of the isolated nanostructures highly depend on their size and morphology including spatial configuration of small groups of the nanoislands. Hence, understanding of nanoisland growth mechanisms would allow one to control the morphology and resonant characteristics of fabricated structure [8]. It is hard to deduce the final configuration of the nanostructures with an analytical approach, especially in case of non-uniform initial silver distribution in glass; therefore, we used a molecular dynamics simulation, which allows simulating nanoislands growth on the base of interparticle interactions calculations.

2. Object and model

An effective technique of the isolated nanostructures fabrication has recently been reported [7, 9]. A soda-lime glass slide is processed in following sequence: an enrichment of glass subsurface layer with silver ions by ion-exchange in $AgNO_3/NaNO_3$ melt, a modulation of silver ions distribution with a profiled anodic electrode ("thermal poling") under DC field and annealing in hydrogen atmosphere resulted in silver clustering in the glass bulk and nanoisland formation on the surface. The mode of annealing and the pattern of the electrode, which in our case is a net of square and rectangular hollows, determine a configuration of nanoisland groups [9]. The silver shifted to the glass bulk in the region of contact glass-electrode ("poled region") cannot reach the surface in the course of annealing contrary to non-poled areas under hollows in the electrode. Thus, the formation of silver nanoislands appears in non-poled regions only. The annealing in reducing atmosphere along with the ion exchange of the glass is a way to fabricate a self-assembled nanoisland film with randomly distributed nanoislands [6]. We used molecular dynamics method to simulate formation of the nanoislands and determine the set of parameters, which provide a qualitative match of simulated configurations with the experimental ones, registered with atomic force microscopy (AFM) and presented in Ref. 10.

3. Molecular Dynamics

Molecular dynamics is a computer simulation method for studying physical movements of atoms and molecules. The particles are allowed to interact for a fixed period giving a view of the dynamical evolution of the system. Trajectories of the atoms are determined by numerical solving of Newton's equations of motion for a system of interacting particles. The forces between the particles and their potential energies are calculated using interatomic potentials. We used molecular dynamics simulator LAMMPS [11] for calculations and atomistic simulation data visualizer OVITO [12] to render the system configuration at a certain time step.

We simulated diffusion of reduced silver atoms along the surface of the glass and their clustering into nanoislands by depositing a fixed flux of N number of neutral atoms on an L x W immobile substrate surface. This is valid for early stages of nanoislands growth, when the majority of reduced silver atoms comes to free surface of the glass, not to already formed nanoislands. All atomic interactions were approximated by a standard 12/6 Lennard-Jones (LJ) potential but its parameters (bond energy, ε, and zero-crossing distance, σ) were determined considering different strength of interaction silver with silver / fused quartz (the poled region of the glass [14,15]) / non-poled glass. We established LJ potential parameters of the substrate atoms interaction as reference value (1, 1) to reduce the number of unknown parameters of the system, and performed all calculations in LJ units system for convenience. Thus, one can restore real bond energy and zero-crossing distance from the scaled values. A Langevin thermostat was applied to the system for temperature control, generic system trajectory was consistent with the microcanonical ensemble. If an atom approached the substrate borderline, it was removed from further computations.

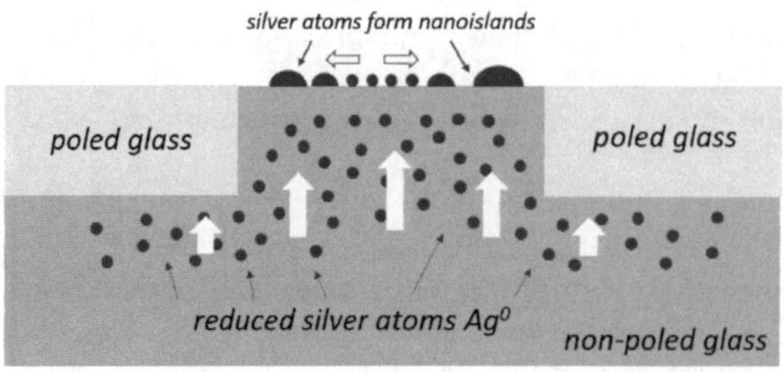

Figure 1. Initial configuration of the modeled system "Ag atoms – poled glass – non-poled glass".

For the simulation of the thermal poling, we limited the area of silver atoms deposition to the L_1 x L_2 non-poled region in the center of the poled substrate (see Figure 1). The flux density corresponded to analytically calculated one appeared after the thermal poling with the profiled electrode [10]. We varied (L_1, L_2) pair to obtain different possible configurations of the isolated nanostructures, such as a nanoisland, a pair, a set of three nanoislands and a "plasmonic molecule" [13].

It should be noted that a typical diameter of an isolated nanoisland is ~100 nm, which corresponds to ~50 million of atoms in case of, e.g. hemispherical nanoisland. Thus, we had to make scaling to reduce computation time and simulate nanoislands consisted of ~1000 atoms only instead of a real number of the atoms.

4. Results and discussion

4.1. Nanoisland structures

The parameters of Ag – Ag and Ag – non-poled glass interactions were established to provide formation of randomly distributed nanoisland film under uniform flux of silver atoms. The image of simulated film is presented in Figure 2 together with the scanning electron microscopy (SEM) image of a typical nanoisland film grown by the out-diffusion technique.

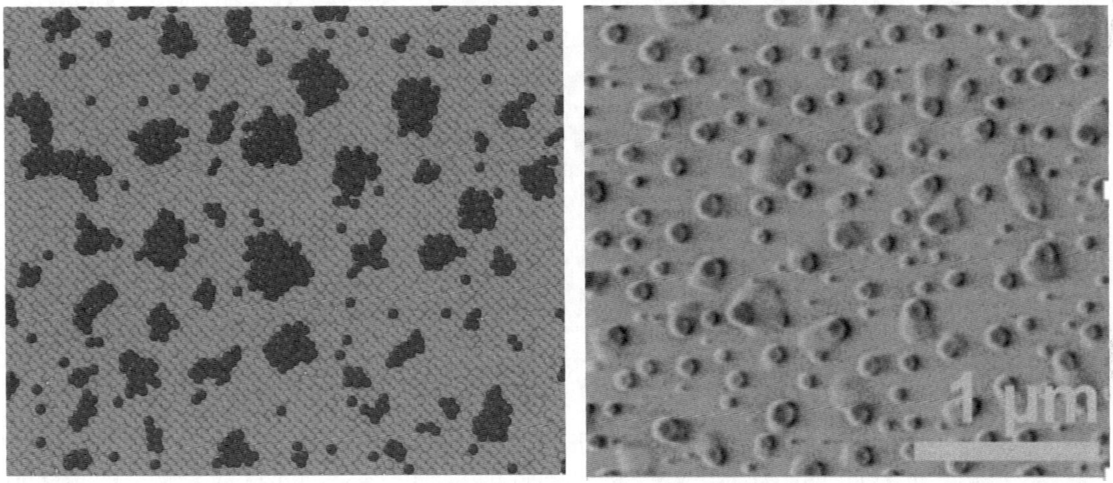

Figure 2. A spatial configuration of a simulated film (left) and SEM image (right) of a nanoisland film grown by out-diffusion technique.

We started with spatially modulated silver flux density according to the analytical model to simulate the formation of isolated nanostructures in course of thermal poling treatment of ion-exchanged glass. However, simple introduction of concentration gradient in the simulation did not result in the appearance of individual nanostructures, whereas the addition of slight difference in surface tension at the interfaces silver – non-poled and silver – poled glass resulted in matching calculated nanoisland configurations with ones experimentally observed. Difference in LJ potential parameters of Ag – poled glass and Ag – non-poled glass interactions is illustrated in Table 1.

Table 1. Lennard-Jones potential parameters for each pair of interacting "atoms" (in LJ units).

	Ag – Ag	Ag – non-poled glass	Ag – poled glass	glass – glass
ε	2.25	2.1	2.0	1
σ	0.95	0.945	0.9	1

The selection of parameters was based on the best statistics of qualitative coincidence of simulated nanostructures and experimental nanoisland configurations. Deviations of ε about 0.1 and σ about 0.01

IOP Conf. Series: Journal of Physics: Conf. Series **929** (2017) 012056 doi:10.1088/1742-6596/929/1/012056

resulted in essential disagreement of these configurations. Unfortunately, direct comparison of the values from Table 1 with literary data is hardly possible because of the shortage of information concerning Ag – glass and Ag – poled glass (fused quartz) interaction potentials. Nevertheless slightly stronger interaction of silver with multicomponent glass than with silica follows from earlier performed diffusion and wetting experiments, and this corresponds to the established relation of LJ parameters.

The set of parameters allowed simulating plasmonic structures, such as an isolated nanoisland, a pair and a set of three nanoislands. Spatial configurations of the nanostructures strongly depend on the aspect ratio of width to height of the hollow in the profiled electrode, and the simulations were in coincidence with the experimental results for the same aspect ratios, as it is shown in Figure 3.

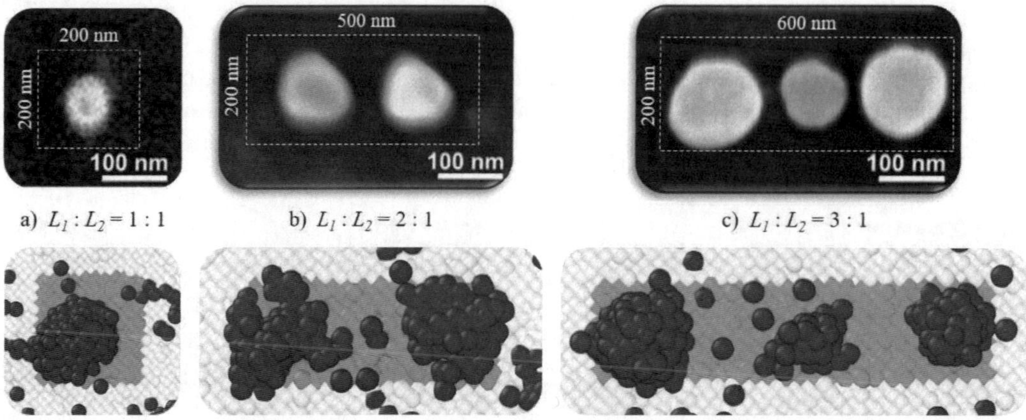

Figure 3. Experimental (top row) and simulated (bottom row) spatial configurations of small group of nanoislands: (a) an isolated nanoisland, (b) a pair of nanoislands and (c) a set of three nanoislands. The dashed outlines correspond to the non-poled regions of the glass.

When the size of the square hollow in the electrode was slightly less than one micron, the nanoparticles tended to grow along the perimeter of the non-poled region on the glass [13]. The same behaviour was observed in the simulations (Figure 4). The atoms that diffused to the glass surface at the very end of the annealing rapidly lost kinetic energy, became low-mobile, and were not able to reach nanoislands placed along the perimeter. These atoms formed smaller clusters in the center of the non-poled glass region.

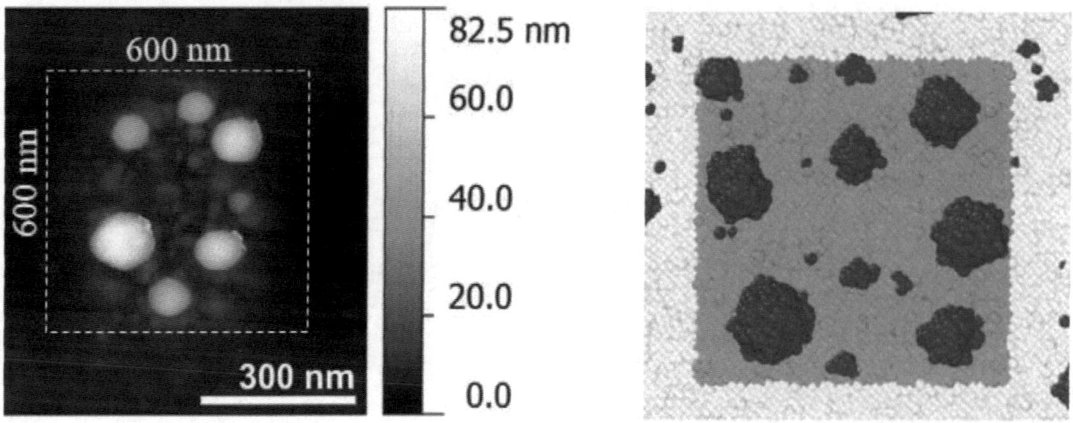

Figure 4. The AFM image (left) and simulated spatial configuration (right) of “plasmonic molecule”.

Statistically 60% of simulations with specified LJ potential parameters but varied initial spatial distribution of silver atoms resulted in the formation of certain configuration (an isolated nanoisland, two and more nanoislands etc.). A similar statistic is observed in the experimental data.

4.2. Dynamical evolution of the system

Molecular dynamics simulation gives one an opportunity to investigate time evolution of the nanoislands formation and growth. During the annealing of an ion-exchanged glass atoms diffuse according to the concentration gradient, hence they move towards the boundaries of the non-poled region of the glass, where clustering into nanoislands occurs (Figure 5a, 5b). While annealed glass is being cooled at the end of the process (the income silver atoms flux is null), individual atoms join the closest nanoparticle along the perimeter of the non-poled region, thus final configuration includes mainly atom clusters and is almost free of single atoms (see Figure 5c). Simulation of the cooling process corresponds to a sample cooling at room temperature after it has been pulled out from the annealing setup. Typically, the annealing treatment lasts 10 minutes and the cooling time is about 5-8 minutes. The optimal simulation cooling duration is about 50% of the annealing time. The slight difference may be explained by the difference in silver atoms fluxes: the flux is immediately reset to zero after the simulation cooling begin, while the flux slowly fades in course of the real cooling.

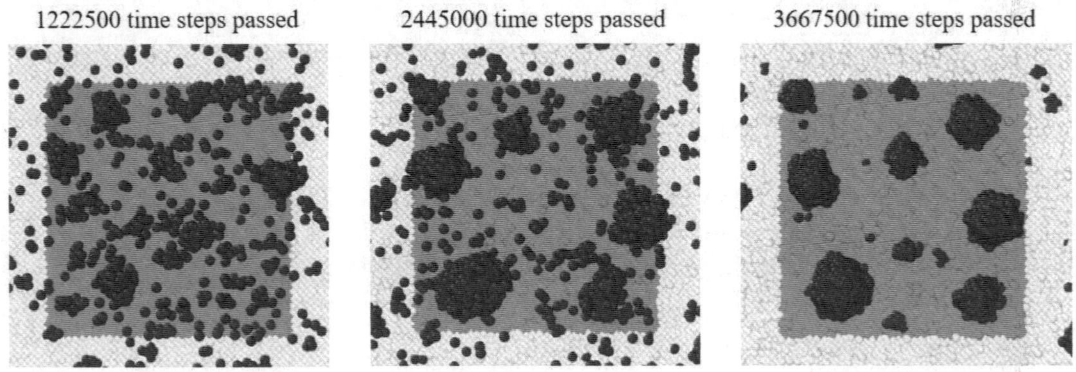

(a) at the middle of the annealing (b) at the end of the annealing (c) after cooling

Figure 5. Dynamical evolution of the simulated system "Ag atoms – poled glass – non-poled glass".

It is worth to note that in the simulation we used number of atoms about five orders of magnitude less than the real quantity of atoms in a nanoparticle. Thus modeled structures, being similar with experimental ones differs with the latter in both temporal and spatial scale.

5. Conclusion

The processes of the formation of silver nanoisland film and their groups under the annealing of an ion-exchanged glass in hydrogen atmosphere were studied and simulated by means of molecular dynamics. Spatial configurations of the nanoisland films grown under a given set of experimental conditions were analyzed, and corresponding parameters of the silver-glass interaction potential, which provide qualitative reproducing of experimentally observed configurations of the nanostructures, were revealed. Temporal evolution of the silver distribution on the surface of the glass during annealing was studied. We studied and simulated specific nanostructures grown on the glass surface after thermal poling with the profiled electrode: the isolated nanoisland, the pair, the set of three nanoislands and the "plasmonic molecule". The simulation showed good qualitative agreement with experimental data.

Acknowledgements

Authors are thankful to their scientific supervisor, Prof. Andrey A. Lipovskii. The work was supported by the Ministry for Education and Science of Russian Federation project #3.2869.2017, AVR also appreciates the Council for grants of the President of the Russian Federation (grant #SP-3509.2016.3).

References

[1] Mie G 1908 *Ann. Physik* IV 25 377-445
[2] Dieringer J A, McFarland A D, Shah N C, Stuart D A, Whitney A V, Yonzon C R, Young M A, Zhang X, Van Duyne R P 2006 *Faraday Discuss.* 132 9-26
[3] Tan W, Shi Z Y, Smith S, Birnbaum D, Kopelman R 1992 *Science* **258** 778–781
[4] Atwater H A, Polman A 2010 *Nature Materials* **9** 205-213
[5] Prieto G, Zečević J, Friedrich H, de Jong K P, de Jongh P E 2013 *Nature Materials* **12** 34–39
[6] Zhurikhina V V, Brunkov P N, Melehin V G, Kaplas T, Svirko Y, Rutckaia V V, Lipovskii A A 2012 *Nanoscale Res Lett.* **7** 676–680
[7] Reduto I V, Chervinskii S D, Kamenskii A N, Karpov D V, Lipovskii A A 2016 Technical Physics Letters 42 93-95
[8] Heisler F, Babich E, Scherbak S, Chervinskii S, Mehedi M, Samusev A, Lipovskii A A 2015 *J. Phys. Chem.* C **119** 26692–26697
[9] Chervinskii S, Reduto I, Kamenskii A, Mukhin I, Lipovskii A A 2016 Farad. Discuss. 186 107–121
[10] Sokolov K, Melehin V, Petrov M, Zhurikhina V, Lipovskii A A 2012 *J. Appl. Phys.* **111** 104307
[11] Plimpton S. 1995 J Comp Phys 117 1-19
[12] Stukowski A. 2010 Modelling Simul. Mater. Sci. Eng. 18 015012
[13] Redkov A, Chervinskii S, Baklanov A, Reduto I, Zhurikhina V, Lipovskii, A 2014 *Nanoscale Res. Lett.* **9** 606-611
[14] Dussauze M, Rodriguez V, Lipovskii A, Petrov M, Smith C, Richardson K, Cardinal T, Fargin E, Kamitsos E I 2010 *J. Phys. Chem.* C **114** 12754–12759
[15] Redkov A V, Melehin V G, Lipovskii A A 2015 J. Phys. Chem. C 119 17298-17307

IOP Conf. Series: Journal of Physics: Conf. Series **929** (2017) 012057 doi:10.1088/1742-6596/929/1/012057

The technique of visualization and evaluation of the emission site distribution for large area field emitters

S V Filippov[1], E O Popov[1,2], A G Kolosko[1,3] and A.V. Shchegolkov[4]

[1]A.F. Ioffe Physico-Technical Institute of Russian Academy of Sciences, ul. Polytechnitscheskaya 26, St.-Petersburg, 194021, Russia
[2]St. Petersburg State Polytechnical University, ul. Polytechnitscheskaya 29, St.-Petersburg, 195251, Russia
[3]The Bonch-Bruevich Saint-Petersburg State University of Telecommunications, pr. Bolshevikov 22, St.-Petersburg, 193232, Russia
[4] Tambov State Technical University, ul. Sovetskaya 106, Tambov, 392000, Russia

s.filippov@mail.ioffe.ru

Abstract. This paper presents a method for online treatment of the distribution pattern of emission sites on the emitter surface, combined with a multichannel registration and processing system of the IVC. Using computerized field emission projector we have studied nanocomposite emitter based on carbon nanotubes "Taunit M" and polystyrene. The distribution pattern of emission sites over the brightness level was obtained as well as the emission area of the sample was determined. Time dependences of the current level, the total brightness and the emission area were registered.

1. Introduction

The development of methods for creating atomic-size directional structures such as carbon nanotubes, graphene sheets and other nanoscaled materials - which have unique physical properties - has led to the production of these materials in industrial scales. Nanocomposite based on abovementioned structures are widely and intensively studied as multi-tip field emitters, which have a potential to be applicated for a wide range of electronic devices (compact X-ray machines, display facilities, mass spectrometers [1-3]).

The uniformity of the emitting surface is a necessary condition for nanocomposite field emitters to operate efficiently. This uniformity can be improved by the technological optimization using widely employed control methods of emissivity such as the evaluation of current-voltage characteristics and registration of the distribution patterns of emission sites [4-6].

At the present time there are several methods for evaluating the surface uniformity of the field cathodes. Using field emission scanning electron microscopy (FE-SEM), high-resolution transmission electron microscope (HRTEM) and scanning electron microscopy (SEM) it is possible to directly observe the surface morphology of the emitter on a nanometer scale [7, 8].

The information about the uniformity of the emission properties is obtained using field emission microscope with a scanning anode (SAFEM) [9], which allows the mapping its emission properties.

However, during the scanning poorly fixed emission sites - for example strongly protruding nanotubes - might follow the scanning electrode that results in an inaccurate pattern of their distribution.

Another method for determining emission uniformity is exposure to the emitter on the resist layer sensitive to electron beam [10]. In this case, however, it is not possible to determine values of the threshold fields of individual regions of the emitter surface. Perhaps the most common method for determining the emission sites density on the field emitter surface is a method of comparision of luminous sites images on a phosphor screen of the field emission projector (FEP). [9, 11, 12]

In some studies, the number of emission sites is directly counted [13-15]. Usually this calculation is performed for nanocomposite structures, which are made in the form of regular arrays of isolated dots. In the case of a thin film field emitter, there are thousands of emission sites and their location on the surface is not ordered, thus the process of emission site counting is highly laborious. Additionally, at a regime of high voltages the entire projection screen glows evenly, thus there are some problems with the resolution of the neighboring sites. Online treatment of images could help solving the abovementioned problems. In the present work we described method of recording and analysis of spatial distribution of emission sites, coupled with the method of registration and online IVC processing.

The nanocomposite MWCNTs / polystyrene was used as a sample for studying field emission properties. Such materials are very promising for the development of the emission electronics [16].

2. Experimental

The nanocomposite was fabricated from a suspension of carbon nanotubes and polystyrene. The suspension was prepared in three steps: 1) polystyrene dissolution in o-xylene; 2) creating a suspension of nanotubes in o-xylene with ultrasonic treatment; 3) mixing of the obtained materials and ultrasonic treatment during 8 hours to increase the uniformity of nanotubes distribution in volume. Thereafter, the suspension was deposited on a polished stainless steel substrate (diameter 10 mm) using a spin coating technique. The concentration of nanotubes in the polymer matrix of the nanocomposite was 10 wt. %.

MWCNT Taunit-M ("NanoTechCentr", TSTU Tambov, Russia) were used used as a filling material. They had the following parameters: length is more than 2 μm, an outer diameter is 8 to 15 nm. The production of carbon nanotubes Taunit-M is described in [17] in details. SEM image of MWCNT powder is shown in figure 1.

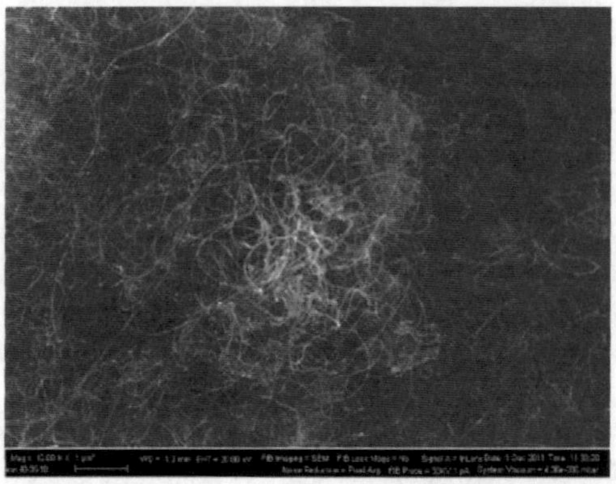

Figure 1. SEM image of MWCNT Taunit-M.

IVC measurements were carried out in a plane-parallel construction, the anode was ITO coated with a phosphor layer as shown in figure 2. The residual pressure in the chamber was ~$5 \cdot 10^{-7}$ torr. Recording and processing of experimental data was carried out by a special program written in LabView 2013 environment using NI-IMAQ for USB Cameras driver software and Vision Development Module 2015.

This experimental setup was assembled in the Ioffe Institute for multi-tip emitter research. Detailed description of the implemented methods for studying the properties of the nanomaterial's field characteristics is described in [18].

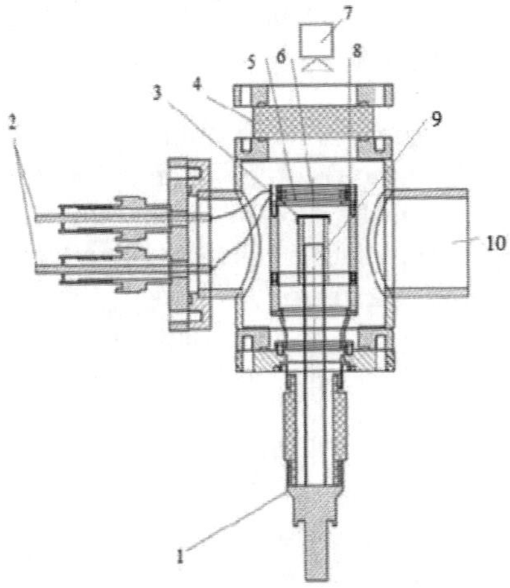

Figure 2. Experimental chamber:
1 - high-voltage input,
2 - inputs of vacuum measurement channels,
3 - emitter (MWCNT / PS),
4 - thick vacuum glass,
5 - ITO glass with a phosphor layer,
6 - chuck for ITO setting,
7 - telephoto USB-microscope (DP-M12 USB eScope),
8 - anode system,
9 - a metal cathode holder,
10 - vacuum pipe to turbomolecular pump.

The capabilities of the system includes: obtaining IVC in the mode of fast scanning half-sine voltage, plotting IVC in the Fowler-Nordheim coordinates, the calculation of the effective parameters of the emitter, such as field enhancement factor β_{eff} and emission area S_{eff}. Further development of a complex research method has led to appearing new features: the registration of distribution pattern of emission sites (DPES), which is displaying on the luminescent projector screen, the calculation of the total brightness of the picture Y and the estimation of its emission area S_Y.

3. Results

The results of the conducted experiments are considered in this section. Figure 3 shows the time dependence of the voltage level applied to the sample and the corresponding emission current in the sample stable operation with I_{max} ~ 100 μA. Registered DPES show that the emission site distribution is constantly changing over time.

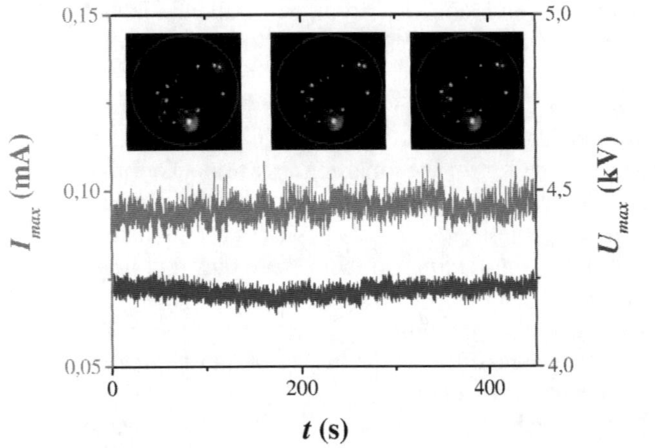

Figure 3. Time dependence of the emission current and voltage for stable current level 100 μA. In the inserts: DPES registered with intervals of 200 s.

The maximum current – which was possible to be received using this measuring system was about 500 μA. Increase of the current led to vacuum discharges, which could damage the luminophore. It is to be noted that this instability is due to insufficient ITO conductance and the quality of the phosphor film, because similar system with a metallic anode and same emitter gave current level that exceed 20 mA. Obviously, the electron emitting streams heat the phosphor coating and lead to its destruction and deterioration of local vacuum in the interelectrode gap that increases the discharge probability.

The emitter operation at three current levels has been studied: 100, 250 and 400 μA.

We carried out determination of the emission area S_Y using online treatment of the DPES in automatic mode with a given level of brightness Y_0 (all pixels of the recorded picture – which brighter than this level - are considered as a part of the emitting surface). For the brightness distribution analysis, the field emission image was converted into gray scales (pixel brightness from 0 to 255). The shape of the emitting surface was in the form of black-and-white pictures.

The setup of a brightness level Y_0 was executed from the analysis of dependence of the number of emission sites N (separate regions of the emitting surface) from the brightness level of their peaks Y_p for an instant DPES at emission current of 400 μA (see fig. 4a).

During the scanning of value Y_p a number of emission sites continuously grew up, but starting from a certain level of brightness (≈ 185), there was a sharp drop in the value of N, due to merger of the most lackluster centers into a single luminous spots. This value we have taken as a threshold for determining the emission area and for further study of its evolution (Y_0 = 185).

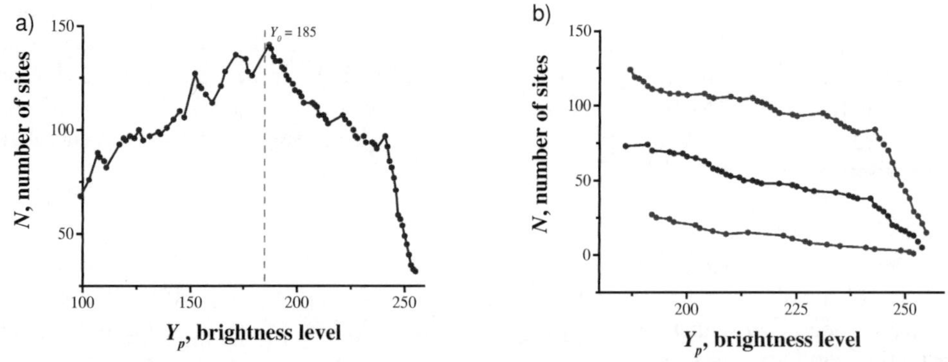

Figure 4. a) Dependence of the number of emission sites N from the brightness level of its peaks Y_p; b) Dependence of the number of emission sites from the brightness level for the instant DPES at three different current levels (400, 250 and 100 μA).

In order to remove emission sites with low brightness from the consideration and avoid impact of the exposure effect on the calculations, a mask was superimposed on the processed image: a brightness of all areas of the surface expelled from the emission area was taken to be equal 0. The formation of the mask was made taking account in the fluctuating nature of the site distribution pattern: surface area - which corresponds to the brightness level $Y > Y_0$ has been accumulated into a single "landscape" by collecting all registered luminescent regions on the picture during 5 minutes with the current level of 400 μA. The result is shown in figure 5.

The luminescent area observed on the projector screen was about 8 % of the total emitter area. The resulting value is the numerical assessment of the degree of emission sites uniformity on the surface of the emitter, as well as degree of its smoothness.

Figure 6a represents measured level of emission current time dependence at three voltage levels. The insets show examples of the recorded IVCs. The achieved stability of the emission current on each step was determined according to the formula $(I_{max} - I_{min}) / I_{mean} \cdot 100\%$ and amounted to: 5,5 % - to 100 μA, 7,6 % - to 250 μA and 12,4 % - for 400 μA. It can be seen that with increasing

IOP Conf. Series: Journal of Physics: Conf. Series **929** (2017) 012057 doi:10.1088/1742-6596/929/1/012057

current level sample stability is falling. It should be related to the inclusion in the emission process new sites and corresponding increase in the adsorption stochastic fluctuations.

A change in the brightness level of field emission image and a change in the value of emission area (fig. 6b) – which is determined by the predetermined level Y_0 - correlate with changes in the fig. 6a. It confirms that it is possible to use the brightness of the individual regions of the cathode in order to determine their own current loading. Figure 4b shows results of scanning of instant DPES using brightness level Y_p (similar to figure 4a) for each of the steps in the figure 6.

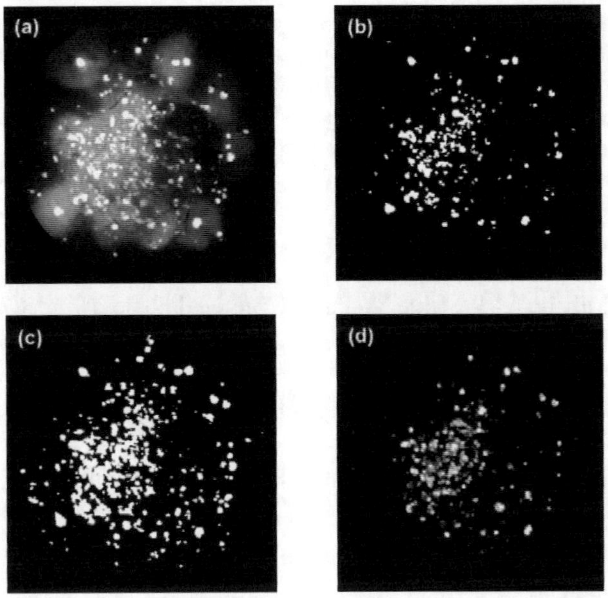

Figure 5. DPES on the cathode surface: a) instant DPES before the treatment, b) black-and-white picture, where the white spots correspond to the brightness above the predetermined $Y > Y_0$, c) DPES obtained by accumulating the instant picture during 5 minutes with the current level of 400 μA, d) instant DPES received after applying the mask.

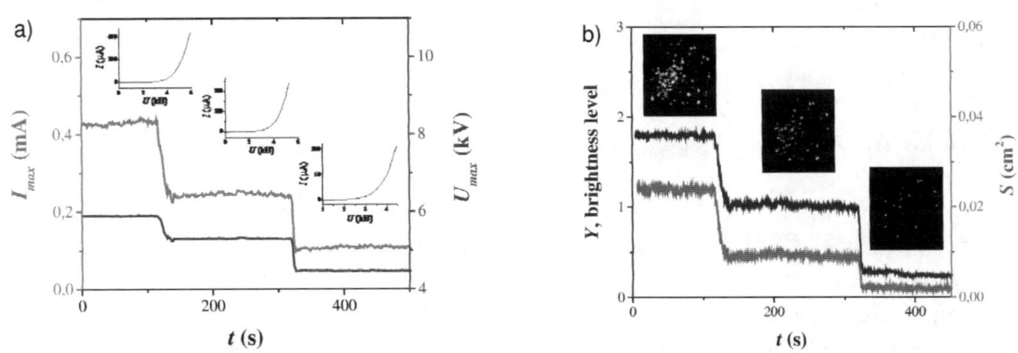

Figure 6. a) Time dependence of the voltage level and the corresponding level of the emission current (steps: 400 uA, 250 μA, 100 μA). The inset shows the instant IVCs for each step; b) Time dependences of the brightness level of image (after applying the mask) and the emission area of this pattern determined for a given brightness level Y_0 = 185. The insets show the instant field emission images, which were used for the calculation of the emission area.

The obtained graphs show that at the emission current level of 250 and of 400 μA distribution of emission sites brightness consists of two sections, which are close to linear form (before and after Y_0 = 240). Possibly that some districts of emitter surface related to high brightness regions have higher temperatures and have no influence of adsorbates, whereas regions with low brightness are influenced by adsorbates on its work function and are responsible for the observed fluctuations of DPES during field emission (see figure 3).

4. Conclusion

The above described technique of multi-channel recording and online data processing of field emission of multi-tip nanocomposite cathodes is a powerful tool for the experimental analysis of the emitting cathode structure and for the defining its relationship with the macroscopic parameters such as total emission current and applied voltage. It allows obtaining the emission sites distribution on the cathode surface for emission brightness levels (which is actually the current loading levels), to estimate the total emission area over luminescent districts of the cathode.

Such numerical analysis of DPES obtained using a field emission projector, still was not carried out. The proposed approach gives great possibilities for further research of large area field emitters.

References

[1] Parmee R J, Collins C M, Milne W I and Cole M T 2015 *Nano Converg.* **2** 1
[2] Kim Y C, Park S H, Lee C S, Chung T W, Cho E, Chung D S and Han I T 2015 *Carbon N. Y.* **91** 304–10
[3] Velaásquez-Garcia L F, Gassend B L P and Akinwande A I 2010 *J. Microelectromechanical Syst.* **19** 484–93
[4] Shiratori Y, Furuichi K, Noda S, Sugime H, Tsuji Y, Zhang Z, Maruyama S and Yamaguchi Y 2008 *Jpn. J. Appl. Phys.* **47** 4780–7
[5] Wang Y Y, Gupta S, Liang M and Nemanich R J 2005 *J. Appl. Phys.* **97**
[6] Xiao X, Ye Y, Zheng L and Guo T 2012 *J. Semicond.* **33** 53004-1-053004–6
[7] Chiu C-C, Tsai T-Y and Tai N-H 2006 *Nanotechnology* **17** 2840–4
[8] Lin P-H, Sie C-L, Chen C-A, Chang H-C, Shih Y-T, Chang H-Y, Su W-J and Lee K-Y 2015 F *Nanoscale Res. Lett.* **10** 297
[9] Nilsson L, Groening O, Groening P, Kuettel O and Schlapbach L 2001 *J. Appl. Phys.* **90** 768–80
[10] Lee J H, Lee S H, Kim W S, Lee H J, Heo J N, Jeong T W, Choi C H, Kim J M, Park J H, Ha J S, Moon J W, Yoo M a., Nam J W, Cho S H, Yoon T I, Kim B S and Choe D H 2005 *J. Vac. Sci. Technol. B Microelectron. Nanom. Struct.* **23** 718
[11] Bocharov G S 2004 *AIP Conf. Proc.* **723** 528–31
[12] Pandey A, Prasad A, Moscatello J P, Engelhard M, Wang C and Yap Y K 2013 *ACS Nano* **7** 117–25
[13] Lysenkov D and Muller G 2005 *Int. J. Nanotechnol.* **2** 239–54
[14] Liu W, Zeng F, Xin L, Zhu C and He Y 2008 *J. Vac. Sci. Technol. B Microelectron. Nanom. Struct.* **26** 32
[15] Jeon J, Goak J, Kim K B and Lee N 2009 *J. of Korean Phys. Soc.* **54** 185–9
[16] Eda G, Emrah Unalan H, Rupesinghe N, Amaratunga G A J and Chhowalla M 2008 *Appl. Phys. Lett.* **93** 2–4
[17] Besperstova G S, Burakova E A, Dyachkova T P and Schegolkov A V 2015 *Fundam. Res. Tech. Sci.* **9** 414–7
[18] Popov E O, Kolos'ko A G, Filippov S V, Romanov P A and Fedichkin I L 2016 *Nanomaterials and Nanostructures - XXI Century* **1** 14-26.

IOP Conf. Series: Journal of Physics: Conf. Series **929** (2017) 012058 doi:10.1088/1742-6596/929/1/012058

Investigation of porous silicon nanopowders functionalized by antibiotic Kanamycin, fluorophore Indocyanine Green

K Bespalova, P A Somov, Yu M Spivak

St. Petersburg Electrotechnical University "LETI", 5 Prof. Popova str., St. Petersburg 197376, Russia

E-mail: kris.alex.bespalova@gmail.com

Abstract. Porous silicon nanopowders for target drug delivery were obtained by electrochemical anodic etching in a hydrofluoric acid solution using the monocrystalline silicon n-type conductivity. Porous silicon powders were obtained by sonification of porous silicon layers. The powders were functionalized by antibiotic Kanamycin and fluorophore Indocyanine Green by the passive adsorption method. The peculiarities of absorption spectra in 190-600 nm region were revealed for functionalized porous silicon powders dispersions in water.

1. Introduction

Currently in modern Material Science is actively grown the field of creating porous materials with controlled parameters (pore diameter, the porosity, structure and thickness of the porous layer, the phase structure on the inner surface of the pores). Porous silicon (por-Si) and por-Si powders are one of the most promising advanced materials in bioengineering (targeted drug delivery, implants, photodynamic therapy, biosensors, etc.)[1-3, 7-9].

A separate area of application of porous silicon in biomedicine currently is the use of porous silicon particles as carrier matrices for drug delivery. Creation of methods of drug delivery is one of the leading trends in modern pharmacology [4-6].

Currently, there are a lot of methods to fix the drug in a porous carrier: method of passive adsorption, electrostatic adsorption, chemical (covalent bonding), and etc. Passive adsorption method in the solution is most commonly used when working with porous silicon based carrier matrix. The method consists in immersing por-Si particles in a predetermined concentration of the drug solution. After the particle pore volume is filled with molecules of drug substance, functionalized particles can be prepared by centrifugation or filtration. The amount of drug substance to be filled is controlled by the concentration of the solution, the time of adsorption, as well as the porosity and pore size of the particles [10].

2. Materials and methods

Por-Si nanopowders were obtained in two steps. In the first step porous silicon nanolayers were obtained by electrochemical anodic etching in a hydrofluoric acid solution from the monocrystalline silicon n-Si(111) with 4.5 Ω·cm resistivity. The choice of process conditions (anodization current density of j = 80 mA / cm2, anodization time of t = 20 min) was based on the experience of previous studies [1]. In the second step por-Si nanolayers were removing by the ultrasonication process. After centrifugation the particles characterized of narrow range of size distributions. Based on SEM data, the particles characterized by the diameter of 300-500 nm. Basically, the particles are rectangular. Figure 1 shows typical size of obtained particles.

For functionalization of such prepared por-Si nanopowders the passive adsorption method was chosen. The nanopowders were immersed into saturated aqueous solution of Kanamycin (or of Indocyanine Green (ICG)) for 24 hours. Absorption spectra were acquired on a PE-5400UV UV–vis spectrophotometer (LLC “Ekohim”).

Figure 1. Morphology of por-Si particles fixed on Si substrate taken by SEM

3.Experiment and results

To reveal the nature of the interaction between the functionalizing substances and particles of porous silicon (control functionalization of porous particles) spectrophotometry method was used. In order to show the result of functionalization, additional absorption spectra were taken: spectrum of the aqueous dispersion por-Si before functionalization, and also spectra of Kanamycin aqueous solution, IGG aqueous solution.

Figure 2 shows absorption spectra of Kanamycin, por-Si aqueous dispersion and aqueous dispersion of functionalized por-Si by Kanamycin. As can be seen from absorption spectra Kanamycin has absorption at 207 nm and 236 nm. Functionalized por-Si by Kanamycin also has absorption in the same ultraviolet region (194 nm and 244 nm) as Kanamycin. So the absorption bands are shifted relative to the solution of Kanamycin, which may indicate the interaction between the porous silicon and Kanamycin molecules.

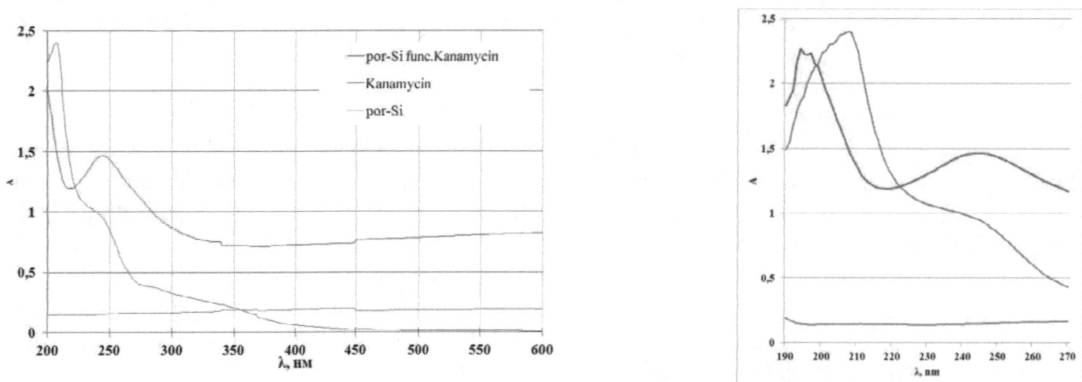

a) b)

Figure 2. Absorption spectra of Kanamycin, por-Si water dispersion and water dispersion of por-Si functionalized by Kanamycin

IOP Conf. Series: Journal of Physics: Conf. Series **929** (2017) 012058 doi:10.1088/1742-6596/929/1/012058

Figure 3 shows absorption spectra of ICG, por-Si aqueous dispersion and aqueous dispersion of functionalized por-Si by ICG. ICG has a number of absorbance peaks: in the near-infrared region with a maximum absorption at 776 nm, at 710 nm and in the ultraviolet region at 213 nm, which corresponds to the typical absorption spectrum of Indocyanine Green. Spectrum of functionalized por-Si by Indocyanine Green has no peaks of absorbance. The main contribution to the signal strength makes the concentration of particles in the solution. That may indicate that the IGG molecules were not fixed on the surface of the porous silicon particles.

Such different nature of the interaction of molecules of Kanamycin and ICG with porous silicon is obviously related to the different composition of the functional groups of analytes. Thus, it is known from previous studies [9-10] that the porous silicon obtained in selected in this work conditions of preparation and post-processing are hydrophilic and contains hydroxyl groups on the surface. Molecules Kanamycin contain amino and hydroxyl functional groups. Therefore, an interaction between a hydrophilic molecule Kanamycin and hydrophilic surface of such prepared porous silicon is energetically favorable, which, in our opinion, is confirmed by absorption spectrum. Furthermore, fastening Kanamycin molecule onto the porous silicon surface by Kanamycin hydroxyl groups is preferred for to the NH_2-group would be in the active form and could provide its therapeutic function. On the other hand, ICG molecules have many hydrophobic CH_2-functional groups, which disadvantageously interact with a hydrophilic porous silicon surface. Therefore, fixing the molecules on the particles of the porous silicon were not observed.

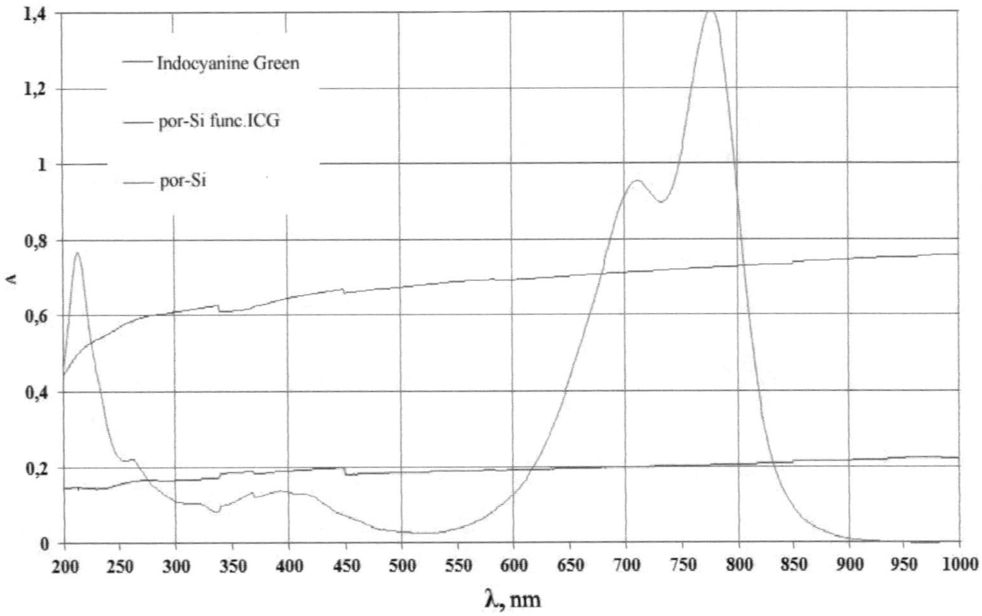

Figure 3. Absorption spectra of ICG, por-Si water dispersion and water dispersion of por-Si functionalized by ICG

4.Conclusion

Thus, Porous Silicon nanopowders were obtained by electrochemical etching. The analysis of the obtained data by scanning electron microscopy showed that the average sizes of obtained particles are 300-500 nm. The porous silicon obtained in selected in this work conditions of preparation and post-processing are hydrophilic and contains hydroxyl groups on the surface. As Kanamycin molecules contain amino and hydroxyl functional groups, interaction with por-Si surface is energetically favorable. ICG molecules have many hydrophobic CH_2-functional groups, which disadvantageously

interact with a hydrophilic porous silicon surface. Therefore, fixing the molecules on the particles of the porous silicon was not observed, which is confirmed, in our opinion, by spectrophotometry data.
According to the results, you can make the following recommendations for the choice of the conditions for obtaining porous matrices carrier on the basis of porous silicon:
- porous silicon particles in the selected conditions are suitable for functionalization of molecules of antibiotic Kanamycin;
- for the functionalization of porous silicon by fluorophore Indocyanine Green adjusting of technological conditions is in need for changing the functional groups on its surface to the hydrophobic. Another line of research may be additional post-processing of porous silicon.
Furthermore, the method of spectrophotometry in the UV range is an appropriate for rapid evaluation of functionalization process of porous silicon particles with target substances.

Acknowledgments
The reported study was this work was supported by project "Preparation and study of porous systems, functionalized nanomaterials, applications in photonics, sensor technology and medicine" (under the Russian Ministry of Education state task № 16.2112.2014/K.

References
[1] Spivak Yu M, Belorus A O, Somov P A, Tulenin S S, Bespalova K A, Moshnikov V A 2015 *Journal of Physics: Conference Series* **643** 012022
[2] Boukherroub R, Wojtyk J T C, Wayner D D M, Lockwood D J 2002 *J. Electrochem. Soc.* **149(2)** H59–H63
[3] Belorus A O, Bespalova K, Spivak Yu M *Conference: 2016 IEEE NW Russia Young Researchers in Electrical and Electronic Engineering Conference (EIConRusNW)* 1-6
[4] Salonena J, Laitinenb L, Kaukonenb A M, Tuuraa J, Björkqvista M, Heikkiläa T, Vähä-Heikkiläa K, Hirvonenb J, Lehtoa V.-P 2005 *J. Control. Release* **108 (2-3)** 362–374
[5] Canham L T 1995 *Advanced Materials* **(7)** 1033-1037
[6] Salonen J, Kaukonen A M, Hirvonen J 2008 *Journal of Pharmaceutical Sciences* **97 (2)** 632–653
[7] Len'shin A S, Kashkarov V M, Spivak Y M, Moshnikov V A 2012 *Glass Physics and Chemistry* 315-321
[8] Anglin E J, Cheng L Y, Freeman W R, Sailor M J 2008 *Adv. Drug Deliv. Rev.* **60 (11)** 1266–1277
[9] Lenshin A S, Kashkarov V M, Seredin P V, Spivak Y M, Moshnikov V A 2011 *Semiconductors* 1183-1188
[10] Spivak Yu M, Mjakin S V, Moshnikov V A, Panov M F, Belorus A O, Bobkov A A 2016 *Journal of Nanomaterials* **(1)** 1-8

IOP Conf. Series: Journal of Physics: Conf. Series **929** (2017) 012059 doi:10.1088/1742-6596/929/1/012059

Screening effect in matrix graphene / SiC planar field emmiters

I L Jityaev[1], A M Svetlichnyi[1], A S Kolomiytsev[1] O A Ageev[1]
[1]Southern Federal University, Institute of Nanotechnologies, Electronics, and Electronic Equipment Engineering, Taganrog, 347922, Russia

E-mail: jityaev.igor@gmail.com

Abstract. The paper describes simulation of matrix field emission nanostructures on the basis of graphene on a semi-insulating silicon carbide. The planar spike-type field emission cathodes were measured. The electric field distribution in an interelectrode gap of the emission structure was obtained. The models take into account the distance between cathode tops. Screening effect condition was detected in planar field emission structure and a way of eliminating was proposed.

1. Introduction

The unique properties of graphene films contribute to their implementation in different areas of nanotechnology and development of electronics [1-6]. Special attention should be paid to the substrate material for the practical application of graphene in nanoelectronics. Silicon carbide is perfect for this purpose. This material has high thermal conductivity, stability in a wide range of temperatures and radiation, the possibility of graphene growth on the entire surface of the wafer by thermal decomposition of silicon carbide in a vacuum [7-13].

A promising direction in the application of graphene films is a field-emission micro and nanoelectronics. [14-17]. The disadvantage of the existing field emission cathodes is the instability of field-emission current due to local overheating, adsorption and desorption of residual gas molecules, ion bombardment of the cathode [18, 19]. Increased resistance to destructive factors is inherent to field emission cathodes based on carbon materials [20].

The stability of field emission cathodes is largely determined by the uniformity of the field strength distribution in the interelectrode gap. Several studies have shown that the emission from the cathode surface is not uniform at high field strengths. Even the "ring effect" is detected on the emission image at pre-breakdown state [21-23]. The nature of this phenomenon is still poorly understood and most likely due to the heterogeneity of the field strength distribution.

One of the factors affecting the emission uniformity is the emergence of the screening effect in matrix field emission structures [24-30]. The emission heterogeneity in matrix structures can lead to varying luminescence intensity of individual sections in the field emission display. Therefore, the identification of factors affecting the field distribution between the cathode and anode in the matrix nanoscale structures is an urgent task of field emission vacuum nanoelectronics.

The object of studying is detection of the screening effect in the matrix nanoscale field emission structures with planar spike-type cathodes on the basis of semi-insulating silicon carbide with graphene films.

2. Design

Spike, blade, film, face and matrix field emission cathodes based on these structures are used to obtain sufficient electric field to the electron emission. Spike-type of field emission cathodes have the greatest curvature of the surface. They allow obtaining high values of the electric field which are required for the start of emission. An important task is the development of energy-efficient nanoscale field emission structures with low threshold field emission [31-33].

Application of the matrix field emission cathodes allows high amperage which is proportional to the number of cathodes in the array. Consequently, it is necessary to have the maximum number of emitters per unit area. However, limiting the number of emitters is determined by the influence of screening effect.

Planar spike-type emitter is used as a single element of the matrix in our work. Previous studies of planar emitters have shown that such a structure has a high-gain and low-threshold voltage [34]. General view of a planar matrix-type field emission structure shown in Figure 1.

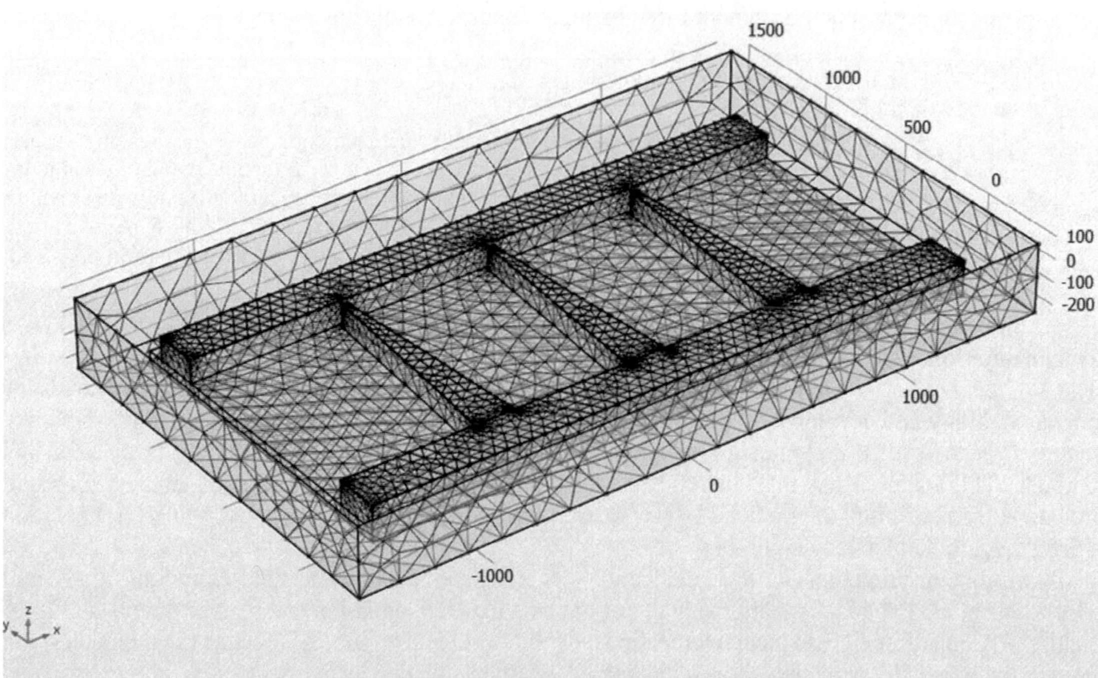

Figure 1. Model of planar matrix field emission structure based on graphene films on a semi-insulating silicon carbide. The dimensions are in nanometers.

3. Simulation and results

Simulation of structural parameters of field emission structures can reduce the labor intensity and reduce the cost of their production and studies [30, 35, 36]. Three-dimensional simulation of the distribution of the electric field strength was performed with the boundary conditions: on the border of the cathode - U_c = const; on the border of the anode - U_a = const; on the border of emission of the cell - $U_e = 0$. Simulation parameters are shown in Table 1. These values are based on experimental models and simulation of single emitters.

The simulation showed that the electric field reaches values of the order of 10^9 V/m at a voltage of 10 V. At the same time, the heterogeneity of the field distribution appears along a top of field emission cathodes (Figure 2).

Table 1. Simulation parameters.

Parameter	Symbol	Value
Distance between the field emission cathode top	L	100 - 1000 nm
Angle of lateral surface relative to the central axis	α	5°
Rounding-off radius of the cathode top	r	10 nm
Number of graphene layers	n	10
Interelectrode distance	d	15 nm
Height of the cathode	h	1 μm
Potential difference	U	10 V

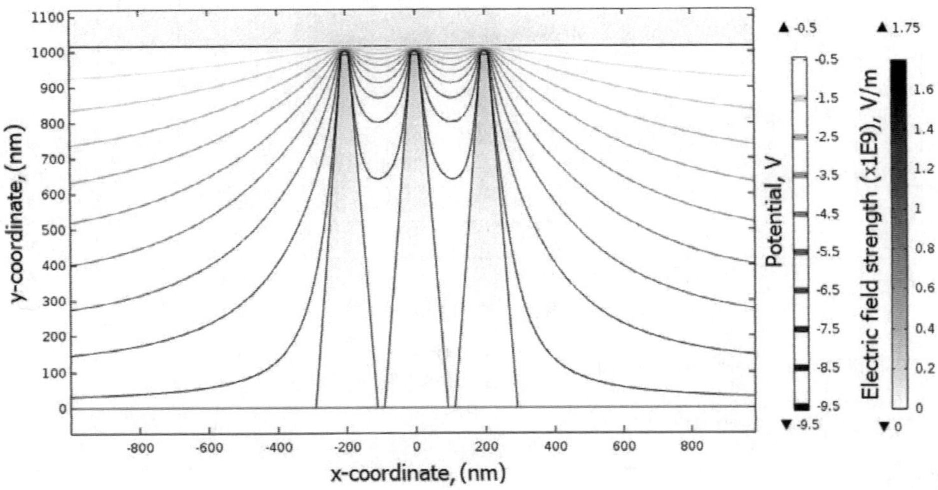

(a)

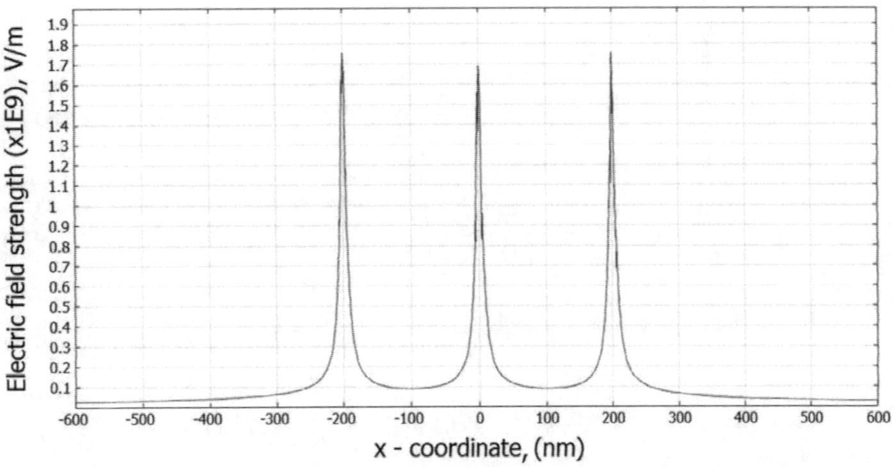

(b)

Figure 2. The screening effect in a planar matrix field emission structure: (a) distribution of the electric field strength at 200 nm distance between the centers of emitters; (b) distribution of the electric field strength along the top of emitters

It was suggested to gradually increase the distance between the centers of emitters for reduction of the screening effect. As a result, it was found that the screening effect was absent in considered structure at a distance between the centers of field emission cathodes of 700 nm or more. Figure 4 shows the distribution of the electric field along the planar field emission cathodes, where screening effect is absent.

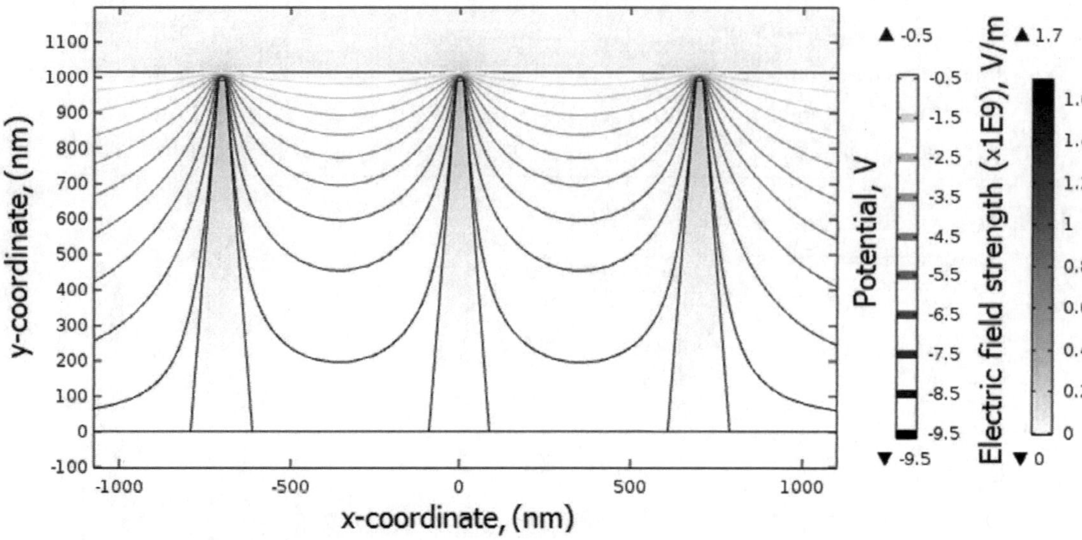

(a)

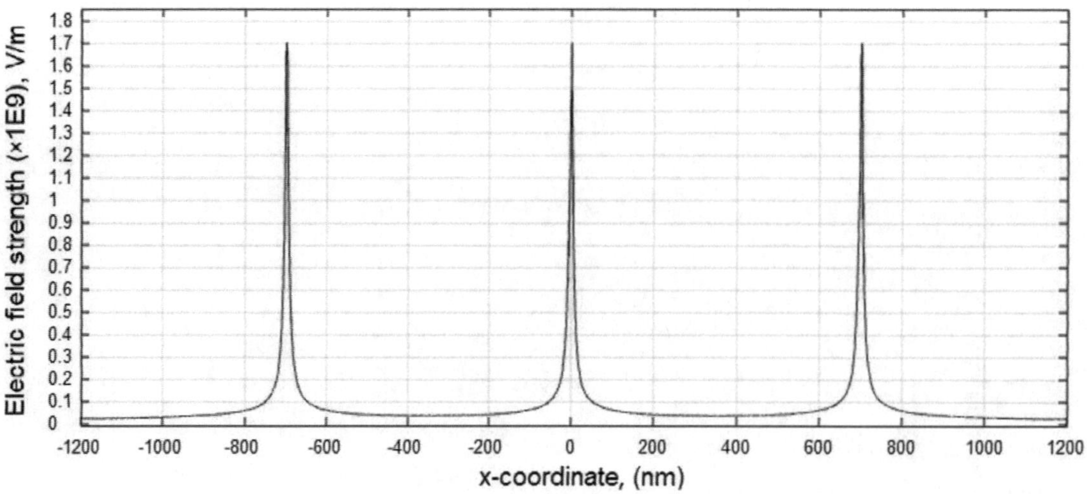

(b)

Figure 3. Absence of the screening effect in a planar matrix field emission structure: (a) distribution of the electric field strength at 700 nm distance between the centers of emitters; (b) distribution of the electric field strength along the top of emitters

4. Conclusions

Simulation allowed us to obtain the distribution of the electric field in the planar matrix field emission structure based on graphene films on a semi-insulating silicon carbide. The value of the field strength corresponds to a spike-type of field emission nanoscale structures [30, 35, 36]. The dependence of electric field strength on the X coordinate showed that the field strength in the edges of emitters exceeds the value of the central emitters at short distance between the emitters. Increasing the distance contributed to the uniformity of the field at each emitter. It was revealed that screening effect is absent when the distance between the emitters is 700 nm and more. These results correlate with the results for the vertical type matrix graphene / SiC field emission nanostructures [30]. However, more distance between the emitters is required for the planar structures. This can be explained by the fact that the lateral surface of the planar structures contributes more to the emission compared to vertical structures. Additional design changes of individual planar emitters can reduce the distance between the emitters while maintaining homogeneity of the field.

Acknowledgments

This work was supported The President of the Russian Federation Grants Council (Grant no. MK-6163.2016.8).

The equipment of the Center for Collective Use "Nanotechnology" and the Research and Education Center "Nanotechnologies" Southern Federal University was used for this study.

References

[1] Novoselov K S, Geim A K, Morozov S V, Jiang D, Katsnelson M I, Grigorieva I V, Dubonos S V and Firsov A A 2005 *Nature* **438** 197
[2] Lee C, Wei X, Kysar J W and Hone J 2008 *Science* **321** 385
[3] Eda G, Unalan H E, Rupesinghe N, Amaratunga G A J and Chhowalla M 2008 *Appl. Phys. Lett.* **93** 233502
[4] Ghosh S, Calizo I, Teweldebrhan D, Pokatilov E P, Nika D L, Balandin A A, Bao W, Miao F and Lau C N 2008 *Applied Physics Letters* **92** 151911
[5] Calderón-Colón X, Geng H, Gao B, An L, Cao G and Zhou O 2009 *Nanotechnology* **20** 325707
[6] Das Sarma S, Adam Sh, Hwang E H and Rossi E 2011 *Reviews of Modern Physics* **83** 407
[7] Lezhenin F F and Gnesin G G 1967 *Soviet Powder Metallurgy and Metal Ceramics* **6** 114
[8] Lebedev A A, Kozlovski V V, Strokan N B, Davydov D V, Ivanov A M, Strel'chuk A M and Yakimova R 2002 *Semiconductors* **36** 1270
[9] Moon J S, Curtis D, Hu M, Wong D, McGuire C, Campbell P M, Jernigan G, Tedesco J L, VanMil B, Myers-Ward R, Eddy C and Gaskill D K 2009 *IEEE Electron Device Letters* **30** 650
[10] Lin Y-M, Dimitrakopoulos C, Jenkins K A, Farmer D B, Chiu H-Y, Grill A and Avouris Ph 2010 *Science* **327** 662
[11] Kim M, Hwang J, Shields V B, Tiwari S, Spencer M G, Lee J-W 2012 *Nanoscale Research Letters* **7** 186
[12] Konakova R V, Kolomys O F, Okhrimenko O B, Strelchuk V V, Volkov E Yu, Grigoriev M N, Svetlichnyi A M and Spiridonov O B 2013 *Semiconductors* **47** 812
[13] Liu X, Chen Y, Sun Ch, Guan M, Zhang Y, Zhang F, Sun G and Zeng Y 2015 *Nanomaterials* **5** 1532
[14] Wu Ch, Li F, Zhang Y and Guo T 2013 *Vacuum* **94** 48
[15] Cai W, Zeng B, Lui J, Guo J, Li N, Chen L and Chen H 2013 *Applied Surface Science* **284** 113
[16] Konakova R V, Okhrimenko O B, Svetlichnyi A M, Ageev O A, Volkov E Yu, Kolomiytsev A S, Jityaev I L and Spiridonov O B 2015 *Semiconductors* **49** 1278
[17] Konakova R V, Okhrimenko O B, Kolomys A F, Strel'chuk V V, Svetlichnyi A M, Ageev O A, Volkov E Yu, Kolomiitsev A S, Zhityaev I L and Spiridonov O B 2016 *Journal of Superhard Materials* **38** 235

[18] Dyke W P, Charbonnier F M, Strayer R W, Floyd R L, Barbour J P and Trolan J K 1960 *Journal of Applied Physics* **31** 790
[19] Todokoro H, Saitou N and Yamamoto Sh 1982 *Japanese Journal of Applied Physics* **21** 1513
[20] Bormashov V S, Baturin A S, Nikolskiy K N, Tchesov R G and Sheshin E P 2007 *Surface and Interface Analysis* **39** 155
[21] Batrakov A V and Proskurovskii D I 1999 *Technical physics letters* **25** 444
[22] Marchand M, Journet C, Adessi Ch and Purcell St T 2009 *Physical Review B* **80** 245425
[23] Zajec B, Nemanic V, Zumer M, Bryan E N and Nemanich R J 2011 *Carbon* **49** 3332
[24] Chen G, Wang W, Peng J, He Ch, Deng Sh, Xu N and Li Zh 2007 *Physical Review B* **76** 195412
[25] Koohsorkhi J, Davoudzadeh N, Mohajerzadeh Sh, Asl Soleimani E and Ghafouri Fard H 2009 *Solid State Phenomena* **154** 83
[26] Smith R C and Silva S R P 2009 *Applied physics letters* **94** 133104
[27] Giubileo F, Bartolomeo A D, Scarfato A, Iemmo L, Bobba F, Passacantando M, Santucci S and Cucolo A M 2009 *Carbon* **47** 1074
[28] Li N, Yan F, Pang Sh, Chen L, Jin D, Xiang W, Zhang D, Dai J and Zeng B 2015 *Journal of the Korean Physical Society* **66** 1186
[29] Cai D and Liu L 2013 *AIP Advances* **3** 122103
[30] Volkov E, Jityaev I and Kolomiitsev A 2015 *IOP Conf. Series: Materials Science and Engineering* **90** 012031
[31] Musa I, Munindrasdasa D A I, Amaratunga G A J and Eccleston W 1998 *Nature* **395** 362
[32] Xu J, Huang X, Li W, Chen K and Xu J 2002 *Journal of Applied Physics* **91** 5434
[33] Fursey G N, Petrick V I and Novikov D V 2009 *Technical Physics* **54** 1048
[34] Jityaev I L, Ageev O A, Svetlichnyi A M, Kolomiytsev A S and Spiridonov O B 2016 *Journal of Physics: Conference Series* **741** 012011
[35] Svetlichnyi A M, Ageev O A, Volkov E Yu, Jityaev I L and Dem'yanenko M V 2015 *Applied Mechanics and Materials* **752-753** 163
[36] Jityaev I L, Ageev O A, Svetlichnyi A M and Spiridonov O B 2016 *AIP Conference Proceedings* **1772** 040010

IOP Conf. Series: Journal of Physics: Conf. Series **929** (2017) 012060 doi:10.1088/1742-6596/929/1/012060

Coulomb explosion of nanosize cylindrical target

P A Golovinski [1,2] **and E A Mikhin**[1]

[1]Physics Research Laboratory, Voronezh State Technical University, Voronezh, 20-Letiya Oktyabrya st., 84, 394006, Russia
[2]Moscow Physical-Technical Institute (State University), Moscow region, Dolgoprudny, Institutskii per., 9, 141700, Russia

E-mail: golovinski@bk.ru

Abstract. We considered the Coulomb explosion of single-component cylindrical nano-clusters. The analytical expression for the energy spectra of ions is obtained, as well as the spectral dependence on the cluster geometrical size is investigated. The laser ionization efficiency for cylindrical and spherical clusters is compared. It is shown that cylindrical targets are preferable in comparison with spherical ones, allowing obtain accelerated ions with higher energy at the same target radius.

1. Introduction

The laser technology development has led to the appearance high-power lasers that generate TW femtosecond pulses, allowing accelerate ions to energies of several MeV [1]. The unique characteristics of ion beams, generated by the ultrashort laser pulses are of great interest due to the possible applications when creating neutron sources, for isotope production and in solving other problems, including technologies of radiation impact on solids and different secondary nuclear reactions. It should be noted that the pulse technology for production micro-beams of protons and neutrons can be used for various medical applications [2].

In interaction laser radiation of ultra-high intensity with nano-clusters and micro-targets the large relativistic gamma factor $\gamma = \sqrt{1 + I\lambda^2 / 2.75 \times 10^{18}}$ ($[\lambda]$–μm, $[I]$–W/cm^2) provides great ponderomotive potential $U_\text{p} = m_\text{e}{}^2(\gamma - 1)$. The ponderomotive force action results in spatial separation electrons and ions, depending on target's composition, shape and size, as well as the laser field parameters. Electrons, leaving the target, have characteristic energy distribution, caused by the action of the electromagnetic wave field. These electrons, moving in a strong field, can participate in the nonlinear Thomson scattering [3], and in the secondary electro-stimulated nuclear reactions [4].

After the laser pulse, nano-target takes uncompensated positive charge, and ions further fly away under the action the Coulomb repulsion. Such expansion of the nanosize clusters is called the Coulomb explosion. Already the experiments with relatively moderate laser field intensity $I \geq 10^{16}$ W/cm^2, ionizing the clusters of Xe, have demonstrated the formation of multiply charged ions up to the charge 40^+ and ion acceleration in the subsequent Coulomb explosion to the energy $\sim 1\,\text{MeV}$ [5]. These experiments stimulated theoretical description the Coulomb explosion and searching its optimal parameters. In particular, impact nonsphericity of clusters on the Coulomb explosion was discussed earlier [6]. It was found that geometry of target significantly influences on the parameters of the Coulomb explosion [7]. An opportunity to reach the maximum ion energy provides formation the ion spot by irradiation thin films using high-power ultrashort laser pulses. On the other hand, investigation of micro sources for various types of nuclear radiation shows that jets of spherical clusters can be very promising for these purposes. At the same time, the cylindrical clusters have an advantage of slow field decrease with distance compared with the spherical targets, and one can expect large final ion energy at the same target diameters. As a consequence of this, we focused on the study of accelerating ions in the Coulomb explosion of the cylindrical targets.

2. Ionization of cylindrical nano-targets

Let us consider cluster ionization within a strong laser field during atomic times, i.e., almost immediately, proceeding from the fact that the degree of ionization is determined by the field strength F. We assume that ionization occurs basically in instants of time when laser field reaches the maximum values. According to this approach [8], we assume that the ionization process follows to Bethe's mechanism of the above-threshold ionization [9], that allows to determine the ionization degree of nano-target. For a spherical target, taking the charge distribution inside of it uniform, one can write an expression for the potential energy of an electron in the cluster field with a charge Q in the form

$$U(r)=\begin{cases}-\dfrac{Q}{2R}\left(3-\dfrac{r^2}{R^2}\right), r\le R,\\ -\dfrac{Q}{r}, r\ge R.\end{cases} \tag{1}$$

Here R is the radius of the target, r is the distance from the center ($e=1$).The ionization energy we take equal the modulus of the electron potential energy $|U(R)|=Q/R$ on the cluster periphery.

In the presence of a laser field polarized in the radial direction, the total potential energy of the system target + field is determined by the expression $U_t(r)=U(r)-Fr$, and the derivative in the maximum is

$$\left.\frac{\mathrm{d}U_t(r)}{\mathrm{d}r}\right|_{r=r_0}=0. \tag{2}$$

Scenario the above-threshold cluster ionization is determined by the equality the binding electron energy and the maximum value of the potential energy. Hence, one can define the ionization threshold as

$$U(r_0)-Fr_0=U(R). \tag{3}$$

Equation (3) gives us the maximum charge Q of the cluster at which the above-threshold ionization is possible and the relevant value for the degree of the cluster ionization $\alpha=Q/ZN$, where Z is the charge of cluster atoms, and N is the total number of atoms. Thus, for spherical cluster ($Q=4FR^2$) with concentration of atoms n the ionization degree is defined by the expression

$$\alpha_s=\frac{3F}{\pi ZRn}. \tag{4}$$

In the same way, we can estimate the ionization degree for cylindrical cluster. The monotonically increasing potential of electron in such a system has the form

$$U(r)=\begin{cases}-\pi\rho\left(R^2-r^2\right), r\le R,\\ 2\pi R^2\rho\ln(r/R), r\ge R,\end{cases} \tag{5}$$

where $\rho = Q/N$ is the bulk charge density. The condition of unstable equilibrium on the top of the potential barrier (2) is now written as

$$\frac{2\tau}{r_0} = F, \tag{6}$$

where $\tau = \pi R^2 \rho$ is linear charge density, i.e., $r_0 = 2\tau / F$.

The potential energy of an electron on the boundary of uniformly charged cylindrical target is taken as zero, so that equation (3) for the above-barrier ionization of cluster is written down as

$$2\tau \ln\left(\frac{2\tau}{FR}\right) - F\frac{2\tau}{F} = 0. \tag{7}$$

By solving equation (7), we obtain the value of the linear charge density of the cylindrical cluster, formed as a result of the above-threshold ionization, and

$$\tau = FR/2. \tag{8}$$

The expression for the ionization degree of the target is

$$\alpha_c = \frac{\tau}{Zn\pi R^2} = \frac{F}{2Zn\pi R}. \tag{9}$$

From the comparison of equations (4) and (9), we can conclude that for spherical and cylindrical targets the ionization degree is described by the same function of the field and the target radius, differing only by a numerical factor, so that

$$\frac{\alpha_s}{\alpha_c} = \frac{3F}{\pi Z R n}\frac{2\pi Z R n}{F} = 6. \tag{10}$$

Table 1 shows the results of calculation the ionization degree of spherical and cylindrical nano-targets with different composition and radius for various laser intensities.

Table 1. The dependence the ionization degree of spherical and cylindrical nano-targets D, Li, C, Kr, Xe, Pt on the radius for different laser intensities.

Ion		D	Li	C	Kr	Xe	Pt
Z		1	3	6	36	54	78
I, W/cm^2	R, nm	α_s/α_c					
10^{17}	1	1.00 / 1.00	1.00 / 0.72	1.00 / 1.00	0.74 / 0.12	0.67 / 0.11	0.13 / 0.02
	10	1.00 / 0.32	0.43 / 0.07	1.00/ 0.17	0.07 / 0.01	0.07 / 0.01	0.01 / 0.00
	20	0.96 / 0.16	0.22 / 0.04	0.52 / 0.09	0.04 / 0.01	0.03 / 0.01	0.01 / 0.00
	30	0.64 / 0.11	0.14 / 0.02	0.35 / 0.06	0.02 / 0.00	0.02 / 0.00	0.00 / 0.00
	40	0.48 / 0.08	0.11 / 0.02	0.26/ 0.04	0.02 / 0.00	0.02 / 0.00	0.00 / 0.00
10^{19}	1	1.00 / 1.00	1.00 / 1.00	1.00 / 1.00	1.00 / 1.00	1.00 / 1.00	1.00 / 0.21
	10	1.00 / 1.00	1.00 / 0.72	1.00 / 1.00	0.74 / 0.12	0.67 / 0.11	0.13 / 0.02
	20	1.00 / 1.00	1.00 / 0.36	1.00 / 0.87	0.37 / 0.06	0.34 / 0.06	0.06 / 0.01
	30	1.00 / 1.00	1.00 / 0.24	1.00 / 0.58	0.24 / 0.04	0.22 / 0.04	0.04 / 0.01
	40	1.00/ 0.80	1.00 / 0.18	1.00 / 0.43	0.19 / 0.03	0.17 / 0.03	0.03 / 0.01
10^{21}	1	1.00 / 1.00	1.00 / 1.00	1.00 / 1.00	1.00 / 1.00	1.00 / 1.00	1.00 / 1.00
	10	1.00 / 1.00	1.00 / 1.00	1.00 / 1.00	1.00 / 1.00	1.00 / 1.00	1.00 / 0.21
	20	1.00 / 1.00	1.00 / 1.00	1.00 / 1.00	1.00 / 0.62	1.00 / 0.56	0.63 / 0.11
	30	1.00 / 1.00	1.00 / 1.00	1.00 / 1.00	1.00 / 0.41	1.00 / 0.37	0.42 / 0.07
	40	1.00 / 1.00	1.00 / 1.00	1.00 / 1.00	1.00 / 0.31	1.00 / 0.28	0.32 / 0.05

IOP Conf. Series: Journal of Physics: Conf. Series **929** (2017) 012060 doi:10.1088/1742-6596/929/1/012060

The data presented in the table correspond to the lower boundary degree of the cluster ionization, because strong laser field provide also other ionization mechanisms, that are comparable in efficiency with the above-threshold ionization, such as the tunneling ionization and the electron impact ionization stimulated by the laser field. Following these estimates, we conclude that the technically accessible laser fields can ionize completely relatively large clusters. In particular, when the laser field intensity is 10^{19} W/cm^2, clusters consisting of light atom elements with a radius up to 10 nm, are ionized completely regardless of their form. It is worth noting that maximum of experimentally accessible intensity greatly exceed this value. Thus lower cylindrical target ionization efficiency is compensated by a significant gain in energy of ions produced in the Coulomb explosion as compared with the spherical target. Therefore, the use of cylindrical clusters is justified when they have sufficiently large length.

3. Coulomb explosion of a uniform cylinder

For further comparison, we will reproduce, at the beginning, the results of the calculation the Coulomb explosion of fully ionized clusters of cold ions, neglecting initial ion velocities. Let's start with a spherical target with a uniform initial density distribution. Since the field strength inside the uniformly charged ball monotonically increases from the center to the periphery, the acceleration of the ions in the outer layers is always greater, and they get the maximum energy at the end of expansion according to the energy conservation law. Thus, the shock waves are not formed in the homogeneous targets. For a spherical layer of thickness $\mathrm{d}r$ the number of particles is

$$\mathrm{d}N = n4\pi r^2\,\mathrm{d}r\,. \tag{11}$$

Energy of ion emitted from such a layer is defined by its potential energy, i.e.,

$$\varepsilon = -ZU(r) = Z^2 n\frac{4\pi r^3}{3}\frac{1}{r} = \frac{\varepsilon_{max}}{R^2}r^2\,, \tag{12}$$

where $\varepsilon_{max} = 4\pi Z^2 nR^2/3$ is the maximum ion energy equal to the potential energy at the cluster boundary. Consequently, the distribution of the emitted ions over the energy is of the form [8]

$$\frac{\mathrm{d}N}{\mathrm{d}\varepsilon} = \frac{\mathrm{d}N/\mathrm{d}r}{\mathrm{d}\varepsilon/\mathrm{d}r} = \frac{3R}{2Z^2}\sqrt{\frac{\varepsilon}{\varepsilon_{max}}}\,,\ \varepsilon < \varepsilon_{max}\,, \tag{13}$$

and the average value of the energy is

$$\varepsilon_{av} = \frac{3}{5}\varepsilon_{max}\,. \tag{14}$$

We turn to the description of the Coulomb explosion of a fully ionized homogeneous cylindrical cluster. Inside it, the electric field increases monotonically in a radial direction as in the spherical cluster which means the absence of shock waves at the expansion. However, the direct use the model of infinite cylinder to obtain the final distribution of the ion energy, after the expansion cylindrical target of finite length, is impossible, because to get the finite energy, one need to take into account the asymptotic behavior of potential at the infinity like the field of a point Coulomb source. Thus, we should propose a model for expanding the target of finite size. To determine the energy of the expanding ions, we replace the finite cylinder by the elongate ellipsoid with a major semi-axis a and small semi-axes R ($a > R$). The volume of the ellipsoid is

$$V = \frac{4\pi}{3}aR^2\,, \tag{15}$$

and the total charge for a uniform density of ions is

$$Q(R) = \frac{4\pi}{3} nZaR^2 . \tag{16}$$

The number of particles in cylindrical layer thickness $\mathrm{d}b$ is equal to

$$\mathrm{d}N = n\frac{8\pi}{3} ab\,\mathrm{d}b . \tag{17}$$

The potential energy (the kinetic energy at infinity) of the charge Z that is located on the surface of the ellipsoid with the current maximum radius b, a constant potential along the surface and the total charge Q is given by the equation [10]

$$U_i(b) = \varepsilon(b) = \frac{ZQ(b)}{2\sqrt{a^2 - b^2}} \ln\frac{a + \sqrt{a^2 - b^2}}{a - \sqrt{a^2 - b^2}} . \tag{18}$$

When $b = a$, the equation (18) becomes exactly the same as for the charged sphere. The differential variation of ion energy $\mathrm{d}\varepsilon$ with a change $\mathrm{d}b$ of the distance from the target axis is given by the expression

$$\mathrm{d}\varepsilon = \frac{2\pi}{3} abn\, Z^2 \left(\left(2a^2 - b^2\right) \ln\left(\frac{a + \sqrt{a^2 - b^2}}{a - \sqrt{a^2 - b^2}} \right) - 2a\sqrt{a^2 - b^2} \right) \left(a^2 - b^2\right)^{-3/2} \mathrm{d}b . \tag{19}$$

The energy distribution for ions after the Coulomb explosion, in the final state can be written as

$$\frac{\mathrm{d}N}{\mathrm{d}\varepsilon} = \frac{\mathrm{d}N/\mathrm{d}b}{\mathrm{d}\varepsilon/\mathrm{d}b} = \frac{4\left(a^2 - b^2\right)^{3/2}}{Z^2} \left(\left(2a^2 - b^2\right) \ln\left(\frac{a + \sqrt{a^2 - b^2}}{a - \sqrt{a^2 - b^2}} \right) - 2a\sqrt{a^2 - b^2} \right)^{-1} . \tag{20}$$

It is not possible to obtain an analytic expression for the function $b(\varepsilon)$ from (18), since equations (18) and (20) give us the spectrum of ions at infinity in parametric form, however, the numerical calculations of the function are simple. The numerical dependences $b(\varepsilon)$ obtained for different values of the parameter a are shown in figure 1.

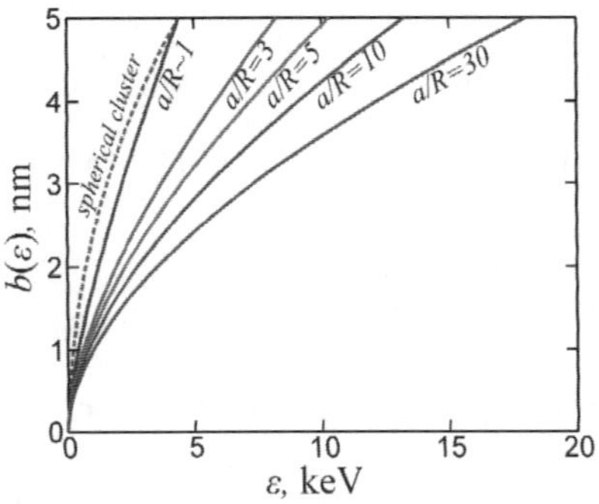

Figure 1. The functional dependence of the initial ion coordinate b and final kinetic energy ε

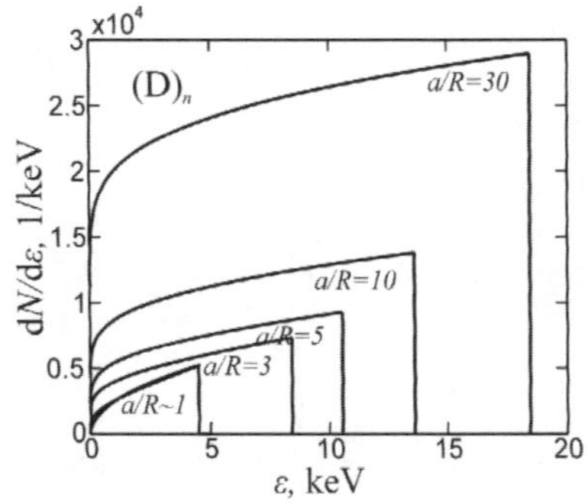

Figure 2. The energy spectrum of the ions after the Coulomb explosion of the deuterium cylindrical target

The dashed line shows the dependence $b(\varepsilon)$ for the spherical cluster. The energy distributions for ions after the Coulomb explosion of fully ionized deuterium clusters with a radius $R = 5\text{nm}$ are shown in figure 2. The ratios between the longitudinal and transverse target sizes are indicated under the curves.

With increasing a, as was expected, we see growth the maximum energy of ions, and for $a/R = 30$ the fourfold gain of energy is achieved compared with the spherical clusters. The ion energy is completely determined by the Coulomb energy of the target and, therefore, depends on the initial cluster charge. With increasing the target radius one can obtain ions with greater energy, but sized effect is restricted by the concomitant reduction in the ionization degree. More effective, in this regard, are the complex targets comprising ions that differ significantly in mass and charge.

4. Conclusions

The quasi-static theoretical model of the Coulomb explosion the cylindrical clusters has allowed us to calculate the energy spectra of ions. The results of numerical simulation showed that for homogeneous cylindrical cluster maximum achievable energy of the ions is much greater than for a spherical cluster of the same radius. The spectrum of expanding target containing deuterium, substantiated the possibility acceleration deuteron to 18 keV if one obtained a uniformly charged cylindrical target with the radius $5\ \mathrm{nm}$ and the length $150\ \mathrm{nm}$. In a collision of deuterons having such energies, we should expect the initiation of nuclear fusion reactions with a significant output of monoenergetic neutrons. Control the laser field intensity and the length of the cylindrical target, clearly, provides the possibility to obtain the required energy deuterons.

As promising for obtaining high-energy ion flows in the Coulomb explosion, the carbon nanotube targets can be considered. Their sizes are optimal for location in the focal spot of the laser field, and the ratio of the length to the radius can reach several hundred units. In addition, the development of nanotechnology opens up broad opportunities for their synthesis. For example, long cylindrical clusters with diameters a few nanometers can be synthesized as carbon nanotubes filled by different compositions [11, 12]. A numerical simulation of the Coulomb explosion of the nanometer-sized carbon-gold nanotubes, containing gold atoms and filled by hydrogen, was carried out [13] showing the formation of highly collimated and almost monoenergetic protons with energies of few MeV.

Further studies suggest simulation the yield of nuclear fusion reactions at counterpropagating expansion of cylindrical targets with different compositions.

Acknowledgments

The work has been supported by the Russian Foundation for Basic Research, agreement No. 16-32-00255.

References

[1] Kryukov P G 2012 *Ultrashort-pulse lasers and their applications* (Dolgoprudniy: INTELLEKT House) (in Russian)
[2] Bychenkov V Yu and Brantov A V 2015 *Eur. Phys. J. Special Topics* **224** 2621
[3] Golovinski P A and Mikhin E A 2011 *Journal of Experimental and Theoretical Physics* **113** 545
[4] Golovinski P A and Mikhin E A 2013 *Physics of Atomic Nuclei* **76** 56
[5] Ditmire T, Tisch J W G, Springate E, Mason M B, Hay N, Marangos J P and Hutchinson M H R 1997 *Phys. Rev. Lett.* **78** 2732
[6] Grech M, Nuter R, Mikaberidze A, Cintio P Di, Gremillet L, Lefebvre E, Saalmann U, Rost J M and Skupin S 2011 *Phys. Rev.* E **84** 056404
[7] Kovalev V F, Popov K I, Bychenkov V Yu and Rozmus W 2007 *Physics of plasmas* **14** 053103
[8] Krainov V P and Smirnov B M 2002 *Phys. Rep.* **370** 273
[9] Bethe H A and Salpeter E E 1957 *J. Quantum Mechanics of One- and Two-Electron* (Berlin: Springer)
[10] Landau L D and Lifshitz E M 1960 *Electrodynamics of Continuous Media: vol. 8* (Pergamon Press)
[11] Eletskii A V 2004 *Phys. Usp.* **47** 1119
[12] Marega R and Bonifazi D 2014 *New J. Chem.* **38** 22
[13] Murakami M and Tanaka M 2013 *Appl. Phys. Lett.* **102** 163101

IOP Conf. Series: Journal of Physics: Conf. Series **929** (2017) 012061 doi:10.1088/1742-6596/929/1/012061

Neutral meson production in Cu+Au collisions at 200 GeV

A. Berdnikov[1], Ya. Berdnikov[1,2], D. Kotov[1,2], P. Radzevich[1], V. Riabov[2] and Sergei Zharko[1]

[1]Peter the Great Saint Petersburg Polytechnic University, 29 Polytechnicheskaya st., Saint-Petersburg, Russia
[2]Petersburg Nuclear Physics Institute, 1 Orlova roscha mcr., Gatchina, Russia

E-mail: zharko@rcf.rhic.bnl.gov

Abstract. Strongly interacting quark-gluon plasma (sQGP) was discovered in central heavy ion collisions at RHIC energies. Jet quenching is one of the sQGP evidence observed in particle yields suppression related to yields measured in elementary nucleon-nucleon collisions. System of Cu+Au collisions is characterized by special nuclear overlap geometry different to other large systems (such as Au+Au). Study of such collision systems will help to estimate sQGP properties with higher accuracy. Neutral mesons such as π^0 give a good opportunity for studying sQGP effects especially jet quenching due to their large production rate. This paper presents results on neutral pion invariant differential spectra and nuclear modification factors measured with PHENIX experiment.

1. Introduction

In 2005 all experiments at Relativistic Heavy Ion Collider, RHIC (BNL, USA) [1] made a statement that strongly interacting Quark-Gluon Plasma (sQGP) had been found in central heavy nuclei collisions (A+A). It is a deconfined and liquid-like state of nuclear matter under extreme conditions of temperature and pressure [2–5].

Jet quenching is one of the evidences for sQGP formation, which manifests itself in suppressed production of high energy hadrons in A+A collisions relative to ones measured in elementary p+p collisions. Partons lose their energy interacting with hot and dense medium, which results in production of hadrons with softened spectra comparing to p+p collisions while direct photons do not interact with the medium, thus their production is not suppressed [6].

In relativistic heavy nuclei collision physics observables are usually measured as a function of collision centrality, which characterize relative geometry of colliding nuclei. In most central collisions with large overlap region centrality is taken equal to 0%. In peripheral collisions where only few nucleons take part in nuclei interactions centrality is about 50-60% (different for different colliding nuclei systems).

Particle suppression in A+A collisions is often studied with so-called nuclear modification factors (R_{AA}) calculated as a function of particle transverse momentum in different centrality A+A collisions:

$$R_{AA}(p_T) = \frac{1}{\langle N_{coll} \rangle} \frac{dN_{AA}(p_T)}{dN_{pp}(p_T)}, \tag{1}$$

where dN_{AA} (dN_{pp}) – hadron invariant differential yields measured in A+A (p+p) collisions $\langle N_{coll}\rangle$ – average number of binary inelastic nucleon-nucleon collisions calculated as a function of centrality by Glauber Monte-Carlo modelling [7]. If nuclear modification factor for hadrons emitted from the medium is equal to unity a collision of two nuclei could be described as a superposition of binary scaled elementary nucleon-nucleon collisions. Otherwise, different from unity R_{AA} suggests that medium effects in nuclei interactions are presented.

Measurements of neutral pion(π^0) yields provide excellent opportunity to probe sQGP formed in A+A collisions and study jet quenching effect. Neutral pions are light mesons consisting of first generation quark (u, d) combinations thus π^0 mesons are a significant part of particles produced in heavy nuclei collisions and their yields could be measured in a wide transverse momentum range with good signal to background ratio.

In 2012 RHIC successfully delivered Cu+Au collisions at $\sqrt{s_{NN}}$ = 200 GeV. This is a first asymmetric heavy nuclear collision system and differs from symmetric ones (Cu+Cu, Au+Au) by special form of nuclear overlap region. The study of particle suppression in such collision system will help to better discriminate between input parameters of various theoretical models and describe partonic energy loss mechanisms more precisely.

2. Experimental setup

PHENIX spectrometer consists of central and forward arms. Forward arms cover $1.2 < |\eta| < 2.2$ range in pseudo-rapidity and full azimuthal angle and provide capabilities for muon measurements [8]. Each central arm (Fig. 1), western and eastern, covers 90 degrees in azimuthal angle and $|\eta| < 0.35$ region in pseudo-rapidity and provides detection of photons, electrons and charged hadrons originating from nuclei collision region [9, 10]. All results shown in the paper were obtained using central arms of the PHENIX detector.

Two beam-beam counters (BBC) [11] are used for z-vertex measurement and centrality classification of nuclei collision events. Also BBC provides the minimum bias (MinBias) trigger. Electromagnetic calorimeter [12] at PHENIX consists of six lead-scintillator sectors (PbSc) and two lead-glass sectors (PbGl) and provides measurements of photon kinematic properties and serves as an effective tool for neutral meson reconstruction. Usage of two different calorimeter types provide independent cross-check of final results within the same experiment. Calorimeter nominal energy resolution is $\delta E/E = 8.1\%/\sqrt{E(GeV)} \pm 2.1\%$ for PbSc and $\delta E/E = 5.9\%/\sqrt{E(GeV)} \pm 0.8\%$ for PbGl. Also electromagnetic calorimeter provides a ERT trigger, which is used for rare high energy events selection.

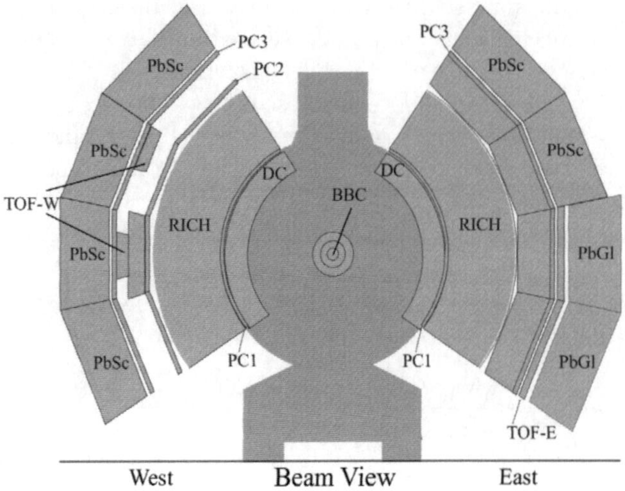

Figure 1. Layout of PHENIX detector central arms for RHIC Run-12 data taking period. Following detectors are presented in the layout: beam-beam counters (BBC), drift chambers (DC), ring-image Cherenkov detectors (RICH), electromagnetic calorimeter sectors(PbSc, PbGl), three layers of pad chambers (PC1, PC2, PC3), time-of-flight system (TOF).

IOP Conf. Series: Journal of Physics: Conf. Series **929** (2017) 012061 doi:10.1088/1742-6596/929/1/012061

3. Data analysis

Neutral pion yields were measured with EMCal by analysing transverse momentum dependent invariant mass distributions of daughter photons in $\pi^0 \longrightarrow \gamma\gamma$ decay channel in different nuclei collision centrality bins. Two data samples are used in the analysis: MinBias ($6 \cdot 10^9$ events) and ERT ($1.8 \cdot 10^{10}$ events). MinBias and ERT data samples were used to measure π^0 meson yields at $p_T < 8$ GeV/c and $p_T > 8$ GeV/c ranges respectively.

Among the analysis details one of the most important is the requirement for each photon from π^0 decay to have energy larger then 0.4 GeV. To obtain better signal-to-background ratio an asymmetry cut was applied on combined photons energies: $\alpha < 0.8$, where $\alpha = |E_{\gamma 1} - E_{\gamma 2}|/(E_{\gamma 1} + E_{\gamma 2})$.

The reconstruction efficiency for neutral pions in acceptance of PHENIX detector was obtained with Geant 3 Monte-Carlo simulation. All event selection cuts were applied into simulation in the same way than in real data. Reconstruction efficiency as a function of transverse momentum was calculated as a ratio of the number of neutral pions reconstructed in simulation to their total number separately for different collision centralities.

Systematic uncertainty of results was obtained by varying different analysis parameters. Main uncertainty comes from possible mismatch of calorimeter absolute energy scale in data and simulation. At high transverse momentum ($p_T > 13$ GeV/c) cluster merging effect gives a significant contribution to the total systematic uncertainty: two photons from high energy π^0 decay have such a narrow opening angle and are reconstructed in electromagnetic calorimeter as one photon, thus π^0 could not be reconstructed. Cluster merging effect is more significant for PbSc subsystem then for PbGl one due to lower sector granularity. Total systematic uncertainty changes from 8% at low p_T to 25% at high p_T for PbSc sectors and from 8% at low p_T to 15% at high p_T for PbGl sectors.

4. Results

Invariant transverse momentum π^0 meson yields as a function of transverse momentum and collision centrality were calculated separately for PbSc and PbGl subsystems according to formula:

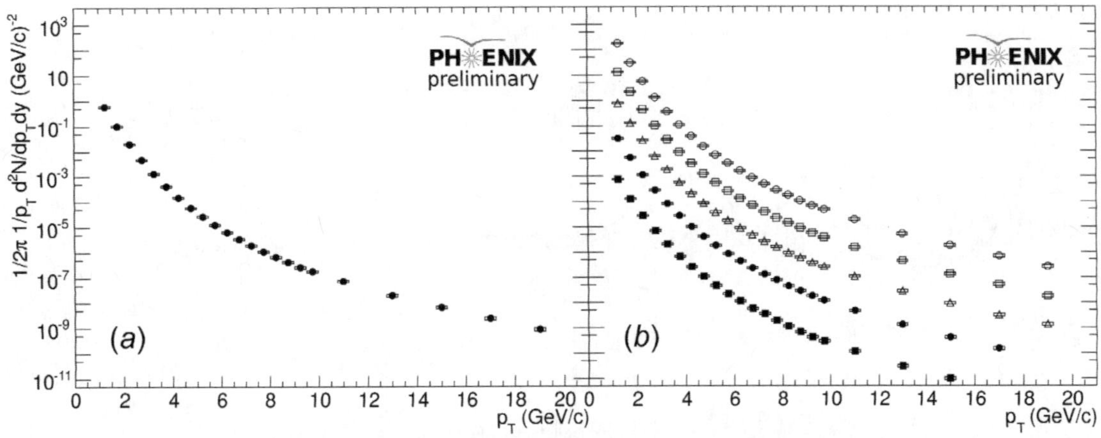

Figure 2. Invariant differential π^0 meson spectra measured in Cu+Au collisions at 200 GeV with no centrality selection (a) and for different centrality bins (b). Curves on the plot (b) are shown for 0-10% (○, multiplied by 10^2), 10-20% (□, multiplied by 10^1), 20-40% (△), 40-60% (●, multiplied by 10^{-1}), 60-90% (■, multiplied by 10^{-2}).

IOP Conf. Series: Journal of Physics: Conf. Series **929** (2017) 012061 doi:10.1088/1742-6596/929/1/012061

$$dN_{AA}(p_T) = \frac{1}{2\pi p_T} \frac{N_{\pi^0}(p_T)}{\Delta p_T N_{events} \epsilon_{acc.\times rec.eff}(p_T) BR}, \quad (2)$$

where N_{π^0} – extracted raw yields, N_{events} – number of analyzed events, $\epsilon_{acc.\times rec.eff}$ – reconstruction efficiency, $BR = 0.998$ – branching ratio for $\pi^0 \to \gamma\gamma$ decay channel.

Resulted yields were obtained by averaging ones measured using PbSc and PbGl subsystems with weights inversely proportional to statistical and systematic uncertainties. Fig. 2 presents invariant differential transverse momentum spectra of neutral pions measured in different centrality Cu+Au collisions. Spectra are measured in a large p_T range: up to 20 GeV/c in central and semi central collisions.

Neutral pion nuclear modification factors were calculated according to formula (1) in different centrality Cu+Au collisions (Fig. 3). Invariant differential spectrum of π^0 mesons measured in p+p collisions at 200 GeV is published in [13]. In central Cu+Au collisions (0-10%) π^0 production is suppressed two times in comparison with π^0 yields measured in elementary p+p collisions. In peripheral Cu+Au collisions (60-90%) a hint on production of neutral pion yields is observed.

Suppression pattern observed for π^0 in Cu+Au collisions at 200 GeV is similar to one observed for reconstructed jets in same collision system [14].

Comparison of π^0 meson nuclear modification factors vs. p_T measured in Cu+Au, Au+Au [15] and Cu+Cu [16] collisions at $\sqrt{s_{NN}} = 200$ GeV and similar numbers of participants is presented

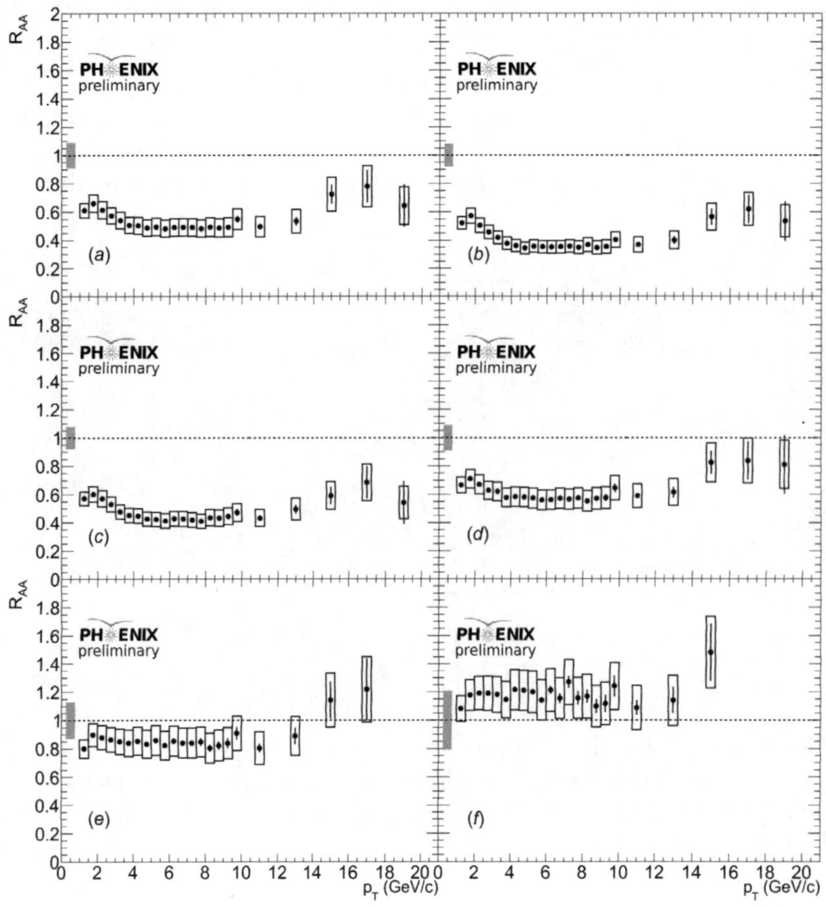

Figure 3. Nuclear modification factors as a function of π^0 mesons transverse momentum measured in 0-10% (a), 10-20% (b), 20-40% (c), 40-60% (d), 60-90% (e) centrality bins and with no centrality selection (f) Cu+Au collisions at 200 GeV. Boxes at markers show transverse momentum dependent systematic uncertainty. Open boxes at unity show scaling uncertainty.

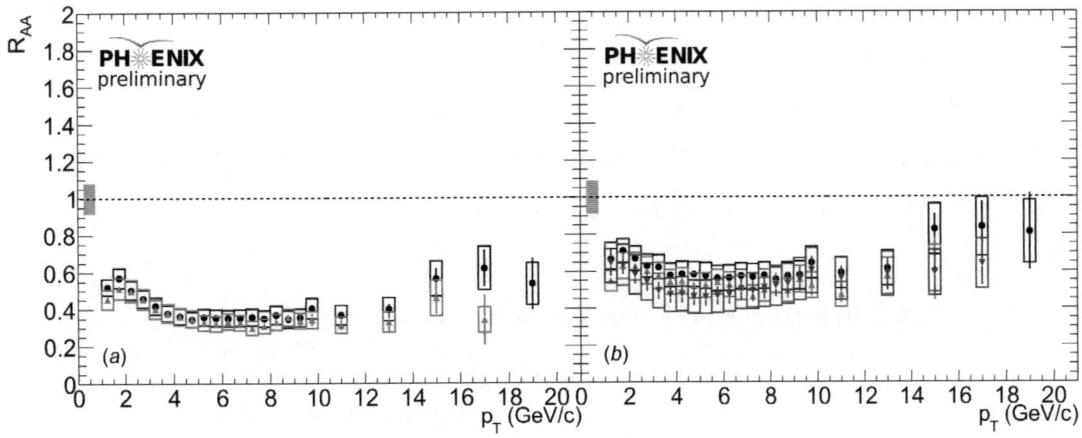

Figure 4. Comparison of π^0 meson nuclear modification factors, measured as a function of transverse momentum in Cu+Au (●), Au+Au (▲) and Cu+Cu (▼) collisions at 200 GeV. Panel (*a*) is for 0-10% Cu+Au with $N_{part} = 177$ and 20-30% Au+Au with $N_{part} = 167$. Panel (*b*) is for 20-40% Cu+Au with $N_{part} = 80.4$, 40-50% Au+Au with $N_{part} = 74.4$ and 0-10% Cu+Cu with $N_{part} = 98.2$.

in Fig. 4 for central and semi-central Cu+Au collisions. One can see that π^0 shows a similar suppression in the whole transverse momentum range. This fact suggests that π^0 suppression level depends on the size of the nuclear overlap region but not on its shape.

5. Conclusion

PHENIX experiment has measured neutral pion invariant differential yields and nuclear modification factors in different centrality Cu+Au collisions at 200 GeV.

In central Cu+Au collisions π^0 meson yields are suppressed approximately by a factor of 2, while in peripheral collisions a hint on enhancement for neutral pion yields is observed. Reconstructed jets measured in Cu+Au collisions at 200 GeV are suppressed in similar way than neutral pion yields. Suppression level of π^0 mesons depends on the size of the medium but not on its shape.

References

[1] Baym G 2002 *Nucl. Phys.* A **698** 23
[2] Arsene I et al 2005 *Nucl. Phys.* A **757** 1-27
[3] Back B et al 2005 *Nucl. Phys.* A **757** 28-101
[4] Adams J et al 2005 *Nucl. Phys.* A **757** 102-83
[5] Adcox K et al 2005 *Nucl. Phys.* A **757** 184-283
[6] Afanasiev S et al 2012 *Phys.Rev.Lett.* **109** 152302
[7] Loizides C 2016 *Phys. Rev.* C **94** 024914
[8] Akikawa H et al 2003 *Nucl. Inst. Meth.* A **499** 537-48
[9] Aizawa M et al 2003 *Nucl. Inst. Meth.* A **499** 508-20
[10] Adcox K et al 2003 *Nucl. Inst. Meth.* A **499** 489-07
[11] Allen M et al 2003 *Nucl. Inst. Meth.* A **499** 549-59
[12] Aphecetche L et al 2003 *Nucl. Inst. Meth.* A **499** 521-36
[13] Adare A et al 2007 *Phys. Rev.* D **76** 051106(R)
[14] Timilsina A 2016 *PHENIX results on reconstructed jets in p+p and Cu+Au collisions* (Kobe: Quark Matter 2015)
[15] Adare A et al 2008 *Phys. Rev. Lett.* **101** 232301
[16] Adare A et al 2008 *Phys. Rev. Lett.* **101** 162301

Determination of the composition of liquids using spectral analysis of the electric discharge radiation

A N Ramazanov[1], D K Kostrin[1] and A A Lisenkov[2]

[1] Department of electronic instruments and devices, Saint Petersburg Electrotechnical University "LETI", 197376, Saint Petersburg, Russia

[2] Institute of Problems of Mechanical Engineering, Russian Academy of Sciences, 199178, Saint Petersburg, Russia

E-mail: ramazanow.alexander@yandex.ru

Abstract. This article considers the reasons stimulating the research and development of a comparatively new method of analyzing the composition of liquids based on the spectral analysis of radiation of an electric discharge. Approaches to creating discharge cells for further analysis are described, including the scheme of the cell developed by the authors. To maintain the discharge has been developed a simple but rather effective electrical scheme with the use of a half-bridge driver. Emission spectra of the deionized and mains water shown in this work allow concluding that the sensitivity of the developed device is sufficient to detect even small concentrations of the substances dissolved in liquid.

1. Introduction

Nowadays the most widely used are chemical and electrochemical methods of the composition analysis of liquids, but they are relatively expensive and need a considerable time for implementation of the analysis. However there are other ways of composition definition that should be mentioned:

- photometric method, which can be used to detect nitrites, phosphates and cyanides;
- gravimetric method for analysis of fats, burnt residues and petroleum products;
- titrimetric method used for the detection of chlorides and dissolved oxygen;
- some others, such as nephelometry, infrared spectrometry and manometric methods.

There is a great interest in researching methods of spectral analysis of the composition of liquids, based on the detection of emission spectra, produced by the substances when heated by an electric discharge. Promising method is the one based on photometric registration of the emission spectra of plasma formed in the channel of an electric discharge [1, 2].

A significant problem with the present methods of the spectral analysis of the liquids composition is the initiation of the electrical discharge, which requires development of a special discharge cell in which as one or both electrodes the fluid itself is used. Such systems have rather small sizes and do not require supplementary gases, which makes them cheaper, more portable and fast compared to the other widely used emission systems.

2. Development of the discharge cell

The initiation of the discharge can be obtained using a variety of the design solutions. For such purposes the following discharge systems can be used:

- system with a liquid electrode (with electrolysis in a glow discharge);

- system with a flowing fluid;
- system with a capillary tube (with discharge in the channel).

Two of the most effective systems are the system with electrolysis in a glow discharge and a system with discharge in the channel. The peculiarity of the first system is that one of the electrodes has a low potential and the second is immersed into the liquid. Space between the electrodes is filled with air or another gas at atmospheric pressure. Initiation of the gas discharge is performed between the metal electrode and the surface of the liquid when the high voltage is applied [3].

There are no requirements to the liquid in this system. The main disadvantage of this design is that the discharge occurs in the air between two electrodes that leads to major difficulties when receiving the spectral characteristics, which in this case requires a high voltage for transmission of the energy to the substances in the water for their heating.

There is also a system design in which both electrodes are immersed into the liquid. Such system contains a capillary tube connecting two reservoirs with the analyzed liquid. Design of the developed discharge cell and the overall scheme of the research setup are shown in figure 1. This discharge cell may be submerged into the test liquid, which allows execution of the analysis in the field.

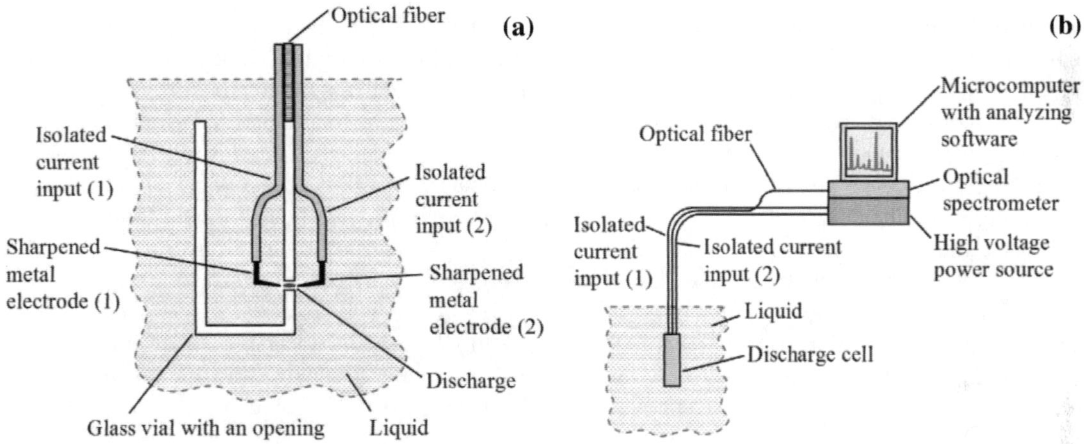

Figure 1. Design of the discharge cell (**a**) and the overall scheme of the research setup (**b**).

The reason for the use of a capillary is that it helps to increase the electric gradient in liquid with limited capacity of the power source, and the capillary contributes to this, as it increases resistance by limiting the formation of the discharge channel [4].

Discharge occurs directly in the liquid filling the capillary, which have a positive impact on the spectrum, as the glow of the electrodes (background spectrum) and also the spectral lines of the air and various substances in the air would not be received by a spectrometer.

Discharge occurring in the described above system is similar to the arc, and it is possible to allocate three stages of the process:

- voltage and current are rising to the moment of ignition of the discharge;
- ignition voltage decreases to the level of maintaining;
- current has no effect on the voltage.

This system has a serious disadvantage – a narrow capillary is polluted if it is used in insufficiently clear liquid, because the combustion substances remain on its walls, which prevents the receiving of the useful spectral data.

3. Development of the high voltage power source

Electric discharge in the system is achieved by applying a pulsed voltage of rectangular form using a high voltage power supply. The resulting radiance of the discharge plasma is received by an optical spectrometer through the wall of a discharge cell.

For excitation of the electric discharge in the described cell has been developed, assembled and studied a rather simple, but efficient scheme of the voltage converter based on a self-oscillating half-bridge driver IR2153 (figure 2).

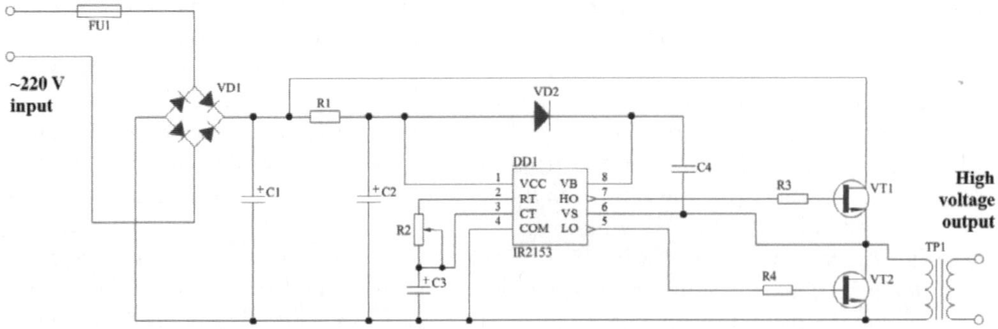

Figure 2. Scheme of the voltage converter based on a half-bridge driver IR2153.

Designed circuit works from the mains with voltage 220 V and frequency 50 Hz. Voltage from the mains is rectified by VD1 and smoothed by the capacitive filters C1 and C2. Used in the experiment IR2153 chip is designed to control the field transistors that operate in a bootstraped mode. The lower output of the C4 capacitor when voltage is applied to the chip begins to alternately switch either to the ground of the scheme or to the supply voltage line. In the moment when transistor VT2 is in the conducting state, the charge of the capacitor C4 begins to pass through the diode VD2. If VT1 is in the conducting state, the power of the upper cascade of the chip is provided due to the energy stored in C4. Use of the potentiometer R2 allows adjusting the frequency in a wide range – from 15 to 50 kHz. The voltage supplied to the transformer TR1 after its passage increases its amplitude up to 15 kV.

Configuring the discharge system is a rather complex task – it is essential to get the maximum efficiency of the excitation of spectral lines corresponding to the contained in the liquid substances, adjusting the mechanical properties of the system: size, shape and arrangement of the electrodes; and electric parameters of the high voltage power source: voltage, current and frequency of the pulses.

4. Results and discussion

In this research for the acquisition of the spectral data the ISM3600 spectrometer and ASpect2010 software were used [5–7]. Both voltage and current were controlled with an oscilloscope using a voltage divider, which reduced the values by 1000 times. Figure 3 shows the spectra obtained for deionized and mains water under the same conditions of the experiment.

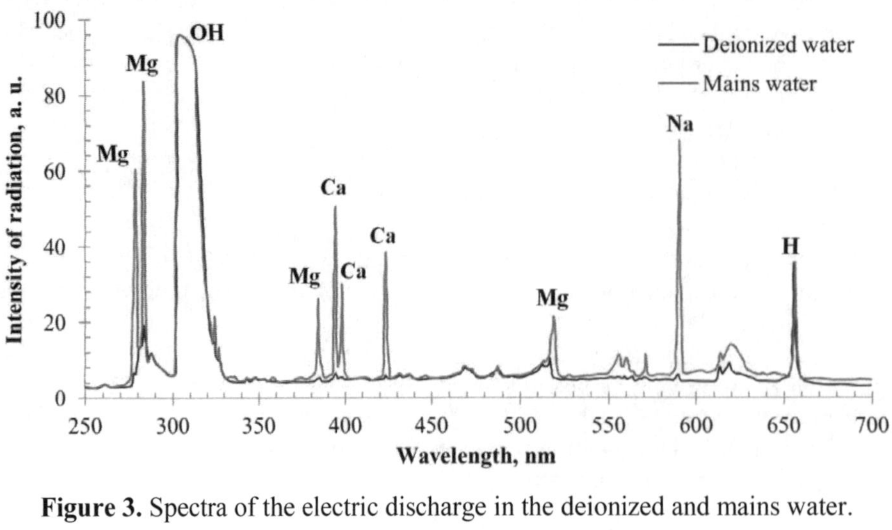

Figure 3. Spectra of the electric discharge in the deionized and mains water.

In both spectra there are lines corresponding to the OH radicals (band in the range 300...320 nm) and hydrogen (656.3 nm). For mains water there are intense lines corresponding to magnesium (279.5, 285.2, 383.8, 518.4 nm), calcium (393.4, 396.8, 422.7 nm) and sodium (589.0 nm). From the spectrum (figure 3) it can be seen that even small concentrations of salts lead to the emergence of rather intense spectral lines corresponding to the dissolved elements. Spectral line 656.3 nm related to the emission of hydrogen atoms can be used for normalization of the obtained spectra.

5. Conclusions

System for analysing the composition of liquids using spectral analysis of the radiation of the discharge in liquid was developed as the result of the conducted research. This device uses a rather simple, but efficient electrical scheme of a high voltage power supply and a specially developed discharge cell. Designing was made bearing in mind the potential field usage of the device. Testing of this system using deionized and mains water showed its high sensitivity for magnesium, calcium and sodium. As for the other chemical elements results are not so promising.

For increasing the efficiency of the system a further research of the more efficient mechanical properties of the system and also electric parameters of the high voltage power source is required. It should be found an optimal scheme of the discharge cell which would allow effective obtaining of the spectral data and also be more durable and less inclined to pollution.

Acknowledgment
The research was partially supported by FASIE.

References

[1] Mezei P and Cserfalvi T 2007 *Applied Spectroscopy Reviews* **42** 573–604
[2] Webb M R and Hieftje G M 2009 *Analytical Chemistry* **3** 862–7
[3] Ramazanov A N, Kostrin D K, Goncharov V D and Lisenkov A A 2016 *Journal of Physics: Conference Series* **729** 012004
[4] Sokolov M A and Brytov I A 2010 *Journal of Analytical Chemistry* **11** 1118–25
[5] Uhov A A, Gerasimov V A, Kostrin D K and Selivanov L M 2014 *Journal of Physics: Conference Series* **567** 012039
[6] Kostrin D K, Uhov A A and Lisenkov A A 2016 *Vakuum in Forschung und Praxis* **3** 34–7
[7] Kostrin D K, Lisenkov A A and Uhov A A 2016 *Journal of Physics: Conference Series* **735** 012055

Investigation of Doppler spectra of laser radiation scattered inside hand skin during occlusion test

I.O. Kozlov, E.A. Zherebtsov, A.I. Zherebtsova, V.V. Dremin, A.V. Dunaev

Biomedical Photonics Instrumentation Group, Scientific-Educational Center of "Biomedical Engineering", Orel State University named after I.S. Turgenev, Orel, Russia

igor57_orel@mail.ru

Abstract. Laser Doppler flowmetry (LDF) is a method widely used in diagnosis of microcirculation diseases. It is well known that information about frequency distribution of Doppler spectrum of the laser radiation scattered by moving red blood cells (RBC) usually disappears after signal processing procedure. Photocurrent's spectrum distribution contains valuable diagnostic information about velocity distribution of the RBC. In this research it is proposed to compute the indexes of microcirculation in the sub-ranges of the Doppler spectrum as well as investigate the frequency distribution of the computed indexes.

1. Introduction

The LDF-method allows to assess the dynamics of a tissue structures blood perfusion by index of microcirculation [1, 2]. This is an integrated assessment which characterizes the general functional state of a local microcirculation system:

$$I_m(\omega) = \int_0^\infty \frac{P(\omega) \cdot d\omega}{i_{dc}}, \qquad (1)$$

where $I_m(\omega)$ – index of microcirculation; ω – frequency of Doppler shift; $P(\omega)$ – amplitude value of power spectrum; i_{dc} – DC component of the photocurrent. According to its physical meaning, index of microcirculation is proportional to the average velocity and concentration of the red blood cells. The time based changes of the registered LDF device perfusion signal contain two main components: constant and variable. The constant component is the average blood perfusion of the selected time interval. The variable component of the signal is caused by physiological factors regulating blood volume and reflects the frequency rhythms of blood flow regulation. Both components are important for the diagnostics of quite a number of diseases. In LDF devices, Doppler shift spectrum is registered in the range of 20 – 24,000 Hz arising after the reflection of radiation from an ensemble of red blood cells moving at different speeds in the range 0.1 – 10 mm/s in small vessels – arterioles, capillaries, and venules in diagnostic volume [3, 4]. Signals from microvessels with different blood flow velocity are mixed in one output time domain signal of perfusion. Consequently, perfusion distribution by Doppler shift frequency is commonly lost during traditional signal processing. Whereas, direct digital processing of received photocurrent is more informative. It was suggested that utilizing occlusion test, cold test and orthostatic test leads to redistribution of perfusion by Doppler shift frequency ranges and this effect may vary in groups of healthy people and people with microcirculatory dysfunction. Thus, redistribution of signal during provocative factors can give significant diagnostic information.

2. Experimental method and equipment

For the designed experimental study appropriate hardware and software were developed. As a laser source, one-mode laser with 1064 nm wavelength was utilized. Si-photodiodes were used to convert detected radiation into photocurrent. Optical fibers were used to deliver radiation to the skin and to collect backscattering light. In the next step, the signal is amplified in a custom electronic board. Analog-to-digital conversion was performed by data acquisition board NI USB 6211. Finally, NI LabVIEW environment installed on PC was implemented for signal processing [5] (Fig. 1). Power spectrum of the signals is computed by classic algorithm based on Fast Fourier Transform.

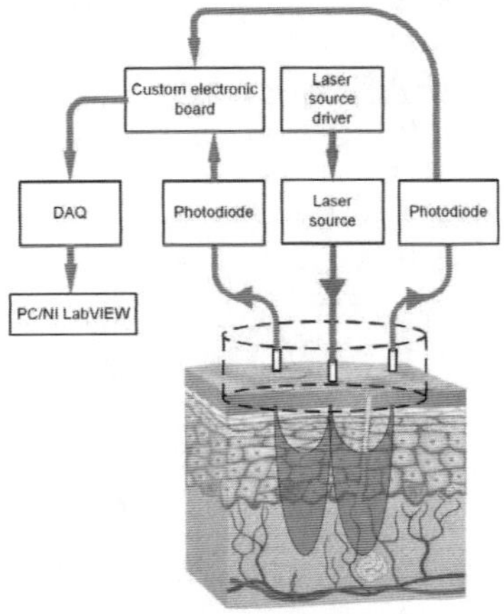

Figure 1 – Scheme of laser Doppler flowmetry device

Applying standard physiological functional tests to a limb results in changing RBC velocity distribution inside the skin. Fourteen experiment of shoulder occlusion were performed in order to record such alternations during the occlusion test [6]. Experiments involved only healthy volunteers. During the experiments simultaneously obtained parts of power spectra from fingers in consecutive frequency ranges 60 – 400 Hz, 400 – 800 Hz, 800 – 1600 Hz, 1600 – 3200 Hz, 3200 – 6400 Hz were also processed. Every experiment was conducted following the protocol: recording of the background level of perfusion (3 min); occlusion test (3 min); post-occlusion recording (5 min). Example of the LDF-gram with integrating in range of Doppler shift 60 – 6400 Hz is presented at Fig. 2. Examples of the LDF-grams with integrating in subranges are presented at Fig. 3.

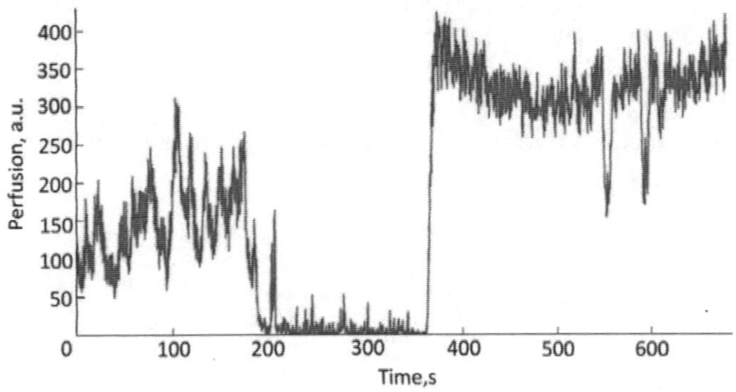

Figure 2 – Example of LDF-gram computed during occlusion test in frequency range 60 – 6400 Hz

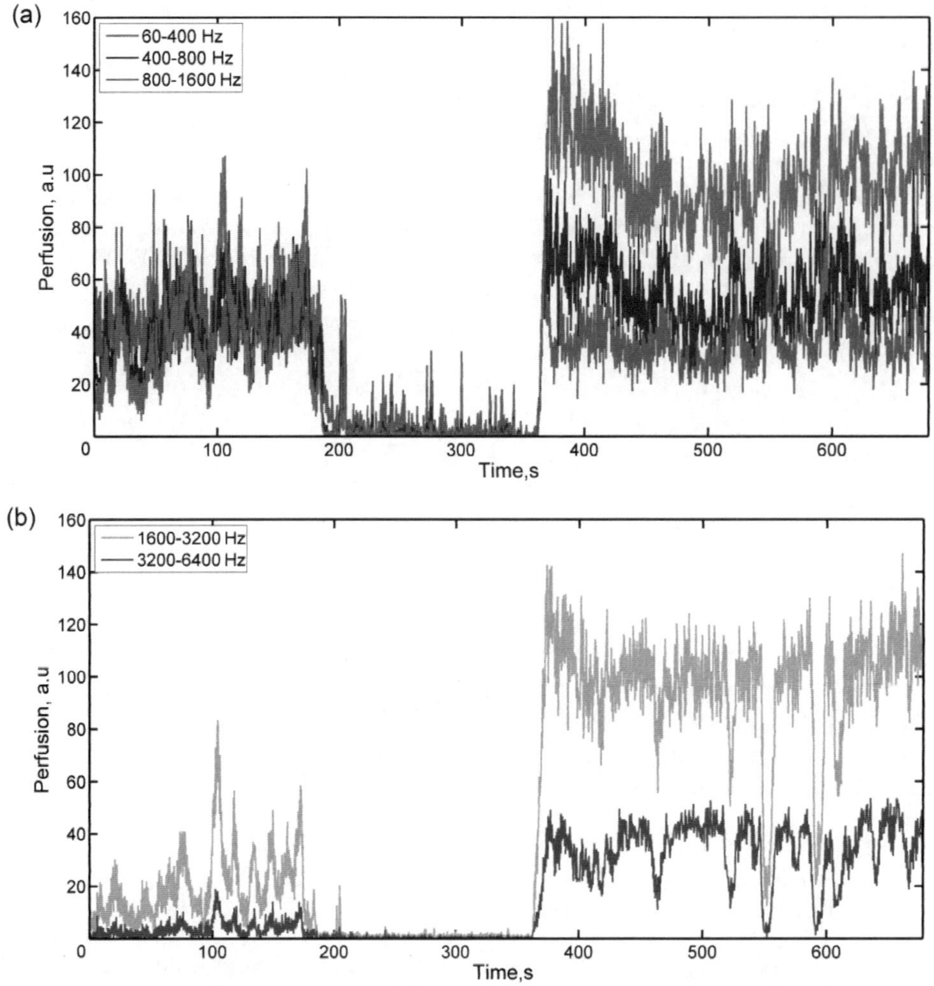

Figure 3 – Example of LDF-grams computed during occlusion test in frequency ranges 60 – 400 Hz, 400 – 800 Hz, 800 – 1600 Hz (a), 1600 – 3200 Hz, 3200 – 6400 Hz (b)

3. Results and discussion

Processing of the obtained experimental data has shown that at the moment of the post-occlusion reactive hyperemia (PORH) Doppler power spectrum undergoes broadening and the spectrum maximum shifts to the high-frequency ranges [7]. This effect can be explained by increasing number of ensembles of RBC with higher velocity in the optical sampling volume of the skin and increasing RBC concentration in low frequency ranges [8, 9]. The main statistical parameters (mean and standard deviation) of perfusion alteration during the occlusion tests were calculated in the selected spectral ranges of power spectrum integration (Fig. 4).

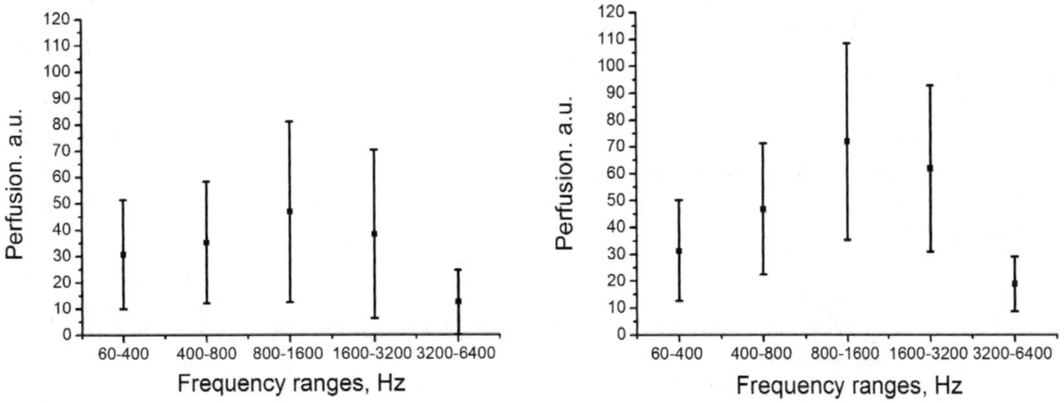

(a) (b)

Figure 4 – Means and standard deviations of perfusion calculated in frequency sub-ranges: (a) pre-occlusion period; (b) post occlusion reactive hyperaemia

In addition, method of visualization of perfusion distribution by frequency of Doppler shift was offered (Fig. 5). These graphs represent values of multiplication of power spectrum and corresponding frequency in color scale according to the following equation:

$$I_m(\omega) = \frac{P(\omega)\cdot\omega}{i_{dc}} \quad (2)$$

Then, all processed perfusion values are colored and presented in time domain.

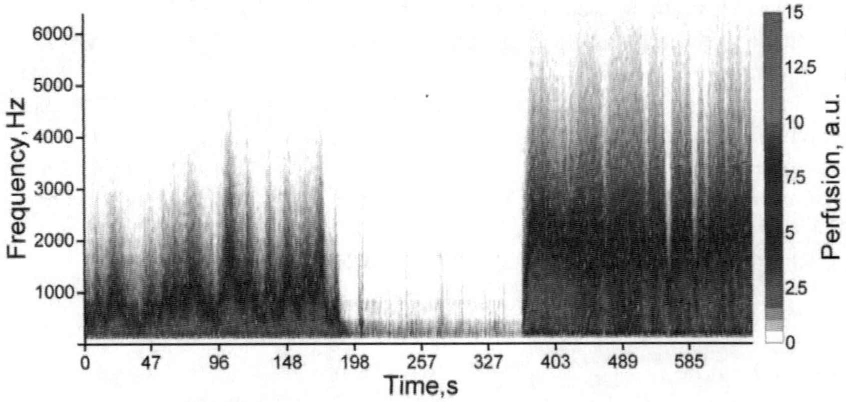

Figure 5 – Visualization of perfusion distribution by frequency of Doppler shift during experiment represented in Fig. 2

This approach can demonstrate registered effect of perfusion redistribution after occlusion. As it presented on the graph, perfusion distribution shifts up to 6000 Hz during post-occlusion reactive hyperemia.

4. Conclusion

As the result of the study we have obtained transient dynamics of perfusion distribution from frequency during such provocative factor as occlusion test. Proposed approach of perfusion distribution visualization can indicate information about concentration and velocity characteristics of blood flow. Further experimental studies in the group of patients with microcirculation diseases are planned. Moreover, identification of microcirculation components such as capillary, arteriole and venule blood flows by analyzing of frequency distribution of Doppler shift is planned. The comparison of the data from the patients with the data from healthy volunteers will allow to substantiate new diagnostic criteria.

Acknowledgements
This work was supported by Russian Foundation for Basic Research (grant №16-32-00662 mol_a) and partially by grant of the President of the Russian Federation for state support of young Russian scientists № MK-7168.2016.8.

References

[1] Bonner R and Nossal R 1981 Model for laser Doppler measurements of blood flow in tissue *Appl. Opt.* **20** 2097-107
[2] Leahy M J and Nilsson G E 2010 Laser Doppler flowmetry for assessment of tissue microcirculation: 30 years to clinical acceptance *Proc. SPIE* **7563** 75630E
[3] Dunaev A V, Zherebtsov E A, Rogatkin D A, Stewart N A, Sokolovski S G and Rafailov E U 2013 Substantiation of medical and technical requirements for noninvasive spectrophotometric diagnostic devices *Journal of Biomedical Optics* **18**
[4] Dunaev A V, Zherebtsov E A, Rogatkin D A, Stewart N A, Sokolovski S G and Rafailov E U 2014 Novel measure for the calibration of laser Doppler flowmetry devices *Proc. SPIE* **8936** 89360D
[5] Hu C-L, Lin Z-S, Chen Y-Y, Lin Y-H and Li M-L 2013 Portable laser Doppler flowmeter for microcirculation detection *Biomedical Engineering Letters* **3** 109-14
[6] Zherebtsova A I, Zherebtsov E A, Dunaev A V, Podmasteryev K V, Pilipenko O V, Krupatkin A I, Khakhicheva L S and Muradyan V F 2015 Study of the functional state of peripheral vessels in fingers of rheumatological patients by means of laser Doppler flowmetry and cutaneous thermometry measurements *Proc. SPIE* **9917** 99170M
[7] Nilsson G E 1984 Signal processor for laser Doppler tissue flowmeters *Medical and Biological Engineering and Computing* **22** 343-8
[8] Fredriksson I, Larsson M and Strömberg T 2010 Model-based quantitative laser Doppler flowmetry in skin *Journal of Biomedical Optics* **15** 057002--12
[9] Jonasson H, Fredriksson I, Larsson M and Strömberg T 2015 *16th Nordic-Baltic Conference on Biomedical Engineering: 16. NBC & 10. MTD 2014 joint conferences. October 14-16, 2014, Gothenburg, Sweden,* ed H Mindedal and M Persson (Cham: Springer International Publishing) pp 52-4

New method for determining concentrations of the mixture components during rapid control

N S Myazin[1], S E Logunov[1], V V Davydov[1, 2, 3], V Yu Rud'[3], N M Grebenikova[1], V V Yushkova[4]

[1]Higher School of applied physics and space technologies, Peter the Great Saint Petersburg Polytechnic University, Saint Petersburg 195251, Russia
[2]Department of photonics and communication lines, The Bonch-Bruevich Saint Petersburg State University of Telecommunications, Saint Petersburg 193232, Russia
[3]All-Russian Research Institute of Phytopathology, Moscow Region 143050, Russia
[4]Saint Petersburg University of Management Technologies and Economics, Saint Petersburg 190103, Russia

myazin.n@list.ru

Abstract. The paper discusses the features of the signal registration of nuclear magnetic resonance in a weak magnetic field from small volume of condensed medium. Method for determining the state of condensed medium in express mode, using measured by nuclear magnetic resonance relaxation time constants T1 and T2, is proposed. We have developed the new method of determining the composition of the investigated mixture and relative concentrations of its components. That method is applicable for mixtures that formed by condensed media similar in both physical structure and chemical composition.

1. Introduction

One of the important directions of condensed media researches is the improvement of existing and development of new methods of rapid control [1–3]. For example, this is especially relevant for environmental monitoring of inaccessible areas of the water bodies and coastal zones, as well as in checking the quality of food products [3–5].

Nowadays requirements for methods of the rapid control became higher. Among them, for example, high measurement accuracy (error not exceeding 1.5%) and universality (method should be applicable for investigation of a large amount of media). However, there is one more criteria, and in most cases, it became primary. Investigation by the rapid control should not change the chemical composition and physical structure of the researched medium. This is required for obtaining confirmation about identified deviations in the sample of condensed medium (this is especially relevant in the customs control or verification of product quality) using the high-resolution spectrometers (e.g., multi-functional X-ray, optical or magneto resonance). Results of the medium researches in these devices provide comprehensive information about its composition. However, these devices are set in stationary laboratories (at significant distance from the place of rapid control conducting). In addition, they are expensive equipment that require special operating conditions and quite expensive for maintenance [1-5]. Therefore, it is desirable to load them only by samples, which shows a deviation from the standard state in order to confirm the results of the rapid control.

Such high requirements significantly restricts the possibility of use of the common rapid control techniques. For example, use of the small-sized optical devices (spectrophotometers, fluorimetric indicators, etc.) for investigation of the liquids with high turbidity require increasing in the laser power

IOP Conf. Series: Journal of Physics: Conf. Series **929** (2017) 012064 doi:10.1088/1742-6596/929/1/012064

[4 6], which leads to the medium heating. Because of an increase in temperature, significant structural changes can occur in the environment, living in biological solutions organisms may die, etc.

The use of nuclear magnetic resonance phenomena for rapid control is one of the solutions to this problem, because researches carried out by nuclear magnetic spectroscopy do not make irreversible changes in the physical structure and chemical composition of the investigated medium.

Nevertheless, use of nuclear magnetic spectroscopy have a number of difficulties. Those difficulties related primarily to the fact that the weight of the magnetic system is the main part of the NMR spectrometer weight. In order to maintain mobility of devices for rapid control their weight should not exceed 4–5 kg. This has led to the fact that the NMR signal is registered in a weak magnetic field with induction $B_0 \approx 100$ mT from small volume of liquid medium [1, 3, 5]. In such circumstances, the effective registration of the NMR signal from liquid medium should be carried out at the protons resonance frequency because protons have the highest sensitivity to the NMR method [3, 4, 7]. In the case of the NMR signal registration at the resonant frequencies of other nuclei signal/noise ratio (S/N) in most cases will be less than 1.3. This S/N value does not allow carrying out measurements with an accuracy higher than 1.5%. This fact eliminates the possibility of the spectrum registration from the investigated medium, as it is implemented in stationary NMR spectrometers [2–4]. Therefore, we developed a method of determining the state of the medium, based on measuring the time of longitudinal T_1 and transverse T_2 relaxation. The articles [3, 5] has showed that we can unmistakably determine the state of the medium only by using results of simultaneous measurements of both relaxation constants.

2. The compact nuclear magnetic spectrometer

To implement the proposed methodology of the rapid control of condensed media we have assembled compact layout of the NMR spectrometer, its structural diagram is shown in figure 1.

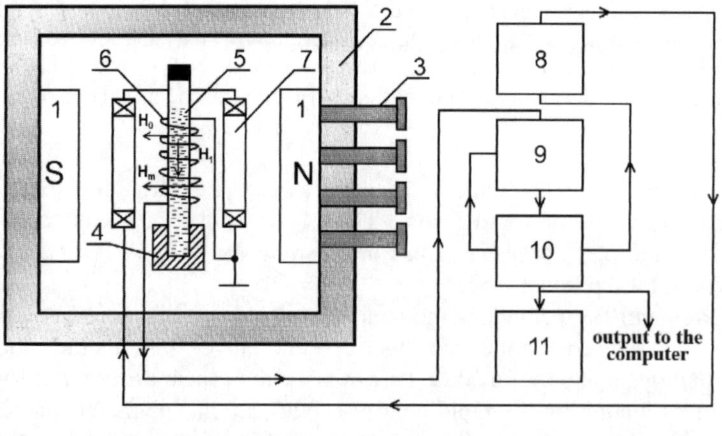

Figure 1 Structural diagram of a small-sized NMR spectrometer: 1 — permanent magnet; 2 — neutral for the placement and alignment of the magnets; 3 — adjusting screws; 4 — locking device for the container with the investigated medium; 5 — container with the investigated medium; 6 — NMR signal registration coil; 7 — modulation coil; 8 — RF generator; 9 — autodyne detector (weak oscillation generator); 10 — processing and control unit; 11 — oscilloscope.

We have designed and manufactured compact magnetic system based on two flat (rectangular) samarium-cobalt magnets 1 (figure 1) with a platinum admixture. In addition, we have designed compact "neutral" 2 with adjusting screws 3 (figure 1) for aligning mutual arrangement of the magnetic system poles. The weight of proposed magnetic system design reduced to 2.8 kg, what corresponds to the requirements of mobility. In our NMR spectrometer design the distance between the magnet poles is 13 mm, magnetic induction $B_0 = 52.8$ mT, inhomogeneity 10^{-3} cm^{-1}, the frequency of the NMR signal registration from protons $f_{NMR} = 2249$ kHz.

Taking into account such conditions, signal registration is possible only with the use of modulation technique. The signals in this case are damped oscillations ("wiggles"). The implementation of pulse methods that are used in stationary NMR spectrometers and relaxometers for registration of the NMR signal is extremely difficult in such small interpolar space [1, 3, 5]. Moreover, compact power sources cannot provide necessary energy for pulse methods for the NMR signal registration even for 20-30 minutes.

For NMR signals registration in a small magnetic system using the modulation technique we designed special probe (figure 2).

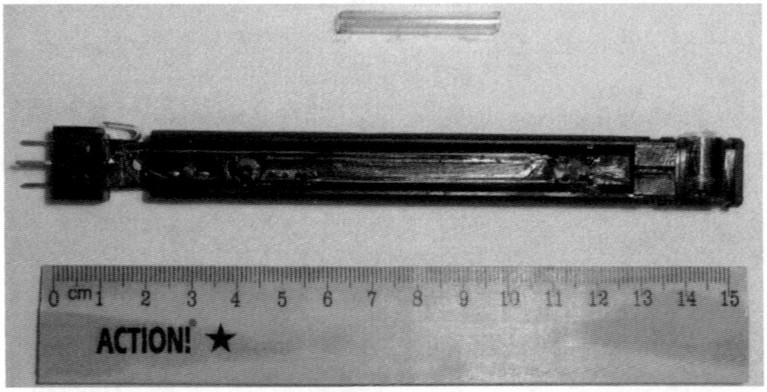

Figure 2 The measuring probe and container for medium

It includes registration coil 6, the modulation coil 7 and device for placing the container (cuvette) (figure 1, 2). The inner cuvette diameter is 3.2 mm; external diameter is 4.8 mm. In high-resolution NMR spectrometers diameters of the cuvette is about 3.0 and 4.6 mm. On figure 3 as example are shown NMR signals from synthetic motor oil ZIC XQ 5W-40 at a temperature T = 291.6 K.

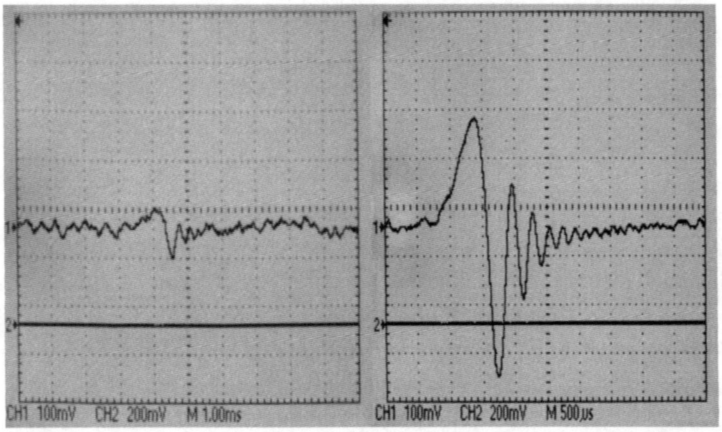

Figure 3 (a, b) Registered NMR signal from the engine oil: (a) — without accumulation; (b) — after accumulation

The analysis of presented in the figure 3 NMR signals shows that we have to use accumulation schemes, because without them it is impossible to be sure that the S/N ratio will be greater than 3.0 which is necessary for the measurement with an accuracy higher than 1.5%

It should also be noted that the use of a probe developed by us allows NMR signal registration from the investigated medium volume V_R less than 0.3 ml. This is important because for precision measurement we have to know sample temperature. If the medium contains in a small volume, after 10–15 s its temperature becomes equal to air temperature, which we can measure using standard instruments with high precision. Use of small container with volume V_R is one of the advantages of the compact NMR spectrometer in comparison with other devices.

3. Methods for determining the relaxation constants

To determine the transverse relaxation time T_2 we use a standard method based on the determination of the effective transverse relaxation time T_2^* from the decay parameters of the line shape envelope of the registered NMR signal [1, 3, 5, 7]. Figure 4 shows an example of the construction envelope of the NMR signal from gel that contains protons.

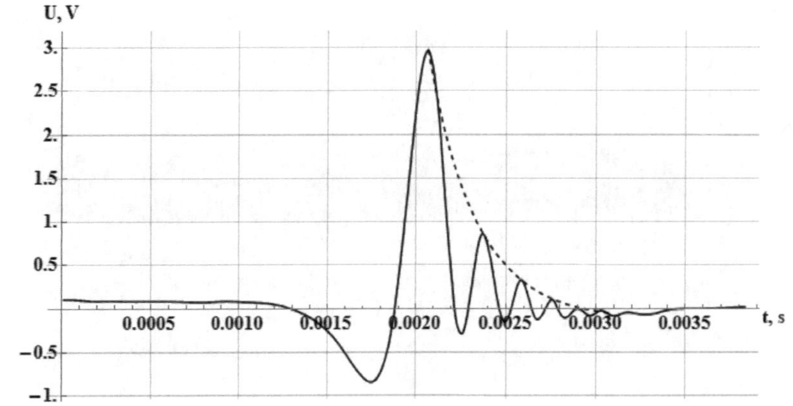

Figure 4 The line shape of the NMR signal registered from the gel at T = 292.3 K

Transverse relaxation time T_2 of investigated medium can be calculated using the ratio:

$$\frac{1}{T_2^*} = \frac{1}{T_2} + \frac{\gamma \Delta \mathrm{H}}{\pi}$$

where ΔH is inhomogeneity of the magnetic field in the placement zone of the NMR signal registration coil.

Almost all successfully used in high-resolution NMR spectrometers standard methods for determining T_1 cannot be implemented in compact NMR spectrometer, except one based on a Giulotto method [8]. But in this method it is necessary to register the shift of the resonance frequency, which in a weak magnetic field is about 0.1–0.2 Hz. Measuring this change in frequency is quite difficult even in the laboratory and almost unrealizable task in field conditions. That is why to determine the longitudinal relaxation time T_1 in a weak magnetic field we have develop the new technique. The amplitude of the registered NMR signal using a modulation method is proportional to the magnetization M of the investigated medium nuclei. Registration of the NMR signal from the medium on a resonant frequency f_{NMR} in a weak magnetic field carried out with the rapid passage through resonance, which corresponds to ratio [7, 8]:

$$H_1 \gg (|\gamma| T_2^*)^{-1}$$

In this case, the investigated medium magnetization M in the registration coil will change as:

$$M = M_0 \left(1 - \frac{1 - \exp\left(-\frac{\tau}{T_1}\right)}{1 + \exp\left(-\frac{\tau}{T_1}\right)} \right) \quad (1)$$

Then, for determination T_1 using (1) we have to measure NMR signal amplitude at two different $\tau = T_m/2$ (T_m – modulation period of field B_0) and solve the system that consist of the equations (1) for various τ. The resulting system of equations is easily transformed into:

IOP Conf. Series: Journal of Physics: Conf. Series **929** (2017) 012064 doi:10.1088/1742-6596/929/1/012064

$$\frac{A_1}{A_2} = \frac{exp\left(-\frac{\tau_1}{T_1}\right) \cdot \left(1 + exp\left(-\frac{\tau_2}{T_1}\right)\right)}{exp\left(-\frac{\tau_2}{T_1}\right) \cdot \left(1 + exp\left(-\frac{\tau_1}{T_1}\right)\right)}, \tag{2}$$

where $A_{1,2}$ — the amplitude of the registered NMR signal for $\tau = \tau_{1,2}$ respectively.

The solution of system (2) for T_1 can obtained both numerically and graphically. To confirm the validity and reliability of used measurement techniques in compact NMR spectrometer we have determined relaxation constants for motor oil (figure 3): $T_1 = 1.021 \pm 0.005$ s and $T_2 = 720.36 \pm 3.59$ μs. Also we have investigate same sample at the same temperature in a stationary NMR relaxometer Minispec mq20 (manufactured by BRUKER), measurement results: $T_1 = 1.0207 \pm 0.0021$ s and $T_2 = 718.94 \pm 1.44$ μs. The values of the relaxation constants matched within the measurement error.

4. The new method for medium components concentrations determining

Recently, during monitoring of the medium state using express methods, often occurs situations when it is desirable to make a decision at place of research, rather than wait for the results of additional studies from a stationary laboratory. This is especially relevant for express control of fuels and other hydrocarbon compounds. To solve this problem using the results of measurements by a small-size NMR spectrometer in the express mode, we developed a new method for processing the detected NMR signal of the medium. New method allows determining the concentration of components in the researched medium if it is formed by substances that are close in both chemical composition and physical structure (for example, a mixture of gasolines, oils or fats). In this case, there is no dissolution of one medium into another, and substances forms a conglomerate. The detected NMR signal from such mixture is the sum signal from each of the mixture components. We proposed a mathematical model that allows us to describe the detected NMR signal from the researched mixtures. The proposed model uses solutions of the Bloch equations in a rotating coordinate system [7, 8]:

$$\begin{gathered}\frac{du}{dt} + \frac{u}{T_2^*} + \Delta\omega \cdot v = 0 \\ \frac{dv}{dt} + \frac{v}{T_2^*} - \Delta\omega \cdot u = -\gamma H_1 M_z \\ \frac{dM_z}{dt} + \frac{M_z}{T_1} - \gamma H_1 v = \frac{M}{T_1}\end{gathered} \tag{3}$$

where $\Delta\omega = \gamma H_0 - \omega$ — field frequency deviation from resonance, $M = \chi_0 H$— magnetization of the investigated medium in the magnetic field of the spectrometer, $v(t), u(t)$ — absorption and dispersion signals.

In case of using a modulation technique to detect the NMR signal in a weak magnetic field, the value of H varies as follows:

$$H = H_0 + H_m sin(\omega_m t), \tag{4}$$

where H_0 — constant magnetic field, H_m — modulation coil field, ω_m — modulation frequency (figure 1).

In this case, in the system of Bloch equations the change in the field frequency deviation from the resonance, with allowance for (4), will have the following dependence:

$$\Delta\omega = \gamma H_0 + \gamma H_m sin(\omega_m t) - \omega \tag{5}$$

One of the features of registering the NMR signal in a weak magnetic field using the modulation technique is that registering should be performed only at the resonance frequency of medium. For most researched media, in the case of detuning the frequency of NMR signal registration ω from the resonance frequency ω_0, the S/N ratio can become less than 1.3. With such value of S/N, the operation of the NMR signal accumulation circuit will be ineffective (without accumulation it is impossible to measure the relaxation constants T_1 and T_2 with the required accuracy) [3, 5, 6]. Therefore, to describe the detected NMR signal using the modulation technique, we can transform equation (5) to the following form:

$$\Delta\omega = \gamma H_m sin(\omega_m t) \tag{6}$$

At the resonant registering frequency $\omega = \omega_0$, taking into account (6), the system of Bloch equations (3) takes the following form:

$$\begin{gathered} u'(t) + \frac{u(t)}{T_2} + \gamma H_m \sin(\omega_m t)\, v(t) = 0 \\ v'(t) + \frac{v(t)}{T_2} - \gamma H_m sin(\omega_m t) u(t) - \gamma H_1 M_z(t) = 0 \\ M_z'(t) + \frac{M_z(t)}{T_1} - \gamma\, v(t)\, H_1 - \frac{M}{T_1} = 0 \end{gathered} \tag{7}$$

where $M = \chi_0 H_0 + \gamma H_m sin(\omega_m t)$ — magnetization of the investigated medium in the magnetic field of the spectrometer, $\chi_0 = NI(I+1)\frac{\mu^2}{3kT}$ — statistical nuclear magnetic susceptibility, N — concentration of paramagnetic particles; μ — particle magnetic moment; k — Boltzmann's constant; T — absolute temperature.

This system of equations is solved for $v(t)$, $u(t)$ and $M_z(t)$ components, taking into account the initial conditions:

$$M_z(0) = \chi_0 H_0;\ u(0) =; v(0) = 0$$

The experimental researches carried out by us showed that the shape of the $G(t)$ line of the registered NMR signal from the substance in a weak magnetic field when tuning the autodyne detector circuit to the maximum of the S/N ratio is described by the following relation:

$$G(t) = \sqrt{Av^2(t) + BU^2(t)}, \tag{8}$$

where $v(t), U(t)$ — absorption and dispersion signals, A, B — coefficients determining the contribution to the NMR signal of the absorption and dispersion signals.

The obtained solution (8), which is in good agreement with the experimental results, made it possible to develop a technique for simulating the detected NMR signal from the mixture, by dividing the received signal into signals from its constituent components. Following relation can represent the shape of the registered NMR signal $G_m(t)$ from the mixture:

$$\begin{gathered} G_m(t) = \sqrt{A_m v_m^2(t) + B_m U_m^2(t)} = \sum_{i=1}^{k} V_i \cdot N_i \sqrt{A_i v_i^2(t) + B_i U_i^2(t)} \\ \sum_{i=1}^{k} V_i = V_r \end{gathered} \tag{9}$$

where $v(t), U(t)$ — absorption and dispersion signals, A, B — coefficients determining the contribution to the NMR signal of the absorption and dispersion signals (m — mixture, i — mixture components).

Since in the experiment all NMR signals, both from the mixture and its components, are formed in the same fields H_0, H_1 and H_m, the solutions obtained for their components $v(t)$ and $U(t)$ from the Bloch equations (7), differ only in the relaxation constants. For the mixture itself, T_1 and T_2 are determined by the NMR signal detected from it. In addition, we know relaxation constants for one of the mixture components, because initially this mixture should be this component in its pure form. The relaxation constants of the remaining components of the mixture, as well as the volumes in which they are contained in the researched medium, are chosen so that (9) is satisfied, the temperature of the researched mixture is known. When relation (9) is fulfilled, through relaxation constants we can determine mixture components, as well as their relative concentrations through their volumes.

IOP Conf. Series: Journal of Physics: Conf. Series **929** (2017) 012064 doi:10.1088/1742-6596/929/1/012064

Figure 5 shows, as an example, the possibility of determining the components and their concentrations in a mixture of two gasolines AI-95 and A 76.

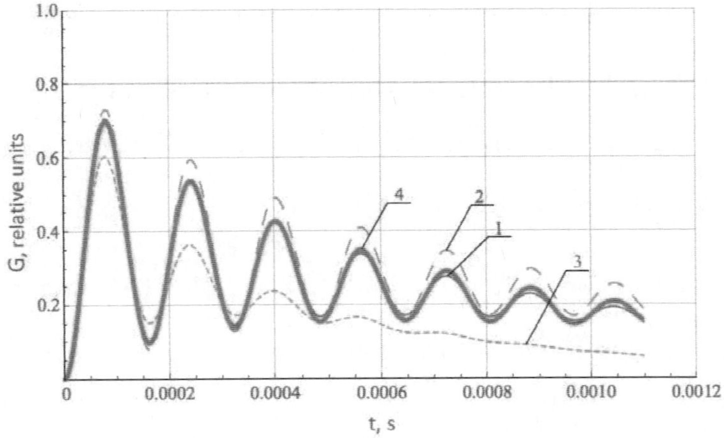

Figure 5 NMR signal line forms. Graph 1 corresponds to the experimental signal from a mixture of gasoline AI-95 and A-76 in the proportion of 75% to 25%. The NMR signal simulation: from pure gasoline AI-95 and A-76 - graphs 2 and 3, mixture of gasoline AI-95 and A-76 in a proportion of 75% to 25% - graph 4.

5. Conclusion

These results demonstrate that the proposed method allows determining composition and ratio between the components during rapid control of the conglomerates. This is greatly extends the functionality of the NMR methods for rapid control. Previously, measurements of T_1 and T_2 gave information only about the presence of deviation from the medium standard state and required its additional study in a stationary laboratory to make a reliable decision on the further use of this medium.

6. References

[1] Karseev A Yu, Vologdin V A, Davydov V V 2015 *Journal of Physics: Conference Series* **643** 012108

[2] Kashaev R S, Gazizov E G 2010 *Journal of Applied Spectroscopy* **77** 347

[3] Davydov V V, Velichko E N, Dudkin V I, Karseev A Yu 2015 *Instruments and Experimental Techniques* **58** 234

[4] Erdogru Y, Manimaran D, Gulluolu M T, Amalanathan M, Hubert Joe I, Yurdakul N 2013 *Optics and Spectroscopy* **113** 573

[5] Davydov V V, Dudkin V I, Karseev A Yu 2014 *Measurement Techniques* **57** 912

[6] Vologdin V A, Davydov V V, Velichko E N 2016 *Journal of Physics: Conference Series* **741** 012095

[7] Leshe A 1963 *Nuclear induction* 684

[8] Chiarotti G, Cristiani G, Giulotto L 1954 *Phys. Rev.* **93** 1241

IOP Conf. Series: Journal of Physics: Conf. Series **929** (2017) 012065 doi:10.1088/1742-6596/929/1/012065

Destructive interference between electric and toroidal dipole moments in TiO_2 cylinders and frustums with coaxial voids

P D Terekhov[1], K V Baryshnikova[1], A B Evlyukhin[1,2] and A S Shalin[1]

[1] ITMO University, 49 Kronversky Ave., 197101, St. Petersburg, Russia
[2] Laser Zentrum Hannover e.V., Hollerithallee, D-30419, Hannover, Germany

E-mail: terekhovpd@gmail.com

Abstract. We demonstrate numerically the possibility of multipole interference in the TiO_2 (titanium dioxide) microcylinders and microfrustums in the wavelength range 210-300 μm. Resonantly strong destructive interference between toroidal and electric dipole contributions to the scattered field is achieved by a geometry tuning. The toroidal and electric dipole mode overlapping at the resonant wavelength with almost total suppression of the total electric dipole moment is achieved.

1. Introduction

Non-radiating current configurations attract attention of physicists and engineers as possible systems for effective cloaking and camouflage [1]. In this work we consider dielectric cylinder with coaxial through hollow under plane wave irradiation. We show that in this system toroidal moment can be effectively induced. Toroidal dipole moment produces non-zero far-field radiation with a similar pattern as an electric dipole moment [2]. Interference of electric and toroidal dipole moments could be in destructive way with total suppression of far fields and reducing of the overall scattering down to the zero. We have researched similar effects for TiO_2 microcylinders in the frequency range 2-3 THz (100-150 μm) in [3]. In this work we continue our study and extend it to microfrustums with coaxial void. We obtain far-field scattering cross-section suppression in the wavelength range 210-300 μm. This frequency range allows us to perform future experiments to prove our researches.

2. Results

2.1. Main formulas

The schematics of the considered system is shown in Fig. 1 for the cylinder and in Fig. 2 for the frustum. Ordinary, usual electric dipole moment is calculated [4] as

$$\mathbf{p} = \int \mathbf{P}(\mathbf{r}')d\mathbf{r}'. \tag{1}$$

where $\mathbf{P}(\mathbf{r}')$ is a polarization vector, $\mathbf{r}'$ is the radius-vector of volume element. According to [4, 5] we calculate multipole moments including toroidal dipole moment via induced polarizations

Figure 1. Cylinder, irradiated by plane wave, which propagates along the cylinder axis. **k** is the wave vector, **E** is the electric field, **H** is the maghetic field.

Figure 2. Frustum, irradiated by plane wave, which propagates along the cylinder axis. **k** is the wave vector, **E** is the electric field, **H** is the maghetic field.

inside scatterer (COMSOL Multiphysics). For example, toroidal dipole moment is determined as [4, 5]

$$\mathbf{T} = \frac{i\omega}{10}\int\{2\mathbf{r}'^2\mathbf{P}(\mathbf{r}') - (\mathbf{r}'\cdot\mathbf{P}(\mathbf{r}'))\mathbf{r}'\}d\mathbf{r}'. \quad (2)$$

The toroidal dipole moment is the third order multipole moment, and one can expect that its contributions rises as the particle size increases with respect to considered wavelength. In our case the particles are large enough in order to support the resonant exitation of the third order multipoles. We calculate the contribution of electric dipole (ED) and toroidal dipole (TD) moments to the scattered field in the far-field as [4, 5]

$$\mathbf{E}_{sca}(\mathbf{r}) \sim \frac{k_0^2}{4\pi\varepsilon_0}\{\mathbf{n}\times\mathbf{P}\times\mathbf{n} + \frac{ik_0\varepsilon_d}{c}\mathbf{n}\times\mathbf{T}\times\mathbf{n}\}, \quad (3)$$

where $\mathbf{r}$ is the radius-vector of an observation point; k_0 is the free-space wave number; ε_d is the relative dielectric permittivity of the surrounding medium (in our case $\varepsilon_d = 1$); ε_0 is the vacuum permittivity; c is the light speed in the vacuum; $\mathbf{n}$ is the unit vector directed along $\mathbf{r}$. Considering several multipole moments up to electric octupole the following expression for the scattering cross-section $\sigma_{\rm sca}$ was obtained (see [4] for details):

$$\begin{aligned}\sigma_{\rm sca} \simeq\ & \frac{k_0^4}{6\pi\varepsilon_0^2|\mathbf{E}_{inc}|^2}|\mathbf{p} + \frac{ik_0\varepsilon_d}{c}\mathbf{T}|^2 + \frac{k_0^4\varepsilon_d}{6\pi\varepsilon_0|\mathbf{E}_{inc}|^2}|\mathbf{m}|^2 \\ & + \frac{k_0^6\varepsilon_d}{720\pi\varepsilon_0^2|\mathbf{E}_{inc}|^2}\sum|Q'_{\alpha\beta}|^2 + \frac{k_0^6\varepsilon_d^2}{80\pi\varepsilon_0|\mathbf{E}_{inc}|^2}\sum|M'_{\alpha\beta}|^2 \\ & + \frac{k_0^8\varepsilon_d^2}{1890\pi\varepsilon_0^2|\mathbf{E}_{inc}|^2}\sum|O'_{\alpha\beta\gamma}|^2 . \end{aligned} \quad (4)$$

where $\mathbf{m}$ is the magnetic dipole moment (MD) of particle; $|\mathbf{p} + \frac{ik_0\varepsilon_d}{c}\mathbf{T}|$ corresponds to the interference of electric dipole and toroidal dipole moments, which can be treated as total electric dipole moment (TED); Q', M' and O' are the electric quadrupole moment tensor (EQ), the magnetic quadrupole moment tensor (MQ) and the tensor of electric octupole moment (OCT) respectively, where $'$ means that matrices of these tensors have been symmetrized and they are traceless. Similar multipole approach up to the second order multipoles has been considered in classical textbooks [6, 7]. The third order multipoles has been taken into account in [5]. This procedure allows to obtain cross-contributions of different tensor moments to scattering equal to zero. Total scattering cross-section was obtained through the integration of Pointing vector over a closed surface in the far zone and the normalization on the incident field intensity [8]. Same multipole decomposition tehnique has been recently used to show so-called super-dipole regime in optical frequency range in [9].

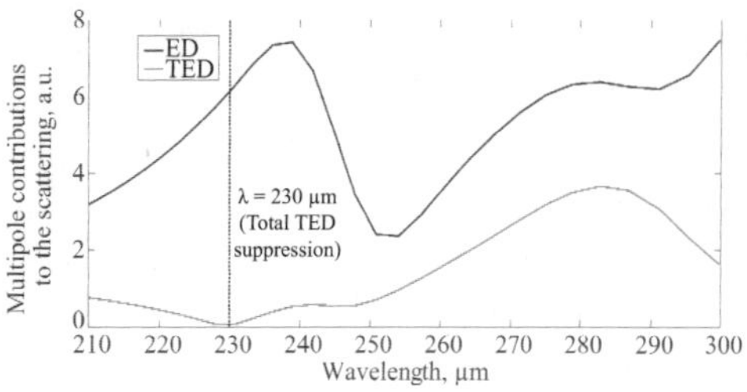

Figure 3. Absolute magnitude of electric dipole moment (blue line) and absolute magnitude of total electric dipole moment (green line) of TiO_2 cylinder, $D = 184.04\ \mu m, H = 27.64\ \mu m$, $D_{core} = 152.5\ \mu m$. The vertical black line corresponds to the wavelength of the almost total TED suppression.

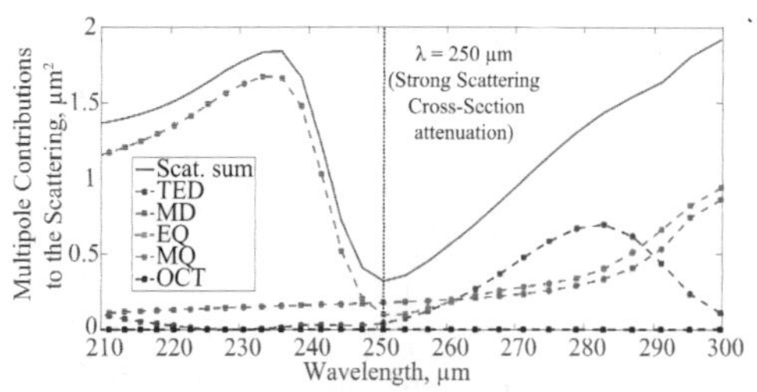

Figure 4. Scattering cross section multipole decomposition spectra for TiO_2 cylinder. $D = 184.04\ \mu m, H = 27.64\ \mu m$, $D_{core} = 152.5\ \mu m$. The dashed blue line corresponds to the TED contribution to the scattering cross-section, the dashed purple line - to the MD contribution, the dashed green line - to the EQ contribution, the dashed red line - the the MQ contribution, the dashed black line - to the OCT contribution and the solid blue line - to the total scattering cross-section.

2.2. Scattering on TiO_2 microdisks with coaxial void

In [3] we have obtained suppression of total electric dipole contribution to the far-field scattering cross section in the frequency range 2-3 THz (100-150 μm). In this work we scale this effect to the wavelength range 210-300 μm. We consider optimized microcilynder parameters and obtain two cases with same coaxial void diameter $D_{core} = 152.5\ \mu m$, same height $H = 27.64\ \mu m$ and different particle diameter $D = 184.04\ \mu m$ for the small cylinder and $D = 215.96\ \mu m$ for the big one. In Fig. 3 and 5 we present ordinary electric dipole moment and total electric dipole moment spectra for small and big cylinders respectively. In both cases we obtain the almost total suppression of total electric dipole moment due to destructive interference of electric and toroidal dipole moments at the wavelengths $\lambda = 230\ \mu m$ for the small cylinder and $\lambda = 264\ \mu m$ for the big one. Wavelengths, corresponding to the almost total suppression, are marked in Fig. 3 and 5 as vertical black lines. In Fig. 4 and 6 we present multipole decomposition of scattering cross section for small and big cylinder respectively. One can observe large magnetic quadrupole contribution. In Fig. 4 one can note significant scattering suppression due to destructive interference of electric and toroidal dipole contributions to the scattering cross-section at the wavelength $\lambda = 250\ \mu m$.

This wavelength is marked in Fig. 4 as vertical black line. Note that at the wavelengths, where TED is equal to zero, the scattering is basically determined by MQ contribution.

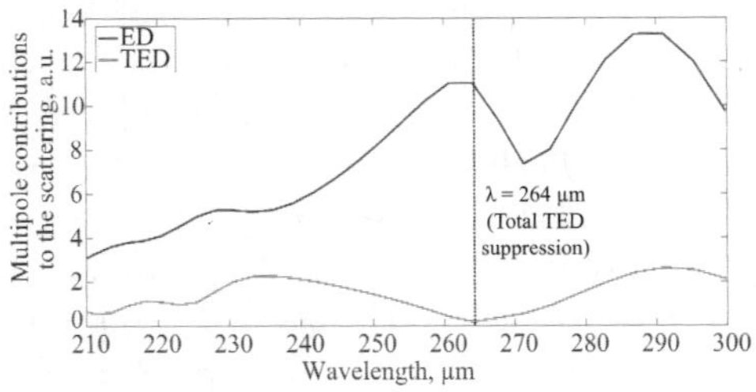

Figure 5. Absolute magnitude of electric dipole moment (blue line) and absolute magnitude of total electric dipole moment (green line) of TiO_2 cylinder, $D = 215.96\ \mu m, H = 27.64\ \mu m$, $D_{core} = 152.5\ \mu m$. The vertical black line corresponds to the wavelength of the total TED suppression.

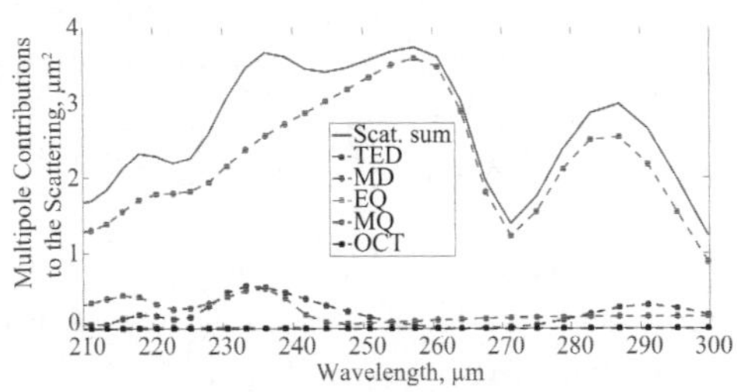

Figure 6. Scattering cross section multipole decomposition spectra for TiO_2 cylinder. $D = 215.96\ \mu m, H = 27.64\ \mu m$, $D_{core} = 152.5\ \mu m$. The dashed blue line corresponds to the TED contribution to the scattering cross-section, the dashed purple line - to the MD contribution, the dashed green line - to the EQ contribution, the dashed red line - the the MQ contribution, the dashed black line - to the OCT contribution and the solid blue line - to the total scattering cross-section.

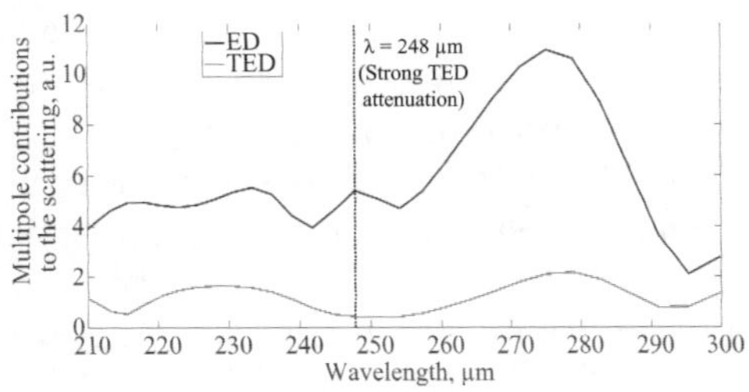

Figure 7. Absolute magnitude of electric dipole moment (blue line) and absolute magnitude of total electric dipole moment (green line) of TiO_2 frustum, $D_{top} = 215.96\ \mu m, D_{bottom} = 184.04\ \mu m$, $H = 27.64\ \mu m, D_{core} = 152.5\ \mu m$. The vertical black line corresponds to the wavelength of the strong TED attenuation.

2.3. Scattering on TiO_2 microfrustum with coaxial void

Frustum microparticles are convenient to fabricate with available equipment as well as cylindrical ones. In this work we consider frustums with coaxial void radius and height, corresponding to the previous subsection. Top frustum radius corresponds to the smaller cylinder, and the bottom radius corresponds to the bigger one. In Fig. 7 we present ordinary electric dipole moment and total electric dipole moment spectra for the TiO_2 microfrustum with $D_{top} = 215.96\ \mu m, D_{bottom} = 184.04\ \mu m, H = 27.64\ \mu m, D_{core} = 152.5\ \mu m$. It is possible to observe the minimum total electric dipole contribution to the scattering cross-section at the wavelength $\lambda = 248\ \mu m$ (marked as the vertical black line in Fig. 7. This wavelength corresponds to average value between wavelengths, corresponding to similar minimum contributions for cylinders ($\lambda = 230\ \mu m$ for the small cylinder and $\lambda = 264\ \mu m$ for the big one). However,

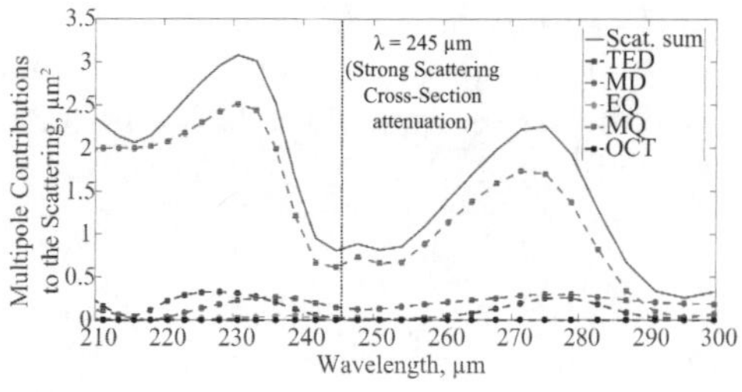

Figure 8. Scattering cross section multipole decomposition spectra for TiO_2 frustum. $D_{top} = 215.96\ \mu m, D_{bottom} = 184.04\ \mu m$, $H = 27.64\ \mu m, D_{core} = 152.5\ \mu m$. The dashed blue line corresponds to the TED contribution to the scattering cross-section, the dashed purple line - to the MD contribution, the dashed green line - to the EQ contribution, the dashed red line - the the MQ contribution, the dashed black line - to the OCT contribution and the solid blue line - to the total scattering cross-section. The vertical black line corresponds to the wavelength of strong scattering cross-section attenuation.

for the frustum microparticle one can observe not total suppression of total electric dipole moment, but its strong attenuation. Multipole decomposition of scattering cross section for considered microfrustum is presented in Fig. 8. We obtain scattering cross-section minimum at the wavelength region, corresponding to the region of total electric dipole attenuation (the vertical black line in Fig. 8). The scattering is basically determined by MQ contribution.

3. Conclusions

In conclusion, TiO_2 microcylinders and microfrustums possess multipole resonant response in the wavelength range 210-300 μm. Toroidal moment has the same contribution to the far-field scattering as electric dipole and interferes with it in destructive manner. Magnitude and resonance frequency of toroidal and electric dipole moments of the cylinder or frustum with coaxial void depends on its aspect ratio and void radius. Destructive interference of electric and toroidal dipole modes can totally destroy total electric dipole contribution to scattering. Total electric dipole moment suppression provided by microfrustum with coaxial void located on the wavelength spectrum at the region, corresponding to the average value of total electric dipole suppression effects, provided with microcylinders with coaxial voids with diameters, corresponding to top and bottom frustum diameters. Strong scattering cross-section attenuation for TiO_2 microfrustum has been obtained. Considered effect of electric dipole radiation suppression can be used in some important applications such as invisibility, optical detection and others.

Acknowledgements

This work was supported, in part, by the Russian Fund for Basic Research within the project 16-52-00112. The analytical calculation of multipolar terms has been supported by the Russian Science Foundation Grant No. 16-12-10287. A.S.S. acknowledges the support of the President of Russian Federation in the frame of Scholarship SP-4248.2016.1 and the support of Ministry of Education and Science of the Russian Federation (GOSZADANIE 2014/190). K.V.B. researching was partially supported by FASIE.

References

[1] Edwards B, Alù A, Silveirinha M G and Engheta N 2009 *Physical Review Letters* **103** 153901
[2] Miroshnichenko A E, Evlyukhin A B, Yu Y F, Bakker R M, Chipouline A, Kuznetsov A I, Lukyanchuk B, Chichkov B N and Kivshar Y S 2015 *Nature communications* **6**

[3] Terekhov P D, Baryshnikova K V, Shalin A S, Evlyukhin A B and Khromova I A 2016 Nonradiating anapole modes of dielectric particles in terahertz range *Days on Diffraction (DD), 2016* (IEEE) pp 406–409
[4] Evlyukhin A B, Fischer T, Reinhardt C and Chichkov B N 2016 *Physical Review B* **94** 205434
[5] Chen J, Ng J, Lin Z and Chan C 2011 *Nature photonics* **5** 531–534
[6] Jackson J D 1975 *Electrodynamics* (Wiley Online Library)
[7] Landau L D and Lifshitz E M 1971 *The classical theory of fields* (Pergamon)
[8] Evlyukhin A B, Reinhardt C, Evlyukhin E and Chichkov B N 2013 *JOSA B* **30** 2589–2598
[9] Terekhov P D, Baryshnikova K V, Shalin A S, Karabchevsky A and Evlyukhin A B 2017 *Optics Letters* **42** 835–838

IOP Conf. Series: Journal of Physics: Conf. Series **929** (2017) 012066 doi:10.1088/1742-6596/929/1/012066

Core-shell Yagi-Uda nanoantenna for highly efficient and directive emission

Olga N. Sergaeva[1,2], **Roman S. Savelev**[1], **Denis G. Baranov**[3,4], **Alexander E. Krasnok**[1,5]

[1]ITMO University, St. Petersburg 197101, Russia
[2]Department of Mechanical and Aerospace Engineering, University of Missouri, Columbia, MO, 65211, USA
[3]Department of Physics, Chalmers University of Technology, 412 96 Gothenburg, Sweden
[4]Moscow Institute of Physics and Technology, 9 Institutskiy per., Dolgoprudny 141700, Russia
[5]Department of Electrical and Computer Engineering, The University of Texas at Austin, Austin, Texas 78712, USA

E-mail: `r.savelev@metalab.com`

Abstract. We study radiation from hybrid Yagi-Uda nanoantennas composed of metal-dielectric core-shell nanoparticles. We show that due to the presence of two types of resonances in each particle at close frequencies the hybrid Yagi-Uda nanoantenna can operate in two different regimes. In the first regime at low frequencies it operates similarly to plasmonic and all-dielectric Yagi-Uda nanoantennas, and it is characterized with highly directive emission in a forward direction. In the second regime at higher frequencies the hybrid nanoantenna can emit with a high directivity in backward direction due to the presence of the hybrid dispersion branch with negative group velocity. Moreover by choosing the appropriate nanoantenna parameters one can achieve the operation regime when due to excitation of dark magnetic dipole modes in nanoantenna high values of directivity and Purcell factor are realized simultaneouslly in extremely narrow frequency range.

1. Introduction

Optical nanoantennas enable enhancement and flexible manipulation of light on the scale much smaller than free space wavelength [1, 2]. Due to this ability, they offer unique opportunities for applications such as optical communications [3], photovoltaics [4], non-classical light emission [5], subwavelength light confinement and enhancement [6], sensing [7], and single-photon sources [8]. A specific type of optical nanoantenna, the Yagi-Uda one, has recently received a widespread attention in the literature [9]. Such nanoantennas consist of several small scatterers and operate similarly to their radio frequency analogues. Yagi-Uda nanoantennas composed of different scatterers, such as core-shell nanoparticles [10], plasmonic nanoparticles [11], high-index dielectric nanoparticles [12] were studied recently. Regardless of the scatterers type, such nanoantennas are characterized by a high directivity in a relatively wide frequency range. Explanation for such behaviour, based on the chain eigenmode analysis was given in Ref. [11] for plasmonic nanoparticles, while it can be also applied to any type of scatterer with a dominant dipole response.

Usually, a Yagi-Uda nanoantenna consists of a reflector and one or several directors. However, as it was pointed out in the Ref. [13], a Yagi-Uda nanoantenna composed of core-

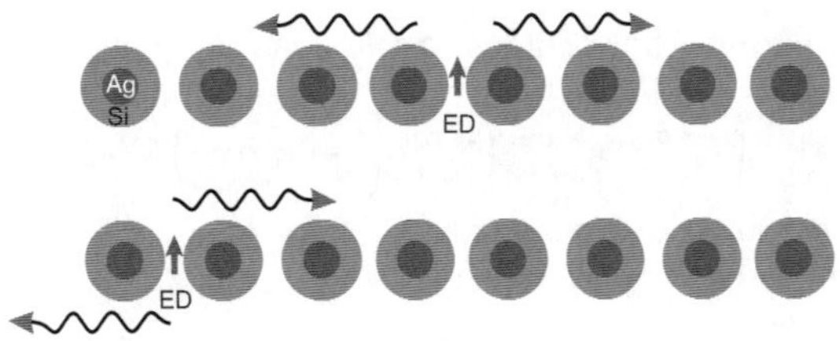

Figure 1. Schematic of the Yagi-Uda nanoantenna – a chain of metal-dielectric core-shell nanoparticles placed with period a. An electric dipole emitter is placed in the center of the chain orthogonally to the chain axis.

shell nanoparticles with both electric dipole (ED) and magnetic dipole (MD) resonant responses at the same frequency does not need a reflector particle, since backward scattering is suppressed automatically, due to Kerker effect [14]. Despite such special feature, core-shell Yagi-Uda nanoantennas still have not been studied in details by now.

Here, we report the results of the theoretical investigation regarding the capabilities of Yagi-Uda nanoantennas composed of Ag-Si core-shell nanoparticles for tailoring the radiation of electric dipole emitters. We show that such hybrid metal-dielectric nanoantenna can operate in two different regimes. In the first regime core-shell nanoantenna operates similarly to the well-known plasmonic and dielectric ones, and it is characterized by very high directivity and low Purcell factor in the wide spectral range at low frequencies. Due to presence of both ED and MD resonances in a single core-shell nanosphere another operation regime is possible. In this regime effective excitation of dark magnetic dipole modes in the chain of core-shell particles by ED source allows one to achieve both high directivity and high Purcell factor in a very narrow frequency range.

2. Electromagnetic properties of core-shell nanoparticles

As an element of the Yagi-Uda nanoantenna, we consider a core-shell nanoparticle with a silver (Ag) core and a silicon (Si) shell. Bi-material core-shell nanoparticles find a broad range of applications in catalysis, nanoelectronics, and biophotonics [15, 16]. They also offer a flexible platform for manipulation of electromagnetic radiation [17, 13]. In contrast to single-material solid particles, the spectral positions of the ED and MD resonances can be tuned almost independently in wide range of frequencies by tuning their geometrical parameters, what makes such particles an advantageous building block for constructing functional nanophotonic devices. Because of the great interest in core-shell particles from the perspective of their unique electromagnetic properties, different methods of their fabrication are being developed [18, 19, 20].

To describe the electromagnetic response of an array of the core-shell nanoparticles under consideration we use the well-known coupled-dipole approximation where each particle is characterized by electric dipole (ED) response (due to core plasmonic particle) and magnetic dipole (MD) response (due to high index dielectric shell). Dipole model for a one-dimensional array of identical particles formulates as follows [21, 22]:

$$\begin{cases} \mathbf{p}_n = \alpha_e(\omega)(\mathbf{E}_n^{(\mathrm{loc})} + \mathbf{E}_n^{(\mathrm{ext})}), \\ \mathbf{m}_n = \alpha_m(\omega)(\mathbf{H}_n^{(\mathrm{loc})} + \mathbf{H}_n^{(\mathrm{ext})}), \end{cases} \tag{1}$$

where $\mathbf{m}_n$ and $\mathbf{p}_n$ are magnetic and electric moments induced in the nth particle ($\propto -i\omega t$), α_m and α_e are magnetic and electric polarizabilities, electric $\mathbf{E}_n^{(\mathrm{ext})}$ and magnetic $\mathbf{H}_n^{(\mathrm{ext})}$ fields at the position of the nth dipole are produced by external source; local electric $\mathbf{E}_n^{(\mathrm{loc})}$ and magnetic

$\mathbf{H}_n^{(\mathrm{loc})}$ fields are produced by all other dipoles in the chain:

$$\begin{cases} \mathbf{E}_n^{(\mathrm{loc})} = \sum\limits_{j\neq n} \left(\widehat{C}_{nj}\mathbf{p}_j - \widehat{G}_{nj}\mathbf{m}_j \right), \\ \mathbf{H}_n^{(\mathrm{loc})} = \sum\limits_{j\neq n} \left(\widehat{C}_{nj}\mathbf{m}_j + \widehat{G}_{nj}\mathbf{p}_j \right), \end{cases} \tag{2}$$

where $\widehat{C}_{nj} = A_{nj}\widehat{I} + B_{nj}(\widehat{\mathbf{r}}_{nj} \otimes \widehat{\mathbf{r}}_{nj})$, $\widehat{G}_{nj} = -D_{nj}\widehat{\mathbf{r}}_{nj} \times \widehat{I}$, $\otimes$ is a dyadic product, $\widehat{I}$ is the unit 3×3 tensor, $\widehat{\mathbf{r}}_{nj}$ is the unit vector in the direction from nth to jth dipole; $A_{nj} = e^{ik_hR_{nj}}\left(\frac{k_h^2}{R_{nj}} - \frac{1}{R_{nj}^3} + \frac{ik_h}{R_{nj}^2}\right)$, $B_{nj} = e^{ik_hR_{nj}}\left(-\frac{k_h^2}{R_{nj}} + \frac{3}{R_{nj}^3} - \frac{3ik_h}{R_{nj}^2}\right)$, $D_{nj} = e^{ik_hR_{nj}}\left(\frac{k_h^2}{R_{nj}} + \frac{ik_h}{R_{nj}^2}\right)$, where R_{nj} is the distance between the nth and jth dipoles, ε_h is the permittivity of the host medium (in our calculations we take $\varepsilon_h = 1$), and $k_h = \sqrt{\varepsilon_h}\omega/c$ is the wavenumber in the host medium, $\omega = 2\pi\nu$, ν is the frequency, and c is the speed of light. Electric and magnetic polarizabilities are defined as $\alpha_e = i\frac{3\varepsilon_h a_1^{sc}}{2k_h^3}$, $\alpha_m = i\frac{3b_1^{sc}}{2k_h^3}$, where scattering coefficients a_1, b_1 can be expressed via the parameters of the layered particle [23].

The Ag-Si particles exhibit resonant dipole responses in near infrared frequency range [17]. Using the radius of the silver core $R_c = 68$ nm and the radius of the silicon shell $R = 225$ nm,

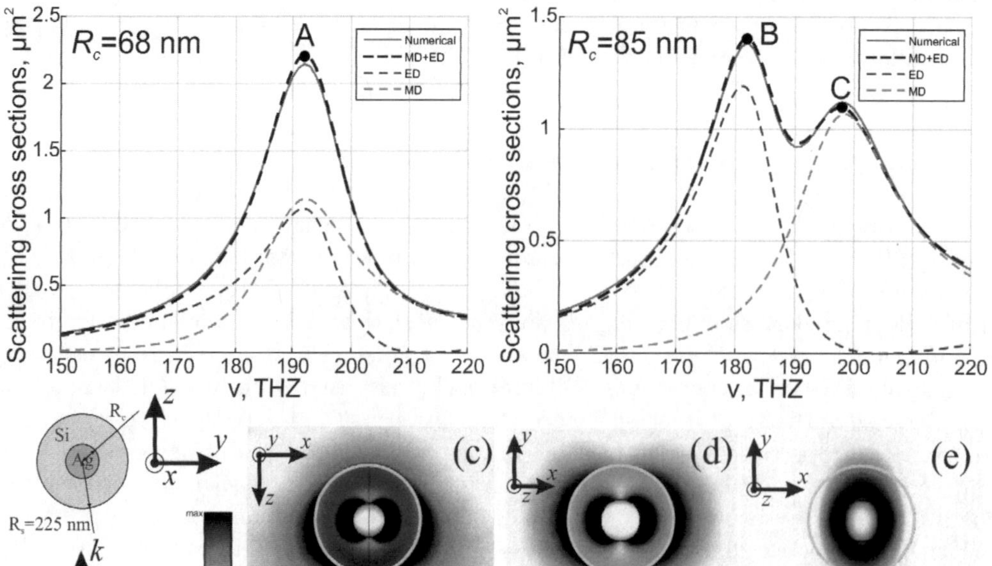

Figure 2. (a,b) Scattering cross-sections (solid red and dashed black curves) for Ag-Si core-shell particles with $R_s = 225$ nm and (a) $R_c = 68$ nm and (b) $R_c = 85$ nm. Dashed black curves are calculated within the framework of the dipole model, and red curves are obtained via full-wave numerical simulation. Partial ED and MD contributions are shown with dashed blue and green curves, respectively. (c-e) Field amplitude distributions at different frequencies: (c) point "A" in (a) (192 THz), (d) point "B" in (b) (182 THz), (e) point "C" in (b) (198 THz), indicating ED+MD, ED and MD resonances, respectively.

the overlap of the ED and MD resonances of the Ag-Si core-shell nanoparticle can be achieved. We assume constant permittivity of silicon $\varepsilon_s = 11.56$, and for the dispersion of silver core we employed Drude model $\varepsilon_c(\nu) = 1 - \nu_p^2/(\nu^2 + i\gamma\nu)$, with plasma frequency $\nu_p = 2180$ THz and collision frequency $\gamma = 4.93$ THz. At the frequency range of interest it very well matches the experimental data given in Ref. [24]. The extinction cross sections of an x-polarized plane wave by core-shell nanoparticle with these parameters along with its partial ED and MD contributions is shown in Figs. 2(a,b) for $R_c = 68$ nm and $R_c = 85$ nm, respectively. The electric field at the resonance frequency in Fig. 2(a) (point "A") combines characteristic field distributions of both ED and MD resonances: namely, strongly enhanced electric field near the core due to the ED plasmonic resonance and slightly enhanced circular distributed electric field due to MD resonance in high-index dielectric shell [see Fig. 2(c)]. Increase of the core radius from 68 nm to 85 nm shifts the ED resonance frequency to ≈182 THz [point "B" in Fig. 2(b)] and MD resonance frequency to ≈198 THz [point "C" in Fig. 2(b)] with corresponding field distributions plotted in Figs. 2(d,e), respectively. This demonstrates the possibility of the efficient tuning the frequencies of the two main resonances of hybrid core-shell particles by simple change of their geometrical parameters.

3. Yagi-Uda core-shell nanoantennas

It is known that far fields produced by magnetic and electric point dipoles cancel each other in backward of forward direction if they have the same amplitude and are oscillating in or out of phase, respectively [14]. Since the backscattering is suppressed automatically in the case of particles with overlapping MD and ED resonances, the Yagi-Uda nanoantenna composed of such elements does not require a reflector [13]. According to this, we place a point electric dipole source (emulating a quantum emitter) oriented orthogonally to the axis of the chain of 8 core-shell nanospheres with $R_c = 68$ nm and $R_s = 225$ nm placed with period $a = 550$ nm. Directivity of a nanoantenna can be calculated within the framework of this model [25]: $D_{\mathbf{max}} = \dfrac{4\pi \cdot \max\left[p(\theta,\varphi)\right]}{P_{rad}}$, where $\max\left[p(\theta,\varphi)\right]$ is the power transmitted in the direction of the main lobe, and $P_{rad} = \int p(\theta,\varphi)d\Omega$ is the total power radiated by a system into the far zone. Frequency dependent directivity for the considered nanoantenna is shown in Fig. 3(a). In order to understand the origin of highly directive emission at low frequencies, following the Koenderink analysis [11] we calculate dispersion properties of an infinite core-shell nanospheres chain. Eigenmodes can be calculated from the analytical closed-form dispersion equations that can be obtained by replacing infinite sums in (2) with analytical functions [26, 27]. In Fig. 3(b) one can observe, that the low frequency chain eigenmode is close to the light line at frequencies $\lesssim$ 170 THz (far from the particle resonances), which corresponds to the high directivity frequency range in Fig. 3(a). Besides the low frequency dispersion branch, due to the presence of two types of resonances there is also a second dispersion branch, which crosses the light line at ≈ 195 THz. For the considered parameters, EDs and MDs amplitudes of the chain eigenmodes are of the same order of magnitude for both branches and consequently there is a strong MD-ED interaction between the neigbour particles. Due to this interaction, the second branch is characterized by a negative group velocity and at the frequency near 195 THz, which corresponds to the observed peak in directivity, the nanoantenna radiates in backward direction [see inset in Fig. 3(a)].

The conventional operation regime of Yagi-Uda nanoantennas at low frequencies is naturally nonresonant. This can be understood from the eigenmode analysis of the considered structure. Eigenfrequencies of the chain of 8 particles with real part below 170 THz [corresponding to the first dispersion branch in Fig. 3(b)], have very large imaginary parts, and thus low frequency operation regime of Yagi-Uda nanoantenna is broadband and nonresonant. Consequently, another important property of nanoantennas, namely Purcell factor (PF), is quite low in this regime [12]. We confirmed it by calculating the PF with the well-known formula [25]

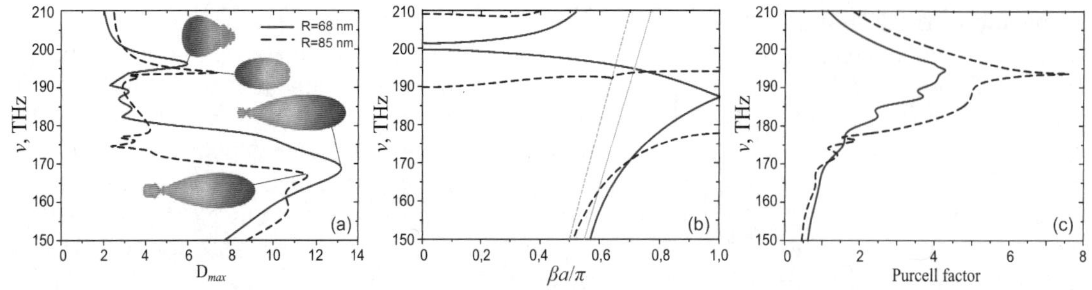

Figure 3. (a) Directivity and (c) Purcell factor for the chain of 8 core-shell nanoparticles with dipole source located on the axis of the chain at the half period from the center of the leftmost particle. (b) Dispersion diagram for the transversely polarized modes of an infinite chain of core-shell nanoparticles; $\beta a/\pi$ is the normalized Bloch wave number and ν is the frequency. Solid blue (dashed black) curves correspond to the following parameters: radius of Ag core $R_c = 68$ nm ($R_c = 85$ nm), radius of silicon shell $R_s = 225$ nm and the period of the chain $a = 550$ nm ($a = 500$ nm).

$F = 1 + \frac{3}{2k^3 d^2}\mathrm{Im}\left[\mathbf{d}^* \cdot \mathbf{E}_s\right]$, where k is the wave number, $\mathbf{d}$ is the dipole moment of the emitter, and $\mathbf{E}_s$ is the scattered electric field at the position of the source. Results of calculations are shown in Fig. 3(c): values of the PF at low frequencies do not exceed 1. The eigenfrequencies of the finite chain that correspond to the second dispersion branch, which is close to the resonances of the single particle, have small imaginary parts and therefore we call this operation regime. In the resonant regime the PF is much higher than in the nonresonant one, but the values of PF are still quite low $\lesssim$10.

In order to increase the PF and consequently the antenna efficiency we employ the method of magnetic dark chain eigenmodes excitation [28]. This method relies on the symmetry matching of the ED source oriented perpendicular to the chain axis and the staggered eigenmode of the chain of MDs, i.e. with MD moments in neighbouring nanoparticles oscillating out-of-phase. However, if the source is placed outside the chain, the modes excitation is not very effective, which is the case, demonstrated in Fig. 3 for the particles with two different radii of the core: in both cases the maximum of the directivity and the PF are quiet low in the resonant operation regime.

According to this, we calculated the PF and directivity for another two different positions of the source: in the center of the chain, and between the first and the second particles. For the case of $R_c = 68$ nm the directivity and the PF in the narrow frequency range corresponding to the resonant operating regime are shown in Figs. 4(a,b). Despite that for these positions of the source the chain eigenmodes are excited more effectively, we observe only a $\approx$4-fold increase in the PF, while the directivity decreases overall.

There are two reasons why increase of the Purcell factor in this case is not very large. First reason is the unique dispersion properties of the chain of hybrid MD+ED dipoles chain near the edge of the Brillouin zone. In the chain of particles with only one dominant dipole response (ED or MD) the group velocity tends to zero and, consequently, the density of photonics states diverges at the *Van Hove singularity* near the edge of the Brillouin zone (in the lossless case) [28]. However, when the particles exhibit both MD and ED resonances at the same frequency, there is no gap between the dispersion branches [see Fig. 3(b)]. Since Purcell factor is proportional to the density of photonics states, very high values of the PF cannot be achieved in the chain of core-shell nanoparticles with the fully spectrally overlapping MD and ED resonances.

Second, while there is a symmetry matching between the magnetic field of electric dipole

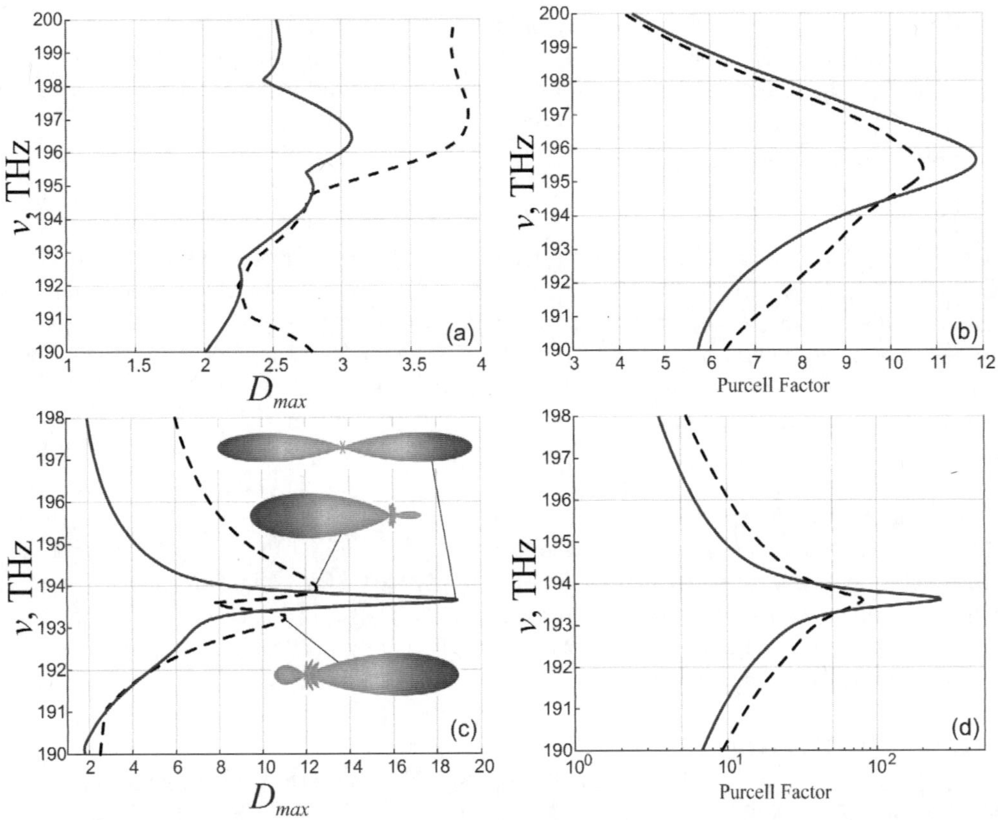

Figure 4. (a,c) Directivity and (b,d) Purcell factor for the chain of 8 core-shell nanoparticles with dipole source located in the middle of the chain (solid blue curves) and between the two leftmost particles (dashed black curves). (a,b) correspond to the $R_c = 68$ nm radius of Ag core, period of the chain is $a = 550$ nm and; (c,d) — to the $R_c = 85$ nm, period of the chain is $a = 500$ nm.

source placed between the particles and MD moments distribution of the staggered chain eigenmode, corresponding electric field and ED moments symmetrically mismatch. Since MD and ED moments in this case are of the same order of magnitude, this also decreases the efficiency of excitation of these modes with electric dipole source.

A different situation is observed for the core-shell particles with $R = 85$ nm. In contrast to the case of $R = 68$ nm, now the second branch is characterized by zero group velocity at the edge of the Brillouin zone, and an infinite local density of optical states [see Fig. 3(b)]. Moreover, the amplitudes of MD moments are substantially larger than amplitudes of ED moments for the second dispersion branch. Therefore, these eigenmodes can be effectively excited with ED source, which allows to achieve both high directivity and high rate of ED source emission in the same frequency range. Note, that since we are interested in the emission of ED source, MD resonance frequency should be higher than ED one. This is so, because the first dispersion branch is characterized with qualitatively the same behaviour for different relative position of resonances and the frequency ranges of high directivity and high Purcell factor do not overlap. While the second dispersion branch can be engineered in different ways.

For the case of $R = 85$ nm, shown in Figs. 4(c,d), we observe that dipole, placed in the center of the chain, emits in both backward and forward directions along the chain with high directivity

≈19 [see insets in Fig. 4(c)], while the PF for this configuration reaches high values up to 270 due to the high density of photonic states at the Van Hove singularity. Emission in a certain direction can be achieved by placing the source between first and second particles [dashed curve in Fig. 4(c)]. In this case the PF is almost one order lower, but source emits predominantly in one direction at certain frequencies [see insets in Fig. 4(c)].

4. Conclusion

To conclude, we have studied emission properties of a dipole source coupled to the Yagi-Uda nanoantenna composed of core-shell nanoparticles. We have demonstrated that the radiation pattern and the emission rate of the nanoantenna fed with electric dipole source strongly depend on the relative position of ED and MD resonance frequencies of a single particle. For most parameters, the hybrid core-shell nanoantenna operates as a conventional Yagi-Uda antenna without reflector, exhibiting highly directive emission in wide frequency range. For certain parameters, almost flat dispersion of the infinite chain of core-shell nanoparticles with inherent high density of photonic states can be realized. In this regime, high values of the Purcell factor and highly directive emission pattern of the nanoantenna can be achieved simultaneously. Finally, while our design of the hybrid nanoantenna is very simple from theoretical point of view, it implies the fabrication of spherical core-shell nanoparticles with high precision, which might be extremely difficult considering current state of technologies. We expect that predicted effects can also be achieved in similar hybrid structures with comparable electromagnetic properties that require simpler fabrication processes [29, 30].

5. ACKNOWLEDGMENTS

This work was supported by RFBR, according to the research projects No. 16-37-60092 mol_a_dk, No. 16-37-60076 mol_a_dk and No. 16-32-00444 mol_a and by Grant from the President of the Russian Federation (MK-381.2017.2).

References

[1] Novotny L and van Hulst N 2011 *Nature Photon.* **5** 83–90
[2] Agio M and Alù A 2013 *Optical antennas* (Cambridge University Press, Cambridge)
[3] Alù A and Engheta N 2010 *Phys. Rev. Lett.* **104**(21) 213902
[4] Atwater H and Polman A 2010 *Nature Materials* **9** 205–213
[5] Maksymov I S, Besbes M, Hugonin J P, Yang J, Beveratos A, Sagnes I, Robert-Philip I and Lalanne P 2010 *Phys. Rev. Lett.* **105**(18) 180502
[6] Stockman M 2004 *Phys. Rev. Lett.* **93** 137404
[7] Liu N, Tang M, Hentschel M, Giessen H and Alivisatos A 2011 *Nature Materials* **10** 631–636
[8] Buckley S, Rivoire K and Vuckovic J 2012 *Reports on Progress in Physics* **75** 126503
[9] Krasnok A, Maksymov I, Denisyuk A, Belov P, Miroshnichenko A, Simovski C and Kivshar Y 2013 *Physics-Uspekhi* **56** 539–564
[10] Li J, Salandrino A and Engheta N 2007 *Phys. Rev. B* **76**(24) 245403
[11] Koenderink A F 2009 *Nano Letters* **9** 4228–4233
[12] Krasnok A E, Miroshnichenko A E, Belov P A and Kivshar Y S 2012 *Opt. Express* **20** 20599–20604
[13] Liu W, Miroshnichenko A, Neshev D and Kivshar Y 2012 *ACS Nano* **6** 5489–5497
[14] Kerker M, Wang D S and Giles C L *J. Opt. Soc. Am.*
[15] Zhong C J and Maye M M 2001 *Advanced Materials* **13** 1507–1511
[16] Ghosh Chaudhuri R and Paria S 2012 *Chemical Reviews* **112** 2373–2433
[17] Paniagua-Domínguez R, López-Tejeira F, Marqués R and Sánchez-Gil J A 2011 *New Journal of Physics* **13** 123017
[18] Ghoshal T, Biswas S and Kar S 2008 *The Journal of Physical Chemistry C* **112** 20138–20142
[19] Zograf G P, Zuev D A, Milichko V A, Mukhin I S, Baranov M A, Ubyivovk E V, Makarov S V and Belov P A *Journal of Physics: Conference Series* **741** 012119
[20] Zograf G P, Rybin M V, Zuev D A, Makarov S V, Belov P A, Lopanitsyna N Y, Kuksin A Y and Starikov S V 2016 *Days on Diffraction (DD)* 460–463
[21] Mulholland G W, Bohren C F and Fuller K A 1994 *Langmuir* **10** 2533–2546

[22] Merchiers O, Moreno F, González F and Saiz J M 2007 *Phys. Rev. A* **76**(4) 043834
[23] Aden A and Kerker M 1951 *Journal of Applied Physics* **22** 1242–1246
[24] Johnson P B and Christy R W 1972 *Phys. Rev. B* **6**(12) 4370–4379
[25] Novotny L and Hecht B 2006 *Principles of Nano-Optics* (Cambridge University Press)
[26] Savelev R, Slobozhanyuk A, Miroshnichenko A, Kivshar Y and Belov P 2014 *Phys. Rev. B* **89** 035435
[27] Shore R A and Yaghjian A D 2012 *Radio Science* **47**
[28] Krasnok A E, Glybovski S B, Petrov M I, Makarov S V, Savelev R S, Belov P A, Simovski C R and Kivshar Y S 2016 *Appl. Phys. Lett.* **108** 211105
[29] Rusak E, Staude I, Decker M, Sautter J, Miroshnichenko A E, Powell D A, Neshev D N and Kivshar Y S 2014 *Appl. Phys. Lett.* **105** 221109
[30] Zuev D A, Makarov S V, Mukhin I S, Milichko V A, Starikov S V, Morozov I A, Shishkin I I, Krasnok A E and Belov P A 2016 *Advanced Materials* **28** 3087–3093

IOP Conf. Series: Journal of Physics: Conf. Series **929** (2017) 012067 doi:10.1088/1742-6596/929/1/012067

Influence of the dynamic Stark effect on long-term frequency stability of a self-oscillating magnetometer with laser-pumped alkali atoms

A A Baranov[1], S V Ermak[1], N K Kulachenkov[2], M V Petrenko[3], E A Sagitov[1] and V V Semenov[1]

[1] Dept. of Quantum electronics, Peter the Great St. Petersburg Polytechnic University, Saint-Petersburg, 195251, Russia
[2] State Research Center of the Russian Federation Concern CSRI Elektropribor, JSC, Saint-Petersburg, 197046, Russia
[3] Physical Gasdynamics Lab, Ioffe Institute, Saint-Petersburg, 194021, Russia

lexusbar@gmail.com

Abstract. This paper presents the results of investigation Stark shift effect influence on the long-term stability of a dual scheme of quantum magnetometers. Such scheme allows suppressing Stark shift components when a certain pumping light polarization is applied. As a result, long-term stability of a quantum sensor increases. However, when low-frequency (LF) and microwave fields are attached to a single vapor cell a coherence circulation in hyperfine structure of alkali atoms takes place. Physical origin of this effect is associated with the so called “dressed” atom theory, when atom is “dressed” by LF field. It yields in multiphoton absorption and resonance frequency shift. First estimates for this shift based on density matrix evolution formalism are provided in the paper.

1. Introduction

One of the most important parameters of the quantum magnetometer with laser pumping is its precision, which is limited by flicker noise at long observation time [1]. This limitation is closely coupled with laser source spectrum instability. Intensity and frequency variations of the pumping light cause Stark shift (or Light Shift effect) of the measured frequency. In alkali atoms such a shift can be described by three components – scalar, vector and tensor [2], which normally act simultaneously that complicates their independent measurement and limits optical pumping regime optimization, when Light shift effect is minimized. In the paper this problem is resolved comparing two Rb^{87} vapor cell magnetometers measurements. One of them was based on a low-frequency (LF) spin oscillator principle, while the second one was built as a passive microwave spectrometer with a resonance frequency lock loop [3].

IOP Conf. Series: Journal of Physics: Conf. Series **929** (2017) 012067 doi:10.1088/1742-6596/929/1/012067

2. Experiment

A simplified scheme of the experimental setup is shown on Figure 1.

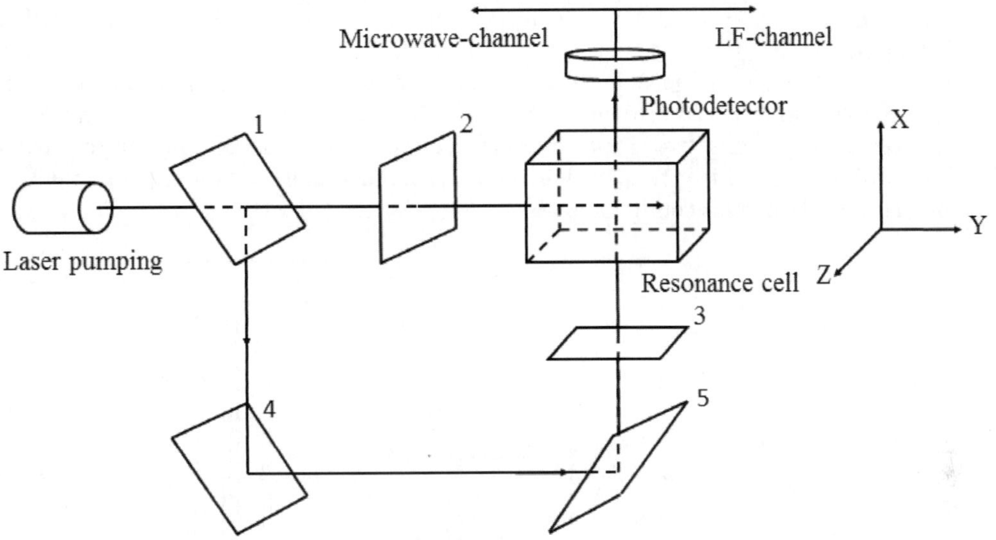

Figure 1. Experimental setup: 1 - beam splitter, 2,3 - circular polarizers, 4,5 - the mirrors. The constant magnetic field is oriented along the 0Y axis, spin oscillator linearly polarized RF field oriented along the axis 0Z.

The most significant feature of the setup is a single optical pumping source and a photodiode detector, which is used for both resonances registration – microwave signal and LF signal of the spin oscillator. Division of the signals was then provided with a LF filter in a microwave channel, which allowed microwave double resonance signal at modulation frequency of a reference oscillator to pass. This detection technique is common for classical radio-frequency (RF) atomic frequency standards. The possibility of a certain light shift component detection in the experiment follows from the fact that on microwave transitions all three components of the light shift act together, whereas on LF transitions of the self-oscillating magnetometer only vector and tensor components exist. This specific feature of the two magnetometers dual scheme allows to determine weighted impact of each light shift component correctly and estimate the influence of their mutual compensation on the long-term stability of the resonance frequency while changing pumping light polarization. Similar estimates were provided when a working magnetic field was measured as an error signal in the microwave channel and as a frequency offset of M_x-magnetometer. After that differential signal of magnetic field variations was processed in terms of Allan deviation. These dependences analysis allows to conclude, that Allan deviation increases with the law $(t^{n-1})^{1/2}$, where n varies within 2,4 – 2,6 due to measurement error.

Experiment showed that self-oscillating magnetometer long-term stability is strongly dependent on the hyperfine structure states coupling, which is induced by microwave transitions. In this case mutual effect of the microwave and LF signals takes place. This effect is not related to magnetic field variations and appears in frequency shifts of the spin oscillator when fixed microwave field intensity shifts are presented. In the work [5] it is noted that physical nature of such frequency shifts is associated with coherence circulation between magnetic sublevels of an alkali atom “dressed” by spin oscillator LF field. Absolute value of this shift strongly depends on magnetic-dipole transition number and reaches its maximum at 0-0 microwave transition, which is commonly used in atomic frequency standards.

In accordance with quantum theory of the “dressed” atom [6], simultaneous action of the LF and microwave fields on alkali atoms yields in multiphoton absorption and reemission of LF field quants, and probing microwave field absorption. The spectrum of microwave transitions between the sublevels of the hyperfine structure takes the form of the central and side groups of satellite lines. The centers of absorption line groups are spaced from the hyperfine transition 0-0 central series frequency by a multiple of the LF field frequency. As an example, arrows on Fig.2 show the possible microwave transitions between the hyperfine structure sublevels of alkali "dressed" atom with a nuclear spin of 3/2 for the center and the first two satellites of the side lines in the case of a significant excess of the radio frequency ω over Larmor frequency ω_0 and line DC and alternating LF and microwave fields orientation (so called $\sigma\pi$ fields configuration).

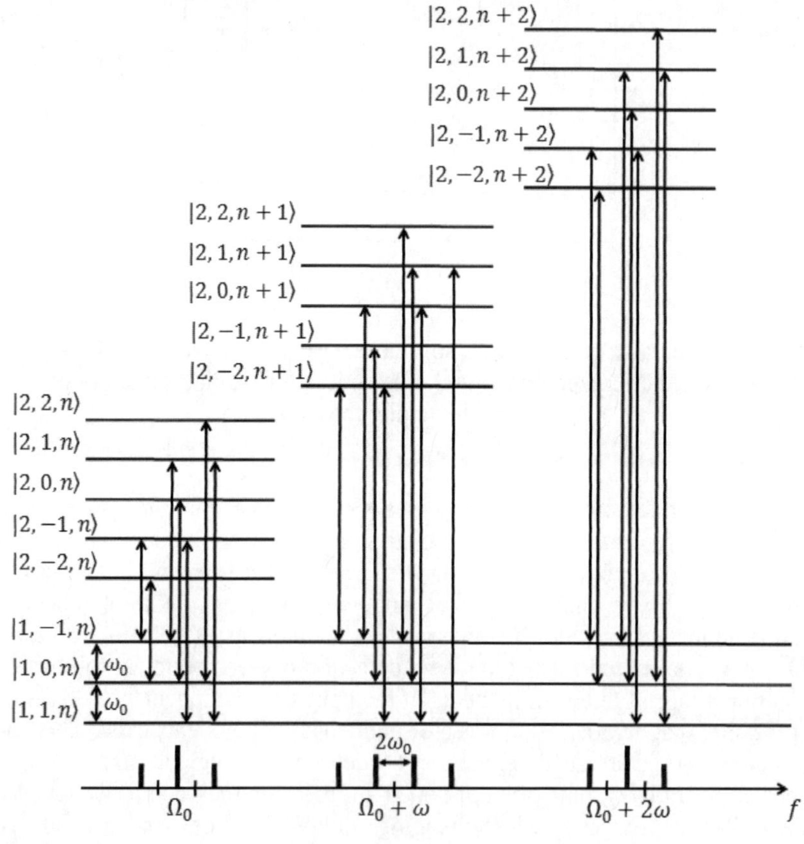

Figure 2. The spectrum of microwave "dressed" alkali atom transitions with a nuclear spin equal to 3/2, $\sigma\pi$ RF field configuration.

In contrast to the embodiment of the microwave spectrum, shown in Figure 2, the experiment with the dual magnetometer scheme corresponds to the resonance case when $\omega \approx \omega_0$. At the same time the spectra of the central and lateral satellites lines series overlap, which results in an increase in the magnetic sublevels with different values of the total angular momentum of the atom coupling. For example, in this case and in accordance with Figure 2 at 0-0 resonance frequency transitions, indicated in Table 1 are induced simultaneously. In the second column of the table probabilities of the corresponding transitions up to the 2nd order Bessel function are indicated.

Table 1. Microwave transitions probabilities, F=1, $m_F \leftrightarrow$ F=2, m_F, $\sigma\pi$ RF fields configuration.

F=1, m_F $\leftrightarrow$ F=2, m_F transitions	Probability of the transition
F=1, m_F = -1 $\leftrightarrow$ F=2, m_F =1	$3/64\ [J_0(\omega_1/\omega) - J_0(3\omega_1/\omega)]^2$
F=1, m_F = 0 $\leftrightarrow$ F=2, m_F =0	$1/64\ [J_0(\omega_1/\omega) + 3J_0(3\omega_1/\omega)]^2$
F=1, m_F = 1 $\leftrightarrow$ F=2, m_F =0	$1/128\ [J_1(\omega_1/\omega) + 3J_1(3\omega_1/\omega)]^2$
F=1, m_F = 0 $\leftrightarrow$ F=2, m_F =1	$3/32\ [J_1(3\omega_1/\omega)]^2$
F=1, m_F = -1$\leftrightarrow$ F=2, m_F =2	$3/256\ [J_1(3\omega_1/\omega) - 3J_1(\omega_1/\omega)]^2$
F=1, m_F = 1$\leftrightarrow$ F=2, m_F =1	$3/64\ [J_2(3\omega_1/\omega) + 3J_2(\omega_1/\omega)]^2$
F=1, m_F = 0$\leftrightarrow$ F=2, m_F =2	$3/128\ [\ J_2(\omega_1/\omega) - 3J_2(\omega_1/\omega)]^2$

The weight contribution of the transitions indicated in Table 1 depends on the population difference between the magnetic sublevels. Laser pumping of the small size resonance cell with antirelaxation wall coating in the magnetometers dual scheme experiment was carried out on the D_2 line of electric-cyclic transition F = 2 $\leftrightarrow$ F^* = 3. For this case when circularly polarized light pumps the atoms of the working substance (Rb^{87}), the system of stationary equations of the density matrix evolution in the balance approximation has the following form:

$$\begin{gathered}
\sigma_{11}=\sigma_{22}=\sigma_{33}=0.125,\\
\sigma_{44}=0.125-0.165\Gamma^*\sigma_{44},\\
\sigma_{55}=0.125+0.075\Gamma^*\sigma_{44}-0.432\Gamma^*\sigma_{55},\\
\sigma_{66}=0.125+0.09\Gamma^*\sigma_{44}+0.324\Gamma^*\sigma_{55}-0.648\Gamma^*\sigma_{66},\\
\sigma_{77}=0.125+0.108\Gamma^*\sigma_{55}+0.576\Gamma^*\sigma_{66}-0.6\Gamma^*\sigma_{77},\\
\sigma_{88}=0.125+0.072\Gamma^*\sigma_{66}+0.6\Gamma^*\sigma_{77},\\
\sum\sigma_{nn}=1,
\end{gathered} \tag{1}$$

where $\Gamma^*$is a ratio of the rates of optical pumping and atoms relaxation processes.

Figure 3 presents the results of diagonal elements of the density matrix calculation in function of the Γ^* parameter.

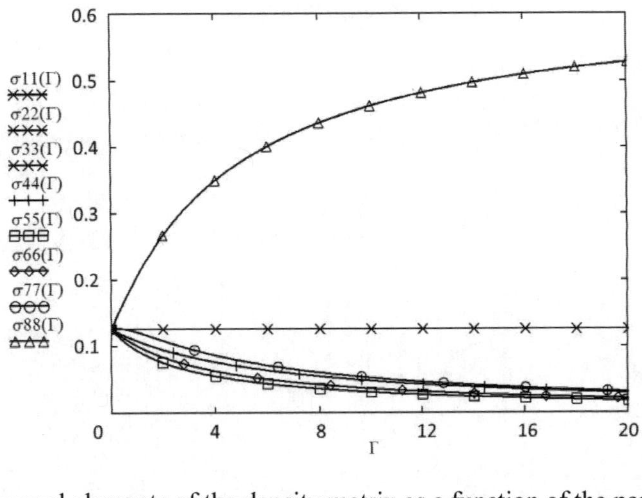

Figure 3. The diagonal elements of the density matrix as a function of the parameter Γ^* for Rb^{87} atoms (F=2$\leftrightarrow$ F^*=3) in case of paraffin-coated cell, $\sigma\pi$ RF field configuration.

The end resonance was observed in the experiment with two magnetometers when the applied fields had σσ configuration with the constant magnetic field orientation perpendicular to the vector of LF and microwave fields. In this case, the probability of the end resonance magnetic dipole transition F = 1, mF = 1 ↔ F = 2, mF = 2 is proportional to $3/64\ [J_2\ (2\omega_1 / \omega)]^2$.

The experimental value of the Bessel functions arguments mentioned above does not exceed 0.1. In this case, taking into account the statistical weight of the corresponding population difference, ratio of the probabilities of inducing magnetic dipole transitions at a frequency of 0-0 and end resonance transitions is of the order of 10^2 in the range of the Γ^* from1 to 5, which has good correlation with experimental data.

3. Conclusions

The analysis carried out in the paper allows to identify further ways of improving the metrological characteristics of quantum devices based on magnetic resonance in optically oriented atoms [7, 8], as well as to expand the functionality of the technique of simultaneous observations of the LF and microwave transitions for different isotopes of the alkali atoms used in atomic clocks and quantum magnetometry.

References

[1] Ermak S V, Zimnitskij P V, Smolin R V, Semenov V V 2014 *Tech. Phys. Lett.* **6** 82.

[2] Happer W, Mathur B S 1967 Effective operator formalism in optical pumping *Phys. Rev.* **163** pp.12-25.

[3] Rihle F 2004 *Frequency Standards: Basics and Applications* (Wiley-VCH Verlag GmbH & Co. KGaA, Weinheim).

[4] Baranov AA, Ermak SV, Sagitov EA, Smolin RV and Semenov VV 2015 *Journal of Experimental and Theoretical Physics* **121** pp. 393-403.

[5] Baranov A A, Ermak S V, Sagitov E A, Smolin R V and Semenov V V 2016 *Tech. Phys. Lett.* **42** 186.

[6] Haroche S 1971 L'atome habillé: une etude theorque et experimentale des propriete physiques d'atomes en interaction avec des photons de radiofrequences *Ann. de Phys.* **6** pp.189-238.

[7] Sagitov E A, Ermak S V, Petrenko M V and Semenov V V 2016 Quantum magnetometers as a base for atomic clock *J. Phys.: Conf. Ser.* 769 012044.

[8] Ermak S V, Petrenko M V and Semenov V V 2016 *Tech. Phys. Lett.* **42** 127.

IOP Conf. Series: Journal of Physics: Conf. Series **929** (2017) 012068 doi:10.1088/1742-6596/929/1/012068

The luminescence properties of nanocrystalline phosphors $Mg_2SiO_4:Eu^{3+}$

A Y Kolomytsev[1], D V Mamonova[2], A A Manshina[3], I E Kolesnikov[4],

[1] Physics Department, Saint Petersburg State University, 3 Ulyanovskaya Street, Peterhof, Saint Petersburg, Russia
[2] Scientific and Technological Institute of Optical Material Science, VNTs S. I. Vavilov State Optical Institute, 36-1 Babushkina, Saint Petersburg, Russia
[3] Institute of Chemistry, Saint Petersburg State University, 26 Universitetskii prospekt, Peterhof, Saint Petersburg, Russia
[4] Centre for Optical and Laser Materials Research, Saint Petersburg State University

Contact infofmation(e-mail): kau210296@mail.ru (Anton Kolomytsev)

Abstract. Nanocrystalline Eu^{3+}-doped Mg_2SiO_4 powders were prepared with combined Pechini-solid phase synthesis. The structural properties were investigated with XRD, SEM and Raman spectroscopy. XRD pattern indicated that $Mg_2SiO_4:Eu^{3+}$ were obtained with formation of other phase: MgO. Raman spectrum revealed good homogeneity and crystallinity of synthesized nanopowders. The luminescence properties were studied with measurement of excitation and emission spectra and decay curves. The effect of Eu^{3+} concentration on $^5D_0$ level lifetime was studied. Most probably, the observed shortening of $^5D_0$ level lifetime with Eu^{3+} concentration is caused by increase of nonradiative process probability.

1. Introduction

The study of inorganic nanomaterials doped with rare earth ions has attracted great interest. Phosphors based on silicates, phosphates, arsenates, chalcogenides doped with rare earth ions despite their widespread use do not fully satisfy the demands of modern technology. For example, forsterite (Mg_2SiO_4) is being currently studied as a potential matrix for bioimplants owing to its superior mechanical properties or as biomarkers for the detection of various diseases owing to its non-toxicity and high resistance to external factors, as well as the relatively low cost of synthesis of this compound [1-2].

Present article is devoted to investigation of nanocrystalline powder $Mg_2SiO_4:Eu^{3+}$synthesized with combined Pechini-solid phase method. We examined structural and luminescence properties by X-ray diffraction (XRD), scanning electron microscopy (SEM), Raman spectroscopy and fluorescence spectroscopy, respectively. A lot of attention is paid to the study of Eu^{3+} doping concentration on the luminescent properties.

2. Experimental

Nanocrystalline powders Mg_2SiO_4 doped with europium were prepared with combined Pechini-solid phase synthesis. Magnesium oxide powder containing a predetermined concentration of rare earth ions is synthesized with Pechini method. For this purpose we prepare magnesium nitrate solution with europium. Then we get citrate complex metals with addition citric acid.

$$Me(NO_3)_3 + 3\ C_6H_8O_7 = [Me(C_6H_8O_7)_3](NO_3)_3\ (Me: Mg, Eu) \quad (1)$$

After that we turn it to polymer gel with etherification reaction:

$$3[Me(C_5H_7O_5COO)_3](NO_3)_3 + C_2H_6O_2 \rightarrow$$

$$\rightarrow MeOOCC_4H_4COOC_2H_2OOCC_2OH(COOC_2H_4OOCC_4H_4COOMe)CCOOMe + (NO_2\uparrow + CO_2\uparrow) + H_2O \quad (2)$$

This gel was calcinated at 850 °C to form $MgO:Eu^{3+}$. The resulting powder is dispersed in ethanol solution. Then we add corresponding quantity of tetraethyl orthosilicate (TEOS) to reach stoichiometric composition of future complex oxide. The slip is dispersed again, and after that TEOS hydrolysis is carried out.

$$SiO_4(C_2H_5)_4 + 2H_2O \rightarrow SiO_2 + 4C_2H_5OH \quad (3)$$

The resulting mixture is dried at 40 °C for a day. Then it is calcinated at high temperature (1000 °C) to form a complex oxide according to the principle of solid phase synthesis.

The described synthesis procedure was used for preparation of powder series with different Eu concentration: 2, 4, 8 and 12 at.%. For luminescence studies the synthesized powders samples of quantity 5 mg were pressed with KBr into pellets.

X-ray phase analysis of the samples was carried out using Bruker «D8 DISCOVER X-ray diffractometer with CuKα -radiation (λ = 1.5406 Å). Electron micrograph images were obtained with Zeiss Merlin scanning electron microscope with the resolution of 1 nm. Raman spectra were investigated on Bruker SENTERRA Raman Microscope with semiconductor laser 488 nm as an excitation source. Photoluminescence spectra of $Mg_2SiO_4:Eu^{3+}$ nanophosphors were recorded with a fluorescence spectrometer Lumina with a Xe lamp (150W power) and the decay curves were obtained with the fluorescence spectrometer Fluorolog-3 with a Xe-flash lamp (150W power).All the measurements were performed at room temperature.

3. Results and discussion

3.1. *Structural properties*

Figure 1 shows the XRD pattern of $Mg_2SiO_4:Eu^{3+}$ 8 at.% nanophosphors synthesized with combined Pechini-solid phase method. This XRD pattern reveals presence of both orthorhombic Mg_2SiO_4 (JCPDS 34-0189) and MgO (JCPDS 77-2364) phase. It should be noted that presence of MgO phase does not affect luminescence properties as far as Eu^{3+} ions replace only Mg^{2+} ions in Mg_2SiO_4 host. Such conclusion can be drawn due to different shape and spectral position of bands at emission spectra of $MgO:Eu^{3+}$ [3]. Partial substitution of Mg^{2+} for Eu^{3+} does not affect the orthorhombic structure of Mg_2SiO_4 phase, indicating that the Eu^{3+} ions locate in the Mg position in the Mg_2SiO_4 crystal lattice.

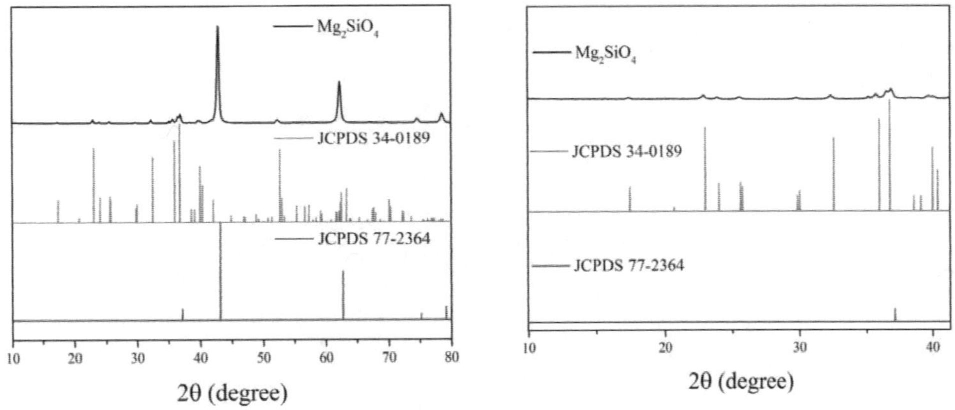

Figure 1. XRD pattern of Mg_2SiO_4:Eu^{3+} 8 at.% phosphor and standard cards for Mg_2SiO_4 (JCPDS 34-0189) and MgO (JCPDS 77-2364)

Figure 2. SEM image of Mg_2SiO_4:Eu^{3+} 8 at.% phosphor

The scanning electron micrograph of the prepared Mg_2SiO_4:Eu^{3+} 8 at.% phosphor is shown in Figure 2. Mg_2SiO_4:Eu^{3+} powder consists of stable agglomerates of nanocrystalline particles with average size about 100 nm.

Raman spectrum of Mg_2SiO_4:Eu^{3+} 8 at.% is presented in Figure 3. The spectrum was measured in spectral region from 80 to 1150 cm^{-1}. The narrow width of Raman lines indicates good homogeneity and crystallinity of synthesized powders that coincide with results of X-ray diffraction and scanning electron microscopy. The Mg^{2+} ions are mainly involved in external vibrations (Mg – SiO_4) in the spectral range ~150 – 400 cm^{-1}, whereas the contributions of internal vibrations of SiO_4 group is significant in the whole frequency region except the low frequency region (0–150 cm^{-1}). The most intense bands are centered at 420, 824, 857, 920, 966 and 1005 cm^{-1}. These vibrations are probably attributed to the SiO_4 stretching mode. It should be noted that Raman spectrum is dominated by line centered at 857 cm^{-1}, whereas band centered at 1005 cm^{-1} has rather small intensity. The obtained results agree well with the theoretical and experimental data from Ref. 4.

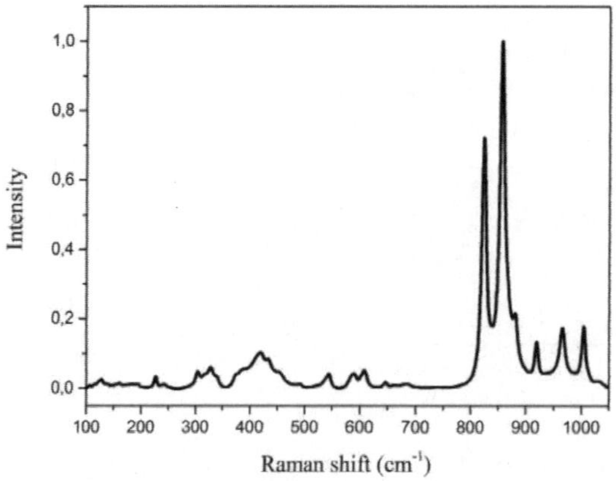

Figure 3. Raman spectrum of $Mg_2SiO_4:Eu^{3+}$ 8 at.% phosphor

3.2. *Luminescence properties*

Figure 4 shows the absorption spectrum of $Mg_2SiO_4:Eu^{3+}$ 8 at.% powder. The wide band in the region of low wavelengths corresponds to absorption of Mg_2SiO_4 host. The narrow lines correspond to the transitions inside Eu^{3+} ion.

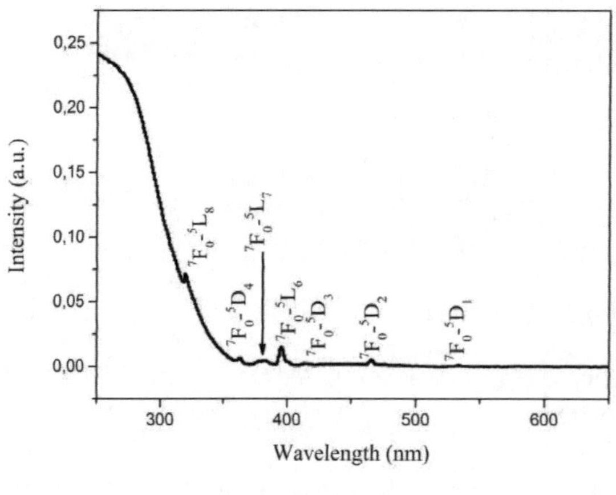

Figure 4. Absorption spectrum of $Mg_2SiO_4:Eu^{3+}$ 8 at.% phosphor

Figure 5 shows the excitation spectrum of $Mg_2SiO_4:Eu^{3+}$ 8 at.% powder for forced electric dipole transition $^5D_0$–$^7F_2$ with maximum at 614 nm. All observed lines are very sharp. As it is known, the most intensive line can be attributed to the $^7F_0$–$^5L_6$ with maximum at 393.5 nm. Other lines are also associated with typical transitions of the Eu^{3+} ion: $^7F_0$–$^5L_8$ (319 nm), $^7F_0$–$^5D_4$ (362 nm), $^7F_0$–$^5L_7$ (379 nm), $^7F_0$–$^5D_3$ (417 nm), $^7F_0$–$^5D_2$ (465 nm), $^7F_0$–$^5D_1$ (527 nm), $^7F_0$–$^5D_0$ (590 nm).

Figure 6 presents emission spectra of the $Mg_2SiO_4:Eu^{3+}$ powders (2, 4, 8, 12 at. %). The spectra consist of bands corresponding to the transitions from the metastable excited state ($^5D_0$) to the ground ^7F_J (J = 0–4) under excitation at 393.5 nm.

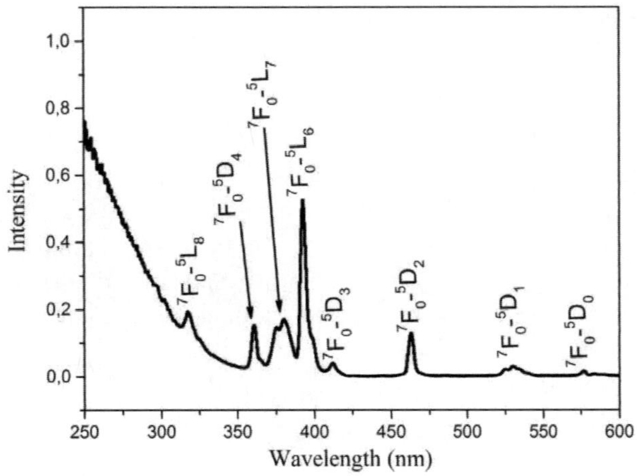

Figure 5. Excitation spectrum of Mg_2SiO_4:Eu^{3+} 8 at.% phosphor

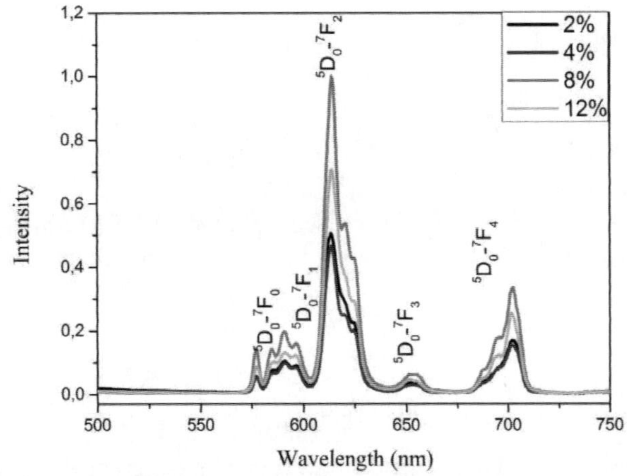

Figure 6. Emission spectra of Mg_2SiO_4:Eu^{3+} phosphors (2, 4, 8, 12 at.%)

Figure 7 shows concentration dependence of the luminescence intensity of the most intensive transition $^5D_0$–$^7F_2$ (614 nm). One can see that luminescence intensity grows with increasing number of Eu^{3+} ions up to 8 at.%. Further increase of Eu^{3+} concentration leads to intensity reduction. This decrease is due to the concentration quenching. Such behavior of the luminescence intensity is typical for rare earth doped materials [5-6].

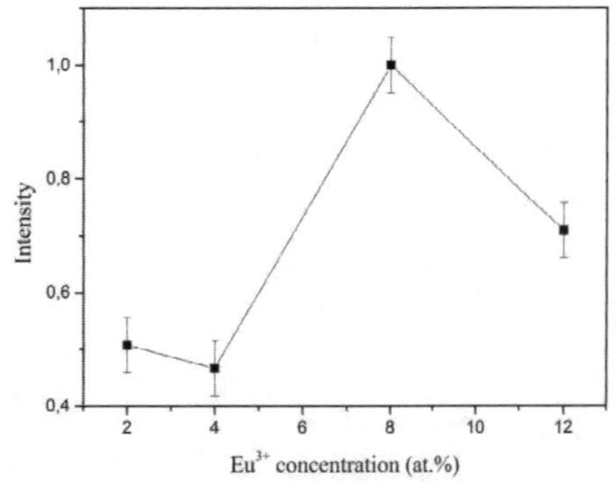

Figure 7. Photoluminescence intensity of Mg_2SiO_4:Eu^{3+} phosphors as a function of doping concentration

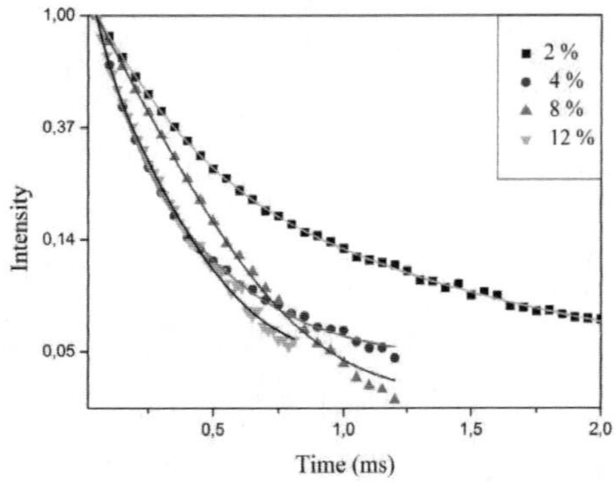

Figure 8. Decay curves of Mg_2SiO_4:Eu^{3+} phosphors (2, 4, 8, 12 at.%)

Figure 8 shows time dependence of the luminescence intensity of Mg_2SiO_4:Eu^{3+} powders (2, 4, 8, 12 at. %). Experimental curves for each compound were approximated with double exponential function.

$$I = A_1 e^{-\frac{t}{\tau_1}} + A_2 e^{-\frac{t}{\tau_2}} \quad (4)$$

Fitting parameters for each sample are presented in Table 1.

Table 1. Fitting parameters for Mg_2SiO_4:Eu^{3+} powders with different Eu^{3+} ions concentration

C(Eu^{3+}), at.%	**A_1**	**τ_1, ms**	**A_2**	**τ_2, ms**
2%	0.21 ±0.01	1.69 ± 0.05	1.00 ±0.01	0.219 ± 0.003
4%	0.18 ± 0.01	0.88 ±0.05	1.32 ± 0.02	0.103 ±0.002
8%	0.11 ± 0.02	0.87 ± 0.09	1.13 ± 0.01	0.203 ±0.003
12%	0.34 ± 0.02	0.42 ± 0.02	1.15 ± 0.02	0.092 ± 0.004

Probably difference from exponential decay can be explained with location of Eu^{3+} ions in two different positions: on the surface and inside the particles [7]. In order to compare lifetimes more convenient, the averaged luminescence lifetimes of Eu^{3+} ions were calculated with obtained approximation parameters using the following formula:

$$\tau = \frac{A_1\tau_1^2 + A_2\tau_2^2}{A_1\tau_1 + A_2\tau_2} \quad (5)$$

Calculated luminescence lifetimes of Eu^{3+} ions for different concentrations are presented in Table 2.

Table 2. Calculated luminescence lifetimes of Eu^{3+} ions for different Eu^{3+} concentrations

Eu^{3+} concentration (at. %)	2	4	8	12
Luminescence lifetime (ms)	1.48	0.46	0.39	0.26

As one can see, the averaged lifetime is decreased along with increase of Eu^{3+} doping concentration. Probably, it can be explained with increase of nonradiative decay rate of Eu^{3+} ions because of decrease of distance between Eu^{3+} ions with growth of doping concentration.

4. Conclusions

The nanocrystalline powders Mg_2SiO_4:Eu^{3+} were prepared with combined Pechini – solid phase synthesis. XRD pattern revealed that synthesized samples contained both Mg_2SiO_4 and MgO phase. XRD and Raman spectroscopy confirmed that synthesized nanocrystalline powders had good homogeneity and crystallinity. The emission and excitation spectra consisted of characteristic lines which can be ascribed to the 4f–4f transitions of Eu^{3+} and the most prominent transition is forced electric dipole $^5D_0$–$^7F_2$ transition (614 nm). Lifetime of excited state $^5D_0$ as a function of doping concentration was investigated. Most probably, the observed decrease of luminescence lifetime can be explained with increase of nonradiative decay rate.

Acknowledgments

This research has been supported by the Russian Foundation of Basic Research (RFBR 16-32-00091) and The Ministry of education and science of the Russian Federation (# 14.604.21.0078, RFMEFI60414X0078). Experimental investigations were carried out in "Center for Optical and Laser materials research", "Research Centre for X-ray Diffraction Studies", "Interdisciplinary Resource Centre for Nanotechnology" (St. Petersburg State University).

References

[1] Kolesnikov, I. E., Tolstikova, D. V., Kurochkin, A. V., Manshina, A. A., & Mikhailov, M. D. (2014). *Eu3+ concentration effect on luminescence properties of YAG: Eu 3+ nanoparticles*. Optical Materials, **37**, 306-310.

[2] Ramesh, S., Yaghoubi, A., Lee, K. S., Chin, K. C., Purbolaksono, J., Hamdi, M., & Hassan, M. A. (2013). *Nanocrystalline forsterite for biomedical applications: synthesis, microstructure and mechanical properties. journal of the mechanical behavior of biomedical materials*, **25**, 63-69.

[3] Devaraja P. B. et al. *MgO: Eu 3+ red nanophosphor: Low temperature synthesis and photoluminescence properties* //Spectrochimica Acta Part A: Molecular and Biomolecular Spectroscopy. – 2014. – **T. 121**. – C. 46-52.

[4] Iishi, K. (1978). *Lattice dynamics of forsterite*. American Mineralogist, 63(11-12), 1198-1208.

[5] Kolesnikov, I. E., Tolstikova, D. V., Kurochkin, A. V., Pulkin, S. A., Manshina, A. A., & Mikhailov, M. D. (2015). *Concentration effect on photoluminescence of Eu 3+-doped nanocrystalline YVO 4*. Journal of Luminescence, **158**, 469-474.

[6] Kolesnikov, I. E., Golyeva, E. V., Kurochkin, A. V., & Mikhailov, M. D. (2016). *Structural and luminescence properties of MgAl2O4:Eu3+ nanopowders*. Journal of Alloys and Compounds, **654**, 32-38.

[7] Kolesnikov, I. E., Povolotskiy, A. V., Mamonova, D. V., Lähderanta, E., Manshina, A. A., & Mikhailov, M. D. (2016). *Photoluminescence properties of Eu 3+ ions in yttrium oxide nanoparticles: defect vs. normal sites*. RSC Advances, **6(80)**, 76533-76541.

IOP Conf. Series: Journal of Physics: Conf. Series **929** (2017) 012069 doi:10.1088/1742-6596/929/1/012069

Application of optical non-invasive methods to diagnose the state of the lower limb tissues in patients with diabetes mellitus

Zharkikh E.V.*, Dremin V.V., Filina M.A., Makovik I.N., Potapova E.V., Zherebtsov E.A., Zherebtsova A.I., Dunaev A.V.

Biomedical Photonics Instrumentation Group, Scientific-Educational Center of "Biomedical engineering", Orel State University named after I.S.Turgenev, 29 Naugorskoe sh., Orel, 302020, Russia

*elena.zharkikh@bmecenter.ru

Abstract. The paper shows the possibility of assessing the functional state of microcirculatory-tissue systems of patients with diabetes mellitus by laser Doppler flowmetry (LDF), diffuse reflectance spectroscopy (DRS) and fluorescence spectroscopy (FS) methods. A review of the existing non-invasive optical technologies used to assess the state of microcirculation and oxygen metabolism in tissues of patients with diabetes is conducted. A series of experimental studies involving 76 patients with diabetes and 46 healthy volunteers was carried out. A wavelet analysis of LDF-grams was used to evaluate the adaptive changes of microcirculation during the temperature tests. The obtained data revealed that the proposed methodology in the form of combined use of several diagnostic technologies (LDF, FS and DRS) allows us to detect the presence or absence of trophic disorders and to evaluate adaptation processes during thermal tests.

1. Introduction

One of the acute problems of modern medicine is the diagnosis of diabetes mellitus (DM) complications. Diagnosis of diabetic microangiopathy and neuropathy is relevant in view of the still unanswered questions about their role in the pathogenesis of diabetic foot syndrome.

It is known that during diabetes, vascular walls and nerve endings suffer damage due to metabolic disorders, lack of oxygen in the tissues and an excess of toxic products of carbohydrate metabolism. As a result, the direct control of these deviations to assess the functional state of patients with DM tissues is the most logical.

Currently, a variety of optical non-invasive methods is used for the study of the microcirculatory-tissue system state.

1.1 Spectrophotometric techniques

The diffuse reflectance spectroscopy (DRS) method has been applied in the assessment of tissue saturation. This method allows to noninvasive continuous measure of changes in the concentration of oxygenated haemoglobin (cHbO2) and deoxygenated haemoglobin (cHb). The level of oxygen saturation of tissues (including disorders in diabetes) is judged by the view of the diffuse reflection spectrum based on different mathematical models or empirical relationships derived for communication the reflection coefficient with a concentration of specified chromophores. As valued characteristics of the skin diffuse reflection spectrum, the slope of the spectrum and the area under the spectrum curve are used most often. One of the diabetes complications is the development of vascular occlusive lesions that may lead to the development of tissue ischaemia. At the same time, a high level of oxyhaemoglobin in the blood can be observed. A number of scientific papers are dedicated to the research status of tissue metabolism in diabetes by diffuse reflectance spectroscopy method. Sujatha *et al* in their work demonstrated a statistically significant decrease in the total haemoglobin concentration in diabetic subjects compared to a control group without diabetes using the diffuse reflectance spectroscopy method [1]. There are study approaches where the relationship of diffuse reflectance coefficients at different wavelengths is selected as a diagnostic parameter. These wavelengths are most often isosbestic points of oxy and deoxyhaemoglobin. The DRS method can be used to study the processes occurring immediately in places of severe metabolic disorders such as tumours or diabetic trophic ulcers.

To assess the state of blood flow and detect possible vascular disorders when applying the method of diffuse reflectance spectroscopy, various stress tests are also used, including venous occlusion and exercises [2, 3]. Diffuse reflectance spectroscopy also is used for assessment of foot ulceration and wound healing processes [1].

Another spectrophotometric technology to evaluate the delivery and tissue oxygen consumption is a hyperspectral imaging technology. The use of hyperspectral technology allows detection of systemic and local microcirculatory disorders in diabetes mellitus [4]. In a number of papers, hyperspectral imaging technology is considered as a tool to noninvasively monitor and predict ulcer formation and healing on the diabetic foot [5]. The paper Nauvong *et al* shows the possibility of using hyperspectral imaging technology to assess cutaneous Hb oxygen saturation under wound healing.

1.2 Optical coherence tomography

As one of the methods to identify possible violations in the state of peripheral blood vessels, an assessment of retinal vessels is applied because they are the first target in diabetes damages. Changes in retinal vascular state reflect a range of pathophysiologic responses to hyperglycaemia, hypertension, inflammation, hypoxia, and endothelial dysfunction, and can predict different diabetes microvascular complications. Disorders of retinal blood vessels may be a marker of the development of neuropathy at an early stage [6]. To assess the state of retinal vessels various methods are used, including Doppler Fourier domain optical coherence tomography. The high-speed Fourier domain OCT enables obtaining of blood velocity and vessel volumes that allows measurements of pulsatile blood flow. Optical coherence tomography also allows measurement of retinal thickness as a marker of disease progression [7].

1.3 Fluorescence techniques

Fluorescence techniques are widely used in the study of disorders associated with the development of diabetes. The fluorescence spectroscopy (FS) method allows recording the fluorescence spectra of endogenous and exogenous skin fluorophores.

An important parameter is the tissue viability is the mitochondrial function. Using indicators of respiratory chain activity, it is possible to predict the death of the cells, to diagnose the state of tissue ischemia, or even to talk about their cancerous activity [8]. One of the assessments of mitochondrial function is the ratio of coenzymes NADH (the reduced form) and FAD (oxidized form). At hyperglycaemic conditions, changes occur in the mitochondrial respiratory chain, resulting in an overall imbalance of NADH/FAD.

Another violation that occurs during prolonged hyperglycaemia and leads to oxidative stress - excessive production of advanced glycation end products (AGE). The accumulation of sugars in the body leads to the glycation of proteins and thus to increase the production of AGEs, which initiate the expression of collagen genes and other proteins of the capillary membrane and skin and lead to the formation of adhesions between the collagen fibres. In diabetic disorders, AGEs accumulate in large amounts in the extracellular matrix. The presence of glycated proteins in the bloodstream may affect many tissues and cells. For example, at glycation of collagen and elastin, fibrosis and atherosclerosis may develop, leading to violations of blood vessel elasticity. A number of papers is devoted to research of AGEs accumulation in tissue by the fluorescence spectroscopy method [9]. Currently to record changes in diabetes related to the accumulation of AGEs various AGE-readers are used [10].

Nowadays one of the promising directions for an effortless and automatic assessment of larger tissue areas in DM is the use of fluorescence imaging technology. This technology allows to evaluate the redox ratio, which characterizes the ratio of coenzyme NADH/FAD in tissues [11]. Furthermore, the use of endogenous fluorescence imaging enables a clear visualization and discrimination of distinct wound regions. AGEs accumulation measurement technology by fluorescence imaging methods are also being developed [12].

1.4 Laser Doppler techniques

For the diagnosis of peripheral blood flow state, laser Doppler flowmetry (LDF) is widely used. This method is highly sensitive and allows measurement of blood flow and detecting of microcirculatory disorders in diagnostic of different microvascular disorders [13]. LDF has frequently been used to measure blood flow in the limbs of diabetic patients. The oscillatory process recorded by this method is a superposition of fluctuations caused by active and passive factors. To diagnose the state of the microcirculation system it is important to determine the contribution of various components of the rhythmic oscillations. One of the most important properties of the microvasculature, according to many researchers, is the reactive hyperaemia that occurs in response to the increased needs of the tissues for oxygen and nutrients. In this regard, different functional stress tests are used for a better assessment of the regulatory mechanisms state. To assess the state of microcirculation in diabetes, orthostatic tests [14], pharmacological tests [15], physical exercises [16] are widely adopted. Local heating test allows to evaluate the response of local regulatory mechanisms to the tissue heating [17]. Numerous studies use LDF for elucidating NO synthase mechanisms for quantification of abnormalities in the endothelium-dependent NO vasodilation and other microvascular research issues.

Similar data can be provided by laser Doppler imaging technique. This technique assesses blood perfusion over large areas and find widespread application in wound healing assessment [18].

1.5 Combined technologies

For a comprehensive diagnosis of the microcirculatory-tissue systems state currently it is promising to apply several diagnostic technologies simultaneously in one diagnostic volume. To assess the state of blood flow and detect possible vascular disorders different combinations of methods are applied. For instance, simultaneous measurements of the intensities of fluorescence and Rayleigh components allow decreasing the age dependency of the separation between diabetic subjects and control group, as well as increasing the overall magnitude of the separation [19]. Also the DRS method can be used for the correction of the data obtained during fluorescence studies[20]. Another example of combined technologies is a lightguide tissue spectrophotometry (O2C). It is a combination of LDF and tissue spectrometry. This method allows measuring both relative blood flow by the laser Doppler technique and haemoglobin oxygenation and haemoglobin amount in tissue by spectrometric techniques [21].

2. Experimental method and equipment

The aim of this work was to evaluate the possibilities of simultaneous application of LDF and the FS methods to identify trophic disorders at earlier stages in the skin of feet of patients with diabetes mellitus. To achieve this goal, the experimental studies were conducted. Local heating probe was chosen as the provocative exposure. Experimental studies included the registration of parameters during 4 of the successive stages. The first stage included registration of a base test of LDF-gram for a 4 minute-period. The second and third stages included registration of a local cold test (t=25 °C) and a local heating test (t=35 °C) for 4 minutes each. The fourth stage included registration of a local heating test (t=42 °C) for 10 minutes. A pair of fluorescence spectra at the excitation source with wavelengths of 365 nm and 450 nm were recorded in the first examination stage to avoid the photobleaching effect.

One study on each foot was 22 min, and indices of both feet recorded successively. Simultaneously with the continuous recording of perfusion a pair of fluorescence spectra under excitation of UV (365 nm) and blue (450 nm) light were recorded at each stage. The light guide probe was brought into contact with the dorsal surface of the foot on a point located on a plateau between the 1st and 2nd metatarsals. All studies were performed in the supine position. After carrying out experiments, LDF-grams of each phase of the study were subjected to adaptive wavelet analysis by LDF 3.0.2.384 software. Experimental studies were carried out by LDF and FS methods using the multifunctional laser noninvasive diagnostic system «LAZMA MC». The «LAZMA-TEST» complex was used to provide thermal effects. Measurements were carried out on 76 patients diagnosed with diabetes and 46 apparently healthy volunteers of approximately the same age. Figure 1 presents schematic views of location of the optical sensor on patient’s foot (a) and the equipment used in the experimental studies (b, c).

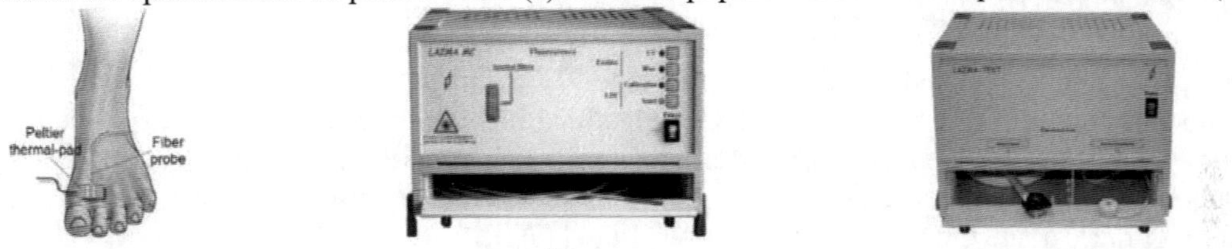

a) b) c)

Figure 1 - Location of the optical sensor on foot (a) and the appearance of the “L ZM MC” complex (b) and the “L ZM -TEST” device (c)

Registration of the blood microcirculation index (I_m) and the I^F_{365} and I^F_{450} fluorescence amplitudes were performed. Before the beginning of each study at the specified point registration of the spectra of skin diffuse reflection was carried out by a compact spectrometer “FLAME” (Ocean Optics, USA). In addition, for patients with ulcers, spectra were recorded directly at ulcers and at one centimetre away from the ulcers (at the intact region).

During the study, the oscillation amplitude values for the 5 main ranges were determined: endothelial (A_e), neurogenic (A_n), myogenic (A_m), respiratory (A_r) and cardiac (A_c). These values were normalized to the standard deviation (σ) and the average value of the index of microcirculation (I_m) were also determined.

3. Results and discussion

The research process revealed that the fluorescence intensity was higher in patients (2.7±0.8 AU, 1.8±0.7 AU) at both excitation wavelengths in comparison to the control group (2.1±0.8 AU, 1.2±0.4 AU). This fact can be explained by the accumulation of advanced glycation end products and formation of glycated proteins, including collagen, which is the largest contributor to the fluorescence spectrum at these wavelengths, in the body of patients with diabetes. At the same time microcirculation index in the group of patients reduced, which is especially pronounced during the local heating tests.

The results obtained in patients and in the control group were compared. Significant differences in some of the above parameters were identified and statistically proved (by One-way ANOVA, $p < 0.05$) for the group of patients relative to the control group.

According to the data of DRS measured, spectra curves of the skin diffuse reflection were analysed. To study the ulcerative processes ratio of the diffuse reflection coefficient, the absorption wavelength of oxyhaemoglobin (540 and 578 nm) was calculated. Erythema coefficient was calculated and analysed according to the method in which area under the curve of the skin optical density in the spectral range of 510-610 nm is estimated.

Table 1 shows the numerical indicators of the study results.

Table 1. The results of 3rd and 4th stages of experimental studies.

Parameters	3 stage		4 stage	
	Control group	Patients	Control group	Patients
I_m, PU	8,68±3,29	6,70±2,65*	20,09±4,46	11,56±3,67*
A_e, PU	0,58±0,31	0,40±0,24*	0,6±0,39	0,42±0,23*
A_n, PU	0,60±0,32	0,38±0,17*	0,5±0,48	0,35±0,17*
A_m, PU	0,49±0,24	0,29±0,17*	0,46±0,36	0,25±0,15*
A_r, PU	0,22±0,15	0,21±0,11	0,33±0,16	0,28±0,17*
A_c, PU	0,36±0,23	0,53±0,25*	1,19±0,48	0,89±0,42*

* - statistical significance of the differences of the indicator values of patients in relation to healthy volunteers with $p<0,05$

The patients have reduced value of nutritive blood flow during local heating (2.42 ± 1.67 PU) in comparison to the control group (4.91 ± 2.85 PU), which indicates redistribution of blood flow in the direction of the shunt component and reducing of capillary perfusion. Reduced values of endothelial oscillations indicate the weakening of the endothelium-dependent vasodilation associated with the secretion of nitric oxide (NO) by vascular endothelium [17].

The lack of growth of the neurogenic oscillation amplitudes in patients shows disorders in the work of sensory peptidergic nerve fibres responsible for the initial stage of vasodilation during local heating.

During the research, the highest blood circulation observed was also revealed in patients with focal disorders. Erythema index for patients without ulcers was higher than that for the control group (17.1 ± 11.7 vs 13.3 ± 8.6), which may indicate the presence of disorders in the peripheral circulation.

4. Conclusions

The proposed methodology allows us not only to detect the presence or absence of trophic disorders, but also to determine possible causes by evaluating adaptation processes during thermal tests and by comparing them with the results obtained in the control group. The use of wavelet analysis for assessing the regulatory mechanisms of peripheral blood flow during heating tests gives the opportunity to explore changes in autoregulation of vascular tone, as well as the regulation of the value of shunt blood flow by sympathetic fibres, which gives an indirect idea of the nervous innervation of blood flow and can indicate the presence of neuropathies.

The combined use of several optics diagnostic technologies allows for a comprehensive approach to the problem of assessing the state of microcirculatory-tissue systems, as well as to identify and predict the development of trophic disorders and diabetic foot syndrome at an earlier stage.

5. Acknowledgments

This work was supported by grant of the President of the Russian Federation for state support of young Russian scientists № MK-7168.2016.8.

References

[1] Anand S, Sujatha N, Narayanamurthy V B, Seshadri V and Poddar R 2014 Diffuse reflectance spectroscopy for monitoring diabetic foot ulcer – A pilot study *Optics and Lasers in Engineering* **53** 1-5

[2] Pichler G, Urlesberger B, Jirak P, Zotter H, Reiterer E, Müller W and Borkenstein M 2004 Reduced Forearm Blood Flow in Children and Adolescents With Type 1 Diabetes (Measured by Near-Infrared Spectroscopy) *Diabetes Care* **27** 1942

[3] Mohler E R, Lech G, Supple G E, Wang H and Chance B 2006 Impaired Exercise-Induced Blood Volume in Type 2 Diabetes With or Without Peripheral Arterial Disease Measured by Continuous-Wave Near-Infrared Spectroscopy *Diabetes Care* **29** 1856

[4] Khaodhiar L, Dinh T, Schomacker K T, Panasyuk S V, Freeman J E, Lew R, Vo T, Panasyuk A A, Lima C, Giurini J M, Lyons T E and Veves A 2007 The Use of Medical Hyperspectral Technology to Evaluate Microcirculatory Changes in Diabetic Foot Ulcers and to Predict Clinical Outcomes *Diabetes Care* **30** 903

[5] Nouvong A, Hoogwerf B, Mohler E, Davis B, Tajaddini A and Medenilla E 2009 Evaluation of Diabetic Foot Ulcer Healing With Hyperspectral Imaging of Oxyhemoglobin and Deoxyhemoglobin *Diabetes Care* **32** 2056

[6] Ikram M K, Cheung C Y, Lorenzi M, Klein R, Jones T L Z and Wong T Y 2013 Retinal Vascular Caliber as a Biomarker for Diabetes Microvascular Complications *Diabetes Care* **36** 750

[7] Zmyslowska A, Fendler W, Niwald A, Ludwikowska-Pawlowska M, Borowiec M, Antosik K, Szadkowska A and Mlynarski W 2015 Retinal Thinning as a Marker of Disease Progression in Patients With Wolfram Syndrome *Diabetes Care* **38** e36

[8] Rafailov I E, Dremin V V, Litvinova K S, Dunaev A V, Sokolovski S G and Rafailov E U 2016 Computational model of bladder tissue based on its measured optical properties *Journal of Biomedical Optics* **21** 025006

[9] Kesavan A, Pachaiappan R, Aruna P R and Ganesan S 2016 Steady state fluorescence spectroscopic characterization of normal and diabetic urine at selective excitation wavelength 280 nm *Proc. SPIE* **9703** 97031H

[10] Gerrits E G, Lutgers H L, Kleefstra N, Graaff R, Groenier K H, Smit A J, Gans R O and Bilo H J 2008 Skin Autofluorescence *Diabetes Care* **31** 517

[11] Quinn K P, Leal E C, Tellechea A, Kafanas A, Auster M E, Veves A and Georgakoudi I 2016 Diabetic Wounds Exhibit Distinct Microstructural and Metabolic Heterogeneity through Label-Free Multiphoton Microscopy *Journal of Investigative Dermatology* **136** 342-4

[12] Larsson M, Favilla R and Strömberg T Assessment of advanced glycated end product accumulation in skin using auto fluorescence multispectral imaging *Computers in Biology and Medicine*

[13] Zherebtsova A I, Zherebtsov E A, Dunaev A V, Podmasteryev K V, Pilipenko O V, Krupatkin A I, Khakhicheva L S and Muradyan V F 2016 Study of the functional state of peripheral vessels in fingers of rheumatological patients by means of laser Doppler flowmetry and cutaneous thermometry measurements *Proc. SPIE* **9917** 99170M

[14] Shahbaz R 2015 Evaluation of Endothelial Dysfunction Using Laser Doppler Flowmetry in Diabetes Mellitus and Peripheral Arterial Occlusive Disease Patients *Journal of Vascular Surgery* **61** 92S

[15] Khan F, Elhadd T A, Greene S A and Belch J J 2000 Impaired skin microvascular function in children, adolescents, and young adults with type 1 diabetes *Diabetes Care* **23** 215

[16] Herriott M T, Colberg S R, Parson H K, Nunnold T and Vinik A I 2004 Effects of 8 Weeks of Flexibility and Resistance Training in Older Adults With Type 2 Diabetes *Diabetes Care* **27** 2988

[17] Zimny S, Dessel F, Ehren M, Pfohl M and Schatz H 2001 Early Detection of Microcirculatory Impairment in Diabetic Patients With Foot at Risk *Diabetes Care* **24** 1810

[18] Krishnan S T M, Quattrini C, Jeziorska M, Malik R A and Rayman G 2007 Neurovascular Factors in Wound Healing in the Foot Skin of Type 2 Diabetic Subjects *Diabetes Care* **30** 3058

[19] Yu N-T, Krantz B S, Eppstein J A, Ignotz K D, Samuels M A, Long J R and Price J 1996 Development of a noninvasive diabetes screening device using the ratio of fluorescence to Rayleigh scattered light *Journal of Biomedical Optics* **1** 280-8

[20] Dremin V V, Zherebtsov E A, Rafailov I E, Vinokurov A Y, Novikova I N, Zherebtsova A I, Litvinova K S and Dunaev A V 2016 The development of attenuation compensation models of fluorescence spectroscopy signals *Proc. SPIE* **9917** 99170Y

[21] Beckert S, Witte M B, Königsrainer A and Coerper S 2004 The Impact of the Micro-Lightguide O2C for the Quantification of Tissue Ischemia in Diabetic Foot Ulcers *Diabetes Care* **27** 2863

IOP Conf. Series: Journal of Physics: Conf. Series **929** (2017) 012070 doi:10.1088/1742-6596/929/1/012070

Investigation of lasers based on coupled waveguides by near-field scanning optical microscopy

Yu S Polubavkina[1], N Yu Gordeev[1,2], A S Payusov[1,2], N V Kryzhanovskaya[1], E I Moiseev[1], F I Zubov[1], S A Mintairov[2], N A Kalyuzhnyy[2], M M Kulagina[1,2], Yu M Shernyakov[1,2], M V Maximov[1], A E Zhukov[1]

[1] St Petersburg Academic University of RAS, 8(3) Khlopina, St Petersburg 194021, Russia

[2] Ioffe Institute, 26 Polytekhnicheskaya, St Petersburg 194021, Russia

Abstract. We have investigated near field intensity distributions of InGaAs/GaAs/AlGaAs lasers possessing broadened waveguides based on coupled large optical cavity structures (CLOC) by scanning near-field optical microscopy (SNOM). The concept allows effective suppressing of the transverse high-order mode lasing. The obtained results can be considered to be the direct proof of pure transverse single-mode emission of the CLOC lasers.

1. Introduction

Many important applications of power semiconductor lasers, such as spectroscopy, frequency converting, surgery, optical pumping sources for solid state lasers or for fiber lasers, require the high optical emission power and the high beam quality[1]-[4]. One of the essential problems limiting the maximum output power of semiconductor lasers is the catastrophic optical mirror damage (COMD). An effective method of reducing the optical power density is an expansion of the efficient mode size in the transverse and lateral directions. It can be reached by simple increasing of the corresponding geometric waveguide size. However, this expansion induces spatial multimode lasing spoiling the far-field patterns and beam quality. Thus a waveguide should be designed in such a way that its thickness does not exceed the boundary condition for high order modes or high order modes are suppressed by gain or loss engineering. The increase of the losses for high order modes is reached by the increase of their overlap with absorbing laser structure layers. The gain selectivity means the adjustment of the active region position so that only the fundamental mode has the high optical confinement factor.

There are many methods of the mode size expansion in the transverse direction. Among them are broadening vertical waveguides based on the photonic crystal [6], super large cavity designs [7], and tapered ridge waveguides [8, 9]. A simple and efficient method of the high order mode suppression was proposed in our recent work [10]. The idea of this approach utilizes resonance coupling in two parallel waveguides with a small distance between them. The waveguides of such lasers are complex and consist of a narrow passive waveguide and broadened waveguide with an active region inside. The effective refractive indices of the eigenmodes of each waveguide have close or equal values to satisfy the phase-matching condition. So the high order mode of the broadened waveguide and the mode of the narrow waveguide are converted into two composite modes (Figure 1) due to the resonance coupling, whereas the fundamental mode of active waveguide does not tunnel into the narrow waveguide. The initial mode intensity is redistributed between the two waveguides and the fundamental mode remains the only lasing mode, thus, the composite modes do not take part in the lasing due to the reduced optical confinement factor (OCF) and increased optical losses resulted from the free-carrier adsorption in the highly doped layers.

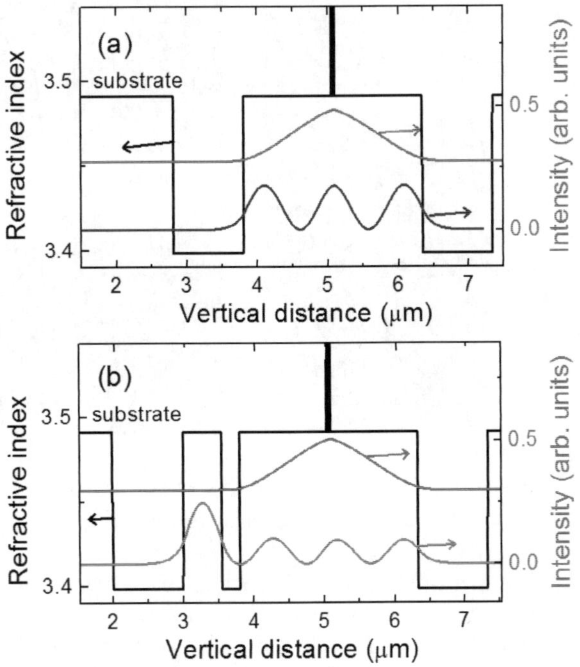

Figure 1. Refractive index profile and simulated intensity distributions of the fundamental (red line), second-order (blue line) and one composite (magenta line) modes for the reference (a) and CLOC (b) lasers.

Accurate and unambiguous information on the laser optical modes can be extracted from the near-field radiation distributions measured directly on the surface of the emitting facet of lasers [12,13] by near-field scanning optical microscopy (SNOM). In this paper, we report on the study of the spatial emission properties of the CLOC and reference lasers investigated by SNOM with the resolution of 100 nm.

2. Experiment and discussion

Both the reference and CLOC laser wafers emitting at the wavelength $\lambda \approx 1.04$ μm were grown by metal-organic chemical vapor deposition (MOCVD) on n-GaAs (100) substrates. The laser structures are much the same as ones investigated earlier [10]. The active GaAs waveguides of the both structures have the thickness of 2.5 μm, just below the cut-off of the fourth-order transverse mode. The CLOC laser structure has an extra 625-nm-thick n-GaAs passive waveguide separated from the active waveguide with 250-nm-thick n-$Al_{0.15}Ga_{0.85}As$ optical barrier. Both wafers were processed into 50-μm stripe lasers. The samples were mounted p-side down on copper heatsinks. Both the CLOC and reference lasers showed the parameters very close to the reported earlier [10].

The near-field intensity distributions at the emitting facet formed after cleaving of the laser wafers, were measured with sub-wavelength spatial resolution using an optical spectrometer (spectral resolution 0.46 nm) integrated with SNOM with an aperture cantilever. In the SNOM regime, we used cantilevers with metal coated pyramidal SiO_2 probes. The aperture of 100 nm was created on the probe ends with the focused ion beam process. Such cantilevers can also be used simultaneously for surface topography imaging with standard atomic force microscopy (AFM) techniques. The radiation penetrating through the nanoaperture was collected with a lens with $NA = 0.5$.

Near-field distribution maps were measured at different currents. At all the excitation levels, the CLOC lasers showed stable single-mode emission. Intensity distributions of the near field patterns of the reference and CLOC lasers measured in pulsed mode at the drive current of 1A are presented in Figure 2. Figures 2a and 2b demonstrate intensity distributions for the reference and CLOC lasers, respectively, in the whole spectrum. The intensity distribution map of the reference laser has a complex pattern manifesting high order mode lasing both in the vertical and lateral directions.

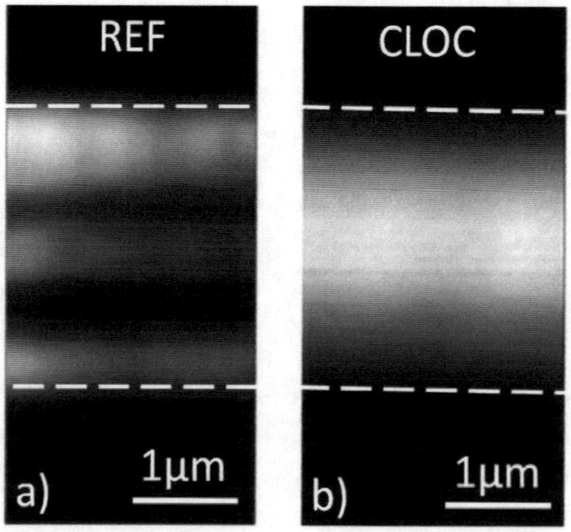

Figure 2. Near-field intensity distribution maps of the reference and CLOC lasers at the current of 1A. White dashed line shows waveguide edges

The intensity distribution map for the CLOC laser has pure transverse single-mode pattern but multi-mode pattern in the lateral direction. The vertical profiles of the intensity distributions are shown in Figure 3. Red lines show simulated intensity distributions of the second-order and fundamental modes for the reference and CLOC lasers, respectively. The measured profiles of the CLOC and reference laser modes are very close to the simulated ones. This fact indicates that the experiment and calculation are in good agreement.

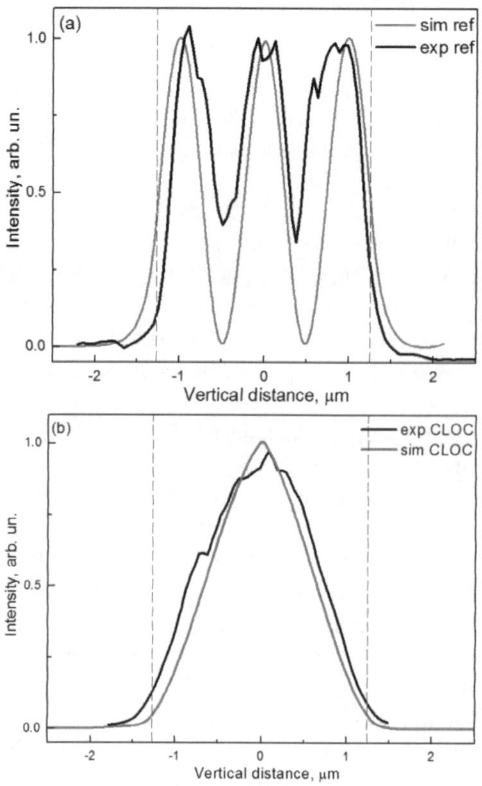

Figure 3. Vertical intensity distribution profiles for the reference and CLOC lasers at the current of 1A. Blue dashed line shows waveguide edges.

3. Conclusions

To conclude, for the first time we have investigated near field intensity distributions of InGaAs/GaAs/AlGaAs lasers based on coupled large optical cavity structures by scanning near-field optical microscopy. We have directly proved pure transverse single-mode lasing having compared the CLOC and reference lasers. The obtained results confirm that the lasers based on Coupled Large Optical Cavities can be the promising for many applications due to their simple design and good beam quality.

Acknowledgements

The authors acknowledge the support from the Russian Foundation for Basic Research (project No 16-29-03279).

References

[1] Inoue Y and Fujikawa S 2000 *IEEE J. Quantum Electron.* **36** 6 751–756
[2] Schulz W and Poprawe R 2000 *IEEE J. Sel. Topics Quantum Electron.* **6** 4 696–705
[3] Brauch U, Loosen P and Opower H 2000 *Topics in Applied Physics* 303–368
[4] Treusch H G and Pandey R 2011 *High-Power Laser Handbook*
[6] Miah M J, Kettler T, Kalosha V P, Posilovic K, Bimberg D H, Pohl J and Weyers M 2015 *IEEE J. Sel. Top. Quantum Electron.* **21** 4900206
[7] Pietrzak A, Crump P, Wenzel H, Erbert G, Bugge F and Trankle G 2011 *IEEE J. Sel. Top. Quantum Electron.* **17** 1715
[8] Kallenbach S et al. 2005 *Proc. SPIE* **5738** 406–415
[9] Kelemen M T et al. 2004 *Proc. SPIE.* **5452** 233–243
[10] Gordeev N Yu, Payusov A S, Shernyakov Yu M, Mintairov S A, Kalyuzhnyy N A, Kulagina M M, Maximov M V 2015 *Opt. Lett.* **40** 2150
[11] Little B E and Huang W P 1995 *Prog. Electromagn. Res.* **10** 217
[12] Lienau Ch, Richter A, Klehr A and Elsaesser 1996 *Appl. Phys. Lett.* **69** (17) 2471
[13] Ankudinov A, Yanul M, Slipchenko S, Shelaev A, Dorozhkin P, Podoskin A and Tarasov I 2014 *Optics Express* **22** 21 26438

IOP Conf. Series: Journal of Physics: Conf. Series **929** (2017) 012071 doi:10.1088/1742-6596/929/1/012071

Plasmonic reflectance anisotropy spectroscopy of metal nanoclusters in a dielectric multilayer. Theory

A V Korotchenkov and V A Kosobukin

Ioffe Physical Technical Institute of the Russian Academy of Sciences,
194021 Saint-Petersburg, Russia

KorotchenkovAlex@gmail.com

Abstract. A theory of plasmonic reflectance anisotropy spectroscopy (RAS) is developed for nanocluster layer at an interface. The model of identical ellipsoidal metal particles occupying the sites of rectangular lattice is used to calculate the effective plasmonic polarizability of nanoparticles. The anisotropic local field due to optically induced dipole plasmons and their interface-conditioned images is taken into account. Within the theory, resonant reflectance anisotropy spectra recently observed for In nanoclusters on InAs surface are explained, the anisotropy being associated with the difference between frequencies of plasmons with orthogonal in-layer polarizations. The frequency difference is treated in terms of anisotropy of the particles shape or/and the layer structure, its sign being opposite for the two types of anisotropy. The plasmonic RAS is concluded to serve as a method for investigating the anisotropy of nanocluster arrays.

1. Introduction

Currently, considerable attention is paid to the optical properties of metal nanocluster arrays. Particularly, reflectance anisotropy spectroscopy (RAS) technique was recently found to reveal the plasmonic anisotropy of metal nanoparticle layers formed on semiconductor surfaces [1–4]. Since the interface does not contribute to RA spectra, it was attributed to inherent macroscopic anisotropy of the nanocluster arrays that is not displayed with conventional methods of diagnostics. Plasmonic nature of the spectral feature observed for In clusters on InAs(001) surface was confirmed by its red shift with increasing nanoparticle size and permittivity of surrounding media [3].

The observed RA spectra were interpreted within the model of regular nanoparticle array with interacting dipole plasmons for In/InAs system [2, 5] and the model of effective medium for In/GaAs [1]. Despite the conceptual difference between these approaches, similar estimations of ellipsoidal indium clusters anisotropy were obtained from the experimental plasmonic reflectance anisotropy spectra. The essential advantage of regular lattice model [5, 6] is the presence of collective plasmon modes, whose properties can be related to the shape of the clusters and the structure of the layer. The effect of anisotropic layer structure is inherently absent in the effective medium model.

This paper presents a theory of plasmonic RAS developed for metal nanoparticle arrays in layered dielectric media. The model of quasi-point dipoles arranged in rectangular lattice at the interface is investigated, the anisotropic local field of dipole plasmons induced in nanoparticles being taken into account. We consider the effects of both nanoparticle shape and the layer structure on the anisotropic optical resonance. Within the model we explain the spectra observed in [2, 3] and fit them to estimate the anisotropy of the In/InAs structures under study.

2. General theory

The paper is aimed at studying the photon-energy dependent RA spectra

$$\frac{\Delta R}{R} = 2\frac{R_x - R_y}{R_x + R_y}. \quad (1)$$

Here, R_α is the reflectivity of the monochromatic linearly polarized wave $\mathbf{E}^{\rm inc}(z,\boldsymbol{\rho}) = \mathbf{e}_\alpha\, \bar{E}^{\rm inc} \exp\left(i\sqrt{\varepsilon_1}k_0 z\right)$ incident normally onto a structure along the z-axis, the polarization subscript α being x or y. Hereafter, we consider the interface $z=0$ between two homogeneous isotropic media, the total permittivity tensor being $\varepsilon^{(0)}_{\alpha\beta}(z,\omega) = \delta_{\alpha\beta}\,\varepsilon^{(0)}(z,\omega)$ with $\varepsilon^{(0)}(z,\omega) = \varepsilon_1$ for transparent medium at $z<0$ and $\varepsilon^{(0)}(z,\omega) = \varepsilon_2(\omega)$ at $z>0$. Nanoparticles are centered at the sites $\boldsymbol{\rho}_{\mathbf{n}} = n_x A_x \cdot \mathbf{e}_x + n_y A_y \cdot \mathbf{e}_y$ of rectangular lattice with the periods A_x and A_y in the plane $z = z_0 = -h < 0$ near the interface (figure 1).

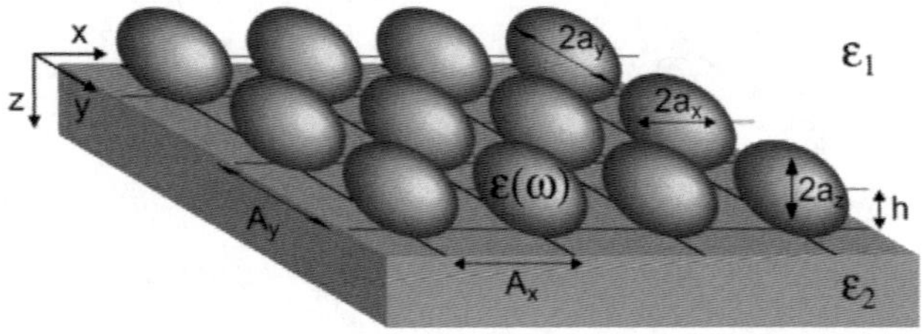

Figure 1. Model of ellipsoidal nanoparticles occupying a rectangular lattice at the interface.

In general [6], the reflectivities R_α are determined by the total electric field $\mathbf{E}(z,\boldsymbol{\rho};\omega)$ obtained from the sequence of equations

$$\sum_\nu \left[\sum_\mu \mathrm{rot}_{\alpha\mu}\,\mathrm{rot}_{\mu\nu} - \varepsilon^{(0)}(z;\omega)k_0^2\delta_{\alpha\nu}\right]\cdot\begin{Bmatrix} E^{(0)}_\nu(z,\boldsymbol{\rho}) \\ D^{(0)}_{\nu\beta}(z,z',\boldsymbol{\rho}-\boldsymbol{\rho}') \\ E_\nu(z,\boldsymbol{\rho}) \end{Bmatrix} = 4\pi k_0^2 \begin{Bmatrix} 0 \\ \delta(\mathbf{r}-\mathbf{r}')\delta_{\alpha\beta} \\ P_\alpha(z,\boldsymbol{\rho}) \end{Bmatrix}, \quad \begin{matrix}(2)\\(3)\\(4)\end{matrix}$$

where $\mathbf{r} = (z,\boldsymbol{\rho})$, $\boldsymbol{\rho} = (x, y)$, $k_0 = \omega/c$, and c is the speed of light. Equations (1) and (2) give the unperturbed (in absence of nanoparticles) electric field $\mathbf{E}^{(0)}$ and tensor Green's function $\hat{D}^{(0)}$, respectively. These solutions of equations (1) and (2) and their Fourier transforms

$$\left\{\mathbf{E}^{(0)}(z;\boldsymbol{\kappa}),\, \hat{G}^{(0)}(z,z';\,\boldsymbol{\kappa})\right\} = \int \mathrm{d}^2\rho \exp(-\mathrm{i}\boldsymbol{\kappa}\cdot\boldsymbol{\rho})\left\{\mathbf{E}^{(0)}(z,\boldsymbol{\rho}),\, \hat{D}^{(0)}(z,z',\boldsymbol{\rho})\right\} \quad (5)$$

satisfy the Maxwell's boundary conditions at $z=0$. In equation (4), the perturbation is provided by the polarization $\mathbf{P}(z,\boldsymbol{\rho};\omega)$ of nanoparticles. Given the components $\chi^{(0)}_{\alpha\alpha}(\omega)$ of diagonal plasmon polarizability tensor for a single nanoparticle, its dipole moment induced by the field $\mathbf{E}$ is $\mathbf{p}_{\mathbf{n},\alpha} = \chi^{(0)}_{\alpha\alpha}(\omega)E_\alpha(\mathbf{r}_\mathbf{n})$. The related polarization of an array of the identical particles is expressed as

$$P_\alpha(z,\boldsymbol{\rho}) = \delta(z-z_0)\sum_\mathbf{n}\delta(\boldsymbol{\rho}-\boldsymbol{\rho}_\mathbf{n})\chi^{(0)}_{\alpha\alpha}(\omega)E_\alpha(z_0,\boldsymbol{\rho}_\mathbf{n}), \quad (6)$$

where the delta function $\delta(z-z_0)\delta(\boldsymbol{\rho}-\boldsymbol{\rho}_\mathbf{n})$ implies that the local plasmons are treated as quasi-point dipoles. Equation (4) with the polarization (6) is rewritten as the equivalent integral equation

IOP Conf. Series: Journal of Physics: Conf. Series **929** (2017) 012071 doi:10.1088/1742-6596/929/1/012071

$$E_\alpha(z,\mathbf{\rho}) = E_\alpha^{(0)}(z,\mathbf{\rho}) + \sum_{\mathbf{n},\beta} D_{\alpha\beta}^{(0)}(z,z_0,\mathbf{\rho}-\mathbf{\rho_n})\chi_{\beta\beta}^{(0)} E_\beta(z_0,\mathbf{\rho_n}). \quad (7)$$

For a regular lattice, the self-consistent solution of equation (7) is obtained using the representations of dipole moment $\mathbf{p}_{\mathbf{n},\alpha} = \chi_{\alpha\beta} E_\beta^{(0)}(\mathbf{r_n}) = \chi_{\alpha\alpha}^{(0)} E_\alpha(\mathbf{r_n})$, where the effective polarizability $\hat{\chi}$ taking account of dipole interactions is introduced for nanoparticles [5, 6]. From equation (7) rewritten for the near field $E_\beta(z_0,\mathbf{\rho_n})$ one obtains the tensor

$$\hat{\chi} = \hat{\chi}^{(0)}\left[\hat{I} - \sum_{\mathbf{n}}{}' \hat{D}^{(0)}(z_0,z_0,\mathbf{\rho_n})\hat{\chi}^{(0)}\right]^{-1}, \quad (8)$$

where prime means exclusion of term with $\mathbf{n}=0$ corresponding to self-interaction of dipoles. On using equation (8) and the representation $\sum_{\mathbf{n}} \exp(-\mathrm{i}\mathbf{\kappa}\cdot\mathbf{\rho_n}) = (2\pi)^2 (A_x A_y)^{-1} \sum_{\mathbf{m}} \delta(\mathbf{\kappa}-\mathbf{b_m})$ equation (7) is transformed to the relation

$$E_\alpha(z,\mathbf{\rho}) = E_\alpha^{(0)}(z,\mathbf{\rho}) + \sum_{\mathbf{m},\beta,\gamma} \mathrm{e}^{\mathrm{i}\mathbf{b_m}\cdot\mathbf{\rho}}\, G_{\alpha\beta}^{(0)}(z,z_0;\mathbf{b_m}) \frac{\chi_{\beta\gamma}}{A_x A_y} E_\gamma^{(0)}(z_0,\mathbf{\rho}) \quad (9)$$

containing summation over the reciprocal lattice vectors $\mathbf{b_m} = 2\pi\left(m_x \mathbf{e}_x A_x^{-1} + m_y \mathbf{e}_y A_y^{-1}\right)$, $\mathbf{m} = (m_x, m_y)$, where m_x and m_y are integers. For short-period lattices $|\mathbf{b}_{\mathbf{m}\neq 0}| > A_\alpha^{-1} >> k_0$, and normally radiated field is given by the only contribution with $\mathbf{b}_{\mathbf{m}=0} = 0$ in equation (9), the other field components vanishing at sub-wavelength distances $|z-z_0| \sim A_\alpha << k_0^{-1}$ from the interface. Then only diagonal tensor components $\chi_{\alpha\alpha}$ and $G_{\alpha\alpha}^{(0)}(z,z_0;0)$ in equation (9) turn to be non-zero, and the polarizations of incident $\mathbf{E}^{\mathrm{inc}}$ and radiated $\mathbf{E}$ waves coincide. As a result, equation (9) gives for reflectivity of light polarized along the α th axis the following expression

$$R_\alpha = \left|r^{(0)} + \Delta r_\alpha\right|^2, \qquad \Delta r_\alpha = \frac{2\pi \mathrm{i} k_0}{\sqrt{\varepsilon_1}} \frac{\chi_{\alpha\alpha}}{A_x A_y}\left(\mathrm{e}^{-\mathrm{i}\sqrt{\varepsilon_1}k_0 h} + r^{(0)}\mathrm{e}^{\mathrm{i}\sqrt{\varepsilon_1}k_0 h}\right)^2, \quad (10)$$

where $r^{(0)} = (\sqrt{\varepsilon_1}-\sqrt{\varepsilon_2})/(\sqrt{\varepsilon_1}+\sqrt{\varepsilon_2})$ is reflectivity of the interface.

3. Plasmonic polarizability of nanoparticles in a layer

Next we investigate the reflectance anisotropy spectrum (1) within the model of identical ellipsoidal nanoparticles (figure 1) whose polarizability components are

$$\chi_{\alpha\alpha}^{(0)} = \frac{a_x a_y a_z}{3} \frac{\varepsilon - \varepsilon_1}{(\varepsilon-\varepsilon_1)N^{(\alpha)} + \varepsilon_1}, \quad (11)$$

the principal axes of the tensor $\hat{\chi}^{(0)}$ are assumed to coincide with the unit vectors $\mathbf{e}_\alpha$ of the lattice. The particles are supposed to be formed of metal with the permittivity $\varepsilon(\omega) = \varepsilon_\infty - \omega_\mathrm{p}^2/(\omega^2 + \mathrm{i}\omega\gamma)$, where ω_p is the plasma frequency and γ^{-1} is the electron relaxation time. The pole of (11) with $\varepsilon(\omega)$ determines the frequency $\omega_\alpha^{(0)} = \omega_\mathrm{p}\sqrt{N^{(\alpha)}/\varepsilon_*^{(\alpha)}}$ of dipole plasmon polarized along α th semi-axis of ellipsoid, $\varepsilon_*^{(\alpha)} = (\varepsilon_\infty - \varepsilon_1) N^{(\alpha)} + \varepsilon_1$. The plasmonic anisotropy is related to the particle shape through the depolarization coefficients $N^{(\alpha)}$ depending on the ratios of semi-axes lengths.

In quasi-static approximation, the tensor $\hat{\chi}$ of effective plasmonic polarizability is obtained from equation (8) on substituting the resonant plasmonic components (11) to give

$$\chi_{\alpha\alpha} = \left[\frac{1}{\chi_{\alpha\alpha}^{(0)}} - \frac{1}{\varepsilon_1 A_\alpha^3} \left(S_\alpha^{(\mathrm{d})} + \frac{\varepsilon_1 - \varepsilon_2}{\varepsilon_1 + \varepsilon_2} S_\alpha^{(\mathrm{i})} \right) \right]^{-1}. \quad (12)$$

The lattice sums

$$S_\alpha^{(\mathrm{d})} = A_\alpha^3 \sum_{\mathbf{n} \neq 0} \frac{3\rho_{\mathbf{n},\alpha}^2 - \rho_{\mathbf{n}}^2}{\rho_{\mathbf{n}}^5}, \quad S_\alpha^{(\mathrm{i})}(h) = -\left(\frac{A_\alpha}{2h} \right)^3 + A_\alpha^3 \sum_{\mathbf{n} \neq 0} \frac{3\rho_{\mathbf{n},\alpha}^2 - (\rho_{\mathbf{n}}^2 + 4h^2)}{(\rho_{\mathbf{n}}^2 + 4h^2)^{5/2}}, \quad (13)$$

corresponding to the fields of in-layer dipoles and their images are derived from the term $\sum_{\mathbf{n}}{}' \hat{D}_{\alpha\alpha}^{(0)}(z_0, z_0, \boldsymbol{\rho}_{\mathbf{n}})$ in which the retardation is neglected.

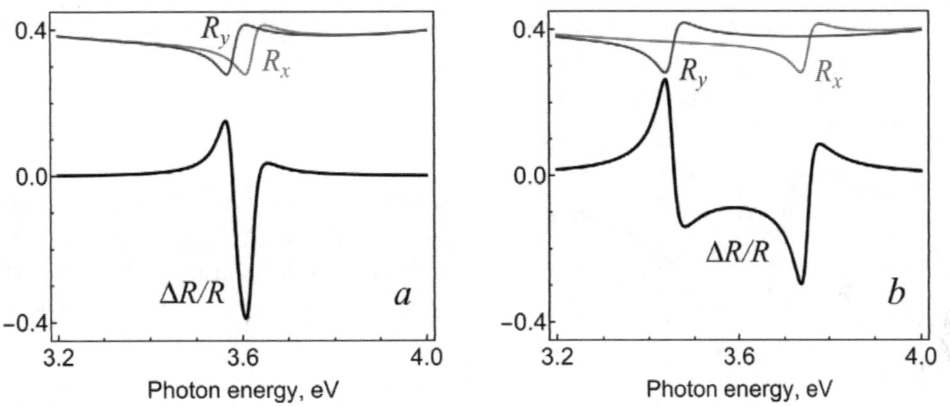

Figure 2. Theoretical spectra $\Delta R/R$ and R_x, R_y calculated, respectively, from equations (1) and (10) with (14) for (a) weak ($\hbar\omega_x = 3.62$ eV, $\hbar\omega_y = 3.58$ eV) and (b) strong ($\hbar\omega_x = 3.75$ eV, $\hbar\omega_y = 3.45$ eV) frequency anisotropy specified by the difference $\hbar|\omega_x - \omega_y|$ at $\hbar\Gamma_{x,y} = 0.04$ eV used as a scale.

The effective polarizability of α-polarized plasmon of a particle (12) can be reduced to the form

$$\frac{k_0}{A_x A_y} \chi_{\alpha\alpha}(\omega) \approx \frac{\Omega_\alpha^2}{\omega_\alpha^2 - \omega^2 - i\omega\Gamma_\alpha}. \quad (14)$$

To interpret, equation (14) is the resonant polarization response of nanoparticles determining the long-wavelength plasmon mode of the layer. The parameters of components $\chi_{xx} \neq \chi_{yy}$ providing theoretical reflectance anisotropy spectra (1), (10) are derived from equation (12) as follows

$$\omega_\alpha = \omega_\alpha^{(0)} \left\{ 1 - \frac{a_x a_y a_z}{3A_\alpha^3} \frac{1}{\varepsilon_*^{(\alpha)} N^{(\alpha)}} \left[S_\alpha^{(\mathrm{d})} + \mathrm{Re} \frac{\varepsilon_1 - \varepsilon_2}{\varepsilon_1 + \varepsilon_2} S_\alpha^{(\mathrm{i})}(h) \right] \right\}^{1/2}, \quad (15)$$

$$\Gamma_\alpha = \gamma + \frac{a_x a_y a_z}{3A_\alpha^3} \frac{\omega_\alpha^{(0)}}{\varepsilon_*^{(\alpha)} N^{(\alpha)}} S_\alpha^{(\mathrm{i})}(h) \,\mathrm{Im} \frac{\varepsilon_1 - \varepsilon_2}{\varepsilon_1 + \varepsilon_2} \,, \tag{16}$$

$$\Omega_\alpha^2 = \frac{\omega_\alpha}{cA_x A_y} \frac{a_x a_y a_z}{3} (\omega_\alpha^{(0)})^2 \frac{\varepsilon_1}{\varepsilon_*^{(\alpha)} N^{(\alpha)}} \,. \tag{17}$$

In the above model, the anisotropy $\omega_x \neq \omega_y$ of frequencies (15) can be attributed to either the particle shape anisotropy $a_x \neq a_y$ or/and the lattice anisotropy $A_x \neq A_y$.

Typical theoretical spectra $\Delta R/R$ for weak and strong anisotropy whose measure is $|\omega_x - \omega_y|/\Gamma$ are compared in figure 2, the two spectra are seen to be qualitatively different. To verify equations (15) - (17) from experiment, we fit measured spectra $\Delta R/R$ of In clusters on InAs(001) surface [3] using formula (14) and considering its parameters as fitting ones. The result of such fitting is shown in figure 3, where only the frequencies $\omega_x \neq \omega_y$ are taken different, and Γ_α and Ω_α are the same for polarizations x and y. The shape of measured (fitted) spectra $\Delta R/R$ (figure 3) is seen to be similar to the weak-anisotropy theoretical spectrum $\Delta R/R$ (figure 2a), despite the difference of Γ. To conclude, this estimation gives evidence of weak anisotropy responsible for spectra $\Delta R/R$ measured in [2, 3], whose representative is the spectrum $\Delta R/R$ in figure 3.

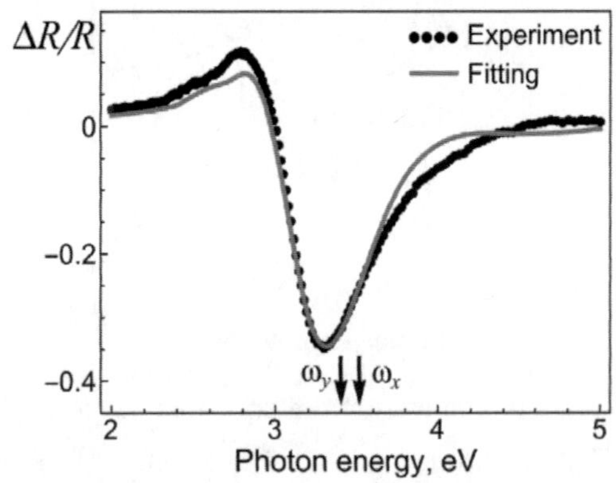

Figure 3. Spectrum $\Delta R/R$ of In nanocluster array on InAs surface (measured in [3], dots), and spectrum (1), (10) fitting the experimental one using function (14), solid line. The plasmon energies $\hbar\omega_y = 3.40$ eV and $\hbar\omega_x = 3.52$ eV are shown with arrows. The other parameters of function (14) are $\hbar\Gamma = 1.17$ eV and $\hbar\Omega = 0.77$ eV for both polarizations.

4. Estimation of layer anisotropy

Given the conclusion that $|\omega_x - \omega_y| << \Gamma$, we introduce the parameters $\omega_\parallel = (\omega_x + \omega_y)/2$, Γ and Ω of hypothetical optically isotropic layer whose small anisotropic deformation provides the fitting $\Delta R/R$ spectrum (figure 3). The calculated parameters (15) - (17) depending on the semi-axes ratio a_z/a for spheroidal particles ($a_x = a_y = a$) and square lattice ($A_x = A_y = A$) are presented in figure 4.

Bearing in mind a small splitting $\Delta\omega_\parallel = \omega_x - \omega_y$, we consider spectra $\Delta R/R$ (figure 3) as anisotropy-conditioned deviation from $\Delta R/R = 0$ corresponding to isotropic layer model with plasmon frequency $\omega_\parallel$. For anisotropic layer of ellipsoidal particles with $a_{x,y} = a \mp \Delta a/2$ in a square lattice (I), or the rectangular lattice $A_{x,y} = A \mp \Delta A/2$ occupied by spheroids (II), the frequencies $\omega_\alpha(a_\alpha, A_\alpha)$ from equation (15) are expanded near a and A up to linear terms in Δa and ΔA as

$$\frac{\omega_x - \omega_y}{\omega_\parallel} = C_{\mathrm{I}} \frac{\Delta a}{a} + C_{\mathrm{II}} \frac{\Delta A}{A} \,. \tag{18}$$

Measured the relative difference $2(\omega_x - \omega_y)/(\omega_x + \omega_y)$, the coefficients $C_{\rm I}$ and $C_{\rm II}$ allow to estimate the necessary deformations $\Delta a/a$ of nanoparticles or $\Delta A/A$ of lattice, respectively [5].

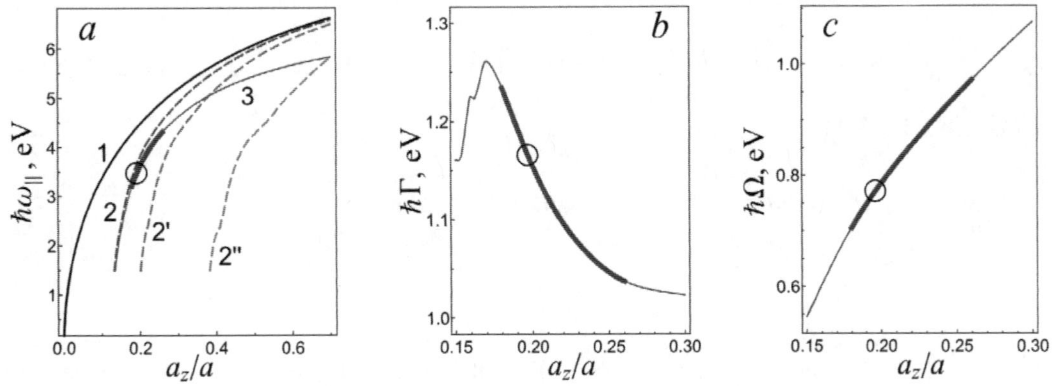

Figure 4. (a) Plasmon frequencies calculated from (15) for a single spheroid (1), for a single spheroid with its image at $h/a_z = 3$ (2), $h/a_z = 2$ (2') and $h/a_z = 1.1$ (2"), and for a layer of spheroids affected by dipoles and their images at $h/a_z = 3$ (3). (b) Decay rate (16) and (c) oscillator strength (17) of plasmons in the layer of spheroids. The parameters are shown depending on semi-axes ratio a_z/a, the condition $R_\alpha < 1$ is met for thick pieces of curves in (a–c). The theoretical values marked by circles at $a_z/a \approx 0.2$ ($a_z = 4$ nm) are very close to fitted ones at $a = 20$ nm and $A = 50$ nm.

With the above parameters of isotropic model, the anisotropy coefficients $C_{\rm I} \approx 1.12$ and $C_{\rm II} \approx -0.28$ are estimated. According to equation (18), the observed frequency splitting $\Delta\omega_{||}/\omega_{||} = 0.12/3.45$ could correspond to the deformation of spheroids $a_y/a_x - 1 \approx 3.1\%$ at $\Delta A = 0$, or deformation of lattice $A_y/A_x - 1 \approx -12.6\%$ at $\Delta a = 0$. Then, the splitting $\Delta\omega_{||} > 0$ could indicate that the particles are flattened and/or the lattice is strained along the x-axis.

5. Conclusion

We have developed a theory of reflectance anisotropy spectra caused by long-wavelength plasmon modes excited in metal nanoparticle layer under the normal incidence of light. The anisotropic plasmons are shown to be responsible for appearance of the resonant reflectance anisotropy spectra which are absent for isotropic layers of nanoparticles. Particularly, within the proposed model the characteristic reflectance anisotropy spectra of indium nanoclusters on InAs surface are interpreted and fitted to estimate the anisotropy of nanoparticle array under study.

The authors are grateful to V L Berkovits for valuable discussion and experimental data provided for fitting.

References

[1] N Esser, A M Frisch, A Roseler, S Schintke, C Goletti, D O Fimland 2003 *Phys. Rev. B* **67** 125306.

[2] V L Berkovits, V A Kosobukin, V P Ulin, A B Gordeeva, V N Petrov 2013 *JETP Letters* **98** 687.

[3] V L Berkovits, V A Kosobukin, V P Ulin, A B Gordeeva, V N Petrov 2015 *Surf. Sci.* **632** L9.

[4] R Verre, K Fleischer, O Ualibek, I V Shvets 2012 *Appl. Phys. Lett.* **100** 031102.

[5] V A Kosobukin, A V Korotchenkov 2016 *Phys. Solid State* **58** 2446.

[6] V A Kosobukin 2012 *Phys. Solid State* **54** 2471.

Coating of Au nanoparticle by Si shell for enhanced local heating

G P Zograf, M I Petrov and S V Makarov

ITMO University, 49 Kronverksky Pr., St. Petersburg, 197101, Russia

E-mail: g.zograf@metalab.ifmo.ru

Abstract. Gold nanoparticles are widely used as a nano-sources of heat, but their properties are governed by skin depth which is usually less than 10 nm. In this work, we propose a mechanism for increasing efficiency of gold nanoparticles heat absorption in optical range. By 'wrapping' Au sphere in Si spherical layer we achieved significantly increased heating the nanoparticle in comparison with the gold filled sphere of the same size. The strong heating is caused by radiative losses suppression and more effective electric field concentration around the gold nanoparticle owing to adding of the Si layer.

1. Introduction

Thermotherapy at nanoscale is an important area of medicine where cancer cells are treated by local hyperthermia [1–3]. Metallic nanoparticles are commonly used as a local source of heat [4]. Indeed, plasmonic nanoparticles are well known for their heat absorption properties and strong degree of electromagnetic field localization, which allow to apply them in wide range of areas, including optics and biomedicine [5]. However, small thickness of a skin depth limits efficiency of plasmonic nanoparticle heating owing to very small volume of effective heat source, i.e. mode volume.

On the other hand, dielectric nanoparticles of high refractive index and low extinction coefficient maintain strong electric and magnetic Mie-resonances [7] in the visible range [8]. In comparison with the plasmonic particles, dielectric ones have much greater filling factors for excited optical modes, being useful for a number of applications [9–11]. In this paper, we are taking advantages of both types of nanoparticles: plasmonic and dielectric ones, for effective localized heating. In particular, by covering gold nanoparticle with spherical layer of silicon we increase the heating temperature of a gold core and extend spectral range of effective heating from near-UV to near-IR region.

2. Model

Our theoretical approach is based on Mie-theory which describes the scattering of an electromagnetic plane wave by a homogeneous sphere. Absorbed power by spherical nanoparticle can be described through its' absorption cross-section C_{abs} and light source intensity I :

$$Q = C_{abs} I \tag{1}$$

Absorption C_{abs} cross-section for spherical nanoparticles and core-shells(layered spheres) are linked with extinction and scattering cross-sections $C_{abs} = C_{ext} - C_{sca}$, whereas C_{ext} and C_{sca}

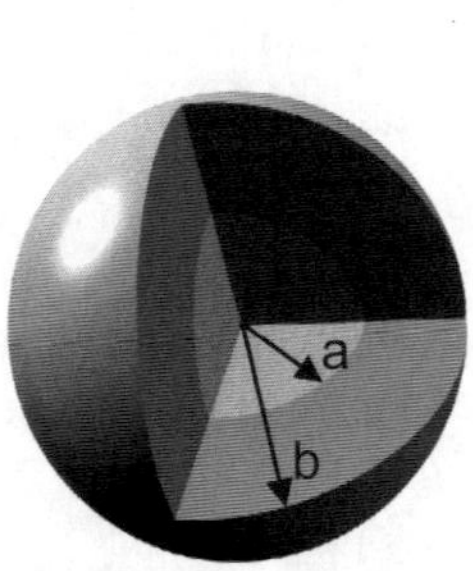

Figure 1. Gold-silicon core-shell concept picture. a - core radius, b - shell radius

are following:

$$C_{sca} = \frac{W_{sca}}{I_i} = \frac{2\pi}{k^2}\sum_{n=1}^{\infty}(2n+1)(|a_n|^2+|b_n|^2)\,, \quad C_{ext} = \frac{W_{ext}}{I_i} = \frac{2\pi}{k^2}\sum_{n=1}^{\infty}(2n+1)Re(a_n+b_n) \quad (2)$$

where W_{sca}, W_{ext} - scattered and extinct energy by core-shell, $k = \frac{2\pi N}{\lambda}$, I_i - incident intensity, a_n and b_n are, so called, scattering coefficients, that describes interaction of nanoparticle with plane wave. For most cases, we can consider same magnetic permeabilities for surrounding medium and a core-shell. These assumptions lead to simplified form of scattering coefficients:

$$a_n = \frac{\psi_n(y)[\psi_n'(m_2y) - A_n\chi_n'(m_2y)] - m_2\psi_n'(y)[\psi_n(m_2y) - A_n\chi_n(m_2y)]}{\xi_n(y)[\psi_n'(m_2y) - A_n\chi_n'(m_2y)] - m_2\xi_n'(y)[\psi_n(m_2y) - A_n\chi_n(m_2y)]} \quad (3)$$

$$b_n = \frac{m_2\psi_n(y)[\psi_n'(m_2y) - B_n\chi_n'(m_2y)] - \psi_n'(y)[\psi_n(m_2y) - B_n\chi_n(m_2y)]}{m_2\xi_n(y)[\psi_n'(m_2y) - B_n\chi_n'(m_2y)] - \xi_n'(y)[\psi_n(m_2y) - B_n\chi_n(m_2y)]} \quad (4)$$

where :

$$A_n = \frac{m_2\psi_n(m_2x)\psi_n'(m_1x) - m_1\psi_n'(m_2x)\psi_n(m_1x)}{m_2\chi_n(m_2x)\psi_n'(m_1x) - m_1\chi_n'(m_2x)\psi_n(m_1x)} \quad (5)$$

$$B_n = \frac{m_2\psi_n(m_1x)\psi_n'(m_2x) - m_1\psi_n(m_2x)\psi_n'(m_1x)}{m_2\chi_n'(m_2x)\psi_n(m_1x) - m_1\psi_n'(m_1x)\chi_n(m_2x)} \quad (6)$$

and $\psi_n(\rho), \chi_n(\rho), \xi_n(\rho)$ - Riccati - Bessel functions, a - core radius, b - shell radius, m_1 and m_2 - relative refractive indices of a sphere and a shell, $x = ka$, $y = kb$. Thus, calculation of heat power Q delivered into core-shell turns out to be an optical problem with a known solution. Meanwhile, in order to define temperature distribution inside and outside a core-shell we need to solve a stationary heat transfer eq. 7 below:

$$\nabla\cdot[\kappa_{in}(\mathbf{r})\nabla T_{in}(\mathbf{r})] = -q(\mathbf{r}) \quad \text{inside}, \quad \nabla\cdot[\kappa_{out}(\mathbf{r})\nabla T_{out}(\mathbf{r})] = 0 \quad \text{outside} \quad (7)$$

Where q power density (in our case is $\frac{Q}{V} = \frac{3Q}{4\pi b^3}$) and $\kappa(\mathbf{r})$ - is thermal conductivity. This leads us to following temperature increment distribution inside and outside core-shell:

$$\Delta T(r)_{in} \approx \Delta T_{CS} \quad r < b\,, \qquad \Delta T(r)_{out} = \Delta T_{CS}\frac{b}{r} \quad r > b \quad (8)$$

$\Delta T_{CS} = \frac{Q}{4\pi\kappa_{out}b}$ - temperature increment in core-shell. This solution is only valid for the case of poorly thermal conductive medium - $\kappa_{in} \gg \kappa_{out}$

IOP Conf. Series: Journal of Physics: Conf. Series **929** (2017) 012072 doi:10.1088/1742-6596/929/1/012072

3. Results and discussions

The above expressions allow us to study scattering and heating behaviour of the core-shell nanoparticles and track their modes evolution with different parameters (e.g. wavelengths, shell and core radii). We consider golden core of 15 nm radius as suitable for most common biomedical applications [baffou review]. Aslo, it is important to study heating properties of nanoparticles in, so called, transparency window of biological tissues (between 700 and 1300 nm). Working in this spectrum range allows to increase light propagation length into biological tissues minimizing absorption and thus heating of healthy cells. Although spherical gold nanoparticles show the best light absorption in visible range between 500 and 600 nm, its' resonant can be shifted into near-IR region by placing golden sphere inside homogeneous spherical silicon layer - shell.

In Fig. 2(a) light scattering by silicon shell with a spherical hole in the centre with a 15 nm radius is presented. The modes evolutions give an insight into the contribution of shell into Au-Si core-shell nanoparticles properties. Particularly, in relatively small nanoparticles the strongest contribution to scattering is given by electric dipole (ED) mode, magnetic dipole (MD) and quadrupole (MQ) modes of the shell.

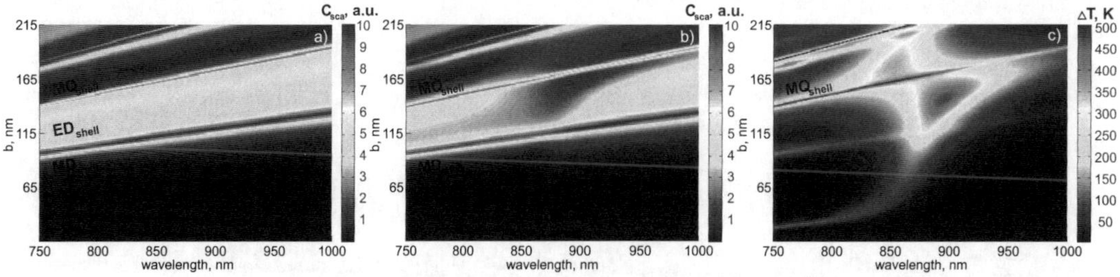

Figure 2. 2D maps for core-shells with fixed core radius of a = 15 nm; x-axis corresponds to incident wavelength, y-axis corresponds to shell radius b, i.e. b = a = 15 nm means appearance of only core. (a) Scattering cross-section of a core-shell with silicon shell and air 'core' in the units normalized by geometrical cross-section; (b) scattering cross-section of gold-silicon core-shell nanoparticle in arbitrary units (normalized by geometrical cross-section); (c) temperature increment of gold-silicon core-shell nanoparticle in degrees of Kelvin irradiated by plane wave with intensity $I_0 = 0.2 mW/\mu m^2$

Putting gold inside the silicon shell, then appears scattering dropping around 900 nm (see Fig. 2(b), where ED of the shell is excited. On the other hand, temperature increase (see Fig. 2(c), which is related to absorption (eq. 1, 8) in the same region shows local peak of heating. This peak corresponds to the core-shell with $b = 150$ nm shell radius. Such configuration demonstrates heating in air up to 500 K (see Fig. 3(a)).

In comparison with spherical gold nanoparticle of the same size, core-shell demonstrates more efficient heating in the transparency region of spectrum up to 10 times (Fig. 3(a)). To evaluate gold core contribution, we have added heating of the silicon shell with excluded gold core. Such comparison of heating improvement possibilities shows the advantage of covering gold sphere by silicon, instead of increasing size of the sphere. Heating properties of gold are limited by skin depth and absorption peak positions, whereas by manipulating shell size we both can control peak position and localize electromagnetic field in the center of the shell, which can dramatically increase temperature. In Fig. 3(b) one can see electric field distribution in the vicinity of the core-shell with considered parameters. Propagation direction of the plane wave is normal to the figure plane and **E**-vector is set to be in x-direction. Typical behaviour of ED resonance both of a shell and a core is manifested in field enhancement on particle/medium boundary in initial electric field vector direction. It means that local dip of scattering noticed in Fig. 2(b)

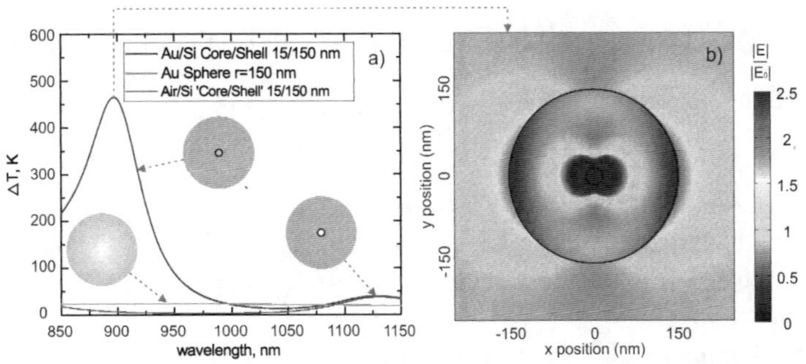

Figure 3. a) Comparison between temperature increments of gold spherical nanoparticle of 150 nm radius, gold-silicon core-shell with a = 15 nm and b = 150 nm and silicon shell with b = 150 nm and excluded core with a = 15 nm b) Electric field magnitude enhancement numerical simulation model at 895 nm wavelength for gold-silicon core-shell with a = 15 nm and b = 150 nm.

can be explained via destructive interference of these two resonant ED modes, which almost completely cancels far-field scattering from the core-shell. Also, the temperature maximum in the same region (Fig. 2(c)) is partially described by the near-field localization and enhancement up to 20 times in the core centre owing to specific mode structure of the shell ED. Therefore, such electric fields overlapping of ED modes in core and shell provides necessary conditions for the most effective heating.

4. Conclusions

In this paper we have shown that optimized gold-silicon core-shell nanoparticle can serve as a effective as nanoscale heater for potential photothermal applications. The origin of the 10fold increased heating efficiency is the reducing of radiative losses and electric field localization around lossy plasmonic core.

5. Acknowledgements

Analitical modeling was financially supported by the Ministry of Education and Science of the Russian Federation (Grant ? 14.Y26.31.0010) and numerical calculations were supported by Russian Science Foundation (Grant 15-19-00172).

6. References

[1] Kennedy L C et al 2011 *Small* vol 7.2 pp 169–183.
[2] Huang X et al 2006 *Journal of the American Chemical Society* vol 128.6 pp 2115–2120.
[3] Atwater H A 2007 *Scientific American* vol 296.4 pp 56–62.
[4] Baffou G and Quidant R 2012 *Laser & Photonics Reviews* vol 7 pp 171–187
[5] Maier S A 2007 *Plasmonics: fundamentals and applications* (Springer Science & Business Media)
[6] Kuznetsov A I, Miroshnichenko A E, Fu Yu H, Zhang J, Luk'yanchuk B 2012 *Scientific Reports* vol 2 492
[7] Mie G 1908 *Annalen der physik* vol 330 pp 377–445
[8] Krasnok A, Makarov S, Petrov M, Savelev R, Belov P and Kivshar Y 2015 *Proc. SPIE* vol 9502 pp 950203
[9] Makarov S, Kudryashov S, Mukhin I, Mozharov A, Milichko V, Krasnok A et al 2015 *Nano letters* vol 15 (9) pp 6187–6192
[10] Dmitriev P A, Baranov D G, Milichko V A, Makarov S V, Mukhin I S et al *Nanoscale* vol 8 (18) pp 9721–9726
[11] Kuznetsov A I et al 2016 *Science* vol 354.6314 aag2472.
[12] Bohren C F and Huffman D R 1983 *Absorption and scattering of light by small particles* (Wiley interscience).

IOP Conf. Series: Journal of Physics: Conf. Series **929** (2017) 012073 doi:10.1088/1742-6596/929/1/012073

Investigation of the methods for optical wavefront parameter manipulation

Tatiana A Vovk, Nikolay V Petrov
Department of Photonics and Optical Information Technology, *ITMO University*
49 Kronverksky Ave, Saint Petersburg 197101, Russia

E-mail: tatiana_vovk@corp.ifmo.ru

Abstract. Here we describe the efficiency of methods for obtaining optical wavefronts with the predetermined parameters, consider the main techniques of the wave field formation, and conduct the analysis of the applicability of these methods to specific problems. We also developed and analyzed the combination of wavefront formation methods. Among the results is the new method of the light beam synthesis based on the combination of the adaptive wavefront optimization approach and the direct prescription of the phase distribution via scalar diffraction theory.

1. Introduction

Formation of optical wave fields with the prescribed complex parameters using digital methods, which can effectively convert energy of laser radiation to the required form, are of particular interest for applications in micro-object manipulation (optical tweezers) [1], controllable thermonuclear synthesis with strong optical pulses [2], and aseptic surgery by the means of laser radiation [3]. Application of new digital methods for controlling light fields allows one to manipulate the field parameters such as amplitude, phase, and degree and kind of polarization. The distinct feature of digital methods is the possibility to create virtual optical devices, whose physical realization is impossible due the absence of manufacturing methods.

From the scalar diffraction theory it is known that the beam amplitude distribution in the arbitrary region of space can be governed by the boundary conditions describing the wave field. Therefore, in order to form the required light field in the given space area one has to find the corresponding amplitude and phase distribution in the wave field modulation plane. It can be implemented using such optical elements as the relief kinoforms (phase holographic platelets) [4], holographic optical elements [5], digital holograms recorded on films [6], and dynamic optical elements with the spatial light modulators (SLMs) [7]. The comprehensive review of the methods for the light field formation and their technological implementations is given in [8].

As of today, numerical calculations of optical fields with the prescribed parameters are conducted using the iteration algorithms, such as Gerchberg-Saxton algorithm [9], and scalar diffraction theory [10]. It should be noted that the employed methods are not suitable for all the existing problems, as their accuracy is not sufficient in some cases. For example, kinoform synthesis is a difficult technological task, and in the most cases, only the structures similar to the kinoforms are synthesized [4]. Suggestion of new methods for obtaining the determined wave field and their further development can potentially solve these problems.

2. Statement of the problem

Numerical simulation of optical wave fields is the common task for optical physics. As a rule, approaches based on geometric optics, iterative algorithms, and scalar diffraction theory are used to solve this problem. In this communication, we study and improve efficiency of the methods for obtaining optical wavefronts with the predetermined parameters. To achieve this, we propose the improvement and optimization of the existing methods for the optical field assignment, as well as the combination of the different methods. We consider and describe the main methods of the wavefront formation. Then we analyze the applicability of these methods to the specific problems.

3. Basic methods of optical field formation

In this section, we consider the basic approaches to formation of the optical fields with the given distribution parameters.

3.1. Direct assignment of the amplitude-phase distribution via scalar diffraction theory

Description of the evolution of the propagating wave field can be performed through the method of scalar diffraction theory. The angular spectrum (AS) determines the wave field amplitude through the each component of the Fourier expansion:

$$U(x,y)=\hat{F}^{-1}\{H(f_x,f_y,l)\cdot\hat{F}[U(x',y')]\}, \tag{1}$$

where $H(f_x,f_y,l)$ is the transfer function on the distance l, given by:

$$H(f_x,f_y,l)=\begin{cases}\exp\left[i\dfrac{2\pi l}{\lambda}\sqrt{1-\lambda^2\cdot(f_x^2+f_y^2)}\right], & f_x^2+f_y^2<\dfrac{1}{\lambda^2}\\ 0, & f_x^2+f_y^2\geq\dfrac{1}{\lambda^2}\end{cases}, \tag{2}$$

Where λ is the wavelength of the propagated optical field, and f_x, f_y are the spatial frequencies along x, y axes, respectively. In practice, the idea of the spatial distribution of the wave field through the AS of plane waves is convenient due to the following reasons:

- There is no additional restrictions on the original description of the wave field through the Rayleigh-Sommerfeld integral;
- Calculation of the diffraction integral can be performed using the fast Fourier transform, which significantly shortens the simulation time.

3.2. Gerchberg-Saxton algorithm

The iterative Gerchberg-Saxton algorithm for the wavefront phase calculations consists of the following stages (see Fig. 1):

1. For the initial phase and intensity distributions defined in the input plane, the complex amplitude in the output plane is calculated;
2. The resulting intensity distribution is then replaced with the necessary intensity distribution, which is required to be formed in the output plane;
3. Numerical calculation of backward beam propagation to the input plane is performed;
4. The resulting intensity distribution is replaced with the necessary intensity distribution in the input plane. Calculated phase distribution is selected as the next approximation.

Then, the iterative process is repeated. The main disadvantage of this method is the inaccuracy of the resulting phase function, associated with the fact that the large number of iterations is required, and it is not always possible due to the time limits.

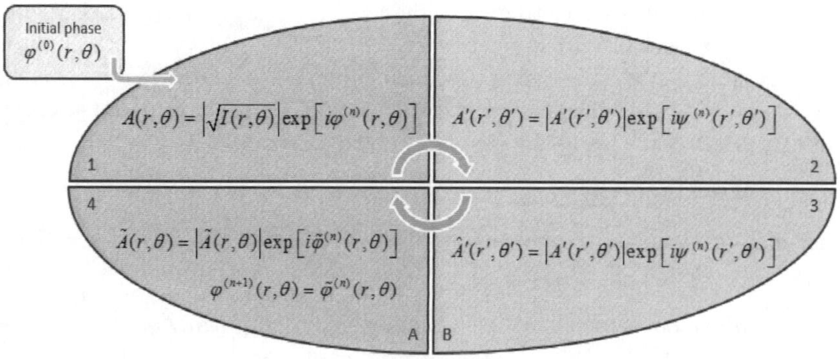

Figure 1. Scheme of the phase distribution reconstruction according to the Gerchberg-Saxton algorithm; *A* and *B* are the input and output planes

3.3. Adaptive optimization of the wavefront (AOWF algorithm)

Initially AOWF algorithm was created to solve the problem of the beam focusing through the scattering media [11]. Plane wavefront passed through the scattering sample, resulting in the randomly distributed intensity pattern on the registration plane. Thereafter, the AOWF algorithm was used converting the wavefront so that the incident beam from the source was focused on the registration plane. Previously, a simple case was considered when the scattering medium was absent [12]. In this case AOWF method formed an array of the optimal phase shift values, which was a discrete lens.

The principle of the algorithm operation can be summarized as follows (see Fig. 2):

1. Plane wavefront falls onto the SLM plane (input plane);
2. Separate segments of the wavefront acquire a predetermined phase delay on SLM pixels;
3. The intensity distribution captured on the registration plane is further numerically compared with the necessary distribution;
4. The computer selects the optimal phase delay in the current segment of SLM.

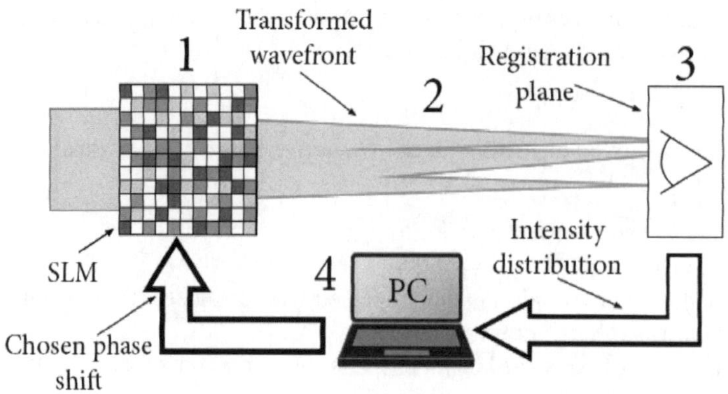

Figure 2. Algorithm of AOWF operation without scattering medium

This process is then repeated for the next segment of SLM. There are many algorithms for segment selection, for example, line-to-line processing of individual SLM segments, binary search, phase mask method, etc. Additional information about these algorithms can be found in [13].

4. Analysis of the applicability of the considered methods

In this section, we compare the proposed methods in the framework of experimental optics problems and tasks.

4.1. Focusing of the beam from the source in an output plane

Consider focusing of the radiation coming from the point source into a single point. This problem is trivial, and the fastest way to solve it is the straightforward usage of scalar diffraction theory, which means consideration of the backward emission of light from the point source to the given plane.

It is still of particular interest to consider the salvation of this problem with the suggested methods, namely the Gerchberg-Saxton algorithm and AOWF, and compare the results with the predictions of scalar diffraction theory. Such a comparison is given in Fig. 3. One can see that with the algorithm execution time being the same, the Gerchberg-Saxton algorithm is the most effective, as it transforms the phase integrally during each iteration, while AOWF consider each segment separately. However, AOWF method results in the closest approximation to the theoretical lens function.

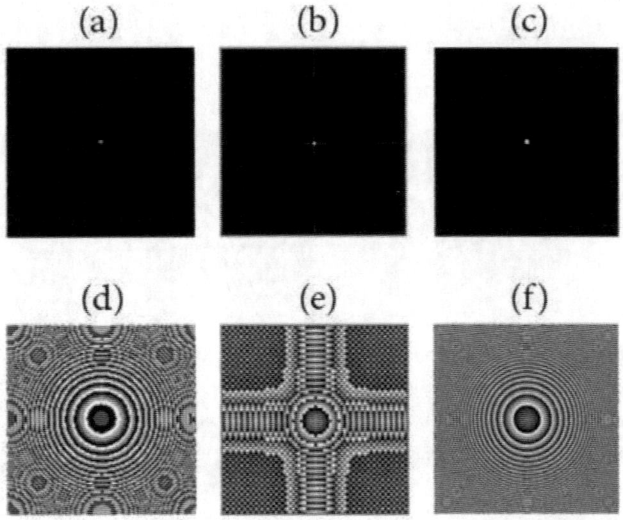

Figure 3. Amplitude distributions ((a) – (c)) in the output plane and phase distributions ((d) – (e)) in the SLM (input) plane after application of AOWF (a, d), Gerchberg-Saxton algorithm (b, e), and using the scalar diffraction theory (c, f)

4.2. Obtaining of a Gaussian beam profile

Modulation of the complex parameters of a laser beam is an important application of the dynamic optical elements created with SLM. Beam manipulation are the major part of the engineering task, and it is of particular importance to be able to obtain an ideal intensity distribution in the beam for the most accurate measurements.

Let us consider an application of the two methods laser beam processing (AOWF method and Gerchberg-Saxton algorithm) and compare the obtained distributions. We assume here that the beam initially have the plane phase distribution and the homogeneous intensity distribution. Results of the calculations are given in Fig. 4.

As is seen from the figure, the intensity distribution that is the closest to the required one is given by the Gerchberg-Saxton algorithm, which, however, results in the additional diffraction maxima, and AOWF algorithm is unable to provide the required distribution.

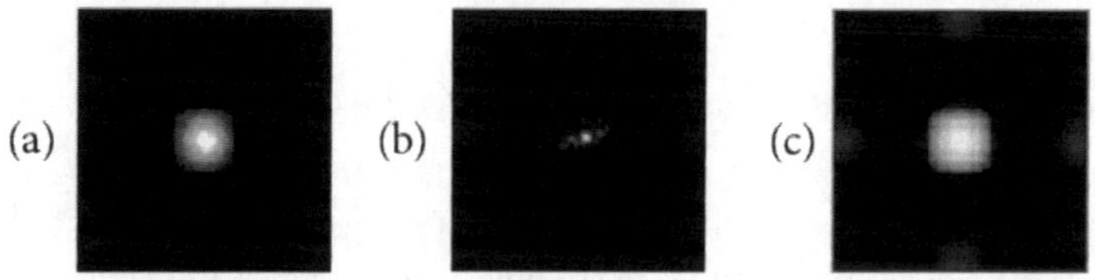

Figure 4. Required intensity distribution of the laser beam (a) and the result of the plane wavefront transformation after application of AOWF (b) and Gerchberg-Saxton algorithm (c)

5. Combination of AOWF and direct assignment of phase distribution

Let us consider the possibility of the combination of classical AOWF method and direct assignment of the phase distribution in a restricted area of SLM plane. A certain set of SLM pixels have a fixed phase shift. Therefore, with this conditions an optical element is synthesized in the SLM plane, consisting of two regions: predetermined phase distribution and optimized phase region. The second region focuses the preset amplitude configuration in the registration plane. The results of this method being applied are shown in Fig. 5.

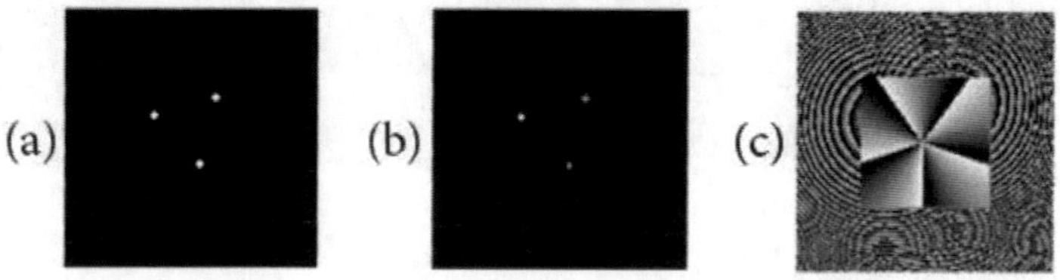

Figure 5. The target image (a), the result of the plane wavefront transformation using combination method (b), and the phase (c) in the SLM plane. Parameters: size of the image 320 × 320 mm, $\lambda = 532$ nm, the phase gradation $i_p = 4$. The size of the fixed singular phase area 160 × 160 mm (topological charge $q = 5$)

Wavefront from the source has been redistributed into the three points [Fig. 5 (b)]. As a result of AOWF configuration we have the phase distribution with the singular condition [Fig. 5 (c)]. This method has the advantage over direct assignment of a phase distribution: in an unfixed region, it is possible to obtain an object of arbitrarily complex configuration without algorithm variation.

Using this combination method beams with unusual properties have been obtained. Consider the propagation of the beam after the registration plane, shown in Fig. 6. The initially focused beam [Fig. 6 (a), (e)] has the spherical phase distributions in the point locations after propagation [Fig. 6 (b), (f)]. Redistributed beam acquires the form of a «donut», and the phase distribution loses its spherical components [Fig. 6 (c), (g)]. Subsequent propagation leads to a complete loss of spherical components and the formation of multiple phase singularities, while «donut» shaped intensity begins to increase in size [Fig. 6 (d), (h)]. It is worth noting that the phase distribution of the spiral beam, which was received during the propagation through the SLM, becomes chiral upon further propagation [Fig. 6 (h)]. This phenomenon is an intrinsic property [14] of spiral beams with a completed intensity distribution [«donut» shaped intensity in Fig. 6 (d)].

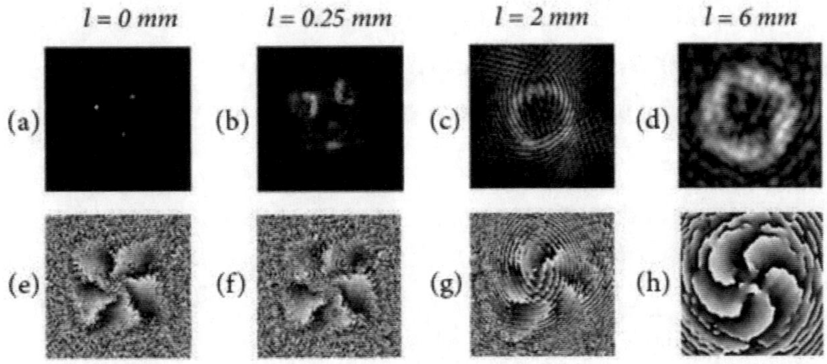

Figure 6. Amplitude [(a) - (d)] and phase [(e) - (h)] distributions of the generated beam after registration plane on a distance *l*

6. Conclusion

In this paper, we have studied the methods for obtaining optical wavefront with the predetermined parameters. Investigated methods have been theoretically described. These methods have been applied to the problems, the solution of which showed their applicability in the various scientific areas. Also, the combination method based on the AOWF algorithm and direct assignment of the phase distribution has been developed and investigated. This method opens up new possibilities for the optical beam synthesis and may result in the development of the new optical modulation algorithms.

7. Acknowledgement

This work is supported by the Grant of the President of Russian Federation No. MK-7068.2016.2.

8. References

[1] Curtis J, et al. 2002 Dynamic holographic optical tweezers *Opt. Commun* **207** 1
[2] Basov N, et al. 1972 Heating of Laser Plasmas for Thermonuclear Fusion *Laser Interaction and Related Plasma Phenomena, Springer US*
[3] Konig K et al. 1999 In Tracellular Nanosurgery With Near Infrared Femtosecond Laser Pulses *Cell Mol Biol* **45** 2
[4] Fishman A 1999 The phase optical elements – kinoforms *Soros Review Journal (Sorosovskij Obrazovatel'nyj Zhurnal)* **12** 1999
[5] Lukin A 2007 Holographic optical elements *J. Opt. Technol.* **74** 1
[6] Perlmutter R 1987 *Optical-digital hologram recording* (USA: pat. 4701006)
[7] Hornbeck L 1991 *Spatial light modulator* (USA: pat. 4956619)
[8] Kim H, et al. 2006 Diffractive optic synthesis and analysis of light fields and recent applications *Jpn. J. Appl. Phys.* **45** 8S
[9] Zalevsky Z, et al 1996 Gerchberg–Saxton algorithm applied in the fractional Fourier or the Fresnel domain *Opt. Lett* **21** 12
[10] Goodman J 2005 *Introduction to Fourier optics* (USA: Roberts and Company Publishers)
[11] Vellekoop I, et al 2007 Focusing coherent light through opaque strongly scattering media *Opt. Lett* **32** 16
[12] Petrov N, et al 2014 *Investigation of interaction of structured illumination with random scattering media* (San Diego: Proc. SPIE)
[13] Vellekoop I, et al 2008 Phase control algorithms for focusing light through turbid media *Opt. Commun* **281** 11
[14] Abramochkin E, Volostnikov V 2004 Spiral light beams *Physics-Uspekhi* **47** 12

IOP Conf. Series: Journal of Physics: Conf. Series **929** (2017) 012074 doi:10.1088/1742-6596/929/1/012074

Optical second harmonic generation and two-photon luminescence in organic microstructures

K. D. Zhdanova[1], N. V. Mitetelo[1], E.A. Mamonov[1], T.V. Murzina[1]

[1] Quantum Electronics division, Department of Physics, Moscow State University, Leninskie Gory, GSP -1, 119991 Moscow, Russia

E-mail: kesha@shg.ru

Abstract. We studied optical and second order nonlinear optical effects in arrays of frustum-shaped microstructures composed of orange-red dye by self-assembling technique.

1. Introduction

Nowadays photonics of micro- and nanosized structures attracts much attention of researchers. This is mostly due to high perspectives for application of such types of structures in optoelectronic devices. Most of the studies are carried out for metal structures, which reveal the enhancement of optical and nonlinear-optical effects enhancement due to local field amplification close to plasmon resonances. This was demonstrated for plasmonic nanostructures, nanoantennas, in plasmonic crystals etc. [1], [2]. Organic nano- and microstructures, being dielectric, are attractive because of a number of unique properties connected with possibility to excite different types of Mie resonances in the visible-near IR spectral range, high and tunable refractive index, a way for the modification of their size and shape.

The easiest method for the composition of organic structures is the application of self-assembly technique, which allows the production of structures with different shapes. This has been demonstrated for the composition of microstructures in the form of empty tubes, hemispheres, spheres etc. Due to a specific design, these structures can be applied as optical cavities, waveguides and microlasers, as was shown in a number of papers [3], [4]. Though microstructures made of organic materials are less stable than those based on inorganic ones, their main feature is a relatively simple way for the manipulation of their linear and nonlinear optical properties by a proper choice of their the chemical compound and composition within a structure. Still a number of nonlinear optical effects were observed for organic microrods, such as two-photon absorption and luminescence, and second harmonic generation [7-9]. The effects were quite pronounced due to a strong localization of the fundamental radiation field inside the microstructures, which increased the light-matter interaction.

In this paper, the second-order nonlinear optical response of an array of nonlinear organic microstructures is studied. We show that an enhancement of the second harmonic generation and two-photon luminescence is attained as compared to homogeneous thin films of the same chemical composition.

2. Samples and experimental setup

2.1. *Samples*

Samples under study are irregular arrays of frustum-shaped organic microstructures fabricated using the self-assembly technique. The details of the samples' fabrication are discussed elsewhere [3]. In brief, the acetonitrile solution of the nonlinear non-centrosymmetric molecules of a red-orange dye, namely, (E)-2-(2-(4-((4-bromophenyl)(phenyl)amino)styryl)-6-tertbutyl-4H-pyran-4-ylidene)malononitrile, is deposited on a glass substrate. The evaporation of the solvent results in the formation of a set of frustum-shaped particles oriented in such a way as is shown in Fig. 1. It is worth noting that such a uniform orientation of

IOP Conf. Series: Journal of Physics: Conf. Series **929** (2017) 012074 doi:10.1088/1742-6596/929/1/012074

frustum-shaped microstructures is rather important for any type of potential applications, and is definitely an advantage of the composition method. The typical upper diameter of a frustum particle is about 2 μm and the opening angle is about 45°. The typical distance between the microstructures is several microns and is larger than their average diameter.

It was shown previously [3] that such a specimen reveals the absorption band at the wavelengths smaller than 500 nm and a luminescence in the spectral range of approximately 550-700 nm.

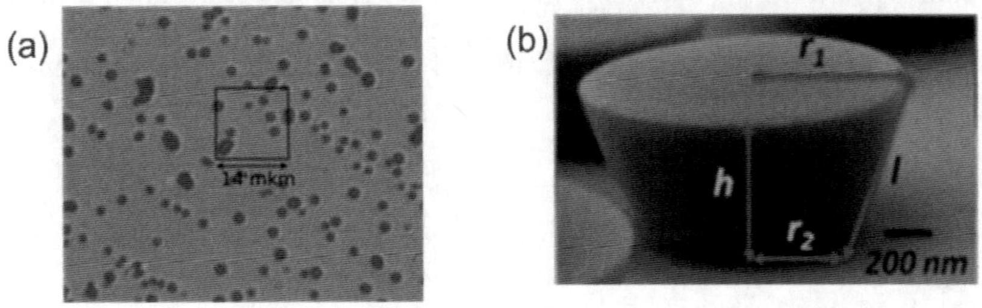

Figure 1. (a) Optical microscopy image of a planar array of frustum-shaped microstructures; (b) Scanning electron image of a single frustum microparticle, the top and bottom radiuses $r_{1,2}$ and the height of the particle h are shown.

2.2 Experimental setup

Prior to the nonlinear-optical studies, linear optical absorption and luminescence spectra were measured when using the diode lasers operating at the wavelengths of 405 and 532 nm, the power of the incident radiation was about 1 mW. Both transmission and reflection schemes were used. For the nolinear-optical measurements, an experimental set-up was used based on a Titanium-Saphire laser tunable in the wavelength range from 740 nm up to 890 nm, with the pulse duration being 100 fs and the pulse repetition rate of 80 MHz. Its average output power was varied from 1 mW up to 200 mW. The scheme of the experimental setup is shown in Figure 2. In order to eliminate the influence of the instability of the laser, the intensity of the nonlinear optical response of a sample was normalized to that attained in the reference channel where a crystalline quartz plate used as a source of the nonlinear optical signal. The spectrum of the nonlinear response, that is the two-photon luminescence (TPL) and second harmonic generation (SHG), was analyzed by a monochromator and detected by the photomultiplier Hamamatsu R4220. The laser beam was focused on the sample into the spot with the diameter of 50 μm. The maximal peak intensity of the fundamental beam was 10 MW/cm^2.

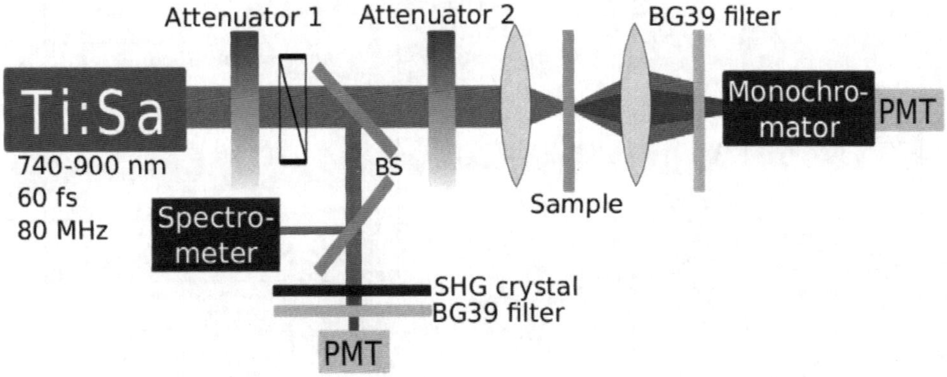

Figure 2. Schematic view of the experimental setup used for the studies of the second-order nonlinear optical effects in organic microstructures.

3. Experimental results and discussion

3.1 Linear-optical measurements

Figure 3 shows the photoluminescence (PL) intensity spectrum emitted by the array of the microstructures under the excitation by the CW laser source of the wavelength of 405 nm and 532 nm, the average power being of 1 mW. This relatively low power level was chosen in order to prevent the destruction of the material of the microparticles. Due to the chosen laser beam spot diameter on the sample several dozens of the microstructures were illuminated simultaneously, all of them having ssmall deviations of their size and shape from the averaged values for the ensemble. Thus the PL did not reveal a set of high-quality peaks associated with the excitation of whispering gallery modes that were observed for individual microparticles [3, 8, 9].

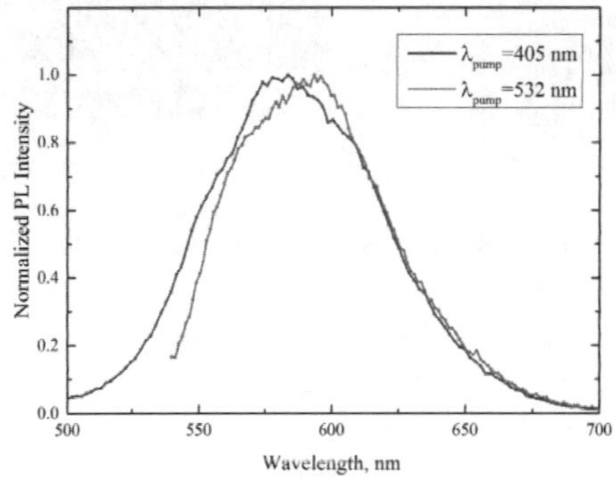

Figure 3. Spectra of the linear (single-photon) luminescence of an array of microstructures obtained for the excitation wavelengths of 405 nm and 532 nm.

Nonlinear optical studies

Typical spectra of the nonlinear optical response measured for various wavelengths of the Ti-sapphire fundamental radiation are shown in Figure 4. The spectra reveal two main features: a wide intensity peak centered at approximately 560 nm, that corresponds to the two-photon excited luminescence, and a smaller in amplitude peak associated with the second harmonic generation. Interestingly, the relative amplitude of this peak depends on the fundamental wavelength.

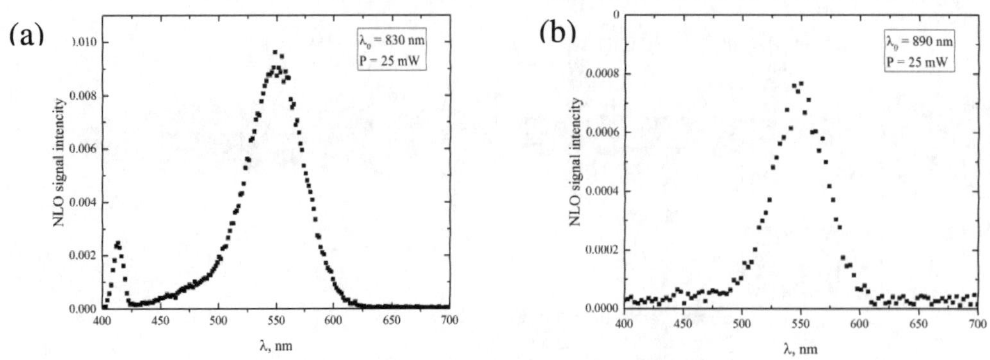

Figure 4. Spectra of the intensity of the nonlinear optical response of arrays of frustum-shaped microstructures measured for the fundamental wavelengths of 830 nm (a) and 890 nm (b), the power of the fundamental beam being $25\ mW$.

In order to study this in more detail, a 3D excitation-emission dependence was measured, which is shown in Figure 5. It can be seen that the highest values of the intensity of the two-photon luminescence is attained closer to the spectral boundaries of the fundamental wavelength tuning. The SHG response, while being much less intensive, also appears and follows the wavelength of the fundamental radiation. It is more pronounced for the longer wavelengths of the fundamental radiation, which is probably connected with the efficiency of the two-photon absorption along with the increase of the SHG escape length. The latter is determined by the absorption spectrum at the SHG wavelengths.

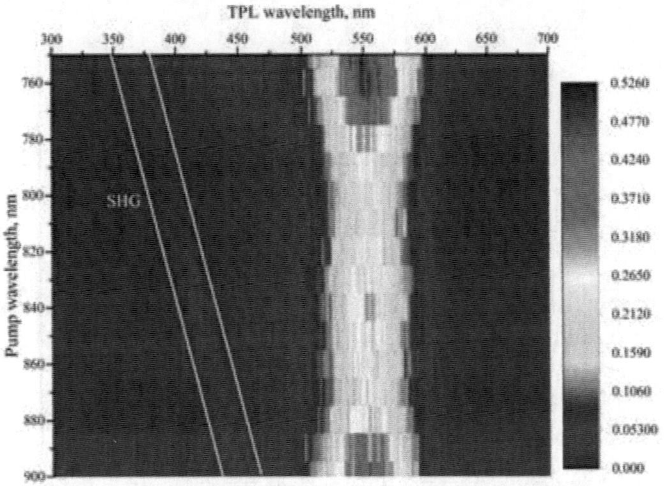

Figure 5. Emission – excitation scattering matrix of the intensity of the noninear optical response of an array of frustum microstructures. The dependence was measured for a constant value of the incident power of the wavelength tuned in the range 750-900 nm.

We performed as well the comparison of the efficiency of the nonlinear optical (NLO) response for a homogeneous orange-red dye of the same chemical composition as was used for composition of the frustum-shaped microstructures. It was found that in the case of microstructures the scattered nonlinear optical signal is about 10 times higher as compared to that from an array of frustum-shaped microstructures. This should be attributed to the local field effects expected for the case of organic microstructures with high value of the refractive index, which brings about strong field localization effects.

An important feature of the interaction of a laser beam with a random array of micron-size structures is the scattering diagram of the nonlinear optical response. In order to study these features more precisely, we measured the nonlinear optical signal (NLO) scattering indicatrix, the corresponding dependencies are shown in Figure 6. Here the averaged NLO intensity is plotted, which contains both the SHG and the two-photon excited luminescence intensity. The incident beam was directed along the normal to the substrate to the sample and the transmitted intensity was registered. The NLO signal is scattered in a wide angular range, with the maxima at approximately ±60°. This is probably connected with the geometry of the microstructures that are oriented in a uniform way. The exact mechanism of this NLO scattering is to be studied further.

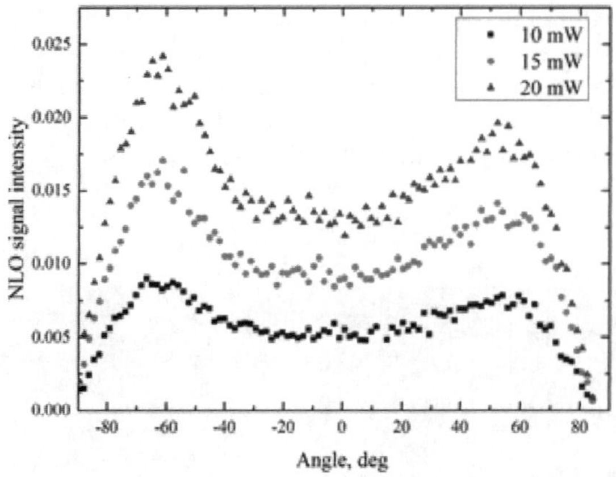

Figure 6. Nonlinear-optical scattering indicatrices measured for an array of frustum microstructures for different power of the incident radiation; zero scattering angle corresponds to the direction of the transmitted beam.

Summing up, we observed that under the excitation by a near-IR femtosecond pulsed radiation the array of micron-size organic frustum-shaped particles exhibits a strong nonlinear optical response, which consists of the two-photon luminescence and optical second harmonic generation scattered over a broad angular range. The emission-excitation spectra shows that the most intense NLO output corresponds to the input wavelengths of 750-770 nm, which is mostly attributed to the highest two-photon absorption in this spectral range. The SHG maximum reveals the opposite dependence, the highest SHG amplitude relatively to the two-photon luminescence was observed for longer wavelengths that shows that in that case larger escape length at the SHG wavelength is the dominant parameter.

Acknowledgenments.

The authors are thankful to our colleagues D. Venkatakrishnarao, YSLV Narayana, R. Chandrasekar from School of Chemistry University of Hyderabad for the composition of the samples. The work was supported by RSCF, grant № 16-42-02024.

References

[1] A.E. Minovich et al., Laser & Photonics Reviews 1–19 (2015)
[2] T. Murzina et al, Phys. Rev B **89**, 121113 (2014).
[3] R. Chandrasekar et al., Adv. Opt. Mater. (2015), DOI: 10.1002/adom.201500362.
[4] Y. S. Zhao et al, J. Am. Chem. Soc., **137**, 62–65 (2014)..
[5] Pal A, Chen S Y, Sen R, Sun T and Grattan K T V. Laser Phys. Lett. **10** (8) 085101 (2013).
[6] Zhao Y S, Xu J, Peng A, Fu H, Ma Y, Jiang L and Yao J Angew. Chem. **120** (38) 7411 (2008).
[7] S. Rosenne, Eran Grinvald, Elijah Shirman et al., Nano Lett., **15**, 7232–7237 (2015).
[8] Y. Duan, C. Ju, G. Yang, et al., Adv. Func. Mat., (2016), DOI: 10.1002/adfm.201602765.
[9] C. Zhang, C.-L. Zou, Y. Yan et al., J. Amer. Chem. Soc., **133**, 7276 (2011).

IOP Conf. Series: Journal of Physics: Conf. Series **929** (2017) 012075 doi:10.1088/1742-6596/929/1/012075

Compensation of the nonuniformity of sensitivity for the multielement infrared photodetector

D K Kostrin, V A Simon, A A Uhov, V A Gerasimov and L M Selivanov
Department of electronic instruments and devices, Saint Petersburg Electrotechnical University "LETI", 197376, Saint Petersburg, Russia

E-mail: dkkostrin@mail.ru

Abstract. Application of spectral devices for the detection of small concentrations of substances presents high demands on the signal dynamic range. Commonly used multielement photodetectors have a substantial nonuniformity of both sensitivity and dark signal that significantly narrows the dynamic range. For correction of such situation it was designed an amplifier with individual correction of gain and offset of a dark signal for each pixel that considerably enhances dynamic range and presents an ability to effectively detect weak signals.

1. Introduction

In optical spectrometers for the registration of radiation in the infrared (IR) region of a spectral range most commonly used are photodiode linear arrays based on InGaAs [1, 2]. Such photodiodes detect radiation in a spectral range 0.9–2.55 μm. These photosensitive arrays use CMOS (complementary metal–oxide–semiconductor) signal processing system with individual amplifiers for each photodiode and a multiplexer for the selection of a channel.

Application of spectral devices for the detection of small concentrations (several ppm) of substances presents high demands on the dynamic range of a signal. It is often necessary to record minor deviations of a signal for the lines of emission or absorption of analyte on the background level of a rather high intensity [3].

2. Development of the amplifier design for compensation of the nonuniformity of sensitivity

Multielement photodetectors of IR region, as a rule, are built using a double-crystal scheme, consisting of a linear array of InGaAs photodiodes and CMOS amplifiers with a control unit (figure 1).

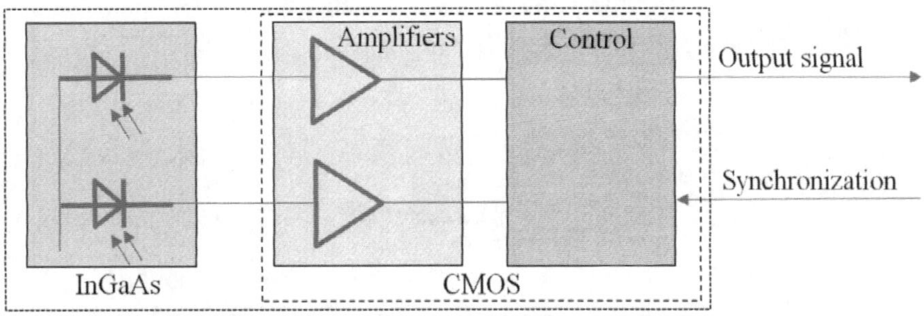

Figure 1. Internal structure of a multielement IR photodetector.

In this design multielement photodetectors have the following disadvantages:

- nonuniformity of the sensitivity of photodetectors can reach 10–20 %;
- photodetectors have a significant nonuniformity of a dark signal, in addition depending on the temperature of a crystal.

Signal obtained using a multielement IR photodetector G9206-256W manufactured by Hamamatsu exposed to uniform illumination is shown in figure 2. It is obvious that with such irregularity of both dark signal and sensitivity the dynamic range of the photodetector is significantly narrowed.

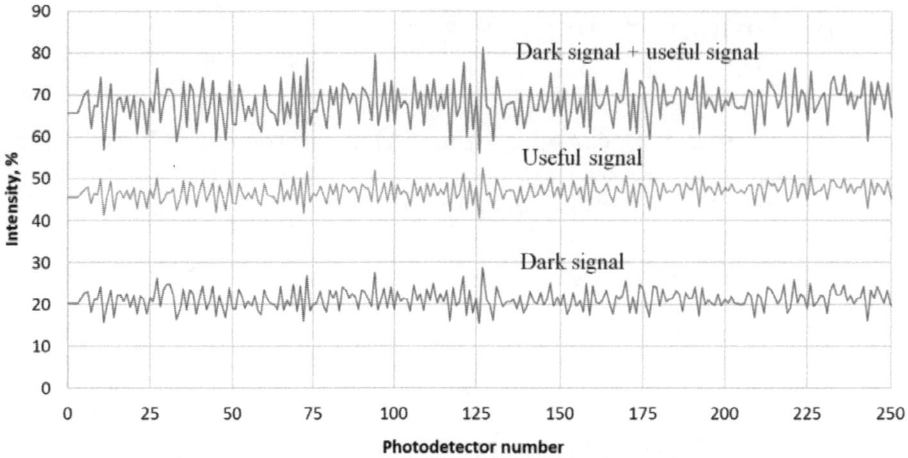

Figure 2. Output signal of a multielement IR photodetector.

Preamplification of the output signal having such a significant nonuniformity leads to the fact that the constant component of a dark signal also increases; consequently the remaining amplitude of a useful signal is noticeably reduced. For example if a photodetector has a dark signal of 20 % compared to the minimum value, the gain ratio of 5 leads to a full saturation of the output signal, making registration of a useful signal using this photodetector unrealizable [4].

For correction of such situation it is possible to presubtract the dark signal level from the output signal before amplification using a differential amplifier. In the case of a multielement IR photodetector such compensation must be carried out individually for each photodiode. Nonuniformity of the sensitivity also should be compensated individually. Thus the structure of the output signal amplification scheme should have two cascades – compensation of the nonuniformity of a dark signal and an adjustable gain of the useful signal (figure 3).

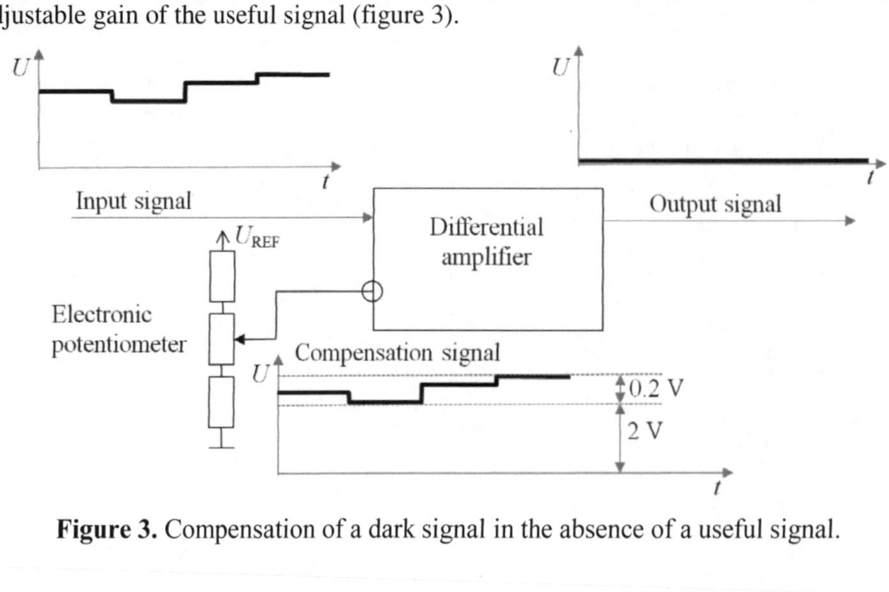

Figure 3. Compensation of a dark signal in the absence of a useful signal.

Amplitude of the useful component of an output signal of a multielement IR photodetector usually does not exceed 2 V, but the signal itself is shifted by a DC (direct current) component on 1.5–2 V. Taking into account nonuniformity of a dark signal of about 10–20 %, the amplitude of the compensation signal reaches 200–400 mV on the background of a constant shift of up to 2 V.

In this case application of common digital-to-analog converters (DAC) becomes inefficient as less than 10 % of their operating range is used. The most convenient in this case is implementation of a so-called "electronic potentiometer". This potentiometer is included into the circuit of a resistive divider that allows to accurately set the upper and lower limits of adjustment, thus ensuring almost 100 % utilization of the working range of adjustment.

Compensation of the nonuniformity of sensitivity requires a software adjustment of gain. Application of standard components, for example programmable gain amplifiers (PGA), does not solve this problem. In these amplifiers the adjustment is made by doubling the gain at each step, meantime in our case an exact change of gain in the range of 20 % is required. The best possible solution will be an amplifier with a digital potentiometer in the feedback circuit (figure 4).

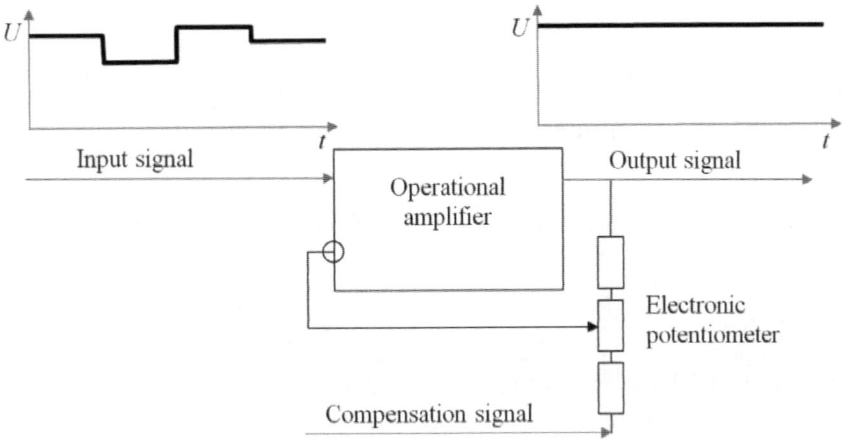

Figure 4. Gain adjustment for the uniform illumination.

As in the case of compensation of a dark signal, an electronic potentiometer is included into the circuit with additional resistors that allow to accurately set the boundaries of the gain control. Compensation signal is also connected to the divider for the subtraction of a DC component.

Signal registration algorithm for one pixel includes the following steps:

- adjustment of the offset for the current pixel in two channels;
- adjustment of the gain for the current pixel in two channels;
- waiting for the sync pulse for the current pixel;
- acquiring data from ADC (analog-to-digital converter) via two channels;
- saving the data in the internal RAM (random-access memory) of the microcontroller.

For the correct operation for this circuit a precalibration is carried out. It consists of several stages:

- darkening of a photodetector and acquisition of an output signal;
- calculation of the correction values of the bias level for each pixel and filling of the appropriate data array in the microcontroller memory;
- application of the offset correction for obtaining of the near zero values of a dark signal;
- registration of a signal under the uniform illumination of a photodetector;
- calculation of the correction values of gain for each pixel and filling of the appropriate data array in the microcontroller memory;
- application of the gain correction for obtaining of similar values of the output signal under the uniform illumination of a photodetector.

Results of the signal detection using the proposed registration system with the correction of both offset and gain are shown in figure 5.

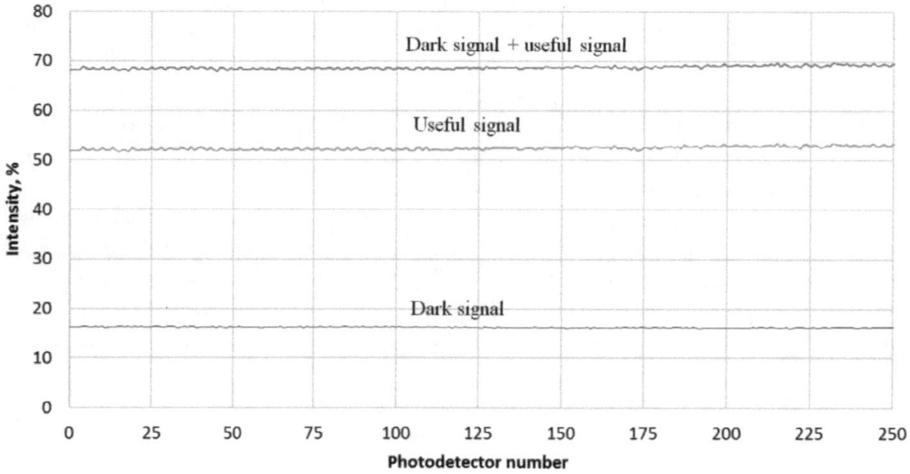

Figure 5. Signal of a multielement IR photodetector with correction of both offset and gain.

3. Conclusions

During the research was developed an amplifier design for a multielement IR photodetector with an individual correction of gain and offset of a dark signal for each pixel. Amplifier, having a minimum cost, significantly enhances the dynamic range and presents an ability to effectively detect weak signals, which is especially important in areas such as Raman spectroscopy. As a practical result of the work it can be mentioned that the use of this sensor and registration system allowed by increasing the spectral resolution in the IR range to improve the reliability of recognition of wood species using a Raman spectroscopy device [5].

References

[1] Balzhiev P E, Bock W J, Eftimov T A, Mikulic P and Arnaudov R 2011 *Proceedings of SPIE* **7753** 77539L

[2] Guoyu L , Tuan Q, Hao Z, Hongwei G, Jian Z, Bo L, Shuzhong Y, Guiyun K and Xiaoyi D 2007 *Applied Optics* **3** 283–6

[3] Vasilevskii A M, Konoplev G A, Svetlov D A, Gerasimov V A, Kostrin D K and Ukhov A A 2014 *Biomedical Engineering* **1** 13–6

[4] Kostrin D K, Lisenkov A A, Uhov A A and Ramazanov A N 2016 *Journal of Physics: Conference Series* **729** 012030

[5] Gerasimov V A, Gurovich A M, Kostrin D K, Selivanov L M, Simon V A, Stuchenkov A B, Paltcev A V and Uhov A A 2016 *Journal of Physics: Conference Series* **741** 012131

Study of electroluminescence of InAs(Sb,P) LED heterostructures

A A Semakova[1,2], N L Bazhenov[2] and K D Mynbaev[1,2]

[1] ITMO University, Saint-Petersburg 197101, Russia
[2] Ioffe Institute, Saint-Petersburg 194021, Russia

t.s.mysterious@mail.ru

Abstract. Electroluminescence of InAs(Sb,P) heterostructures grown on InAs substrates was studied in the temperature range T=4.2–300 K. At low temperatures (T=4.2–50 K), stimulated emission was observed. This effect was due to optical resonator, which was formed between the lower face of the LED chip with the solid metal contact and the upper face with semiconductor/air interface. The emission became spontaneous at higher temperatures due to the effect of CHHS Auger recombination process, when the energy of a recombining electron-hole pair was transferred to another hole with the latter transitioning to the spin-orbit-splitted band. The results obtained show that structures based on InAs(Sb,P) are a promising material for fabrication of vertical-emitting mid-infrared lasers.

1. Introduction

It is known that characteristic optical absorption bands of many important chemicals (CH_4, CO_2, NO_2, H_2S, CO, etc.) are located in the mid-infrared wavelength range (MWIR, wavelengths λ 2 to 6 μm) [1]. Sensors of these chemicals are of great demand in industry, in environmental control systems and in medicine. The most promising are MWIR sensors based on light-emitting diodes (LEDs) and semiconductor photodetectors, and it is worthwhile to fabricate light emitters and photodetectors on the basis of the same materials, as it greatly simplifies manufacturing processes.

To enhance the efficiency of optoelectronic devices, it is important to understand and control processes that govern absorption and/or emission of light in real structures. For that task, it is worth studying their operation not only at working temperatures, but also at lower temperatures. In this case, effects can be observed that may enable one to identify the mechanisms of physical processes in the devices in more detail. Within the frames of such an approach, we have studied electroluminescence (EL) of InAs(Sb,P) LED heterostructures in a wide temperature range, from 4.2 K up to 300 K.

2. Experimental details

The heterostructures were grown with the use of metal-organic chemical vapour deposition at Microsensor Technology, LLC, using synthesis method described elsewhere [2]. For all the heterostructures, a strongly sulphur-doped InAs substrate (n-type conductivity, electron concentration at 77 K $n_{77} \approx (1\text{-}2)\times10^{18}$ cm^{-3}) was used. An active layer of the heterostructures was made of InAs or InAsSb. This layer was not intentionally doped and had electron concentration $n_{77} \approx 10^{16}$ cm^{-3}, presumably due to background donors. A typical thickness of the active layer was 2 to 3 μm. On top of the active layer, a p-type InAsSbP barrier layer was grown. This layer was doped with zinc and had hole concentration at 77 K $p_{77} \approx 2\times10^{18}$ cm^{-3}. Parameters of the heterostructures are given in Table 1.

Table 1. Parameters of the studied heterostructures.

Structure type	Active layer	Barrier layer
A	InAs	$InAs_{0.15}Sb_{0.31}P_{0.54}$
B	$InAs_{0.94}Sb_{0.06}$	$InAs_{0.40}Sb_{0.20}P_{0.40}$
C	$InAs_{0.93}Sb_{0.07}$	$InAs_{0.70}Sb_{0.10}P_{0.20}$
D	$InAs_{0.91}Sb_{0.09}$	$InAs_{0.48}Sb_{0.18}P_{0.34}$

LED chips with 380×380 μm size were fabricated with the use of standard photolithography and wet chemical etching. Electrical contacts were based on a multi-layer Cr–Au–Ni–Au composition. A 'flip-chip' technology was used, so a non-transparent solid contact was placed on the top epitaxial layer, while a ring-type contact with 35 μm thickness and 200 μm internal diameter was placed on the InAs substrate. The emission was collected from the side of the substrate, as the latter was transparent for the light emitted by the active layer due to the strong doping. For the measurements, the chips were placed on TO-18 holders. EL spectra were recorded under pulse excitation (frequency 1 kHz, pulse duration 1 μs) with the use of computer-controlled installation employing a grating monochromator and a lock-in amplifier. InSb photodiode was used as a detector.

3. Experimental results

Figure 1(a) shows normalized EL spectra of two heterostructures at *T*=300 K, which is a typical working temperature for this kind of LEDs. Figure 1(b) shows normalized EL spectra for three types of studied heterostructures at *T*=4.2 K at various driving currents *I*. The heterostructure of type *A* had the active region made of pure InAs. The heterostructure of type *C* had the active region made of InAsSb with InSb molar fraction 0.07 (see Table 1). In Fig. 1 (a) one can observe a red-shift of the maximum of the spectrum (curve *2* vs. curve *1*) due to the introduction of InSb, which has the energy bandgap value E_g smaller than that of InAs, and some broadening of the EL line typical of alloys as compared to binary compounds. The full-width at half-maximum (FWHM) of the spectrum at 300 K is 50 meV for the heterostructure of type *C* vs. 35 meV for the heterostructure of type *A*.

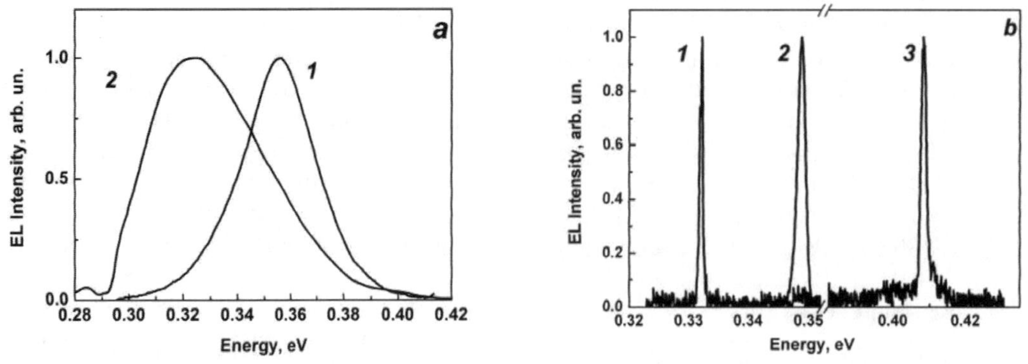

Figure 1. Normalized EL spectra of the heterostructures of type *A* (curve *1*) and type *C* (curve *2*) at *T*=300 K at driving current *I*=1 A (a), and spectra of the heterostructures of type *C* (*I*=0.6 A, curve *1*), type *B* (*I*=0.4 A, curve *2*) and type *A* (*I*=1.2 A, curve *3*) at 4.2 K (b).

As can be seen in Fig. 1(b), the spectra recorded at 4.2 K differ in shape and FWHM values from those recorded at 300 K. For example, for the structure of type A, at the high-energy side of the barely visible wide emission band with FWHM 20 meV, which would correspond to the narrowing of the initial bandwidth of 35 meV with the temperature decreasing, there appeared a second narrow line with FWHM of 2 meV. A study of the dependences of these bands on *I* showed that with the latter increasing from 0.1 up to 0.6 A, the intensity of the wide band increased insignificantly, while that of the narrow band increased substantially. For the structures of types *B, C,* and *D* at the temperatures close to 4.2 K, we also observed quite narrow emission bands. With increasing of the current, the integral EL intensity was increasing linearly, while the peak maximum was blue-shifting, presumably reflecting the filling of the conduction band and lifting of the Fermi quasi-level for electrons. EL spectra of the structures of type *C* at *T*=77 K previously demonstrated a clearly defined modal structure [3]. A similar structure of EL band was observed for some of the structures of type *D* at 4.2 K (not shown).

With temperature increasing, for all the structures we observed that narrow EL peaks transformed to broad ones: for example, for the structure of type *C*, the FWHM of the main emission peak increased from 7 meV to 400 meV. The exact transition point depended on the type of structure. Figure 2 shows the energy of the EL peaks E_{EL} vs. temperature for structures of types *A* and *C*. All the spectra were recorded at the same driving current *I*=0.6 A. It is seen that for the structure of type A at $T \geq 70$ K the energy of the emission peak corresponds to the value of E_g of the active layer calculated in accordance with the data presented by Vurgaftman *et al.* [4]. At lower temperatures, when the peak of EL was narrow, the energy of the peak was lower than E_g. Similar dependences of E_{EL} on the temperature were observed for structures of type *C*. In this case, however, at higher temperatures (typically, at *T*>120 K) E_{EL} values were larger than those of calculated E_g by ~20 meV. At *T*<100 K, depending on the type of the heterostructure, the values of E_{EL} were close to calculated E_g or smaller.

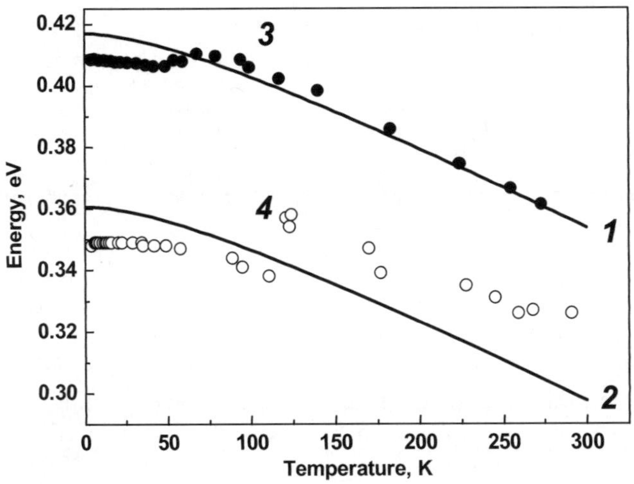

Figure 2. Calculated temperature dependences of E_g for InAs (straight curve *1*) and $InAs_{0.93}Sb_{0.07}$ (curve 2) and experimentally determined position of EL peaks for heterostructures with corresponding active layer composition (points): InAs (type *A*) (*3*) and $InAs_{0.93}Sb_{0.07}$ (type *C*) (*4*).

4. Discussion

The analysis of the obtained results shows that at low temperatures (4.2–50 K), in the studied structures conditions for stimulated emission held true. It is worth noting that in LED structures with the active layer made of InAs, a similar effect was first reported on by Matveev *et al.* [5] (however, there, in contrast to the present work, double heterostructures were studied). Stimulated emission from structures with InAsSb-based active layer has been observed so far only in specially designed laser structures: in particular, one can mention strip-waveguide lasers based on double heterostructures with high symmetrical barriers [6–8]. In our case, resonators were not fabricated specially; the cleft edges of the LED chips were not mirror-like either. The inter-mode spacing in the EL spectra of 6 nm [3] allowed for assessing the length of the resonator as 280 μm using a simple method described by Grebenshchikova *et al.* [8]. This value appeared to be much closer to the thickness of the LED chip (~300 μm) rather than to the distance between the cleft edges of the chip (380 μm). Also, it appeared that EL signal in the stimulated emission mode coming from the surface of the LED chips was much stronger than that collected from the chip edges. Thus, we concluded that in our structures the optical resonator was formed normal to the growth plane, most probably, between the surface of the chip with a gold Ohmic contact and the substrate, which was polished chemically in the process of making LEDs.

The experimental data showed that transition from stimulated to spontaneous emission occurred at different temperatures (50 to 100 K) for different types of heterostructures (see Figure 2). Table 2 shows that actually this transition was not related to a particular temperature, but rather, to the energy of the bandgap: the transition occurred when the energy of the 'effective' bandgap coincided with the energy of the spin-orbit splitting Δ_{SO} in the material of the active layer (Δ_{SO} was calculated on the basis of the data by Vurgaftman *et al.* [4]). It was natural to suggest that with temperature increasing and E_g decreasing, we observed a resonant 'switch-on' of the CHHS Auger process, when the energy of recombining electron-hole pair was transferred to a hole transitioning to the spin-orbit-splitted band. With further temperature increasing and E_g continuing to decrease, the resonance condition $E_g=\Delta_{SO}$, that was responsible for quenching of the effect of stimulated emission, disappeared, but the emission remained spontaneous. This was obviously due to the influence of Auger processes of other types, which suppressed amplification: likely, the dominating process was that with the participation of two electrons and a heavy hole with excitation of electron in more energetic state (CHCC), or that involving two heavy holes and an electron, with the conversion of a heavy hole into a light hole (CHHL).

Table 2. Calculated values of the energy of spin-orbit splitting and experimentally determined energies of the quenching of the stimulated emission.

Structure type	Δ_{SO}, eV	'Quenching' energy, eV
A (InAs)	0.37-0.41	0.41
B, *C*, *D* (InAsSb)	0.33-0.37	0.34

5. Conclusion

When studying electroluminescence of InAs(Sb,P) heterostructures, at low temperatures (4.2–100 K), stimulated emission from the structures was observed. This effect, however, was quenched with the temperature increasing due to the resonant 'switch-on' of the CHHS Auger process. Despite this, it appears as InAs(Sb,P) heterostructures are promising for fabrication of the vertical-emitting MWIR lasers, since even under the influence of Auger recombination it is possible to obtain stimulated emission with minimum requirements for optical resonators. To increase the operating temperature of the devices, a different chemical composition of active layer may be used, which would allow to shift the $E_g=\Delta_{SO}$ resonance condition to higher temperatures.

References

[1] Krier A, Ed. 2006 *Mid-Infrared Semiconductor Optoelectronics*. Springer Series in Optical Sciences, Vol. 118 (Springer: Berlin)
[2] Sopanen M, Koljonen T, Lipsanen H and Tuomi T 1994 *J. Cryst. Growth* **145** 492
[3] Mynbaev K D, Bazhenov N L, Semakova A A, Mikhailova M P, Stoyanov N D, Kizhaev S S, Molchanov S S, Astakhova A P, Chernyaev A V, Lipsanen H and Bougrov V E 2017 *Semiconductors* **51** 239
[4] Vurgaftman I, Meyer J R and Ram-Mohan L R 2001 *J. Appl. Phys.* **89** 5815
[5] Matveev B, Zotova N, Il'inskaya N, Karandashev S and Stus' N 2005 *Phys. Stat. Sol. C* **2** 927
[6] Lane B and Razeghi M 2000 *J. Cryst. Growth* **221** 679
[7] Astakhova A P, Bez'yazychnaya T V, Burov L I, Gorbatsevich A S, Ryabtsev A G, Ryabtsev G I, Shchemelev M A, Yakovlev Yu P 2008 *Semiconductors* **42** 228
[8] Grebenshchikova E A, Zotova N V, Kizhaev S S, Molchanov S S and Yakovlev Yu P 2001 *Tech. Phys.* **46** 1125

IOP Conf. Series: Journal of Physics: Conf. Series **929** (2017) 012077 doi:10.1088/1742-6596/929/1/012077

Edge-emitting lasers based on coupled large optical cavity with high beam stability

A Serin[1,2], N Gordeev[2,3], A Payusov[2,3], Y Shernyakov[2,3], N Kalyuzhnyy[2], S Mintairov[2] and M Maximov[2,3]

[1] Peter the Great St. Petersburg Polytechnic University, Polytechnicheskaya, 29, 195251, Saint Petersburg, Russia
[2] Ioffe Institute, Saint Petersburg, Russia
[3] St. Petersburg Academic University, Saint Petersburg, Russia

E-mail: spbgate21@gmail.com

Abstract. In this paper we present a study on temperature and current stability of far-field patterns of lasers based on the coupled large optical cavity (CLOC) concept. Previously it has been shown that the CLOC structures allows effective suppressing of high-order mode lasing in broadened waveguides. For the first time we report on transverse single-mode emission from the CLOC lasers with 4.8 μm thick waveguide. Using broadened waveguide allowed us to reduce the divergence of the far-field patterns down to 14° in continuous-wave (CW) regime. Far-field patterns proved to be insensitive to current and temperature changes.

1. Introduction

Nowadays high-power semiconductor lasers with high-quality beam are desired for a large number of applications, such as material processing, solid-state laser pumping, frequency conversion, medical surgery etc. [1] because of their small size, efficiency, and reliability. Most designs of such devices utilize broadened waveguides in order to reduce the output beam divergence and to lower the power density on the laser facets. The latter is critical to avoid catastrophic optical mirror degradation (COMD) [2]. One obstacle that makes it difficult to expand the waveguide is the appearance of high-order transverse modes, which negatively affects the device characteristics such as the beam quality and divergence. A number of approaches has been developed in order to eliminate high-order mode oscillations in thick waveguides of high-power semiconductor lasers [3-7]. However, along with high-order mode suppression, many of these approaches either too sensitive to changes of the refractive indices of the waveguide layers [3-6] or tend to have significant overlap of the fundamental mode with the highly doped cladding layers, which leads to additional optical loss [7]. Some of the approaches are free of these negative effects, but they could be used to eliminate only one high-order vertical mode [8].

Recently novel waveguide design that allows selective eliminating of high-order modes in broadened waveguides was suggested, and transverse single mode lasing from 2.5μm thick waveguide was demonstrated [9]. This new concept is based on coupled optical waveguides and thus named as coupled large optical cavity (CLOC). In this paper, we present a study of lasing parameters of the advanced CLOC lasers having two passive waveguides. Also, stability of far-field pattern of CLOC based lasers with temperature was investigated for the first time. It is shown that the CLOC laser beam is robust against overheat, typically found in high-power lasers.

IOP Conf. Series: Journal of Physics: Conf. Series **929** (2017) 012077 doi:10.1088/1742-6596/929/1/012077

2. Experiment

The laser wafer (figure 1) was grown by metal-organic vapor phase epitaxy on *n*-GaAs (100) substrate. An $Al_{0.1}Ga_{0.9}As$ active waveguide (thickness 4.8μm) incorporating two InGaAs quantum wells (wavelength 1.04μm) and two thin GaAs passive waveguides with the thicknesses of 310 nm and 375 nm were sandwiched between two p- and n-doped $Al_{0.2}Ga_{0.8}As$ claddings. The passive waveguides were separated from the active waveguide by 400 nm $Al_{0.2}Ga_{0.8}As$ layers. All the layers except the active waveguide were doped to the level of $2\times10^{18}cm^{-3}$. The active region was shifted toward the p-side to lower series resistance and optical loss on free holes accumulated in the waveguide. In order to minimize the resistance, 2μm of the active waveguide from the n-side was n-doped up to the concentration of $5\times10^{16}cm^{-3}$.

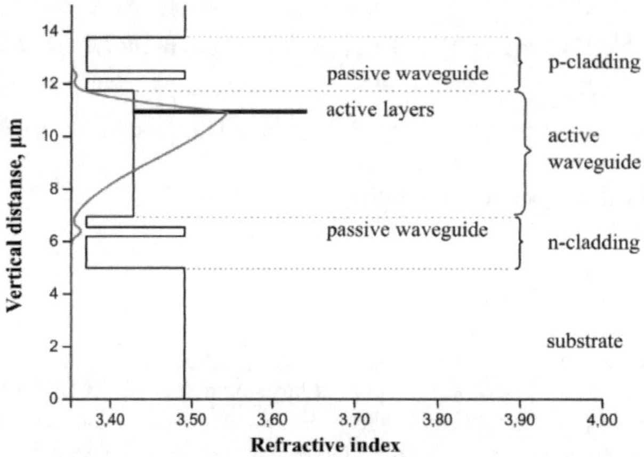

Figure 1. Refractive index profile of the CLOC laser (black line) and simulated fundamental mode intensity profile (red line).

Besides the fundamental mode in active waveguide, its first and second eigenmodes had rather large optical confinement factors (OCF) due to the waveguide thickness and shifted active region. Therefore, two passive waveguides where designed to select the first and second modes. Figure 1 shows refractive index profile and simulated intensity profile for the fundamental mode in the vertical direction. One can see that the fundamental mode penetrates into the additional passive waveguides, which may cause some additional optical loss.

The laser wafer was processed into 50-μm-wide shallow-mesa ridge lasers using etching through the p-contact and partly through the p-cladding layers. Devices with different cavity length where obtained by cleaving the processed wafer. Laser parameters were measured both in pulse (750 ns, 5 kHz) and CW regimes. Samples where mounted p-side down on copper heat sinks to reduce overheat. A Peltier cooler was used to control the temperature of the mounted samples during experiments.

3. Results and Discussion

All tested devices emitted on the fundamental vertical mode, which was confirmed by the far-field measurements. The dependence of the inverse differential efficiency on the cavity length allowed us to estimate the internal quantum efficiency 80% and the internal loss, which was 5.5 cm^{-1}. Threshold current density was 1 kA/cm^2 for the 2 mm long devices. High threshold current density resulted from the reduced fundamental mode OCF. We attribute rather high internal losses to the free carrier absorption in the partially doped waveguide and to the possible diffusion of zinc from the p-doped layers into the p-side of the active waveguide. Figure 2 shows measured L-I curve for the mounted 1.5 mm CLOC laser in the CW regime. Maximum output power of 1.8 W was limited by thermal roll-over.

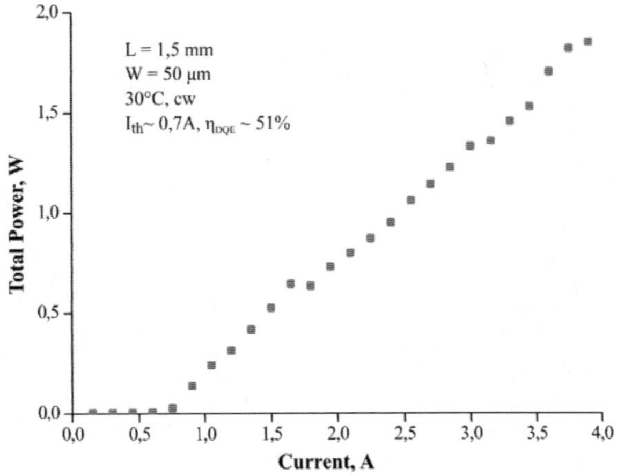

Figure 2. L-I characteristic of the CLOC laser measured in the CW regime.

Figure 3(a) shows far-field patterns of the 1.5 mm device obtained in the CW regime at different driving currents. One can see that single-lobe patterns remain stable throughout entire pumping current range. The obtained results are in good agreement with the simulation (inset in figure 3(a)). Full width at half maximum (FWHM) versus CW current is plotted in figure 3(b) for 1.5 mm and 2 mm devices. Divergence of 12-14° is very close to the diffraction limit for the 4.8μm emitting aperture. Thus, we attribute the small fringes at ±20 degrees to the fundamental mode penetration into the additional passive waveguides rather than to the presence of high-order modes. Further optimization of the structure may eliminate this effect.

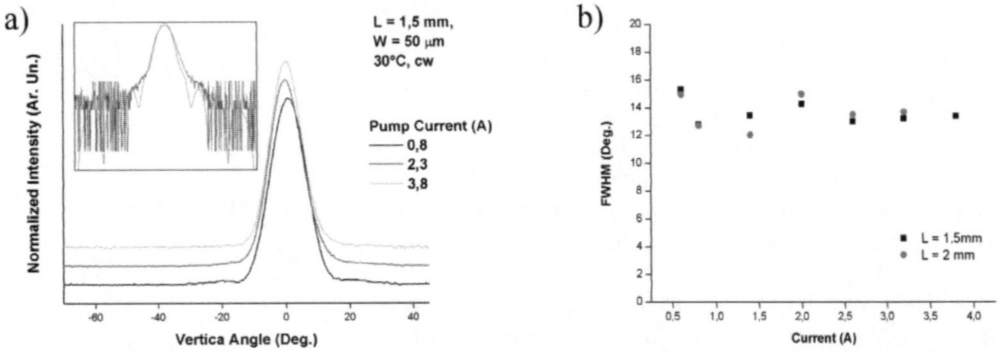

Figure 3. Far-field patterns (a) and FWHM (b) of CLOC lasers, obtained in CW regime at different currents. The comparison of calculated and measured far-field patterns is shown on the inset of (a) in semilogarithmic scale.

Far-field patterns measured in the pulsed regime where used to analyze temperature behavior of the lasing mode. Driving current was 10-15% over the threshold and had been adjusted according to the temperature dependence of threshold current temperature dependence for each device. Figure 4(a) shows measured far-field patterns at different temperatures. The FWHM dependence on the temperature is plotted in figure 4(b). Current and temperature changes do not cause any pronounced modifications of the beam divergence.

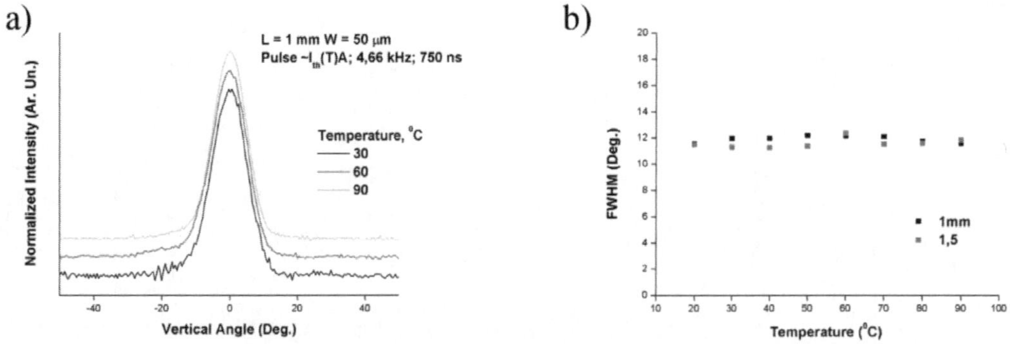

Figure 4. Far-field patterns (a) and FWHM (b) of the CLOC lasers obtained in pulsed regime at different temperatures.

4. Conclusions

We have demonstrated for the first time transverse single-mode operation of the CLOC lasers possessing 4.8 μm thick waveguide with two additional passive waveguides. The CLOC structure allowed obtaining current and temperature insensitive narrow far-field patterns with the diffraction limited divergence of 14 degrees and the CW output power of 1.8 W. We believe that the CLOC structures is a simple, universal, and efficient solution that could be used to advance characteristics of high power diode lasers.

Acknowledgment

The authors acknowledge the support from the Russian Foundation for Basic Research (project No 16-29-03279) and the Ministry of Higher Education and Science of the Russian Federation (project No 3.9787.2017/8.9).

References

[1] W. Schultz *et all* (2000) IEEE J. Sel. Top. Quant. Elect. 6(4) 696
[2] D. Botez (1999) Appl. Phys. Lett. 74 3102
[3] Md Miach *et* all (2015) IEEE J. Sel. Top. Quantum Electron. 21(6) 4900206
[4] G. Lin *et all* (1996) IEEE Photonics Technol. Lett. 8 1588
[5] A. Pietrzak *et all* (2011) IEEE J. Sel. Top. Quantum Electron. 17(6) 1715
[6] B.S. Ryvkin *et all* (2013) J. of Appl. Phys. 114 013104
[7] V.A. Shchukin *et all* (2011) IEEE Quantum Electron. 47(7) 1014
[8] N.A. Pikhtin *et all* (2004) Electronics Letters 40(22) 1413
[9] N.Yu. Gordeev *et all* (2015) Optics Letters 40(9) 2150

IOP Conf. Series: Journal of Physics: Conf. Series **929** (2017) 012078 doi:10.1088/1742-6596/929/1/012078

The luminescent structures based on colloidal quantum dots CdZnSeS/ZnS

I I Mikhailov, L I Kozlovich, P O Tadtaev, T D Lebedeva, A V Solomonov, S A Tarasov,

Saint Petersburg Electrotechnical University «LETI» (ETU)

iimihalov@gmail.com

Abstract. Phosphor mixture based on colloidal quantum dots (CQDs) has been developed. Devices based on nitride LEDs and phosphor mixtures of CQDs have been created. CRI value of the devices exceeded 95.

1. Introduction
In the last decade optoelectronic devices based on colloidal quantum dots are being actively developed [1]. CQDs are nanoscale particles [2] that are placed in the solution and that retain their state as a result of passivation properties of surfactants. Such nanoscale particles can be prepared by using relatively inexpensive colloidal synthesis techniques. These techniques do not require complex equipment [3-8]. In this case, properties of colloidal quantum dots depend on the initial materials and technological parameters of the synthesis. The most important property of CQDs as quantum-sized particles is the dependence of their energetic parameters from geometric dimensions. This allows to adjust characteristics of CQDs by changing technological parameters of the synthesis [9-11]. One of the promising semiconductor devices are QD-based WLEDs [12] and LEDs coated with phosphor mixtures of the CQDs with the colour rendering index (CRI) close to 100. During this study, we created three prototypes of such devices. We based them on semiconductor light-emitting diodes with wavelengths of 290 nm, 360 nm and 405 nm.

2. Description of investigated CQDs
We used three compositions of CQDs to create light-emitting structures. First composition was based on cadmium sulphide (CdS) CQDs with peak wavelengths of 380, 400, 420, 440, 460 and 480 nm. Second composition was based on CdZnSeS solid solution, covered with wide-band shell of ZnS, with peak wavelengths of 500 and 550 nm. Third composition was based on CdSe, also covered with wide-band shell of ZnS, with peak wavelengths of 600, 650 and 680 nm.

3. Calculation of phosphors. The interaction between CQDs.
Before creating luminescent structures, we studied interaction of colloidal quantum dots mixed in a solution and after deposition on the substrates of different types. Interaction between the CQDs in the mixture, in most cases, led to an increase in the intensity of the longwave part of the spectrum and to a decrease of the intensity in the shortwave part of spectrum. We studied these interactions in detail. Parameters of interactions were determined for the colloidal quantum dots based on cadmium sulphide, cadmium selenide and solid solutions thereof. Obtained interaction parameters were in a form of an array of calibration coefficients, taking into account the effect of the substrate and the method of coating.

IOP Conf. Series: Journal of Physics: Conf. Series **929** (2017) 012078 doi:10.1088/1742-6596/929/1/012078

Calibration coefficients were entered into further described algorithm and software for calculation of the spectral characteristics of luminescent structures. It is shown that the strongest interaction is manifested for the non-enveloped structures. In some cases, this interaction was so strong, that it almost entirely suppressed their luminescence as a result of agglomeration and radiation. Developed technique prevented agglomeration through the use of polymer matrix, layer deposition of various types of particles arrays and their subsequent annealing.

Quantum dots of different sizes can be used to construct the spectrum of arbitrary form. Spectrum of the solar radiation at the earth's surface is one of the most important spectra for the analysis of the colour characteristics. This spectrum can be simulated using a popular approach, which essentially is an approximation of the solar spectrum to spectrum of the reference blackbody at a given temperature. The simulation showed that the spectrum with a CRI of 99.5 can be obtained from mixtures based on ten samples of CQDs different sizes.

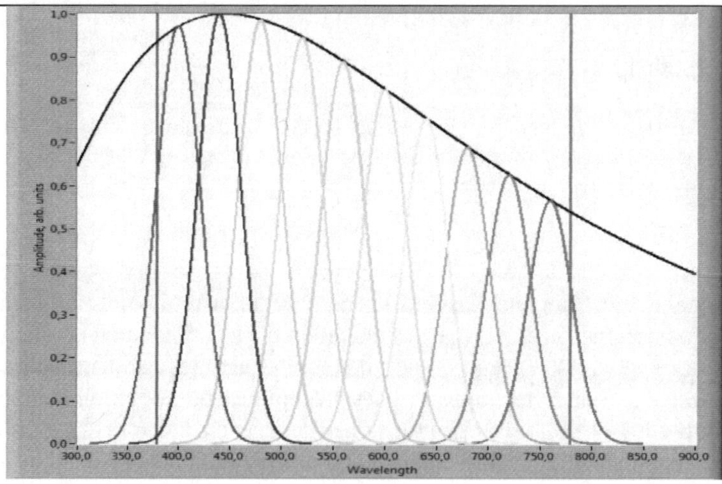

Figure 1. Simulation of the standard spectrum (blackbody at Tc = 6504K) using CQDs.

4. Technology of creation of phosphors

We used the following particle deposition methods: drip coating method, spin-coating, spray atomization (microdroplet method) or soaking. Next, a layer of deposited quantum dots was dried or annealed. Most frequently used combination was drip deposition and centrifugation, followed by annealing in vacuum at temperatures of 30 - 150°C for 3 - 30 minutes depending on the concentration and composition of CQDs. To obtain desired density and distribution of colloidal quantum dots on the substrate surface, these procedures were carried out repeatedly.

Selected CQDs samples had their required wavelength of the radiation in the visible range. Spectra outside the visible range were not considered. Two parameters controlled the intensity at the corresponding wavelength: concentration in the initial solution of CQDs and quantum yield of the corresponding sample. The concentration of CdS and CdSe CQDs in the initial solution was 5 mg/ml and 100 mg/ml, respectively. Therefore, CdSe CQDs samples were diluted to appropriate values. The quantum yield for the CdS and CdSe CQDs was 10% and 15-80%, respectively. The resulting mixture was deposited on a glass substrate, including quartz, and annealed for complete evaporation of the solvent. In some cases, the operation was repeated in order to increase the density of the particles to the desired values.

Thus, we were able to develop luminescent emitting devices based on CQDs of metal chalcogenides by choosing the appropriate structure, taking into account calibration coefficients and making appropriate modelling. In this research we used the reference light sources of devices with colour temperatures of 6504 K and 5503 K. The source of characteristics corresponded to solar radiation spectrum and took into account the absorption of the earth's atmosphere. Created devices were based on multi-component mixtures of cadmium sulphide and cadmium selenide CQDs, and solid solutions thereof, deposited on the substrate.

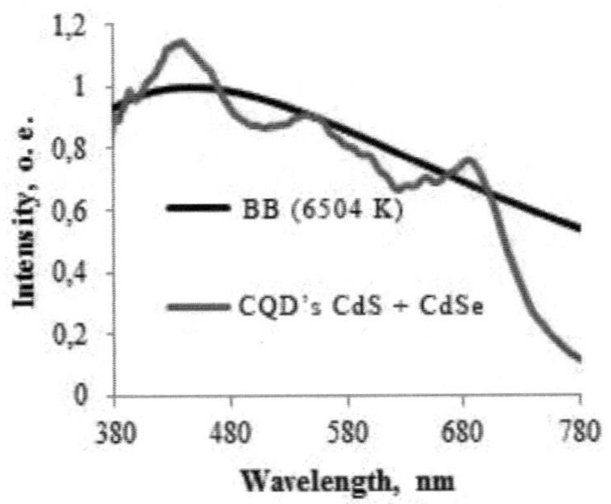

Figure 2. The luminescence spectrum of emitter based on the CQDs with a colour temperature of 6504 K

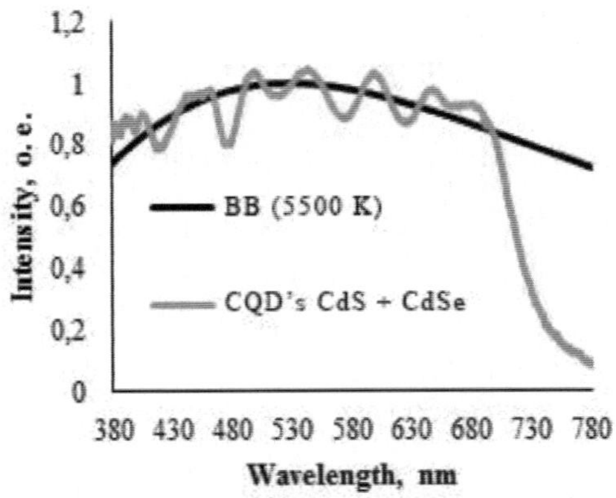

Figure 3. The luminescence spectrum of emitter based on the CQDs with a colour temperature of 5500 K

Luminescent emitting devices with high colour rendering index were created based on the established structures (Figure 4). Devices were a system of LEDs with luminophore placed in a single package. LEDs were used as emitter with different luminescence wavelengths in the ultraviolet and violet region (290, 360, 405 nm). It was best to use an ultraviolet emitter with wavelength of 290 nm because its radiation does not fall into the visible range and does not distort the shape of the spectrum.

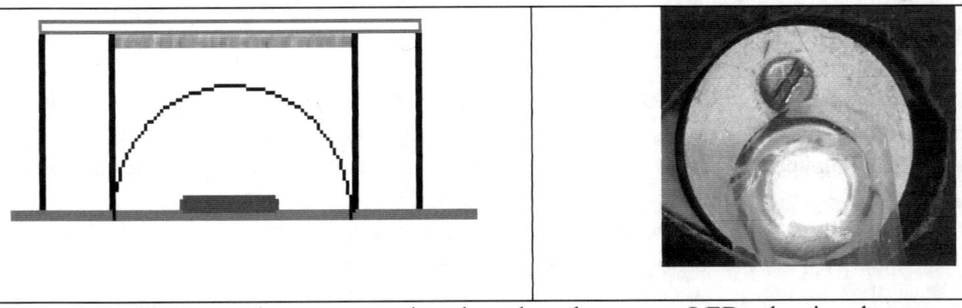

Figure 4. Fluorescent emitter based on the system LED + luminophore

We searched for optimal design for the structure, converted the largest part of the pump radiation into visible light, and selected the best thickness of the luminescent layer and concentration of the CQDs. The shape of the spectral characteristics of the emitting devices was fully compatible with the spectra of luminescent structures and depended insignificantly on the current in the LED. We used LEDs with radiation wavelengths of 360 and 405 nm to reduce the cost of devices. The main difference in the spectra for these emitters in the short-end of the range was in an additional component associated with partial penetration through the substrate of the LED radiation.

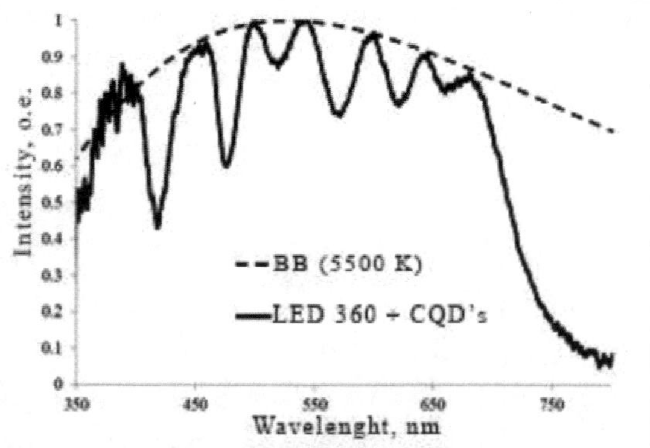

Figure 5. The emission spectrum of the luminescent device based on LED, with a peak wavelength of 360 nm

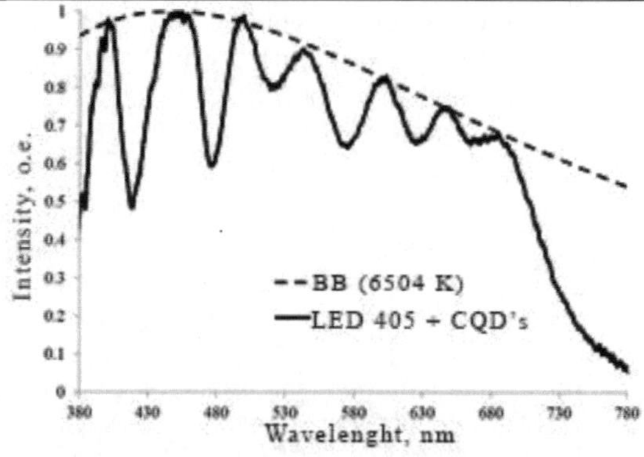

Figure 6. The emission spectrum of the luminescent device based on LED, with a peak wavelength of 405 nm

High temperature stability characteristics were achieved through the use of quantum dots in reemitting layers. The possibility of implementing almost any form of spectral characteristics of the luminescence devices is especially important. In many cases, the process is very complicated when using luminophores based on other materials.

As a result, we created the reference light sources devices with colour temperatures of 6504 K and 5503 K. The source of the characteristics was is correspondence with the solar radiation spectrum, and took into account absorption of the Earth's atmosphere. Method of creating such devices with almost any desired shape of the spectral characteristics in the visible spectrum range was developed.

5. Conclusion

Created luminescent structures showed efficiency of using CQDs to create light-emitting structures with spectral characteristics of given shape. We showed that the use of a mixture of three or four types of CQDs allowed to obtain a high colour rendering index.

We estimated the parameters that had influence on interaction CQDs when mixed and deposited on the substrate. We established that the blending CQDs with the shell and without the shell in liquid form leads to aggregation and suppression of luminescence. The solution of this problem was to deposit layers of mixtures of different types of CQDs. During creation of light-emitting structures we deposited first narrow-band and then wide-band samples of quantum dots.

Created light-emitting structure corresponded to the three standards: the spectrum of the sun on the earth's surface, blackbody at 5500 K and blackbody at 6504 K. These structures had a colour rendering index CRI = 95 ... 97.

The work was supported by the Ministry of Education of the Russian under project of the state task in the field of scientific activity, project number 16.1750.2017/4.6.

References

[1] I I Mikhailov, S A Tarasov, I A Lamkin, P O Tadtaev, L I Kozlovich, A V Solomonov, E M Stepanov. Optical parameters of ITO/TPD/Alq3/Al luminescent structures, containing arrays of CdSe/ZnS colloidal quantum dots// Journal of Physics: Conference Series 741 (2016) 012103

[2] Murray, C.B. Synthesis and characterization of nearly monodisperse CdE (E = sulfur, selenium, tellurium) semiconductor nanocrystallites / C. B. Murray, D. J. Norris, M. G. Bawendi // J. Am. Chem. Soc. – 1993 – Vol.115 – P.8706.

[3] Mikhailov II, Tarasov SA, Presnyakova AV, Romanovsky DS Investigation of the evolution of photoluminescence spectra as a method for analyzing the growth processes of colloidal quantum dots // Izvestiya of Saint-Petersburg State Electrotechnical University LETI. - 2013. T. 5. P. 22-28.

[4] I.I.Mikhailov, S.A.Tarasov, A.V.Solomonov, L.B.Matyushkin, D.S.Mazing. The investigation of

the luminescence properties of colloidal quantum dots based on cadmium chalcogenides // Functional materials, Vol.21, No.2, 2014, c. 142-145.

[5] S.A. Tarasov, O.A. Aleksandrova, A.I. Maksimov, E.V. Maraeva, L.B. Matyushkin, E.A. Men'kovich, V.A. Moshnikov, S.F. Musikhin. Study of the Self-Organization Processes in Lead Sulfide Quantum Dots // Semiconductors, 2014, Vol. 48, No. 13, pp. 1729–1731.A reference

[6] D. S. Mazing, L. B. Matyushkin, O. A. Aleksandrova, I. I. Mikhailov, V. A. Moshnikov and S. A. Tarasov. Synthesis of cadmium selenide colloidal quantum dots in aquatic medium // Journal of Physics: Conference Series 572 (2014) 012028

[7] I. I. Mikhailov, S. A. Tarasov, A. V. Solomonov, O. A. Aleksandrova, L. B. Matyushkin, D. S. Mazing. The study of CdSe colloidal quantum dots synthesized in aqueous and organic media // Journal of Physics: Conference Series 572 (2014) 012029

[8] A. Antipov, etc. Luminescence of colloidal CdSe/ZnS nanoparticles: high sensitivity to solvent phase transitions // Nanoscale Res Lett. 2011; 6(1): 142

[9] I. I. Mikhailov, S. A. Tarasov, I. A. Lamkin, M. Y. Andreev, A. V. Solomonov. The luminescence properties of colloidal quantum dots // Journal of Physics: Conference Series 643 (2015) 012080

[10] Tarasov SA, Aleksandrova OA, Lamkin IA, Maksimov AI, Maraeva EV, Mikhailov II, Moshnikov VA, Musikhin SF , Nalimova SS, Permyakov NV, Spivak Yu.M., Travkin PG "Luminescent properties of systems" porous silicon-containing matrices - PbS nanoparticles "// Izvestiya Vysshikh Uchebnykh Zavedenii. Electronics. 2014. No. 4 (108). Pp. 21-26.

[11] S.A. Tarasov, I.I. Mikhailov, E.M. Stepanov, D.V. Vylegzhanin, MO Gurevich, LI Kozlovich, AV Korneeva, PO Tadtaev, A.V. Solomonov. Luminescent properties of structures containing arrays of colloidal quantum dots CdS, CdSe / ZnS, and CdZnSeS / ZnS // Nano- and microsystem technique, 2016. Vol. 18, No. 9. p. 558-562.

[12] Su L., etc. Recent Progress in Quantum Dot Based White Light-Emitting Devices // Top Curr Chem (J). 2016 Aug; 374(4):42.

IOP Conf. Series: Journal of Physics: Conf. Series **929** (2017) 012079 doi:10.1088/1742-6596/929/1/012079

Calculation of Quantum Efficiency of Photodetectors Based on A3B5 Metal-Solid Solutions Contacts

V S Zemtsovskaya, S A Tarasov, I A Lamkin, A V Solomonov
Saint-Petersburg Electrotechnical University "LETI", Prof. Popova 5, St. Petersburg 197376, Russia

E-mail: Zemtsovskaya.vera@yandex.ru

Abstract. In our investigation, we have modelled the quantum efficiency of the photodetecting structures based on the p-n junction and metal-semiconductor contacts. The calculation has been performed based on the solution of the continuity equation, taking into account the impact of the surface recombination. In most cases, this process has an adverse effect on the parameters of instruments; in particular, in photodetectors, this effect should reduce sensitivity. However, in our investigation it is shown that the effect of the surface recombination may be used as a useful factor for developing selective photodetectors. The effect of the surface recombination process will significantly increase with an increase of the absorption coefficient, since in this case more photons will be absorbed at the surface. Since there is a significant dependence of the absorption coefficient on the incident photon energy in semiconductors, the effect of the surface recombination on the photodetector sensitivity in the long-wave and short-wave regions of the spectral range will be significantly different.

1. Introduction

Photodetectors based on metal-semiconductor contacts are widely used in modern electronics. Such devices offer high sensitivity and enhanced response rate [1]. Furthermore, relatively simple methods of the vacuum deposition of the metal layers are used in their creation, which has a beneficial effect on their cost. The key characteristics of photodetectors are their quantum efficiency and spectral characteristics of sensitivity [2]. In many cases, the calculations of these characteristics are limited to consideration of dependence of the absorption coefficient for a semiconductor. At the same time, in case of the structures based on metal-semiconductor contact, it is necessary to consider a number of additional factors, in particular, the surface condition, the above-barrier transport processes, et cetera [3].

Consideration of the processes of the surface recombination has been a specific feature of the modelling performed. The surface recombination is understood as a dramatic increase in the probability of the electron-hole interaction processes caused by the presence of a higher concentration of defects at the outer boundary of a semiconductor crystal or near the heterointerfaces inside the multilayer structure. In most cases, this process has an adverse effect on the parameters of devices. In field effect transistors, the effect of the surface recombination leads to a decrease in the amount of the charge carriers passing through the channel, which deteriorates the transfer coefficients and other characteristics of a device. In photodetectors, this effect should reduce photoresponse, since the electron-hole pairs generated by a flux of photons will recombine near the surface of a semiconductor until they reach ohmic contacts. Nevertheless, in our investigation it is shown that the effect of the surface recombination may be used as a useful factor enabling photodetectors with selective spectral characteristics, without using external optical filters. In this case, sensitivity in the spectrum maximum S_{max} slightly decreases with a decrease of the spectral characteristics width in the half maximum up to the values of $\Delta\lambda_{0.5}$ = 40 nm or less. Development of selective photodiodes is one of the main challenges of modern optics [4]. The use of such effects is relevant to develop solar-blind and visible-blind photodiodes [5, 6].

2. Experimental results and discussions

Let us consider the photodetective structure based on the p-n-junction. Let this structure be illuminated from the side of the p-region by a flux of photons Φ_0 with energy $\hbar\omega$ that is greater or equal to the band gap width of a semiconductor E_g. Let us take the p-region thickness of width d as large enough for surface defects not to affect the processes in the space charge region of width w. In this structure, the incident light will give rise to generation of the electron-hole pairs practically in the full depth thereof. In this case, the coordinate x corresponding to the geometric position of the photon absorption point will be determined by the Bouguer-Lambert Law, whereby the radiation intensity will decrease exponentially with the passing of light through a semiconductor. The concentration of the nonequilibrium charge carriers in the structure will change in a similar way. We can calculate the distribution of this concentration based on the solution of the continuity equation. In the case under consideration, this solution becomes:

$$n_p(x) = n_{p0} + \tau_n \left(1 - \frac{e^{-\frac{x}{L_n}} s\tau_n}{L_n + s\tau_n} \right) \Phi_0 \alpha e^{-\alpha x}, \tag{1}$$

where τ_n and L_n are the lifetime and diffusion length for the minority charge carriers; in this case, for the electrons in the p-region.

The key factors that determine the distribution of the generated carriers in the structure depth are the absorption coefficient α and the surface recombination rate s. Let us consider the distribution for the case of a thick p-layer (d > 0.1 mm). It is obvious that with an increase of the absorption coefficient, the depth of the light penetration decreases. This means that for small values of α, the light will pass deep into the structure, while for large values of α, almost all the photons will be absorbed at the very surface.

If there is significant surface recombination ($s > 10^5$ m/s), the concentration of the nonequilibrium carriers at the semiconductor surface will decrease. However, the impact of this effect for the curve with the large absorption coefficient will be much higher, since in this case much more carriers will emerge at the very surface. The quantum yield of such a photodetector, and, hence, its sensitivity will generally correspond to the areas under the curves. In semiconductors, there is a significant dependence of the absorption coefficient on the incident photon energy, which is particularly evident near the absorption edge. This makes it possible to talk about the different impact of the surface recombination rate on the photodetector sensitivity in the long-wave and short-wave regions of the spectral range.

IOP Conf. Series: Journal of Physics: Conf. Series **929** (2017) 012079 doi:10.1088/1742-6596/929/1/012079

To verify this fact, we have modelled the effect of the absorption coefficient on the quantum yield of a photodetector considering the effect of the surface recombination. The modelling was based on the solution of the continuity equation for the p- and n-regions, as well as for the space-charge region. Let us consider each of these regions in more detail.

P-region. The concentration of the photogenerated minority charge carriers for the p-region may be determined from the following equation

$$\frac{d^2(\Delta n)}{dx^2} - \frac{\Delta n}{L_n^2} = -\frac{q\Phi_0 \cdot \alpha \exp(-\alpha x)}{D_n}. \tag{2}$$

In our investigation, boundary conditions were determined in such a way as to take account of the effect of the surface recombination on the current transport processes in the photodetective structures:

$$s \cdot \Delta n\big|_{x=0} = D_n \frac{d(\Delta n)}{dx}\Big|_{x=0}, \qquad \Delta n\big|_{x=w} = 0. \tag{3}$$

Then the current density will be determined by the following expression

$$J_{\phi.n.} = q\Phi_0 \frac{\alpha L_n}{\alpha^2 L_n^2 - 1}\left[\frac{\alpha L_p + \frac{L_n}{L_s} - e^{-\alpha w}\left(\frac{L_n}{L_s} ch\left(\frac{w}{L_n}\right) + sh\left(\frac{w}{L_n}\right)\right)}{\frac{L_p}{L_s} ch\left(\frac{w}{L_n}\right) + sh\left(\frac{w}{L_n}\right)} - \alpha L_n e^{-\alpha w}\right] \tag{4}$$

Here we have used a connection between the surface recombination rate and the diffusion coefficient

$$s = \frac{D_n}{L_s}, \tag{5}$$

that allowed to simplify the equation record.

Space-charge region. In this case, the photocurrent is determined by the generation in the region from the point w to the point d

$$J_{\phi.p\text{-}n} = -q\int_w^d G(x)dx \tag{6}$$

whereby we obtain the expression for the photocurrent density in the p-n-junction:

$$J_{\phi.p\text{-}n} = q\Phi_0\left[\left[1 - \exp(-\alpha d)\right]\exp(-\alpha w)\right] \tag{7}$$

N-region. The concentration of the photogenerated minority charge carriers for the n-region may be determined from the following equation

$$\frac{d^2(\Delta p)}{dx^2} - \frac{\Delta p}{L_p^2} = -\frac{q\Phi_0 \cdot \alpha \exp(-\alpha x)}{D_p} \tag{8}$$

with the use of boundary conditions

$$\Delta p\big|_{x=w+d} = 0, \qquad \Delta p_{x\to\infty} \to 0 \tag{9}$$

Then the current density will be determined by the following expression

$$J_{\text{ф.р.}} = -qD_p \frac{d(\Delta p)}{dx}\Big|_{x=w+d} = q\Phi_0 \frac{\alpha L_p}{1+\alpha L_p} \exp[-(w+d)\alpha] \tag{10}$$

As it is known, the quantum efficiency of a photodiode may be determined from the value of the photocurrent arising when its sensitive area is illuminated by radiation of power P_{opt}, according to the formula (1.3). Then dependence of the quantum efficiency on the absorption coefficient in the structure based on the p-n-junction may be calculated from the sum of the photocurrents specified above for the considered parts of the structure:

$$\eta(\alpha) = e^{-\alpha w}(1-e^{-\alpha d}) + \frac{\alpha L_p}{1+\alpha L_p} e^{-(w+d)\alpha} +$$

$$+\frac{\alpha L_n}{\alpha^2 L_n^2 - 1}\left[\frac{\alpha L_p + \frac{L_n}{L_s} - e^{-\alpha w}\left(\frac{L_n}{L_s} ch\left(\frac{w}{L_n}\right) + sh\left(\frac{w}{L_n}\right)\right)}{\frac{L_p}{L_s} ch\left(\frac{w}{L_n}\right) + sh\left(\frac{w}{L_n}\right)} - \alpha L_n e^{-\alpha w}\right] \tag{11}$$

The results of the modelling of the spectral characteristics of the photosensitive structures based on the p-n-junctions at different surface recombination rates show that by increasing the absorption coefficient up to 100 cm^{-1}, the quantum yield rapidly increase, and a change of the surface recombination rate has almost no effect on this increase. At the same time, with an increase of the absorption coefficient up to the values of $\alpha = 10^5$ cm^{-1}, an increase of the surface recombination rate leads to a decrease of the quantum efficiency by two orders of magnitude. It is important that sensitivity in the maximum of the spectral characteristics does not decrease so significantly (no more than by 20 %).

The difference in a change of the quantum efficiency in various parts of the spectral characteristics is more evidently shown in Figure 1. It is seen that sensitivity in the maximum practically does not change, and short-wavelength sensitivity drops by almost 100 times.

Thus, modelling has shown that it is possible to achieve the selectivity effect in the structures based on the p-n-junctions by affecting the semiconductor surface and by changing the recombination processes occurring therein.

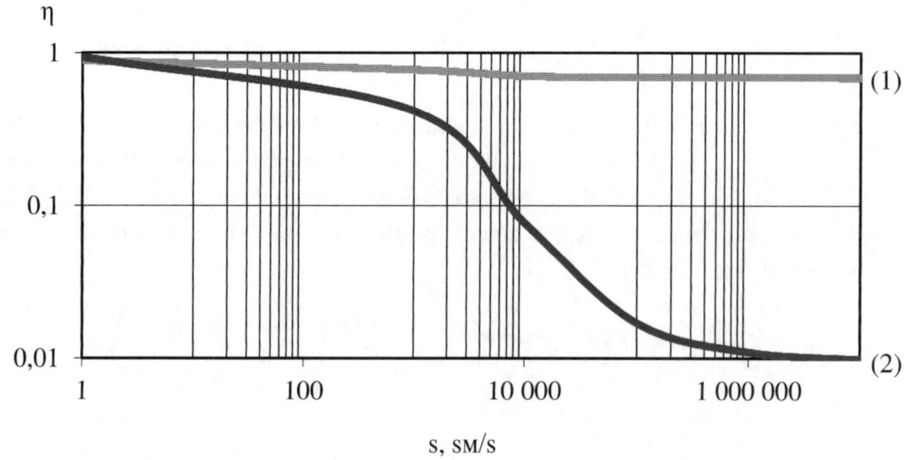

Figure 1. Dependence of the quantum yield on the surface recombination rate in the short-wave part of the spectral characteristics (1) and in its maximum (2) for a photodetector based on the p-n-junction.

We can increase the effect of the surface recombination on the characteristics of a photodetector in case we use a rectifying metal-semiconductor contact in the basis of the photodetector structure. Let us consider a device based on the Schottky barrier n-type metal-semiconductor. In this case, the structure is changed as follows: instead of the p-region, there will be the metal layer in the structure, which is in direct contact with the surface of the n-type semiconductor. In this case, the surface recombination will have an impact on the processes occurring in the space charge region, i.e. where effective separation of the electron-hole pairs should occur. However, at a relatively low dopant concentration in a semiconductor, the space charge region may be quite extensive, and there will be a difference between the effect of the surface recombination on the long-wave and short-wave parts of the spectral characteristics, which is similar with the structure based on the p-n-junction.

In order to test the hypotheses specified above we have modelled dependence of the quantum yield on the absorption coefficient in the structures based on the Schottky barrier. Illumination of the structure was carried through the upper metal contact. The thickness of the metal layer was taken as low enough to consider the metal film semi-transparent with the minimal absorption therein. In this case, the contribution of the effect of the above-barrier transport may be considered insignificant. The modelling was performed based on the solution of the continuity equation.

In the structures based on the Schottky barriers, it is convenient to consider the calculation of the current transport processes in terms of the drift component J_{dr} and diffusion component J_{dif} of the total current:

$$J_{total} = J_{dr} + J_{dif} \tag{12}$$

The diffusion current component is determined similarly to the current in the n-region of the structures based on the p-n-junction:

$$J_{dif} = q\Phi_0 \left(\frac{\alpha L_p}{1+\alpha L_p} \left(e^{-\alpha w} \right) \right) \tag{13}$$

The drift current component is calculated from the equation that describes the photogeneration processes of the charge carriers in the space charge region. But in the case of the Schottky barriers, we should take into account the presence of the surface states at the metal-semiconductor boundary and the recombination related therewith:

IOP Conf. Series: Journal of Physics: Conf. Series **929** (2017) 012079 doi:10.1088/1742-6596/929/1/012079

$$G(x) = \Phi_0 \alpha e^{-\alpha x}\left(1 - \frac{S\tau}{L_p + \tau S} e^{-\frac{x}{L_p}}\right) \quad (14)$$

Then the drift current component may be determined from the following equation:

$$J_{dr} = -q\int_0^w G(x)dx = q\Phi_0\left(1 - e^{-\alpha w} + \frac{S\tau}{L_p + \tau S}\frac{\alpha L_p}{1 + \alpha L_p}\left(e^{-w\left(\frac{1+\alpha L_p}{L_p}\right)} - 1\right)\right) \quad (15)$$

Then, summing the expressions for the drift and diffusion current, it is possible to determine the total current in the Schottky barriers:

$$J_{total} = q\Phi_0\left(1 - e^{-\alpha w} + \frac{S\tau}{L_p + \tau S}\frac{\alpha L_p}{1 + \alpha L_p}\left(e^{-w\left(\frac{1+\alpha L_p}{L_p}\right)} - 1\right) + \left(\frac{\alpha L_p}{1 + \alpha L_p}\left(e^{-\alpha w}\right)\right)\right) \quad (16)$$

Then, we may determine dependence of the quantum efficiency on the absorption coefficient in the structure based on the Schottky barrier.

Taking into account the effect of the surface recombination rate, the quantum efficiency in the structure based on the Schottky barrier is determined by the following expression:

$$\eta = \left(1 - e^{-\alpha w} + \frac{s\tau}{L_p + \tau s}\frac{\alpha L_p}{1 + \alpha L_p}\left(e^{-w\left(\frac{1+\alpha L_p}{L_p}\right)} - 1\right) + \left(\frac{\alpha L_p}{1 + \alpha L_p}\left(e^{-\alpha w}\right)\right)\right) \quad (17)$$

The results of the modelling of the spectral characteristics of the photosensitive structures based on the Schottky barriers at different surface recombination rates are generally similar to those for the structures based on the p-n-junction, but the effects described above are more evident. In particular, a drop of sensitivity increases both in the maximum and in the short-wave part of the spectral characteristics. However, sensitivity in the maximum decreases just by 1.5 - 2 times, while the short-wavelength sensitivity decreases by 2 orders of magnitude at the absorption coefficient of $\alpha = 3 \cdot 10^4$ cm^{-1} and the half-width of the spectral characteristics at the same values of the surface recombination rate is significantly smaller.

When considering the results of the modelling, we should also consider that it is very difficult to get the surface recombination rates that are smaller than s = 10^2 m/s in real structures. Therefore, a decrease of sensitivity in these structures should be counted from the specified curve. In this case, a drop of sensitivity in the maximum of the spectral characteristics is no longer seen as so significant (no more than 10 %), while its decrease in the short-wavelength part of the spectrum still amounts to 50-100 times (Figure 2).

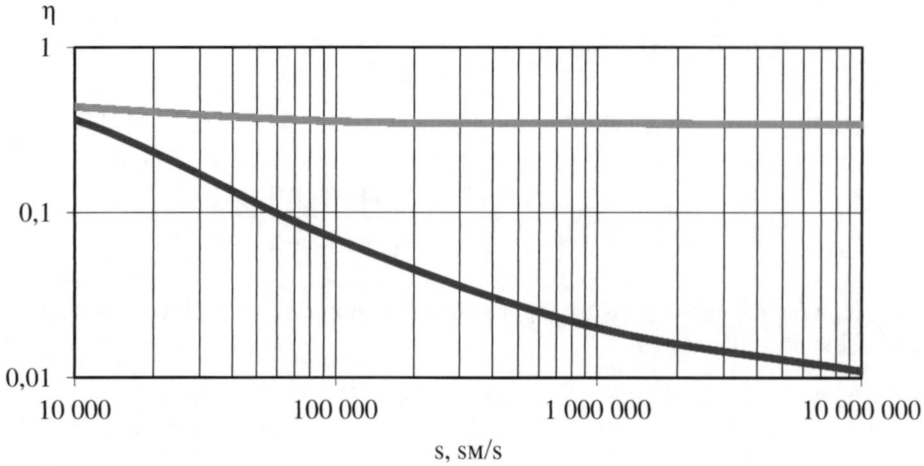

Figure 2. Dependence of the quantum yield on the surface recombination rate in the short-wave part of the spectral characteristics (1) and in its maximum (2) for a photodetector based on the p-n-junction.

Thus, it is shown that control of the surface recombination allows to change the sensitivity range of photodetectors.

3. Conclusion

It is established that when modelling the quantum efficiency of the photodetective structures based on the p-n junction and metal-semiconductor, it is necessary to take into account the effect of the surface states. It is shown that an increase in the surface recombination in the photodetective structures reduces the sensitivity range. In photodiodes based on the p-n-junction, at an increase of the absorption coefficient up to the values of $\alpha = 10^5$ cm^{-1}, an increase of the surface recombination rate leads to a decrease of the quantum efficiency by two orders of magnitude. It is important that sensitivity in the maximum of the spectral characteristics does not decrease so significantly (no more than by 20 %). In photodiodes based on metal-solid solutions contacts, a drop of sensitivity increases both in the maximum and in the short-wave part of the spectral characteristics. However, photosensitivity in the maximum decreases just by 1.5 - 2 times, while the short-wavelength sensitivity decreases by 2 orders of magnitude at the absorption coefficient of $\alpha = 3 \cdot 10^4$ cm^{-1}, and the half-width of the spectral characteristics at the same values of the surface recombination rate is significantly smaller.

4. Acknowledgment

The study was supported by the Ministry of Education of the Russian Federation within the framework of the state project in the field of scientific activity, project number 16.1307.2014K.

5. References

[1] Lamkin I, Tarasov S 2013 *J. of Physics: Conference Series* **461** 012025

[2] Xiaodong Wang, Weida Hu, Ming Pan, Liwei Hou, Wei Xie, Jintong Xu, Xiangyang Li, Xiaoshuang Chen, Wei Lu 2014 *J. of Applied Physics* **115** 013103

[3] Tarasov S A, Andreev M Y, Lamkin I A, Solomonov A V 2016 *J. of Physics: Conference Series* **741** 012085

[4] Lamkin I A, Andreev M Y, Tarasov S A, Solomonov A V, Kurin S Yu 2015 *Proceedings of the 2015 IEEE North West Russia Section Young Researchers in Electrical and Electronic Engineering Conference 2 – 4 February* 30

[5] Zhipeng Zhang, Holger von Wenckstern, Jörg Lenzner, Michael Lorenz, Marius Grundmann 2016 *Applied Phys. Let.* **108** 123503
[6] Lamkin I A, Tarasov S A, Petrov A A, Menkovich E A, Solomonov A V, Kurin S Yu 2014 J. of Physics: Conference Series **572** 012063

Spectral characteristics of InP photocathode with a surface grid electrode

N S Myazin[1], K J Smirnov[2,3], V V Davydov[1,3], S E Logunov[1]

[1]Department of Quantum Electronics, Peter the Great Saint Petersburg Polytechnic University, Saint Petersburg 195251, Russia
[2]OJSC "NRI "Electron", Saint Petersburg, Russia
[3]Department of photonics and communication lines, The Bonch-Bruevich Saint Petersburg State University of Telecommunications, Saint Petersburg 193232, Russia

myazin.n@list.ru

Abstract. Results of indium phosphide structures research, showing the possibility of using it in the near inferred range (NIR) photocathodes of InP / InGaAs, are represented. An optimal method of obtaining the atomically clean indium phosphide surface was suggested. The activation process of indium phosphide was shown and its spectral characteristics were given. Researches of the photoemission dependence of the structure with surface grid electrode upon different bias voltages were carried out.

1. Introduction

The creation of a photocathode, which operates in the NIR range, also known as the transferred electron photocathode, was proposed by Bell [1]. Such photocathodes can move the long-wavelength threshold of photoemission by increasing the internal energy of the photoelectrons. Their structure has been realized on the basis of heteropair indium phosphide – solid solution of indium – gallium arsenide. The InGaAs forbidden band is narrow enough to detect near-infrared range. However, this narrow–band material ("absorber") cannot provide by itself the negative electron affinity (NEA) on the surface, which is a determinant factor for obtaining high quantum efficiency for such structures. However, InGaAs can be perfectly matched to the InP crystal lattice, which is a conductor of electrons (emitter) generated in the absorber. The indium phosphide is an A_3B_5 semiconductor which allows getting on its surface the effective state of the NEA. Semiconductor photocathodes activated to negative electron affinity (NEA) serve as electron beam sources for many applications. InP photocathode is an emitter with InP/InGaAs heterostructure. These heterostructures allows getting greater quantum efficiency as compared to other semiconductor materials in case of detection reflective laser beam in near infrared range [2]. That is why the purpose of the work is to study the indium phosphide purification methods to obtain the high efficiency photocathode, and a study of its persistence.

2. Purifying of InP surface

The structures based on the indium phosphide are characterized by low thermal resistance and heating over with the temperature of 593 K lead to the degradation of these structures, so that the typical cleaning methods for metals and thermally stable materials cannot be used. Sulfur passivation leads to stable surface termination, but the chemisorbed sulfur atoms cannot be removed completely by thermal annealing at temperatures below the decomposition temperature of InP. Process of obtaining a clean InP surface is rather difficult, especially pre–vacuum chemical cleaning step. The treatment with HCl-C_3H_8O and subsequent vacuum annealing is useful method for obtaining the atomically clean GaAs surfaces, another semiconductor of the A_3B_5 group [3]. However, etching with these chemical compounds is ineffective for InP surfaces.

Therefore, we use the efficient three-step method of chemical etching combined with a thermal cleaning in vacuum to obtain the atomic clean surface of InP [4, 5]. Samples of InP (111): Zn, with concentration $p=1\cdot10^{18}cm^{-3}$ were used for studies. Sample of InP structure with surface grid is shown in figure 1.

Figure 1. InP structure with surface grid

Etching time was increased, as compared to [4], and the mode was adjusted to increase in rate of the removing of a thick natural oxide layer by the use of ultrasound in the etching process. Besides that the intermediate etching step was added: a weak solution of sulfuric acid was used. Strong acid solutions with HCl or H_2SO_4, which are used in third step of chemical cleaning, are able to remove the surface oxide and leave the InP surface passivated with elemental P which is, in turn, terminated with H [4]. This yields a hydrophobic surface and allows using lower temperatures during annealing. All steps of chemical cleaning are shown in table 1:

Table 1. Optimized three-step method of chemical cleaning

	Solution	Concentration
1	H_2SO_4:H_2O_2:H_2O	4:1:100
2	H_2SO_4:H_2O	1:10
3	H_2SO_4:H_2O	1:1

Protective layer formed on InP surface are also allows to carry etched structures from chemical laboratory bench to vacuum installation in atmosphere. This is an important achievement in terms of surface oxidation and uncontrolled contamination of surfaces. We studied the effect of etching on the titanium grid and the possibility of etching agents use without degradation of the grid structure. As a result, we found that etching is equally effective for two different acid solutions HCl and H_2SO_4, however, in case of HCl titanium grid is exposed to chemical degradation.

3. Activation of InP structures

The ultra-high vacuum installation was used to conduct the research in a vacuum, which could reach the level of $1\cdot10^{-9}$ Pa. Structural diagram of installation is shown in figure 2.

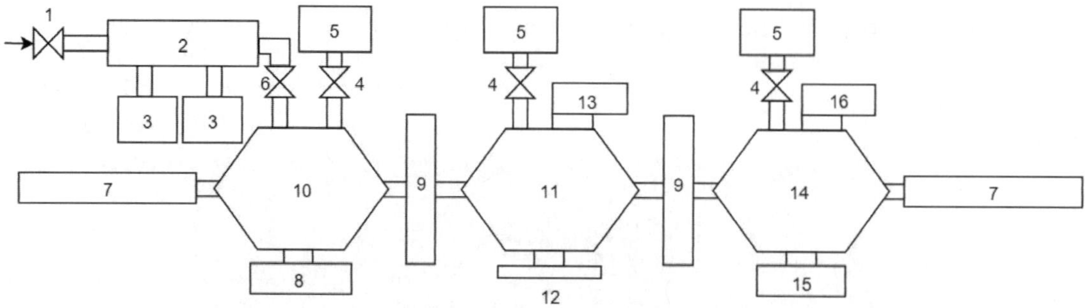

Figure 2. Structural diagram of the ultra-high vacuum installation: 1 – nitrogen valve, 2 – trunk-line, 3 – molecular sieve foreline trap, 4 –ion pump's valve, 5 –ion pump, 6 –trunk-line valve, 7 – manipulator, 8 – entry gate, 9 – gate valve, 10 – loading chamber, 11 – press and housing preparation chamber, 12 – special vacuum press, 13 – viewport, 14 – activation chamber, 15 – sources of Cs and O_2, 16 – sapphire viewport

As a result of the etching process the hydrophobic surface InP structure was obtained [4]. After heating in a vacuum chamber at a temperature of about 573 K, the net InP atomic surface was received, that's ready for the activation with Cs and O_2, which allows to form effective state of NEA on InP photocathode surface.

The method of the activation structure called «yo-yo» [6], based on a continuous simultaneous supply of cesium and oxygen was used. The process for obtaining the maximum photoemission current was about 90 minutes. The progress of activation is presented in figure 3.

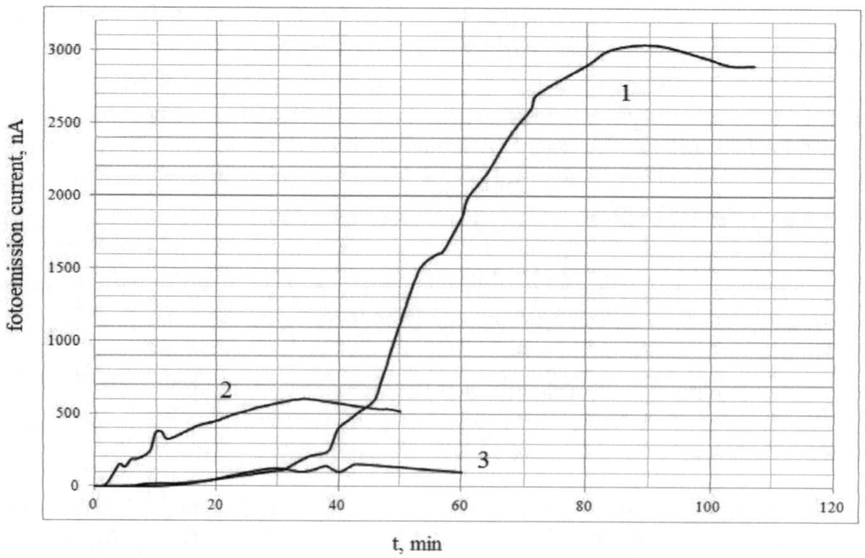

Figure 3. Activation of structure etched by 1. H_2SO_4. 2. C_3H_8O+Br. 3. C_3H_8O+HCl

Research was conducted on InP structures which etched by solutions of C_3H_8O+Br, C_3H_8O+HCl and our new three stage method of chemical cleaning with solutions of H_2SO_4. After graphical analysis, we found that with new method of chemical cleaning the photoemission current increase in 6 times as compared to previous experiments. In addition to pure samples were also studied samples with the grid electrode, used to create the photocathode with the transferred field. The dependence of the photoemission current on the applied bias voltage is shown in figure 4.

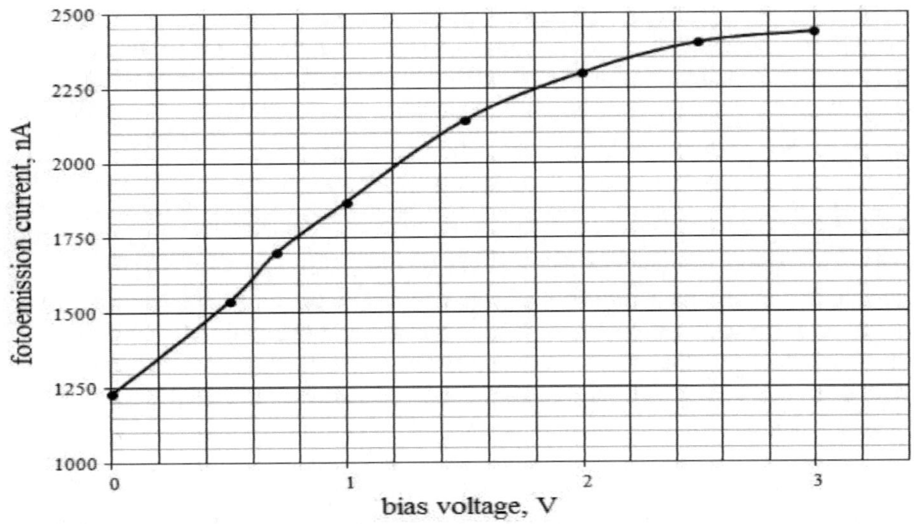

Figure 4. The dependence of photoemission current on the applied bias voltage

Measurements were carried out in range of 500 to 950 nm. Based on figure 3, quantum efficiency (QE) is calculated using the ratio:

$$QE = \frac{I \cdot h \cdot c}{P \cdot e \cdot \lambda} \cdot 100\% \qquad (1)$$

where I — the photoemission current, P — the monochromator radiation power at wavelength λ, h — the Planck's constant, c — the speed of light e — the elementary charge.

Calculation of the dependence between QE and λ using (1) are represented in figure 5.

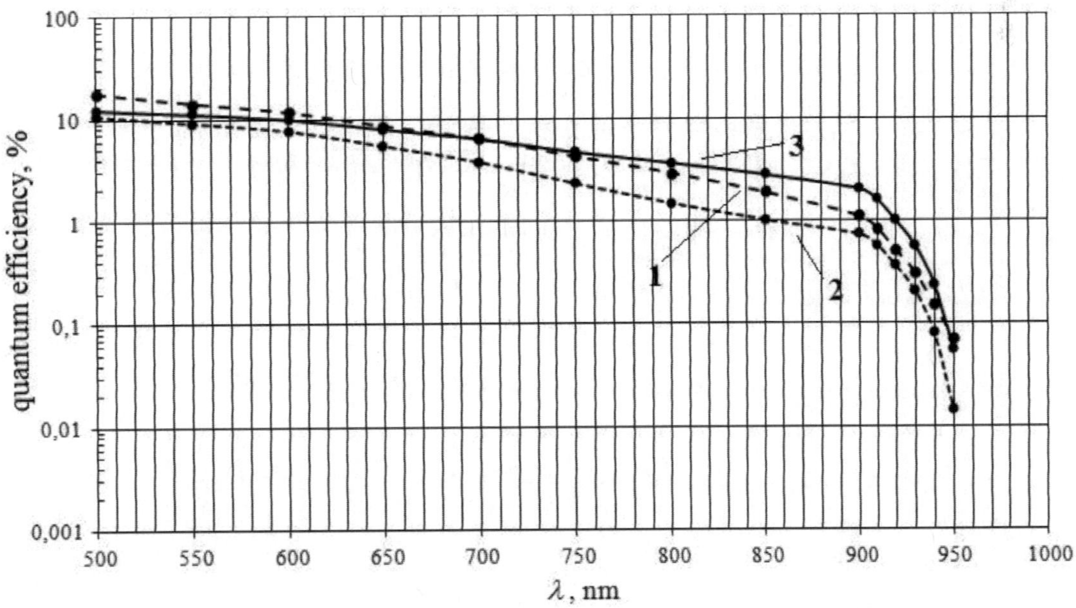

Figure 5. Spectral characteristics of InP:Zn (111) photocathodes, p=1·10^{18}cm^{-3}. 1 – without grid; 2 – with grid on U_r= 0 V; 3 – with grid on U_r= 1.2

The analysis of the obtained experimental results on figure 5 allows us to make the following conclusions. The quantum efficiency is greater than 10% at a wavelength of 630 nm, and more than 1% for 900 nm on indium phosphide. In [5] the QE at 632 nm was in the range 9-12% and spectral characteristics were not present. The photolithographic grid stamped on of InP, has demonstrated not only the overall photoemission gain increase, but also provided the greater sensitivity at longer wavelengths when a voltage was applied to the grid with the help of the pressure contact. For example at the wavelength of 650 nm the QE increases by 46%, and by 177% at 900 nm. The resulting dark current was small (less than 0.05 nA at U_r=3V), but at the applied voltage exceeding 2V the photoemission current has saturation, which makes the voltage increase weakly effective.

As the result of our experiments we got that spectral sensitivity of InP samples, which was etched by solutions of C_3H_8O+Br and C_3H_8O+HCl, was lower than 1% at the wavelengths from 500 to 700 nm. Another important fact is that the value of quantum efficiency of structures was irregularly on the surface of InP. The difference of quantum efficiency between various points, which was illuminated with monochromator radiation beam, was more than 100%. This indicates that surface of the InP structure is not purified. For structures etched by our method, difference of quantum efficiency in various points is on the level of 10-15%.

Persistence of InP photocathode is determined by the vacuum level in the activation chamber. QE of the structure is decreases by 50% in 24 hours on the vacuum lower than $1x10^{-8}$ Pa, however it can be recovered by inclusion of the cesium source.

4. Conclusion

Obtained results of the researches confirm the validity of the usage of our method for purifying InP structures. Based on the experimental results analysis, further ways of improvements current technologies for obtaining photocathode on the basis of InP/InGaAs heterostructure were provided. These heterostructures work in near infrared range and could be used in devices for special purposes, where high processing speed of detection reflected radiation from the object is needed. At present time solid-state analogs of these devices cannot provide the same processing speed.

5. References

[1] Bell R L US 1974 Patent N23958 143
[2] Ermolaev A N, Krishpents G P, Davydov V V, Vysoczkiy M G *2016 J. of Phys: Conference Series* **741** 012171.
[3] Tereshchenko O E, Chikichev S I, Terekhov A S 1999 *Appl. Surface Science* 142 75–80
[4] Sun Y, Liu Z, Machuca F, Pianetta P, Spicer W 2005 *SLAC-PUB*-11018
[5] Sun Y, Liu Z, Pianetta P 2007 *SLAC-PUB*-12710
[6] Chanlek N, Herbert J, Jones R, Jones L, Middleman K, Militsyn B 2014 *J. Phys. D: Appl. Phys.* **47** 055110

IOP Conf. Series: Journal of Physics: Conf. Series **929** (2017) 012081 doi:10.1088/1742-6596/929/1/012081

Molecular-beam epitaxy of 7-8 μm range quantum-cascade laser heterostructures

A V Babichev [1,2], D V Denisov [1], A V Filimonov [1], V N Nevedomsky [2], A S Kurochkin [1], A G Gladyshev [1], L Ya Karachinsky [1,2,3], G S Sokolovskii [2], I I Novikov [1,2,3], A Bousseksou [4], A Yu Egorov [1,3*]

[1]Connector Optics LLC, 194292 St. Petersburg, Russia
[2]Ioffe Institute, 194021 St. Petersburg, Russia
[3]ITMO University, 197101, St. Petersburg, Russia
[4]Centre for Nanoscience and Nanotechnology (C2N Orsay), CNRS UMR9001, Univ. Paris Sud, Univ. Paris Saclay, 91405, Orsay, France

E-mail: anton.egorov@connector-optics.com

Abstract. The method of molecular beam epitaxy demonstrates the possibility to create high quality heterostructures of quantum cascade lasers in a spectral range of 7-8 μm containing 50 quantum cascades in an active region. Design based on the principle of two-phonon resonant scattering is used. X-ray diffraction and transmission electron microscopy experiments confirm high structural properties of the created heterostructures, e.g. the identity of the composition and thickness of epitaxial layers in all 50 cascades. Edge-emitting lasers based on the grown heterostructure demonstrate lasing with threshold current density of 2.8 kA/cm^2 at a temperature of 78 K.

1. Introduction

Quantum-cascade lasers (QCL) emitting in a spectral range of 7-8 μm can effectively be used for remote gas analysis and various medical applications [1]. In particular, lasers with a wavelength of 7.5 μm are used for detection of SO_x [2] and trinitrotoluene [3]. The application of laser radiation with wavelength near 7.8 μm provides the detection of methane concentration [4], and can be used for remote monitoring of leaks in pipelines and at chemical plants.

The implementation of QCL can be done in different ways. The results of studies of QCL in the range of 7-8 μm published so far were based on structures with two- [5] or three-phonon resonance scattering [6], on superlattice [7], on structures with transitions between a bound state and continuum, based on the strained heteropairs [8] and on non-resonance escape of charge carriers [9].

Basic technologies to create QCL heterostructures are metal-organic vapour-phase epitaxy (MOVPE) and molecular beam epitaxy (MBE). MOVPE allows one to create heterostructures with a greater growth speed. MBE allows implementation of lasers with higher quantum efficiency [10]. Additional improvement of quantum efficiency of QCL is possible to achieve by implementing multi-cascade QCL heterostructures containing up to 100 cascades [11]. Manufacturing of such multi-cascade QCL heterostructures requires maintaining of high identity of chemical composition and cascade layer thickness throughout the whole long process of epitaxy [12]. The utilization of industrial

IOP Conf. Series: Journal of Physics: Conf. Series **929** (2017) 012081 doi:10.1088/1742-6596/929/1/012081

MBE system satisfies these requirements, which was demonstrated by a number of research groups [13-15]. The use of state of the art multi-substrate industrial MBE systems such as Riber 49-7000, Veeco Gen 200-2000 and Oxford VG Semicon V80H-V150 substantially increases the overall yield.

2. Experimental

QCL heterostructures were grown by Connector Optics LLC using MBE system - Riber 49, equipped with a solid-state source of arsenic of cracker type and ABI 1000 cells for gallium and indium metals. ABI 1000 source has a cylindrical crucible and a dual-zone heater and specially designed to ensure high stability of metal flow during a long epitaxy process. ABI 1000 requires no temperature correction while it is emptied, e.g. during the depletion of the metal loaded in the cells. To create QCL heterostructure on InP substrate of (001) orientation doped with sulphur up to 3×10^{18} cm^{-3} level, a 2 µm buffer layer was grown with the level of Si doping of 1×10^{17} cm^{-3}. The next layer of $In_{0.53}Ga_{0.47}As$ having the thickness of 0.5 µm was doped up to level 5×10^{16} cm^{-3}. Next, the heterostructure active area was grown. The layout of the single quantum cascade is presented in Table 1. To obtain photons of 0.025±0.001 eV energy we implemented the design, which provides the transition from the top- to the lower-level of 5.3 nm quantum well (Layer 17, Table 1). Depletion of the lower-level was achieved by means of two-phonon resonance scattering with emission of two longitudinal optical (LO) phonons [5]. Here with the electron from the lower-level of the first quantum well (Layer 17, Table 1) is scattered emitting LO-phonon to another lower discrete level of the neighboring quantum well (Layer 15, Table 1), from which it scatters again emitting LO-phonon to another lower discrete level of the following quantum well (Layer 13, Table 1) and then it is released into the injector. To ensure the scattering with LO-phonon emission, the energy gap between levels in neighboring quantum wells should be close to LO-phonon energy, which is 0.035 eV in this case. The final QCL heterostructure includes 50 such quantum cascades (QC) and is formed using of broadband, $Al_{0.48}In_{0.52}As$, and narrowband, $In_{0.53}Ga_{0.47}As$, solid alloys, matched to the lattice parameter of InP substrate.

After growing the 50 QCs we consequently formed the layers of $In_{0.53}Ga_{0.47}As$ with n-doping level of 1×10^{17} cm^{-3} and 1×10^{19} cm^{-3}, having the thickness of 0.1 µm and 0.02 µm, respectively. The level of layer doping was calibrated on the test epitaxial layers of $In_{0.53}Ga_{0.47}As$ and InP. The precise calibration of the chemical composition of epitaxial layers and growth speeds of $In_{0.53}Ga_{0.47}As$ and $Al_{0.48}In_{0.52}As$ solid alloys was carried out by growing of a special test heterostructure based on 10 pairs of alternating layers of $Al_{0.48}In_{0.52}As$ /$In_{0.53}Ga_{0.47}As$, each of 10 nm thick. All samples were grown under the same equivalent pressure of arsenic flow (As4) equal to 1×10^{-5} Torr and the same growth speed of 0.2 nm/s. QCs were grown at a substrate temperature of 480°C. During the growing of the doped layers of $In_{0.53}Ga_{0.47}As$ and InP, the epitaxy temperature was lowered to 450-460°C. To ensure better uniformity of thickness of epitaxial layers in all QCs during their epitaxy, we increased rotation speed of the substrate holder up to 24 rpm. The rotation speed was chosen in such a way that during the growth of the thinnest layer in the quantum cascade (layer $Al_{0.48}In_{0.52}As$ of 1 nm thickness) the substrate holder have been made two full rotations.

3. Results and discussion

The thickness of QCL heterostructure layers was controlled using transmission electronic microscopy (TEM) and X-ray diffraction. TEM measurements were made using the transmission electron microscope JEM2100F. The samples in cross-sectional geometry were prepared using a standard technique by plane cleavage of the structure (1-10) and their further polishing [16]. In the final stage the samples were narrowed by Ar+ ion etching with energies of 3–4 keV. The images were made at the acceleration voltage of 200 kV. To define the interfaces between the layers, we plotted the image intensity profile, which was orthogonal to the layer interfaces. The averaging width was 20 nm. The location of the edges between the layers was taken by the half-height of transitional contrast from one layer to the other. The cross-sectional TEM image of a QCL heterostructure is shown in Figure 1. The total thickness of one period was equal to 532 Å. The thickness of the cascade according the TEM measurements and taking into account full errors, was 532±50 Å.

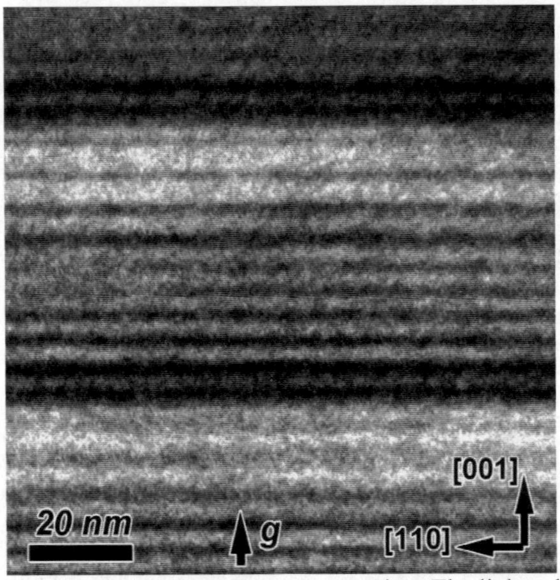

Figure 1. TEM image of a QCL heterostructure cross-section. The lighter layers correspond to InGaAs, darker layers to AlInAs.

Figure 2 shows the calculated (according to specification) and experimental X-ray diffraction curves of a QCL heterostructure. The X-ray diffraction measurements with high resolution were made in the vicinity of a symmetric reflex (004) of InP using PANalytical X'PertPro diffractometer in a parallel X-ray beam geometry. The 6 kW X-ray source was represented by a tube with a rotating copper anode (λ=0.15406 nm). FWHM of the primary beam was below 12″, which was provided by using a fourfold Ge (220) monochromator. Figure 2 shows characteristic peaks of X-ray diffraction from InP substrate and periodic QCL heterostructure. There is a complete match between the zero peak of the satellite structure and the position of the peak of InP substrate. This fact proves the exact correspondence between the chemical composition of the epitaxial layers and the values in the growth specification. In addition, this figure shows 13 satellites, resulting from the periodic structure of quantum cascades. Satellites, located to the left of the substrate peak, have FWHM of 26±5″, and the ones located to the right are slightly widened. When calculating the total thickness of the quantum heterostructure cascade, we took into account the angular position of the peaks with clear maxima. The calculated thickness of the QC for the experimental curve was 526±7 Å. Thus, there was an exact matching between the thickness of the quantum cascade measured in experiment with the cascade thickness determined using the specification (see Table 1). Furthermore, it should be noted that satellites have a distinct shape with one maximum. In the case of non-optimized parameters of the epitaxial process, the splitting of the satellite peaks could appear and several maxima could emerge, which is associated with fluctuations of the composition and thickness of epitaxial layers in different heterostructure cascades [16]. This result indicates a significantly better flow stability of Group III elements when using ABI 1000 cell with a cylindrical crucible, comparing to a general ABN 700 cell with a conical crucible [16].

IOP Conf. Series: Journal of Physics: Conf. Series **929** (2017) 012081 doi:10.1088/1742-6596/929/1/012081

To prove the optical quality of the grown QCL heterostructure and to study its optical properties, we processed the laser diodes in ridge geometry with width of 20 microns. The lasing spectrum is shown in the Figure 3a. The maximum of stimulated radiation intensity corresponds to a wavelength of 7.6 µm. A typical linewidth of lasing spectra (FWHM) is about 2.8 nm. Typical light-current and voltage-current characteristics of the QCL laser diode measured at a temperature of 78 K are shown in Figure 3b. Threshold current of 0.9 A (which corresponds to the threshold current density of 2.8 kA/cm^2) was obtained. The clear diode type voltage-current dependence without parasitic current flow channel was observed.

Table 1. Design of a QCL cascade.

No	Layer material	Layer thickness according to specification (Å)	Layer thickness determined by TEM analysis (Å)
1	$In_{0.53}Ga_{0.47}As$	24	24
2	$Al_{0.48}In_{0.52}As$	24	21
3	$In_{0.53}Ga_{0.47}As$	26	24
4	$Al_{0.48}In_{0.52}As$	21	22
5	$In_{0.53}Ga_{0.47}As$	26	24
6	$Al_{0.48}In_{0.52}As$	18	25
7	$In_{0.53}Ga_{0.47}As$	27	24
8	$Al_{0.48}In_{0.52}As$	16	31
9	$In_{0.53}Ga_{0.47}As$	29	26
10	$Al_{0.48}In_{0.52}As$	17	24
11	$In_{0.53}Ga_{0.47}As$	31	31
12	$Al_{0.48}In_{0.52}As$	25	18
13	$In_{0.53}Ga_{0.47}As$	44	43
14	$Al_{0.48}In_{0.52}As$	12	19
15	$In_{0.53}Ga_{0.47}As$	52	42
16	$Al_{0.48}In_{0.52}As$	12	25
17	$In_{0.53}Ga_{0.47}As$	53	31
18	$Al_{0.48}In_{0.52}As$	10	26
19	$In_{0.53}Ga_{0.47}As$	17	25
20	$Al_{0.48}In_{0.52}As$	43	27
Total thickness of the cascade		527	532

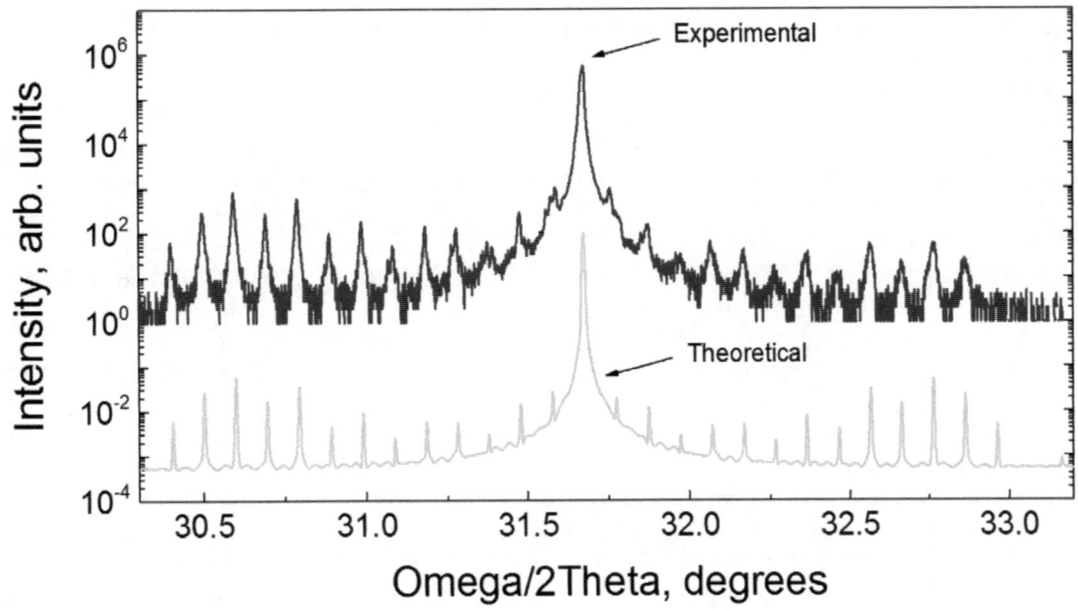

Figure 2. X-ray diffraction curves of QCL heterostructure. The experimental curve is shown in the top, the calculated counterpart is shown in the bottom.

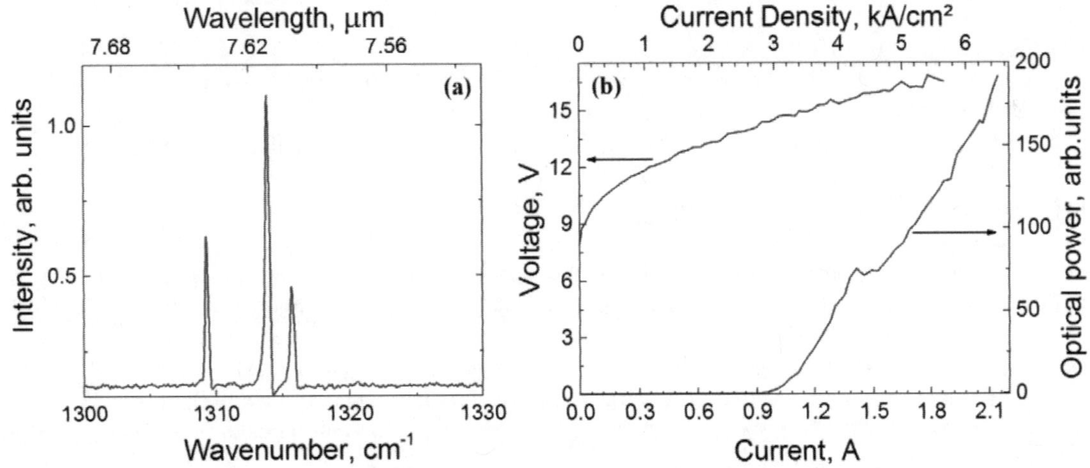

Figure 3. (a) The lasing spectrum measured at 78 K; (b) voltage-current curve (left vertical axis) and light-current curve (right vertical axis) of a QCL laser ridge measured at a temperature of 78 K.

4. Conclusions

We have studied the parameters of the epitaxy for 7-8 μm optical range multi-cascade QCL heterostructures grown using the industrial multi-substrate MBE system - Riber 49 equipped with ABI 1000 cells. To achieve QCL lasing in the specified spectral range we used the cascade design based on two-phonon resonant scattering. Using X-ray diffraction and transmission electron microscopy, we investigated structural properties of QCL heterostructures and confirmed their high structural perfection, e.g. the identity of their compositions and thicknesses of epitaxial layers in all 50 cascades and full compliance with the numerical specification. The heterostructures were used for manufacturing of stripe lasers with 20 μm stripe width. Laser ridges demonstrated the lasing at a wavelength of 7.6 μm and threshold current density of 2.8 kA/cm^2 at a temperature of 78 K.

5. Acknowledgment

The reported study was partially supported by RFBR research project № 16-29-03289. Structural characterization was made on the equipment of the Federal Joint Research Centre «Material science and characterization in advanced technology» (Ioffe Institute, St. Petersburg, Russia).

References

[1] Kim S S, Young C, Vidakovic B, Gabram-Mendola S G, Bayer C W and Mizaikoff B 2010 *IEEE Sensors J.* **10** 145-58
[2] Yu J S, Slivken S and Razeghi M 2010 *Semicond. Sci. Technol.* **25** 125015
[3] Pushkarsky M B, Dunayevskiy I G, Prasanna M, Tsekoun A G, Go R and Patel C K N 2006 *Proc. Nat. Acad. Sci.* **103** 19630-34
[4] Darvish S R, Zhang W, Evans A, Yu J S, Slivken S and Razeghi M 2006 *Appl. Phys. Lett.* **89** 251119
[5] Xu G, Moreau V, Chassagneux Y, Bousseksou A, Colombelli R, Patriarche G, Beaudoin G and Sagnes I 2009 *Appl. Phys. Lett.* **94** 221101
[6] Maulini R, Lyakh A, Tsekoun A and Patel C K N 2011 *Opt. Express.* **19** 17203-11
[7] Gmachl C, Tredicucci A, Capasso F, Hutchinson A L, Sivco D L, Baillargeon J N and Cho A Y 1998 *Appl. Phys. Lett.* **72** 3130-32
[8] Leavitt R P, Bradshaw J L, Lascola K M, Meissner G P, Micalizzi F, Towner F J and Pham J T 2010 *Opt. Eng.* **49** 111109
[9] Troccoli M, Lyakh A, Fan J, Wang X, Maulini R, Tsekoun A G, Go R and Patel C K N 2013 *Opt. Mater. Express.* **3** 1546-60
[10] Troccoli M 2015 *IEEE J. Sel. Top. Quantum Electron.* **21** 61-7
[11] Gmachl C, Capasso F, Tredicucci A, Sivxo D L, Kohler R, Hutchinson A L and Cho A Y 1999 *IEEE J. Sel. Top. Quantum Electron.* **5** 808-16
[12] Fedosenko O, Chashnikova M, Machulik S, Kischkat J, Klinkmüller M, Aleksandrova A, Monastyrskyi G, Semtsiv M P and Masselink W T 2011 *J. Cryst. Growth* **323** 484-7
[13] Babichev A V *et al* 2016 *Semiconductors* **50** 1299-03
[14] Garcia M, Vermersch F J, Marcadet X, Bansropun S, Carras M, Wilk A, Chaix C and Sirtori C 2006 *Proc. SPIE* (San Jose, CA, USA) p 613304
[15] Han Y J, Freeman J R, Isac N, Colombelli R, Dean P, Davies A G and Linfield E H 2016 *Opt. Express* **24** 28583-93
[16] Egorov A Y, Brunkov P N, Nikitina E V, Pirogov E V, Sobolev, M S, Lazarenko A A, Baidakova M V, Kirilenko D A and Konnikov S G 2014 *Semiconductors* **48** 1600-04

The concept for realization of quantum-cascade lasers emitting at 7.5 µm wavelength

I I Novikov[1,2], A V Babichev[1,2], V E Bugrov[1], A G Gladyshev[2], L Ya Karachinsky[1,2], E S Kolodeznyi[1], A S Kurochkin[1], A V Savelyev[1,3], G S Sokolovskii[1,4,5], A Yu Egorov[1,2*]

[1] ITMO University, 197101, St. Petersburg, Russia
[2] Connector Optics LLC, 194292, St. Petersburg, Russia
[3] St. Petersburg Academic University, 194021, St. Petersburg, Russia
[4] Ioffe Institute, 194021, St. Petersburg, Russia
[5] Peter the Great St. Petersburg Polytechnic University, 195251, St. Petersburg, Russia

E-mail: anton.egorov@connector-optics.com

Abstract We consider the advantages and disadvantages of various designs of waveguide for heterostructures of quantum cascade lasers (QCL) in a spectral region of 7.5 µm. Based on a numerical calculation we make a comparison of light wave distribution in QCL waveguides with different designs. We demonstrate the benefits of practical QCL realization with an extended five-layered waveguide formed by introducing extra layers of InGaAs, which allows to modify the spatial distribution of the light wave and get the rectangular shape of the spatial distribution of light wave intensity in the laser active area.

1. Introduction

The operation principle of quantum cascade lasers (QCL) is based on intersubband electronic transitions between quantum states of semiconductor nanoheterostructure [1-2]. The wavelength of the radiation emitted by an active QCL area (by a set of quantum wells) is mainly determined by the thickness of semiconductor layers and not by their bandgap. Utilization of effects of resonant tunneling and miniband conductivity allows one to create an injector area and interconnect active zones. The combination of the injector and the active zone is called a cascade. Thus, alternating between active and injector areas [2-11], one can create a QCL heterostructures containing large number of connected cascades. This cascaded laser structure provides a relatively simple way for scaling of the output optical power of a single emitter. An electron, passing through each cascade, emits a photon. During its passage through the all QCL cascades, one electron emits multiple photons. The larger number of cascades in the QCL, the greater number of photons, one electron emits during its passage.

Hence, increasing number of cascades should lead to increasing of the optical output power of the device and to decreasing of the threshold current. The number of cascades which is required for realization of an efficient QCL is typically 30 or more. However, this is true only in the case of perfect identity of all cascades [12-14]. In practice, technical difficulties of maintaining stable parameters during the epitaxial process for a long period of time, that is needed for growing of a QCL nanoheterostructure, may cause changes in the thickness and composition of epitaxial layers that shifts the cascade amplification spectrum. As a result, the potential gain may be lost.

For creation of QCL with lasing wavelength 7.5 μm it is advisable to use $In_{0.53}Ga_{0.47}As$ and $In_{0.52}Al_{0.48}As$ solid alloys. Mole fraction of indium (0.53 and 0.52) in these solid alloys is chosen in that way that the lattice parameter of the layer solid alloy matches to the lattice parameter of InP substrate. One of the most important physical parameter of semiconductor layers comprising QCL heterostructures is their refractive index, which contains real and imaginary parts. The refractive index depends on a selected material and wavelength, as well as on the doping level. For n-type doping of InP-matched solid alloys, silicon is commonly used. The influence of doping on the refractive index of the material can be calculated using Drude-Lorentz approach [15], where plasma frequency is defined by the expression:

$$\omega_p = \sqrt{\frac{n_e q^2}{\varepsilon_0 \varepsilon_\infty m_e}} \tag{1}$$

where key parameter n_e is the conduction electron density, q is the elementary charge, m_e is the electron mass, ϵ_0 is the permittivity of free space, ϵ_∞ is the high-frequency permittivity, which is actually equal to the local level of doping. Then the total electric susceptibility can be expressed as:

$$\varepsilon = (n + ik)^2 = \epsilon_\infty \left(1 - \frac{\omega_P^2}{\omega(\omega + i/\tau)}\right) \tag{2}$$

where n is the real refractive index, k is the extinction coefficient, ω is the frequency, $1/\tau$ is the scattering rate.

The calculated values of real and imaginary parts of refractive indexes of $In_{0.53}Ga_{0.47}As$ and $In_{0.52}Al_{0.48}As$ solid alloys and InP substrate for different levels of n-type doping are shown in Table 1.

Table 1. calculated values of real and imaginary parts of refractive indexes of $In_{0.53}Ga_{0.47}As$, $In_{0.52}Al_{0.48}As$ semiconductor solid alloys and InP substrate for different levels of n-type doping.

	$In_{0.53}Ga_{0.47}As$		$In_{0.52}Al_{0.48}As$		InP	
Doping level $\times 10^{16}$ (cm^{-3})	n	k $\times 10^{-3}$	n	k $\times 10^{-3}$	n	k $\times 10^{-3}$
1.0	3.375	0.45	3.147	0.27	3.071	0.28
5.0	3.369	2.27	3.143	1.29	3.067	1.42
10	3.362	4.52	3.139	2.58	3.062	2.84
50	3.300	23.1	3.104	13.0	3.024	14.4
100	3.222	47.4	3.061	26.4	2.976	29.2
500	2.526	302	2.687	150	2.559	170

The presented data shows that with an increase of doping levels the real part of the refractive index decreases and the imaginary part increases. An important consequence of the increasing of the imaginary part of the refractive index is that it causes in increasing of optical losses due to an increase of radiation absorption on free charge carriers with increasing of material doping.

During the formation of heterostructure active region consisting of alternating layers of solid alloys of $In_{0.53}Ga_{0.47}As$ and $In_{0.52}Al_{0.48}As$, we have a significant energy gap in a conduction band equal to 520 meV, which enables the creation of QCL active area comprising $In_{.53}Ga_{0.47}As$ quantum wells and $In_{0.52}Al_{0.48}As$ barriers. At the same time, the characteristic values of the real part of the average refractive index of the active region and the injector are usually in the range 3.27-3.29 [16]. The real part of the refractive index of $In_{0.53}Ga_{0.47}As$ layers is equal to 3.375 [15], of $In_{0.52}Al_{0.48}As$ layers is 3.147 [17], and of InP substrate is 3.071 (the shown values relate to low doping level).

The QCL number of cascades selected when designing the heterostructures determines the thickness of the light-emitting laser area and is equal to the product of the total thickness of one cascade by a number of cascades. Due to a smaller refractive index, InP and InAlAs materials are used as the claddings of the waveguide. Quantum cascades, which consist of alternating layers of InGaAs and InAlAs, form the core of the waveguide. In each QCL cascade electrons relax from the excited state of the quantum well to the ground state, in the result photons are emitted. One electron, while passing "h" cascades, can emit "h" photons. For this reason, the QCL threshold current density must decrease with the increasing number of cascades. Additionally, an important parameter is the overlap of the light-emitting region, which is also the gain region, with the light wave in the waveguide. A part of the light wave is located in the waveguide core and the other part penetrates into the claddings of the waveguide. With increasing thickness of waveguide core, the fraction of the light wave located in it increases (confinement factor increases) and the threshold current decreases. With further increase of waveguide thickness due to increasing the number of cascades, the confinement factor tends to unity. A further increase in the number of cascades will not lead to a reduction in threshold current density, since confinement factor has already reached its maximum value. For each individual cascade, the confinement factor will even decrease, since the light wave is redistributed between the greater number of cascades and the peak intensity of the light wave in the waveguide reduces. Furthermore, the efficiency of side cascades is quite low because of their low interaction with light wave.

2. Calculation results and discussion

The comparison of the light wave distribution in the narrow and broad waveguides is shown in Figures 1 and 2. The boundary conditions were used according [18]. As the bottom cladding, in these QCL designs we used InP substrate of n-type doped up to the level of 3.0×10^{18} cm^{-3}, as the top cladding we used two layers of InAlAs having thicknesses 2.0 μm and 1.0 μm and doped with silicon up to levels of 1.0×10^{17} cm^{-3} and 1.0×10^{18} cm^{-3}, respectively. Waveguide consists of 50 (or 99) quantum cascades with an average refractive index of 3.291. The confinement factors are 70% and 93%, and the optical waveguide losses due to the penetration of light waves into doped regions are equal to 90 cm^{-1} and 22 cm^{-1} for 50 and 99 cascades, respectively. Thus, one can see advantages of a broad waveguide.

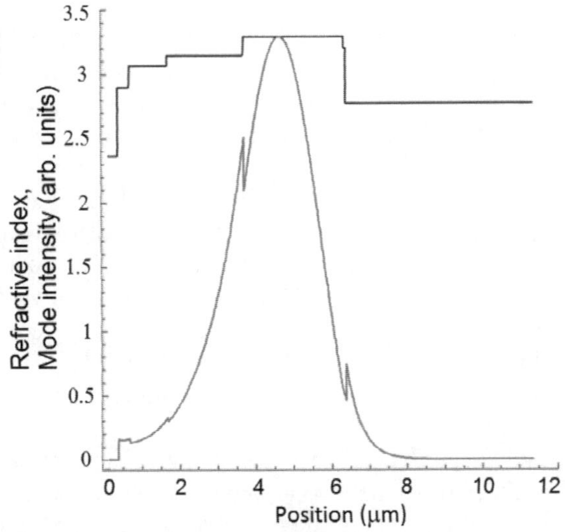

Figure 1. Distribution of the refractive index and 7.5 μm light wave intensity in QCL heterostructure with a narrow waveguide (50 cascades) on InP substrate.

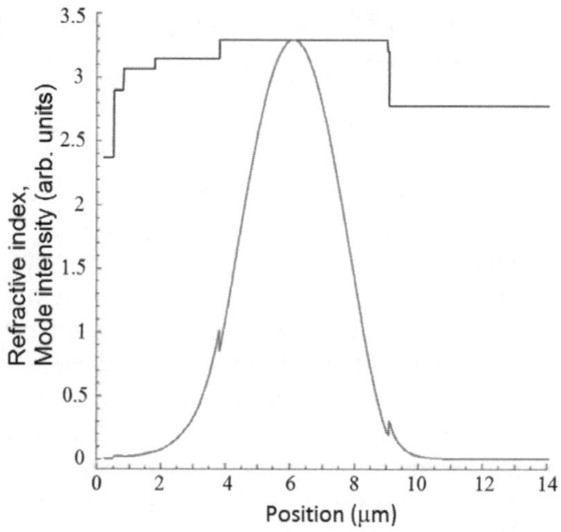

Figure 2. Distribution of the refractive index and 7.5 μm light wave intensity in QCL heterostructure with a narrow waveguide (99 cascades) on InP substrate.

However, one should keep in mind that the increase in the number of cascades results in a proportional increase in the operation voltage of the device and corresponding heating. Low efficiency of the side cascades should also be noted.

In practice, when constructing quantum-cascade lasers, for extension of the waveguide and reduction of optical losses one can use additional InGaAs layers, which have even greater refractive index compared with quantum cascades.

The use of such layers enables one to extend the waveguide and substantially modify the spatial distribution of the light wave. As a result, one can get not only an extended waveguide with reduced number of cascades, but equally efficiently use all the cascades due to forming of a rectangular profile of optical mode in the gain area of laser heterostructure. Utilizing of extended five-layered optical waveguide with a rectangular optical mode profile not only allows one to achieve maximum QCL quantum efficiency values at a fixed number of cascades, but also to increase a differential optical gain of active medium. In this case, one advantageously postpones the moment, when the gain saturation takes place, which is observed at higher pumping levels. Naturally, the gain saturation should primarily occur in the parts of the active medium, where the optical mode has its maximum. In an extended optical waveguide with a rectangular optical mode profile, the intensity of the optical mode is substantially the same throughout the whole gain region and the peak value of the optical mode intensity is decreased. Therefore, the start of gain saturation will occur at higher total optical power or at a higher pumping current. The calculations show that the start of gain saturation in the narrow optical waveguide (Figure 1) will occur when the optical power is by half lower compared with the optical waveguide with rectangular mode profile (Figure 3). Another advantage of the extended optical waveguide with rectangular optical mode profile is that it provides lower laser beam divergence, which in turn substantially facilitates focusing of laser radiation. The distribution of refractive indexes in QCL heterostructure with the extended five-layered waveguide, as well as the spatial distribution of the light wave intensity are shown in Figure 3.

The waveguide includes 50 quantum cascades with an average refractive index 3.291. Furthermore, InGaAs layers doped with silicon up to the level of 5.0×10^{16} cm^{-3} are additionally introduced into the waveguide. The optimal thickness of InGaAs layers, that is necessary to obtain a rectangular spatial distribution of the light wave in the laser active region, is equal to 2.17 μm from the side of InP and 1.67 μm from the side of InAlAs layers. The doping levels and thickness of all other layers for heterostructures shown in Figures 1, 2 and 3 are identical.

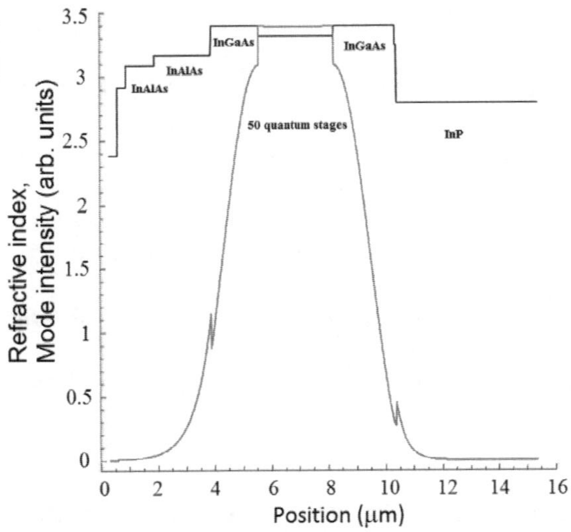

Figure 3. Distribution of the refractive index and 7.5 μm light wave intensity in QCL heterostructure with extended five-layered waveguide on InP substrate. The gain region includes 50 cascades.

The confinement factor for presented five-layered waveguide is 51%, and waveguide losses due to the penetration of light waves in the doped InGaAs layers and claddings are 36 cm^{-1}, this value is twice lower than for the case shown in Figure 1. At the same time this waveguide has all the advantages described above.

3. Conclusions

It has been shown that the use of QCL with extended five-layered waveguide formed by introducing extra layers of InGaAs with higher refractive index compared with quantum cascades, allows, on the one hand, to expand the waveguide core and reduce the proportion of the light wave penetrating the claddings of the waveguide and, on the other hand, to achieve significant modification of the spatial distribution of the light wave in the waveguide. By introducing sufficiently thick InGaAs layers on both sides of the light-emitting region, it is possible to obtain a rectangular spatial distribution of the intensity of the light wave in the active region of the laser (in quantum cascades) and to achieve high quantum efficiency of all cascades.

4. Acknowledgment

This work was supported by the Ministry of Education and Science of the Russian Federation in the framework of the Federal target program "Research and development in priority areas of Russian science and technology complex for 2014 – 2020", code 2016-14-579-0009, agreement No. 14.578.21.0204, unique ID: RFMEFI57816X0204.

References

[1] Faist J, Capasso, F, Sivco D L, Sirtori C, Hutchinson A L and Cho A Y 1994 *Science* **264** 553–56

[2] Faist J, Capasso F, Sirtori C, Sivco D L, Hutchinson A L and Cho A Y 1995 *Appl. Phys. Lett.* **66** 538–40

[3] Beck M, Hofstetter D, Aellen T, Faist J, Oesterle U, Ilegems M, Gini E and Melchior H 2002 *Science* **295** 301–05

[4] Wang Q J, Pflügl C, Diehl L, Capasso F, Edamura T, Furuta S, Yamanishi M and Kan H 2009 *Appl. Phys. Lett.* **94** 011103

IOP Conf. Series: Journal of Physics: Conf. Series **929** (2017) 012082 doi:10.1088/1742-6596/929/1/012082

[5] Tredicucci A, Capasso F, Gmachl C, Sivco D L, Hutchinson A L and Cho A Y 1998 *Appl. Phys. Lett.* **73** 2101–03
[6] Fujita K, Furuta S, Sugiyama A, Ochiai T, Edamura T, Akikusa, N., Yamanishi M and Kan, H. 2007 *Appl. Phys. Lett.* **91** 141121
[7] Lyakh A, Maulini R, Tsekoun A, Go R, Pflügl C, Diehl L, Wang Q. J., Capasso F and Patel C K N 2009 *Appl. Phys. Lett.* **95** 141113
[8] Yamanishi M, Fujita K, Edamura T and Kan H 2008 *Opt. Express* **16** 20748–58
[9] Fujita K, Yamanishi M, Furuta S, Sugiyama A and Edamura T 2012 *Appl. Phys. Lett.* **101** 181111
[10] Liu P Q, Hoffman A J, Escarra M D, Franz K J, Khurgin J B, Dikmelik Y, Wang X, Fan J Y and Gmachl C F 2010 *Nat. Photonics* **4** 95–8
[11] Yao Y, Charles W O, Tsai T, Chen J, Wysocki G and Gmachl C F 2010 *Appl. Phys. Lett.* **96** 211106
[12] Gmachi C, Capasso F, Tredicucci A, Sivxo D L, Kohler R, Hutchinson A L and Cho A Y 1999 *IEEE J. Sel. Top. Quantum Electron.* **5** 808–16
[13] Fedosenko O, Chashnikova M, Machulik S, Kischkat J, Klinkmüller M, Aleksandrova A, Monastyrskyi G, Semtsiv M P and Masselink W T 2011 J. Cryst. Growth 323 484–7
[14] Babichev A V *et al* 2016 *Semiconductors* **50** 1299–03
[15] Dressel M and Grüner G 2003 *Electrodynamics Of Solids* (Cambridge: Cambridge University Press)
[16] Bousseksou, A, Chassagneux Y, Coudevylle J R, Colombelli R, Sirtori C, Patriarche G, Beaudoin G and Sagnes I 2009 *Appl. Phys. Lett.* **95** 091105
[17] Sirtori C, Gmachl C, Capasso F, Faist J, Sivco D L, Hutchinson A L and Cho A Y *Opt. Lett.* **23** 1366–68
[18] Faist J 2013 *Quantum cascade lasers* (Oxford: Oxford University Press) p 93

IOP Conf. Series: Journal of Physics: Conf. Series **929** (2017) 012083 doi:10.1088/1742-6596/929/1/012083

Simulating of solid-state electrocaloric cooler based on multi-layered ferroelectric capacitor structures

A V Es'kov[1,2], P Yu Belyavskiy[1,2], A S Anokhin[2], O V Pakhomov[2], A A Semenov[1], A A Nikitin[1,2], V A Krylov[2], M T Bui[2]

[1]Department of Physical Electronics and Technology, St. Petersburg Electrotechnical University, St. Petersburg 197376, Russia
[2]International Laboratory "MultiferrLab", ITMO University, St. Petersburg 197101, Russia

E-mail: pbeliavskiy@gmail.com

Abstract. On the basis of the phenomenological theory of Landau was simulated electrocaloric response in ferroelectric. The obtained data were used for computer simulating of cooling of layered structure

Due to the decreasing size and increasing power of electronic components the problem of cooling and temperature control is one of the hottest in the modern electronics and electrical engineering. The effective heat transformation could be implemented in the solid state structures using electrocaloric effect [1, 2]. The main technical difficulty standing in the way of developing a small-sized high-performance solid-state cooler on electrocaloric effect, is the need to use heat switches for heat removal process from the cooling object [3-5]. Earlier, in works [6-9], principle and thermodynamic cycle of the electrocaloric cooler without the use of thermal switches have been proposed. The principle of operation is based on the effect of the difference values of electrocaloric effect in ferroelectric materials with polarization and depolarization of the sample [10].

In previous works [11, 12] we experimentally investigated electrocaloric response in bulk ferroelectric samples based on a solid solution of barium strontium titanate and demostrated the difference in the electrocaloric responses for polarization and depolarization. However, this information is not enough for correct computer simulation of solid-state cooler based on electrocaloric effect without the use of heat switches. For additional parameters, such as the dependence of electric polarization on the temperature and the electric field and dependence of heat capacity on the temperature and the electric field we used different mathematical transformations of a phenomenological expression (1) for the free energy G as a Taylor expansion in the order parameter P written by L. Landau [13].

$$\Delta G(P,T) = a\cdot(T-T_0)\cdot P^2 + b\cdot P^4 + c\cdot P^6 - E\cdot P, \quad (1)$$

where P is electrical polarization; T – temperature; E – electric field; a, b & c – modeling coefficients. Coefficients a, b & c were selected to best meet the experimental data obtained earlier. With the use of the expression (1), we can find the values of the electric polarization at different values of the temperature and the electric field by solving the equations (2) and selecting only those solutions that meet the minimum ΔG.

$$\frac{\partial \Delta G(P,T)}{\partial P} = 0. \tag{2}$$

Figure 1 presents the dependence of electric polarization versus temperature and electric field. Figure 2 presents the first derivative of electric polarization *dP(T, E)/dT* versus temperature and electric field. Anomalous behavior of the heat capacity C_a near the phase transition can be obtained by equation (3):

$$C_a(T,E) = -T\frac{\partial \Delta G(T,E)}{\partial T}. \tag{3}$$

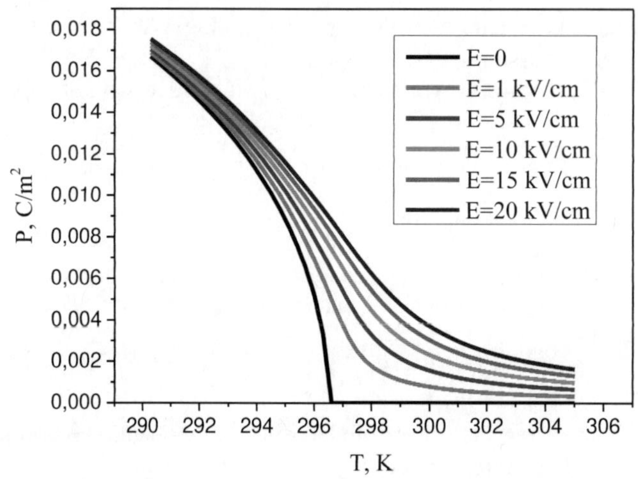

Figure 1. The dependence of electric polarization versus temperature and electric field.

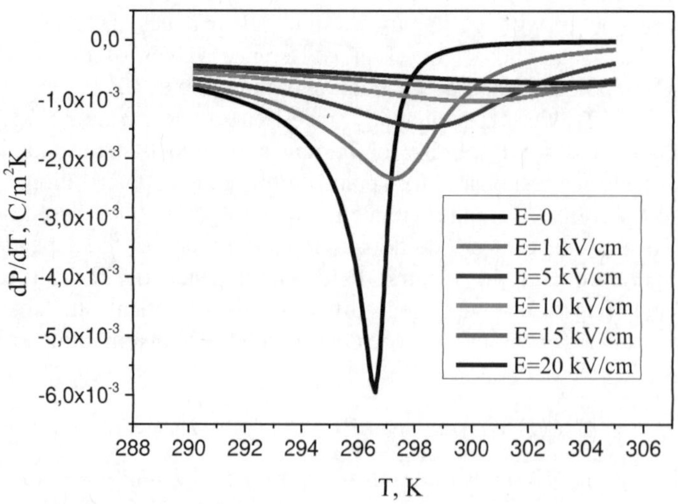

Figure 2. The first derivative of electric polarization on temperature versus temperature and electric field.

Figure 3 presents the dependence of anomalous heat capacity versus temperature and electric field. Data on the electric polarization and anomalous behavior of specific heat obtained before we can use to calculate the value of electrocaloric response due to polarization and depolarization. The response is calculated by solving differential equation (4):

$$\rho \cdot (C(T) + C_a(T,E)) \cdot \frac{\partial T}{\partial t} = -T \frac{\partial P(T,E)}{\partial T} \frac{\partial E}{\partial t}, \tag{4}$$

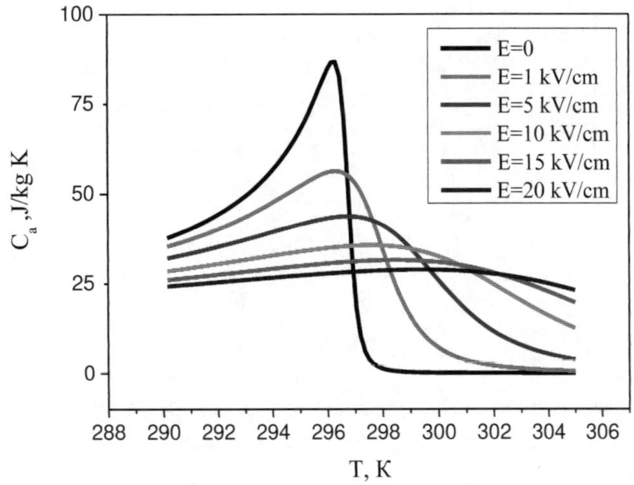

Figure 3. The dependence of anomalous heat capacity versus temperature and electric field.

where C - heat capacity, ρ – density of ferroelectric. Figure 4 presents the dependence of the electrocaloric response due to polarization and depolarization at the value of electric field 20 kV/cm. As we can see, we have a reasonable agreement between the theoretical models and experimental data in the required temperature range. Experimental data were obtained on bulk ceramic samples of a solid solution of strontium barium titanate with Curie temperature T_C=296.7K ($0.6BaTiO_3$-$0.4SrTiO_3$). Samples were prepared according to standard ceramic high-temperature technology and were disks with a diameter of 10mm and a thickness of 0.5mm. The electrocaloric response during polarization and depolarization was recorded by a microthermocouple. The magnitude of the applied electric field was 20kV/cm. The measurements were carried out only in the region of interest to us, namely, above the Curie temperature, where the electrocaloric response during depolarization is greater than the electrocaloric response during polarization. As can be seen from the approximation, the maximum of the electrocaloric responce under polarization and depolarization is shifted with respect to the Curie temperature. The maximum of the electrocaloric responce during depolarization is above the Curie temperature, correspondingly the maximum of the electrocaloric responce under polarization is below the Curie temperature. Which is in good agreement with the theory.In the area of 299 K the difference in the electrocaloric responses for polarization and depolarization is 4 mK for theoretical data and 5 mK for experiment.

IOP Conf. Series: Journal of Physics: Conf. Series **929** (2017) 012083 doi:10.1088/1742-6596/929/1/012083

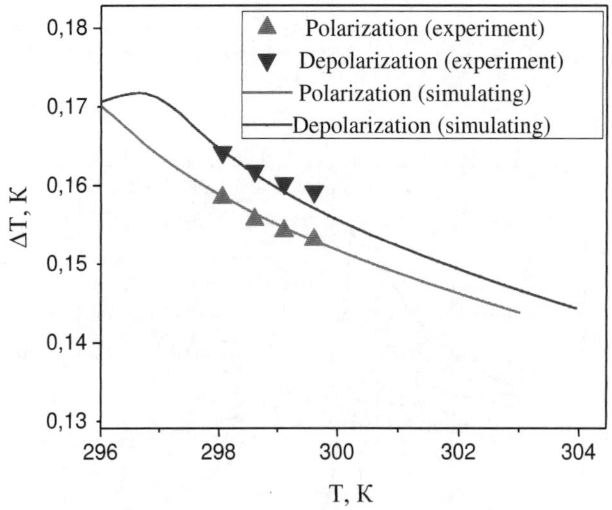

Figure 4. Dependence of the electrocaloric response due to polarization and depolarization at the value of electric field 20 kV/cm.

The above theoretical model we used for modeling layered structure consisting of two capacitive elements. The structure chosen for simulation simulates the heat distribution between the two thin ferroelectric layers in a multilayer film capacitor when supplying the control signal to them (Figure 5). The layer thickness was chosen to be 10 microns. Periodic square pulses of the electric field were supplied to the ferroelectric layers with different frequencies. The pulses for second layer were shifted by a half period. The pulses were cause periodic heating (cooling) of the electrocaloric elements and a redistribution of the temperature field along the structure. We will investigate a one—dimensional model under the assumption that the temperature changes only along the x axis. Thermal conductivity $\lambda(x)$ are assumed to be constant for each layer according to his material. In this case, the temperature distribution $T(x, t)$ along the line could be founding by solving the heat conduction equation (5):

$$\rho\cdot(C(T)+C_a(T,E))\frac{\partial T}{\partial t}=\frac{\partial}{\partial x}\lambda(x)\frac{\partial T}{\partial x}-T\frac{\partial P(T,E)}{\partial T}\frac{\partial E}{\partial t}, \tag{5}$$

which satisfies the initial and boundary conditions:

$$T(x,0)=T_0; \qquad -\lambda\frac{\partial T}{\partial x}\bigg|_{x=0,L}=0. \tag{6}$$

IOP Conf. Series: Journal of Physics: Conf. Series **929** (2017) 012083 doi:10.1088/1742-6596/929/1/012083

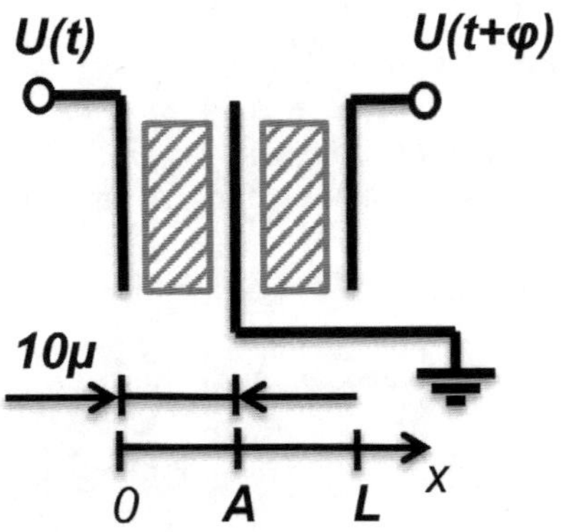

Figure 5. Schematic model of ferroelectric layers and electric connections.

In simulation temperature has been selected for the initial temperature is 299 K. The difference of electrocaloric responses for polarization and depolarization and the phase shift of control signals leads to a decrease in temperature between the layers. Figure 6 presents the results of the computer simulation of the temperature variation in the point A between the ferroelectric layers for pulses with frequencies 500 Hz, 1 kHz, 2 kHz and 5k Hz. As can be seen from the figure the velocity of cooling is 0.3 K/sec for 500 Hz, 0.97 K/sec for 1 kHz, 2.7 K/sec for 2 kHz and 9 K/sec for 5 kHz. Volumetric cooling capacity will be correspondingly 0.9 W/cm^3, 2.9 W/cm^3, 8.1 W/cm^3 and 27 W/cm^3. As we can see, although the cooling power increases with increasing frequency, but in reality will increase fundamental loss in the ferroelectric layers, which are not accounted for in the above model, what ultimately will limit the increase of cooling capacity.

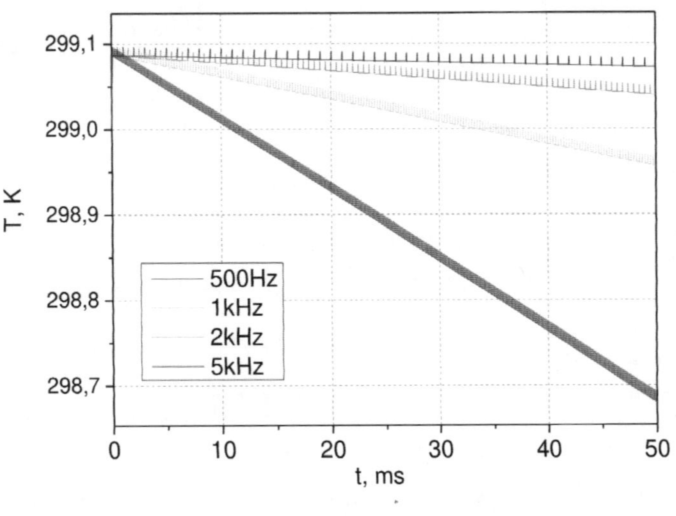

Figure 6. Dynamics of change in the temperature between the ferroelectric layers in point A.

Acknowledgments
The work was supported by RFBR (Grant № 15-32-20357 mol_a_ved) and by Ministry of Education and Science of Russian Federation (Project "Goszadanie"). A part of work was financially supported by the Government of Russian Federation, grant 074-U01. The research was partially supported by FASIE.

IOP Conf. Series: Journal of Physics: Conf. Series **929** (2017) 012083 doi:10.1088/1742-6596/929/1/012083

References

[1] Valant M 2012 *Prog. in Mater. Sci.* **57(6)** 980-1009

[2] Ozbolt M, Kitanovski A, Tusek J, Poredos A 2014 *Int. J. of Refrigeration* **40** 174-188

[3] Jia Y and Ju Y S 2012 *App. Phys. Lett.* **100(24)** 242901

[4] Gu H, Qian X, Li X, Craven B, Zhu W., Cheng A, and Zhang Q M 2013 *App. Phys. Lett.* **102(12)** 122904

[5] Guo D, Gao J, Yu Y J, Santhanam S, Slippey A, Fedder G K, Yao S C 2014 *Inter. J. of Heat and Mass Transfer* **72** 559-564

[6] Karmanenko S F, Pakhomov O V, Prudan A M, Starkov A S, Es'kov A V 2007 *J. Europ. Ceramic Soc* **27** 3109

[7] Es'kov A V, Karmanenko S F, Pakhomov O V, Starkov A S 2009 *Phys. of Solid State* **51(8)** 1574-1577

[8] Starkov A S, Karmanenko S F, Semenov A A, Es'kov A V, Semikin D, Hagberg J 2009 *Phys. of Solid State* **51(7)** 1510-1514

[9] Pakhomov O V, Karmanenko S F, Semenov A A, Starkov A S, Es'kov A V 2010 *Tech. Phys.* **55(8)** 1155-1160

[10] Karmanenko S, Semenov A, Dedyk A, Es'kov A, Ivanov A, Beliavskiy P, Pakhomov O 2014 New approaches to electrocaloric-based multilayer cooling *Eng. Mater.* **34** 183-223

[11] Es'kov A V, Belyavskii P Yu, Anokhin A S, Pakhomov O V, Semenov A A, Myl'nikov I L, Nikitin A A, Bui M T, Cherkasskii M A, Plotnikov V V 2016 *Tech. Phys.* **61(7)** 1112-1114

[12] Es'Kov A V, Belyavskiy P Y, Anokhin A S, Semenov A A, Pakhomov O V, Mylnikov I L, Nikitin A A, Krylov V A 2016 *J. of Phys.: Conf. Ser.* **741** 012163

[13] Landau L D 1937 *Zh. Eksp. Teor. Fiz.* **7** 19–32

IOP Conf. Series: Journal of Physics: Conf. Series **929** (2017) 012084 doi:10.1088/1742-6596/929/1/012084

Spectrum tuning in multi-junction solar cells measurements

S.A. Bogomolova, M.Z. Shvarts, N. Kh. Timoshina

Ioffe Institute, 26 Polytechnicheskaya, St.Petersburg, 194021, Russia

s_bogomolova@mail.ru

Abstract. This paper presents the technique of spectrum tuning in the multi-lamp solar simulator's workspace and accuracy (uncertainty) evaluation for measurement results of spectral irradiance in order to determine its suitability for GaInP/Ga(In)As/Ge multi-junction solar cells testing has been performed.

1. Introduction

At the research and development stage of multi-layer nanoheterostructures for multi-junction (MJ) GaInP/Ga(In)As/Ge solar cells (SCs), it is necessary to ensure matching the photocurrents for GaInP and Ga (In)As subcells of the entire MJ SC. For getting the optimum value of the MJ SC efficiency such matching should be at a maximum photocurrent, which could be achieved in the selected structure of GaInP/Ga(In)As/Ge MJ SC. It is known that the achievement of complete matching is extremely challenging, both in architecture (the choice of semiconductor materials of the multilayer structure) and technological (terms and conditions of the growth of the semiconductor layers in the formation of multi-junction structure) reasons. In measuring I-V characteristics of MJ SC on solar simulators (SSs), the photocurrent mismatch may arise by two reasons:

- non-optimal spectral response of the subcells, which results in the difference between values of photocurrents at a preset spectral irradiance formed by SS;
- dissimilarity of the spectral irradiance generated by SS from the standard one [1] leads to the subcells' photocurrent mismatch even for a MJ SC with initially photocurrent matched subcells.

In the paper, a procedure for tuning the spectral irradiance in a multi-lamp SS has been proposed, and accuracy estimation for photovoltaic parameters in measuring GaInP/Ga(In)As/Ge MJ SC has been performed at the condition of "blue-red ratio" variation in a spectral irradiance of the pulsed SS.

2. Technical and metrological characteristics of the pulsed solar simulator

The SS under consideration ensures both forming of irradiance with characteristics according to class A in correspondence with [1] and dynamic correction of the spectral irradiance. The picture of the pulsed SS is in figure 1. The irradiance source of SS is constructed on the base of two closely located xenon lamps, which allows tuning the "blue-red ratio" in the spectral irradiance due to the principle of optical absorption of irradiance of one lamp in the gas discharge plasma of another one. Each lamp generates a light pulse with a flat part duration of 1 ms. During this time, both the measurement of the spectral irradiance in the place of a SC location (by means of a high-speed spectroradiometer) and the record of SC I-V characteristic (by an analog-digital converter in supplying sweep voltage to a SC) can be performed. The time-dependences of the flash illumination intensity, voltage applied to a solar cell and current flowing through the SC are presented in figure 2.

Figure 1. Dual-lamp high power solar simulator (5000 X)

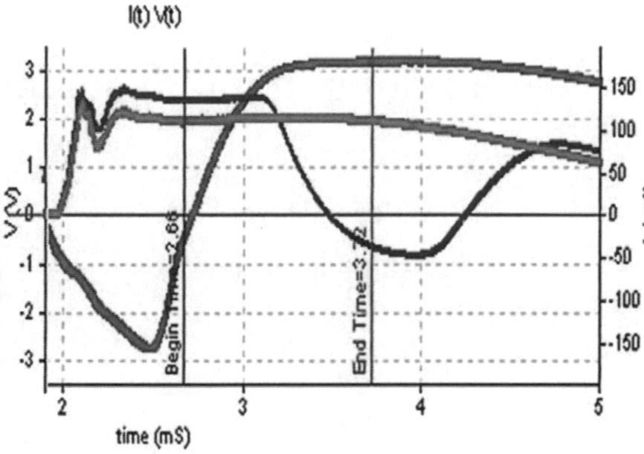

Figure 2. Time-dependences of the flash illumination intensity (red); voltage applied to a SC (green) and current flowing through the SC (blue).

In varying supply voltage on the lamps the gradual variation of the irradiance intensity in three spectral ranges is ensured, that allow adjusting precisely the spectral irradiance with simultaneously maintaining the value of the spectral mismatch within acceptable limits in regard to class A [1].

The basic technical and metrological characteristics of the pulsed SS are presented in table 1.

Table 1. Basic technical and metrological characteristics of the pulsed SS

Characteristic	Value
Flat part duration of irradiance pulse (at deviations not greater than ± 2%), ms	1.0 – 1.2
Voltage applied to the SC from an active load, V	+/- 12
Current being registered, A	0 – 10
Accuracy in measuring voltage and current, %	± 0.2
Duration of recording the dark and light I-V characteristics, ms	1
Duration of measuring one pair of "current-voltage" values, μs	8 – 10
Irradiance (spectrum AM1.5D), W/m^2	1000

3. Procedure for spectrum tuning in measuring I-V characteristics of MJ SCs

The theoretical basis of the proposed procedure is a well-known fact about the high sensitivity of the fill factor (FF) for I-V characteristic to photocurrent matching of the subcells: it reaches the minimum in the state of matched photocurrents [2-4]. The procedure for tuning the spectral irradiance in measuring I-V characteristics of MJ SC on SS contains the following steps:

1) Measurements of the spectral irradiance should be performed in varying voltage on simulator lamps (in varying the "blue-red ratio" in the spectrum) with using a high-speed spectroradiometer.

2) Plots of dependencies of real values and confidence interval for photocurrent density of MJ SC's subcells on the lamps voltages should be formed. The following expression should be used in calculating the current density:

$$J_{ph} = \int_{\lambda 2}^{\lambda 1} E(\lambda) \cdot SR(\lambda) \cdot d\lambda, \qquad (1)$$

where *E(λ)* – spectral irradiance of the SS; *SR(λ)* – dependence of the spectral response on wavelength λ for subcells of MJ SC.

3) Range of voltages on the lamps confined by the projections of cross-section points of the confidence interval's lines for photocurrent density of MJ SC's subcells should be determined. This range determines the permissible variation of the "blue-red ratio" in the simulator's spectrum, at which the required subcells' photocurrent matching or mismatching is ensured.

4) Plot of the dependence of the MJ SC efficiency on voltage on the lamps should be formed, and the expanded uncertainty caused by the ambiguity in choosing voltages on the lamps ("blue-red ratio" in the simulator's spectrum) should be estimated.

At the initial step of the procedure realization, measurements of the spectral irradiance in varying voltage on the simulator's lamps (varying the "blue-red ratio" in the simulator's spectrum) were carried out (figure 3) with using high-speed spectroradiometer "Flash Measurement System UV-VIS-IR", for which the traceability of spectral irradiance measurement results to SI units was ensured. The main element of the spectroradiometer's construction is a multi-element photodetector (CCD matrix), cells of which detect spectral lines of the incident irradiance converting them into a current signal.

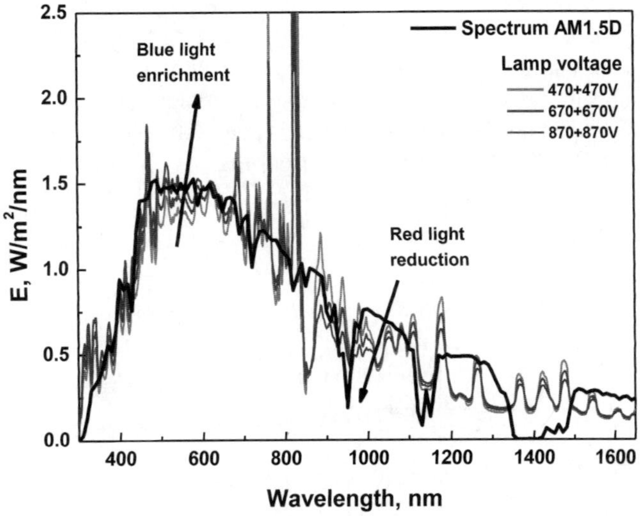

Figure 3. Spectral irradiance in the simulator's workspace at changing voltage on the lamps (a variation of "blue-red ratio" in the simulator's spectrum) and standard spectrum AM 1.5

The spectral irradiance data produced by spectroradiometer could be determined as follows [5]:

$$Y_{SS}(\lambda_i) = \frac{Y_{SS,i}}{R_i \cdot k_L(\lambda_i) \cdot k_T(\lambda_i)}, \qquad (2)$$

where $Y_{SS,i}$ – signal of the *i*-th pixel of the CCD matrix in measuring the spectral irradiance; R_i – spectral photoresponce of the *i*-th pixel of the CCD matrix being determined in calibrating the spectroradiometer sensitivity to the spectral irradiance; $k_L(\lambda_i)$ – correction factor allowing for the spectroradiometer signal nonlinearity; $k_T(\lambda_i)$ – correction factor allowing for the effect of temperature on the spectral irradiance measurement result.

The uncertainty budget of the spectral irradiance measurement result is presented in table 2.

At the next step, spectral dependences of the external quantum yield (spectral response) of the MJ SC's subcells were recorded on an installation elaborated in the Laboratory for Photovoltaic Converters of the Ioffe Institute [4]. The principle of these measurements bases on comparing photocurrent values of the tested and reference SCs [7].

Table 2. Uncertainty budget of the spectral irradiance measurement result (type *A* and type *B* evaluation methods [6])

Value X_i	Standard uncertainty $u(x_i)$	Probability distribution	Sensitivity coefficient, c_i	Contribution to the standard uncertainty $u_i(y)$, %
$Y_{SS,i}$	0.600 %	Gaussian	1.0	0.11
R_i	0.520 %	Gaussian	1.0	0.27
$k_L(\lambda_i)$	0.580 %	Rectangular	1.0	0.012
$k_T(\lambda_i)$	0.576 °C	Rectangular	0.05 %/°C	0.029
$Y_{SS}(\lambda_i)$				0.28

As a result of statistical analysis of measurement data, standard uncertainties (type *A* [6]) were obtained, which characterize similarity of results in measuring the spectral response at a definite photocurrent signal: u_{SR} = 0.18 %.

On the base of experimental dependencies SR(λ) and E(λ), estimates of the photocurrent densities for the subcells of MJ SC were accounted in correspondence with the expression (1), and the dependencies of the subcells' photocurrent densities on voltage of the lamps were constructed (figure 4). The spectral dependences of the external quantum yield (spectral response) of the GaInP/Ga(In)As/Ge MJ SC's subcells are depicted in figure 5.

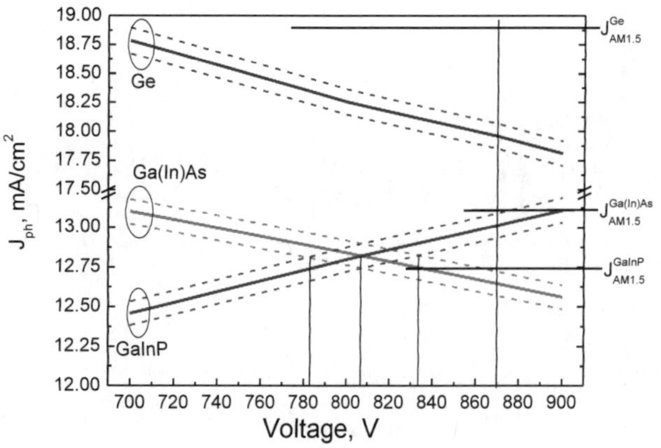

Figure 4. Dependence photocurrent density values (J_{ph}) on voltage on the lamps (dashed lines indicate the boundaries of the confidence intervals for photocurrent density values; $J^{GaInP}_{AM1.5}$, $J^{GaAs}_{AM1.5}$, $J^{Ge}_{AM1.5}$ are the subcells' photocurrent density values at standard spectrum AM 1.5D)

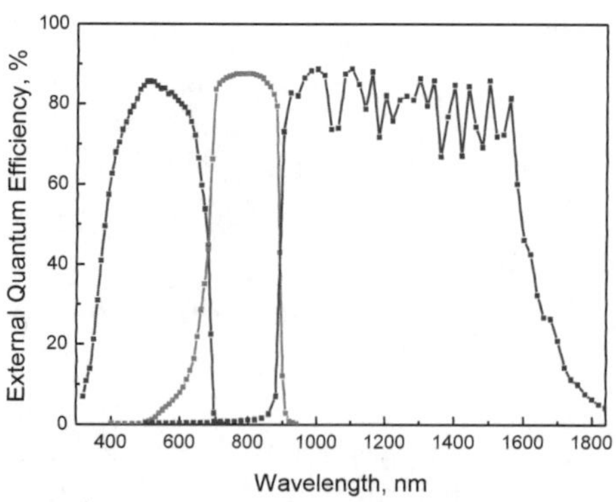

Figure 5. Spectral dependence of External Quantum Efficiency for GaInP/Ga(In)As/Ge MJ SC's subcells

Then positions of confidence interval's boundaries for subcells' photocurrent density were determined by the results of calculation of the combined expanded uncertainty in correspondence with the expression:

$$U_{J_{ph}} = \pm 2\sqrt{u_{E(\lambda)}^2 + u_{SR}^2}, \quad (3)$$

where u_{SR} – standard uncertainty of the subcell's spectral response; $u_{E(\lambda)}$ – standard uncertainty of the spectral irradiance for the multi-lamp SS.

The estimated combined expanded uncertainty is $U_{Jph} = \pm$ 0,66 %. The calculated boundaries of the confidence intervals are shown in figure 4. Projections of the cross-section points of the confidence interval's lines have allowed determining the range of lamps voltages (variation of the "blue-red ratio" in the spectral irradiance of SS), at which a designed or required subcells' photocurrent mismatch was ensured with conforming accuracy.

At the final step of the procedure realization, measurements of the GaInP/Ga(In)As/Ge MJ SC output photovoltaic parameters have been carried out in varying "blue-red ratio" in the simulator's spectrum, and dependence of the FF versus lamp's voltage were constructed (figure 6). At every change of the "blue-red ratio" in the spectral irradiance, a control of the irradiance in the simulator's workspace by the GaAs monitor cell was performed.

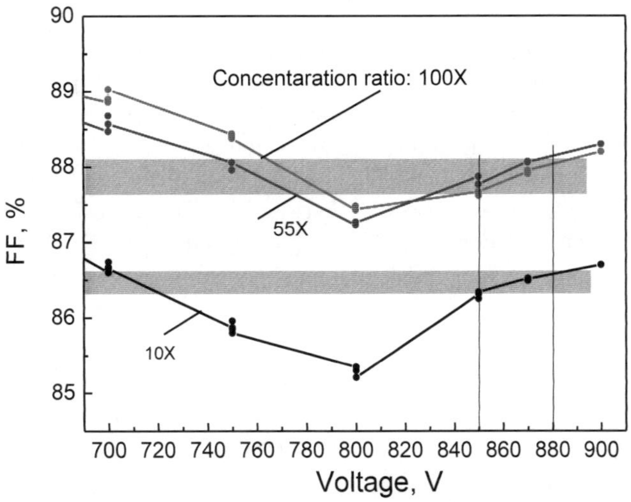

Figure 6. Dependence of the fill factor on voltage on the simulator's lamps

4. Conclusion

Calculation results have shown that for considered type of MJ SC the equalization of photocurrents in GaInP- GaAs subcells pair is expected to be at spectral irradiance formed by dual-lamp pulsed SS at approximately 810 V on the lamps (figure 4).

On the basis of experimental I-V recording the minimum value of the fill factor, indicating the photocurrent matching condition, is reached at lamp voltage of 800 V (figure 6). Since both obtained values are within the tolerance of the lamp voltage variations at accurate tuning the spectral irradiance (0 – 40) V. So proposed technique of spectral irradiance tuning allows reaching the required photocurrents mismatch for MJ SC with high precision.

The presented example shown that for MJ SC under study the photocurrent mismatch corresponding to AM1.5D standard test condition is proposed to be at (870 ± 20) V on the solar simulator's lamps and, thus, the accurate cell efficiency could be determined through I-V recording at this simulator's spectrum.

IOP Conf. Series: Journal of Physics: Conf. Series **929** (2017) 012084 doi:10.1088/1742-6596/929/1/012084

Acknowledgments
The reported study was funded by RFBR according to the research project № 16-29-03216.

The authors wish to thank Prof. V. M. Andreev for support of this work. Special thanks to V.R. Larionov, D. A. Malevskiy and P. V. Pokrovskiy for simulator development.

References
[1] IEC 60904-9:2007. Photovoltaic devices. Pt. 9: Solar simulator performance requirements. 2nd edn.
[2] Adelhelm R, Berger D 2003 Requirements for a large area solar simulator regarding the measurement of MJ solar cells *Photovoltaic Energy Conversion: Proc. 3rd World Conf. (Osaka)* pp 821-824
[3] McMahon W E e. a. 2008 Fill factor as a probe of current-matching for GaInP2/GaAs tandem cells in a concentrator system during outdoor operation *Prog. Photovolt: Res. Appl.* vol 16. pp 213-224
[4] Rumyantsev V D, Andreev V M, Larionov V R, Malevskiy D A, Shvarts M Z 2007 Indoor characterization of multijunction concentrator cells under flash illumination with variable spectrum *Solar Concentrators for the Generation of Electricity or Hydrogen: Proc. 4th Int. Conf.* (El Escorial) pp 277-280
[5] Sperfeld P, Dzafic D, Plag F, Haas F, Kröber C, Albert H, Nevas S, Pape S 2012 Usability of compact array spectroradiometers for the traceable classification of pulsed solar simulator *Proc. 27th European Photovoltaic Solar Energy Conference and Exhibition* pp 3060-65.
[6] ISO/IEC Guide 98-3: 2008 Uncertainty of measurement: Part 3. Guide to the expression of uncertainty in measurement (GUM: 1995).
[7] IEC 60904-8:2014. Photovoltaic devices - Part 8: Measurement of spectral responsivity of a photovoltaic (PV) device. 3rd edn.

IOP Conf. Series: Journal of Physics: Conf. Series **929** (2017) 012085 doi:10.1088/1742-6596/929/1/012085

Mechanism of domain wall interaction with electric field in iron garnet films

A. S. Sergeev

Faculty of Physics, M. V. Lomonosov Moscow State University, Moscow, 119991 Russia

E-mail: a.sergeev@physics.msu.ru

Abstract. Domain walls in ferrimagnetic iron garnet films move under the influence of the applied static electric field and thus show the magnetoelectric behavior. We study theoretically two mechanisms proposed to describe this phenomenon. First one is based upon the inhomogeneous magnetoelectric interaction and the second one relies on the change of magnetic anisotropy parameters caused by electric field. We show that neither mechanism can be excluded by the general qualitative arguments.

1. Introduction

Physical quantities of electric and magnetic nature behave differently under space inversion and time reversal operations. Electric field (polarization) can be associated with charge displacements, so it changes sign under space inversion and remains intact under time reversal. Magnetic field (magnetization), which is classically related to the circular motion of point charge shows the opposite behavior.

Maxwell equations relate quantities of different symmetry by means of spatial and temporal derivatives. At first sight it seems that the static coupling between electricity and magnetism is impossible. While this is true in vacuum, solid state systems provide numerous counter examples. These are known as *magnetoelectric* (ME) materials.

The simplest kind of ME effect is called linear and homogeneous. If the crystal cell breaks both inversion symmetry (by the crystal structure) and time-reversal symmetries (due to magnetic order), it is possible to have ME term f_{ME} in the free energy expansion given by $f_{ME} = \alpha_{ij} E_i H_j$. A well-known example of the material with ME coupling of this type is Cr_2O_3, which has been being studied since 1960s. Interestingly, some of the microscopic details of the ME interaction in this compound still remain unclear [1].

Another way to obtain ME coupling in theory is to use the term that includes magnetic parameter twice together with the electric one and the spatial derivative. Indeed, such a combination is invariant under both the inversion and time-reversal symmetries, and thus can be present as a part of the free energy. For instance, one can arrange these quantities in the following manner:

$$f^i_{ME} = \gamma_{ijkl} P_i M_j \partial_k M_l. \quad (1)$$

This combination corresponds to the case of *inhomogeneous* ME effect [2] . Electric polarization may appear as a consequence of the presence of the spatial spin structure that breaks inversion symmetry.

One can also consider the term with the spatial derivative acting on the electric parameter (external electric field in this case)[3]:

$$f^a_{ME} = \beta_{ijkl} M_i M_j \partial_k E_l \tag{2}$$

which can be thought as the change of magnetic anisotropy parameters under the influence of the spatially inhomogeneous electric field.

Experimentally it was found that domain walls in ferrimagnetic iron garnet films bend and shift out of the equilibrium position when electric field created by sharp non-magnetic electrode is applied [4]. This situation includes non-uniform magnetization distribution and highly inhomogeneous electric field as well, so in principle both f^i_{ME} and f^a_{ME} contributions can play a certain role [3].

The aim of the present work is to determine which mechanism underlies the magnetoelectric interaction in the iron garnet films. We begin with brief review of experimental results in Section 2 and then analyze proposed mechanisms in Section 3.

2. Magnetoelectric behavior of domain walls in iron garnet films

Equilibrium, room temperature state of magnetization vector distribution in the iron garnet films studied in [4] is stripe domain structure, characterized by domain width of approximately 20μm. Static electric field created by sharp (tip radius is about 5μm) non-magnetic electrode acts on the nearby domain wall differently depending on the sign of the applied voltage. Domain walls attract to the electrode if the voltage is positive and repel otherwise. Total amount of the wall displacement near the electrode is approximately 1 μm at the voltage value of $500V$.

In the modified setting, additional in-plane magnetic field is applied which deforms domain wall micromagnetic structure but doesn't destroy domains. Magnetic field is directed perpendicular to the domain walls [5]. This changes domain wall behavior significantly. First, the displacement increases up to tens of micrometers and involves several domain walls. Secondly, the direction of displacement depends on the direction of magnetic field. Thirdly, the direction of the shift changes from one wall to another.

One should also note that the presence of the magnetoelectric properties of the domain walls relies heavily on the crystallographic orientation of the film. These effects were only observed in the films with "low-symmetry" substrate orientation such as (210) or (110), and never in (111) films. Details of the experiment can be found elsewhere [6].

3. Possible mechanisms of ME interaction

3.1. Inhomogeneous ME effect

For the crystal of cubic symmetry the coupling of the type (1) takes the following form [7]:

$$\vec{P} = \gamma\chi_e[(\vec{M}\cdot\nabla)\vec{M} - \vec{M}(\nabla\cdot\vec{M})] \tag{3}$$

where χ_e is the dielectric susceptibility.

It follows that the domain wall can possess the bulk electric polarization and associated surface electric charges, depending on its magnetization vector distribution. For instance, domain wall of the Bloch type in which magnetization vector rotates in the plane of the wall does not break inversion symmetry and thus makes the appearance of polarization impossible (Eq. (3) gives $\vec{P} = 0$ in this case).

In order to calculate the force of the electric field acting on the domain wall we take the derivative of the energy of the interaction $w(\vec{M}, \vec{E}) = -\vec{P}(\vec{M}) \cdot \vec{E}$ with respect to the position x_0 of the domain wall:

$$F_x = -\frac{\partial}{\partial x_0}\left(\int_a^b w(\vec{M}, \vec{E})\, dx\right) \tag{4}$$

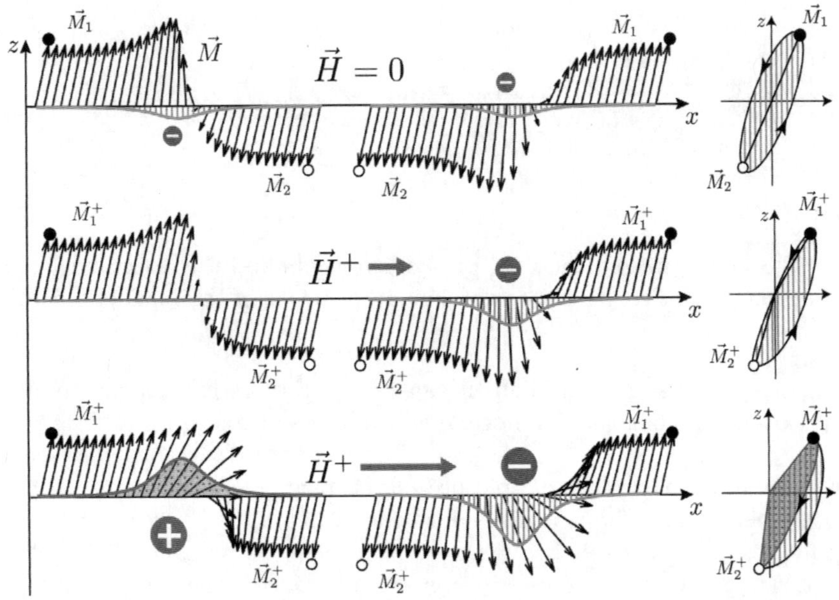

Figure 1. Typical magnetization vector distributions for two neighboring domain walls in the absence (upper panel) and in the presence of the medium (middle panel) and strong (lower panel) magnetic field. Trajectories of the magnetization vector in the (x, z) plane for both domain walls are shown on the right.

where x axis is directed perpendicular to the wall's plane and a, b stand for the coordinates of the middle points of the neighboring domains. Here we assume that displacement of the domain wall occurs without change of its micromagnetic structure. Our model is one-dimensional which is justified by the fact that effective field of magnetic anisotropy in the films under consideration is stronger then magnetostatic stray fields and suppresses spatial variation of magnetization in the plane of the domain wall (y, z). In three dimensions, the force F_x is to be thought of as the pressure on the domain wall's surface.

We suppose that variation of the electric field on scale of the domain wall's width is negligibly small. Then the last expression takes the form:

$$F_x = \left(\int_{DW} \gamma \chi_e [(\vec{M}\nabla)\vec{M} - \vec{M}(\nabla\vec{M})]\, dx \right) \left(\frac{\partial \vec{E}}{\partial x} \right)_{x=x_0}, \tag{5}$$

where DW means region containing domain wall.

Linear density of the bound electric charge, or the z-component of the vector appearing as a result of the integration here has a simple visual interpretation. It can be shown that this quantity equals the area swept out by the magnetization vector in the (x, z) plane (Fig. 1, right). For the domain wall of Bloch type this area equals zero. Magnetic anisotropy of the films determines complex form of the magnetization vector distribution. We are interested, however, only in the area, which is a topological property in a sense that it does not depend on the details of the geometry of $\vec{M}(x)$.

Thus the force depends on the electric field direction and on the micromagnetic structure of the domain wall. Once the latter is deformed by the external magnetic field, the response of the domain wall to the electric field changes together with its electric polarization.

One can assume that initially all the domain walls have uniform sense of rotation. It is reasonable because the inversion symmetry of the sample is broken by the substrate and by the

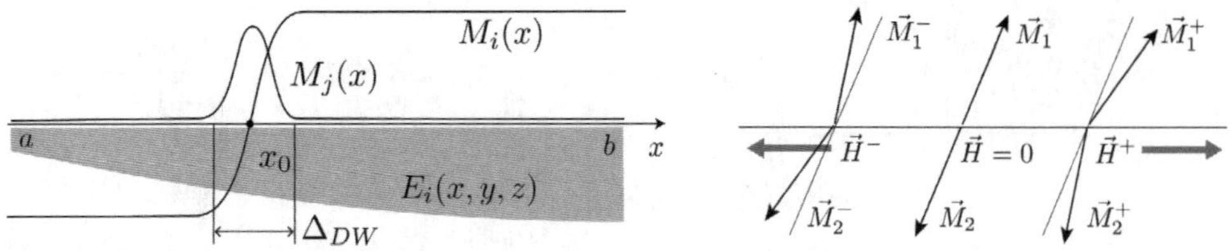

Figure 2. Profiles of the magnetization vector components and the electric field (left) and the tilting of the magnetzation vector in the domains in presence of the magnetic field (right).

very process of the epitaxial growth. Thus it can lead to the presence of the "built-in" uniform background polarization along the film normal which forces all the walls to have the same sense of rotation.

With this assumption the plausible picture of the magnetization vector path deformation gives the correct behavior of the domain wall in the presence of the magnetic filed. As shown in Fig. 1, applying of the magnetic field leads eventually to the relatively large electric charges of opposite sign at the neighboring walls (lower panel). During the process, charge of the one of the domain walls will pass through zero value (middle panel) — the fact that is observed experimentally as the vanishing of the displacement of the wall at some value of the magnetic field $\vec{H}$.

3.2. Change of the anisotropy parameters

Now we consider the energy of the from given by the Eq. (2) as a cause of the force acting on the domain wall. As in the previous case, we differentiate the interaction energy with respect to the position of the domain wall:

$$F_x^{kl} = -\frac{\partial}{\partial x_0}\left(\int_a^b \beta_{kl}(\vec{M}(x))\,\frac{\partial E_l}{\partial x_k}\,dx\right) \tag{6}$$

where we denote $\beta_{kl}(\vec{M}) \equiv \beta_{ijkl}M_iM_j$.

If one takes into account only those kl components that lead to a reasonable force direction (the force does not depend on the side the electrode approaches the wall from etc.), we are left with two components, F_x^{zz} and F_x^{xz}.

Physically, change of the anisotropy constants creates potential landscape for any magnetic texture which contains magnetization vector that does not lie in the easy plane or the easy axis (see Fig. 2). Since in the present case this change can be caused by the inhomogeneous electric field, one can expect that any magnetic texture will be attracted to, or repelled from the electrode without any reference to the magnetization vector distribution. This explains the behavior of the domain walls in the absence of the external magnetic field and can be described by the following equation:

$$F_x^{zz} = -\left[\int_{DW} \beta_{zz}(\vec{M}(x))\,dx\right]\frac{\partial}{\partial x}\left(\frac{\partial E_z}{\partial z}\right)_{x=x_0}. \tag{7}$$

Since the expression is quadratic in M_i, domains do not contribute to the force, because one has $\vec{M}_1 = -\vec{M}_2$ for the magnetization of the nearby domains.

However, in the presence of the external magnetic field, magnetization direction in domains is tilted and the last equality does not hold. Instead, one has $\vec{M}_1^+ = -\vec{M}_2^-$ and $\vec{M}_2^+ = -\vec{M}_1^-$, which can be deduced from the fact that anisotropy energy is quadratic in magnetization vector components (for the notation $\vec{M}^{\pm}$, see figures). Thus domain wall displacement can lower the energy of the system because of the change of the volumes of the domains divided by this wall.

In the case of $\vec{H} = \vec{H}^{+}$ we have:

$$F_x^{xz} = (\beta_{xz}(\vec{M}_2^{+}) - \beta_{xz}(\vec{M}_1^{+})) \left(\frac{\partial E_z}{\partial x}\right)_{x=x_0} \tag{8}$$

One can see that the last expression changes sign both on the reversal of the direction of $\vec{H}$ (i.e. $\vec{M}^{+} \leftrightarrow \vec{M}^{-}$) and for neighboring domain walls ($\vec{M}_1 \leftrightarrow \vec{M}_2$). It follows that the hypothesis of the electric-field-induced change of the anisotropy constants allows one to explain, at lest in principle, all the experimental results.

4. Conclusion

Domain walls in the iron garnet films show the rich variety of magnetoelectric phenomena, but it is not clear which mechanism underlies this behavior, even at the phenomenological level. We have considered two possible mechanisms, inhomogeneous magnetoelectric interaction and changing of the magnetic anisotropy constants under the influence of the electric field. It is shown that both mechanisms can, in principle, lead to the observed behavior of the domain walls.

The first mechanism based on the inhomogeneous ME interaction is sensitive to the domain wall internal micromagnetic structure and requires that all the walls have the same chirality in the state with no magnetic field applied. The second one works without any assumptions apart from that the values of $\beta_{zz}(\vec{M})$ and $\beta_{xz}(\vec{M})$ are significantly large.

To outline directions of the further study, one may note that our setting was very general. For instance, it did not include important information about crystallographic orientation of the film. This may put some restrictions on the particular components of the material tensors describing both mechanisms.

Acknowledgments

Author thanks A. P. Pyatakov, T. B. Kosykh and E. P. Nikolaeva for fruitful discussions. Support of RFBR grant No. 16-32-00840 mol_a is gratefully acknowledged.

References

[1] Malashevich A, Coh S, Souza I and Vanderbilt D 2012 *Physical Review* B **86** 1–7 (*Preprint* arXiv:1207.5873v2)
[2] Bar'yakhtar V G, L'vov V A and Yablonskii D A 1983 *JETP Lett.* **37** 673
[3] Arzamastseva G V, Balbashov A M, Lisovskii F V, Mansvetova E G, Temiryazev A G and Temiryazeva M P 2015 *Journal of Experimental and Theoretical Physics* **120** 687–701
[4] Logginov A S, Meshkov G A, Nikolaev A V, Nikolaeva E P, Pyatakov A P and Zvezdin A K 2008 *Applied Physics Letters* **93** 182510
[5] Pyatakov A P, Sechin D A, Sergeev A S, Nikolaev A V, Nikolaeva E P, Logginov A S and Zvezdin A K 2011 *EPL (Europhysics Letters)* **93** 17001
[6] Pyatakov A P, Sergeev A S, Nikolaeva E P, Kosykh T B, Nikolaev A V, Zvezdin K A and Zvezdin A K 2015 *Phys. Usp.* **58** 981–992
[7] Mostovoy M 2006 *Physical Review Letters* **96** 067601

IOP Conf. Series: Journal of Physics: Conf. Series **929** (2017) 012086 doi:10.1088/1742-6596/929/1/012086

Magnonic crystals based on ferrite and multiferroic periodic structures

A V Drozdovskiy, I A Ustinova, A A Nikitin, V V Vitko, A B Ustinov

Department of Physical Electronics and Technology, St. Petersburg Electrotechnical University, 197376 St. Petersburg, Russia

E-mail: drozdovskiy@gmail.com

Abstract. A comparison of spectrum characteristics of ferrite and multiferroic magnonic crystals has been made for the first time. It is shown that ferrite-ferroelectric magnonic crystal band gaps shift in the low frequencies against ferrite magnonic crystal. Theoretical and experimental results are in good agreement.

1. Introduction

Recent years an increased interest for investigation of spin waves propagating in spatially periodic magnetic film waveguide (the so-called magnonic crystals) takes place [1,2]. A feature of the spin wave spectrum of such structures is the presence of band gaps that appear due to Bragg reflection.

Such structures can be made of ferrite films [3,4] or of ferrite-ferroelectric structures [5,6

]. In the latter case, their dispersion characteristics can be controlled by both magnetic and electric field. It is assumed that such structures can become the basis of the functional microwave (MW) devices of the new generation. At the same time, a survey of the literature shows that the ferrite-ferroelectric periodical structures were studied not enough.

The aim of the present work is an investigation of the performance characteristics of the multiferroic magnonic crystals and their comparison with ferrite-film magnonic crystals.

2. Theoretical investigation

Investigated ferrite-ferroelectric structure is shown in figure 1. It consists of an yttrium iron garnet (YIG) film. A ferroelectric slab can be placed on the top of a periodic magnetic structure. Ferroelectric slab is assumed to be in a paraelectric phase with a high dielectric constant. This assumption is correct for barium strontium titanate (BST) ceramic of the composition $Ba_{0.5}Sr_{0.5}TiO_3$ at the room temperature. Spin waves (SWs) can propagate in the single ferrite magnonic crystal. Hybrid spin-electromagnetic waves (SEWs) propagate in the layered periodic ferrite-ferroelectric structure.

Consider the development of the electrodynamic model of the structure. The multiferroic structure is assumed to be magnetized by uniform magnetic field H along the z axis, and the waves runs in the direction perpendicular to the direction of the x axis (Fig. 1). This field orientation corresponds to propagation of quasi-surface SWs or SEWs [5] with the maximum of the field distribution occurring on the upper interface. We do not take into account existence of the dielectric substrate because it has a low dielectric constant about 10.

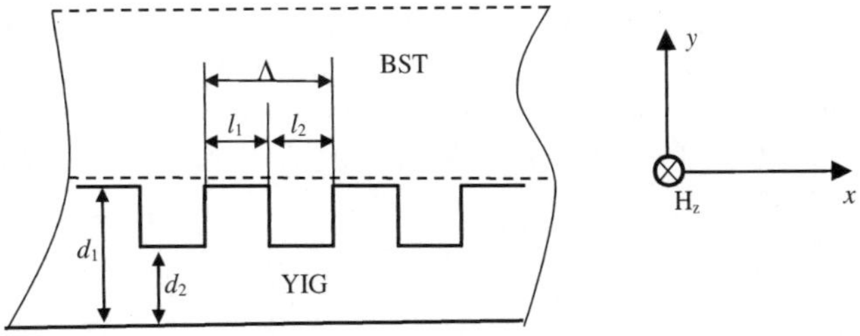

Figure 1. Schematic representation of the magnonic crystal based on ferrite-ferroelectric structure.

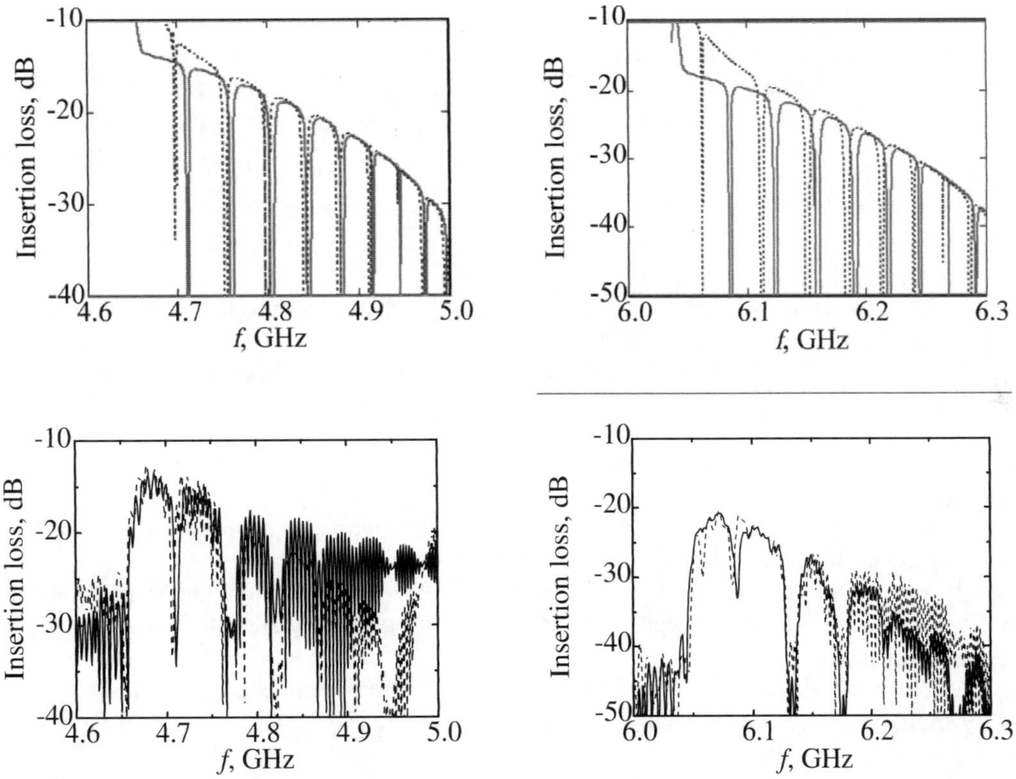

Figure 2. The theoretical (top) and experimental (bottom) transmission characteristics of magnonic crystal (solid line) and multiferroic magnonic crystal (dotted line). Left graphs correspond $H=1001$ Oe, right graphs correspond to $H=1450$ Oe.

The theoretical model is developed in two stages. In the first stage we derive the dispersion equation and calculate the dispersion characteristics of the SWs in free ferromagnetic film [8] and the SEWs in the homogeneous ferrite-ferroelectric waveguiding structure [7]. As a second step we calculate the dispersion relation of the SEWs in periodic ferrite-ferroelectric structure. We used the following formula obtained by the method of coupled waves [9]:

$$\cos(K\Lambda) = \cos(k_1 l_1)\cos(k_2 l_2) - \frac{k_1^2 + k_2^2}{2k_1 k_2} \cdot \sin(k_1 l_1)\sin(k_2 l_2), \quad (1)$$

where K is a Bloch wave vector, Λ is a period of the structure, k_1 и k_2 are wave numbers of SEWs propagating in thin and thick sections of the structure with d_1 and d_2, respectively, l_1 and l_2 are the distances passed by the wave in the thin and thick sections of the structure (see Figure 1). Finally, the amplitude-frequency characteristics (AFC) of the multiferroic periodic structure are calculated by following formula:

$$H = 20\lg(e^{-\alpha x}), \quad (2)$$

where $\alpha = \left|\frac{dk(\omega)}{dH}\right|\Delta H + \varepsilon_d \left|\frac{dk(\omega)}{d\varepsilon_d}\right|\tan\delta$ is a spatial damping decrement of the SWs and SEWs, correspondingly [10], ΔH is the half-width of the curve of the ferromagnetic resonance, $\tan\delta$ is a tangent of dielectric losses of the ferroelectric layer.

Upper graphs in the figure 2 demonstrate theoretical results. The calculation was carried out for the ferrite layer having parameters corresponding to single-crystal YIG film: $4\pi M_s = 1760\,\mathrm{G}$, d_1=5.7 μm, d_2=4.8 μm, $\varepsilon_f = 14$, $\Delta H = 0.5\,\mathrm{Oe}$. Parameters of the ferroelectric layer corresponds to the ceramic solid-state solution of $Ba_{0.5}Sr_{0.5}TiO_3$: $\varepsilon_d = 1200$, L_d=1000 μm, $\tan\delta = 10^{-2}$. Period of the structure Λ was equal 400 μm. The width of the groove l_2 was 50 μm. It is seen that the attachment of the ferroelectric slab to the surface of the YIG film magnonic crystal leads to the frequency shift of the band gaps. The increase of magnetic field leads to increase of the shift value. This is due to the existing of the region of a strong dispersion of the SEWs which corresponds to the maximum hybridization between the spin waves in the ferrite film and the electromagnetic waves in the ferroelectric slab.

3. Experiment

To verify the theoretical results, we carried out an experiment with a magnonic crystal formed from a YIG film by chemical etching. The fabricated sample of the magnonic crystal had a width of 2 mm and a length of 60 mm. The YIG film with thickness d_1=5.8 μm and saturation magnetization of 1750 G was grown by liquid-phase epitaxy on a gadolinium-gallium garnet (GGG) substrate. The ferromagnetic resonance line-width ΔH was 0.5 Oe. To avoid possible unwanted reflected spin waves, the edges of the MC were cut at an angle less than 45^0. The experimental periodic structure had a geometry similar to that in the calculations, namely a period Λ of 400 μm, a groove length of 50 μm, and a groove depth of 1.1 μm.

A delay line structure was used for the measurements. Spin waves were excited and received using short-circuited input and output microstrip antennas having a length of 2 mm and a width of 50 μm. The experimental device was placed between poles of an electric magnet in a uniform magnetic field. The field was directed within the plane of the magnonic crystal along the antennas.

The distance between the antennas was 8 mm. Thus, 20 periods of the structure under study were situated along the SW propagation path. Delay line design allows the slab of BST overlay on the entire region of the magnonic crystal. A Rodhe & Schwarz ZVA-40 vector network analyzer was used to measure the amplitude-frequency characteristics of the sample under study.

The experimental results are shown in the bottom graphs in the fig. 2. As is seen from the figure, the presence of the periodicity leads to the formation of band gaps both in magnonic crystal and in multiferroic structure. It can be readily seen that the theory sufficiently well describes the characteristic features of the stop bands, namely, their frequency position. Our numerical simulations did not take into account the SW excitation ant reception losses. Therefore, the theoretical curve shows a loss which is smaller than that observed in the experiment. The results are in good agreement with previous theoretical calculation [11, 12].

We used dispersion equation of spin-electromagnetic waves propagating in infinite in plane ferrite-ferroelectric structure. At the same time the experimental structure had a width of 2 mm. Discrepancies between the frequency shift of the stop bands in the experimental and numerical data could be explained by this assumption.

4. Conclusions

In conclusion, amplitude-frequency characteristics of ferrite and multiferroic magnonic crystals has been studied. It is shown that ferrite-ferroelectric magnonic crystal band gaps shifts in the low frequencies against ferrite magnonic crystal. The value of the frequency shift depends on magnetic field. Thus, magnonic crystals based on the planar multiferroic structure consisting of ferrite and ferroelectric layers can find a variety of applications.

Acknowledgement

The theoretical work was supported in part by Ministry of Education and Science of Russian Federation (Project "Goszadanie"). The experimental work was supported by Russian Science Foundation (Grant# 14-12-01296-P).

References

[1] Krawczyk M and Grundler D 2014 *J. Phys.: Condens. Matter* **26** 123202
[2] Nikitov S A, Kalyabin D V, Lisenkov I V, Slavin A N, Barabanenkov Yu N, Osokin S A, Sadovnikov A V, Beginin E N, Morozova M A, Sharaevsky Yu P, Filimonov Yu A, Khivintsev Yu V, Vysotsky S L, Sakharov V K and Pavlov E S 2015 *Physics-Uspekhi* **58**
[3] Vopson M M 2015 *Critical Reviews in Solid State and Materials Sciences*. **40** 223-50
[4] Sun N X and Srinivasan G 2012 *Spin* **2** 1240004
[5] Demidov V E, Kalinikos B A and Edenhofer E 2002 *J. Appl. Phys.* **91** 10007
[6] Ustinov A B and Kalinikos B A 2014 *Tech. Phys. Lett.* **40**. 568-70
[7] Ustinova I A, Nikitin A A and Ustinov A B 2016 *Tech. Phys.* 61. 473-476
[8] Damon R W and Eshbach J R 1961 *J. Phys. Chem. Sol.* **19**. 308-320
[9] Yariv A and Yeh P 1984 *Optical waves in crystals* (New York: Wiley)
[10] Ustinov A B, Kalinikos B A, Tiberkevich V S, Slavin A N and Srinivasan G 2008 *J. Appl. Phys.* **103** 063908
[11] Morozova M A, Grishin S V, Sadovnikov A V, Romanenko D V, Sharaevskii Yu P, Nikitov S A 2015 *IEEE Trans. on Magn.* **51(11)** 2802504
[12] Drozdovskii A V, Nikitin A A, Ustinov A B, Kalinikos B A 2014 *Tech. Phys.* **59(7)** 1032-1035

IOP Conf. Series: Journal of Physics: Conf. Series **929** (2017) 012087 doi:10.1088/1742-6596/929/1/012087

Entanglement between two Rydberg atoms induced by a thermal field

T S Mastyugina and E K Bashkirov

Department of General and Theoretical Physics, Samara National Research University, 34, Moskovskoye shosse, Samara, 443086, Russia

E-mail: tatyana_mastyugina@mail.ru

Abstract. We investigated two Rydberg atoms successively passing a vacuum or a thermal cavity taking into account the detuning. The atoms was assumed to be initially prepared in the Bell types entangled atomic states. Calculating the negativity we investigated the dynamics of atom-atom entanglement both for the vacuum and the thermal field. The special features of negativity behavior have been studied comprehensively foe small and large values of detunings. For thermal field and small detunings we established the effect of sudden death and birth of entanglement.

1. Introduction

Entanglement is not only one of the most surprising features of quantum theory, but also provides an important resource for various quantum information processes such as quantum information, quantum communication, and quantum cryptography. Therefore, great efforts have been made to investigate entanglement characterization, entanglement control, and entanglement production in different systems. Much attention is given to a system containing two two-level atoms, since they can represent two qubits, the building blocks of the quantum gates that are essential to implement quantum protocols in quantum information processing. Two-atom entangled states have been demonstrated experimentally using Rydberg atoms, cold trap ions, impurity spins in solids, superconducting circuits and cavity quantum electrodynamics schemes (cavity QED) [1]. Cavity QED with Rydberg atoms has been a favorite tool to checking the foundations of quantum mechanics including entanglement. Rydberg atoms was proposed a two decade ago to implement quantum gates between neutral atoms. The theoretical and experimental aspects of Rydberg-mediated quantum information processing received for the first decade are reviewed by M. Saffman and T. Walker [2]. The newest review of quantum computation with neutral atom qubits are presented by M. Saffman an coauthors [3, 4]. A brief overview of the current state of the experimental research on the development of the element base of quantum computers with qubits based on single neutral atoms trapped in optical traps are made by I. Ryabtsev and coauthors [5].

The entanglement between two initially independent Rydberg atoms successively passing a vacuum cavity have been demonstrated by S. Haroche et all. [6]. The entanglement procedure involves the resonant coupling, one by one, of the atoms to a high Q microwave superconducting cavity. The atoms, prepared in circular Rydberg states, exchange a single photon in the cavity and become entangled by this indirect interaction. Later S. Haroche and coauthors demonstrated the multiparticle entanglement for two atoms and a single-photon cavity mode [7]. A number of theoretical works have been devoted to investigations of entanglement properties of two atoms passing through cavities one after another, thereafter interacting with a various state radiation fields inside the cavities [8]-[13]. In our paper [14] we examined the dynamics of entanglement of two initially entangled Rydberg atoms successively passing a thermal cavity with different velocities. In this paper we continued this investigations taking into account the detuning between the cavity mode and the atom frequency. The dynamics and transfer of entanglement of two atoms interacting with quantum field of a cavity in the presence of detuning has been investigated in a number of papers [8, 15-19]. Some of these papers deal with the atoms simultaneously interacting with common cavity field. In other papers most attention has been concentrated on the entanglement dynamics of atoms passing consecutively through a cavity [8, 18, 19]. B. Gosh et al. [8] considered the model with effective two-photon interaction between atom and cavity field taking into account detuning and Stark shift. They found that though the two-atom entanglement generally diminishes with the increase of the two-photon detuning and the Stark shift, it is possible to sustain the entanglement over a range of interaction times by making the detuning and the Stark shift compensate each other. Consideration is being given to atoms prepared in their upper states before they enter the cavity. D. Gonta and S. Fritzsche [18] proposed a scheme to generate an entangled state between two four-level atoms that interact effectively by means of a detuned optical cavity and a laser beam that acts perpendicularly to the cavity axis. It is shown how the degree of entanglement for two atoms passing through the cavity can be controlled by manipulating their velocity and the (initial) distance between the atoms. Calderderón and coauthors [19] considered the sequential passage of two Rubidium atoms with effective two-photon transitions, prepared in specific Rydberg states, through a high-Q cavity, initially prepared in its vacuum state. They shown that, choosing a particular interaction time, the two-photon detuning can be used to control the atomic entanglement. We concerned our attention to investigation of the entanglement properties of two initially entangled Rydberg atoms passing through a single-mode cavity with a thermal field. The main goal of the present paper is to investigate the influence of detuning and thermal photons on entanglement between Rydberg atoms.

2. Model and negativity calculations

The physical system under consideration consists of two separate identical two-level Rydberg atoms passing through a cavity one after another and interacting with the cavity field. With the exception of the initial atomic state and cavity temperature, the parameters of the considered model are assumed to be the same as in experiment [6]. In [6] the author used two circular Rydberg states of Rb with principal quantum number 51 or 50 before crossing the cavity. The relevant cavity mode has is slightly off resonant with the transition at $\omega_0 = 51.1$ GHz between exited and ground states (detuning $\delta = 170$ kHz). Exact resonance between the atoms and the cavity is obtained by switching on a small electric field across the cavity mirrors while each atom is crossing a cavity. The Rabi frequency $\Omega/2\pi$ of the atom transition resonant with the cavity mode was 48 kHz. One can easily calculate that $\delta \approx 1.2g$, where g is the atom-field coupling constant. The cavity was cooled to 0.6 K, i.e. the mean thermal photon number was negligible. The cavity photon damping time was 112 ms, much shorter than the interval between two experimental sequences. In our theoretical consideration we took into account the detuning and thermal photons. We also considered the entangled initial atomic states.

The Hamiltonian of the joint "one atom+field" system with the dipole and rotating wave approximation can be written as

$$H = (1/2)\hbar\omega_0\,\sigma_i^z + \hbar\omega\, a^+a + \hbar g\,(a^+\sigma^- + a\sigma^+), \tag{1}$$

where $(1/2)\sigma^z$ is the inversion operator, $\sigma^+ = |+\rangle\langle-|$, and $\sigma^- = |-\rangle\langle+|$ are the transition operators between the excited $|+\rangle$ and the ground $|-\rangle$ states, a^+ and a are the creation and the annihilation operators of photons of the cavity mode and ω is the frequency of the cavity mode. We introduce the detuning as $\delta = \omega_0 - \omega$.

The evolution operator corresponding to the Hamiltonian (1) is

$$U_{A_i}(t) = e^{\frac{\imath\omega_0 t}{2}}|0,-_i\rangle\langle 0,-_i| + \sum_{n=0}^{\infty} e^{-\imath\omega(n+\frac{1}{2})t}\{A_n|n+1,-_i\rangle\langle n+1,-_i| + B_n|n,+_i\rangle\langle n,+_i| +$$

$$+C_n\left(|n+1,-_i\rangle\langle n,+_i| + |n,+_i\rangle\langle n+1,-_i|\right)\}.$$

Here

$$A_n = \cos(\Delta_n t/2) + \frac{\imath\delta}{\Delta_n}\sin(\Delta_n t/2),\ B_n = \cos(\Delta_n t/2) - \frac{\imath\delta}{\Delta_n}\sin(\Delta_n t/2),\ C_n = \frac{\imath\Omega_n}{\Delta_n}\sin(\Delta_n t/2),$$

where $\Delta_n = \sqrt{\delta^2 + \Omega_n^2}$, $\Omega_n = 2g\sqrt{n+1}$ and $|n\rangle$ is the Fock state for the cavity mode.

Suppose that before the first atom enters the cavity the two atoms have been prepared in Bell-type entangled states of the form

$$|\Psi(0)\rangle = \cos\Theta|+,-\rangle + \sin\Theta|-,+\rangle \tag{2}$$

or

$$|\Psi(0)\rangle = \cos\Theta|+,+\rangle + \sin\Theta|-,-\rangle, \tag{3}$$

where Θ defines the degree of initial atomic entanglement $(0 \le \Theta \le \pi/2)$, and one-mode cavity field is in a thermal state $\rho_F(0) = \sum_n p_n|n\rangle\langle n|$, where the probabilities $p_n = \bar{n}^n/(1+\bar{n})^{n+1}$. Here $\bar{n}$ is the mean photon number in the cavity mode $\bar{n} = (\exp[\hbar\omega/k_B T] - 1]^{-1}$, k_B is the Boltzmann constant and T is the equilibrium temperature of the cavity mirrors.

Solving the evolution equation we derived the density matrix of considered system for time moment τ when the first atom leaves the cavity. The considered density matrix on the other hand is the initial state of the system prior to entering into the cavity of the second atom. We deduced the density matrix of the whole system at the moment t when the second atom leaves the cavity. Taking a partial trace over the field variables we obtained from the density matrix of the whole system the reduced atomic density operator $\rho_{at}(t,\tau)$ in the two-atom basis $|+,+\rangle$, $|+,-\rangle$, $|-,+\rangle$, $|-,-\rangle$ both for entangled state (2) and (3).

For two-qubit system described by the density operator $\rho_{at}(t,\tau)$, a measure of entanglement or negativity can be defined in terms of the negative eigenvalues μ_i^- of partial transpose of the reduced atomic density matrix $\rho_{at}^{(T_1)}$ [20, 21]. The negativity is $\varepsilon = -2\sum_i \mu_i^-$. When $\varepsilon = 0$ two qubits are separable and $\varepsilon > 0$ means the atom-atom entanglement. The case $\varepsilon = 1$ indicates maximum entanglement.

Using the exact expressions for reduced atomic density matrix $\rho_{at}(t,\tau)$ we derived the partial transpositions and calculated the exact formulae for negativity both for state (2) and (3). We do not give the explicit form of negativities, as they are too cumbersome. The results of numerical calculations of negativity for entangled initial atomic state (2) and (3) are shown in Figs. 1-5. The curves were obtained under the assumption that $\tau = t/2$ as in experiment [6].

3. Results and discussion

In numerical calculations we have turned our attention to the effects of the detuning both for the vacuum and the thermal field. Without loss of generality, we derived that the detuning is necessary and meaningful. Fig. 1 illustrates the dynamics of negativity as a function of a scaled time gt for separable atomic state $|+,-\rangle$. The detuning sharply reduces the maximum degree of

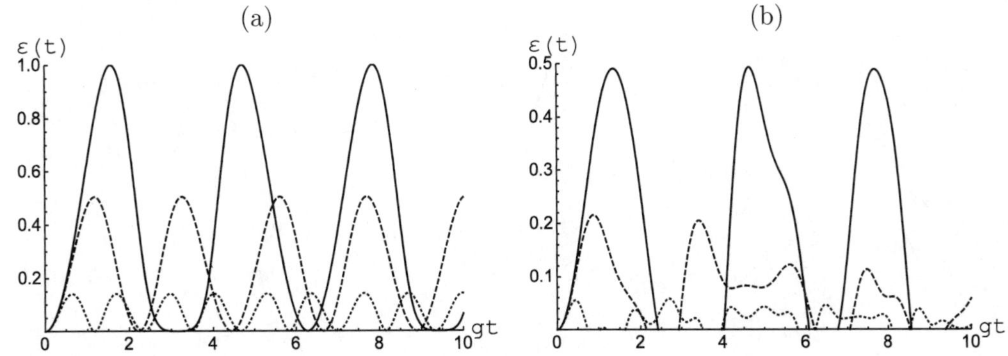

Figure 1. The negativity as a function of a scaled time gt for separable atomic state $|+,-\rangle$ with $\delta = 0$ (solid) and $\delta = 5$ (dashed). The mean photon number $\bar{n} = 0$ (a) and $\bar{n} = 0.5$ (b).

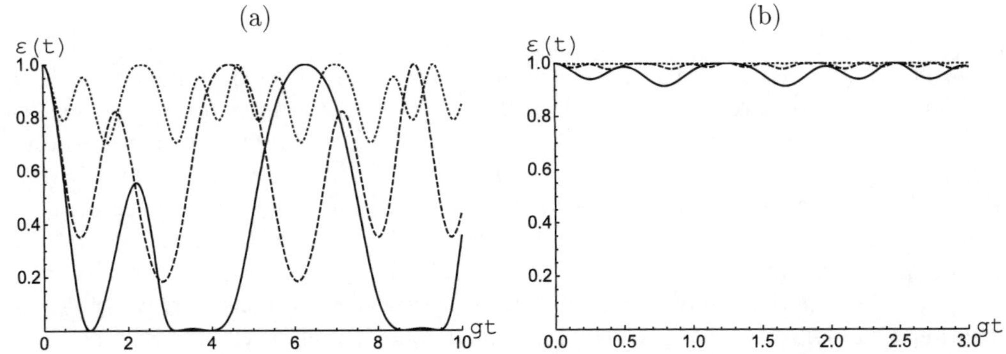

Figure 2. The negativity as a function of a scaled time gt for entangled initial state (2) with $\Theta = \pi/4$ and a vacuum cavity field. The detuning $\delta = 0$ (solid), $\delta = 2g$ (dashed) and $\delta = 5g$ (dotted) (a) and $\delta = 10g$ (solid), $\delta = 20g$ (dashed) and $\delta = 50g$ (dotted) (b).

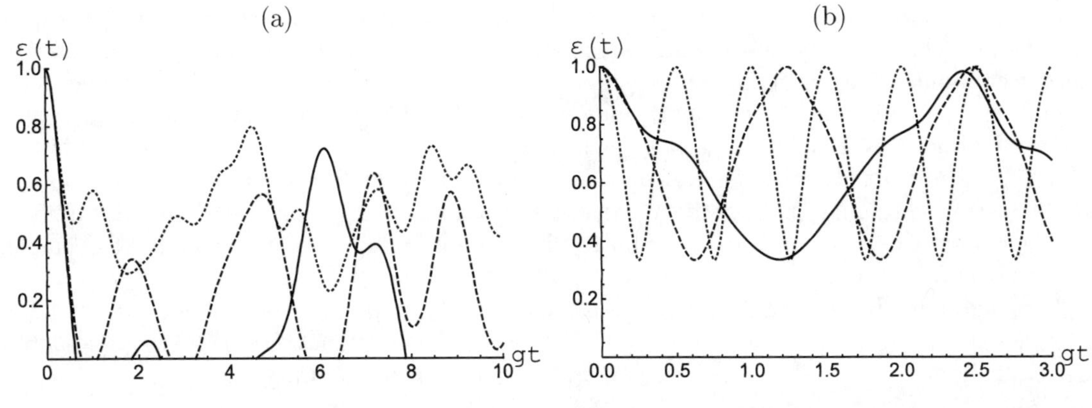

Figure 3. The negativity as a function of a scaled time gt for entangled initial state (2) with $\Theta = \pi/4$ and a thermal cavity field with $\bar{n} = 0.5$. The detuning $\delta = 0$ (solid), $\delta = 2g$ (dashed) and $\delta = 5g$ (dotted) (a) and $\delta = 10g$ (solid), $\delta = 20g$ (dashed) and $\delta = 50g$ (dotted) (b).

entanglement both for the vacuum and the thermal field. The above effects were noted firstly for initial separable atomic state $|+,+\rangle$ [8]. For vacuum state results agree closely with experiment.

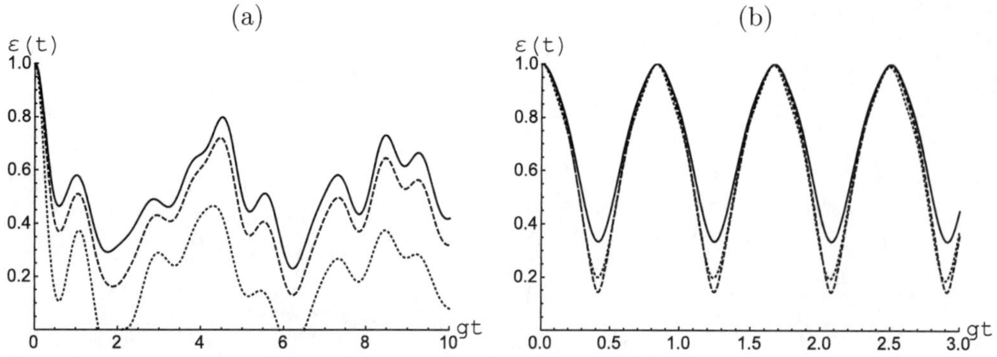

Figure 4. The negativity as a function of a scaled time gt for entangled initial state (2) with $\Theta = \pi/4$ and detuning $\delta = 5$ (a) and $\delta = 50$ (b). The mean photon number $\bar{n} = 0.5$ (solid), $\bar{n} = 0.7$ (dashed) and $\bar{n} = 1.5$ (dotted).

S. Haroche et al. [6] noted that atom-cavity coupling was negligible when detunig ($\delta \approx 1.2g$) takes place. When the atom-field detuning is large enough, there is no energy exchange between the cavity and atom. The field, which acts as a medium, is virtually excited during the atom-atom coupling process. In the considered case the interaction between atoms through virtual media does't lead to entanglement. Note that for atoms simultaneously interact with common field the virtually excited cavities may produce the maximally entangled two-atom states [22].

In Fig. 2 we plot the negativity as a function of a scaled time gt for entangled state (2) and a vacuum field. Fig. 2 shows that small detinings leads to decreasing of the negativity oscillations and stabilization of the entanglement. For large detunings the virtually excited medium doesn't destroy the initial atomic quantum correlations or entanglement. Fig. 3 shows the same for thermal field. Fig 3(a) shows that for slight detunings the entanglement behavior is very similar to that for a vacuum field, but the time periodicity destroys because system has a set of Rabi frequencies. One can see from Fig. 3(a) that for a thermal cavity field and small detunings the sudden death and birth of entanglement takes place. Figs. 3(b) demonstrates that for large detunings the amplitude of entanglement oscillations are scarcely affected by detuning. This result is more or less to be expected from the fact that the quantum correlations of atoms are affected by the medium involving not only the virtual but the real photons of thermal field. The frequencies of oscillations increase with increasing the Rabi frequencies $\Omega_n \approx \delta$ ($\delta \gg g$ and $\bar{n} \sim 1$). In Fig. 4 we plot the negativity as a function of a scaled time gt for entangled initial state (2) and different values of mean photon numbers with detuning $\delta = 5g$ (a) and $\delta = 30g$ (b). For small detunings the negativity decreases with increasing of the mean photon number. For large detunings and relatively large mean photon numbers ($\bar{n} > 1$) the negativity behavior is scarcely affected by $\bar{n}$.

The negativity behavior for entangled initial state (3) is qualitatively similar to that for entangled initial state (2). In Fig. 4 we show the negativity as a function of a scaled time gt for thermal field and atomic state (3) for small and large values of detunings.

4. Conclusions

We used the negativity to study the entanglement of the system of two initially entangled Rydberg atoms successively passing a thermal cavity taking into account the detuning. We derived the exact expressions for the reduced atomic density matrix and calculated the negativity formulae for different Bell's initial atomic state and thermal cavity field. We investigated the entanglement turning our attention to the role of detuning in entanglement behavior. Our

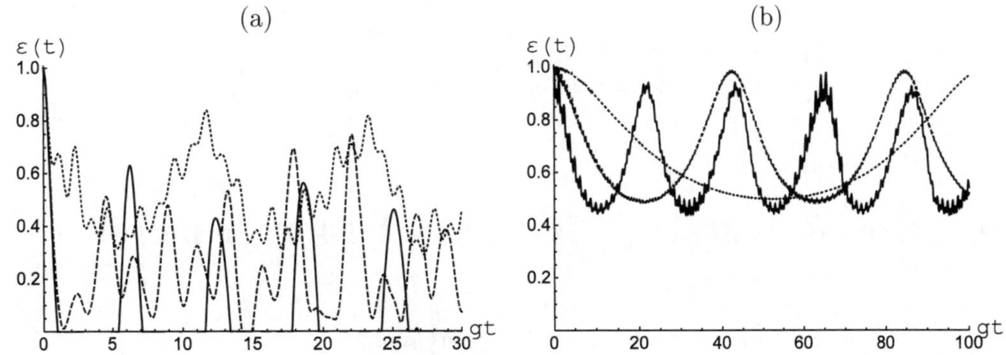

Figure 5. The negativity as a function of a scaled time gt for entangled initial state (3) with $\Theta = \pi/4$ and a thermal cavity field with $\bar{n} = 0.5$. The detuning $\delta = 0$ (solid), $\delta = 2g$ (dashed) and $\delta = 5g$ (dotted) (a) and $\delta = 10g$ (solid), $\delta = 20g$ (dashed) and $\delta = 50g$ (dotted) (b).

numerical results reveal that for small detunings the presence of detuning leads to decreasing of the entanglement amplitude oscillations and stabilization of the degree of entanglement both for the vacuum and for the thermal field. For vacuum field and large detunings the initial entanglement ceases to vary in time. For thermal field and large detunings the entanglement amplitude oscillations not decrease with detuning increasing. For thermal field and small detunings the effect of sudden death and birth of entanglement takes place. For large detunings such effect vanishes. For separable initial atomic states the detuning reduces the maximum degree of entanglement both for the vacuum and the thermal field. These results show that the atom-atom entanglement can be controlled by changing the system parameters, such as the mean photon numbers and detuning.

References

[1] Buluta I, Ashhab S and Nori F 2011 *Rep. Prog. Phys.* **74** 104401
[2] Saffman M and Walker T G 2010 *Rev. Mod. Phys.* **82** 2313
[3] Saffman M 2016 *J. Phys. B: At. Mol. Opt. Phys.* **49** 202001
[4] Beterov I I et al. 2014 *Laser Phys.* **24** 074013
[5] Ryabtsev I I et al. 2017 *Russian Microelectr.* **46** 109
[6] Hagley E et al. 1997 *Phys. Rev. Lett.*. **79** 1
[7] Rauschenbeutel A. et al. 2000 *Science* **288** 2024
[8] Ghosh B, Majumdar A S and Nayak N 2008 *J. Phys.* B **41** 065503
[9] Liao Q, Fang G, Wang Y, Ahmad M A and Liu S 2010 *Chin. Phys. Lett.* **8** 1191
[10] Liao Q , Fang G, Ahmad M A and Liu S 2011 *Optics Commun.* **384** 201
[11] Yan X Q 2009 *Chaos Solit. Fract.* **41** 1645
[12] Bashkirov E K and Nikiforova Y A 2012 *Computer Opt.* **36** 468
[13] Bougouffa S and Ficek Z 2012 Phys. Scr. **T147** 014005
[14] Bashkirov E K and Mastuygin M S 2016 *J. Phys.: Conf. Ser.* **735** 012026
[15] Zhang B 2010 *Opt. Comm.* **283** 4676
[16] Sebawe Abdalla M, Khalil E M, Obada A.S.-F, Peřina J and Křepelka J 2015 *Eur. Phys. J. Plus* **130** 227
[17] Liao 2016 *Commun. Theor. Phys.* **65** 684
[18] Gonta D and Fritzsche S 2009 *J. Phys. B: At. Mol. Opt. Phys.* **42** 145508
[19] Calderderón O, Joya M R and Fonseca Romero K M 2016 *J. Phys.: Conf. Ser.* **687** 012073
[20] Peres A 1996 *Phys. Rev. Lett.* **77** 1413
[21] Horodecki R, Horodecki M and Horodecki P 1996 *Phys. Lett.* A **223** 333
[22] Zheng S B and Guo G C 2000 *Phys. Rev. Lett.* **85** 2392

IOP Conf. Series: Journal of Physics: Conf. Series **929** (2017) 012088 doi:10.1088/1742-6596/929/1/012088

Quantum size effect in single-crystalline bismuth nanorods

E A Sedov[1], K –P Riikonen[2] and K Yu Arutyunov[1, 3]

[1] National Research University Higher School of Economics, Moscow, 101000, Russia
[2] Nano Science Center, Department of Physics, University of Jyväskylä, PB 35, 40014, Jyväskylä, Finland
[3] P.L. Kapitza Institute for Physical Problems RAS, 119334, Moscow, Russia

E-mail: Sedov1993@yandex.ru

Abstract. In a metal sample, where at least one of the dimensions is comparable with the de Broglie wavelength of conduction electrons, the quantum size effects (QSE) should be observed. QSEs manifest themselves as non-monotonic dependencies of various material properties as function of relevant dimension. QSE should be particularly noticeable in materials with charge carrier(s) effective mass less than the free electron mass. Bismuth is one of the most suitable semi-metal to observe QSE due to small effective masses and small the Fermi energy. However, bismuth has a high anisotropic energy spectrum. Hence to observe QSE which can be interpreted with reasonable accuracy, it is mandatory to fabricate single-crystal nanostructure with known orientation of crystallographic axes. In this paper several short bismuth nanowires (nanorods) were investigated, and oscillating dependence of electric resistance on effective cross section was found. Theoretical calculations provide a reasonable agreement with experiment. The quantum-size phenomena are important for operation of a wide spectrum of nanolelectronic devices.

1. Introduction

The laws of classical physics might describe well the behaviour of macroscopic objects, but if dimension(s) are reduced down to certain scales, one might observe qualitatively different quantum phenomena. The origin of the QSE is related to the text-book problem of energy level quantization of a particle with mass m confined inside a potential well with characteristic dimension a: $E(n) = n\hbar^2\pi^2/2ma^2$, $n = 1,2,3 ....$ QSEs affect various electronic properties of metals: electric conductivity, Hall and thermoelectric coefficients, reflectivity, etc. [1]. However in a typical metal with effective masses m^* comparable to free electron mass m_0 QSEs manifest at scales of about 1 nm. While in materials with $m^*<<m$, like antimony, bismuth or their alloys [2], the QSEs can be observed at scales ~ 100 nm achievable by modern nanofabrication.

In bismuth there are two types of charges: 'heavy' T-holes with $m^h \approx m_0$ and 'light' L-electrons with $m^e << m_0$ (Fig. 1a). Heavy T-holes fix the energy of the Fermi level and, as a first approximation, one might neglect the energy quantization effect for T-holes. While for L-electrons one might expect the manifestation of QSE (Fig. 1b). As the energy spectrum of bismuth is strongly anisotropic, the expected phenomena should dependent on particular orientation of the sample with respect to its crystallographic axes.

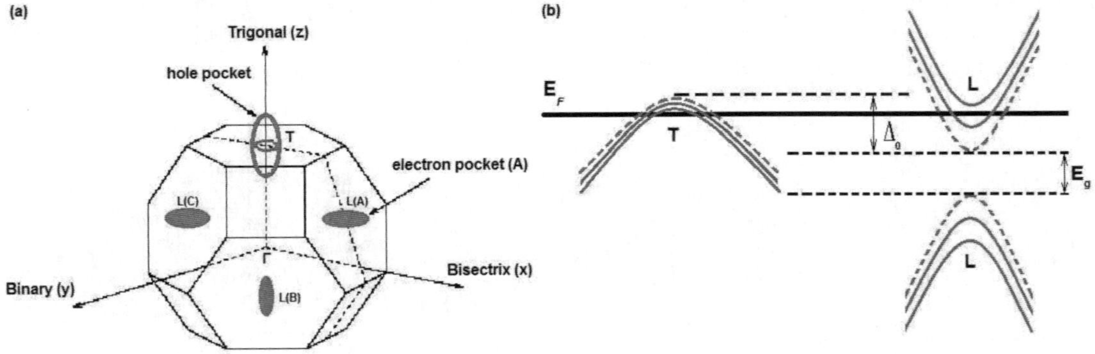

Fig. 1. (a) Brillouin zone and Fermi surface of bulk bismuth with three 'light' L-electron and one 'heavy' T-hole pockets. (b) Energy spectrum of bismuth. Dashed lines stand for bulk, while solid lines correspond to splitting of energy levels due to QSE.

First observations of QSE refer to early 1960-th including both experimental [3] and theory [4] studies. Later, other low-dimensional structures were investigated: microcylindres [5], nanowires [6, 7] and point contacts [8, 9]. Unfortunately, experiments with bundles of nanowires [6, 7] can provide only qualitative results because the contributions of individual nanowires are averaged. The experiments realized with scanning tunneling microscope (STM) technique [8, 9] demonstrated clear size-dependent change of electrical conductivity. However, the results of STM experiments [8, 9] are difficult to interpret, because the crystallographic orientation and the shape of an STM-pulled point contacts are not clearly defined.

In this work we experimentally demonstrate that electric resistivity of single-crystalline bismuth nanorods non-monotonically varies with reduction of cross section. The size dependence of resistance oscillates and can be interpreted in terms of relevant theoretical model [10].

2. Theory

The model [10] assumes parabolic E($\boldsymbol{k}$) dependence for bismuth spectrum [11, 12] both for T-holes and L-electrons with the corresponding effective mass diagonal elements (in units of free-electron mass) for holes m_x^h=m_y^h=0.059, m_z^h=0.634 and electrons m_x^e=0.00139, m_y^e=0.291, m_z^e=0.0071 (Fig. 1a). The energy gap is $E_g = 13.7 \pm 0.1\, meV$ in L-point and the band overlap energy is $\Delta = 38\, meV$ (Fig. 1b).

Model [10] solves Boltzmann kinetic equation for bismuth nanowire of rectangular cross section w*t, where w is width and t is thickness. Following Ref. 10 one obtains the oscillatory dependence of electrical resistivity vs. cross section:

$$R^e = \left(\frac{2}{\pi} \frac{q_e^2}{\hbar} \frac{1}{\varrho} \frac{\mu_x^e}{m_z^e} \sum_{m,n=1}^{[r_w],[r_t]} \frac{(2\hbar/V_0)^2 \sqrt{U_{mn}^e}}{\sum_{m',n'=1}^{[r_w],[r_t]} \Lambda_{m'n'}^{mn} / \sqrt{U_{m'n'}^e}} \right)^{-1} \quad (1)$$

where $U_{mn}^e \equiv 1 - (m/r_w)^2 + (\mu_y^e/\mu_x^e)[1 - (n/r_t)^2]$; $\Lambda_{m'n'}^{mn} \equiv (2 + \delta_{mm'})(2 + \delta_{nn'})$, here δ_{ab} is Kronecker delta function; $\varrho \equiv N/\Omega$ is the value of the carriers density. Parameters $r_w \equiv w/w_0$ and $r_t \equiv t/t_0$ are width and thickness of nanowire, normalized by the relevant dimensions $w_0 = \hbar\pi/(2M_x\Delta_x)^{1/2}$ and $t_0 = \hbar\pi/(2M_y\Delta_y)^{1/2}$ corresponding to metal-to-insulator transition: when the gap opens between the lowest L-electron and the highest T-hole bands (Fig. 1b). Here $M_{x,y} \equiv (m_{x,y}^e m_{x,y}^h)/(m_{x,y}^e + m_{x,y}^h)$ and $\mu_{x,y}^e \equiv (\Delta_{x,y} m_{x,y}^h)/(m_{x,y}^e + m_{x,y}^h)^{-1}$, $\Delta_{x,y}$ is the energy bands overlap due to confinement of charge carriers in corresponding direction x or y. The outer summation accounts for contribution of the different (m, n) sub-bands. Formally, to calculate the total electric conductivity one should also consider the T-hole contribution. However, as for certain particular directions (e.g. along the bisectrix axis) the T-hole masses are significantly larger than for L-electrons, one can neglect the QSE for holes. The observation simplifies the calculation allowing one to consider the hole contribution to integral electric resistivity as an additive size-independent term.

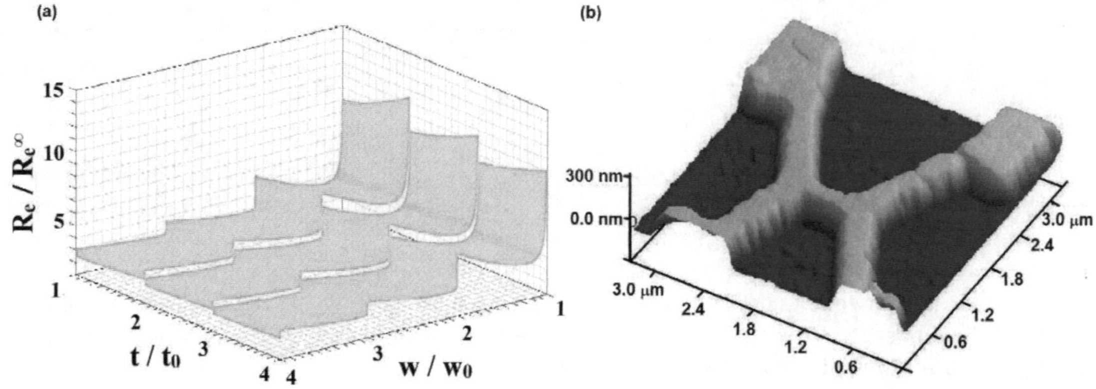

Fig. 2. (a) Calculated electron resistivity R_e, normalized by the bulk value R_e^∞, for bismuth nanowire of rectangular cross section $w*t$ cut along bisectrix axis as function of thickness t and width w in units of threshold values t_0 and w_0, corresponding to metal-to-insulator transition. (b) AFM image of typical bismuth nanostructure on mica substrate after several sessions of ion milling.

Obviously for material with highly anisotropic energy spectrum as bismuth, the electric resistivity $R_e(w,t)$ depends on orientation of the sample relative to crystallographic axes. If the sample is cut along high-symmetry crystallographic axis, Eq. (1) provides rather simple $R_e(w,t)$ dependence (Fig. 2a). If one cuts the sample along an arbitrary direction, the QSE in all three non-equivalent L-pockets of the Fermi surface will result in a complicated $R_e(w,t)$ dependence with secondary maxima and minima. It should be noted that a particular experimentally observed R_e(cross section) dependence is affected not only by the effective diameter $(w*t)^{1/2}$, but also by the actual values of width w and thickness t, representing a 'slice' of three-dimensional $R_e(w,t)$ function (Fig. 2a).

3. Experimental

In order for detect the quantum size effect we fabricated several single crystal bismuth nanorods with typical dimensions: length 0.8-1 μm, width 0.3-0.4 μm and thickness 0.2-0.3 μm. All nanostructures were designed for measuring resistance in four probe configuration (Fig. 2b).

For QSE studies several approaches have been proposed [5, 7, 13, 14]. In this paper we used a standard lift-off e-beam lithography on high speed (~1 nm/s at 10^{-6} mbar) pure bismuth deposited film on mica substrate heated to 140 ^{0}C. Atomic force (AFM) and scanning electron (SEM) microscope analyses revealed polycrystalline structure of the films with the grain size up to 1.5 μm. X-ray analysis of co-deposited wide films showed that the trigonal axis in each grain is perpendicular to the substrate plane. Contrary to Ref. [14], where long bismuth nanowires containg multiple grains connected in series, our lithographically fabricated structures (the 'nanorods') were made deliberately rather short to enable the very central part of the sample (the 'body') be a single crystal (Fig. 2b).

To observe the QSE in between the sessions of resistivity measurements we used low-energy Ar^+ directional ion milling to reduce the effective diameter of the nanorods [15]. The method permits progressive removal of material from the surface just by few atomic layers. At the utilized acceleration energies < 1 keV the penetration depth of Ar^+ ions inside the metal matrix is minimal < 2 nm being comparable to the thickness of chemically corrupted layer at the very surface [16]. The technique enables the consistent reduction of a nanostructure cross section $(w*t)^{1/2}$ down to the sub-50 nm scales

[17]. AFM analysis was used to evaluate the dimensions of the structures after each session of ion milling [16].

4. Results and discussion

Between sessions of ion beam milling we measured the resistance *R* of bismuth nanorods as function of their transverse dimensions: width *w* and thickness *t*. We examined several samples and all of them demonstrated non-monotonic $R_e(w,t)$ dependencies (Fig. 3). In the thinnest structures approaching $(w*t)^{1/2} \sim 50$ nm we have observed a pronounced increase of electric resistance, which can be associated with QSE-mediated metal-to-semiconductor transition [17]. However, even the wider samples at scales $(w*t)^{1/2} < 200$ nm demonstrate peaks of electric resistivity (= dips of conductivity), corresponding to crossing of the size-quantized energy bands and the Fermi level (Fig. 1b).

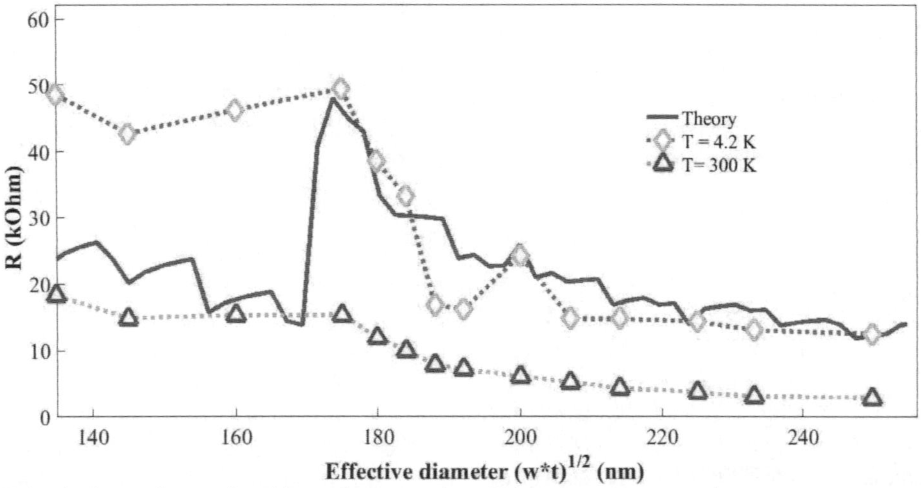

Fig. 3. R(w,t) dependence for bismuth nanowire. Triangles correspond to room temperature and squares stand for T=4.2 K data. Dashed lines, connecting experimental points, are guides for eye. Solid line is the best theory fit using Eq. (1).

In Fig. 3 one can trace a certain correlation between the calculation based on model [10] and experiment, but the agreement is far from being perfect. Presumably the mismatch originates from uncertainty in orientation of the sample axis (= current lines) against the crystallographic axes of the central grain, forming the 'body' of the sample. Though for electric measurements we have used 4-probe configuration, the relatively short length of the nanorod compared to width and thickness, does not allow to consider the measuring probes as non-invasive and, hence, the current lines are not strictly parallel to each other. The best-fit theory dependence in Fig. 3 considers the simplified arrangement: the bisectrix axis in the central grain ('body') is parallel to the current lines. Then the two main peaks originate from L-electrons with minimal effective mass $m^e_x \approx 0.00139 m_0$. Small peaks are contributions from two other L-electron pockets with larger effective masses. Heavy T-holes contribute to electric conductivity with much less pronounced size dependence due to their large effective mass $m^h >> m^e$. Formally better agreement between experiment and theory can be achieved if to consider the contributions of several single crystalline grains with mismatched binary and orientations (e.g. the central grain and the two neighbors from the probes) connected in series. However as the X-ray structural analyses of each particular nanostructure could not be performed (we do not have such facilities) the 'improvement' would be speculative. One should also notice that reduction of temperature down to T=4.2 K makes the experimental $R_e(w,t)$ dependencies more 'sharp'. The observation reflects the obvious fact that finite temperature T broadens each size-quantized energy level by $k_B T$ making the crossing with Fermi level 'washed out'.

5. Conclusions

We fabricated short bismuth's single-crystalline nanorods and measured their resistivity as function of effective cross section $(w*t)^{1/2}$. The R(w,t) data demonstrated clear non-monotonic dependence, which is consistent with model [10] that takes into account the quantum size effects in bismuth nanowires with rectangular cross section. The agreement between experiment and theory is not perfect, presumably, due to difficulty to fabricate a truly single-crystalline one-dimensional nanostructure with well defined orientation against crystallographic axes. Nevertheless our observations can be considered to be in a semi-quantitative (or at least - qualitative) agreement with theory predictions.

The results suggest that production of the next generation of ultra-small nanoelectronic devices should take into account the quantum size effects, as it is a universal phenomena and should affect basically all electronic properties of such tiny objects [1].

Acknowledgements

The article was prepared within the framework of the Academic Fund Program at the National Research University Higher School of Economics (HSE) in 2016 (grant No. 16-05-0029) and supported within the framework of a subsidy granted to the HSE by the Government of the Russian Federation for the implementation of the Global Competitiveness Program.

References

[1] Tringides M C, Jatochowski M and Bauer E 2007 *Phys. Today* ***60*** *50-54*
[2] Lutski V N, 1970 *Phys. Stat. Sol.* ***1*** *199-220*
[3] Ogrin Yu V, Lutskii V N and Elinson M I 1966 *Zh. Exper. Teor. Fiz. Pisma.* ***3*** *114-118*
[4] Sandomirskii V 1967 *BSov. Phys. JETP* ***25*** *101-106*
[5] Nikolaeva A, Huber T, Konopko L, Tsurkan A 2010 *J. Low. Temp. Phys.* ***158*** *530-535*
[6] Sun X, Zhang Z and Dresselhaus M S 1999 *Appl.Phys. Lett.* ***74*** *4005-4007*
[7] Heremans J et. al 2000 *Phys. Rev. B* ***61*** *2921-2930*
[8] Costa-Krämer J L Garcia N and Olin H 1997 *Phys.Rev. Lett.* ***78*** *4990-4993*
[9] Rodrigo J G García-Martín A Sáenz J J and Vieira S 2002 *Phys. Rev. Lett.* ***88*** *246801-1 - 246801-4*
[10] Farhangfar S 2006 *Phys. Rev. B* ***74*** *205318-1 -205318-5*
[11] Lax B and Mavroides J G 1960 *Solid State Physics* (Academic Press, New York)
[12] Zhang Z et. al 2000 *Phys. Rev. B* ***61*** *4850-4861*
[13] Arutyunov K Yu Ryynanen T V Pekola J P and Pavolotski A B 2001 *Phys. Rev. B* ***63*** *092506-1 - 092506-4*
[14] Farhangfar S 2007 *Phys. Rev. B* ***76*** *205437-1 - 205437-4*
[15] Savolainen M et. al 2004 *Appl. Phys. A* ***79*** *1769 -1773*
[16] Zgirski M et. al 2008 *Nanotechnology* ***19*** *055301-1 -055301-6*
[17] Sedov E A Riikonen K-P and Arutyunov K Yu 2017 *to be published elsewhere*

IOP Conf. Series: Journal of Physics: Conf. Series **929** (2017) 012089 doi:10.1088/1742-6596/929/1/012089

Effect of carrier spectra nonsphericity and subband mixing on CHHS Auger process rate in deep quantum wells

N V Pavlov[1], **G G Zegrya**[1]

1 Ioffe Institute, 194021, Russian Federation, St Petersburg, Politekhnicheskaya st., 26

E-mail: pavlovnv@mail.ru

Abstract. The Kane's equations are written and solved with taking into account nonsphericity of the ***kp*** Hamiltonian. Charge carrier energy spectra and wave functions are obtained and analized. Subbands of dimensional quantization in the AlSb/InAs$_{0.84}$Sb$_{0.16}$/AlSb system are calculated with taking subband mixing into account. CHHS Auger process coefficient in the AlSb/InAs$_{0.84}$Sb$_{0.16}$/AlSb system is calculated with and without taking energy spectra nonsphericity and subband mixing into account. It is obtained that energy spectra nonsphericity and subband mixing lead to Auger matrix element and CHHS Auger process rate decrease.

1. Introduction

Improving the quantum efficiency of optoelectronic emitters mid-infrared range at room temperature is one of the most actual problems of semiconductor microelectronics. This is because in this wavelength range are: atmospheric transparency window, the absorption peaks of many toxic industrial gases, and the maximum absorption of the human body tissues radiation. Thus, the semiconductor emitters of this range can be used in telecommunication and security systems, in location and night vision systems, as component of gas analyzer structures and also as a so-called optical scalpel for photodynamic therapy (selective destruction of cancer cells) and laser therapy (contributing to the acceleration of wound healing).

One of the most promising ways of solving this problem is using semiconductor heterostructures with deep quantum wells based on InAsSb compounds as the active region. These structures have the highest values of the electron density and the overlap integral between the states of the valence band and the conduction band. The low value of the band gap in these structures is the cause of significant nonparabolicity of charge carrier energy spectra even for the ground states, which in turn leads to significant corrections to the values of the radiative and Auger recombination rates. In this paper we considered the CHHS Auger process which involves an electron and two heavy holes, with transition of one of the holes to the spin-orbit splitting zone

The purpose of present investigation is to calculate the CHHS Auger process coefficient in the AlSb/InAs$_{0.84}$Sb$_{0.16}$/AlSb with taking energy spectra nonsphericity and subband mixing into account.

2. Basic Equations

We used 8×8 ***kp*** Hamiltonian [1] that takes interaction with the higher bands into account up to quadratic terms by the wave vector but neglects the relativistic linear terms and the term with heavy electron mass.

The Kane equations near the Γ-point are given by following expressions:

$$(E_C - E)u = -iP\boldsymbol{k}\cdot\boldsymbol{v},$$

$$(E_V - \delta - E - \frac{\hbar^2}{2m_0}(\gamma_1 - 2\gamma_2))\boldsymbol{v} + i\delta[\sigma\times\boldsymbol{v}] - \frac{3\hbar^2}{m_0}\gamma_3\boldsymbol{k}(\boldsymbol{k}\cdot\boldsymbol{v}) + \frac{3\hbar^2}{m_0}(\gamma_3-\gamma_2)\begin{pmatrix} k_x^2 v_x \\ k_y^2 v_y \\ k_z^2 v_z \end{pmatrix} = -iP\boldsymbol{k}u \quad (1)$$

Here E_C and E_V are the energies of conduction and valence band edges, P is the Kane matrix element, $\delta=\Delta_{so}/3$ is the spin-orbit splitting constant, m_0 is the free electron mass, γ_1, γ_2 and γ_3 are the generalized Luttinger parameters, $\boldsymbol{\sigma}$ is the Pauli matrices, u and $\boldsymbol{v}$ are the spinor components of wave function with s- and p-symmetry, respectively, $\boldsymbol{k}$ is the wave vector and E is the energy.

If $\gamma_2 = \gamma_3$,that corresponding the spherical approximation, equations (1) transform to the common used ones [2].

The energy spectrum and the wave function of heavy holes, calculated from (1), are:

$$\left(E+\frac{\hbar^2k^2}{2m_h}\right)^3 - C_h k^2\left(E+\frac{\hbar^2k^2}{2m_h}\right)^2 + \left\{C_h^2(k_x^2k_y^2+k_y^2k_z^2+k_z^2k_x^2) - 3\delta^2\right\}\left(E+\frac{\hbar^2k^2}{2m_h}\right)$$
$$-C_h^3k_x^2k_y^2k_z^2 + 2C_hk^2\delta + 2\delta^3 = 0, \quad (2)$$

$$\boldsymbol{v} = H_1\begin{pmatrix} q\cos k_h x\xi \\ -ik_h\sin k_h x\xi \\ \delta\left(E+\frac{\hbar^2k^2}{2m_h}\right)^{-1}(-k_h\sin k_h x\xi + q\cos k_h x\eta) \end{pmatrix}$$

$$+H_2\begin{pmatrix} q\sin k_h x\eta \\ ik_h\cos k_h x\eta \\ \delta\left(E+\frac{\hbar^2k^2}{2m_h}\right)^{-1}(-k_h\cos k_h x\xi - q\sin k_h x\eta) \end{pmatrix}. \quad (3)$$

Where $m_h = \frac{m_0}{\gamma_1 - 2\gamma_2}$, $\xi = \frac{1}{\sqrt{2}}\begin{pmatrix}1\\-1\end{pmatrix}$, $\eta = \frac{1}{\sqrt{2}}\begin{pmatrix}1\\1\end{pmatrix}$, $C_h = \frac{3(\gamma_3-\gamma_2)\hbar^2}{m_0}$, H_1 and H_2 are the normalization coefficients. We can see that heavy hole effective mass on the (100) direction is $m_{hh} = \frac{m_0}{\gamma_1+2\gamma_2-4\gamma_3}$ and differs from m_h. We note that wave function (3) differs from similar one in spherical approximation because division $\left(E+\frac{\hbar^2k^2}{2m_h}\right)^{-1}\delta$ is not unity.

Energy spectra and wave function of electrons, light and spin-orbit holes are more complex. If k_z=0 they are given by following equations:

$$E - E_g - \delta = -Pf_3((1+f_2^2)k^2 + 2Ck_x^2q^2) \quad (4)$$

$$u = A_1\cos kx\eta + A_2\sin kx\xi$$

$$\boldsymbol{v} = A_1 f_3 \begin{pmatrix} iqg_t \cos k_x x\xi + g_y \sin k_x x\eta \\ -iqg_x \cos k_x x\eta + g_t \sin k_x x\xi \\ f_2 \left\{ (g_t - ig_y)\sin k_x x\xi + q(ig_x - g_t)\cos k_x x\eta \right\} \end{pmatrix}$$

$$+ A_2 f_3 \begin{pmatrix} iqg_t \sin k_x x\eta - g_y \cos k_x x\xi \\ -iqg_x \sin k_x x\xi - g_y \cos k_x x\eta \\ f_2 \left\{ (g_t - ig_y)\cos k_x x\eta + q(g_t - ig_x)\sin k_x x\xi \right\} \end{pmatrix}, \quad (5)$$

where $q = k_y$, $g_x = 1 + f_2^2 + Ck_x^2$, $g_y = 1 + f_2^2 + Cq^2$, $g_t = if_2^2 - f_2$, $f_3 = \dfrac{f_1}{g_x g_y + g_t^2}$,

$$f_2 = -i\delta\left(E + \frac{\hbar^2 k^2}{2m_h}\right)^{-1}, f_1 = -\left(E + \frac{\hbar^2 k^2}{2m_h}\right)^{-1}\left\{P + \frac{3\gamma_3 \hbar^2 (E_g + \delta - E)}{m_0 P}\right\}, \quad C = -C_h\left(E + \frac{\hbar^2 k^2}{2m_h}\right)^{-1},$$

A1 and A2 are the normalization coefficients.

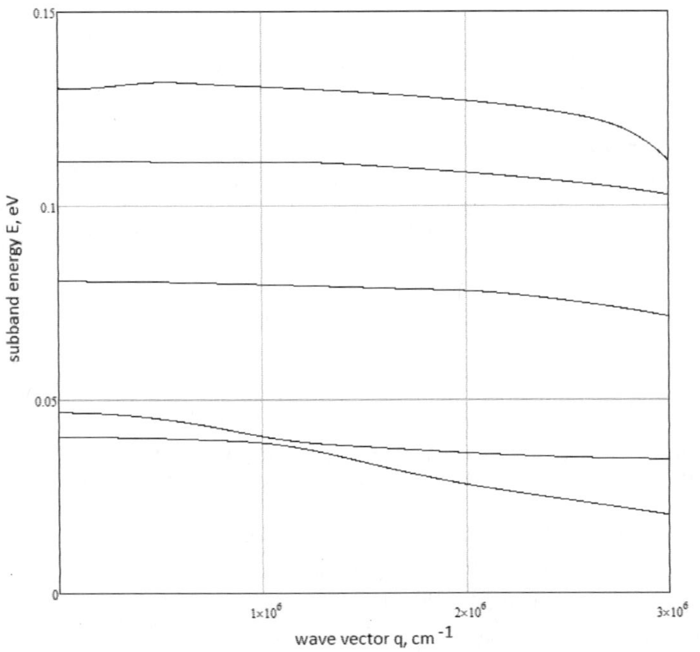

Figure 1. Subband energies in $AlSb/InAs_{0.84}Sb_{0.16}/AlSb$ quantum well with width $a = 100A$.

.

We can derive the boundary contitions by integrating equations (1) near the interface. Thus we obtain:

$$v_-^+ = 0, \left.\frac{\partial v_y}{\partial x}\right|_-^+ = 0, \left.\frac{\partial v_z}{\partial x}\right|_-^+ = 0, \frac{\hbar^2(6\gamma_3 + \gamma_1 - 4\gamma_2)}{2m_0}\left.\frac{\partial v_x}{\partial x}\right|_-^+ = \left(P + \frac{3\gamma_3\hbar^2(E_g + \delta - E)}{m_0 P}\right)u_-^+ \quad (6)$$

In Fig.1 the results of numerical calculation of subband energies in $AlSb/InAs_{0.84}Sb_{0.16}/AlSb$ quantum well with width $a = 100A$. We can observe the regions with negative effective mass in first and second subbands and also interaction between fourth and fifth subbands. Thus we can conclude that there are sufficient subband mixing effect in $AlSb/InAs_{0.84}Sb_{0.16}/AlSb$ deep quantum wells.

3. Auger matrix element and CHHS Auger process coefficient

Auger matrix element can be calculated from the following formula [2]:

$$M = \frac{4\pi e^2}{\kappa_0}\frac{1}{2q}\int_{-\infty}^{+\infty}\int_{-\frac{a}{2}}^{\frac{a}{2}} \overline{\psi_4(x_1)}\,\overline{\psi_3(x_2)}\psi_1(x_1)\psi_2(x_2)\exp(-q|x_1 - x_2|)dx_1 dx_2 \qquad (7)$$

Where ψ_1 is the wave function of electron, ψ_2 and ψ_3 are the wave functions of two heavy holes and ψ_4 is the wave function of spin-orbit split hole. The bar above ψ denotes the complex conjugation.
We can write matrix element M as the sum of two terms:

$M = M_1 + M_2$

where

$$M_1 = \frac{4\pi e^2}{\kappa_0(k_4^2 + q^2)}\left(F(\frac{a}{2})\int_{-\frac{a}{2}}^{\frac{a}{2}} \exp(qx_2)\overline{\psi_3(x_2)}\psi_2(x_2)dx_2 - F(-\frac{a}{2})\int_{-\frac{a}{2}}^{\frac{a}{2}} \exp(-qx_2)\overline{\psi_3(x_2)}\psi_2(x_2)dx_2 \right),$$

$$M_2 = \frac{4\pi e^2}{\kappa_0(k_4^2 + q^2)}\int_{-\frac{a}{2}}^{\frac{a}{2}} \overline{\psi_4(x_2)}\,\overline{\psi_3(x_2)}\psi_2(x_2)\psi_1(x_2)dx_2 \qquad (8)$$

M_1 corresponds to the thresholdless prosess, M_2 corresponds to the quasithreshold process. Here $F(x) = \frac{V_c}{E_g}\exp(-qx)\overline{\psi_4(x)}\psi_1(x)$, $q = |\boldsymbol{q}_1 - \boldsymbol{q}_4| = |\boldsymbol{q}_2 - \boldsymbol{q}_3|$.

Matrix elements M_1 and M_2 are proportional to overlap integrals $I_{14} = \int_{-\frac{a}{2}}^{\frac{a}{2}} \overline{\psi_4(x)}\psi_1(x)dx$ and $I_{23} = \int_{-\frac{a}{2}}^{\frac{a}{2}} \overline{\psi_3(x)}\psi_2(x)dx$. We can conclude that the subband mixing leads to the CH and HS overlap integrals decrease for the ground heavy hole state decrease because normalization coefficient H_1 which makes the largest contribution to these integrals becomes much smaller.
CHHS Auger process coefficients can be derived with the following formulas:

$$C_1 \approx \frac{2\pi^2 e^4}{\kappa_0^2}\frac{V_c}{E_g}\frac{k_c^2\kappa_c^2}{(k_c^2+\kappa_c^2)^2}\frac{F\left(\frac{\Delta}{E_g}\right)}{a^2(a+2\kappa_c^{-1})}\frac{\hbar^3}{m_{so}^2(\Delta - E_g)^3}\left\langle \frac{k_{h1}^2 k_{h2}^2 q_{h1}^2 (q_{h1}^2 + q_{h2}^2)}{(k_{h1}^2 + q_{h1}^2)^3 (k_{h2}^2 + q_{h2}^2)} \right\rangle, \qquad (9)$$

$$C_2 \approx \frac{\pi^2 e^4}{4\kappa_0^2}\frac{E_c}{E_g}\frac{k_c^2\kappa_c^2}{(k_c^2+\kappa_c^2)^2}\frac{F\left(\frac{\Delta}{E_g}\right)}{a^2(a+2\kappa_c^{-1})}\frac{\hbar^3}{m_{so}^2(\Delta - E_g)^3}$$

$$\times\left\langle \frac{1-cos(k_{so}-k_{h1}-k_{h2}-k_c)a}{(k_{so}-k_{h1}-k_{h2}-k_c)^2}\frac{q_{h2}^2\left\{(k_{so}^2+k_{h1}^2)q_{h1}^2 + q_{h2}^2 k_{h1}^2 + 2k_{h1}^2(\boldsymbol{q}_{h1}\cdot\boldsymbol{q}_{h2}) + [\boldsymbol{q}_{h1}\times\boldsymbol{q}_{h2}]^2\right\}}{k_{so}(k_{h1}^2+q_{h1}^2)(k_{h2}^2+q_{h2}^2)} \right\rangle, \qquad (10)$$

where $F(x)$ is weakly varying function in order of unity and angle brackets denotes averaging on the distribution function.

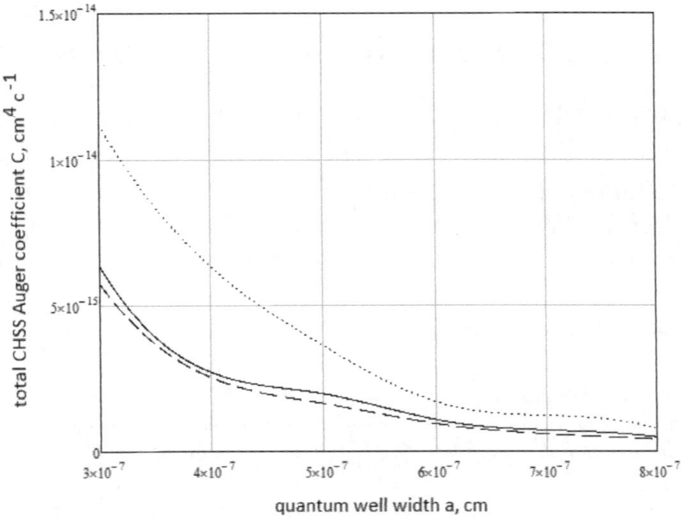

Figure 2. Dependence of total CHHS Auger recombination coefficient in $AlSb/InAs_{0.84}Sb_{0.16}/AlSb$ system on quantum well width.

Dependence of total CHHS Auger recombination coefficient in $AlSb/InAs_{0.84}Sb_{0.16}/AlSb$ system on quantum well width is shown in Fig.2. Result of calculation from formulas (9-10) with taking energy spectra nonsphericity and subband mixing into account is shown by solid line, result of numerical calculation from formula (7) is shown by dashed line and result of calculation from formulas (9-10) without taking energy spectra nonsphericity and subband mixing into account is shown by dotted line. We can see that taking energy spectra nonsphericity and subband mixing into account leads to decrease total CHHS Auger recombination coefficient.

4. Conclusions

This paper we investigated the effect of charge carrier energy spectra nonsphericity and the valence band subband mixing on the value of CHHS Auger recombination coefficient in the $AlSb/InAs_{0.84}Sb_{0.16}/AlSb$ system. The Kane equations are written from the nonspherical ***kp*** Hamiltonian. Charge carrier energy spectra and wave functions are obtained. It is shown that the subband mixing leads to CH and HS overlap integral and, therefore, Auger matrix element decrease. It is also shown that energy spectra nonsphericity and subband mixing leads to decrease total CHHS Auger recombination coefficient.

5. Aknowledgements

This work is supported in part by RFBR grant № 16-08-01130-A.

References

[1] Zegrya G G 1997 *Mid-infrared strained diode lasers* (Antimonide-Related Strained Layer Heterostructures) ed Manasreh M O (Gordon & Breach Science Publishers, Amsterdam) p 273
[2] Zegrya G G and Polkovnikov A S 1998 *J. Exp. Theor. Phys.* **86** 815

IOP Conf. Series: Journal of Physics: Conf. Series **929** (2017) 012090 doi:10.1088/1742-6596/929/1/012090

Bleaching compensation in GaAs/AlGaAs quantum wells by above-barrier illumination

I. A. Solovev, Yu. V. Kapitonov, V. G. Davydov, Yu. P. Efimov, S. A. Eliseev, V. V. Petrov and V. V. Ovsyankin

Saint Petersburg State University, Ulyanovskaya 1, Petrodvorets, St. Petersburg, 198504, Russia

E-mail: st024004@student.spbu.ru

Abstract. We report an experimental study of the nonlinear response of single GaAs/AlGaAs quantum well. It was shown that bleaching effect manifested as reversible exciton spectral lines broadening can be suppressed by additional above-barrier illumination.

1. Introduction

One of the most promising alternatives to current silicon-based electronics is the information processing in purely optical way [1–3]. Exciton and excitonic complexes in A3B5 nanostructures strongly interact with light due to the relatively large oscillator strength and could serve as a physical model of two-level system that draws special attention in the area of optical computing. However bleaching effect manifested as reversible exciton spectral line broadening limits usefulness of GaAs/AlGaAs structures in informational photonics.

Earlier it was shown that at relatively weak and intermediate intensities exciton bleaching is observed while spectrally integrated peak intensity in the absorption spectrum remains the same [4]. The main mechanism of bleaching is said to be broadening of the exciton line induced by scattering of excitons on carriers and exciton-carriers complexes. Carriers captured in QW reveal long-lived kinetics in microsecond scale [5]. Moreover PL signal from electron hole pair could be clearly detected up to tens of second after excitation due to non-exponential decay of carrier concentration [6, 7]. Bleaching effect could be reduced by lowering of Al percentage in $Al_xGa_{1-x}As$ QW barriers down to few percent [8].

Carriers could be observed even in nominally undoped quantum wells (QWs) grown by Molecular Beam Epitaxy (MBE) due to the unintentional doping. Presence of carriers in QW leads to the formation of a peak in the photoluminescence (PL) spectrum at energies below exciton recombination line [9] . This peak corresponds to the exction-carrier bound state – trion. Additional optical pump could lead either to the buildup or to the reduction of charge carriers concentration in QW depending on the pump spectral position relative to the band gap [10]. Alterations in carriers concentration or even in their sign [11] may be revealed by analysis of opposing changes in trion and exciton peak amplitudes in PL.

PL spectroscopy is generally used for detection of such charged exciton complexes [5, 9–11]. Still it could not give the full picture of the exciton dynamics due to the difficulty of the factorization of the exciton resonance width. We propose the Brewster angle reflection

IOP Conf. Series: Journal of Physics: Conf. Series **929** (2017) 012090 doi:10.1088/1742-6596/929/1/012090

spectroscopy [8, 12, 13] as a tool enabling independent measurement of the radiative width and nonradiative broadening of the exciton resonance. In this work we present a joint study of exciton resonance bleaching in single QW by means of PL and reflection spectroscopies and suggest possible means to control and eventually suppress this parasitic effect.

2. Experiment

The sample E296 grown by MBE contains nominally undoped single 12 nm GaAs QW with $Al_{0.28}Ga_{0.72}As$ barriers. We use Ti:Sapphire laser as a broadband probe for reflection spectra measurement. The sample was kept at 5 K in a closed-loop helium cryostate and spectra were detected by a spectrometer equipped with a CCD detector. An additional tuneable cw-laser, 650 nm and 532 nm lasers (refered further as red and green respectively) are used as pumping light sources for studying of light-induced bleaching under different excitation conditions.

The reflection spectrum is well described by the Lorentz curve [14, 15] when light is linearly polarized perpendicularly to the sample surface and angle of incidence coincides with Brewster angle:

$$K_R(\omega) = \frac{\Gamma_R^2}{(\omega - \omega_0)^2 + (\Gamma_R + \Gamma_{NR})^2}, \tag{1}$$

where ω_0 – frequency of exciton resonance, Γ_R – radiative width, Γ_{NR} – nonradiative broadening.

3. Results and discussion

In the case of resonant pump excitation to the light-hole exciton resonance (LH-exc) a bleaching of the reflection at the heavy-hole exciton resonance (HH-exc) was observed [8, 12]. This effect has monotonous dependency on the pump intensity and is related to the scattering of excitons on excessive pump generated carriers accumulated in the QW. Observation of electron-related shake-up lines in magneto-PL in the case of LH-exc excitation proved negative sign of carriers and unintentional n-doping of the sample [16].

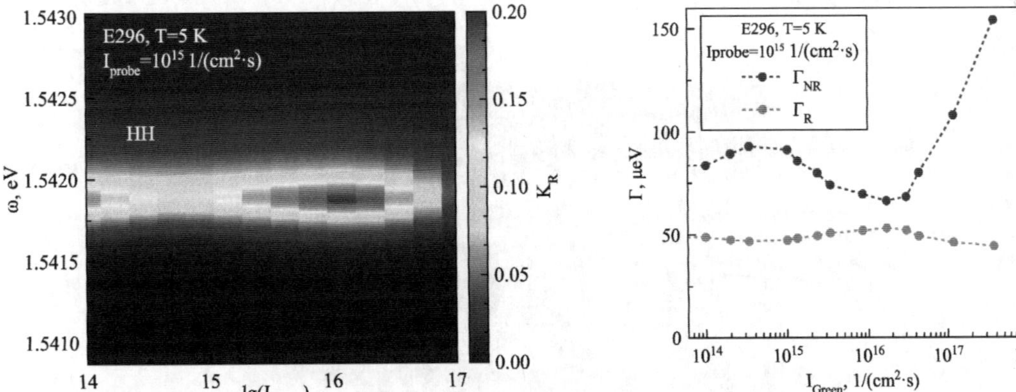

Figure 1. Dependence of $K_R(\omega)$ (a), Γ_R and Γ_{NR} (b) on above-barrier pump intensity I_{Green}.

On the contrary non-resonant above-barrier pump (λ= 532 nm) leads to the non-trivial behaviour of $K_R(\omega)$ (Fig.1). Parameters Γ_R and Γ_{NR} were extracted from reflection spectra using (1). At weak pump intensity reflection spectrum is similar to the unpumped case and broadens with rising pump intensity. But starting from I_{Green} = 10^{15} 1/(cm^2 s) the trend becomes opposite. Intermediate intensity pumping reduces bleaching – Γ_{NR} and $K_R(\omega_0)$ reach

IOP Conf. Series: Journal of Physics: Conf. Series **929** (2017) 012090 doi:10.1088/1742-6596/929/1/012090

its minimum and maximum respectively at $I_{Green} = 10^{16}$ 1/(cm ^{2}s). More intensive pump causes bleaching again.

Γ_R remains constant in the whole range of pump intensities, so $K_R(\omega)$ behaviour is determined by Γ_{NR} changes. We attribute non-monotonous Γ_{NR} dependence on I_{Green} to the change of the charge carriers density in QW. Above-barrier pump creates positively charged carriers flow into QW that recombine with resident electrons at $I_{Green} = 10^{16}$ 1/(cm ^{2}s). This situation is similar to one analyzed in [9–11], however we detect two reversing point with the monotonous rise of I_{Green}.

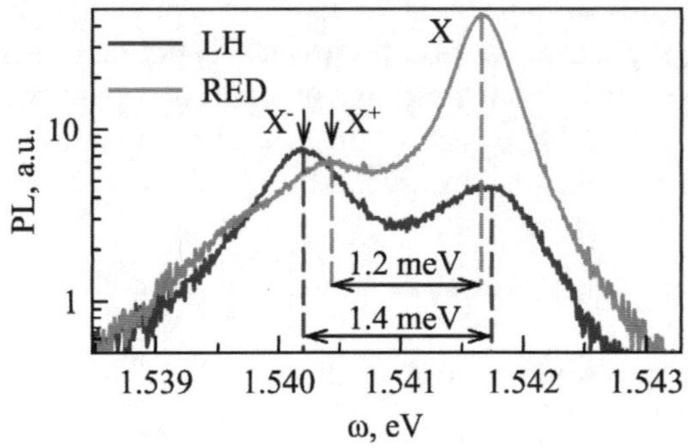

Figure 2. PL spectra at resonant (blue curve) and non-resonant (red curve) pumping.

Change of the sign of charge carriers in QW was further proved by PL spectroscopy. Fig 2 illustrates PL spectra at LH-exc and non-resonant above-barrier pumping (λ= 650 nm). In both cases at energies below HH-exc resonance new lines emerge. These lines correspond to the luminescence from charged exciton complexes – negative X^- and positive trions X^+ in the case of resonant and nonresonant above-barrier excitation respectively. Difference in trion signs is also proved by difference of trion binding energies (Δ_{X^-}= 1.4 meV and Δ_{X^+} = 1.2 meV).

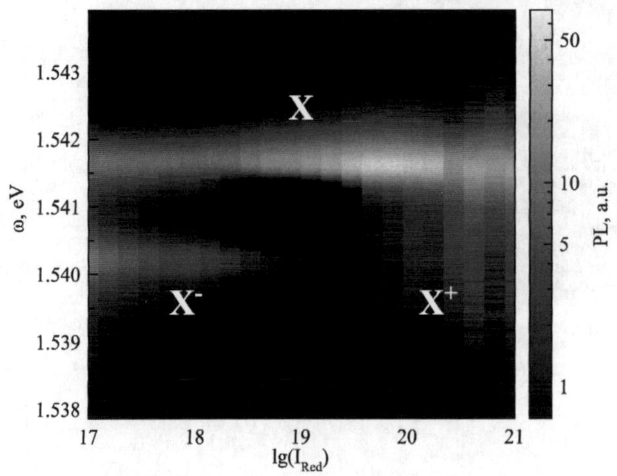

Figure 3. Dependence of PL spectrum on above-barrier pump intensity I_{Red}.

By fine tuning of above-barrier pumping intensity we can observe change from negative trion state to the positive one through the point of full charge compensation in QW (Fig. 3) as in [11].

At compensation point PL signal from neutral exciton state (X) reaches its maximum and PL signal from trions disappears. Quenching of X^- line is a result of compensation of resident electrons in QW. Further accumulation of holes in QW causes buildup of the peak corresponding to the recombination of X^+ trions.

4. Conclusion

By means of PL and reflection spectroscopy the bleaching effect in GaAs/AlGaAs QW in the presence of above-barrier illumination was investigated. The independence of the radiative width Γ_R on the pump intensity was found. At certain above-barrier illumination intensity nonradiative broadening Γ_{NR} reaches its minimum corresponding to the compensation of exceeding charge carriers within QW. As a result above-barrier illumination could be used for control of carriers sign as well as their full compensation.

5. Acknowledgments

The reported study was partially supported by RFBR, research project № 14-29-07247 nnio_a and № 16-29-03115 ofi_m. The work of I.A.S. was supported by the German-Russian Interdisciplinary Science Center (G-RISC) funded by the German Federal Foreign Office via the German Academic Exchange Service (DAAD). This work was carried out on the equipment of the SPbU Resource Center "Nanophotonics" (photon.spbu.ru).

References

[1] Allen L and Eberly J H 1975 *Optical Resonance and Two-Level Atoms*, (New-York: Wiley)
[2] Gerlovin I Ya et al 2000 *Nanotechnology* **11** 383-86
[3] Gerlovin I Ya et al 2000 *J. of Luminescence* 8789 421-22
[4] Wake D R et al 1992 B **46** 20 13452-60
[5] Kochiev M V et al 2015 *Jetp Lett.* **101** 183-88
[6] Gilinsky A M and Zhuravlev K S 2001 *Appl. Phys. Lett.* **79** 3455
[7] Reshchikov M A et al 2003 *Appl. Phys. Lett.* **83** 266
[8] Solovev I A et al 2015 *J. Phys.: Conf. Series* **643** 012085
[9] Osborne J L et al 1996 *Phys. Rev.* B **53** 13002-10
[10] Sibeldin N N et al 2001 *Nanotechnology* **12** 591–96
[11] Volkov O V et al 1998 *Jetp Lett.* **67** 744-50
[12] Poltavtsev S V and Stroganov B V 2010 *Phys. of the Solid State* **52** 1899-905
[13] Poltavtsev S V et al 2014 *Solid State Comm.* **199** 47-51
[14] Benedict M G and Trifonov E D 1988 *Phys. Rev* A **38** 2854
[15] Ivchenko E L 2004 *Optical Spectroscopy of Semiconductor Nanostructures* (Berlin: Springer)
[16] Shapochkin P Yu et al 2015 *23rd Int. Symp. Nanostructures: Physics and Technology* (St. Petersburg: St. Petersburg Academic University).

IOP Conf. Series: Journal of Physics: Conf. Series **929** (2017) 012091 doi:10.1088/1742-6596/929/1/012091

Plasmonic enhancement of photoluminescence from cadmium sulfide and lead sulfide quantum dots

L Matyushkin and A Percova

Department of micro- and nanoelectronics, St. Petersburg Electrotechnical University LETI, Prof. Popov str. 5, St. Petersburg 197376, Russia

E-mail: leva.matyushkin@gmail.com, anastasijapercova@gmail.com

Abstract. Plasmonic core/shell nanoparticles Ag/SiO_2 with different shell thicknesses were synthesized. Localized plasmon resonance was investigated by UV-vis spectroscopy. Arranged layers of plasmonic nanoparticles and quantum dots were obtained on a quartz substrate using a spin-coating technique. Plasmon-enhanced photoluminescence excited by 405 nm laser corresponding to the plasmon resonance was studied on CdS and PbS quantum dots structures. The possibilities of twofold and fourfold intensity increase are demonstrated for CdS and PbS quantum dots respectively.

1. Introduction

In metallic nanoparticles, where the electrons are confined in the three dimensions, the electron density oscillations (plasmons) induce an electric field around the nanoparticle that can be much larger than the incident light one [1]. Such plasmonic particles open the possibility to amplify, concentrate and manipulate electromagnetic field at the nanoscale and enhance light absorption and luminescence of organic fluorophores [2, 3, 4] or inorganic quantum dots (QDs) [5, 6, 7] located near the metallic nanoparticles. Moreover, the metallic particle can modify the ratio of radiative and non-radiative QD decay rates and alter its quantum efficiency [8].

However, luminescence can be strongly quenched when a distance between plasmonic and emitting particles is small [9]. One of the reasons is that a charge carrier from an electron-hole pair photoexcited in an emitting particle can tunnel and be captured by metallic nanoparticle, so radiative recombination won't happen (Figure 1, the leftmost piece). In this case, the luminescence is undesirably reduced in comparison with a system consisting only of radiating particles. To avoid this effect, it is necessary to suppress tunneling by creating a barrier between the particles. Such a barrier is usually made by a dielectric [10] or polyelectrolyte [5] spacer between plasmonic and emitting layers. The spacer thickness is typically ones or tens of nanometers, and there is some optimum thickness of the layer whereby maximal enhancement is observed. On the one hand, spacer thickness must be thick enough to suppress charge carriers tunneling, on the other hand, spacer thickness should be thin enough so the metal particle electromagnetic field can effectively impact the emitting particle.

In this paper, we propose instead a specially created layer between metal and semiconductor nanoparticles layers, use metal/dielectric nanoparticles containing a spacer as a shell of the structure. Possible ways to specify the distance between plasmonic and emitting particles in such way are shown in Figure 1. For instance, the metallic nanoparticle can be covered by SiO_2

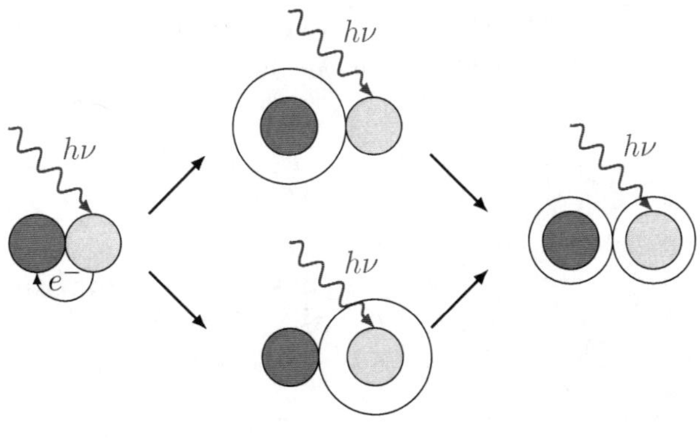

Figure 1. The scheme of luminescence quenching problem and options for creating a spacer between a metal nanoparticle and a quantum dot

shell while emitting nanocrystals can be covered by wide-bandgap semiconductor material shell to obtain type-I heterojunction nanostructures. In the most general case, the spacer may be distributed between the plasmon and quantum dot nanoparticles, so both types of particles can be coated. In this article emitting particles are semiconductor nanocrystals of cadmium sulfide CdS and lead sulfide PbS, covered with non-conducting oleic acid molecules. Plasmon particles are Ag/SiO_2 — silver core particles coated by silica layer of different thickness. All of these particles types are obtained by colloid chemistry techniques.

2. Experiment

Synthesis of 20 nm in diameter monodisperse silver nanoparticles was produced by the known procedure [11]. For the coating of silver nanoparticles by silica, it was applied a modified Stober technique, described earlier for gold nanoparticles [12]. Different shell thickness h_S was achieved using different tetraethoxysilane (TEOS, silica precursor) concentrations. The following formula was derived of geometric and chemical reasons to determine TEOS amount V_P needed to add into a given volume of the sol V_{sol} of silver core particles to obtain specified h_S:

$$V_P = \frac{\eta_C}{\eta_S} \cdot \frac{M_C M_P}{M_S} \cdot \frac{\rho_S}{\rho_C \rho_P} \cdot \frac{V_{CS} - V_C}{V_C} \cdot C_C \cdot V_{sol}, \tag{1}$$

where p, c, s are indices correspond to the precursor shell, shell material and the core material, respectively, η — chemical yields of reactions for the formation of core and shell, M — molar masses, ρ — densities, d_C — core diameter, C_C — concentration of core nanoparticles. The average diameter of the silver core particles d_C was calculated by following linear equation [13]:

$$d_C = \frac{\lambda_{max} - 382.6}{1.18} \tag{2}$$

The results obtained from formula (1) are in agreement with atomic force microscopy data for Ag/SiO_2 nanoparticles deposited separately on an atomically smooth silicon surface.

Absorption spectra of Ag/SiO_2 particles are shown in Figure 2. As it is evident from spectra, shell-covered particles retain their monodispersity. The thicker a shell, the more red-shifted plasmon resonance peak due to the environment dielectric constant alteration (Figure 2, *b*). With such thin shells environment, the dielectric permittivity can be considered as combined dielectric permittivity of silicon dioxide and an aqueous medium. Particles were synthesised with the same core diameter d_C = 20 nm and different shell thickness h_S = 15, 25, 35 and 40

IOP Conf. Series: Journal of Physics: Conf. Series **929** (2017) 012091 doi:10.1088/1742-6596/929/1/012091

nm. These shell thicknesses were chosen because we expected the maximum enhancement when the spacer is less than 50 nm due to our preliminary work.

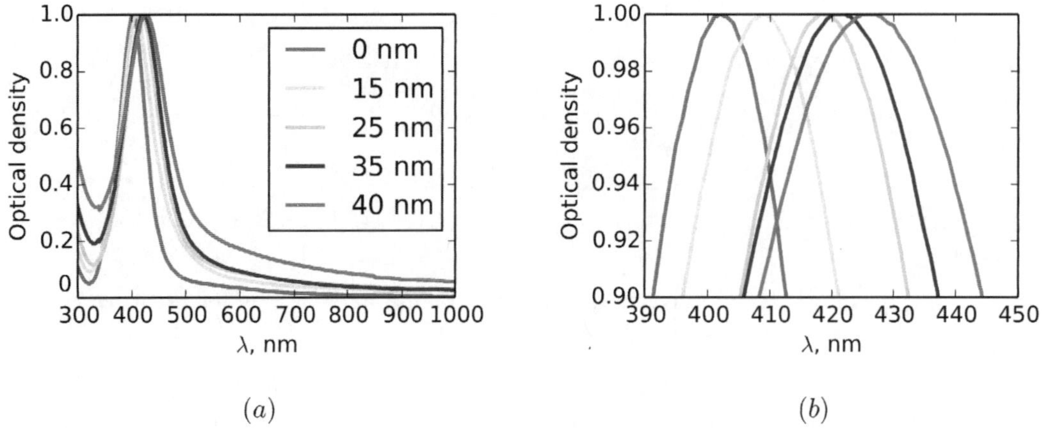

Figure 2. (a) Absorption spectra of Ag/SiO_2 particles with different shell thickness and (b) the same spectra on a larger scale near the plasmon resonance maxima (T = 300 K)

Materials of quantum dots were selected as significantly differ in their optical properties, but rather similar in chemical nature. Absorption and photoluminescence (PL) spectra of colloidal quantum dot solutions are shown in Figure 3.

The PL maximum is Stokes-shifted from the first exciton peak in the absorption spectrum. In the case of CdS QDs besides narrow fundamental luminescence band, the wide PL band is observed in the long-wavelength region, related to unpassivated surface states in the core material, which also display radiative recombination. Spectral characteristics shape and peak positions stay the same for quantum dots, applied on the substrate.

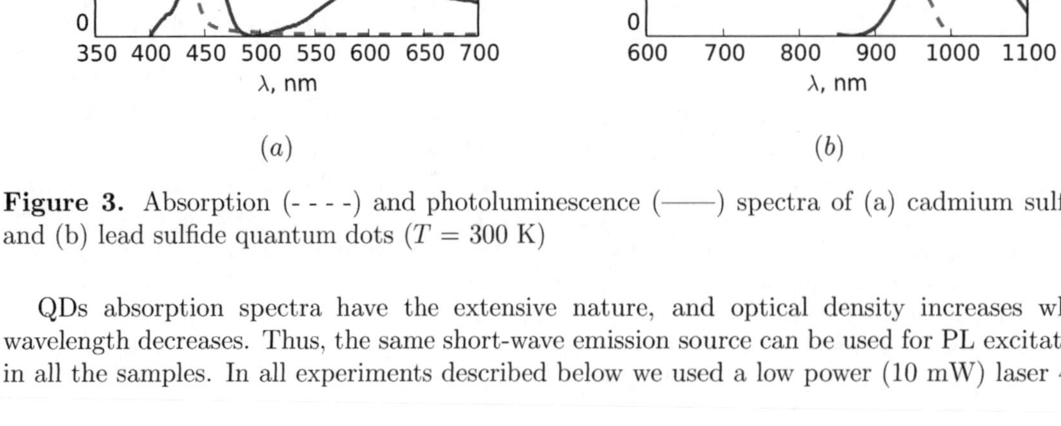

Figure 3. Absorption (- - - -) and photoluminescence (——) spectra of (a) cadmium sulfide and (b) lead sulfide quantum dots (T = 300 K)

QDs absorption spectra have the extensive nature, and optical density increases when wavelength decreases. Thus, the same short-wave emission source can be used for PL excitation in all the samples. In all experiments described below we used a low power (10 mW) laser 405 nm, corresponding to the simultaneous excitation of the plasmon resonance in Ag/SiO_2 particles and efficient PL excitation of semiconductor nanocrystals.

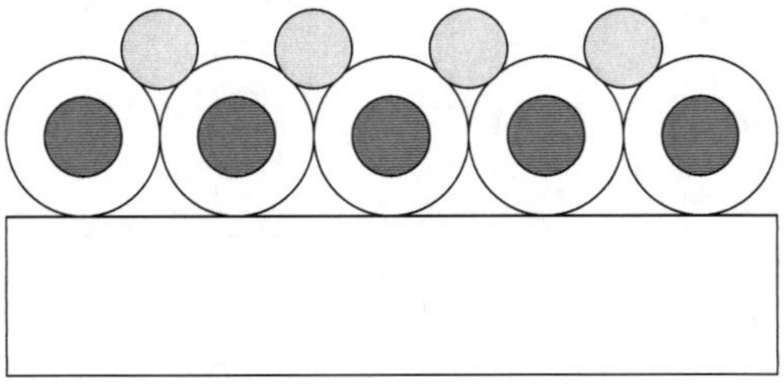

Figure 4. The scheme of structures for study of the plasmon luminescence enhancement

The general view of structures produced for the study of plasmon-enhanced luminescence is in Figure 4, where Ag/SiO_2 — nanoparticles with different shell thickness (0 — no shell, 15, 25, 35 and 40 nm) and QDs — quantum dots of different material (CdS or PbS). For obtaining these structures the following procedure was carried out: 20 μl of plasmonic nanoparticles aqueous solution was deposited on a quartz substrate by spin-coating. Then the substrate was placed in a furnace at 200°C for 5 min. to remove a solvent. To achieve a continuous uniform layer of plasmonic particles described procedure was repeated 5 times. Then the procedure was repeated for QDs solutions, but the solvent removal was carried out at 100°C temperature. Control samples were layers of quantum dots on quartz glass substrate without the underlayer plasmonic particles.

The obtained structures further were investigated by PL spectroscopy. The excitation of the samples PL was carried out by 405 nm laser from above through a thin layer of quantum dots. The spectra were measured by fluorescence spectrometer based on a diffraction grating and a silicon photodiode. Laser beam falls at 45° angle, but excited PL from QDs is normal to the input slit of monochromator. Polarization of incident and emitted light was not studied in this paper.

3. Results and discussion

For describing luminescence enhancement let's preliminarily introduce the concept of the enhancement factor:

$$K = \frac{\max(I(\lambda))}{\max(I_0(\lambda))} \tag{3}$$

where $I(\lambda)$ and $I_0(\lambda)$ — structure photoluminescence spectra in the presence and absence of plasmonic particles correspondingly. Photoluminescence spectra of CdS and PbS structures in compose with different shell thickness Ag/SiO_2 nanoparticles are shown in Figure 5.

As it can be seen from CdS photoluminescence spectra luminescence associated with defect states also enhances. The largest enhancement factors (Figure 5, right) $K = 1.9$ for CdS and $K = 4.2$ for PbS are observed at $h_S = 35$ nm for both types of QDs. However, it should be taken into account that the thickness of the shells had discrete values and the optimum thickness can vary for different types of particles. Shapes of enhancement factor curves are also similar for both types of particles, despite their different PL wavelength region. CdS QDs show low quantum efficiency compared to PbS QDs and substantially their PL is determined by defect states on the surface, so we see that wide PL band related to those states is strongly enhanced. The sharp nature of the amplification curve is most likely due to the fact that both tunneling effects and plasmon field amplification effects are nonlinearly dependent on the distance. As

IOP Conf. Series: Journal of Physics: Conf. Series **929** (2017) 012091 doi:10.1088/1742-6596/929/1/012091

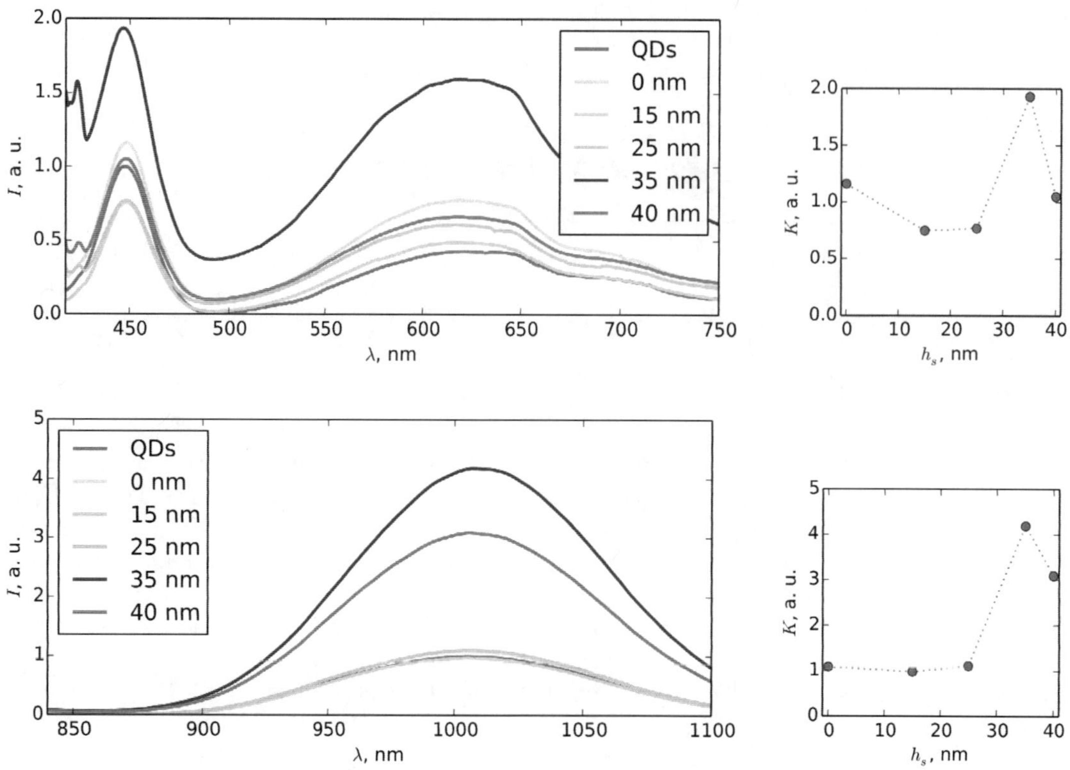

Figure 5. The photoluminescence spectra of CdS (top) and PbS QDs in the presence of plasmonic particles with different shell thickness. Right plots illustrate the dependence of the enhancement factor on the shell thickness

follows from the absorption spectra of semiconductor and metallic nanoparticles (Figures 2 and 3), the techniques used to synthesize nanoparticles make it possible to obtain monodisperse nanocrystals, so we do not observe broadening of the gain characteristics that could be observed with an appreciable dispersion of particle diameter and shell thickness.

We assume that maximizing the effect on some optimal thickness is explained not only by considerations described in the introduction (the competition the tunneling of charge carriers and enhancing the electromagnetic field) but also by factors such as the total change of light scattering when using thicker shells (Figure 2). Some impact can be affected by changing the diameter ratio plasmonic nanoparticles and quantum dots. In addition, quantum dots of cadmium sulfide and lead sulfide have significantly different dielectric constants, which as we noted earlier affects the plasmon resonance of particles. This diversity of effects requires a clear theoretical consideration, but in this short paper, we only show the experimental fact of the enhancement effect itself.

4. Conclusions

The effect of the plasmon enhancement of photoluminescence was investigated for CdS and PbS nanoparticles in the presence of plasmon nanoparticles Ag/SiO_2 with different thickness of the dielectric coating (15 – 40 nm). Increasing of PL intensity for all the studied QDs is observed at some optimum spacer thickness, which is consistent with the results of the literature review.

The greatest gain for investigated photoluminescence quantum dots are observed at $h_S \approx 35$ nm with maximal enhancement coefficients: $K = 4.2$ for PbS and $K = 1.9$ for CdS quantum dots.

This approach, when instead of a system of metal particles layer/insulator layer, we used the composite particles of metal/insulator simplifies the process of creating structures with quantum dots, where it required the plasmonic enhancement of the absorption of light or luminescence. Our research shows that this enhancement can be very significant but it depends on the quantum dots, their optical properties, and quantum efficiencies.

References

[1] Garcia M 2011 *Journal of Physics D: Applied Physics* **44** 283001
[2] Tam F, Goodrich G P, Johnson B R and Halas N J 2007 *Nano letters* **7** 496–501
[3] Chen Y, Munechika K and Ginger D S 2007 *Nano letters* **7** 690–696
[4] Chowdhury M H, Ray K, Gray S K, Pond J and Lakowicz J R 2009 *Analytical chemistry* **81** 1397–1403
[5] Kulakovich O, Strekal N, Yaroshevich A, Maskevich S, Gaponenko S, Nabiev I, Woggon U and Artemyev M 2002 *Nano Letters* **2** 1449–1452
[6] Munechika K, Chen Y, Tillack A F, Kulkarni A P, Plante I J L, Munro A M and Ginger D S 2010 *Nano letters* **10** 2598–2603
[7] Wu J, Lee S, Reddy V, Manasreh M, Weaver B, Yakes M, Furrow C, Kunets V P, Benamara M and Salamo G 2011 *Materials Letters* **65** 3605–3608
[8] Geddes C D and Lakowicz J R 2002 *Journal of Fluorescence* **12** 121–129
[9] Dubertret B, Calame M and Libchaber A J 2001 *Nature biotechnology* **19** 365–370
[10] Okamoto K, Niki I, Shvartser A, Narukawa Y, Mukai T and Scherer A 2004 *Nature materials* **3** 601–605
[11] Li H, Xia H, Wang D and Tao X 2013 *Langmuir* **29** 5074–5079
[12] Mine E, Yamada A, Kobayashi Y, Konno M and Liz-Marzán L M 2003 *Journal of colloid and interface science* **264** 385–390
[13] Van Hoonacker A and Englebienne P 2006 *Current Nanoscience* **2** 359–371

IOP Conf. Series: Journal of Physics: Conf. Series **929** (2017) 012092 doi:10.1088/1742-6596/929/1/012092

Assessing the impact of the discharge conditions on the accommodation coefficient of electrons in plasma of helium and neon in narrow extended tubes

A S Kiselev and E A Smirnov

Department of electronic instruments and devices, Saint Petersburg Electrotechnical University "LETI", 197376, Saint Petersburg, Russia

E-mail: alex.kiselev.epu@gmail.com

Abstract. In this work current-voltage characteristics of a glow discharge in neon and helium in long discharge tubes of various diameters at different pressures are investigated. Dependence of the longitudinal gradient of the plasma potential in the positive column of the pressure for different diameter of the discharge tube is found. Depending on the accommodation coefficient of electrons in various diameters, pressures and gas filling discharge gaps are obtained. The possibility of using described techniques for the calculation of parameters of the plasma glow discharge laser is shown.

Gas-discharge device parameters are determined by current-voltage characteristic (CVC) of the discharge gap. CVC provides understanding of the state of the active medium, including the extent of atmospheric gases leakage [1]. Plasma impedance affects the current discharge circuit dynamic stability [2]. The impedance to a large extent depends on the values of the static and differential resistances. In addition, the type of CVC has significant influence on the dynamic properties of the discharge. Current-voltage characteristic view is determined largely by the behavior of the longitudinal potential gradient. The longitudinal potential gradient E_z in the positive column (PC) glow discharge is proportional to the square root of the accommodation electrons coefficient χ. Accommodation coefficient of electrons is called the average proportion of the total electron energy W_e that is lost at various collisions with gas particles. These losses are related to heating and gas emission due to elastic and inelastic processes, energy transfer to walls. The longitudinal potential gradient E_z in the positive column is associated with the coefficient of accommodation by the following expression [3]:

$$E_z = \frac{1,5kT_e\sqrt{\chi}}{e\lambda_e}, \tag{1}$$

where T_e – the electron temperature, $\lambda_e = \lambda_{e0} / p$ – the mean free path of electrons in the gas, p – gas pressure.

Function $\chi = f(T_e)$ can be found taking into account shares of energy lost by electrons in elastic and inelastic collisions:

$$\chi = \chi^* + \chi_a + \chi_w, \tag{2}$$

where $\chi^* = 2m_e/M$ – is the average fraction of the energy lost by electron in the elastic collision; χ_a – average fraction of energy loss due to the excitation; χ_w – share of energy losses at the walls.

The coefficient of accommodation is a complex function of the discharge conditions. Rigorous calculation χ for the discharge conditions in the narrow long tubes is difficult due to the diversity and complexity of the processes occurring in the plasma. Experimental data about χ in pure gases there are in the literature [3, 4]. They reflect the relationship $\chi = f(W_e)$ for a very narrow range of discharge conditions. Using values χ when calculating of the reference literature is difficult.

Maquette with the four discharge tubes with an internal diameter of 1.1, 2, 3.4, 5.2 mm used in the experiments, with movable anodes. Anodes are moved by magnetic drive within the change length PS 0.1 ... 0.25 m. Maquette was connected to a vacuum pumping and filling system, allowing to vary the gas filling and the operating pressure. The characteristics of the discharge in pure helium and neon are investigated in this study. In the study was an attempt to clarify the accommodation coefficient values in order to increase the accuracy of calculations of the positive column of plasma parameters. The idea χ refinement based on the use of semi-experimental approach to the problem. The essence of the approach is as follows. Initially information about the behavior of E_z in a wide range of discharge conditions based on the experimental current-voltage characteristics was obtained (figures 1, 2).

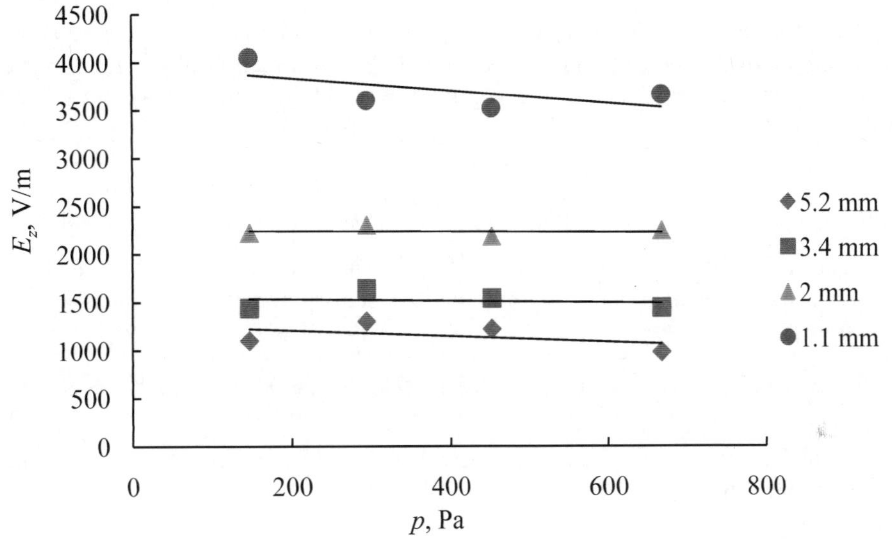

Figure 1. Experimental dependence of $E_z = f(p, d)$ for a neon

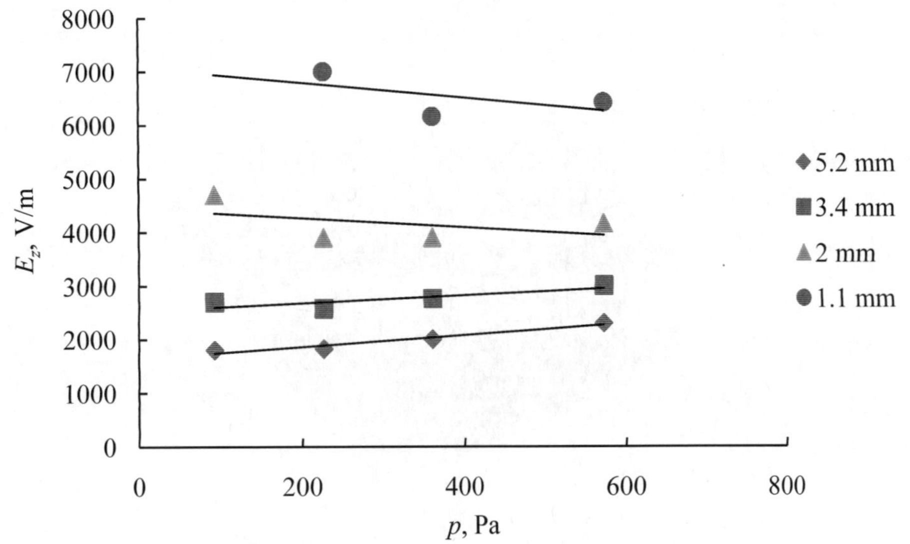

Figure 2. Experimental dependence of $E_z = f(p, d)$ for a helium

Next, the electron temperature using the developed computer program calculated T_e [5]. Changes ranges of pressure p, of the discharge channel diameter d, of the gas mix correspond to the experimental conditions. Method of calculation T_e is based on the decisions transcendent ionization balance equation. The plasma is contained in a cylindrical tube with an internal radius R. Recombination in the gas volume is negligible, and charge neutralization occurs mainly due to ambipolar diffusion with subsequent recombination on the walls. The frequency of ionization ν_i with considering the expression for the coefficient of ambipolar diffusion is defined as [3]

$$\nu_{\mathrm{i}} = \frac{b_{\mathrm{i}} k (2{,}405)^2 T_{\mathrm{e}}}{R^2 e}. \tag{3}$$

where $b_{\mathrm{i}} = b_{\mathrm{i0}} / p$ – mobility of ions.
On the other hand the average frequency ionization is recorded as

$$\nu_{\mathrm{i}} = n_{\mathrm{a}} \int_0^{\infty} q_{\mathrm{i}}(U) \upsilon(U) f(U) dU, \tag{4}$$

where n_{a} – is the concentration of atoms; q_{i} – ionization cross section; $f(U)$ – Maxwell distribution function for electrons over energies. Approximations of the cross section of ionization for atoms and molecules were assumed linear. Thus, after integrating, was obtained an expression for the frequency of ionization:

$$\nu_{\mathrm{i}} = n_{\alpha} \left(\frac{8kT_{\mathrm{e}}}{\pi m_{\mathrm{e}}} \right)^{0,5} \cdot \alpha_{\mathrm{i}} \exp\left(\frac{-eU_{\mathrm{i}}}{kT_{\mathrm{e}}} \right) \left(U_{\mathrm{i}} + \frac{2kT_{\mathrm{e}}}{e} \right). \tag{5}$$

The transcendental equation (5) is applicable for the single component active media of the gas discharge lasers. For multicomponent gas mixtures the sum of ionization frequencies of all gases within the mixture should be used:

$$\sum_j \nu_{\mathrm{i}}(j) = \frac{(2{,}405)^2 b_{\mathrm{i}} k T_{\mathrm{e}}}{R^2 e} = \frac{23{,}1 b_{\mathrm{i}} k T_{\mathrm{e}}}{e d^2}.$$

On the basis of the analytic relationship (1) between E_z and T_e and equating the calculated of the longitudinal gradient of the experimental value was obtained the family dependencies $\chi = f(p, d,$ kind of gas) (figures 3, 4).

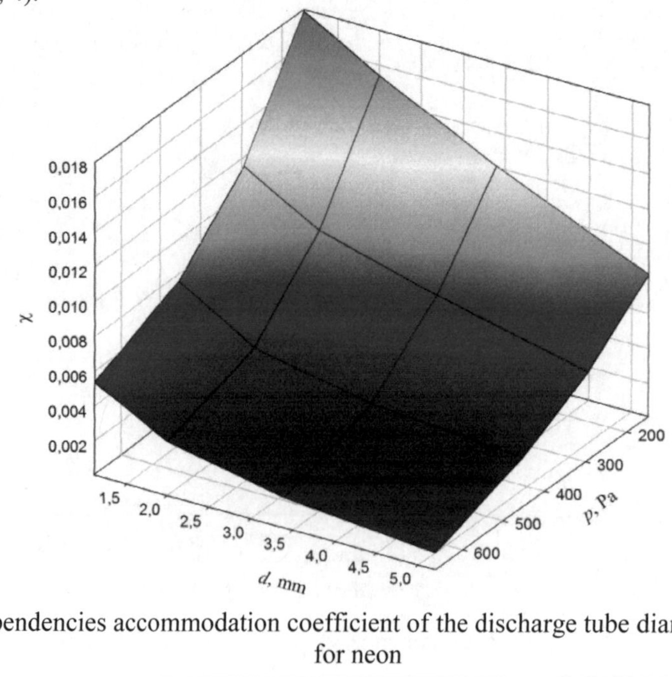

Figure 3. Dependencies accommodation coefficient of the discharge tube diameter and pressure for neon

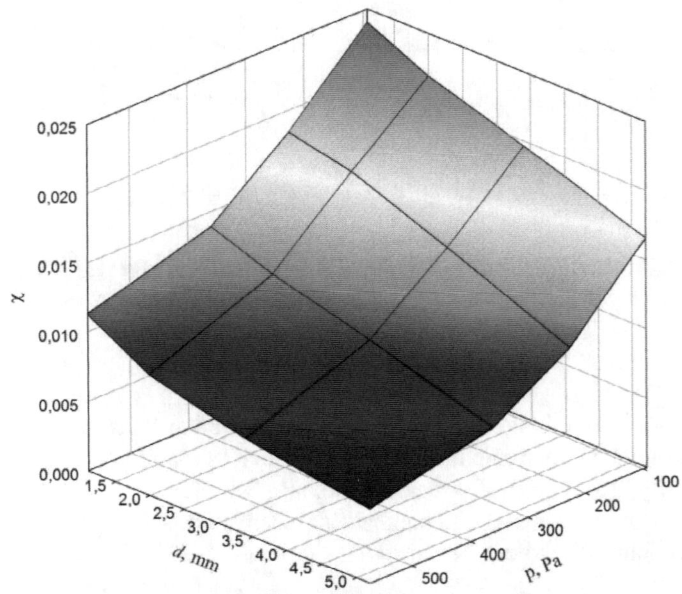

Figure 4. Dependencies accommodation coefficient of the discharge tube diameter and pressure for helium

The admissibility of a semi-empirical approach to assessing χ confirmed by the coincidence with an accuracy of 10 ... 15% with the results obtained for the pure helium and neon, with those known from the literature [3]. Satisfactory agreement between the experimental and reference data makes it possible to draw a conclusion about the applicability of methods for determining the accommodation coefficient for the glow discharge lasers. In further studies we intend to extend the ranges of changes of the discharge conditions and gas mixtures, including typical for glow discharge lasers, such as He-Ne and CO2 lasers.

References

[1] Smirnov E A, Morozov S S and Timoshichev A A 2008 *Petersburg Journal of Electronics* **2–3** 39–48
[2] Smirnov E A 2005 *Proceeding of Saint Petersburg Electrotechnical University "LETI"* **2** 43–8
[3] Granovsky V L 1971 *Electrical current in gas* (Moscow: Science)
[4] Levitsky S M 1964 *Collection of problems and calculations on physical electronics* (Kiev: Publishing House of Kiev University)
[5] Kiselev A S and Smirnov E A 2014 *Vacuum technique and technology* **1** 56–9

Plasma of a vacuum-arc discharge for obtaining carbon-based coatings

M I Pikus[1], A A Lisenkov[2], D K Kostrin[1] and S A Trifonov[1]
[1] Department of electronic instruments and devices, Saint Petersburg Electrotechnical University "LETI", 197376, Saint Petersburg, Russia
[2] Institute of Problems of Mechanical Engineering, Russian Academy of Sciences, 199178, Saint Petersburg, Russia

E-mail: mar.pikus@mail.ru, lisran@yandex.ru

Abstract. Formation of the carbon films and coatings of different structural modifications can be efficiently achieved by sputtering graphite in a vacuum-arc plasma source. In this case the plasma flux is shaped in a form of a current-carrying jet with fairly distinct lateral borders. Spectral analysis shows that this plasma flux contains positively charged, excited and neutral carbon particles. In this paper is shown that for a technological cycle of deposition of a carbon-based coating it is possible to form a sublayer of the substrate material carbide, providing a possibility for the later growth of a well-formed coating.

1. Introduction

Carbon materials and coatings are used in various fields of science and technology. The unique ability of carbon is the possibility of receiving mixed electronic structures, along with the basic *sp*^{3}-, *sp*^{2}- and *sp*-configurations. Obtaining of the carbon material by sputtering presents an ability of acquiring an infinite number of its condensed forms [1]. Technological methods and conditions of film deposition determine the physical properties of the received materials.

For obtaining materials consisting of carbon atoms the heat treatment of various carbonaceous substances in both the condensed and gas phases is necessary [2]. From the gas phase it is possible to receive soot, fullerenes, pyrocarbon (pyrographite) and diamonds. If soot and fullerenes are the result of processes occurring in the reaction volume, the pyrocarbon and diamonds are formed by deposition of carbon atoms on a solid substrate.

Among the materials built from carbon atoms a special place holds so-called active carbon. It can be represented as a structure made from the condensed carbon atoms, forming a certain pattern. Its distinction from the other known compounds is a different degree of intra- and interlayer disordering of atoms. In this structure, the carbon has a so-called pore space, the volume and pore size of which is determined by the size of the primary crystallites, the nature of their packing and their relative orientation.

2. Features of the coating deposition from a vacuum-arc discharge

Significant progress in the formation of carbon films and coatings of different structural modifications was achieved by sputtering graphite in a vacuum-arc plasma source [3, 4]. The nature of the particle deposition from a plasma flux generated by a vacuum-arc discharge depends on their energy, which determines not only the adhesion of the formed coating to the substrate, but also the structure and composition of a film and the presence of defects.

When particles contact the surface, their interaction occurs, depending on the energy of the chemical bonds between the atoms in the treated surface, the surface electric fields arising due to the asymmetry of the crystal lattice on the surface, lattice constants and the substrate temperature [5].

If ions interacting with the surface have so high energy that completely free substrate atoms from the chemical bonds, the physical sputtering of the surface is observed. During ion implantation, the ions penetrate into the crystal lattice and remain in it [6]. Upon the surface impact deposited particles are relived from their energy and transferred to the adsorbed state. Further, the process of condensation on the substrate surface is determined by a sufficiently large diffusion mobility of the particles.

In the formation of the coating there are three types of adhesion: mechanical – due to the van der Waals forces; chemical – occurs with the formation of new phases at the interface, its strength is higher, the greater the chemical affinity between the coating and the substrate materials; diffuse – occurs when the substrate temperature is such that there are the processes of materials counter-diffusion.

The quality of the formed coating and the possibility of the textured coating growth are determined by the conditions of condensation, and for each pair condensate–substrate, at a given deposition rate, there is a temperature at which the growth of a crystal-oriented film is observed.

Thus, the process of the coating formation from the plasma of a vacuum-arc discharge includes two stages: heating of the machined part and modification of the structural and phase composition of the surface; condensation of the working substance from the plasma flux. The flow of each process is determined by the energy of deposited ions and the flux density to the machined part [7].

3. Results and discussion

As a coating material in this work was used carbon obtained by a thermal decomposition of a high-density graphite MPG-6 [8] in the plasma of a vacuum-arc discharge, existing in vapors of the eroded cathode. Spraying was carried out from the cathode spot region, which as a heat source on the cathode surface can be modeled in a form of a circle. As movement speed of the cathode spot on the graphite cathode is rather small, significant roles in the spraying process have the temperatures of the cathode and in the region of the cathode spot [9].

Sprayed particles form a directed plasma flux (figure 1) including three groups of particles: charged (ions and electrons that initiate chemical reactions in a working volume), excited and neutral (initial substances, atoms, radicals, intermediate products of the reactions).

Figure 1. Cathode spot on the working surface of a graphite cathode.

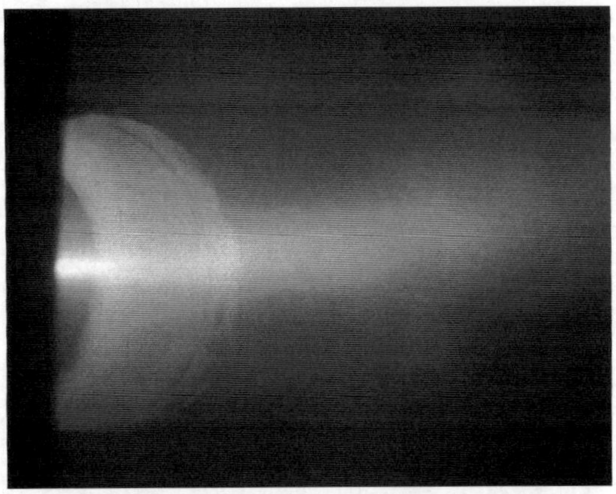

Figure 2. Current-carrying plasma jet providing the shorting of an arc discharge in vacuum.

The role of the plasma electron component in the initiation of the reactions is crucial. Distinctive feature of sputtering a graphite cathode in vacuum (10^{-3} Pa and below) is that the plasma flux providing shorting of the arc discharge from the cathode to the anode is in a form of a current-carrying jet with fairly distinct lateral borders (figure 2). The jet in each time point is arbitrarily shorted to the walls of a vacuum chamber, acting as the anode. It is assumed that the kinetic pressure of the plasma $p = n_e T_e + n_i T_i$ at the flux boundary is balanced by a transverse compression of the cathode plasma jet by an intrinsic magnetic field.

When graphite is sputtered by a cathode spot of a vacuum-arc discharge (p = $7.8 \cdot 10^{-3}$ Pa, I_{arc} = 80 A), in the plasma flux are recorded positively charged (CII, CIII), excited (C*) and neutral (C) carbon particles, as well as molecules having a linear structure, the carbon material conglomerates and the composite particles resulting from the connection of several particles (figure 3).

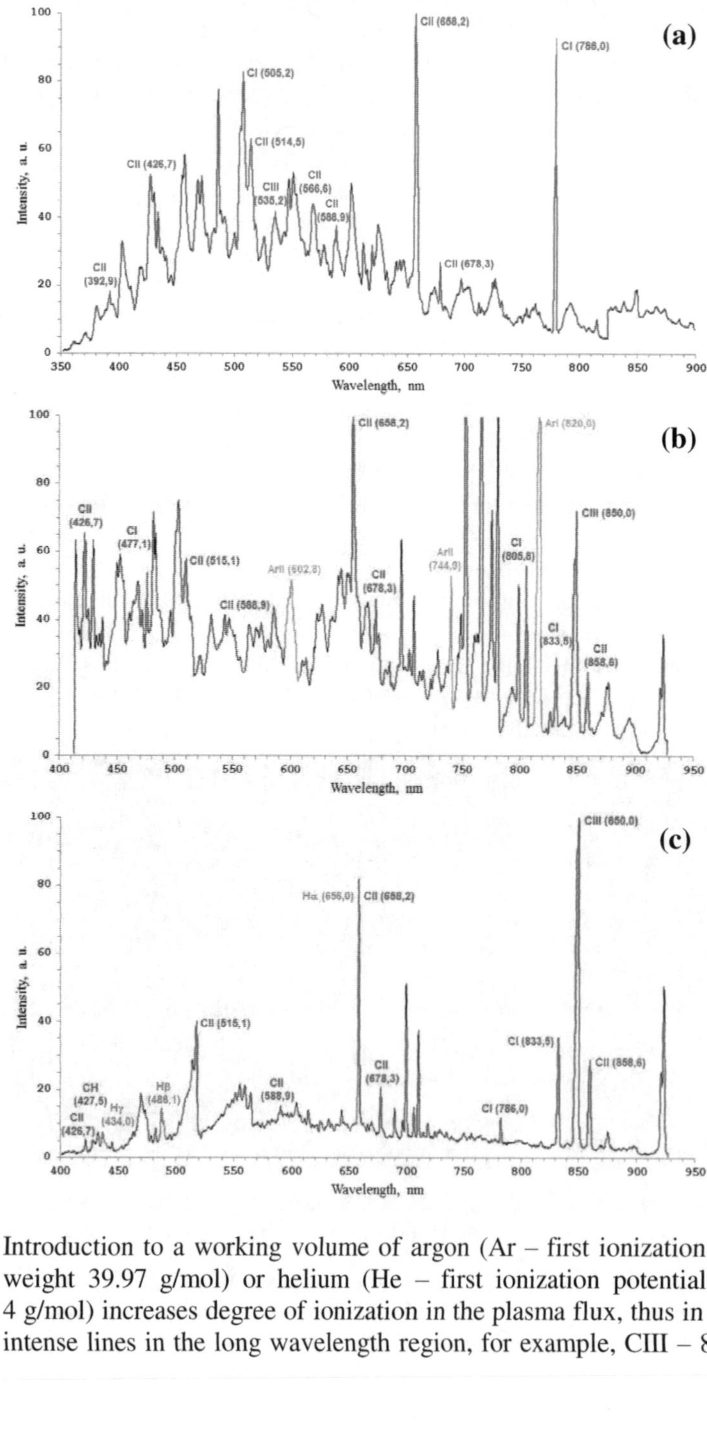

Figure 3. Decoding of the plasma flux spectrogram when spraying: **(a)** – graphite cathode (MPG-6); **(b)** – graphite cathode with the addition of argon; **(c)** – graphite cathode with the addition of benzene vapor. Plasma flow was studied using optical spectrometer, developed on the basis of a modified one-dimensional CCD array TCD1304 in Saint Petersburg Electrotechnical University "LETI". Control of the analyzer is carried out on the emission spectrum of a mercury-quartz low pressure lamp. Spectrometer allows determining the composition of the generated plasma flux in the wavelength range from 220 to 950 nm with optical resolution of 1.5 nm.

Introduction to a working volume of argon (Ar – first ionization potential U_{i1} = 15.75 V, molecular weight 39.97 g/mol) or helium (He – first ionization potential U_{i1} = 24.47 V, molecular weight 4 g/mol) increases degree of ionization in the plasma flux, thus in the emission spectrum are observed intense lines in the long wavelength region, for example, CIII – 850.0 nm. Also on the spectrograms are observed the lines of Balmer series of atomic hydrogen (H_α and H_β), as well as the emission spectra of C_2 (Swan band) and CH (Gere band) molecules.

If carbon conglomerates have a straight trajectory and move with the thermal velocity, the charged particles due to the presence of an external magnetic field are derived from the arc and, while having complex trajectories, move along the field lines.

Modification of the substrate material surface (in this case molybdenum substrate is used) is achieved through the alignment of the following physical processes: sputtering of a graphite cathode by a cathode spot of a vacuum-arc discharge, formation, transportation and extraction of the charged component of the plasma flux, acceleration (bias voltage set at the sample is varied from –100 to –1000 V) and deposition of the positive carbon ions on the refractory base (substrate temperature range – from 300 to 900 K).

When the equilibrium condition between the receiving of the charged carbon particles (dn_C/dt) on the treated surface and the process of material transfer into the depth of the substrate (dn_{dif}/dt) is met, for the multicomponent systems the advantageous is not the coexistence of elements, but the coexistence of the chemical compounds [10]. In this case, simultaneously with the saturation of the surface layer with carbon (diameters of carbon and molybdenum atoms are respectively 0.15 and 0.28 nm) on the substrate takes place the reaction of the formation of a surface carbide compound: $Mo_{sub} + 2C \rightarrow Mo_{sub}C + C$.

The solubility of carbon in molybdenum in the temperature range 300...1070 K is determined approximately as 0.3 % by weight and at temperatures above 1770 K dramatically increases. Therefore, further penetration of carbon into the surface region is accompanied by the formation in it of a bulk molybdenum carbide (Mo_2C) having a lattice of a hexagonal structure Bh. In all X-ray diffraction patterns peaks of the base – molybdenum and the lines of a well-formed Mo_2C with preferential orientations (101), (100) and (002), as well as (102), (110) and (103) are observed. From the achieved lines of Mo_2C may be marked the lines of maximum intensity: (102) – 2.4595 (ASTM – 2.442) and (103) – 2.2794 (ASTM – 2.292) (figure 4).

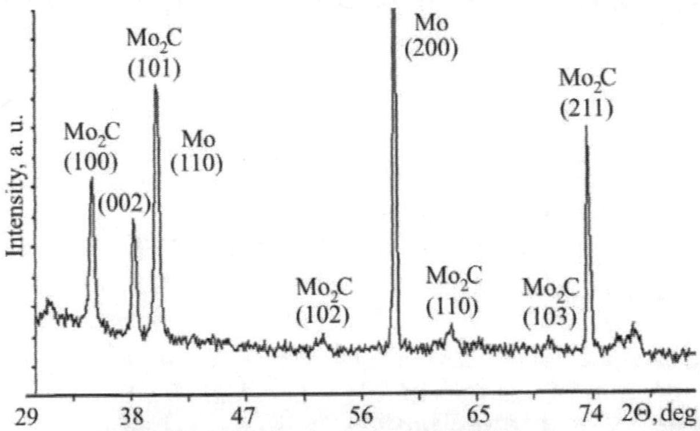

Figure 4. X-ray diffraction pattern of molybdenum carbide (Mo_2C) on a molybdenum substrate ($Cu_{K\alpha}$ radiation).

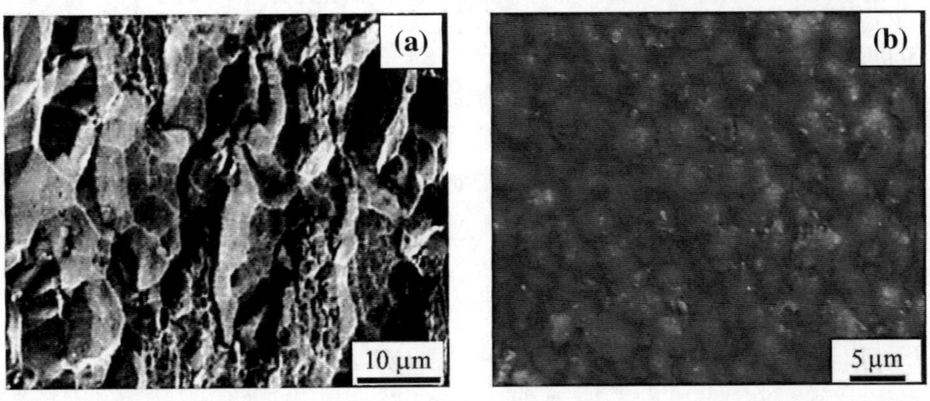

Figure 5. Carbon coating on a molybdenum substrate **(a)** with the emerging system of crystallites **(b)**.

Thickness of formed carbide is determined by the substrate temperature and the energy of the deposited particles. For the studied samples the layer thickness is 2...3 μm. The increase in the processing time or amplification of the flux of charged carbon particles ($dn_C/dt > dn_{dif}/dt$) leads to a formation on the surface of the carbon coating (figures 5, 6), the structure of which is determined by the growth and the mutual orientation of the primary crystallites, the nature and degree of intra- and interlayer ordering of atoms. In turn, the size of the crystallites (figure 5(b)) is determined by the number of initial nuclei, which is greater, the higher is the temperature. In this case, the higher is the temperature the smaller is the size of the crystallites.

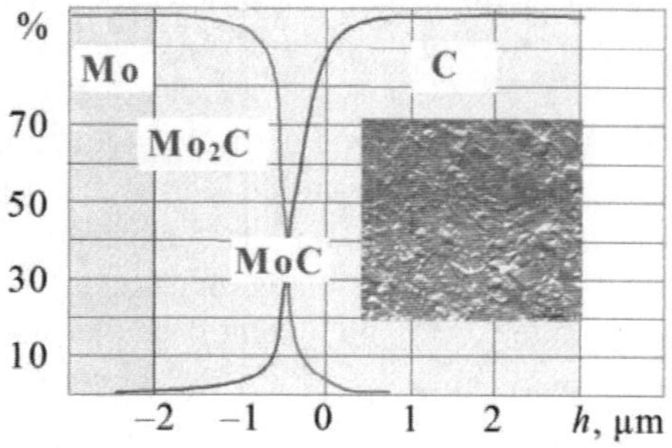

Figure 6. Distribution of the elements across the thickness (results are obtained using X-ray energy dispersive microanalysis system QUANTAX, defining all the elements, starting from Be (4) and allowing to obtain information on the elemental composition with a spatial resolution of 1 μm).

4. Conclusion

Thus, it is shown that stimulation of the growth processes and the ability to control the structure of the various carbon phases is achieved by using ions of plasma flux of a vacuum-arc discharge generated during sputtering of a graphite cathode. In one technological cycle of deposition of a coating based on carbon, it is possible to form sequentially: a sublayer of the carbide substrate material (Me_2C–MeC) for the future growth of a well-formed coating (figure 5). The presence of the intermediate layer with the agreed properties of the substrate–coating reduces tension and increases the adhesion strength (figure 5).

References

[1] Delhaes P 2001 *Graphite and Precursors* (Amsterdam: Gordon and Breach Science Publishers)
[2] Presser V, Heon M and Gogotsi Y 2011 *Advanced Functional Materials* **5** 810–33
[3] Kostrin D K and Lisenkov A A 2016 *Materials Science Forum* **843** 278–83
[4] Bystrov Yu A, Vetrov N Z, Lisenkov A A and Kostrin D K 2014 *Vakuum in Forschung und Praxis* **5** 19–23
[5] Barchenko V T, Vetrov N Z and Lisenkov A A 2013 *Technological vacuum-arc plasma sources* (Saint Petersburg: Saint Petersburg Electrotechnical University)
[6] Bystrov Yu A, Vetrov N Z and Lisenkov A A 2011 *Technical Physics Letters* **8** 707–9
[7] Frolov V I, Lisenkov A A and Barchenko V T 2010 *Physical basis of the use of low-temperature plasma* (Saint Petersburg: Saint Petersburg Polytechnic University)
[8] Stankus S V, Savchenko I V, Agadzhanov A Sh, Yatsuk O S and Zhmurikov E I 2013 *High Temperature* **2** 179–82
[9] Bystrov Yu A, Kostrin D K, Lisenkov A A and Vetrov N Z 2015 *Vakuum in Forschung und Praxis* **2** 22–5
[10] Bystrov Yu A, Vetrov N Z and Lisenkov A A 2013 *Technical Physics Letters* **10** 914–6

On prediction of turbulent heat transfer in the framework of Improved Delayed Detached Eddy Simulation with Wall Functions

M S Gritskevich[1], A V Garbaruk[1]

[1]Peter the Great St. Petersburg Polytechnic University, 195251, St. Petersburg, Russia

E-mail: gritckevich@gmail.com

Abstract. An investigation of Improved Delayed Detached Eddy Simulation (IDDES) with Δy^+-insensitive Near Wall Treatment (NWT) for prediction of turbulent heat transfer is performed in the framework of general-purpose CFD code ANSYS-Fluent for both fully attached and separated flows. The obtained results show that the considered approach is capable to predict the mean and RMS temperature with sufficient accuracy on the grids with the wall-adjacent step of about 1% of the velocity boundary layer thickness.

1. Introduction

Nowadays, hybrid RANS-LES methods (HRLM) are very promising for prediction of turbulent heat transfer at high Reynolds numbers typical for industrial applications [1,2], where the conventional RANS (Reynolds Averaged Navier-Stockes) models are not sufficiently accurate or informative, while LES (Large Eddy Simulation) is still beyond the capabilities of modern computers. Particularly, HRLM are widely employed on grids with the sufficient wall-normal resolution to integrate the governing equations down to the viscous sublayer to calculate the temperature pulsations [3–7], which are required for example for the thermal fatigue prediction [8]. However, generation of such grids could be a challenging task for complex industrial flows and, therefore, using of Δy^+-insensitive near wall formulations is required for more precise wall boundary conditions specification.

It should be mentioned that a comprehensive assessment of HRLM with WF has been recently performed in [9] for Shear Stress Transport based Improved Delayed Detached Eddy Simulation (SST-IDDES) [10] with the Δy^+-insensitive Near Wall Treatment (NWT) [11]. Particularly, it has been shown that using of grids with the wall-adjacent step of about 1% of the velocity boundary layer thickness allows accurate prediction of the mean and RMS velocity for a wide range of fully attached and separated flows.

However, it is not evident that the wall-normal grid requirements [9] are sufficient for prediction of the mean and RMS temperature fields since the temperature boundary layer thickness could be noticeably smaller than the velocity one depending on the considered Prandtl number. Therefore, the principal objective of this work is to estimate the accuracy of SST-IDDES with NWT for flows with turbulent heat transfer.

2. Assessment of SST-IDDES with NWT

Three test cases namely the periodic channel flow [12], the backward facing step [13], and the thermal mixing in the T-Junction [14] are considered to estimate the accuracy of SST-IDDES with

IOP Conf. Series: Journal of Physics: Conf. Series **929** (2017) 012094 doi:10.1088/1742-6596/929/1/012094

NWT for both attached flows and for flows with separation and reattachment. The employed computational setups are exactly the same as those in [9]. Particularly, two grids are considered throughout the paper. Mesh 1 utilizes the steps of 10% and 5% of the velocity boundary layer thickness in the streamwise and spanwise directions respectively, while the wall-normal step is changing from its minimum value corresponding to $\Delta y_w^+<1$ up to 5% of the velocity boundary layer thickness in the core-flow. Mesh 2 is exactly the same as Mesh 1 with the only difference for the wall-adjacent step, which is about 1% of the velocity boundary layer thickness.

It is worthwhile mentioning, that all simulations within the paper are performed in the framework of the general purpose CFD code ANSYS-Fluent [15]. Finally, the inlet content required for simulation of spatially developing flows is introduced with the use of Synthetic Turbulence Generator (STG) [16].

2.1. Periodic channel flow

The simulations of the periodic channel flow [12] (Figure 1) are conducted at the Reynolds number $Re_\tau=\rho\cdot u_\tau\cdot h/\mu=4200$ (u_τ is the friction velocity, $h=0.5\cdot H$ is the channel half-height, ρ is the density, and μ is the dynamic viscosity) and at three Prandtl numbers $Pr=\mu\cdot C_p/\lambda=0.07$, 0.7, and 7.0 ($C_p$ is the specific heat capacity and λ is the thermal conductivity) corresponding to liquid metals, air, and water. The lower and upper channel walls are kept at constant temperature of T_0 and $T_0+\Delta T$ respectively.

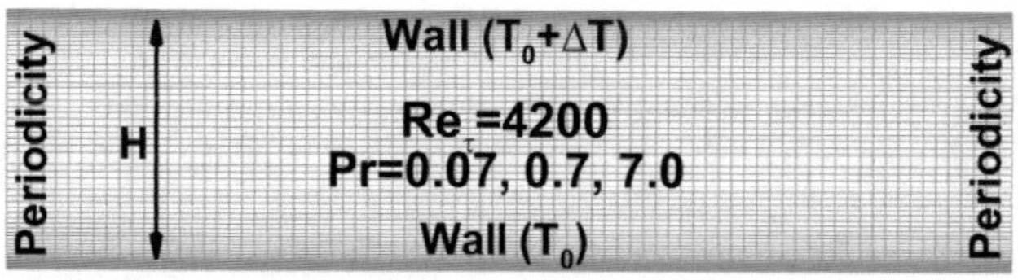

Figure 1. Computational setup for the periodic channel flow

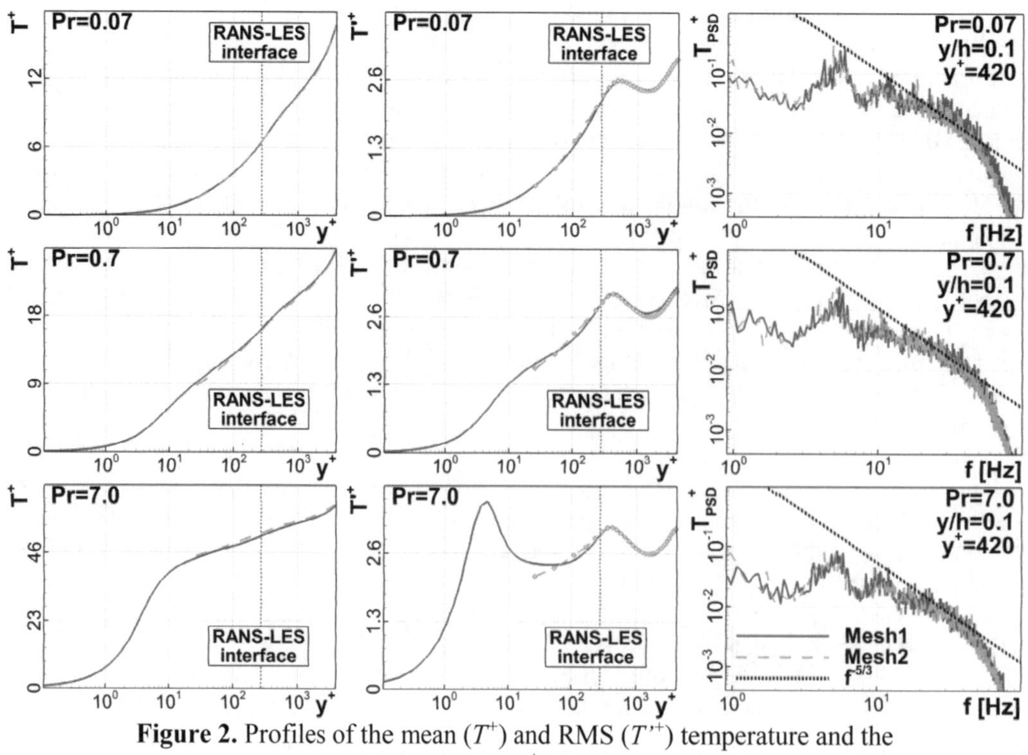

Figure 2. Profiles of the mean (T^+) and RMS (T'^+) temperature and the temperature power spectral density ($T_{PSD}{}^+$) for the considered Prandtl numbers

To be able to estimate the grid effect on the heat transfer prediction the profiles of the mean and RMS temperature as well as the temperature power spectral density are shown in Figure 2. As seen, for all the considered Prandtl numbers Mesh 1 and Mesh 2 yield almost identical results. It should be noted, however, that Mesh 2 is not capable to capture a peak of the RMS temperature near the wall at Pr=7.0, which is observed on Mesh 1. However, it is still not clear if this peak could be properly predicted on Mesh 1 since in this region SST-IDDES is operating in the RANS mode due to the insufficient grid resolution and, therefore, a comparison of the RMS temperature on Mesh 1 with the DNS data is required, which is not available from [12].

Summing up, virtually identical solutions are obtained on Mesh 1 and Mesh 2 indicating that the grid with the wall-adjacent step of about 1% of the velocity boundary layer thickness is sufficient at small and moderate Prandtl numbers. At the same time, for large Prandtl numbers, for which the temperature boundary layer is typically thinner than the velocity one, some peculiarities of the heat transfer in the vicinity of the wall could be missing on Mesh 2, while in the core-flow both grids still yield almost identical solutions.

2.2. Backward facing step

The backward facing step test case (Figure 3) has been experimentally studied in [13] at $Re=\rho\cdot U_0\cdot H/\mu=28000$ (U_0 is the inlet bulk velocity, H is the step height, ρ is the density, and μ is the dynamic viscosity) and Pr=0.7 (the definition of Pr is the same as for the periodic channel flow). The lower wall is heated with the constant flux of Q_0, while the adiabatic condition is preserved elsewhere.

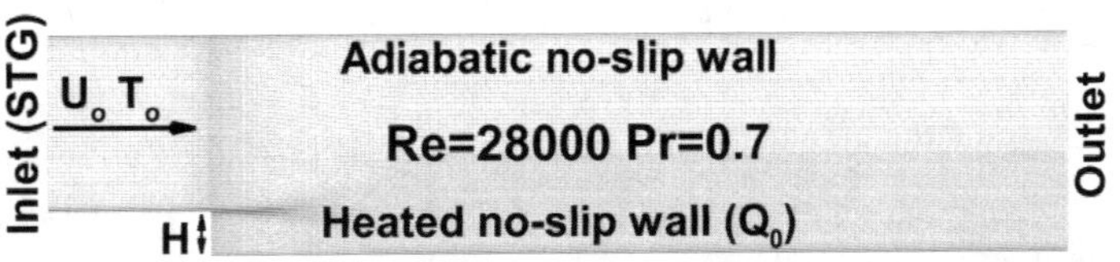

Figure 3. Computational setup for the backward facing step flow

To assess the performance of SST-IDDES with NWT the wall distributions are firstly shown in Figure 4. As seen, Mesh 2 yields higher Stanton number near the reattachment point and further downstream comparing to those on Mesh 1 (better agreement with the experimental data on Mesh 2 is likely due to the error cancelation). Consistently, the mean temperature is slightly lower near the reattachment point and further downstream and slightly higher within the separation zone. At the same time, the RMS temperature distributions are almost identical on the considered grids in the most part of the domain except the small region in the vicinity of the step, where Mesh 2 yields slightly higher RMS temperature.

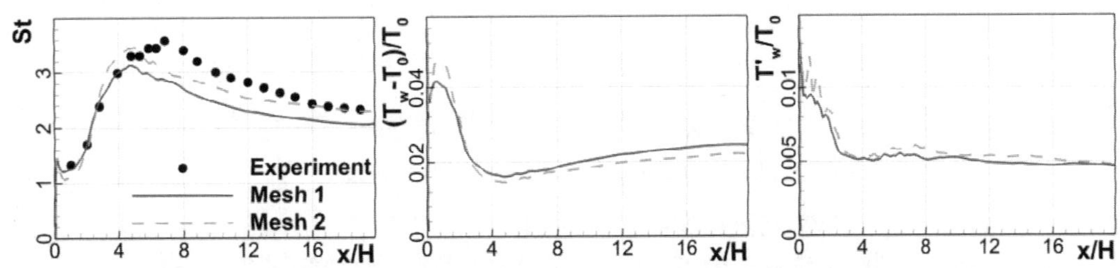

Figure 4. Distributions of the Stanton number (St) and of the mean ($(T_w-T_0)/T_0$) and RMS (T'_w/T_0) temperature along the bottom heated wall

Similar conclusions could be drawn for the mean and RMS temperature profiles (Figure 5). Particularly, almost identical solutions are obtained on Mesh 1 and Mesh 2, however, in the vicinity of the reattachment point Mesh 2 seems to be insufficiently fine to properly capture the temperature gradients in several near wall cells.

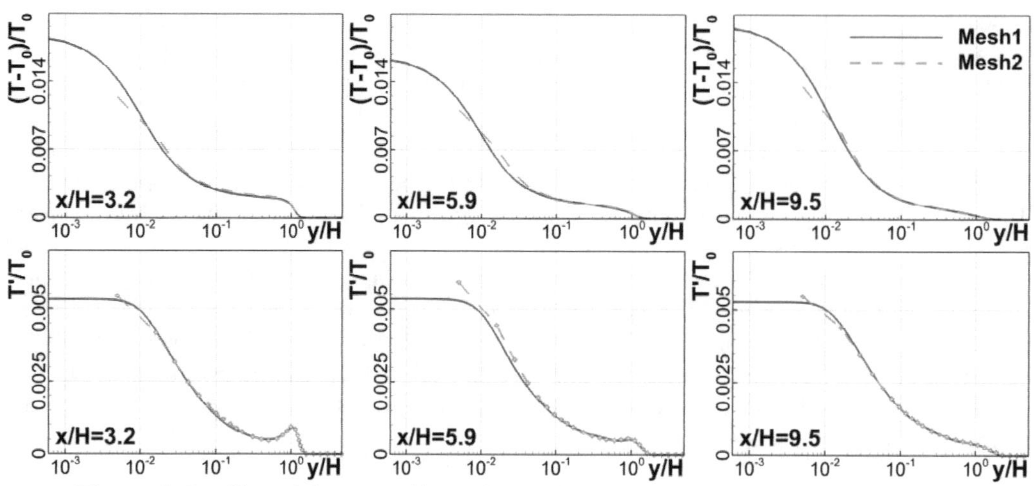

Figure 5. Profiles of the mean ($(T\text{-}T_0)/T_0$) and RMS (T'/T_0) temperature at x/H=3.2, 5.9, and 9.5

Thus, SST-IDDES with NWT is able to provide a reasonably precise results on the considered grids not only for the attached but also for separated flows. However, coarsening of the grid in wall-normal direction results in slightly lower accuracy in the vicinity of the stagnation point likely due to the insufficient wall-normal grid resolution.

2.3. Thermal mixing in the T-Junction

The experimental setup for the thermal mixing in the T-junction [14] consists of vertical and horizontal pipes with inner diameters of $0.7{\cdot}D$ and D (Figure 6). The bulk velocity is kept constant throughout the experiment and is equal to $1.3{\cdot}U_0$ and U_0 in the vertical and horizontal pipe respectively. The temperature of the water is T_0 in the vertical pipe and $T_0\text{-}\Delta T$ in the horizontal one with the adiabatic condition preserved on both horizontal and vertical pipes. These parameters correspond to $Re=\rho{\cdot}U_0{\cdot}D/\mu$=100000 ($\rho$ is the density and μ is the dynamic viscosity) and to Pr=7 (the definition is similar to those in sections 2.1 and 2.2).

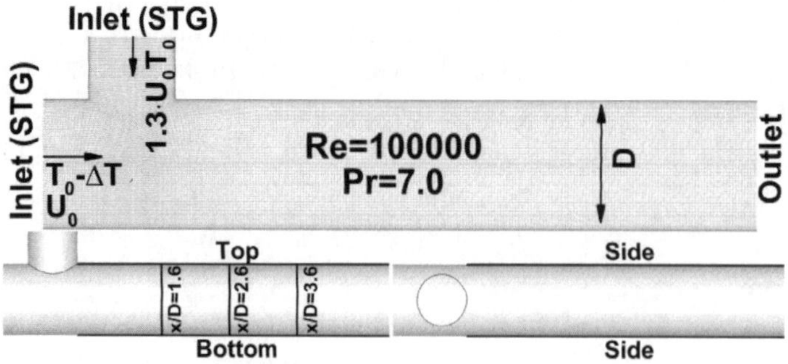

Figure 6. Computational setup for the T-Junction flow

As seen from Figure 7, Mesh 1 and Mesh 2 provide almost identical mean and RMS temperature distributions along the horizontal pipe walls (the corresponding sections are depicted in Figure 6), which agree fairly well with the experimental data [14]. Particularly the peak of the RMS temperature, which is very important for the thermal fatigue predictions, is properly captured everywhere in the domain. Consistently, the profiles of the mean and RMS temperature are almost the same at all the considered vertical sections (Figure 8).

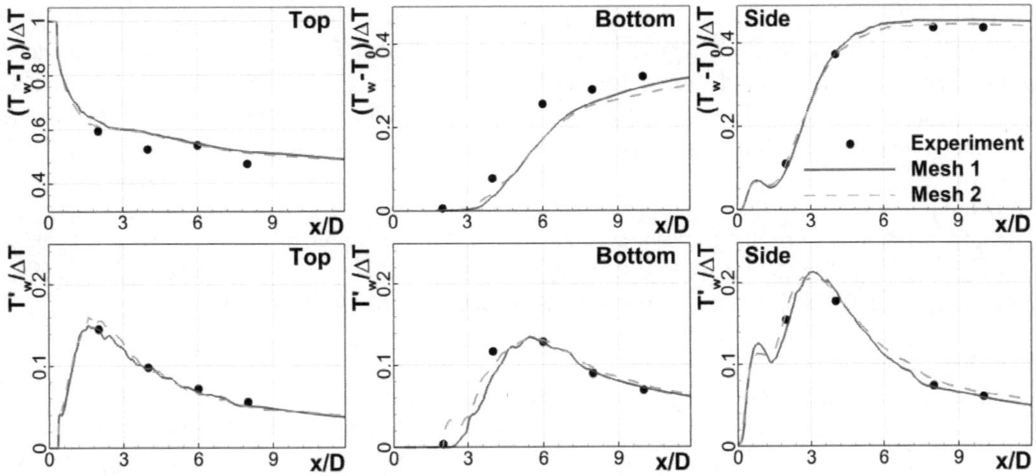

Figure 7. Distribution of the mean ($(T_w-T_0)/\Delta T$) and RMS ($T'_w/\Delta T$) temperature along the horizontal pipe

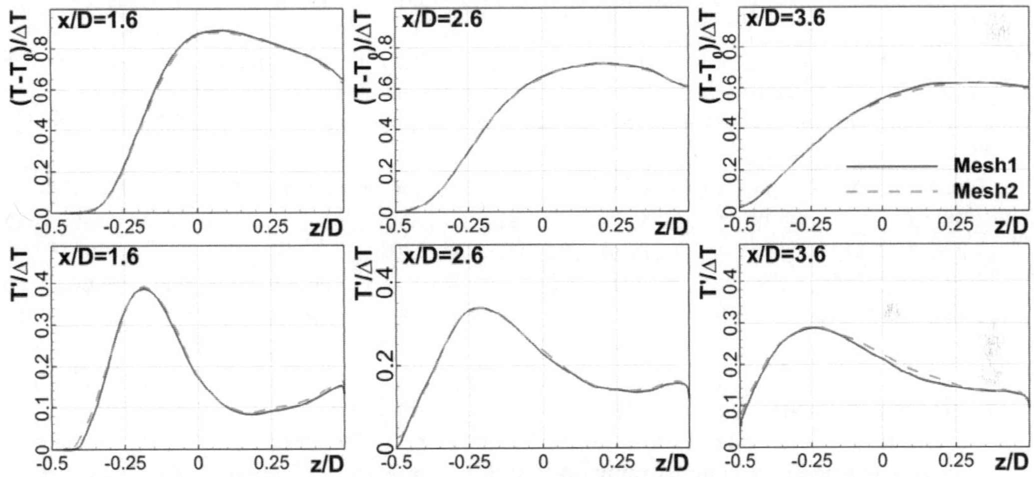

Figure 8. Profiles of the mean ($(T_w-T_0)/\Delta T$) and RMS ($T'/\Delta T$) temperature at x/D=1.6, 2.6, and 3.6

Summing up, the obtained results justify that SST-IDDES with NWT is capable to predict the heat transfer in case of three-dimensional separation and reattachment on grids with the wall-adjacent step of 1% of the velocity boundary layer thickness.

3. Conclusions

An assessment of SST-IDDES with NWT for the heat transfer prediction has been performed for several attached and separated flows.

It has been shown, that for small and moderate Prandtl numbers using of grids with the wall-adjacent step of about 1% of the velocity boundary layer thickness is sufficient to achieve almost identical mean and RMS temperature profiles with those on the conventional grids with $\Delta y^+_w<1$. At the same time, for large Prandtl numbers, for which the temperature boundary layer is typically thinner than the velocity one, using of such coarsened grids results in the missing RMS temperature peak in the vicinity of the wall, whereas almost identical profiles are still obtained in the core-flow. Nevertheless, taking into account that SST-IDDES is operating in the RANS mode in that region it is not clear if the RMS temperature peak could be properly captured even on conventionally fine wall-normal grids.

Finally, it should be mentioned, that within the current work the energy equation is solved as a passive scalar and, thus, the flow physics is fully separated from the peculiarities of the heat transfer. However, when the energy and momentum equations are strongly coupled (e.g. for compressible flows) the abovementioned effect of high Prandtl numbers become more important and, therefore, the temperature boundary layer thickness should be also taken into account for the grid generation.

Acknowledgement
The computations within the paper are conducted with the support of Saint-Petersburg Branch of the Joint Supercomputer Center of the Russian academy of sciences. The authors acknowledge the financial support from the Russian Science Foundation (grant No. 14-11-00060).

References
[1] Spalart P R 2009 Detached eddy simulation *Annual Review of Fluid Mechanics* **41** 181–202
[2] Menter F R, Schütze J and Gritskevich M S 2012 Global vs. zonal approaches in hybrid RANS-LES turbulence modelling *Progress in Hybrid RANS-LES Modelling, Notes on Numerical Fluid Mechanics and Multidisciplinary Design Volume 117* pp 15–28
[3] Viswanathan A K and Tafti D K 2006 Detached eddy simulation of turbulent flow and heat transfer in a two-pass internal cooling duct *International Journal of Heat and Fluid Flow* **27** 1–20
[4] Kubacki S and Dick E 2011 Hybrid RANS/LES of flow and heat transfer in round impinging jets *International Journal of Heat and Fluid Flow* **32** 631–51
[5] Yin Z and Durbin P A Passive Scalar Transport Modeling for Hybrid RANS/LES Simulation *Flow Turbulence and Combustion*
[6] Smirnov E M, Smirnovsky A A, Schur N A, Zaitsev D K and Smirnov P E 2016 Numerical simulation of turbulent heat transfer past a backward-facing step: 2D/3D RANS versus IDDES solutions *Journal of Physics: Conference Series* **745** 32016
[7] Gritskevich M S, Garbaruk A V and Menter F R 2014 Computation of wall bounded flows with heat transfer in the framework of SRS approaches *Journal of Physics: Conference Series* **572** 12057
[8] Smith B L, Mahaffy J H and Angele K 2013 A CFD benchmarking exercise based on flow mixing in a T-junction *Nuclear Engineering and Design* **264** 80–8
[9] Gritskevich M S, Garbaruk A V and Menter F R A Comprehensive Study of Improved Delayed Detached Eddy Simulation with Wall Functions *Flow Turbulence and Combustion*
[10] Gritskevich M S, Garbaruk A V, Schütze J and Menter F R 2012 Development of DDES and IDDES formulations for the k-ω shear stress transport model *Flow Turbulence and Combustion* **88** 431–49
[11] Esch T and Menter F R 2003 Heat transfer prediction based on two-equation turbulence models with advanced wall treatment *Turbulence Heat an Mass Transfer 4* pp 1–8
[12] Hoyas S and Jimenez J 2008 Reynolds number effects on the Reynolds-stress budgets in turbulent channels *Physics of Fluids* **20** 101511
[13] Vogel J C and Eaton J K 1985 Combined heat transfer and fluid dynamic measurements downstream of a backward-facing step *ASME Journal of Heat Transfer* **107** 922–929
[14] Odemark Y, Green T M, Angele K, Westin J, Alavyoon F and Lundström S 2009 High-cycle thermal fatigue in mixing tees: new large-eddy simulations validated against new data obtained by PIV in the Vattenfall experiment *ASME Proceedings* **ICONE17-75** 775–85
[15] Mathur S R and Murthy J Y 1997 A pressure-based method for unstructured meshes *Numerical Heat Transfer* **32** 195–215
[16] Shur M L, Spalart P R, Strelets M K and Travin A K 2014 Synthetic Turbulence Generators for RANS-LES Interfaces in Zonal Simulations of Aerodynamic and Aeroacoustic Problems *Flow, Turbulence and Combustion* **93** 63–92

The influence of interatomic collisions on the angular distribution of Cs atomic flow from microchannel

A Yu Rumyantsev[1], M V Petrenko[1,2], S A Poniaev[2,3], Yu A Shustrov[2], V P Kochegarov[1] and D L Schennikov[1]

[1] Russian Institute of Radionavigation and Time, 120 Obukhovskoy Oborony prospect, St. Petersburg, 192012, Russia
[2] Ioffe Institute, 26 Politekhnicheskaya Street, St. Petersburg, 194021, Russia
[3] Mozhaisky Military Space Academy, 13 Zhdanovskaya Street, St. Petersburg, 197198, Russia

E-mail: m.petrenko@mail.ioffe.ru , s.poniaev@gmail.com

Abstract. Beams of alkali atoms are typically used in precise quantum devices, such as atomic beam frequency standards. In order to know more exactly parameters of atomic beam tubes and frequency standards (frequency instability), it is necessary to investigate the propagation of Cs atoms after the channel exit. In this work theoretical calculations of angular distribution of Cs atomic flow from a microchannel are carried out. The Monte-Carlo direct numerical simulation of Cs atom propagation is also performed to estimate the influence of interatomic collisions on the atomic flow.

1. Introduction

Alkali atoms are typically used in precise quantum devices, such as frequency standards based on alkali vapour cells [1,2] or atomic beam tubes [3], and quantum magnetometers with optical pumping [4,5]. Beams of Cs atoms with temperature in the range of 100 – 130ºC are used in atomic beam frequency standards. These beams outflow from microchannels with typical length-to-diameter ratios of 50 – 200. In order to know more exactly parameters of atomic beam tubes and frequency standards (frequency instability), it is necessary to investigate the propagation of Cs atoms after the channel exit. Nowadays theoretical calculations are widely used to describe Cs atomic flow [3], but the influence of interatomic collisions on the atomic flow is not taken into account in these calculations.

In this work theoretical calculations of angular distribution of Cs atomic flow from a microchannel are carried out. The Monte-Carlo direct numerical simulation of Cs atom propagation is also performed to estimate the influence of interatomic collisions on the atomic flow.

2. Results of theoretical calculations

The propagation of Cs atoms is investigated by using a microchannel with the following parameters: channel length $L = 1$ cm, channel diameter $d = 50$, 100 and 200 μm. The scheme of the flow is presented in figure 1. The input volume contains Cs vapor with temperature $T = 120$ °C, so the free path of Cs atoms takes on a value $\lambda = 0.066$ cm since the cross-section of Cs atom interatomic collisions is $\sigma = 2.35\times10^{-13}$ cm^2 [6]. The pressure of Cs vapor in the input reservoir takes on a value

IOP Conf. Series: Journal of Physics: Conf. Series **929** (2017) 012095 doi:10.1088/1742-6596/929/1/012095

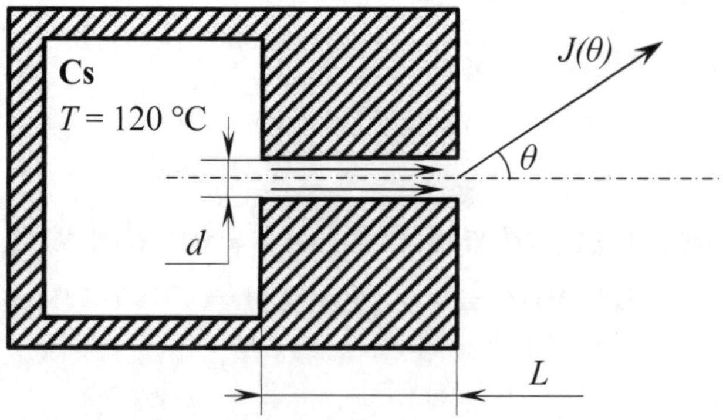

Figure 1. Scheme of Cs atom flow through microchannel.

P_{Cs} =1.85×10^{-3} Torr, and Knudsen number is $Kn_L = \lambda/L = 0.066$. After the propagation through the microchannel Cs atoms flow into vacuum and the angular distribution of the atomic flow is described by the function $J(\theta)$.

It is necessary to note that the increase in the channel length to the values that exceed the free path of Cs atoms in the input volume leads to an essential influence of interatomic collisions on the atom propagation inside the channel and on the angular distribution of the atomic flow after the channel. In our work we used the theoretical model [7] in which the atom-wall interaction obeys the law of diffuse reflection and the atom scattering due to interatomic collisions is taken into account.

The angular distribution of Cs atomic flow at the microchannel exit is defined by

$$J(\theta) = \frac{\dot{N}}{\pi K} j(\theta), \tag{1}$$

where $J(\theta)$ is the intensity of the atomic flow in atoms·s^{-1}·Sr^{-1} at a polar angle θ from the tube axis, $\dot{N}$ is the total flow rate through the channel, K is the transmission probability and $j(\theta)$ is the angular distribution function. Since the gas in the input volume is Maxwellian, the total atomic flow rate through the channel is defined by the Cs vapor characteristics in the input reservoir and by the channel parameters

$$\dot{N} = K\left(\frac{n\bar{v}}{4}\right)\left(\frac{\pi d^2}{4}\right), \tag{2}$$

where n is the number density of Cs atoms in the source, and $\bar{v}$ is the mean speed of the atoms. Since the integral of $J(\theta)$ over the front hemisphere must be equal to the total flow rate, K and $j(\theta)$ are related by

$$K = 2\int_0^{\pi/2} j(\theta)\sin\theta\, d\theta. \tag{3}$$

The calculation of the angular distribution of the atomic flow from different microchannels was carried out with and without taking into account interatomic collisions. The calculation was performed by using the expressions for the function $j(\theta)$ which are defined in [7,8]. The parameters of the calculation in the collisionless case corresponding to the calculation performed in [3,9] were chosen.

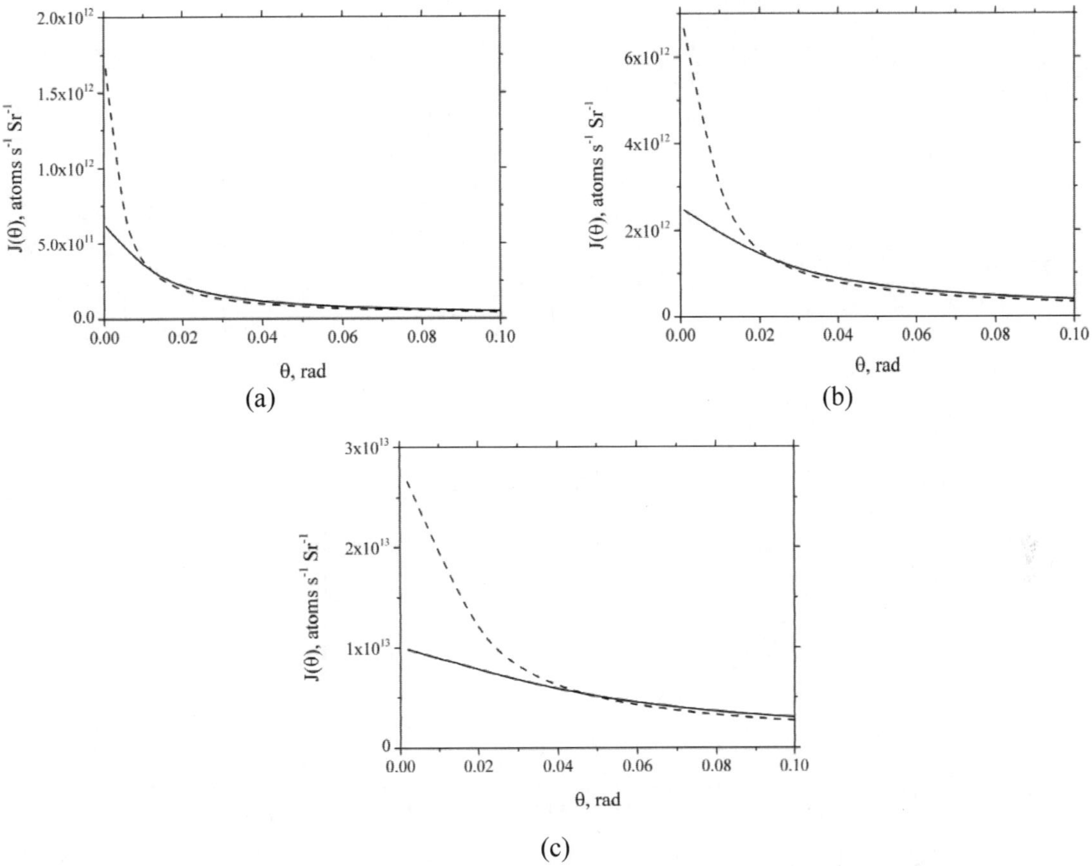

Figure 2. Angular distribution of Cs atomic flow from microchannel. The results of theoretical calculation. (a) channel diameter $d = 50$ μm, (b) $d = 100$ μm, (c) $d = 200$ μm.

The results of the calculation are shown in figure 2. In the graphs the solid lines show the angular distributions in which interatomic collisions are taken into account, and the dashed line is for the collisionless case. It follows from the results obtained in the calculations that the consideration of interatomic collisions in the description of the propagation of atoms leads to a reduction in the atomic flow in the axial (central) part of the atomic flux in 2.5 – 3 times depending on the channel geometry. The angular distribution becomes flatter and exceeds the distribution obtained without taking into account interatomic collisions for large angles θ.

3. Results of numerical simulation

The numerical simulation was performed by the DSMC method [10] using the OpenFOAM software package [11]. The calculations were performed for collision and collisionless flows. The variable hard sphere (VHS) model was employed [12,13] for simulation of collision flows.

The interaction of the sample particles with the channel walls was calculated in the diffuse approximation, i.e., as a reflection with adhesion. The direction of the reflected particles was determined by the cosine law, and their speeds were randomly chosen on the basis of the maxwellian thermal distribution of velocities at the temperature equal to the temperature of the channel walls. In our simulation the temperature of the channel wall was taken to be 120 °C.

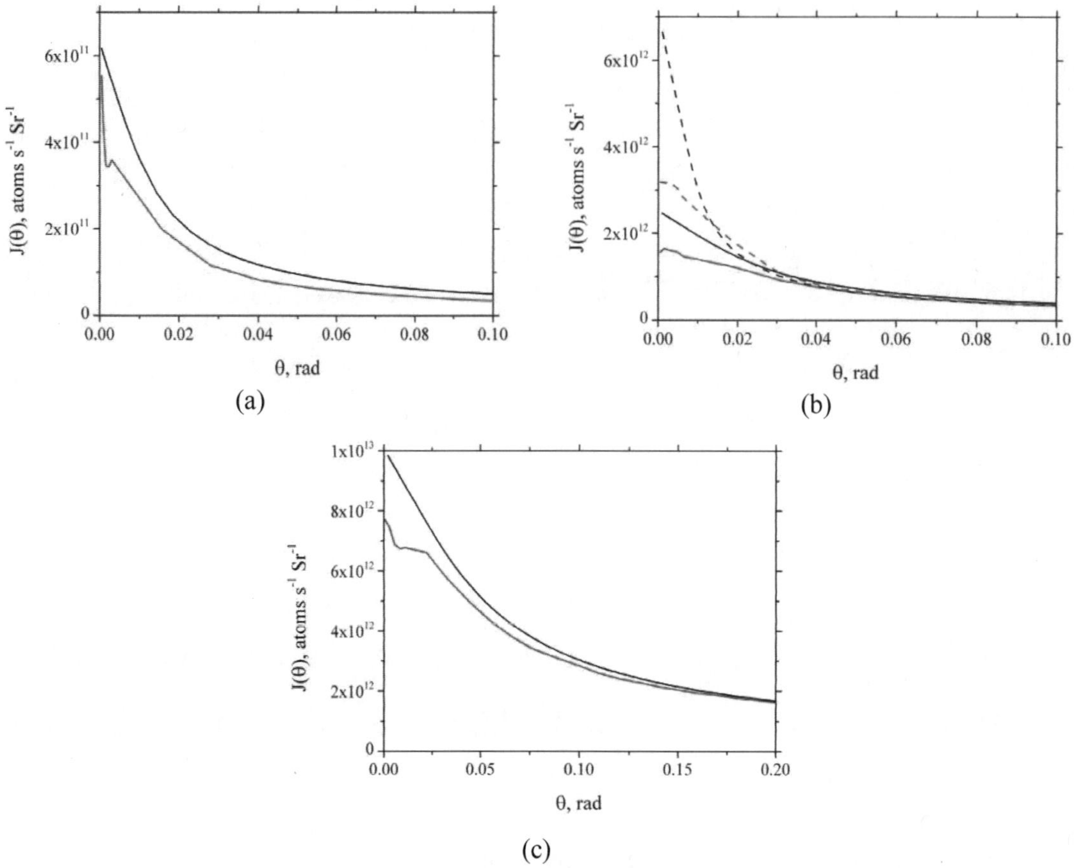

Figure 3. Angular distribution of Cs atomic flow from microchannel. The results of numerical simulation. (a) channel diameter $d = 50$ μm, (b) $d = 100$ μm, (c) $d = 200$ μm.

The angular distribution of the Cs atom flow was calculated by using the number of sample particles passing through a unit area at angle θ. The number of sample particles was time-averaged over the period from 0.05 s to 0.1 s from the flow start. The results are shown in figure 3. The red curve show results of numerical simulation, the black curve show results of theoretical calculations for comparison, the solid lines are angular distributions obtained with interatomic collisions taken into account, the dashed lines were obtained for the collisionless case. It can be seen from figure 3 that the results of numerical simulation are in good agreement with the results of theoretical calculations. The behavior of the angular distributions in the axial region of atomic beam obtained numerically can be explained by the statistical errors, taking into account the limited number of sample particles passing through the unit area in the axial region. These flows of atoms obtained numerically are quantitatively lower than the theoretical values.

Table 1 shows the values of total flow rate through the channel in case of interatomic collisions are taken into account. The values $\dot{N}_{th}$ and $\dot{N}_{num}$ represent the flow of atoms through the channel, calculated by analytical formulas and by numerical simulation respectively.

Table 1. Total flow rate through the channel. Results of theoretical and numerical calculations in the case of collision flow.

d/L	$\dot{N}_{th}$, atoms s^{-1}	$\dot{N}_{num}$, atoms s^{-1}	$\dot{N}_{th} / \dot{N}_{num}$
0.005	3.675×10^{10}	2.172×10^{10}	1.69
0.01	2.918×10^{11}	2.08×10^{11}	1.40
0.02	2.297×10^{12}	1.70×10^{12}	1.35

4. Conclusion

The values of the total flow rate of atoms from the channel obtained in numerical simulation are lower than the theoretical rates in 1.35 – 1.69 times, depending on the channel geometry. The difference can be explained by using a fairly simple calculation model of diffuse law of interaction of atoms with the channel walls. Further investigations are necessary to model a more complex interaction of the particles with the walls, in particular taking into account the direction and speed of the particles before the interaction.

Results of numerical simulation are in good agreement with theoretical calculations. This agreement allows the use of analytical formulas in the description of the spread of the atoms from the channel in the case of a simple geometry of the system, for example in the case of a single cylindrical channel. In the case of more complex geometrical systems, such as a channel with a square cross-section or multichannel collimator, the result obtained by using analytical formulas can have a significant divergence with the results of numerical simulation and with the real physical process.

The angular distributions of Cs atomic flow obtained in our work by using numerical simulation and analytical formulas show a significant influence of interatomic collisions inside a channel on the atomic flow. If interatomic collisions are taken into account, a significant decrease (in 2.5 – 3 times) of the atomic flow rate in the axial (central) region of the beam results. Since the central region of the atomic beam is used in the atomic beam frequency standards, it is necessary to consider interatomic collisions when describing the propagation of cesium atoms from microchannels.

References

[1] Ermak S V, Petrenko M V and Semenov V V 2016 *Tech. Phys. Lett.* **42** 127
[2] Fedorov M I, Ermak S V, Petrenko M V, Pyatyshev E N and Semenov V V 2016 *J. Phys.: Conf. Ser.* **769** 012046
[3] Pimenov A V and Pleshanov S A 2010 *Electronnaya Tekhnika. SVCH Tekhnika* **507** 16 *(in russian)*
[4] Baranov A A, Ermak S V, Sagitov E A, Smolin R V and Semenov V V 2016 *Tech. Phys. Lett.* **42** 186
[5] Sagitov E A, Ermak S V, Petrenko M V and Semenov V V 2016 *J. Phys.: Conf.Ser.* **769** 012044
[6] Vanier J and Audoine C 1989 *The quantum physics of atomic frequency standards* (Philadelphia: IOP Publishung Ltd) p 453
[7] Olander D R and Kruger V 1970 *J. Appl. Phys.* **41** 2769
[8] Lucas C B 2014 *Atomic and Molecular Beams. Production and Collimation* (New York: CRC Press) p 211
[9] Beijerinck H C W and Verster N F 1975 *J. Appl. Phys.* **46** 2083
[10] Bird G A 1994 *Molecular gas dynamics and the direct simulation of gas flows* (Oxford: Clarendon)
[11] OpenFOAM®, www.openfoam.com
[12] Bird G A 1978 *Annu. Rev. Fluid Mech.* **10** 11
[13] Bird G A 1983 *Phys. Fluids* **26** 3222

Investigation of electrode erosion products of alternating current plasma torch operating on mixture of steam, carbon dioxide, methane and carbon tetrachloride

J. A. Kuchina[1], V.E. Kuznetsov[1], D.I. Subbotin[1,2], V.E. Popov[1], E.O. Serba[1], J.D. Dudnik[1], A.I. Litvyakova[1,2], I.A. Cherepkova[2], A.V. Surov[1]

[1]Institute for Electrophysics and Electric Power of RAS
[2]SPSIT, Saint-Petersburg State Institute of Technology

Abstract. The most important unit of the plasma torch mainly subjected to the thermal load of the electric arc is electrode. Tungsten, hafnium, yttrium, aluminum, copper, iron and their alloys are used as electrode materials depending on the plasma-forming gas. Copper-based alloys can be rationally used for gases containing oxygen. The experiments were performed on the three-phase alternating current vortex plasma torch with power of 120 kW working on a mixture of steam, carbon dioxide, methane and carbon tetrachloride for the following flow rates 2.9 g/s, 2.9 g/s, 1.21 g/s and 2.44 g/s respectively. The material of erosion of copper rod electrodes was sampled from water-cooled walls of electric arc plasma torch channels made of stainless steel. The sample was divided into magnetized and non-magnetic parts. According to the energy-dispersive elemental analysis the products of electrode erosion contain oxides and chlorides of copper and iron, as well as metallic copper and iron.

1. Introduction

Plasma-chemical techniques can be used for a wide range of plasma-chemical processes: municipal solid waste gasification [1], natural gas reforming [2], toxic waste processing [3]. Conventional combustion does not provide high temperature for a long time, therefore, the thermal plasma can perform these conditions. In addition all these processes require powerful plasma torches with long life time and in most cases it is determined by the life time of electrodes.
A number of methods are used to reduce erosion of electrodes: choice of the optimal electrode material, shielding gases [4], intensive water cooling, electric current reducing at high voltage [5], rotation of the arc attachment point by magnetic [6] and hydrodynamic forces [7]. High-voltage AC plasma torches may have a higher life time of the electrodes, as lower electric current causing intense heating is required for the same power. The IEE RAS has developed a wide range of AC plasma torches with rod and hollow electrodes and vortex stabilization of the electric arc [8], which provides high energy efficiency.

Erosion products can participate in a plasma-chemical process, for example, as a catalyst in certain reactions [9] (including side reactions). Therefore it is important to know their composition, especially during the decomposition of organochlorine compounds because they can form dioxins. There is a reason to believe that chlorine and the erosion products form safe substances [10], which increases the decomposition degree of toxic compounds.

2. Experimental section

2.1. Laboratory facility

Study of electrode erosion products was conducted on a laboratory test facility (Figure 1) consisting of the plasma torch, plasma chemical reactor, gases supply and sampling systems.

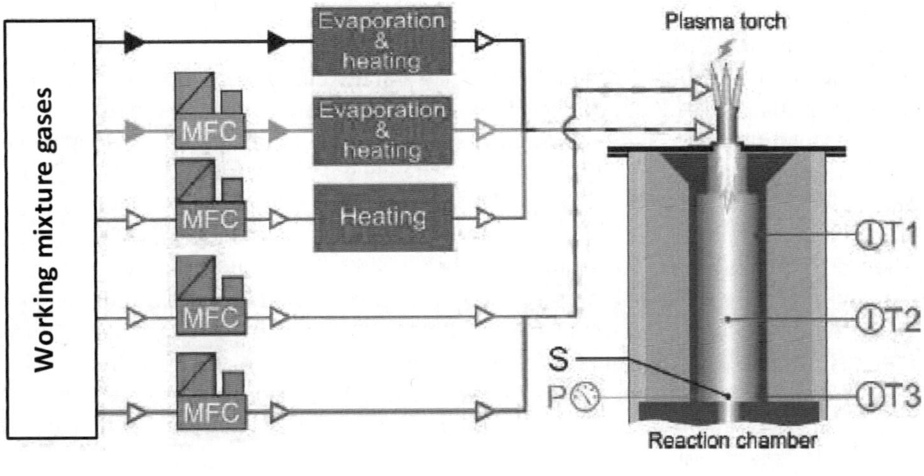

Figure 1. Schematic diagram of the laboratory test bench

S – sampling point; P – pressure gauge; T1 – T3 – thermocouples; MFC – mass flow controller

The reactor is equipped with a high-temperature thermocouples and sampling probe. The used plasma generator is a high voltage AC plasma torch with rod electrodes [4] made of copper. Electric arcs are burning in separate electric arc channels and are closed on each other in the external space. The electric arc channel has two zones for plasma-forming gas supply: electrode zone and electric arc burning zone. Therefore chemically more active compounds will not virtually interact with the electrode material. In the presented schematic diagram carbon dioxide and small amount of methane are fed into the electrode zone. Carbon dioxide is used as a shielding gas and methane is used for production of the desired reaction composition and required power. The model chemical system in operation includes mixture of methane, carbon dioxide, steam and carbon tetrachloride. This mixture is chosen to study the decomposition of organochlorine compounds by high temperature steam reforming and hydrocracking:

$$CCl4+1.049H2O+0.427CO2+0,476CH4=1.903CO+4HCl$$

IOP Conf. Series: Journal of Physics: Conf. Series **929** (2017) 012096 doi:10.1088/1742-6596/929/1/012096

Carbon tetrachloride flow rate was reduced 10 times compared to the theoretical one to ensure safe conditions of the experiment and to prevent possible formation of highly toxic substances.

2.2. Sampling and analysis

Electrode erosion products were collected from the water cooled stainless steel walls and separated into magnetic and non-magnetic part by a permanent magnet. Elemental analysis was carried out using a scanning electron microscope (Quanta 200 FEG) on the aluminum substrate under the following conditions: voltage - 20 kV, vacuum, room temperature, image scanning speed - 91.19 s, working distance - 0.11 mm. Product gas was sampled from the reactor through the sampling probe, then condensate was separated and product gas composition was determined by mass spectrometer Cirrus 300. Hydrogen chloride content in the condensate was determined by acid-base titration.

3. Results and discussions

Flow rate of plasma forming media: CH_4 - 1.21 g/s, CO_2 - 2.90 g/s, H_2O - 2.90 g/s CCl_4 - 2.44 g/s. The effective value of the arc voltage drop is 1390 V, an electric current - 51.2 A, plasma torch power - 116.7 kW. As a result of carbon tetrachloride processing by steam, methane and carbon dioxide plasma, only hydrogen chloride is found among the gaseous chlorine-containing compounds. The composition of the product gas is the following: CH_4 - 1.84% vol., H_2 - 32.46%, CO_2 - 33.62%, CO - 12.97%, HCl - 19.13%. Elemental analysis of the magnetic part of erosion products of the copper electrode shows that the magnetic sample is composed of oxygen (42.27 mol.%), copper (34.59 mol.%), carbon (9.32 mol.%), iron (6.88 mol.%), chlorine (4.09 mol.%) and small amounts of chrome (1.74 mol.%), nickel (0.81 mol.%) and silicon (0.29 mol.%). Elemental analysis of the non-magnetic part of erosion products of copper electrode shows that non-magnetic sample consists of copper (46.29 mol.%) carbon (31.12 mol.%), oxygen (19.63 mol.%), iron (1.09 mol.%) and as well as small amount of chlorine (1.09 mol.%), nickel (0.50 mol.%), chrome (0.37 mol.%). The presence of chromium, nickel and silicon in these samples is acconted for the presence of these elements in the material of the plasma torch case. Iron is presented in two parts of the sample, presumably indicating the presence of magnetic and non-magnetic iron oxides. Images of the sample with magnification of 2500 (Fig. 2, 3) were obtained by scanning electron microscope, which shows that the particles of the samples have different sizes and morphology. Probably, the magnetic part of the sample corresponds to copper ferrite ($CuFe_2O_4$) , which is formed in the high-temperature zone by the thermal steam plasma from the copper electrodes and the iron of the plasma torch case. Particles obtained by the co-precipitation method at high temperature have an ordered magnetic structure [11]. This material is used as a catalyst for the conversion of carbon monoxide by steam [12]. This plasma method can be effective for the production of this catalyst.

Figure 2. Micrograph of the magnetic part of the sample

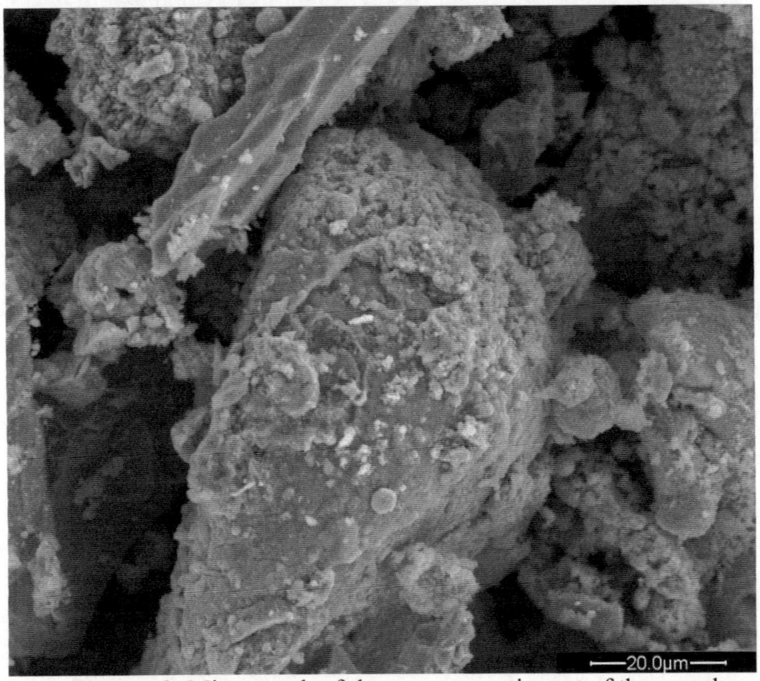

Figure 3. Micrograph of the non-magnetic part of the sample

IOP Conf. Series: Journal of Physics: Conf. Series **929** (2017) 012096 doi:10.1088/1742-6596/929/1/012096

4. Conclusion

Under the action of the electric arc the electrodes is decomposed and the electrode material reacts with hot gas. Thus, metal particles carried out by the plasma stream react with chlorine with formation of corresponding chlorides. Perhaps this is one of the reasons of organochlorine compounds absence in the reaction products. Remaining chlorine reacts with hydrogen to form hydrogen chloride. In our opinion, the chemical reaction of the carbon tetrachloride decomposition is facilitated by the fact that copper ferrite is found in erosion products which is an active catalyst for the conversion of carbon monoxide by steam. The products of erosion of electrodes can be effectively used for carrying out of the plasma-chemical process, which can significantly increase the efficiency of plasma processing methods and reduce the amount of waste.

References

[1] Evangelisti S, Tagliaferri C, Clift R, Lettieri P, Taylor R and Chapman C 2015 *Waste Management* **43** 485–96
[2] Rutberg P, Kuznetsov V, Popov V, Popov S, Surov A, Subbotin D and Bratsev A 2015 *Applied Energy* **148** 159–168
[3] Pan X, Yan J, Xie Z 2013 *Journal of Environmental Sciences* **25 #7** 1362-1367
[4] Rutberg Ph G, Nakonechny Gh V, Pavlov A V, Popov S D, Serba E O, Surov A V 2015 *J. Phys. D: Appl. Phys.* **48 # 24** 245204 – 245216
[5] Rutberg Ph G, Safronov A A, Popov S D, Surov A V and Nakonechny Gh V 2005 *Plasma Phys. Control. Fusion* **47** 1681–96
[6] Surov A V, Popov S D, Serba E O, Nakonechny Gh V, Spodobin V A, Ovchinnikov R V, Kumkova I I and Shabalin S A 2012 *J. Phys.: Conf. Ser.* 406 012007.
[7] Fulcheri L, Fabry F, Takali S and Rohani V 2015 *Plasma Chem. Plasma Proc.* **35** 565–85
[8] Rutberg Ph G, Kuznetsov V A, Serba E O, Popov S D, Surov A V, Nakonechny Gh V and Nikonov A V 2013 *Applied Energy* **108** 505–14
[9] Vinogradov S E, Kuznetsov V E, Gorynin I V, Oryschenko A S, Rutberg Ph G, Rybin V V, Safronov A A, Shekalov V I, Shiryaev V N 2006 *Voprosy materialovedeniya* **45 #1** 1-7
[10] Rubio S, Quintero M, Rodero A 2011 *J. Hazard. Mater.* **186** 820–826
[11] Pandya P B, Joshi H H, Kulkarni R G 1991 *Journal of Materials Science Letters* **8** 474–476
[12] Estrella M, Barrio L, Zhou G, Wang X, Wang Q, Wen W, Hanson J C, Frenkel A I, Rodriguez J A 2009 *J. Phys. Chem.* **113** 14411–14417

Numerical simulation of natural convection in wedge-shaped domain with isothermal free surface

N A Yaroslavtseva and N G Ivanov

Department of Fluid Dynamics, Combustion and Heat Transfer,
Peter the Great St.Petersburg Polytechnic University,
Polytechnicheskaya str. 29, 195251, St.Petersburg, Russia

E-mail: nadya_yar94@mail.ru

Abstract. The contribution deals with 2D laminar unsteady natural convection in a wedge-shaped reservoir model induced by the isothermal surface heating of a water basin being colder than surrounding atmosphere. The problem formulation considered corresponds to large-scale convection development during a cloudy day when the solar radiation impact is negligible. Numerical simulation was performed using an in-house Navier-Stokes code SINF. The focus of the paper is on the accurate resolution of the initial period of the convective circulation pattern development. The dependence of the predicted convective structures on the computational domain size as well as on the boundary condition at the free surface is analysed. The influence of geometry on the buoyancy-induced flow formation is discussed.

1. Introduction

Natural convection induced by heating or cooling of the shallow nearshore zone with non-uniform depth plays an important role in transportation of nutrients, pollutants and chemical species. Started in 1980s, numerous field experiments in lakes or coastal waters showed stable large-scale flows forming in this zone. These flows could have an impact on the pollutants residence time in the nearshore regions. For example, experiments by James and Barko performed in 1988 [1] examined nighttime convective circulation patterns; the effect of the patterns on the transport of phosphorus between littoral and pelagic zones was significant. Later contributions also pointed to the effects of the nearshore convection, e.g., measurements by Niemann et al. [2] allowed to suppose that the convective circulation is the origin of the reef-borne material transport to the deeper strata of the Red Sea.

The origin and development of convective circulation patterns detected in the field experiments have been studied also in many contributions operating with scaling analysis and numerical simulation, see, e.g., the reviews in [3, 4]. Many papers reported the analysis data for test problems with imposed horizontal temperature gradient that drives the flow. Apart from this scenario, other contributions considered different mechanisms of horizontally uniform thermal forcing when the geometry plays the key role in the formation of the convective flows. Thus, Lei and Patterson [5] carried out an analysis of the unsteady natural convection in a triangular domain induced by the absorption of solar radiation. The horizontally uniform internal heating source due to radiation was set in this case, and the principal driving force for the circulation came from the heat exchange on the sloping bottom. For the same triangular domain, Lei and Patterson [6] investigated also the case of nighttime cooling through the water surface, and a heat flux condition quantified the heat loss. Later Bednarz et al. [7] considered natural convection in a wedge-shaped domain for periodic thermal forcing corresponding to daytime heating and nighttime cooling at the water surface. Numerical simulation in [7] has modeled the flow development relevant to cloudy atmospheric conditions without any influence of the solar radiation and the flow is driven by the ambient temperature changes.

Bednarz et al. [8, 9] confirmed experimentally development of convection in a wedge-shaped reservoir model. In [8] the experimental model was cooled from above, while in [9] the quasi-steady natural convection flow in the domain subject to periodic thermal forcing at the water surface was considered. Yu et al. [4] performed a scaling analysis for the case of the iso-flux surface heating with the focus on the horizontal conduction effects. On the contrary, Mao et al. [3] considered the isothermal surface heating that corresponds to the overcast day conditions when the air temperature is higher than the water body temperature and the solar radiation impact is negligible. In a rectangular reservoir, an isothermal warm surface results in a stable thermal stratification without any motion, so that the simplified isothermal surface heating model points to pure effects of the wedge shape.

The current study considers the wedge-shaped reservoir model with the same surface boundary condition as was adopted in [3]. Numerical simulation of two-dimensional (2D) transient natural convection induced by isothermal surface heating was performed for various computational domains, with the focus on the accurate resolution of the initial period of the convective circulation pattern development.

2. Problem definition and computational aspects

The 2D wedge-shaped reservoir model considered is shown in figure 1, x is the distance from the shoreline and y is the vertical position, the origin of the coordinate system is set at the shoreline. The flat free surface approximation is used. The water depth varies with the offshore distance until the maximum depth of the reservoir, h, is reached. The bottom slope $A = \mathrm{tg}\alpha$ is one of the parameters of the problem. Two cases with the bottom slope values of $A = 0.1$ and $A = 0.2$ are considered in the current paper.

To the right of the region with the sloping bottom the water depth is uniform. To make sure that the length of the region with the horizontal bottom is enough to avoid the influence of boundary BC on the nearshore convection, a special study of the effect of the distance FC on the solution was performed, FC length values of $10h$ and $15h$ were considered. To avoid the singularity problem near the shoreline, an auxiliary vertical boundary AD is introduced, AD = $0.1h$, and small triangle to the left of AD is excluded from the computational domain.

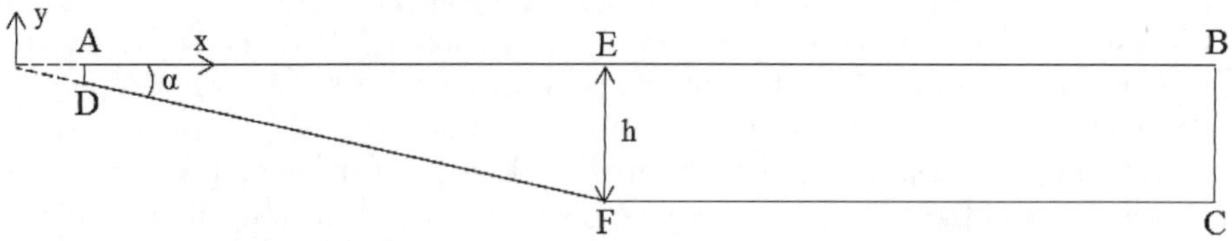

Figure 1. Computational domain and coordinate system

Development of convective circulation patterns from the initial stationary isothermal state is modeled, and the initial temperature is assumed as the reference temperature, T_{ref}. Starting from the time instant $t = 0$, heating is imposed at the free surface, and the boundary is treated as isothermal with the uniform temperature T_{fs}. The reference temperature difference is defined as $\Delta T_{\mathrm{ref}} = T_{\mathrm{fs}} - T_{\mathrm{ref}}$.

The buoyancy-induced transient flow is simulated using the 2D Navier-Stokes and energy equations. The incompressible fluid with constant physical properties is considered. The buoyancy effects are represented with the Boussinesq approximation. Assuming that the buoyancy velocity $V_b = (g\beta\Delta T_{\mathrm{ref}}h^3)^{1/2}$ is the velocity scale, here g is the gravity acceleration and β is the thermal expansion coefficient, and the ratio h/V_b is the time scale, the governing equations in the non-dimensional form are written as:

$$\frac{\partial u}{\partial x}+\frac{\partial v}{\partial y}=0$$
$$\frac{\partial u}{\partial t}+u\frac{\partial u}{\partial x}+v\frac{\partial u}{\partial y}=-\frac{\partial p}{\partial x}+\sqrt{Pr/Ra}\,\Delta u \quad ,$$
$$\frac{\partial v}{\partial t}+u\frac{\partial v}{\partial x}+v\frac{\partial v}{\partial y}=-\frac{\partial p}{\partial y}+\sqrt{Pr/Ra}\,\Delta v+T$$
$$\frac{\partial T}{\partial t}+u\frac{\partial T}{\partial x}+v\frac{\partial T}{\partial y}=\frac{1}{\sqrt{Pr\,Ra}}\Delta T$$

where u and v are the horizontal and vertical velocity components, respectively, p is the pressure and T is the temperature. The equations comprise two non-dimensional parameters: the Rayleigh number, $Ra = g\beta\Delta T_{ref}h^3/\nu a$, and the Prandtl number, $\Pr = \nu/a$, here ν is the kinematic viscosity and a is the thermal diffusivity. The current contribution considers one set of these parameters: $Pr = 7$, $Ra = 10^5$.

The initial conditions are $u = 0$, $v = 0$, $T = 0$. The boundary conditions are as follows. The sloping and flat sections of the bottom are treated as adiabatic walls with the no-slip boundary conditions, so that $u = 0$, $v = 0$, $\partial T/\partial n = 0$. The same boundary conditions are set at the auxiliary boundary AD. For the isothermal flat free surface AB two cases were considered: (1) the stress free condition, so that $\partial u/\partial y = 0$, $v = 0$, $T = 1$, and (2) the no-slip condition, so that $u = 0$, $v = 0$, $T = 1$. The conditions at the distant boundary BC that connects the domain considered with the larger central part of the reservoir were varied. First, an open boundary condition was set at the line BC, with an approximation of the uniform pressure distribution over the boundary; in case of inflow the temperature value $T = 0$ is set. Second, a symmetry boundary conditions were set at the line BC, so that $u = 0$, $\partial v/\partial x = 0$, $\partial T/\partial n = 0$.

The computations were performed using the structured version of the in-house 3D steady/unsteady Navier-Stokes code SINF (Supersonic to INcompressible Flows) being under development since 1993 [10-12]. This version of the code is based on the second-order finite-volume spatial discretization using the cell-centered variable arrangement and body-fitted block-structured grids [10]. The artificial compressibility (AC) technique and/or the SIMPLEC method are used for the pressure-velocity coupling in the case of incompressible fluid flows (see [11] for details). The viscous fluxes are evaluated on the basis of central differences. Weighted corrections are calculated on the basis of the Rhie and Chow approach. Three-layer second-order scheme is implemented for physical time stepping. The code is parallelized based on the MPI standard and the domain decomposition strategy according to the grid block structure [12].

In the current contribution, the QUICK scheme was applied to compute the convective fluxes. The SIMPLEC technique was used in more stable case of the no-slip surface condition. For the cases with the stress-free surface condition the coupled AC/SIMPLEC algorithm was activated.

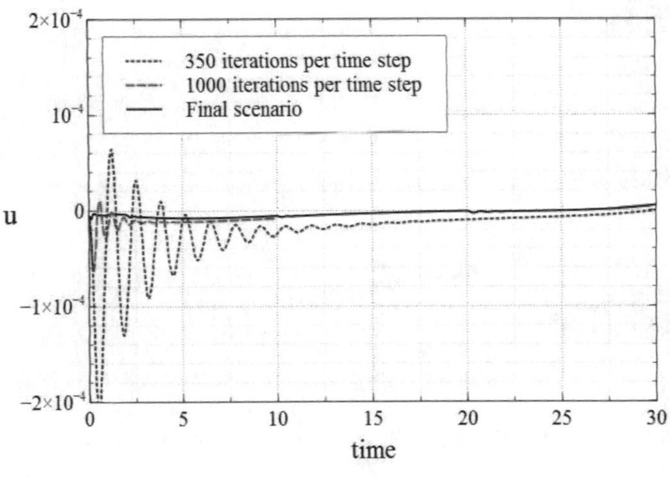

Figure 2. The effect of solution convergence at each time step on the velocity evolution at the monitoring point located at $x = 10.16$, $y = -0.36$; final scenario is: 5000 iterations per time step from the start to the 2nd time units; 2500 iterations from the 3rd to the 10th time unit, 1500 iterations from the 11th to the 20th time unit, 750 iterations from the 21st to the 50th time unit, then 500 iterations

The uniform meshes were used for the computations. A mesh dependency study was performed for the case of FC = 10h using seven successively refined grids, the finest grid size was more than 40 thousand cells. The solutions obtained with the mesh of 451×75 nodes (33,300 cells) were accepted as mesh-independent, and all the data presented in the paper were computed using spatial resolution of such quality. The time step of 0.05 time units was chosen after several time step dependency tests. Finally, special attention was paid on the effect of solution convergence at each time step. To provide sufficient convergence during initial period of surface heating, large amount of iterations at each time step was necessary (see figure 2). Later, when the temperature gradients are reduced, it is possible to use less number of iterations, the final scenario of the iteration variation is given in figure 2.

3. Results and discussion

Formation of the buoyancy-induced flow in the region with the sloping bottom is illustrated in figure 3 where temperature and velocity fields are shown for $A = 0.1$ at three successive time instants. A zone with pronounced temperature gradient is formed near the heated free surface. It interacts with the inclined bottom at low x-values, and slight thermal forcing is generated in the horizontal direction driving a convective pattern with clockwise water motion. The intensity and extension of this pattern grow in time significantly during the period illustrated, though typical velocity values in the pattern are much less than the scale V_b value, and even at $t = 30$ do not exceed 1% of the buoyancy velocity.

Figure 4 gives quantitative information on time evolution of the convective pattern near the shoreline: the plot shows the maximum value of x-velocity that is extracted at each time instant from the vertical line $x = 2$. The velocity values given are very low as the monitoring line is close to the left boundary of the convective pattern. The motion intensity in this region grows during the initial period of about 20 time units, and then velocity values decrease gradually. The results computed correspond to the data obtained previously by Mao et al. [3] also shown in the figure. Some difference between the current results and the data from [3] could be attributed to the auxiliary vertical boundary AD introduced in the current study. Note that to compare the velocity values, the data from [3] were rescaled: $t = t_{[3]}(Ra{\cdot}Pr)^{1/2}$, $u = u_{[3]}(Ra{\cdot}Pr)^{-1/2}$.

As mentioned above, a special test case with the no-slip condition at AB was computed. If the no-slip boundary condition is set at the free surface, convection behavior is qualitatively the same as in case of the stress-free condition: the size of the convective cell is comparable in both the cases as the streamline patterns for the no-slip condition (not shown) looks like the cells given in figure 3 a,c,e. However, as expected, the velocity values in the fluid body are approximately two times lower in case of zero velocities at the surface, see blue rhombs in figure 4.

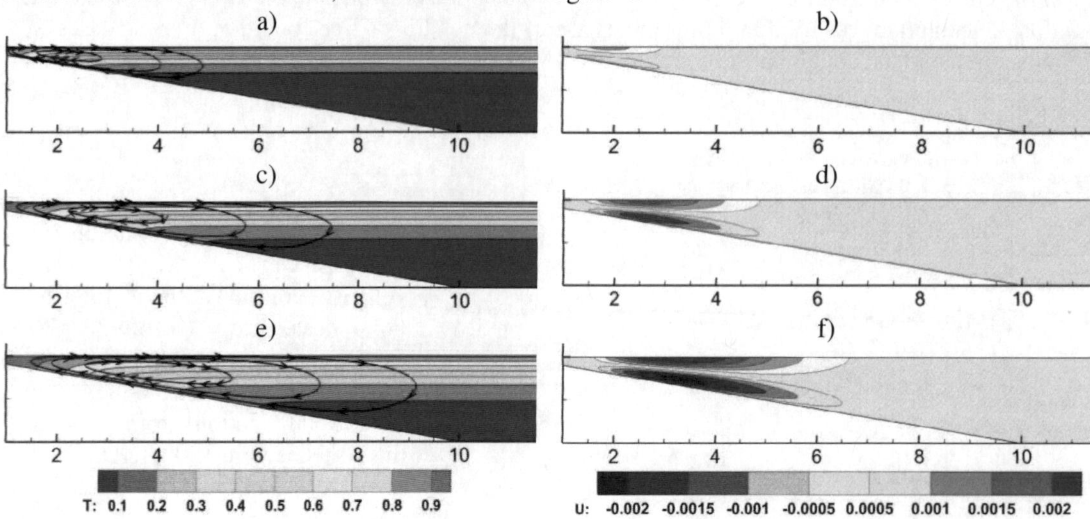

Figure 3. Temperature fields with streamlines (left) and x-velocity component distributions (right) at (a, b) $t = 10$, (c, d) $t = 20$, and (e, f) $t = 30$; $A = 0.1$

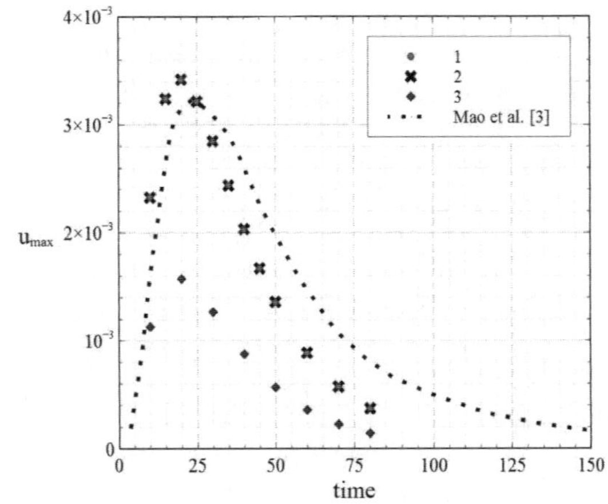

Figure 4. Evolution of the maximum value of x-velocity at $x = 2$; (1) – stress-free surface boundary, symmetry condition at BC; (2) – stress-free surface boundary, pressure condition at BC; (3) – no-slip surface boundary, pressure condition at BC; the computed results are compared with the data from [3]; $A = 0.1$

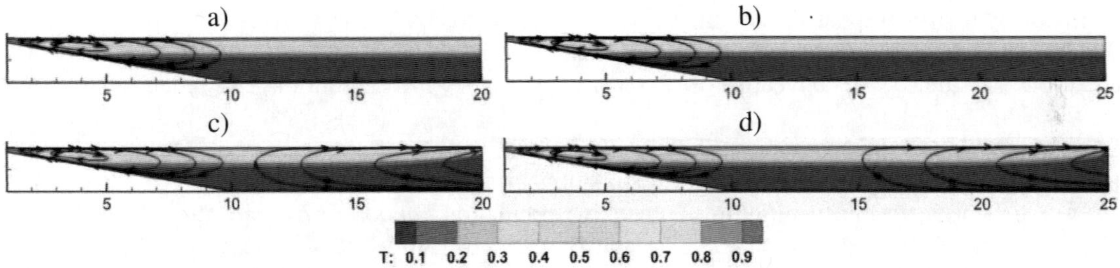

Figure 5. Temperature fields with streamlines for (a, c) FC = 10h and (b, d) FC = 15h: (a, b) symmetry and (c, d) pressure condition at the distant boundary BC; t = 30

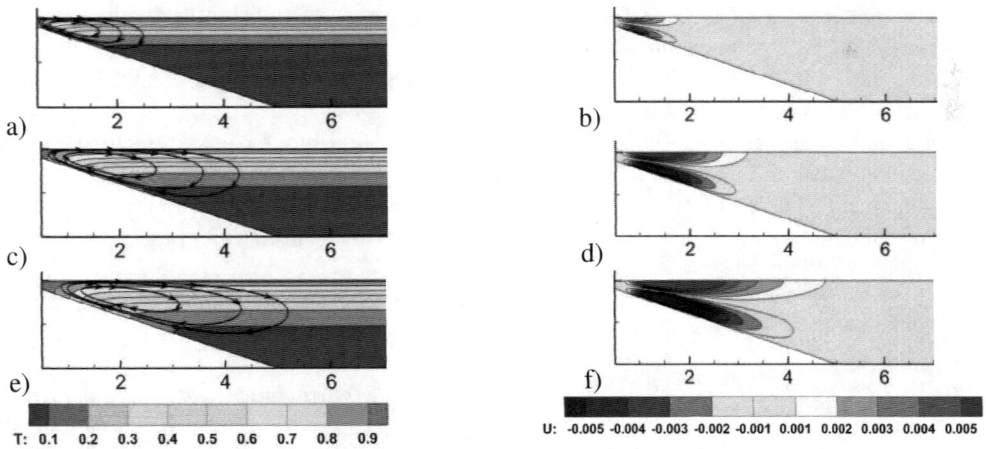

Figure 6. Temperature fields with streamlines (left) and x-velocity component distributions (right) at (a, b) t = 10, (c, d) t = 20, (e, f) t = 30; A = 0.2

Figure 4 illustrates also the effect of the condition at the distant boundary BC on the convective pattern in the nearshore region. In reality the nearshore zone is connected with a much larger central part of the reservoir, and the convective pattern could be influenced by the processes far from the shoreline. The current model problem does not consider these long-distance processes, and it is supposed that the open boundary does not affect the flow in the region with the sloping bottom. Figure 4 gives an evidence that at $x = 2$ the solutions with the symmetry and the pressure conditions are the same at least till the 80th time unit. However, as it is visible from figure 5, if the pressure boundary condition is set, the second convective structure is generated near this boundary that was detected also in [3]. Remarkably, at $t = 30$ considered in figure 5 two convective cells do not interact. If the length of the region with the uniform water depth is increased, this convective structure is shifted to the right (compare figures 5c and 5d). It could be concluded that the length of the region with the horizontal bottom of $10h$ is enough to avoid the influence of the right boundary on the near shore convection.

Finally, the influence of bottom slope value on the convective cell formation is studied. A comparison of figure 6 where the results for $A = 0.2$ are given with similar plots presented in figure 3 for $A = 0.1$ allows to conclude that the convective cell size increases with the bottom slope reduction. However, for less bottom slope the intensity of the convective pattern is lower than for higher A. The reason for it is in higher value of horizontal temperature gradient in case of larger bottom slope.

4. Conclusions

Natural convection in a 2D wedge-shaped reservoir model induced by isothermal surface heating at the water surface was simulated numerically using an in-house Navier-Stokes code SINF. Transient large-scale convective pattern with clockwise water motion was detected in the region with the sloping bottom, typical velocities did not exceed 1% of the buoyancy velocity. It was shown that the length of the region with the horizontal bottom of $10h$ is enough to avoid the influence of the right open boundary on the nearshore convection. The influence of the bottom slope value on the convective cell formation was studied: the convective cell intensity decreases with the bottom slope reduction.

References

[1] James W F and Barko J W 1991 Estimation of phosphorus exchange between littoral and pelagic zones during nighttime convective circulation *Limnol. Oceanogr.* **36** pp 179–87

[2] Niemann H, Richter C, Jonkers H M and Badran M I 2004 Red sea gravity currents cascade near-reef phytoplankton to the twilight zone *Mar. Ecol.-Prog. Ser.* **269** pp 91–99

[3] Mao Y, Lei C and Patterson J C 2012 Unsteady nearshore natural convection induced by constant isothermal surface heating *J. Fluid Mech.* **707** pp 342–68

[4] Yu P, Patterson J C and Lei C 2015 A scaling analysis of transient natural convection in a reservoir model induced by iso-flux heating *J. Fluid Mech.* **764** pp 219–49

[5] Lei C and Patterson J C 2002 Unsteady natural convection in a triangular enclosure induced by absorption of radiation *J. Fluid Mech.* **460** pp 181–209

[6] Lei C and Patterson J C 2005 Unsteady natural convection in a triangular enclosure induced by surface cooling *Intl J. Heat Fluid Flow* **26** pp 307–21

[7] Bednarz T P, Lei C and Patterson J C 2009 Unsteady natural convection induced by diurnal temperature changes in a reservoir with slowly varying bottom topography *Int. J. of Thermal Sciences* **48** pp 1932–42

[8] Bednarz T P, Lei C and Patterson J C 2008 An experimental study of unsteady natural convection in a reservoir model cooled from the water surface *Exp. Therm. Fluid Sci.* **32** pp 844–56

[9] Bednarz T P, Lei C and Patterson J C 2009 An experimental study of unsteady natural convection in a reservoir model subject to periodic thermal forcing using combined PIV and PIT techniques *Exp. Fluids* **47** pp 107–17

[10] Smirnov E M 2000 Solving the full Navier-Stokes equations for very-long-duct flows using the artificial compressibility method *Proc. of ECCOMAS 2000 - European Congress on Comp. Methods in Appl. Sciences and Engin. (Barcelona, Spain, 11 - 14 September 2000)* pp 1–17

[11] Smirnov E M and Zaitsev D K 2004 Computations of internal flows using an artificial-compressibility solver enhanced with an elliptic pressure-correction procedure *Proc. of ECCOMAS 2004 - European Congress on Comp. Methods in Appl. Sciences and Engin. (Jyväskylä, Finland, 24 - 28 July 2004)* pp 1–13

[12] Zaitsev D K, Shchur N A and Smirnov E M 2009 Numerical simulation of 3D turbulent flows around bodies subjected to vortex-induced and forced vibration *Lecture Notes in Computational Science and Engineering* **67** pp 347–54

IOP Conf. Series: Journal of Physics: Conf. Series **929** (2017) 012098 doi:10.1088/1742-6596/929/1/012098

Numerical study of MHD supersonic flow control

A.I. Ryakhovskiy[1,2] and A.A. Schmidt[1]

[1] Plasma Physics and Astrophysics Department, Ioffe Institute, 194021, 26 Polytechnicheskaya, Saint-Petersburg, Russian Federation
[2] Institute of Applied Mathematics and Mechanics, Saint-Petersburg Polytechnical University, 195251, 29 Polytechnicheskaya, Saint-Petersburg, Russian Federation

e-mail: alexey.i.ryakhovskiy@mail.ioffe.ru

Abstract. Supersonic MHD flow around a blunted body with a constant external magnetic field has been simulated for a number of geometries as well as a range of the flow parameters. Solvers based on Balbas-Tadmor MHD schemes and HLLC-Roe Godunov-type method have been developed within the OpenFOAM framework. The stability of the solution varies depending on the intensity of magnetic interaction The obtained solutions show the potential of MHD flow control and provide insights into for the development of the flow control system. The analysis of the results proves the applicability of numerical schemes, that are being used in the solvers. A number of ways to improve both the mathematical model of the process and the developed solvers are proposed.

1. Introduction

The recent challenges in the aerospace industry have caused the resurgence of research in flow control concepts. One of such concepts in the magnetohydrodynamic (MHD) flow control. It is based on the idea of altering the shock wave structure by using the interaction between charged particles in the flow and a magnetic field. The field, that is generated by either a coil with a current or a magnet, affects the charged particles in an incoming gas flow through the ponderomotive force and joule heating. This effect can be used to reduce aerodynamic heating on an atmospheric entry capsule or even to change its' descent trajectory [1]. Another possible application is controlling the pressure position jump within the engines of hypersonic scramjet during the changing of altitude or regime of the flight. The development of such a system will require both theoretical and experimental research into the aspects of this interaction. Computational fluid dynamics (CFD) can be an invaluable instrument to the research supplementing limited experimental capabilities. In the recent decades, there have been several studies concerning MHD flow control for different regimes [2],[3]. Most of them deal with the hypersonic flow, since it is characteristic for both descent capsules and scramjets. We aim to develop a comprehensive model of the phenomenon based on a free CFD software to aid the design of MHD flow control system. In this paper, we deal primarily with supersonic flow. The applications developed and the results obtained in this study will serve as a springboard towards a comprehensive open-source simulation tool for MHD hypersonic flow.

2. Physical description

We consider a supersonic flow around a blunted body of a shape characteristic for a re-entry capsule. Three different cases are considered. Case geometries are cylindrically symmetrical in order to reduce required computational power. First case geometry simulates the shape of "Soyuz" spaceship's atmospheric return capsule, while the second one is shaped similar to the analogous capsule of NASA's Phoenix spacecraft.

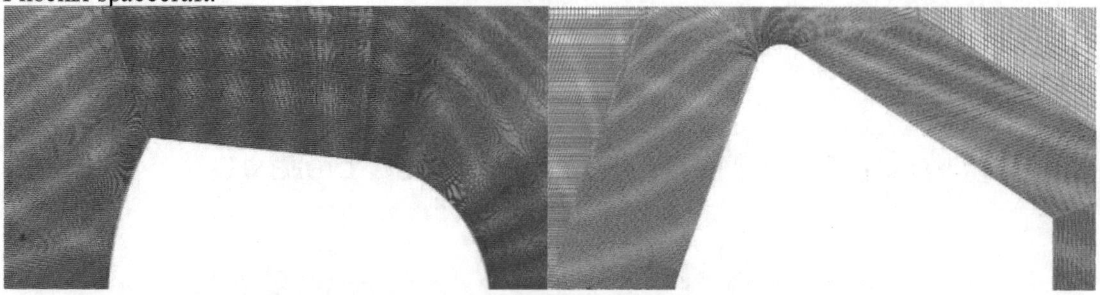

Figure 1. Sample of mesh for case geometries. 1 and 2

In the first two cases magnetic field is generated by a coil placed inside the model of the each craft at the conjunction between the forebody and middle body. Typical configuration of magnetic field is shown on the picture.

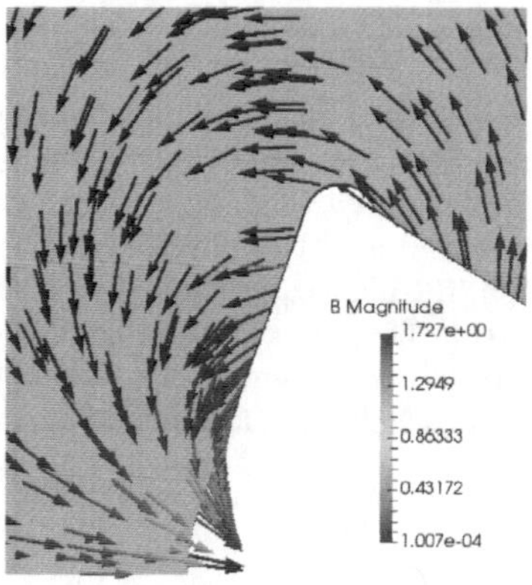

Figure 2. Magnetic field configuration for the case 2

Finally, third case represents an experiment conducted in Ioffe Institute [2]. The setup consists of a cylinder with a cone on top. It has an electrode core, surrounded by a magnetic core, as well as a ring electrode at the base of its conical top. The electrodes are plugged to a power source, so that a discharge between them ionizes the gas in front of the body. By raising the conductivity of the gas the intensity of interaction between the incoming flow and generated magnetic field is therefore increased.

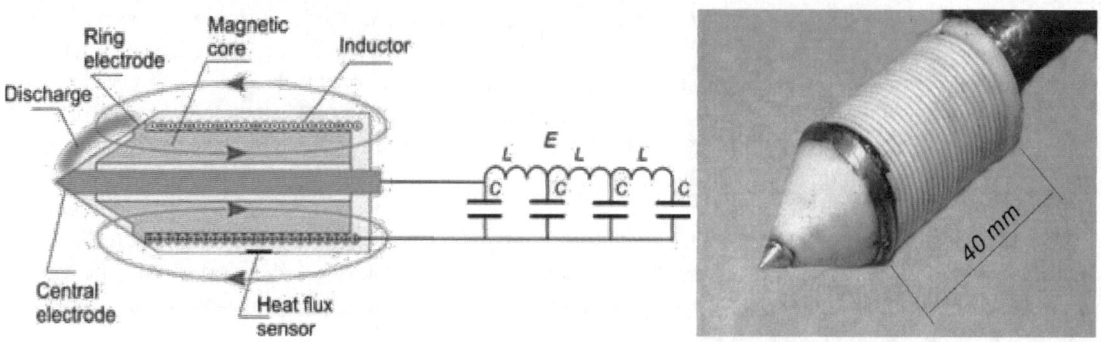

Figure 3. Scheme (left) and a photo of an experimental setup that is represented by case 3.

Hydrodynamic parameters of the flow are listed in the table below. The gas in consideration is air. The intensity of MHD interaction is characterized by Stuart number which can range from 0 to 0.5.

IOP Conf. Series: Journal of Physics: Conf. Series **929** (2017) 012098 doi:10.1088/1742-6596/929/1/012098

Table 1. The values of characteristic dimensionless quantities of the studied flow

Number	Value
M (Mach	2-6
Re (Reynolds)	~10^5
Re_{mag} (magnetic Reynolds)	~0.1
Kn (Knudsen)	$10^{-3} – 10^{-5}$
S (Stuart)	0 - 0.5
Ha (Hartmann)	~100

3. Mathematical model

Our simulation uses magnetohydrodynamic model, which is widely used in similar problems [3]. Almost all of the flow's parameters are well within the range of applicability of the continuous representation. It should, however be noted, that in a real case Knudsen number can locally fall below the threshold of model's applicability. We have tried to numerically determine where such a breakdown can occur.

The system of MHD equations consists of a set of conservation laws as well as magnetic field equation. However, in this study we only consider cases with a constant magnetic field so the latter can be dropped. The effect of magnetic interaction therefore is described by adding ponderomotive force term into the equation of motion and joule heating term into energy conservation equation. Also, due to low magnetic Reynolds number we can disregard self-induced electrical field. Using these assumptions and the generalized Ohm law we can rewrite the governing system of equations as follows.

$$\begin{gathered}\frac{\partial \rho}{\partial t} + \nabla \cdot (\rho \mathbf{v}) = 0, \\ \frac{\partial}{\partial t}(\rho \mathbf{v}) + \rho(\mathbf{v}\nabla)\mathbf{v} = \nabla P + \boldsymbol{j} \times \boldsymbol{B}, \\ \frac{\partial}{\partial t}\left(\frac{1}{2}\rho v^2 + \rho e\right) = \nabla \cdot [P\mathbf{v} + \lambda_\epsilon \nabla T] + \boldsymbol{j} \cdot \boldsymbol{E}, \\ j = \sigma(\boldsymbol{E} + \mathbf{v} \times \boldsymbol{B}), \qquad \boldsymbol{B} = \boldsymbol{B}_0(x, y, z)\end{gathered} \tag{1}$$

4. Numerical method

The studied flow is highly compressible, which gives a density-based approach a significant advantage in shock resolution accuracy. We use *rhoCentralFoam,* a density-based solver from OpenFOAM standard toolkit, as a starting point for our MHD solver. *rhoCentralFoam* implements Kurganov-Tadmor difference scheme [4], that is reliable at resolving steep pressure gradients, associated with shock wave formation. In addition Kurganov-Tadmor scheme has a MHD modification, proposed by Balbas and Tadmor [5]. This scheme for the system of MHD equations provides a guideline into how to modify *rhoCentralFoam* solver according to our needs.

Another solver for MHD supersonic flow has been developed based on *dbnsTurbFoam* from foam-extend fork of OpenFOAM. It is a Godunov-type coupled density-based solver utilizing Rusanov flux. We used a positive conservative method based on HLL and Roe methods [6] to add MHD functionality to *dbnsTurbFoam.*

To achieve the convergence of the method we start with a constant distribution for the temperature, pressure and velocity field and solve the problem without MHD interaction. After the solution settles to a stationary regime, we start iterations with a magnetic field and continue them until another stationary flow regime is achieved. To keep the computations stable, time step had to be limited by a local cell Courant number threshold of 0.9.

To resolve the flow separation that can occur behind the body we have used *Spalart-Allmaras IDDES* turbulence model. We can take liberty with choosing the turbulence parameters since the turbulence effect is negligible in front of the body and the region, where it is mode substantial, is not of a particular interest to us in the scope of this study.

5. Analisys of the results

Numerical results for supersonic flow around blunted body without MHD interaction can be validated using theoretical values for the shock wave pressure jump [7] :

$$P_2 = \frac{P_1}{\gamma + 1}[2\gamma M^2 - (\gamma - 1)] \tag{2}$$

where P_1 is the free-stream flow pressure and P_2 is the pressure behind the shockwave.

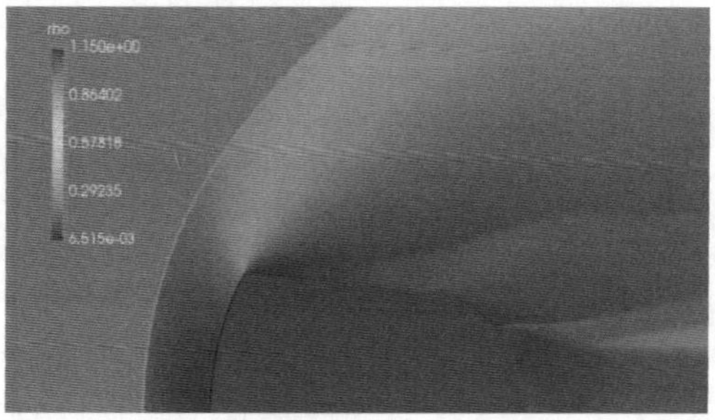

Figure 4. Numerically obrained density distribution for case 1

Table 2. Comparison between free-stream flow pressure numerically obtained pressure behind the shockwave

Mach	P_1 (10^3 Pa)	P_2 (10^3Pa)
2		135
4	30	555
6		1255

The comparison of experimental and numerical shock fronts show agreement despite the fact that conductivity in numerical model was set to be constant, which isn't the case in the real experiment.

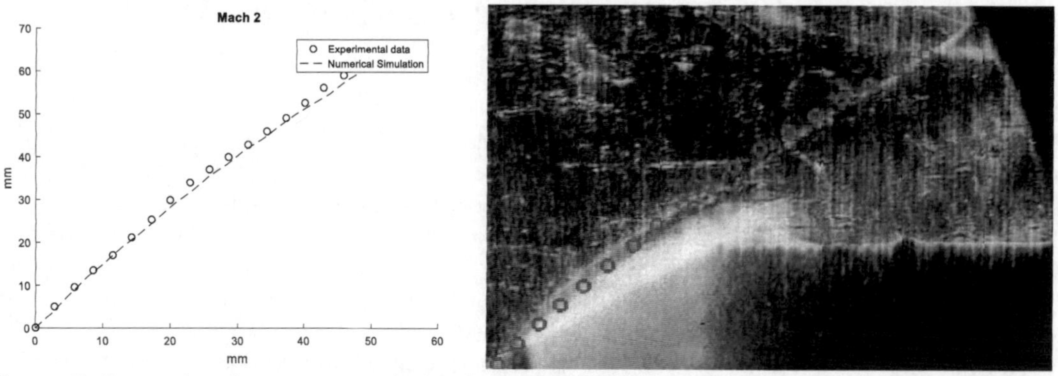

Figure 5. Comparison between numerical and experimental shock from (left) and a shadowgram of experimental shock front (right)

Graphs below present numericaly obtained profiles of temperature on the generatrix of case 2 body with and without magnetic field for Mach 2, 4 and 6 flows. Characteristic magnetic field induction in these cases is B_c = 1.5T, which makes Stuart number 0.5, 0.3, 0.1 respectively. The zone around 0.6m

IOP Conf. Series: Journal of Physics: Conf. Series **929** (2017) 012098 doi:10.1088/1742-6596/929/1/012098

mark on the generatrix corresponds to the maximum of magnetic field. Numerically obtained temperature graphs show significant dip in this region. The source of this dip is obviously the magnetic interaction between the generated field and the incoming ionized gas flow. The dip becomes extreme with lower Mach/high Stuart flows. This can signify a breakdown of continuous model for the extremely low density flow. This fact presents a challenge that will require significant reworking of the model to fully overcome.

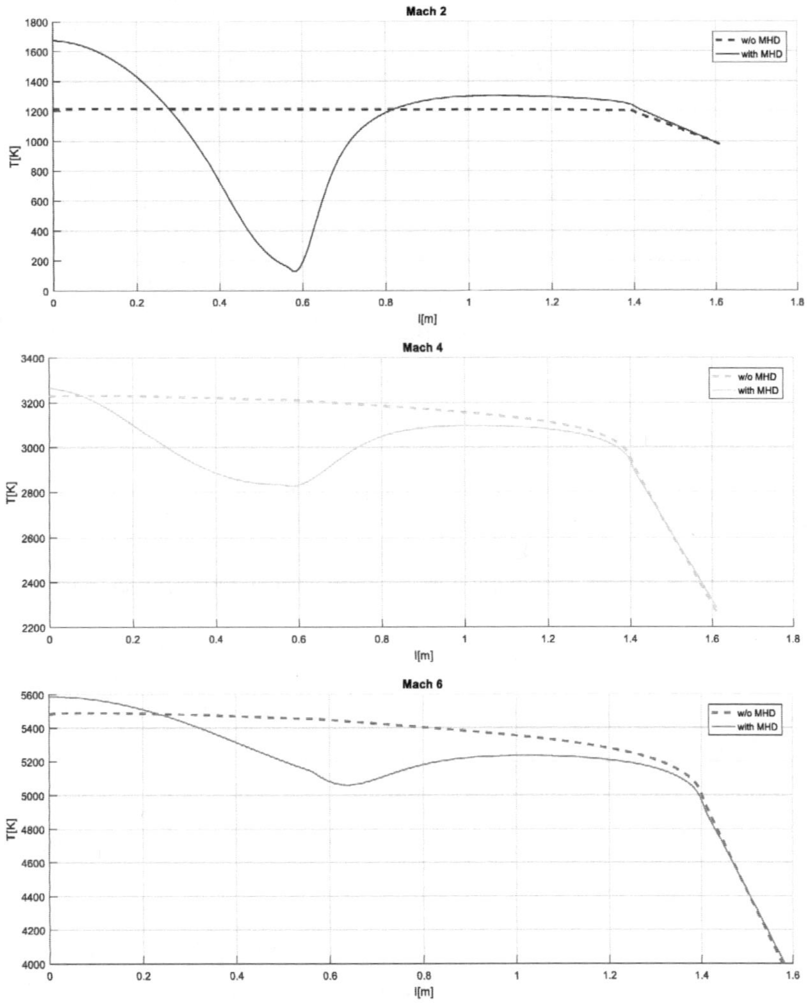

(a)
(b)
(c)

Figure 6. Comparison between temperature profiles on the generatrix of the body with (solid) and without (dashed) MHD interaction for Mach 2(a), 4(b) and 6(c) flows (B_c.= 1.5T).

Both central difference and Godunov-type solver proved stable for the flows with Stuart number ranging from 0 to 0.5. Instabilities for higher Stuart number can be attributed to low-density effects.

6. Conclusion

The aforementioned results show the potential of MHD flow control to significantly reduce the surface temperature of supersonic vehicle if the magnetic field is strong enough. However, our simulation lacks proper models for such important aspects as radiational heat transfer and ionization. It also should be modified to include multi-species capabilities integral for the modeling of hypersonic flows, for which the concept of MHD control is more relevant. Such a modification can also bring resolution to the problem of low-density effects.

The developed solvers have shown the applicability of both approaches to our problem. Both solvers can be used to resolve MHD supersonic flows with Stuart number ranging from 0 to 0.5.

Other question that remain open include determining necessary energy input for ionizing the gas to sufficient degree as well as choosing the optimal configuration of the magnetic field. Answering these question is going to require enhancing the model as well as additional numerical experimentation.

References

[1] Kawamura M, Nagata Y, Katsurayama H, Otsu H, Yamada K and Abe T 2013 Magnetoaerodynamic Force on a Magnetized Body in a Partially Ionized Flow *Journal of Spacecraft and Rockets* vol *50*(2), pp 347-351

[2] Bityurin, V., Bocharov, A., Lineberry, J. 2005 MHD flow control in hypersonic flight. *AIAA/CIRA 13th International Space Planes and Hypersonics Systems and Technologies Conference* p. 3225.

[3] Fujino, T., Yoshino, T., & Ishikawa, M. 2008. Numerical Analysis of Reentry Trajectory Coupled with Magnetohydrodynamics Flow Control. *Journal of Spacecraft and Rockets*, *45*(5), pp 911-920.

[4] Bobashev S V, Erofeev A V, Lapushkina, T A, Mende N, Poniaev S, Sakharov V and Van Wie D 2006. Recent results on MHD flow control at Ioffe institute. *14th AIAA/AHI Space Planes and Hypersonic Systems and Technologies Conference* p 8012

[5] Ryakhovskiy A I, Schmidt A A 2016 MHD supersonic flow control: OpenFOAM simulation *Trudy ISP RAN /Proc. ISP RAS* vol 28(1) pp 197-206

[6] Kurganov A and Tadmor E 2000 New high-resolution central schemes for nonlinear conservation laws and convection–diffusion equations *Journal of Computational Physics* vol 160(1) pp 241-282

[7] Balbás J, Tadmor E and Wu C C 2004 Non-oscillatory central schemes for one-and two-dimensional MHD equations *Journal of Computational Physics* vol 201(1) pp 261-285

[8] Janhunen, P. (2000). A positive conservative method for magnetohydrodynamics based on HLL and Roe methods. *Journal of Computational Physics* vol 160(2) pp 649-661

[9] Landau L D and Lifshitz E M 1986 *Theoretical Physics* vol 6 Hydrodynamics p 342

IOP Conf. Series: Journal of Physics: Conf. Series **929** (2017) 012099 doi:10.1088/1742-6596/929/1/012099

An automatic hybrid numerical scheme for global RANS-LES approaches

E K Guseva, A V Garbaruk, M Kh Strelets

Peter the Great St. Petersburg Polytechnic University, 29 Polytechnicheskaya Str., 195251, St. Petersburg, Russia

E-mail: katia.guseva@inbox.ru

Abstract. A new automatic high-order hybrid central-difference/upwind scheme is developed for the finite-volume approximation of the inviscid fluxes within global hybrid RANS-LES approaches. Performance of the scheme is illustrated by examples of computations of three types of flow: a flow with massive separation with the use of Delayed Detached-Eddy Simulation (DDES), a flow with separation and reattachment with the use of DDES with shear-layer adapted subgrid length scale, and a fully attached flow with the use of Improved DDES (IDDES).

1. Introduction

Nowadays hybrid RANS-LES approaches to turbulence modeling are considered as a compromise between the high quality of Large Eddy Simulation (LES) in separated flow regions and low computational costs of the methods based on the Reynolds Averaged Navier-Stokes equations (RANS) in the near-wall regions of the flow. These methods are powerfull tools of computational fluid dynamics and can be used in a wide range of applications. However their performance strongly depends upon the choice of discretization scheme for the governing equations.

In the zonal RANS-LES approaches that choice is dictated by their zonal nature: in LES regions the use of low-dissipative schemes is essential to ensure an acceptable resolution of small turbulent scales, while in RANS sub-domains more dissipative schemes should be used to avoid non-physical oscillations of the solution on the coarse, highly distorted grids typically used in these subdomains.

In the global hybrid approaches the RANS and LES sub-domains are defined automatically, and the position of the RANS-LES interface depends not only on the distance to the nearest wall and grid size but also on the solution. As a result, for global approaches the use of automatic weighted schemes is the most appropriate technique.

For the Detached Eddy Simulation (DES, [1]) such a scheme was proposed in the work [2] where it was demonstrated that for the "natural" DES applications (massively separated flows) this scheme ensures both high resolution of turbulent structures and stability (later on, this conclusion was supported in a large number of numerical studies). However, in the present work this scheme is shown to be sub-optimal in the framework of enhanced modifications of DES, namely, IDDES [3] and DDES with the shear-layer-adapted (SLA) subgrid scale [4], and a new hybrid scheme is proposed which performs equally well within all these DES versions. The scheme was tested in the framework of three types of simulations. First, in order to make sure that for the natural DES application area it performs not worse than the scheme [2], the scheme was applied for DDES of a massively separated flow. After that, a comparison of performance of the proposed scheme and scheme [2] was carried out for a flow with separation and reattachment in the framework of DDES combined with SLA definition of the subgrid length-scale and for a fully attached flow in the framework of the IDDES.

2. Formulation of automatic hybrid scheme

The idea of the proposed scheme is similar to that of the scheme [2]. In particular, it consists of the use of a weighted central-difference (CD) and upwind-biased (UB) schemes with a solution-dependent weight function. However the specific UB scheme and the weight function proposed in the present study differ from those of the scheme [2]. In particular, the weight function results in an effectively CD scheme not only in the detached flow areas, as in [2], but also in the attached boundary layers with resolved turbulent structures. In the inviscid (irrotational) areas of the flow and in the boundary layers with no turbulent content (RANS zone), the weight function, just as the weight function of [2], reduces the scheme to an UB one, but the UB scheme chosen in the present work is the so called BCD scheme [5] rather than fifth or third order upwind schemes, as in [2]. The BCD scheme works as an upwind or a central-difference scheme depending on whether the convection boundedness criterion is satisfied or not. Thus, the proposed scheme for the inviscid fluxes reads as:

$$F = (1-\sigma)F_{CD} + \sigma F_{BCD} \tag{1}$$

Here F_{CD} and F_{BCD} are the inviscid fluxes approximated with the 4th order centred scheme and with the BCD scheme respectively, the latter presenting a combination of the 3rd order upwind and 4th order centered schemes, and σ is the weight function defined as follows:

$$\sigma = \max\left(f_{inv}, f_{\nu t}, f_{2D\,BL}\right) \tag{2}$$

The functions f_{inv}, $f_{\nu t}$, $f_{2D\,BL}$ are aimed at identifying different regions of the flow where the more stable BCD scheme should be used. The first one, f_{inv}, serves for identification of the inviscid regions of the flow and is defined as follows:

$$f_{inv} = 1 - \tanh\left(B^4\right), \tag{3}$$

where $B = C_{H3}\Omega\max(S,\Omega)/\max\left\{\left(S^2+\Omega^2\right)/2, \Omega^2_{\min}\right\}$ is a function borrowed from the original hybrid scheme [2], S and Ω are magnitudes of the mean strain and vorticity, $C_{H3} = 2.0$, $\Omega_{\min} = \tau^{-1}$ and τ is a characteristic convective time. The functionality of f_{inv} is presented in Figure 1a, where the field of this function for the hump flow (see section 3.3 below) is shown.

The two other functions in the relation (2), $f_{\nu t}$ and $f_{2D\,BL}$, ensure treatment of the attached boundary layers with no turbulent content with the BCD scheme and the boundary layers with resolved turbulent structures with the CD scheme. The $f_{\nu t}$ function is defined as follows:

$$f_{\nu t} = 1 - \tanh\left\{\left(2\frac{\nu_{t\,LES}}{\nu_t}\right)^2\right\}, \tag{4}$$

where $\nu_{t\,LES} = \left(\min\{0.2\Delta_{\max}, 0.41 d_w\}\right)^2 S$ is the eddy viscosity estimated using the Smagorinsky model with a subgrid scale definition $\Delta_{\max}$. This function activates the BCD scheme if the eddy viscosity predicted by the hybrid approach is higher than that predicted by LES.

The $f_{2D\,BL}$ function is defined by the following relation:

$$f_{2D\,BL} = \tanh\left(8r_{dt}^2\right)\left(1 - F_{KH}(<\mathrm{VTM}>)\right), \tag{5}$$

where r_{dt} is an argument of the DDES shielding function [6] and the $F_{KH}(<VTM>)$ function identifies 2D flow regions [4]:

$$F_{KH}(<\mathrm{VTM}>) = \max\{F_{KH}^{\min}, \min\{F_{KH}^{\max}, F_{KH}^{\min} + \frac{F_{KH}^{\max} - F_{KH}^{\min}}{a_2 - a_1}(<\mathrm{VTM}> - a_1)\}\}.$$

Here $\mathrm{VTM} \equiv \dfrac{\sqrt{6}\,|(\hat{\mathbf{S}}\cdot\boldsymbol{\omega})\times\boldsymbol{\omega}|}{\omega^2\sqrt{3tr(\hat{\mathbf{S}}^2)-[tr(\hat{\mathbf{S}})]^2}}$, angular brackets denote averaging over the current and closest neighboring cells, and the empirical constants are: $F_{KH}^{\max}=1.0$, $F_{KH}^{\min}=0.0$, $a_1 = 0.005$ и $a_2 = 0.01$.

The function equal to $\max\{f_{vt}, f_{2D\_BL}\}$ ensures approximation of the inviscid fluxes in the attached boundary layer with no turbulent content ($x < 0.6c$) with the BCD scheme (Figure 1b).

Finally, a snapshot of the resulting weight function σ in the hump flow is presented in Figure 1c. One can see that with this weight function the scheme (1) performs as the BCD scheme in the irrotational flow region and in the boundary layer with no turbulent content and as the centred scheme in the separated region and in the reattached boundary layer with resolved turbulent structures.

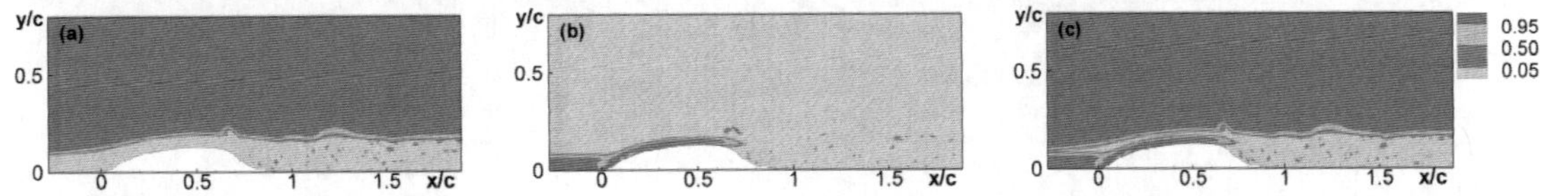

Figure 1. Snapshots of the f_{inv} (a), $\max(f_{vt}, f_{2D\,BL})$ (b), and resulting weigh function σ (c) from the simulation of the hump flow (see section 3.3 below)

3. Results and discussion

In this section computational set-ups are presented along with results of simulations of three test-cases chosen for evaluation of performance of the proposed scheme. The scheme was implemented in the incompressible branch of the NTS code, details of which are outlined in [7].

3.1. DES of massively separated flow

The first considered flow is that past the NACA 0021 airfoil at 60 degrees angle of attack studied in the experiment [8]. This is an example of massively separated flow, for which the DES approach was originally proposed and validated. The experimental Reynolds number based on the chord length, c, and free stream velocity, U_0, was equal to $2.7\cdot10^5$. Computational domain and "O-type" grid in an XY-plane used in the simulations are shown in Figure 2. The radius of the domain is equal to 15c and the grid has 211 cells in the azimuthal and 140 cells in the radial direction. It is refined in the vicinity of the leading and trailing edges of the airfoil where enhanced resolution is needed and is clustered towards the airfoil surface so that the first near-wall step in wall units is less than 1.0. The spanwise size of the domain was set equal to the value of 4c. The 3D grid was generated by extruding of the XY-grid in the spanwise direction with a constant step equal to 0.02c. As a result, the 3D grid contained nearly $5.5\cdot10^6$ cells. The time step in the simulations was equal to $0.02c/U_0$, which ensures the CFL number is less than 1.0 in the major part of the domain. The simulations were performed using the SA-based DDES.

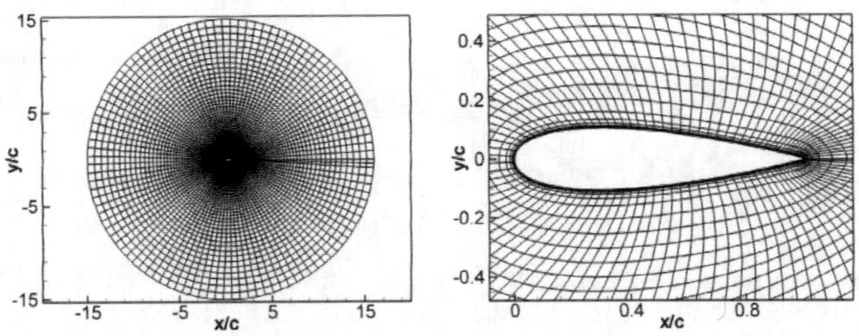

Figure 2. Computational domain and grid in XY plane (every other grid line is shown)

Results of the simulations with the current scheme are compared with those obtained with the hybrid scheme [2]. As can be seen from vorticity magnitude snaphots presented on Figure 3, the proposed scheme provides somewhat faster shear layer roll up than the scheme [2]. This is explained by the different behavior of schemes in the separation region and in the close vicinity of the leading and trailing edges of the airfoil, where the weight of CD in the new scheme turns out to be higher than that in scheme [2] (Figure 4). Other than that, the weight of CD in the current scheme in the near-wall region within the recirculation bubble is higher than in scheme [2]. As a result, the mean pressure coefficient predicted by the proposed scheme is a bit closer to experimental data than the one predicted by scheme [2] (Figure 5). Note, however, that with the proposed scheme, slightly non-monotonic behavior of the pressure is observed near the edge of the airfoil.

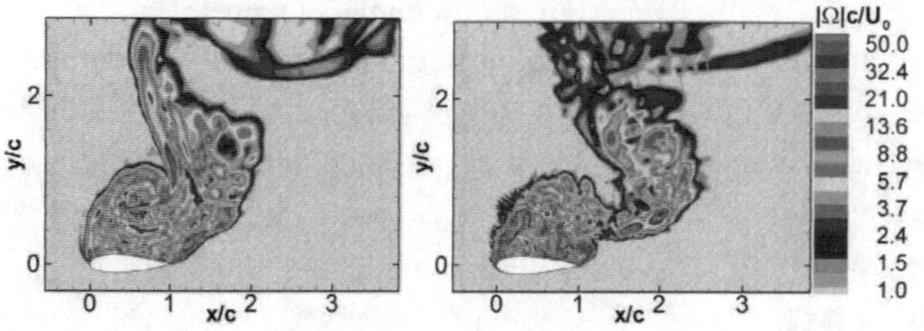

Figure 3. Vorticity snapshots from DDES carried out with the hybrid scheme [2] (left frame) and hybrid scheme (1) (right frame)

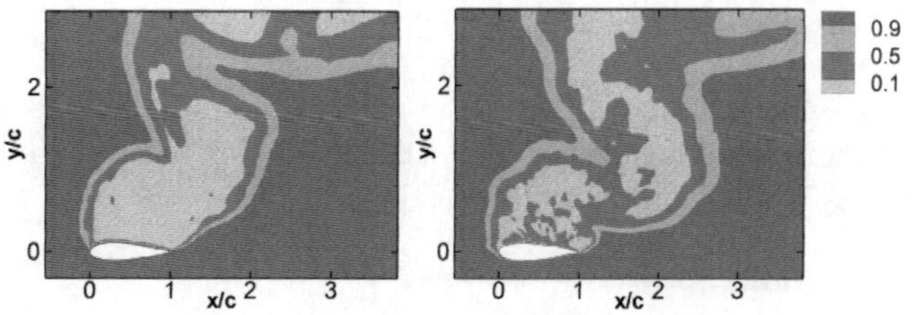

Figure 4. Snapshots of the weight function of scheme [2] (left frame) and of scheme (1) (right frame)

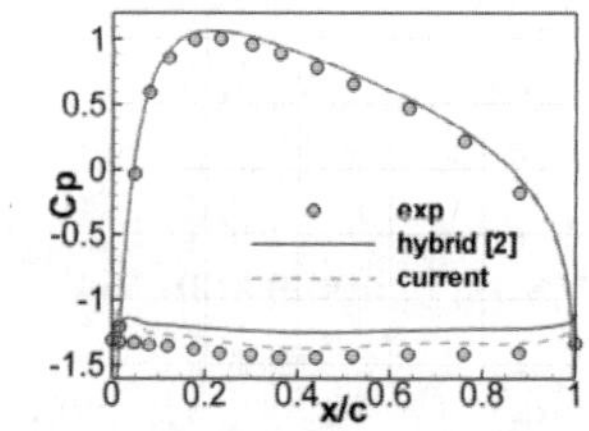

Figure 5. Pressure coefficient distribution

3.2. Shear-layer adapted DDES of a flow over the wall-mounted hump

This flow (the so called 2D NASA wall-mounted hump) was studied in the experiments [9]. The Reynolds number based on the hump length c and reference velocity U_0 (maximum free stream velocity at the inlet of the domain) in the experiment was equal to $9.36{\cdot}10^5$, while the height of the hump, h, was equal to $0.128c$. The computational domain and the grid in the XY-plane used in the simulations are shown in Figure 6. The grid has 510 and 126 cells in the x- and y-directions, respectively. The size of the domain in the homogeneous z direction is $0.4c$, and the grid-step Δz is uniform and equal to $1.0{\cdot}10^{-2}c$. Thus, the total grid size is about 2.5 Million. The time step in the simulations is equal to $2{\cdot}10^{-3}c/U_0$ which ensures the CFL number is less than 1.0. Simulations were performed using the SA-based DDES with shear-layer-adapted subgrid scale [4].

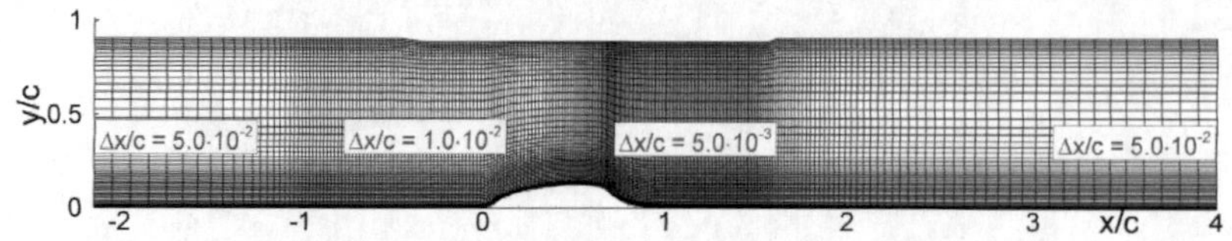

Figure 6. Computational domain and grid in XY-plane (every other grid line is shown)

Again, results of the simulations with the hybrid scheme (1) are compared with those obtained with the use of the scheme [2].

As seen in Figure 7, the weight functions of the two schemes in these simulations turn out to be rather different. In particular, in the scheme [2] the weight of the upwind scheme is close to 1.0 in the near-wall region both in the recirculation zone and in the reattached boundary layer. In contrast to this, in the scheme (1) these regions are treated with the pure CD scheme (Figure 7). As a result, using scheme [2] leads to a bit slower roll-up of the separated shear layer, to a noticeable under-prediction of the skin friction in the recirculation zones and in the reattached boundary layer, and to a strong under-prediction of the resolved Reynolds stresses (see Figure 8). All these lead to a somewhat better agreement of the predictions obtained with the use of scheme (1) with the experimental data.

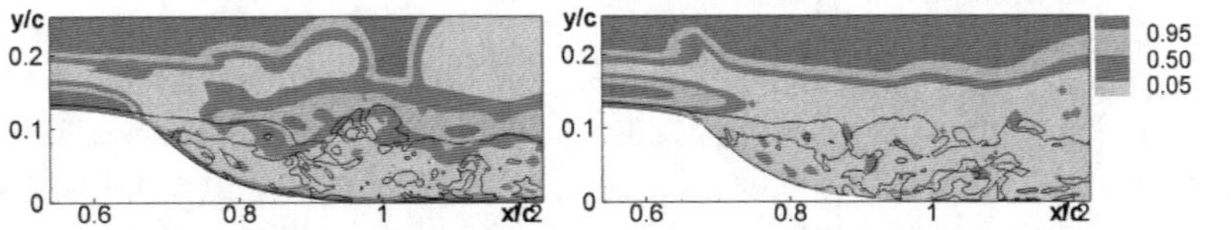

Figure 7. Snapshots of the weight function of scheme [2] (left frame) and of scheme (1) (right frame). The black line corresponds to non-dimensional vorticity magnitude equal to 20

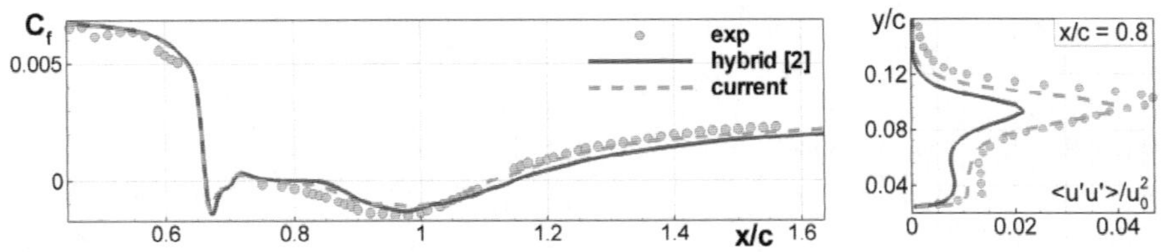

Figure 8. Distribution of the mean skin friction coefficient (left frame) and profile of resolved normal stress at $x/c = 0.8$ (right frame)

3.3. IDDES of the wall bounded flow

Zero pressure gradient boundary layer over the flat plate is widely used for validation of turbulence models, both RANS and WMLES. In the present work this test case is used to evaluate performance of the proposed scheme within the framework of IDDES in the WMLES mode.

The Reynolds number based on the momentum thickness of the incoming boundary layer was equal to 1000. The computational domain has a size of $L_x = 25\delta_0$, $L_y = 10\delta_0$, and $L_z = 3\delta_0$ in the streamwise, wall-normal and spanwise directions respectively. The grid $N_x \times N_y \times N_z = 251\times71\times61$ was uniform in the x- and z-directions ($\Delta x = 0.1\delta_0$, $\Delta z = 0.05\delta_0$) and clustered near the wall in the wall-normal direction with $y_{min} = 0.0025\delta_0$. The corresponding steps in wall units are about $\Delta x^+ = 40$, $\Delta z^+ = 20$, and $\Delta y^+_{min} = 1.0$. The total size of the grid results in $1.1\cdot10^6$ cells. The incoming turbulent content was created with the use of a Synthetic Turbulence Generator [11].

Results of the simulations are presented in Figure 9.

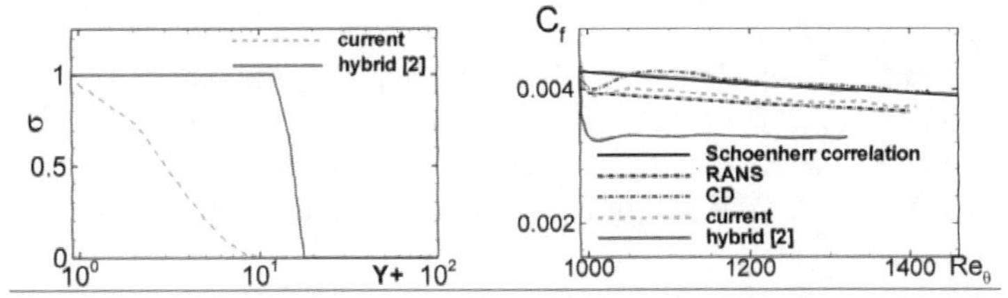

Figure 9. Profile of the weight function at $x = 0.2$m (left frame) and streamwise distribution of the mean skin friction coefficient in the flat plate boundary layer (right frame)

The figure reveals a drastic difference between the weight functions of the scheme [2] and of the current scheme in this flow. One can see that scheme [2] treats the inner part of the boundary layer up to the distance to the wall about 10 wall units with the upwind scheme, while the current scheme treats this region with virtually CD scheme. As a result, the latter ensures much higher accuracy of the mean flow prediction. This is illustrated by the comparison of the skin-friction coefficient distributions computed with the use of the two schemes with the similar distribution predicted by the pure 4th order centred scheme (see right frame in Figure 9).

Conclusions

A new hybrid scheme for global RANS-LES approaches is developed and validated in the framework of different DES-like approaches. It is shown that within DDES of the massively separated flow over NACA 0021 airfoil, the scheme performs equally well as the scheme [2]. For a flow with separation and reattachment (NACA wall-mounted hump) computed with the use of DDES combined with shear-layer-adapted definition of the subgrid length-scale, the proposed scheme is superior to the scheme [2]. A major advantage of the proposed scheme over scheme [2] is observed in a fully attached flow (zero pressure gradient boundary layer) treated by IDDES in the wall-modelled LES mode.

Acknowledgments

The results of the work were obtained using computational resources of Peter the Great Sainte-Petersburg Polytechnic University Supercomputing Center (http://www.spbstu.ru).

References

[1] Spalart P R, Jou W-H, Strelets M and Allmaras S R 1997 Comments on the feasibility of LES for wings, and on a hybrid RANS/LES approach *Proceedings of first AFOSR international conference on DNS/LES* pp. 137–147

[2] Travin A, Shur M, Strelets M and Spalart P R 2002 Physical and numerical upgrades in the detached-eddy simulation of complex turbulent flows *Adv. LES Complex Flows Proc. Euromech Colloq. 412* pp. 239–254

[3] Shur M L, P. R. Spalart, M. K. Strelets, and Travin A K 2008 A hybrid RANS-LES approach with delayed-DES and wall-modelled LES capabilities *Int. J. Heat Fluid Flow* vol. **29** no. 6 pp. 1638–1649

[4] Shur M L, Spalart P R, Strelets M K and Travin A K 2015 An Enhanced Version of DES with Rapid Transition from RANS to les in Separated Flows *Flow, Turbul. Combust.* vol. **95** no. 4 pp. 709–737

[5] Jasak H, Weller H G and Gosman A D 1999 High resolution NVD differencing scheme for arbitrarily unstructured meshes *Int. J. Numer. Methods Fluids* vol. **31** no. 2 pp. 431–449

[6] Spalart P R, Deck S, Shur M L, Squires K D, Strelets M K and Travin A 2006 A New Version of Detached-eddy Simulation, Resistant to Ambiguous Grid Densities *Theor. Comput. Fluid Dyn.* vol. **20** no. 3 pp. 181–195

[7] Shur M L, Strelets M K and Travin A K 2004 High-Order Implicit Multi-Block Navier-Stokes Code: Ten-Year Experience of Application to RANS/DES/LES/DNS of Turbulence *http://cfd.spbstu.ru/agarbaruk/c/document_library/DLFE-42505.pdf*

[8] Swalwell K, Sheridan J and Melbourne W 2003 Frequency Analysis of Surface Pressures on an Airfoil After Stall *21st AIAA Applied Aerodynamics Conference* no. June pp. 1–8.

[9] Greenblatt D, Paschal K B, Yao C-S, Harris J, Schaeffler N W and Washburn A E 2005 A Separation Control CFD Validation Test Case Part 2 - Zero Efflux Oscillatory Blowing *AIAA J.* pp. 1–24

[10] Shur M L, Spalart P R, Strelets M K and Travin A K 2014 Synthetic turbulence generators for RANS-LES interfaces in zonal simulations of aerodynamic and aeroacoustic problems *Flow, Turbul. Combust.* vol. **93** no. 1 pp. 63–92

Laboratory simulation of Hot Jupiters atmosphere expansion

P A Prokopov, I F Shaikhislamov, Yu P Zakharov, V G Posukh, A V Melekhov, E L Boyarintsev, A G Ponomarenko

Institute of Laser Physic (ILP) SB RAS, Novosibirsk, Av. Lavrentyeva 15 b, 630090, Russia

paprok312@gmail.com

Abstract. Hot Jupiters (HJ) are exoplanets, gas giants with low orbits (≤ 0.1 a.u.). The stellar X-ray and ultraviolet (XUV) radiation energy deposition result in heating ionization and the consequent expansion of planetary atmosphere. Expansion of upper atmosphere under certain conditions could be so large that the majority of light atmospheric constituents overcome the gravitational binding and escape from the planet in a form of hydrodynamic wind. Besides interaction of two counter-streaming plasma flows (stellar wind and ionized upper layers of planet atmosphere), each of this flows interact with planetary magnetic field. In such complex situation laboratory simulation can provide data that can't be obtained by computer simulation or observation. Experiment was carried out on KI-1 facility: high-vacuum chamber 5m long, 1.2 m in diameter with pressure ~ 10^{-6} Torr. Magnetic dipole with two attached laser targets played the role of a planet, and background plasma from θ-pinch used for simulation of stellar wind. As a result, data on a behavior of plasma density and magnetic field were obtained. The novel phenomenon was registered: magnetic field is transferred by the cloud of laser plasma, which was not observed before in experiments or calculations.

1. Introduction

Observation and study of planets outside the Solar System is a popular and rapidly developing field of modern physics [1]. Significant portion of discovered planets are close-orbits (≤0.1 a.u.) gas giants ($0{,}2M_J<M_P<8M_J$), so-called Hot Jupiters. Upper layers of atmosphere are extremely heated by intense X-ray and ultraviolet radiation (XUV) of a host star. Expansion of ionized upper layers of atmosphere can overcome the gravitational binding and escape from the planet surface in a form of hydrodynamic wind (similar to the solar wind). This plasma stream interacts with solar wind and planetary magnetic field.

In such complex situation (counter-streaming plasma flows with a presence of magnetic field) laboratory simulation can provide new information, which cannot be obtained by computer simulations or observation of exoplanets. This work represents the description and results of experiment, which simulates all principal aspects of Hot Jupiters: relatively uniform solar wind, planetary magnetic field and plasma stream originates from inside of this magnetic field.

2. Experimental setup

Simulation experiment was carried out on KI-1 facility [2] of ILP SB RAS (figure 1). High-vacuum chamber 5 m long and 1.2 m in diameter with pressure in working mode ~10^{-6} Torr. Solenoid, which covers almost the entire outer surface of the chamber, created magnetic field B_0 inside the chamber with magnitude ~20 G at its axis. Chamber was filled with H^+ background plasma (BP) from θ-pinch plasma-source with density ~ $0.5 \cdot 10^{13}$ cm^{-3}, represents the stellar wind. Magnetic dipole with magnetic moment $\mu = 7.5 \cdot 10^5$ G/cm^3 and a fall of time ~250 μs, oriented perpendicular to the chamber axis simulates the exoplanet with magnetic field. To simulate plasma flow from upper layers of planetary atmosphere, two polyethylene laser targets were attached to its surface. CO_2 laser beam with specific form (~70 ns peak and ~1μs tail with energy ~150 J for each) was focused symmetrically on each target into a spots with diameter ~2 cm to produce a laser plasma cloud (LP).

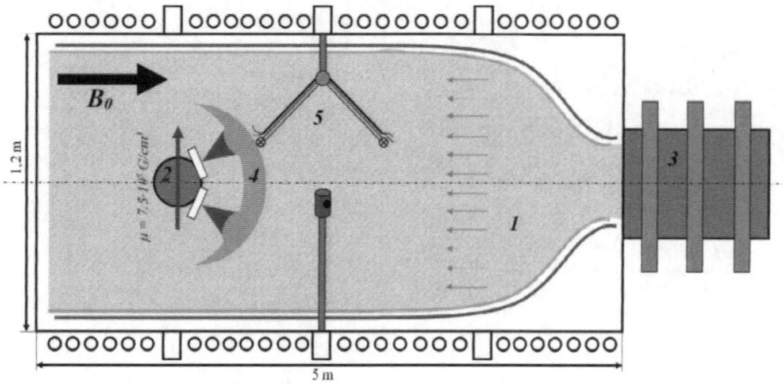

Figure 1. Scheme of experimental setup. High-vacuum chamber 5 m long and 1.2 m in diameter. External magnetic field B_0 ~ 20 G
1 – background plasma
2 – magnetic dipole
3 – θ-pinch
4 – laser plasma
5 – diagnostic probes

Each cloud propagated along the normal to laser targets surface. At certain distance from the target these two clouds of plasma merged into a single cloud, which represents planetary wind. Figure 2 is a picture from image converter that illustrates expansion of laser plasma in a presence of a dipole magnetic field inside the background plasma flow. Coordinate system: X axis directed along the symmetry axis of the vacuum chamber, Z axis oriented vertically along the magnetic moment of dipole, and corresponding Y axis.

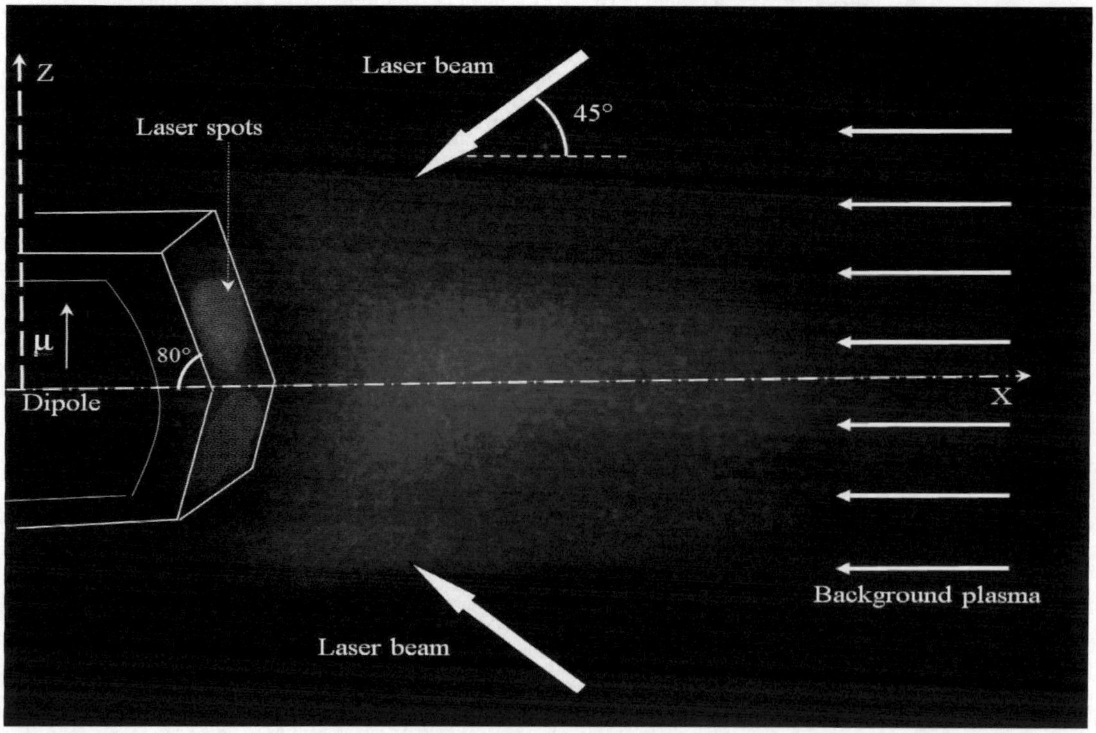

Figure 2. Image converter picture of laser plasma expansion. Contours of the dipole, direction of laser beams and BP flow, coordinate axis are marked on the figure.

Two Langmuir probes with attached three-dimensional magnetic probes at a distance 40÷90 cm from the target were used for plasma diagnostic. Three-dimensional magnetic probe is a three small coils made of tungsten wire, oriented in three orthogonal directions with effective area ~ 1 cm^2. There was also Langmuir probe and directed ion collector [3, 4] at a distance 207 cm from the target (90 cm from θ-pinch aperture).

3. Experimental results

LP consisted for the most part of ions H^+ and C^{4+} in approximately equally proportion. The front of LP propagates along the chamber axis at the speed ~170 km/s in a cone with half-angle ~30°. A total number of electrons per solid angle, measured far from the target (where both clouds of LP already merged), was about $N_{e,\Omega} \approx 2.2\cdot10^{18}$ sr^{-1}, and the kinetic energy of ions was about $W_\Omega \approx 32$ J/sr. Typical form of LP signal is presented on figure 3, hereafter time t = 0 corresponds to the laser pulse. LP density had a three bunches, divided by 550 ns between the peaks (interval at the moment of LP generation; time interval between the peaks increases during the LP propagation, and reaches 3 – 5 μs at the moment of registration). Second peak was the biggest one, and had a twice smaller speed, than first one.

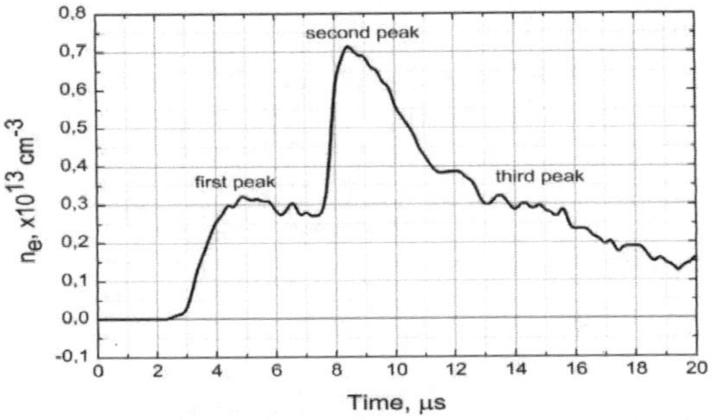

Figure 3. Typical signal of LP density in vacuum in absence of magnetic field, measured by Langmuir probe at a distance 67 cm from laser target. It can be seen that typical LP cloud consist of three peaks of different amplitude

Figure 4 represents density distribution of BP at two different distances from the target. X = 207 cm is near to θ-pinch aperture, so plasma density there is several times greater than at X = 67 cm. During its propagation from the source, the density of BP decrease by a cubic law, which is confirmed by measurements, presented on figure 5.

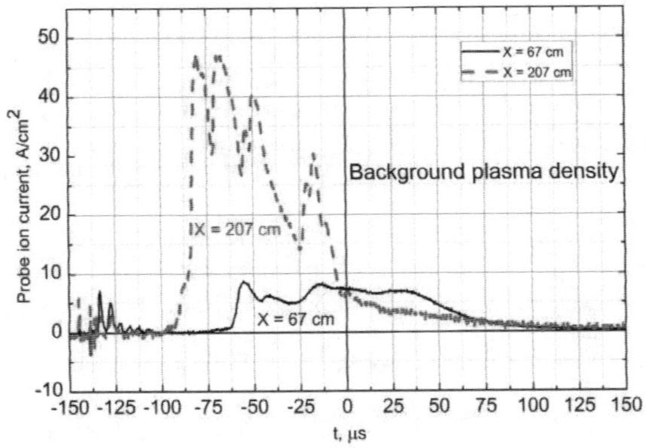

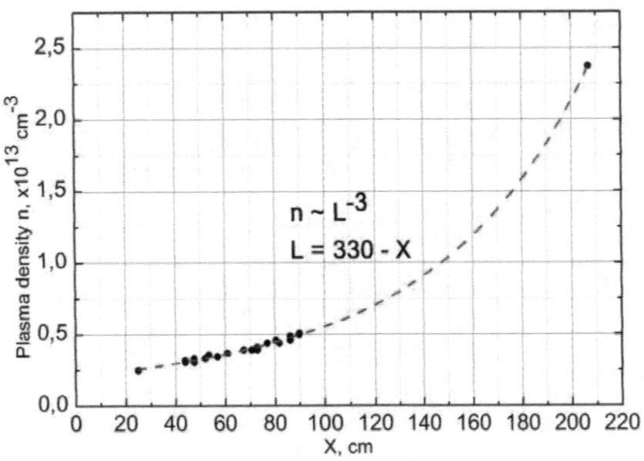

Figure 4. Signal of unperturbed background plasma without magnetic dipole, measured by Langmuir probes at distances 67 and 207 cm. Ion probe current is proportional to the plasma density

Figure 5. Spatial distribution of BP density along the X axis (black dots). Dashed line show cubic law.

Figure 6 illustrate interaction between LP and BP. Solid black line shows density distribution of unperturbed BP. Dashed line density distribution in case of LP propagates inside the BP, a sharp jump of concentration corresponds to arrival of LP to the probe. Grey line represents the data obtained from directed ion collector; its construction allows us to detect ions, whose speed vectors are close to one direction. Here collector was oriented to detect only ions of BP, so the gray line coincidence with the black up to the time of arrival of perturbation caused by LP. After this moment collector shows partial displacement of BP. All measurements, presented on figure 6, were made at a distance 67 cm from the laser target.

Despite the expulsion of BP the interaction between LP and BP is rather weak. An evidence of a weak interaction between LP and BP is well demonstrated on figure 7, which illustrate the plasma density behavior in different modes (LP in vacuum; LP in BP; LP in BP with dipole magnetic field), at a distance 67 cm (figure 7a) and 207 cm (figure 7b) from the target.

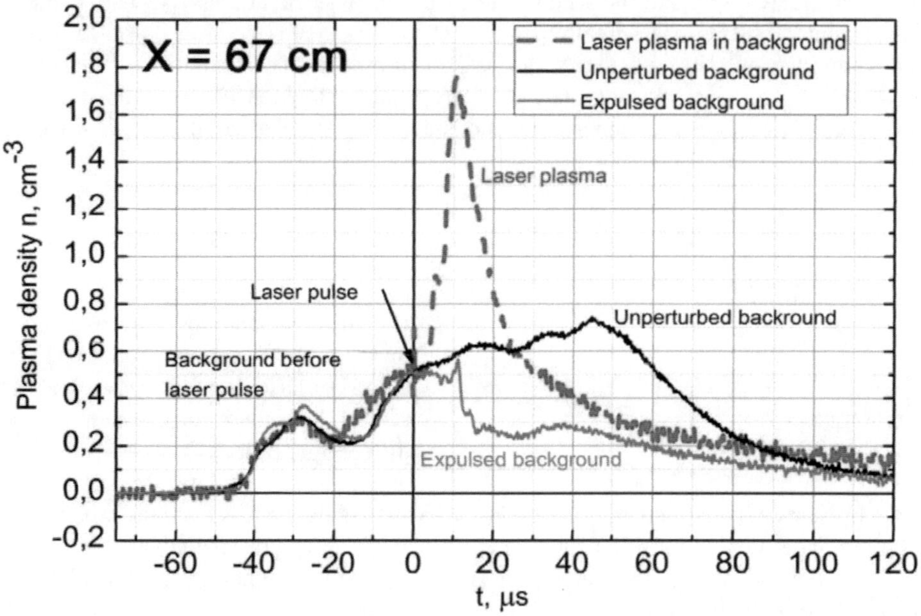

Figure 6. Demonstration of interaction of LP and BP. Solid black line represents unperturbed background. Dashed line – laser plasma in background, sharp peak, caused by laser plasma, can be seen. Grey line allows to see the expulsion of BP, it represents data from directed ion collector, which is oriented toward θ-pinch aperture and doesn't register LP ions.

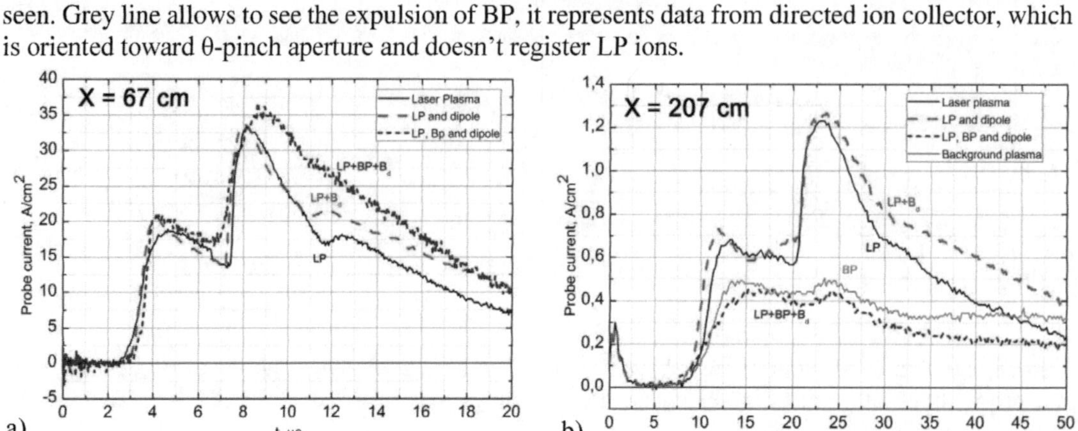

Figure 7. Ion probe current in different modes at distances 67 cm (a), 207 cm (b). LP (solid black) – laser plasma in vacuum in absence of magnetic field; LP+B_d (dash) – laser plasma and dipole field; LP+BP+B_d (short dash) – laser plasma in background plasma with dipole; BP (solid grey) – background plasma without dipole.

Data obtained at a distance 67 cm shows a little difference in different modes, significant difference appears only at a distance 207 cm, where presence of BP causes almost 3 times reduction of LP density. Measurements in absence and in presence of dipole magnetic field give similar results, which demonstrate, that interaction takes place due to Coulomb collisions regardless the presence of magnetic field.

Let's consider the behavior of magnetic field. Magnetic moment of the dipole is directed along the Z axis, so B_Z is largest component of magnetic field. Figure 8 illustrates the behavior of B_Z in three modes: LP with BP without dipole; LP with dipole without BP; LP with BP and dipole. In the first two cases the variation of magnetic field is not higher than 2÷3 G. In the first case it could be an irregular magnetic field carried by θ-pinch plasma and swept by LP. Magnitude of magnetic field in second case correspond to undisturbed dipole magnetic field, calculated by $B = \mu/x^3$. In the third case, when dipole and θ-pinch are on, variation of magnetic field is about 23 G which is much larger than in the other two cases. This is a novel experimental result, generation of such powerful magnetic field by counter-streaming plasma flows has not been observed before. Magnetic field at such significant distance from dipole should be displaced by BP, which is confirmed in our previous experiments.

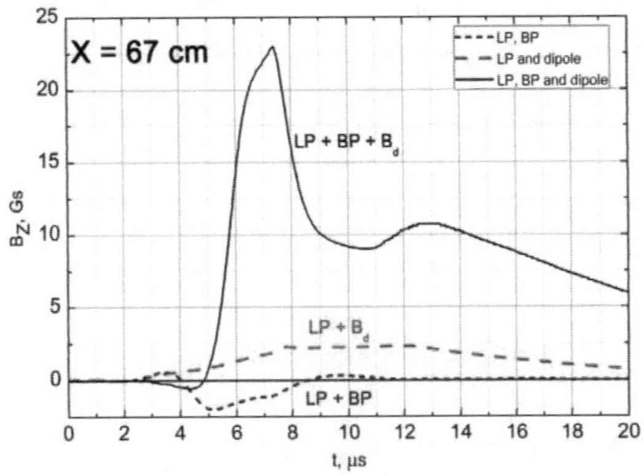

Figure 8. Perturbations of magnetic field at X=67 cm in different experimental modes. LP+BP (solid) – laser plasma in background plasma; LP+B_d (dash) – laser plasma in vacuum with dipole; LP+BP+B_d (short dash) – laser plasma in background plasma with dipole

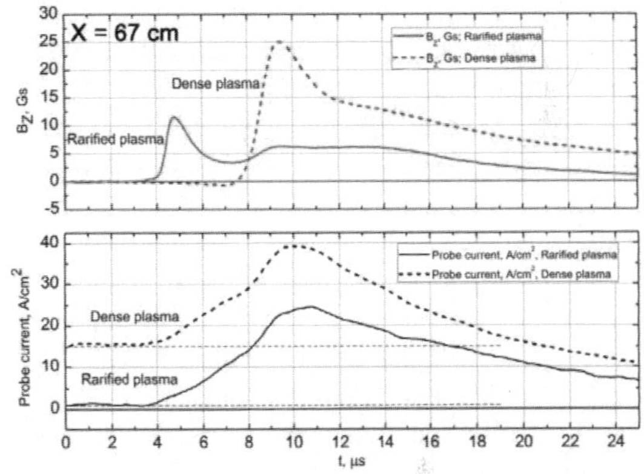

Figure 9. Measurements of ion probe current and magnetic field in rarified (solid) and dense (dash) BP at X = 67 cm. Stronger perturbations of magnetic field are observed in dense plasma.

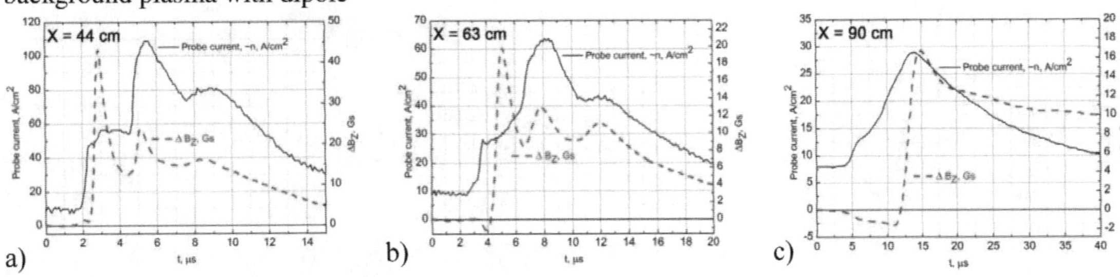

Figure 10. Correlation of ion probe current (solid) and magnetic field (dash), measured at different distances. At X = 44 cm (a) three peaks of density and magnetic field can be seen (dipole magnetic field at this distance ~9 G); at X = 63 (b) first peak of density has fallen but magnetic field still has three peaks (dipole magnetic field at this distance ~3 G); at X = 90 cm (c) only the second peak of density and magnetic field can be seen (dipole magnetic field at this distance ~1 G).

Obviously, this magnetic field is not generated only by interaction of LP and BP, because significant magnitude is observed only when dipole is switched on, which means that magnetic field of 23 G at a distance 44 cm from dipole was generated near the dipole and was transferred at that distance by LP. Figure 10 illustrates in more detail, that magnetic field is transferred by LP, simultaneous measurements of plasma density and variation of magnetic field shows that peaks of magnetic field correlated with density peaks at each three distances.

Another important result is dependence on the plasma density of the magnetic field. The comparison of magnetic fields in rarified and dense BP is presented on figure 9. In rarified plasma the magnitude of magnetic field is smaller but it corresponds to first peak of LP, whereas in dense plasma the twice greater magnetic field transferred by second peak of LP.

4. Discussion and conclusions

Summing up the results of experiment, weak interaction of LP and BP, due to the Coulomb collisions, takes place, and the measurements with directed ion collector shows that BP is expulsed by LP. Perturbations of B_Z component of magnetic field about 23 G were detected at significant distance from dipole, when dipole is switched on and in presence of θ-pinch plasma.

Since a strong magnetic field is observed only in mode, when dipole is switched on and LP propagates in BP, a possible explanation of a presence of such strong magnetic field at this distance, is the transportation of magnetic field by LP away from dipole. The approximate possible scenario of this mechanism is described below.

BP ions are captured by magnetic field of dipole when they reached a certain distance. BP ions create a mini-magnetosphere around dipole with frozen magnetic field. After a laser pulse expanding cloud of LP expulse ions of BP with frozen magnetic field due to Coulomb collisions. Magnetic field is carried by electrons, which move with ions to keep plasma quasi-neutral. Perturbations of B_Z corresponds the peaks of LP density in area close to dipole, when LP density is greater than BP density. But at larger distances, where BP ions are dominant, dynamic of magnetic field is governed by BP density. For this reason, the first peak of LP and corresponding peak of magnetic field are not observed already at a distance >70 cm.

Illustration of two stages of this process is presented on fig. The first stage is formation of magnetosphere before the laser pulse and creation of LP. The second stage is after the laser pulse, expanding LP carries with it ions of BP with magnetic field at significant distance, where magnetic field is approximately zero in unperturbed background.

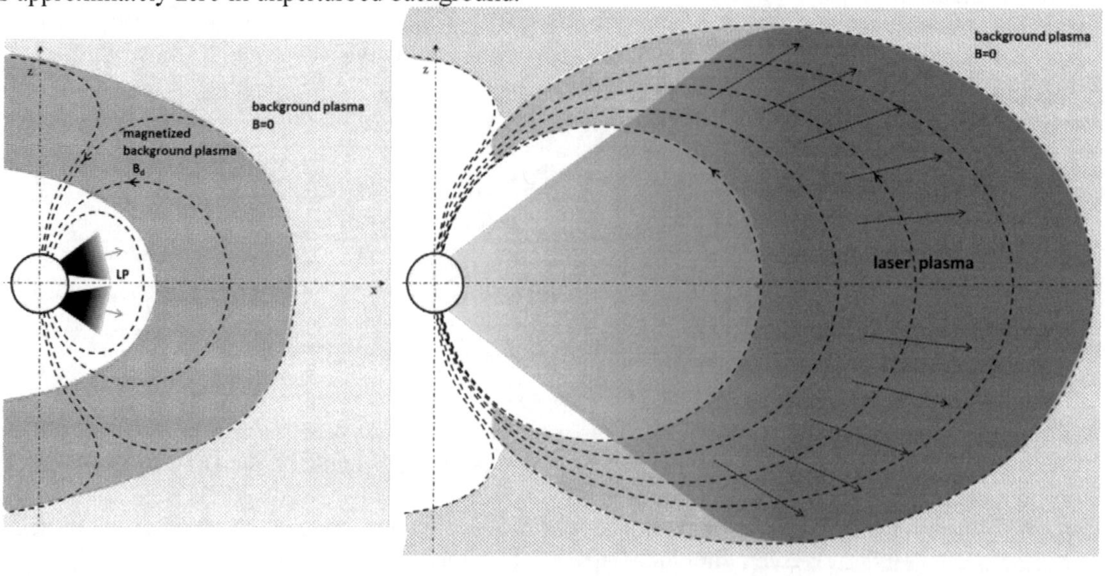

Figure 11. Schematic illustration of expansion of magnetosphere. On the left picture BP ions are captured by dipole magnetic field and create mini magnetosphere. On the right picture expanding LP takes away BP ions and magnetic field.

Earlier experiments at KI-1 facility [5, 6, 7] showed, that dipole magnetic field is expelled by background almost everywhere inside the vacuum chamber. In uniform magnetized background the cloud of super-Alfvenic LP decelerates and generates non-linear magnetosonic wave [8] or shockwave [9] (if there is enough energy). The present conditions principally differ from previous experiments, in the present case magnetized background creates the mini-magnetosphere, and magnetic field is localized in a small shell. LP generated inside this magnetosphere expands and takes away from dipole background ions with magnetic field at significant distance.

References

[1] Shaikhislamov I F, Khodachenko M L, Sasunov Y L, Lammer H, Kislyakova K G and Erkaev N V 2014 Atmosphere expansion and mass loss of close-orbit giant exoplanets heated by stellar XUV. I. Modeling of hydrodynamic escape of upper atmospheric material *Astrophys. J.* **795** 132

[2] Zakharov Yu 2003 Collisionless laboratory astrophysics with lasers, *Plas sc. IEEE Transact. on plas. sc* **31** 6 pp 1243-1251

[3] Ponomarenko A G, Antonov V M, Posukh V G, Melekhov A V, Boyarintsev E L, Afanasyev D M and Yurkov R N 2004 Simulation of non-stationary processes in the solar wind and its impact on the Earth magnetosphere *Report on MinPromNauka project Investigation of Solar activity and its magnifications in near Earth space and atmosphere* part III (in Russian)

[4] Shaikhislamov I F, Zakharov Yu P, Posukh V G, Melekhov A V, Antonov V M, Boyarintsev E L and Ponomarenko A G 2014 Laboratory model of magnetosphere created by strong plasma perturbation with frozen-in magnetic field Plasma *Plasma Phys. Control. Fusion* **56** 125007

[5] Ponomarenko A G, Antonov V M, Posukh V G, Melekhov A V, Boyarintsev E L, Afanasyev D M and Yurkov R N 2004 Simulation of non-stationary processes in the solar wind and its impact on the Earth magnetosphere *Report on MinPromNauka project Investigation of Solar activity and its magnifications in near Earth space and atmosphere* part III (in Russian)

[6] Ponomarenko A G, Zakharov Yu P, Antonov V M,Boyarintsev E L, Melekhov A V, Posukh V G, Shaikhislamov I F and Vchivkov K V 2008 Simulation of strong magnetospheric disturbances in laser-produced plasma experiments *Plasma Phys. Control. Fusion* **50** 074015

[7] Shaikhislamov I F, Antonov V M, Zakharov Yu P, Boyarintsev E L, Melekhov A V, Posukh V G and Ponomarenko A G 2009 Laboratory simulation of field aligned currents in an experiment on laser-produced plasma interacting with a magnetic dipole *Plasma Phys. Control. Fusion* **51** 105005

[8] Shaikhislamov I F, Zakharov Yu P, Posukh V G, Melekhov A V, Boyarintsev E L, Ponomarenko A G and Terekhin V A 2015 Experimental study of collisionless super-Alfvénic interaction of interpenetrating plasma flows *Plasma Physics Reports* **41** pp 399-407

[9] Zakharov Yu P, Ponomarenko A G, Tishchenko V N, et al 2016 Generation of laser plasma bunches with a high efficiency of energy concentration for laboratory simulation of collisionless shock waves in magnetized cosmic plasma *Quantum Electronics* **46** 5 pp 399–405

Acknowledgments

This work was supported by SB RAS Research Program (project II.10.1.4 N 01201374303), Russian Fund for Basic Research grants 14-29-06036 and 16-52-14006, the RAS Presidium Program: “Extreme laser radiation: Physics and fundamental applications”.

IOP Conf. Series: Journal of Physics: Conf. Series **929** (2017) 012101 doi:10.1088/1742-6596/929/1/012101

Comparative analysis of transition models at different farfield turbulence intensities

A S Stabnikov and A V Garbaruk,

Peter the Great St. Petersburg Polytechnic University. Russia, 195251, St. Petersburg, Polytechnicheskaya, 29

an.stabnikov@gmail.com

Abstract. Four differential models for laminar-turbulent transition prediction were tested with different freestream turbulence levels on two cases: turbulent boundary layer and flow around NACA0021 airfoil. Computational results show that none of the considered models are able to predict transition in a wide range of freestream conditions. SA e^N model is applicable only for natural transition scenario with low freestream turbulence intensity and predicts fully turbulent solutions in other cases. Among γ-based models SST γ-Re_θ and SST γ have shown better agreement with experiment than SA γ.

1. Introduction

Though most of engineering flows are turbulent, they often contain laminar areas. The most important example is the boundary layer, initial part of which is laminar due to low local Reynolds number. Downstream, with growth of the Reynolds number, the boundary layer becomes turbulent. Since laminar and turbulent boundary layers develop differently it is very important to predict location of laminar-turbulent transition precisely. Unfortunately, it is impossible in the frame of traditional turbulence models for Reynolds-Averaged Navier-Stokes equations closure [1]. This circumstance has led to creation of special models for prediction of laminar-turbulent transition which can be considered as an extension of traditional RANS models.

The location and type of laminar-turbulent transition in the boundary layer are dependent on different factors, with the most significant of them being the freestream turbulence level $Tu = 100\% \cdot \overline{u'^2} / U_0$, where $\overline{u'^2}$ is the intensity of turbulent fluctuations and U_0 is the freestream velocity. At low turbulence levels (Tu<0.8%) natural transition is observed whereas bypass transition is typical for higher turbulence levels. Sometimes, in flows with adverse pressure gradient and low freestream turbulence levels (often takes place on wings and airfoils), another type of transition takes place - the so-called "bubble" transition, which occurs when a laminar boundary layer detaches from the surface, instantly becoming turbulent and reattaching.

Sensitivity of the transition type and location to turbulence level is very difficult to model. The present work aims to check the ability of different transition models to catch this effect correctly.

2. Considered models

Four differential transition models were chosen for testing. Two of them, SST γ-Re_θ [2] and SST γ [3], are based on the SST [4] turbulence model. Both are developed to predict all types of transition and use an additional transport equation for turbulent intermittency γ to control the turbulent kinetic energy production. These models are widely used and are implemented in a range of commercial CFD codes.

The third model, SA-γ of Coder and Maughmer [5], is an example of applying the γ approach to Spalart–Allmaras one-equation turbulence model [6]. Intermittency equation in this model was derived by adaptation of the SST γ-Re_θ model equation.

The last considered model, SA e^N of Coder [7], contains a differential equation for the growth factor N with source terms based on the Drela [8] realization of e^9 method [9] which was initially developed for natural transition prediction. For extension of this method to higher *Tu* levels, Mack [10] correction was used. It proposes a way of changing the magnitude of the critical disturbance growth rate factor from constant value, equal to 9, to one depending on the turbulence level.

3. Test cases

This study addresses two test cases: flat plate boundary layer and flow around NACA0021 airfoil. For both cases experimental data with different turbulence intensities is available.

All the computations were carried out using the NTS CFD code [11]. The code uses a finite volume method and structured multiblock overlapping grids (the so-called “Chimera” technology). Convective fluxes in the equations for an incompressible fluid were approximated with the third order Rogers-Kwak upwind scheme [12] and in the turbulence models equations with a second order upwind scheme. All of the presented solutions are fully converged and grid-independent.

3.1. Flat plate transitional boundary layer

Experiments with transitional boundary layers at different turbulent intensities and pressure gradients (T3 series) were conducted in Rolls Royce [13]. Three cases with no pressure gradient from these series were considered in the present work, namely T3A- with 0.87 % freestream intensity, T3A with 3.3% and T3B with 6.2%. Even at the lowest intensity bypass transition takes place.

The computational domain contains a freeflow region upstream of the plate’s leading edge to minimize the pressure gradient effect. The boundary conditions are as follows: uniform velocity profile was set at the inlet boundary, constant pressure at outlet and upper boundaries, no-slip wall at the plate. Turbulence characteristics at the start of the plate were chosen to coincide with experimental values. A computational grid of 305x146 cells was used.

Experimental and computational skin friction coefficients distribution for all cases and models are presented on figure 1. One can see that there is a significant difference in the transition location between models and it is mostly pronounced for the lowest turbulence intensity.

The best agreement with experimental data is obtained using SST γ-Re_θ and SST γ models, which is not surprising since they were calibrated using T3 series experiments. The SST γ results are slightly closer to experimental in T3A and T3B cases while in T3A- both models predict an early transition, SST γ-Re_θ being much closer to experiment.

Results of SA-based models are in less consistency with experimental data than of the SST-based models for all considered cases: SA γ predicts too early transition in all instances, while SA e^N predicts the earliest transition in case of T3A- (fig. 1a), and in cases T3A (fig. 1b) and T3B (fig. 1c) with higher turbulence intensity the solution is fully turbulent with no observable transition at all. This result is predictable and is in agreement with the nature of e^N method. It can be concluded that SA e^N model can only be used for low turbulence intensities when natural transition is observed, despite the Mack critical N-factor correction [10].

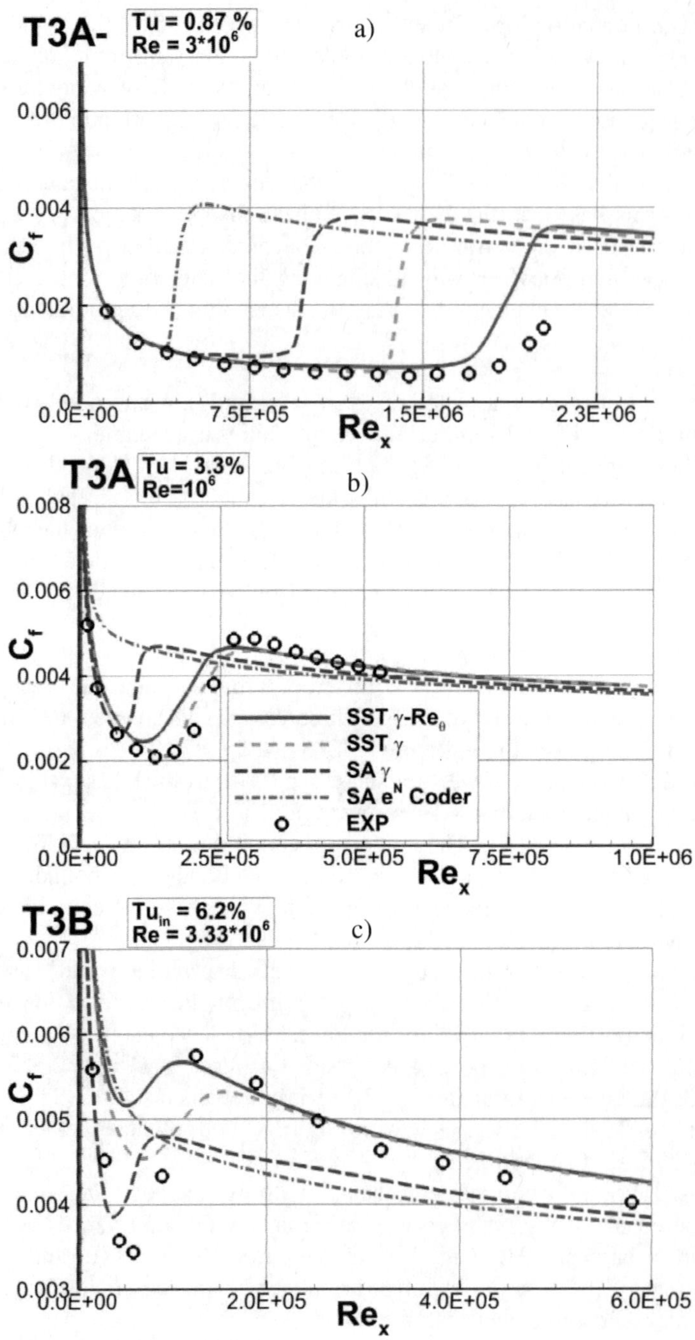

Fig. 1. Comparison of experimental skin friction coefficient with results of computations of three considered cases: T3A- (a), T3A (b), T3B (c)

3.2. NACA0021 Airfoil

Symmetrical airfoil NACA0021 with relative thickness of 21% was considered under conditions that were used in the experiment [14]. The computations are carried out at Reynolds number equal to

$2.7 \cdot 10^5$, based on the chord and freestream velocity and three different levels of turbulence intensity: 0.6%, 3% and 6%. Angles of attack from 0° to 30° were considered.

Computations were performed on a structured 593x200 cells C-grid in a 40C x 40C computational domain, where C is the airfoil chord.

Dependence of lift and drag coefficients on the angle of attack for the low turbulence level (Tu=0.6%) is shown on Figure 2. For the attached flows (angles of attack less than 10°) all of the considered models' results are close to each other and are in good agreement with experimental data. The difference between solutions for higher angles of attack is mostly connected with underlying turbulence model (SST or SA) than with transition prediction approach.

Results of SST γ-Re_θ and SST γ models are close and in better agreement with experimental data than SA based models' results. The SST γ leads to fully detached flow regime (boundary layer at suction side separating from the leading edge of the airfoil) at lower angles of attack. Despite the different transition modelling approach both SA based models also show similar results. The fully detached flow appears at higher attack angles than in the experiment, which leads to 15-20% difference in maximum lift coefficient. The good agreement with experimental data at high angles of attack (22-30°) seems to be a coincidence and is most likely caused by error cancellation.

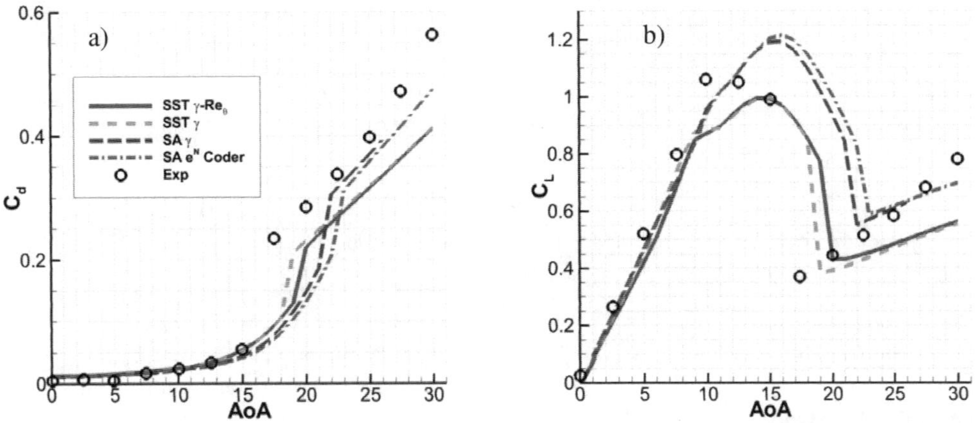

Fig. 2. Comparison of drag (a) and lift (b) coefficients for NACA0021 airfoil with experimental data

For the investigation of turbulence intensity effect the 15° angle of attack flow was considered. Skin friction coefficient distributions (Fig. 3) show that this effect strongly depends on the specific transition modelling approach rather than baseline turbulence model. For the low freestream turbulence intensity (*Tu*=0.6%) all models predict similar flow structure (Fig. 3a): bubble transition on the suction part of the airfoil and laminar flow on the pressure side. However, the bubble size is approximately two times longer for all γ-based models than for the SA e^N model.

With increasing turbulence intensity a significant difference occurs between solutions obtained with different models. In the case of *Tu*=3% (Fig. 3b) SA e^N model shows fully turbulent behavior on both sides. This fact confirms once more that e^N-based methods are inapplicable at high turbulence intensities. On the pressure side both SST-based models predict laminar flow whereas SA γ predicts transition at about ¾ of the chord. The type of transition on the suction side depends on the model: bubble transition is obtained with SST γ and bypass transition - with SST γ-Re_θ and SA γ. Further increase of turbulence intensity (*Tu*=6%) results in transition on pressure side for all considered models (Fig. 3c), while the type of transition on the suction side is similar to that at *Tu*=3%.

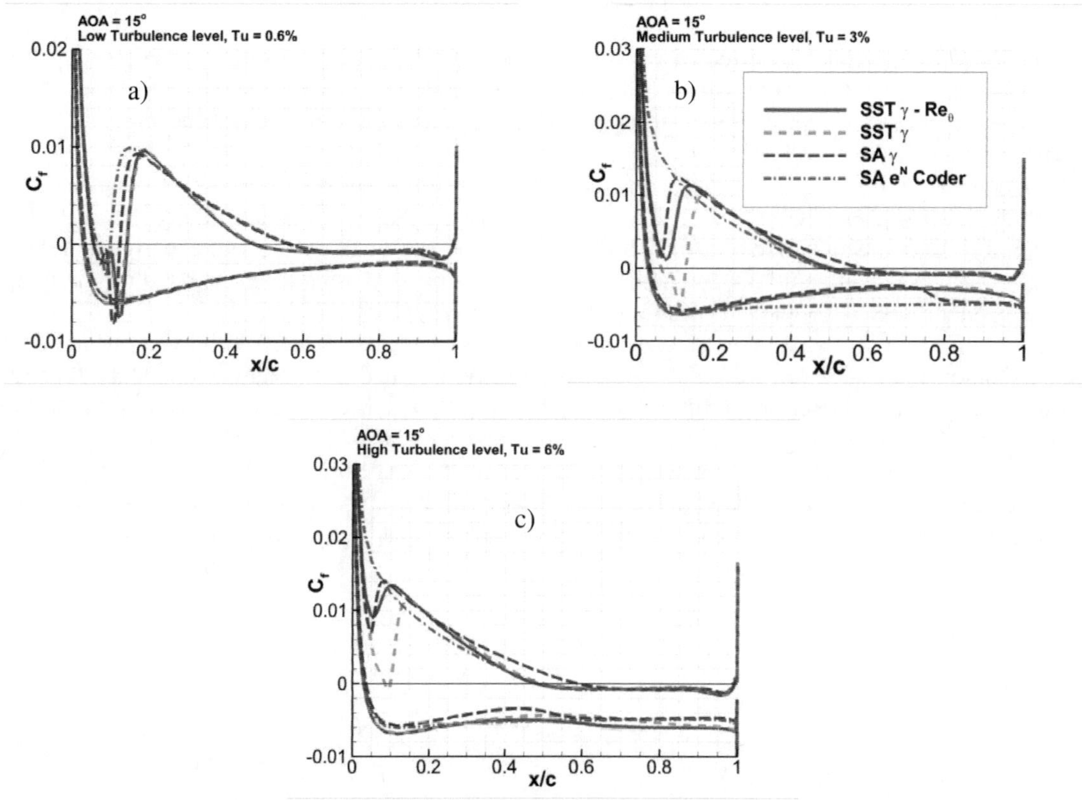

Fig. 3. Skin friction coefficient distribution for different turbulent intensities: a – 0.6%, b – 3%, c – 6%

Effect of turbulence intensity on the drag C_d and lift C_L coefficients is shown on Figure 4. One can see that none of the models are able to predict rapid drag coefficient growth with turbulence level increasing (Fig. 4a). Only SST-γ predicts monotonous growth for the lift coefficient, which is consistent with experiments (Fig. 4b), but the slope is approximately three times lower than in the experiment. Other models predict quantitatively wrong dependence: lift coefficient is decreasing for the SA-based models and not monotonous for SST γ-$Re_θ$.

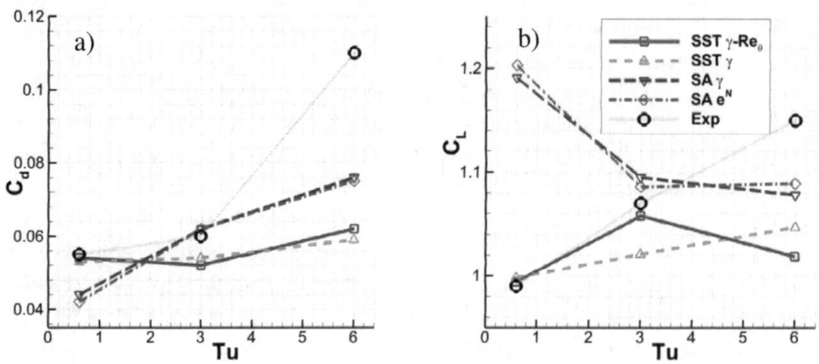

Fig. 4. Dependence of skin friction (a) and drag coefficients (b) on turbulence level for all studied models

4. Conclusions

The main conclusion from the present work is that existing transition models are far from perfection. None of the considered models are able to adequately predict transition in a wide range of freestream turbulence intensities. SA e^N model is applicable only for cases with low *Tu* values where natural transition is observed, while at high *Tu* it is inclined to predict fully turbulent solutions.

SST γ-Re_θ and SST γ have shown better agreement with experimental results and thus surpass the SA γ model. However, it should be noted that SST-based models are more complex and often lead to poorer convergence and higher computational cost than SA-based models.

References

[1] C.L. Rumsey, P.R. Spalart. 2008. Turbulence Model Behavior in Low Reynolds Number Regions of Aerodynamic Flowfields. 38th AIAA Fluid Dynamics Conference and Exhibit. *AIAA-2008-4403.*

[2] F.R. Menter, R.B. Langtry. 2009. Correlation-Based Transition Modelling for Unstructured Parallelized Computational Fluid Dynamics Codes. *AIAA J.* 47(12), 2984-2906.

[3] F.R. Menter, P. E. Smirnov, Tao Liu, R. Avancha. 2015. A One-Equation Local Correlation-Based Transition Model. *Flow Turbulence Combust.* DOI 10.1007/s10494-015-9622-4

[4] F. R. Menter. 1993. Zonal Two Equation k-ω Turbulence Models for Aerodynamic Flows. *AIAA Paper 93-2906.*

[5] J.M. Coder, M. D. Maughmer. 2012. One-Equation Transition Closure for Eddy-Viscosity Turbulence Models in CFD. *AIAA Paper 2012-0672.*

[6] P. R. Spalart, S. R. Allmaras. 1992. A one-equation turbulence model for aerodynamic flows. *AIAA Paper 92-0439.*

[7] J.M. Coder, M. D. Maughmer. 2014 Computational Fluid Dynamics Compatible Transition Modelling using an Amplification Factor Transport Equation. *AIAA Journal.*

[8] M. Drela, M. B. Gilest. 1986. Viscous-Inviscid Analysis of Transonic and Low Reynolds Number Airfoils. *AIAA Journal* Vol. 25, No. 10 1347- 1354

[9] A. M. O. Smith, N. Gamberoni. 1956. Transition, Pressure Gradient and Stability Theory. Technical Report ES-26388, Douglas Aircraft Company.

[10] L. M. M. Mack. 1975. Aerodynamic Analysis Requiring Advanced Computers. Paper nr. 4. Technical Report SP-347, NASA.

[11] M. K. Strelets. 2001. Detached Eddy Simulation of Massively Separated Flows. *AIAA Paper 2001-0879.*

[12] S. E. Rogers, D. Kwak. 1991. An Upwind Differencing Scheme for the Incompressible Navier-Stokes Equations. *Applied Numerical Mathematicis,* 8, 43–64.

[13] A. M. Savill. 1993. Some Recent Progress in the Turbulence Modelling of By-Pass Transition. Near-Wall Turbulent Flows, edited by R. M. C. So, C. G. Speziale, and B. E. Launder, *Elsevier*, NewYork, 1993, p. 829.

[14] K.E. Swalwell. 2005. The effect of turbulence on stall of horizontal axis wind turbines. PhD thesis, Monash University, Australia

Non-linear correction for the k-ω SST turbulence model

A A Matyushenko, A V Garbaruk
Peter the Great St.Petersburg Polytechnic University
Russia, Saint-Petersburg, 195251, Politechnicheskaya, 29

E-mail: alexey.matyushenko@gmail.com

Abstract. Non-linear correction for the k-ω SST model based on the WJ-BSL-EARSM model was developed and tested for a wide range of flows. For the basic turbulence flows the corrected model (SST-NL) results are close to those of the linear SST model and surpass WJ-BSL-EARSM ones. At the same time non-linear correction is capable to predict secondary motion in rectangle channels and prevents false corner separation for flows around a wing-body junction as well as the WJ-BSL-EARSM model.

1. Introduction

Despite well-known fundamental restrictions of the Linear Eddy-Viscosity RANS turbulence models (LEVM), they are still most widely used in industrial and aerospace applications. Such popularity of these models is due to the existence of time-proved and numerically stable models for then. Reynolds Stress Models (RSM), which in principle present a more general class of models with a wider range of applicability, could be considered as an alternative of LEVM. The advantages of RSM are revealed most significantly for turbulent flows with secondary motion induced by Reynolds stress anisotropy. One of the impressive examples of such flows is a flow around a wing-body junction where LEVM predict false corner flow separation and only RSM are able to capture correct flow topology. However for relatively simple flows accuracy of RSM is less than accuracy of the best linear models. Other than that the worse robustness and convergence of equations prevent wide spreading of these models.

One of the best turbulence LEVM in terms of robustness, computational cost and accuracy for simple flows is the k-ω SST model [1]. However this model is not able to prevent false corner separation. Therefore the aim of the present work is to resolve this issue and develop a non-linear correction for the SST model (SST-NL).

2. Non-linear correction for the SST model

The SST model contains two transport equations for turbulent kinetic energy, k, and specific dissipation rate, ω. In the original SST model the Reynolds stress tensor is computed using the Boussinesq relation

$$\tau_{ij} = \frac{2}{3}k\delta_{ij} - 2\nu_t S_{ij} \tag{1}$$

where $\nu_t = \dfrac{a_1 k}{max(a_1\omega, SF_2)}$ is the eddy viscosity, S_{ij} - strain rate tensor, $S = \sqrt{2S_{ij}S_{ij}}$ - strain rate invariant, F_2 - SST function, and $a_1 = 0.31$.

The proposed correction affects only the Reynolds stress tensor computation and consists of addition of non-linear terms to relation (1).The «constitutive relation» in the SST-NL reads as follows:

$$\tau_{ij} = \frac{2}{3}k\delta_{ij} - \overbrace{2\nu_t S_{ij}}^{\substack{LINEAR\\TERM}} - \overbrace{\frac{3}{2}k\left[\frac{1}{Q}\left(S^*_{ik}\Omega^*_{kj} - \Omega^*_{ik}S^*_{kj}\right) + \frac{N}{Q_1}\left(S^*_{ik}\Omega^*_{kl}\Omega^*_{lj} + \Omega^*_{ik}\Omega^*_{kl}S^*_{lj} - \frac{2}{3}IV\delta_{ij} - II_{\Omega}S^*_{ij}\right)\right]}^{NON-LINEAR\ \ TERMS} \quad (2)$$

where $S^*_{ij} = \frac{1}{2C_\mu\omega}\left(\frac{\partial u_i}{\partial x_j} + \frac{\partial u_j}{\partial x_i} - \frac{2}{3}\frac{\partial u_k}{\partial x_k}\delta_{ij}\right)$ and $\Omega^*_{ij} = \frac{1}{2C_\mu\omega}\left(\frac{\partial u_i}{\partial x_j} - \frac{\partial u_j}{\partial x_i}\right)$ are the non-dimensional strain rate and vorticity tensors and other quantities are defined as:

$$N = C_1' + \frac{9}{4}\sqrt{2C_\mu II_S}, \quad Q = \frac{\left(N^2 - 2II_\Omega\right)}{A_1}, \quad Q_1 = \frac{Q}{6}\left(2N^2 - II_\Omega\right),$$

$$II_S = S_{mn}S_{nm}, \quad II_\Omega = \Omega_{mn}\Omega_{nm}, \quad IV = S_{mn}\Omega_{nk}\Omega_{km} .$$

Note, that the non-linear terms in (2) are based on the WJ-BSL-EARSM [2] which is a modification of the WJ EARSM [3].

3. Application Test Cases

This section presents a comparative assessment of the SST-NL model with «its parents», SST and WJ-BSL-EARSM, and consists of three main parts. The first part presents results of computations of basic two dimensional turbulence flows including wall-bounded flows and flows with turbulent boundary layer separation. The second shows comparison of the models for rectangle channels where secondary motion in the sharp corners takes place. Finally an example of a flow with a wing-body junction is discussed in the third part.

The simulations of all considered flows were carried out using ANSYS FLUENT and the proposed correction was implemented using UDF (User Defined Functions). The Coupled method is used for pressure-velocity coupling. The inviscid fluxes in the momentum and turbulence equations are approximated with the use of the Second Order Upwind scheme. For pressure the «Standard» interpolation (weighted interpolation based on central coefficients) is utilized and the gradients are approximated with the use of cell based Green-Gauss method. For all test cases computational meshes ensure grid-independent solution and convergence of iterations was achieved.

3.1. Basic turbulence flows

The goal of these computations is to check the influence of the suggested modification on the accuracy of simple turbulent flow prediction. Three tasks were considered: zero pressure gradient boundary layer, backward facing step and axisymmetric diffuser.

The first case, Turbulent Boundary Layer (TBL) was experimentally investigated by Wieghardt and Tillmann [4] and was included into the AFOSR-IFP Stanford database of turbulent flows [5]. The corresponding incompressible flow was computed at the Reynolds number based on inlet velocity and length of the flat plate Re=10^7.

An experimental investigation of the second case, an incompressible flow downstream of a Backward-Facing Step (BFS), was carried out by Vogel and Eaton [6]. The Reynolds number based on the maximum inlet velocity and the step height H is equal to Re = $2.8{\cdot}10^4$. Inlet boundary is placed 3.8H upstream, and the outlet boundary is located approximately 40H downstream from the step. In order to reproduce the experimental inflow conditions in the simulations, a precursor computation of the flow in the plane channel was carried out, and the inflow conditions were extracted from this computation at the cross-section where the experimental boundary layer displacement thickness have been matched.

The last test case, CS0 diffuser by Driver [7], presents an example of shallow separation from a smooth wall. The test case geometry consists of an axisymmetric diffuser caused by adverse pressure gradient with an internally mounted cylinder along the centerline. The Reynolds number based on the maximal inlet velocity and the inner cylinder diameter D is equal to $2.8\cdot 10^5$.

Figure 1 shows schemes of the considered flows and the skin friction coefficient distribution for the considered cases. One can see that for the TBL flow all considered models are almost identical but for separated flows (BFS and CS0) the difference is significant. Namely, the linear and non-linear SST models are close to each other and are in better agreement with experimental data than the WJ-BSL-EARSM model, which predicts less separation for the backward-facing step or even no separation for the CS0 diffuser.

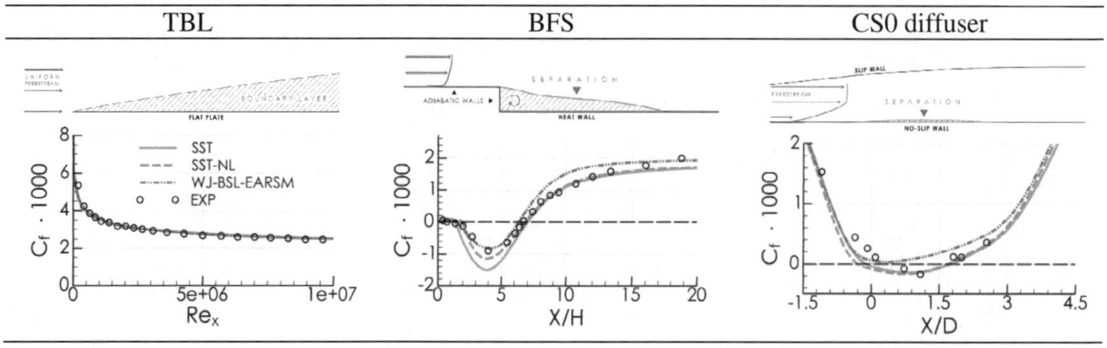

Figure 1. Schemes of the considered flows (upper row) and skin friction coefficient distribution for the TBL, backward-facing step and CS0 diffuser

3.2. Flows in rectangular ducts

Flows in rectangle ducts are examples of the essential influence of Reynolds stress anisotropy which initiates secondary flow motion. In some cases the secondary flow can result in a strong change in a flow structure and this effect is beyond the capabilities of linear eddy viscosity models to predict.

3.2.1. Fully developed flow in square duct

The flow is considered under the conditions of the DNS [8] at a Reynolds number based on the averaged friction velocity and the channel width *H* of 1200. The computations are carried out in a «2.5D mode», which includes momentum equations for all the three velocity components but assumes that their streamwise derivatives are zero. A constant streamwise pressure gradient is adjusted to meet the DNS Reynolds number. At the solid walls, the no-slip boundary conditions are imposed.

The streamlines at XY section show that SST-NL is capable of predicting secondary flow as well as WJ-BSL-EARSM while the SST model fails (see Figure 2). Lateral velocity at diagonal line (X=Y) predicted by SST-NL is slightly closer to the DNS data than the WJ-BSL-EARSM solution (Figure 3a) and the SST model as well as other LEVM predicts no secondary motion. As a result streamwise velocity predicted by SST-NL is in better agreement with DNS data than the results of the SST model (Figure 3b).

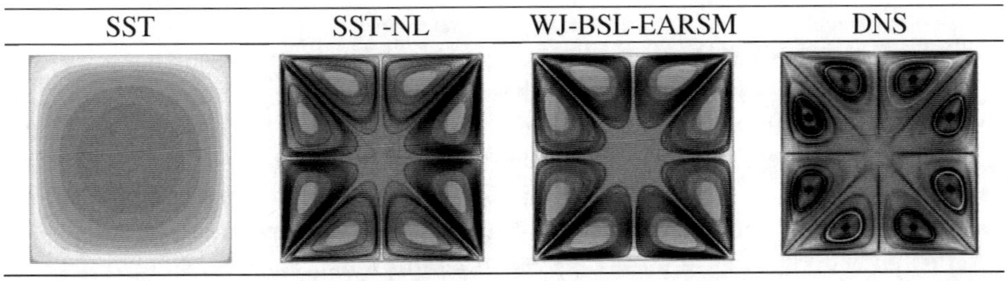

Figure 2. Comparison of streamline velocity contours and secondary motion with the DNS data in the fully developed square duct flow.

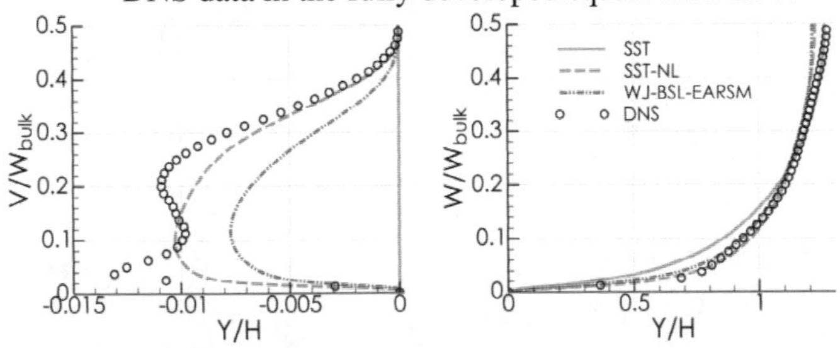

Figure 3. Lateral and steamwise velocity profiles on the diagonal of the square duct.

3.2.2. Flow in a rectangular diffuser

Flow in the rectangular diffuser (Figure 4a) [9,10] is characterized by strong separation which is very sensitive to turbulence modelling. The Reynolds number based on the bulk velocity and height of the duct, H, is equal to 10^4. In accordance with the experimental setup, the inlet flow is considered as fully developed flow. The velocity and turbulence characteristics in the inlet plane are specified using precursor computation of the developed rectangular channel flow with the same turbulence model as for the main computation. The no-slip boundary conditions are specified at the diffuser walls.

The pressure coefficient $C_P = \left(p - p_{ref}\right)/\left(\frac{1}{2}\rho u_b^2\right)$ at the centerline of the bottom flat wall is presented at Figure 4b (p_{ref} is the pressure at the first experimental point; u_b is a bulk velocity in the inlet channel). One can see that the results of the SST-NL and WJ-BSL-EARSM are close to each other and better agree with the experimental data than the linear SST model.

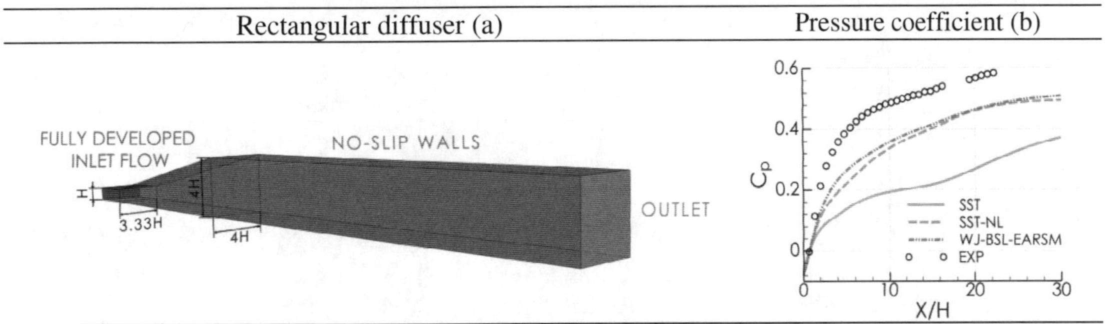

Figure 4. Rectangular diffuser (a) and comparison of computed and measured pressure distributions along centerline of the bottom wall (b).

3.3. Flow around a wing in wind tunnel with no-slip side walls.

The flow around an A-Airfoil wing mounted between sidewalls was experimentally investigated in the ONERA F2 wind tunnel [11] at a Reynolds number based on the wing chord C of $2.1\cdot10^6$. No corner separation from the wing-sidewall junction was observed at AoA $\leq 16°$ in the experiment while LEVM models tend to predict this separation. The incompressible flow was considered at AoA from 13° to 16°. The height of the tunnel ($L_y = 3C$) and wing span ($L_z = 2.3C$) correspond to experimental wind tunnel parameters (see Figure 5). Uniform velocity and turbulent characteristics are specified at the inlet boundary. Top and bottom tunnel walls are modelled as slip walls. The beginning of the no-slip side walls ($2.25C$ upstream from the leading edge) ensures the same boundary layer thickness on the side wall near the leading edge as it was observed in the experiment. No-slip boundary conditions are specified on the airfoil surface.

Streamlines over the wing suction side show flow structure and separation size near wing-body junction (Figure 6). The SST model predicts corner separation for all considered angles of attack. At AoA=13° this separation is very small but at AoA≥13° it exceeds half of the tunnel width. For both non-linear models the corner separation is absent. Moreover one can see that with increasing AoA the separation appears near the wing center and then spread on the entire wing suction side.

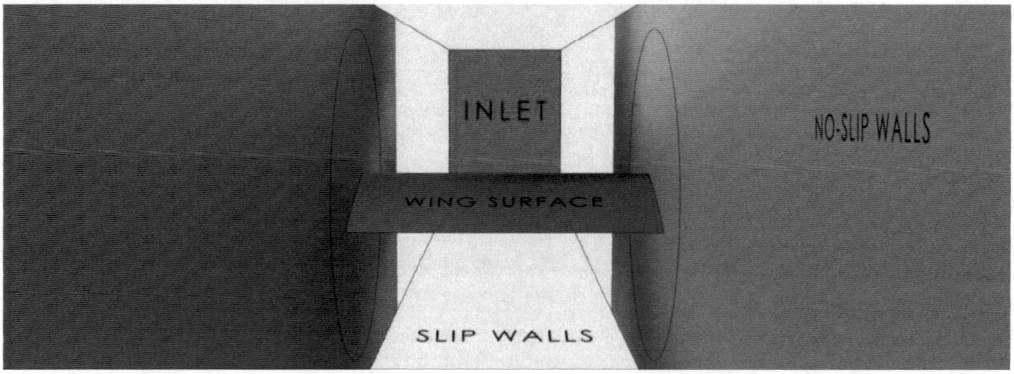

Figure 5. Computational domain used for the A-Airfoil test case.

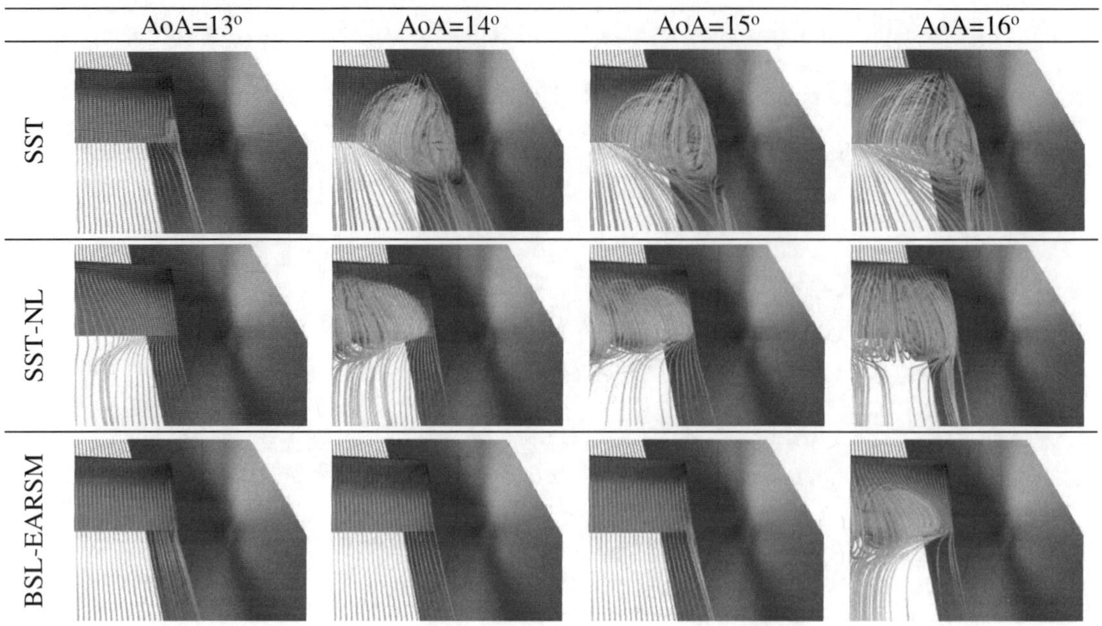

Figure 6. Visualization of streamlines near wing-sidewall junction.

4. Conclusions

A non-linear correction for the k-ω SST model was developed and the corrected model (SST-NL) was examined. Tests show that the SST-NL predicts basic turbulence flows as well as the SST model and surpasses the WJ-BSL-EARSM model. For flows in rectangle ducts, where secondary flow leads to a strong changes in the flow, results of the SST-NL and WJ-BSL-EARSM are close to each other and are in better agreement with the experimental data than the linear SST model results. Finally, the SST-NL model as well as WJ-BSL-EARSM model prevents false corner separation for flow around a wing-body junction. So, for the considered flows SST-NL exceeds both «parents», namely SST and WJ-BSL-EARSM.

Acknowledgements

The results of the work were obtained using computational resources of Peter the Great Saint-Petersburg Polytechnic University Supercomputing Center (http://www.spbstu.ru).

References

[1] Menter, F. R., Kuntz, M. & Langtry, R., (2003). Ten Years of Industrial Experience with the SST Turbulence Model.

[2] Menter F. R., Garbaruk A. V. & Egorov Y., (2009). Explicit Algebraic Reynolds Stress Models for Anisotropic Wall-Bounded Flows. Proceedings of EUCASS – 3rd European Conference for Aero-Space Sciences.

[3] Wallin S., Johansson A.V. (2000). An explicit algebraic Reynolds stress model for incompressible and compressible turbulent flows. J. Fluid Mech., v.403, No.1, pp.89-132.

[4] Wieghardt, K. and Tillmann, W., (1951). On the turbulent friction layer for rising pressure. NACA TM 1314.

[5] Coles, D.E., and Hirst, E.A., Eds, (1969). Computation of turbulent boundary layers- 1968 AFOSR-IFP-Stanford Conference. Vol. 2, Stanford University, CA.

[6] Vogel, J.C. and Eaton, J.K., (1985). Combined Heat Transfer and Fluid Dynamic Measurements Downstream of a Backward-Facing Step, Vol. 107, Journal of Heat Transfer, pp. 922 – 929.

[7] Driver, D.M., (1991). Reynolds shear stress measurements in a separated boundary layer flow. AIAA-91-1787.

[8] Raiesi, H., Piomelli, U., Pollard, A. (2011). Evaluation of Turbulence Models Using Direct Numerical and Large-Eddy Simulation Data. Journal of Fluids Engineering, v.133, No.2, 021203 (10 pp.).

[9] Cherry, E. M., Elkins, C. J., Eaton, J. K. (2008). Geometric sensitivity of three-dimensional separated flows. International Journal of Heat and Fluid Flow, v.29, No.3, pp.803–811.

[10] Cherry, E. M., Elkins, C. J., Eaton, J. K., (2009). Pressure measurements in a three-dimensional separated diffuser. International Journal of Heat and Fluid Flow, v.30, No.1, pp.1–2.

[11] Gleyzes, C., Capbern, P., (2003). Experimental study of two AIRBUS/ONERA airfoils in near stall conditions. Part I: Boundary layers. Aerospace Science and Technology, vol. 7, pp. 439–449.

Analytical and numerical solutions to Stefan problem in model of the glaciation dynamics of the multilayer cylinder in sea water

V V Mikova[1], **G I Kurbatova**[2], **N N Ermolaeva**[3,4]

Faculty of Applied Mathematics and Control Processes, Saint-Petersburg State University, Universitetskii prospekt 35, Petergof, Saint Petersburg, Russia

E-mail: n.ermolaeva@spbu.ru

Abstract. A mathematical model of the glaciation dynamics of the multilayer cylinder is suggested. The model reflects the specific feature of glaciation in salt water. The choice of effective model parameters and thermal characteristics of the increasing ice is considered. The calculations of problem using front-tracking finite difference method and using analytical solution found in quasi-stationary approximation are presented.

1. Introduction

Many papers have studied the sea-ice formation, for example, [1]–[6]. Extensive studies on the physical and structural properties of sea ice were carried out [7]–[11]. This work continues research of the sea-gas pipelines in the northern seas, which began in "Models of sea gas-pipelines"[12]. The nonstationary problem of the gas-pipeline glaciation, which has multilayer cylindrical wall in sea-water is considered. In contrast to the freshwater, in the salt water the inflow of salt into the water layer adjacent to the front of glaciation occurs. It increases the density of water and causes the additional convective flows. The aggregate effect results in increasing of total heat flux from water to the front of glaciation. This mechanism is taken into account by introduction in the Stefan condition of the additional heat flux $\alpha\frac{dy}{dt}$ from the adjacent water layer, that is proportional to the rate of ice growth $\frac{dy}{dt}$. α is the effective parameter of the model having dimensions of the volumetric heat source, in the fresh water $\alpha = 0$. The selection technique of average salinity s_i of growing ice from which depend all thermal and physical characteristics of sea ice, and the selection of effective parameter α are explained below.

2. Mathematical model

The model of the cylindrical gas pipeline glaciation with two coating layers in the salt water is written as follows:

$$\rho_1 c_1 \frac{\partial T_1}{\partial t} = \lambda_1 L(T_1), \quad r \in (R, R_1), \quad t > t_0; \tag{1}$$

$$t = t_0, \quad T_1(r) = T_1^0(r); \tag{2}$$

[4] Present address: Faculty of Applied Mathematics and Control Processes, Saint-Petersburg State University, Universitetskii prospekt 35, Petergof, Saint Petersburg, Russia.

$$t > t_0, \quad r = R: \quad T_1 = T_0(t); \tag{3}$$

$$r = R_1: \quad T_1 = T_2, \quad \lambda_1 \frac{\partial T_1}{\partial r} = \lambda_2 \frac{\partial T_2}{\partial r}; \tag{4}$$

$$\rho_2 c_2 \frac{\partial T_2}{\partial t} = \lambda_2 L(T_2), \quad r \in (R_1, R_2), \quad t > t_0; \tag{5}$$

$$t = t_0, \quad T_2(r) = T_2^0(r); \tag{6}$$

$$t > t_0, \quad r = R_2: \quad T_2 = T_i, \quad \lambda_2 \frac{\partial T_2}{\partial r} = \lambda_i \frac{\partial T_i}{\partial r}; \tag{7}$$

$$\rho_i c_i \frac{\partial T_i}{\partial t} = \lambda_i L(T_i), \quad r \in (R_2, R_2 + y(t)), \quad t > t_0; \tag{8}$$

$$t = t_0, \quad y = y_0, \quad T_i(r) = T_i^0(r); \tag{9}$$

$$t > t_0, \quad r = R_2 + y(t): T_i = T_*; \tag{10}$$

$$\lambda_i \frac{\partial T_i}{\partial r}\bigg|_{R_2+y(t)} - q = \gamma \rho_i \frac{\mathrm{d}y}{\mathrm{d}t}, \quad q = q_w + \alpha \frac{\mathrm{d}y}{\mathrm{d}t}, \quad t > t_0. \tag{11}$$

We use the following designations: $L = \frac{1}{r}\frac{\partial}{\partial r}\left(r\frac{\partial}{\partial r}\right)$ is the Laplace operator in cylindrical coordinate system (r, z, φ) under conditions $\frac{\partial}{\partial z} = \frac{\partial}{\partial \varphi} = 0$; λ_k, ρ_k, c_k $T_k = T_k(r,t)$ are the thermal conductivity [W/m.K], the density [kg/m^3], the specific heat [kJ/kg.K] and the temperature distribution [K] in k-th layer, indices $k = 1$, 2, i correspond to the following regions: 1 to the first layer of pipeline wall (steel), 2 to the second layer of pipeline wall (concrete), i to the ice layer; R is the inner radius of the gas-pipeline [m], R_1, R_2 are the outer radii of the pipeline wall layers [m]; $T_0(t)$ is the temperature distribution in the ice layer at the initial moment of time. γ is the latent heat (of fusion) for sea ice [kJ/kg]; q_w is the radial component of the heat flux vector from seawater to glaciation front; $r = R_2 + y(t)$ is the coordinate of the glaciation front; $y = y(t)$ is the ice thickness [m] upon the surface of pipeline at the time moment t; y_0 is the ice thickness [m] at the initial moment of time t_0. By $r = R_2 + y_0$ following equalities hold $T_i^0(R_2 + y_0, t_0) = T_*$. Model can be extended to a larger number of coating layers. The equation (11) is the modified Stefan condition, which covers the additional heat flux from the water. In model the parameters λ_i, c_i, ρ_i, T_*, γ, α, q_w are considered to be constant.

3. Selection of parameters

For selecting parameters of model, which are in agreement with the known experimental data, was used following technique. The values of the effective parameter α^0, of the range of the ice temperature changes and of the ice salinity s_i^0 were set. The values of λ_i, c_i, ρ_i, γ were calculated using known experimental data given in view of the empirical formulas and of the tables in [13,14].

The other parameters of model q_w, T_*, ρ_1, c_1, λ_1, ρ_2, c_2, λ_2, R, R_1, R_2 were assumed to known and unchanged. Time dependence of the ice growth rate $W = \frac{dy}{dt}$ is calculated using the given below algorithm to numerical solution of system (1)–(11). The average salinity s_i [$^0/_{00}$] of growing ice was calculated using semiempirical formula by Tsurikov [13]:

$$s_i = S_w \frac{7\sqrt{W}}{7\sqrt{W} + 10.30}, \tag{12}$$

S_w is the salinity of sea water [$^0/_{00}$], $W = \frac{dy}{dt}$ is the rate of ice growth [mm/h]. The sea water salinity was chosen to be $S_w = 35$ $^0/_{00}$. The ice salinity s_i values were found on the Tsurikov formula (12) using dependence $W = \frac{dy}{dt}$ calculated from model. The mean value of salinity

$< s_i >$ is compared to the selected value s_i^0. If the difference between $< s_i >$ and s_i^0 was over $2\ {}^0/_{00}$, the procedure selecting effective parameters s_i and α was repeated according to the same scheme with new s_i^0 and α^0.

The following correlated parameters of model were found according to the technique described above:

$$\begin{gathered} q_w = 30.4447\,\mathrm{W/m^2}, \quad T_* = 271.236\,\mathrm{K}, \quad T_0 = (T_* - 6)\ \mathrm{K}, \\ R = 0.51\,\mathrm{m},\ \rho_1 = 10^4\,\mathrm{kg/m^3},\ \rho_2 = 2300\,\mathrm{kg/m^3},\ c_1 = 450\,\mathrm{J/(kg\cdot K)}, \\ c_2 = 924\,\mathrm{J/(kg\cdot K)},\ \lambda_1 = 24\,\mathrm{W/(m\cdot K)},\ \lambda_2 = 1.7\,\mathrm{W/(m\cdot K)}, \\ c_i = 2100\,\mathrm{J/(kg\cdot K)},\ \lambda_i = 2.15\,\mathrm{W/(m\cdot K)},\ \alpha = 330048.81\,\mathrm{kJ/kg}, \\ \delta_1 = 0.04\,\mathrm{m},\ \delta_2 = 0.12\,\mathrm{m},\ \gamma = 303000\,\mathrm{J/kg},\ \rho_i = 931\,\mathrm{kg/m^3}. \end{gathered} \tag{13}$$

4. Numerical solution algorithm

If the temperature upon the outer surface of the gas pipeline is lower than the sea water-ice transition temperature, the pipeline glaciation may occur. The thickness of the ice layer increases in time, and the ice growth rate converges to zero. At an equilibrium state the ice thickness y_* reaches the maximum (steady) value and the modified Stefan condition (11) passes to the equality: $\lambda_i \frac{\mathrm{d}T_i}{\mathrm{d}r}\Big|_{R_2+y_*} - q_w = 0$. This equality together with the stationary heat transfer equations in heat-insulation layers and with other conditions of model enables to calculate the maximum thickness of the ice layer y_* from the transcendental equation:

$$\frac{\lambda_i(T_* - T_0)}{(R_2 + y_*)(\frac{\lambda_4}{\lambda_1}\ln\frac{R_1}{R} + \frac{\lambda_4}{\lambda_2}\ln\frac{R_2}{R_1} + \ln\frac{R_2+y_*}{R_2})} - q_w = 0. \tag{14}$$

For the numerical solution of this Stefan-type problem can be used a method with the explicit tracking of moving surface. We use the method of Douglas and Gallie [15], [16], which is an iterative finite difference method, with variable time steps. In this approach the time step size τ_{n+1} is variable and at the $(n+1)$-th temporal level it is determined so that the ice thickness increased on constant value h during this time step. The value h is the space grid step size. The convergence of this numerical method was proved for a Stefan-type problem in [15], [16]. The time step size in the null approximation $\tau^{(0)}$ at the $(n+1)$-th temporal level is set to equal previous step size τ_n at the n-th temporal level: $\tau^{(0)} = \tau_n$.

The calculation on model begins at the moment of time t_0, when necessary and sufficient conditions are fulfilled for beginning glaciation of the outer surface of pipeline:

$$T_2^0(R_2) < T_*, \quad \lambda_2 \frac{\partial T_2}{\partial r}\Big|_{R_2} > q_w.$$

The first step τ_1 was chosen from the difference version of equation (11):

$$\tau_1 = \frac{(\gamma\rho_i + \alpha)h^2}{(\lambda_i(T_* - T_2^0[N2]) - q_w\,h)},$$

$T_2^0[N2]$ is the temperature in second insulation-layer at boundary with ice at initial time moment, $N2$ is the array length of temperature of second insulation-layer. Arrays of the temperatures in two insulation-layers and in ice layer at the s-the iteration is calculated using the implicit finite difference scheme with step size τ^s.

The step size τ^{s+1} at the $(s+1)$-th iteration at the $(n+1)$-th temporal level is calculated using obtained array of the temperature T_i^s in the ice layer according to the formula:

$$\tau^{s+1} = \tau^s + (\gamma\rho_i + \alpha)(h - h^s)/(q_{II} - q_w), \; q_{II} = \lambda_i(T_i^s[1] - T_i^s[0])/h,$$

$$h^s = \tau^s(q_I - q_w)/(\gamma\rho_i + \alpha), \quad q_I = \lambda_i(T_i^s[n+1] - T_i^s[n])/h,$$

$(n+1)$ is the array length of the temperature in the ice layer at the $(n+1)$-th temporal level. The iterative process is terminated if the inequality $|\tau^{s+1} - \tau^s| \leq \varepsilon$ holds for a specified small quantity ε.

By this algorithm the time step size τ, the overall glaciation time t, the ice growth rate W and the temperature distributions are calculated for every thickness of the ice layer y.

5. Results

As an example, in Table 1 we present the calculation results of the beginning of the gas pipeline glaciation in seawater with parameter values (13) by the presented algorithm.

Table 1.

y, cm.	0.3	0.5	1	3	5	7	9	11	12.5
t, h.	13.60	20.624	43.500	139.220	264.170	423.300	633.021	916.34	1219.44
τ, h.	3.935	4.033	4.287	5.471	7.018	9.120	12.132	16.811	22.472
W, mm/h.	0.254	0.248	0.233	0.183	0.142	0.110	0.082	0.060	0.040

For parameter values (13) the maximum ice thickness y_* is equal to 18.2378 cm.

The ice salinity calculation $s_i(W)$ on values $W(t)$ from the Table 1 using Tsurikov formula (12) is presented in Table 2.

Table 2.

W, mm/h.	0.254	0.248	0.233	0.183	0.142	0.110	0.08	0.06
$s_4(W)$, ‰	8.93	8.85	8.645	7.88	7.135	6.44	5.70	5.0

As follows in table 2 the mean salinity $< s_i > (W)$ of the growth ice, calculating by the formula (12) is equal to 7.3 $^0/_{00}$, which confirms that the set parameters (13) is correlated.

It is possible to show that model at quasi-stationary approximation has the analytical solution when the ratio of the ice layer thickness to the outer radius of the gas pipeline is small: $y/R_2 \ll 1$.The analytical solution in this approximation is

$$t - t_0 = -\frac{(y - y_0)}{a} + b \ln\left(\frac{R_2 - cy_0}{R_2 - cy}\right), \tag{15}$$

$$a = \frac{q_w \, t_x}{r_x \, (\gamma\rho_i + \alpha)}, \qquad b = \frac{\lambda_i \, (T_* - T_0) \, (\gamma\rho_i + \alpha)}{q_w^2 \, t_x (1 + d)},$$

$$c = \left(\frac{\lambda_i (T_* - T_0)}{r_x R_2 (1+d) q_w} - \frac{d}{1+d} \right)^{-1}, \; d = \frac{\lambda_i}{\lambda_1} \ln \frac{R_1}{R} + \frac{\lambda_i}{\lambda_2} \ln \frac{R_2}{R_1},$$

the values t, t_0, y, y_0, a, b, c, d, R_2 in the equation (15) are dimensionless, r_x, t_x are the characteristic values of variables r and t respectively.

In Table 3 we present the comparing the results of the numerical modeling (Table 1) and the calculation using analytical solution (15) with the initial conditions $t_0 = 20.624$ h., $y_0 = 0.5$ cm.

Table 3.

y, cm.	1	3	5	7	9	11
t, h.	43.50	139.22	264.17	423.30	633.02	916.34
$\tilde{t}$, h.	41.49	138.67	261.92	418.73	620.28	884.51

As follows in table 3 the ice layer with the thickness of 1 cm is formed a faster in accordance with the analytical solution, which is based on the quasi-stationary temperature distribution. The calculations of nonstationary problems suggest that at the very beginning of the processes the temperature distributions in the layers are different from the quasi-stationary ones. This is reflected in the fact that the values t and $\tilde{t}$ are differ for $y = 1$ cm. Further, with a reasonable degree of accuracy the heat processes can be considered the quasi-stationary, which is confirmed by the values t, $\tilde{t}$ close to each other in domain $y = 3$ cm. Distinctions between numerical and analytical solutions increases for $y \geq 5$ cm. This is due to the fact that, the ice layer while growing, less and less holds the condition $y \ll R_2$. Consequently, as follows in table 3 if that the conditions using to obtain analytical solution are satisfied, then the analytical and numerical solutions to the glaciation dynamic problem are coincident. The calculation by the presented algorithm of the numerical solution of different variants of the gas pipeline glaciation shown that the process of glaciation can estimated as stationary for defined thickness of ice for many sets of parameters and the thickness of ice can calculate using analytical solution obtained (15).

6. Conclusion

The paper offered the modification of Stefan condition, which allows to take into account the feature of glaciation in salt water. The algorithm of the numerical solution of the glaciation model of multilayer cylindrical wall in sea-water and the selection procedure of the average thermal characteristics of the growing ice were suggested. The analytical solution of the problem was given for the approximation of the thin layer. The presented comparing the numerical and analytical solution confirms reliability and acceptable accuracy of the suggested numerical solution to the nonlinear system of equations.

References

[1] Lepparanta M 1993 A review of analytical models of sea-ice growth *Atmosph.-Ocean* **31:1** 23-138
[2] Pehovich A I 1983 *Fundamentals of hygro-ice-thermics* (Leningrad: Energoatomizdat) p 200 in Russian
[3] Anderson D L 1961 Growth rate of sea ice *J. of Glaciol.* **3** 1170-2
[4] Andreev O M 2007 Parametrization of the vertical salinity distribution of annual sea ice for thermodynamic modeling in the Arctic *Probl. of the Arctic and Antarc.* **75** 99-105 in Russian
[5] Gupta S C 2003 *The Classical Stefan Problem: Basic Concepts, Modelling and Analysis* (Amsterdam: Elsevier) p 385
[6] Ebert E E 1993 An intermediate one-dimensional thermodynamic sea ice model for investigating ice-atmosphere interactions *J. Geophys. Res.* **98** 10085-109

[7] Weber J E 1977 Heat and salt transfer associated with formation of sea-ice *Tellus* **29** 151-60
[8] Cox G F 1974 Salinity variations in sea ice *J.Clacoil.* **23** 109-20
[9] Fukusako S 1990 Thermophysical properties of ice, snow and sea ice *Int. J. of Thermophy.* **11** 353-72
[10] Weeks W F 1998 *Physics of ice covered seas* vol 1, ed M Leppäranta (Helsinki: Helsinki University Press) pp 25-104
[11] Cottier F, Eicken H and Wadhams P 1999 Linkages between salinity and brine channel distribution in young sea ice *J. of Geophy. Res.* **104** 15859-71
[12] Kurbatova G I, Popov E A, Filippov B V, Filippov V B and Filippov K B 2005 *Model Offshore Pipelines* (St. Petersburg: St. Petersburg State University) p 156 in Russian
[13] Sazonov K E 2007 *Materials Science. Material properties. Test methods. Ice and snow* (St. Petersburg: RSHU) p 195 in Russian
[14] Doronin Yu P and Kheysin D E 1975 *Sea Ice* (Leningrad: Gidrometeoizdat) p 320 in Russian
[15] Douglas J and Gallie T M 1955 On the numerical integration of a parabolic differential equation subject to a moving boundary condition *Duke Math. J.* **22** 557-70
[16] Vasilyev F P 1963 The method of finite differences to solve phase Stefan problem *Comp. Math. and Math. Phys. J.* **3** 861-73 in Russian

IOP Conf. Series: Journal of Physics: Conf. Series **929** (2017) 012104 doi:10.1088/1742-6596/929/1/012104

The influence of thermal properties of anisotropic thermoelement substrate to the temperature distribution and the value of thermal EMF

P.A. Popov

194021, Ioffe Institute, St. Petersburg, Russia

pavel.popov@mail.ioffe.ru

Abstract. Temperature and the electric potential distributions in anisotropic thermoelements situated on substrates with different thermal properties are calculated. The influence of the geometric dimensions and thermophysical properties of the thermoelement and the substrate material to the electrical signal of the sensor is investigated. It is shown that at the initial stage of heating the influence of the thermal conductivity anisotropy of the thermoelement is negligible. On further heating the thermal properties of the substrate significantly affect to the temperature and potential distribution. Using a sensors with long thermoelements situated on the substrate with high thermal diffusivity can greatly simplify the processing of the measurement results.

1. Introduction

Traditionally, anisotropic thermoelements are used as thermoelectric power generators and cooling devices [1, 2]. Usually, in such devices, the temperature distribution is known and the subjects of the study are features of thermal EMF generation. Recently, thermoelements were used as sensitive elements of the heat sensors. In this case it is necessary to solve the inverse problem – to determine the heat flux based on electric signal of the sensor obtained in the experiment. In the general case it is difficult to do because of the two-dimensional distribution of temperature due to anisotropy of the thermal conductivity. However, for the same reason, under certain thermal properties of the substrate it can be achieved close to the one-dimensional temperature distribution. It greatly simplifies the processing of the output signal of this type of sensors [5]. In this paper we analyze the influence of thermal properties of substrates on the temperature distribution in the anisotropic thermoelement and the value of thermal EMF in the case of heat sensors.

The heat sensor is a battery of series-connected bismuth anisotropic thermoelements of length $1 \div 10\, mm$ fixed on mica substrate [3, 4]. They have a fast response, high sensitivity and noise immunity, and are successfully used in various thermophysical experiments [5–7].

2. Mathematical model

Consider the model of the heat sensor that consists of an rectangular anisotropic thermoelement (1) of length l and thickness h, placed on a nonconducting substrate (2) of thickness s (Fig. 1). The heat flux q_h passes through the working surface of the thermoelement $y = h$. The side surfaces $x = 0, x = l$ of the thermoelement and substrate are thermal insulated. On the back surface of the substrate $y = s$ occurs the convective heat exchange with the environment $q_s = \alpha_s(T_s - T_e)$, where α_s – the heat transfer coefficient, T_s and T_e – the temperature of the substrate and the environment respectively. Electrical insulation boundary condition $j_n = 0$ is set on all the surfaces of thermoelement. All physical properties of the sensor and the substrate are constant.

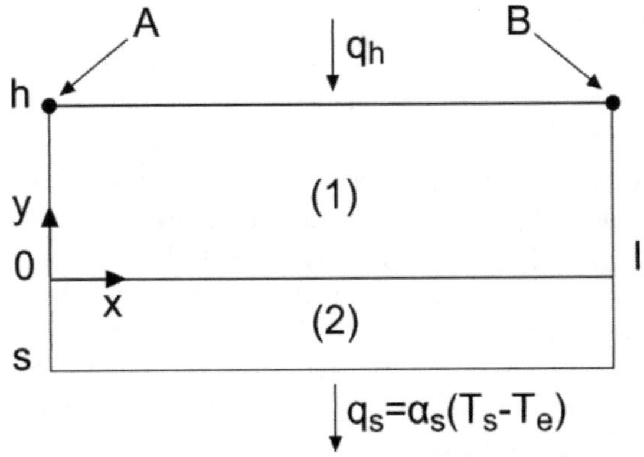

Figure 1. Computational domain: (1) – anisotropic thermoelement, (2) – substrate.

Further we consider two versions of the heat sensor model. In the first version (full model) it takes into account the anisotropy of the thermoelement material properties and the temperature $T(x,y)$ and the potential $\varphi(x,y)$ are found by solving the system of equations [2]:

$$C\rho\frac{\partial T}{\partial t} = \mathrm{div}\mathbf{q}$$

$$\mathrm{div}\mathbf{j} = 0$$

where $\mathbf{q} = -\lambda\nabla T$, $\mathbf{j} = -\sigma\nabla\zeta - \sigma\alpha\nabla T$ – heat flux and current density, T – temperature, ζ – electrochemical potential, λ, σ, α – tensor of thermal conductivity, electrical conductivity and Seebeck coefficient. The numerical solution of this the system is carried out in the COMSOL Multiphysics. The voltage $\Delta\varphi$ is determined between A and B points located on the working surface of the thermoelement $y = h$. The experimental data obtained in a stationary calibration of gradient heat flux sensor (GHFS) [3] were used to verify the model of the heat sensor. Sensor was a battery of 10 connected in series bismuth thermoelements of length $l = 2.2\,mm$ and thickness $h = 0.2\,mm$ mounted on a mica substrate. Heat flux $q_h = 100\ kW/m^2$ passed through the working surface of the sensor. The output signal of the sensor was $U = 6.8\,mV$. According to the calculation results corresponding to the conditions of this experiment, the voltage between points A and B was $\Delta\varphi_{AB} = 0.43\,mV$, and the full voltage of the battery was $\Delta\varphi = 4.3\,mV$. Difference between the calculation results and the experimental data does not exceed 35%.

In the second version of the model (simplified model) it takes into account only the anisotropy of the Seebeck coefficient while thermal conductivity of the material is isotropic ($\lambda = \lambda_{yy}$). The temperature distribution $T = T(y)$ is obtained from the solution of the one-dimensional heat transfer equation, and the voltage between points A and B is calculated by Thomson formula $\Delta\varphi_T = \alpha_{12}\Delta T\frac{l}{h}$, i.e. considered only transverse component of the thermal EMF [1].

3. Calculation parameters

In further calculations we consider the typical bismuth thermoelements of length $l = 2, 5, 10\ mm$ and thickness $h = 0.2\ mm$ $(l/h = 10, 25, 50)$ which are located on substrates of mica and copper (thermal diffusivity is $a = \frac{\lambda}{C\rho} \sim 10^{-7}\,m^2/s$ and $a \sim 10^{-4}\,m^2/s$ respectively). Thermal conductivity is equal $\lambda = \begin{bmatrix} 7{,}6 & -1{,}2 \\ -1{,}2 & 8{,}3 \end{bmatrix}, W/m\cdot K$, electrical conductivity is equal $\sigma = \begin{bmatrix} 7{,}9 & -0{,}9 \\ -0{,}9 & 8{,}5 \end{bmatrix} \cdot 10^5, 1/Ohm\cdot m$, Seebeck coefficient is equal $\alpha = \begin{bmatrix} -72{,}8 & 23{,}9 \\ 23{,}9 & -87{,}2 \end{bmatrix} \cdot 10^{-6}, V/K$.

IOP Conf. Series: Journal of Physics: Conf. Series **929** (2017) 012104 doi:10.1088/1742-6596/929/1/012104

Heat flux $q_h = 100\ kW/m^2$ passes through the working surface of the sensor $y = h$. At initial state temperature of anisotropic thermoelement and substrate is $T = T_e$. Termination condition of the calculation is the equality of heat fluxes $q_s = q_h$. The difference between voltage $\Delta\varphi$ calculated using the full model and $\Delta\varphi_T$ obtained using the simplified model allows us to evaluate the applicability of the one-dimensional model for calculating the required temperature difference $\Delta T = T_h - T_0$.

4. Calculation results

Fig. 2 shows the voltages $\Delta\varphi$ and $\Delta\varphi_T$ for two ranges of time. At the initial stage of heating the temperature of thermoelement back surface T_0 remains unchanged. For the considered thermoelements that time amounts to $\approx 0.5\ ms$ [8]. During which time the results of the calculation for both models are close enough (figures (a) and (b)). That means at this stage the influence of thermal conductivity anisotropy is negligible and temperature distribution is almost one-dimensional. The difference between the curves in the case of a short thermoelement is explained by the comparable contribution of the longitudinal and transverse components. In the case of a long thermoelement, the contribution of only a transverse component is decisive.

On further heating the thermal properties of the substrate material is beginning to affect to the temperature distribution in the thermoelement and to the thermal EMF. In the case of mica substrate the temperature field becomes a two-dimensional, and the longitudinal component of thermal EMF significantly increases. In the case of copper substrate the temperature distribution remains close to a one-dimensional. For short thermoelements $\Delta\varphi$ differs from $\Delta\varphi_T$ for both variants of the substrate material, and in the case of a long thermoelements the voltage curves practically coincide. At that stage the voltage of thermoelements with copper substrate is greater for all aspect ratios l/h.

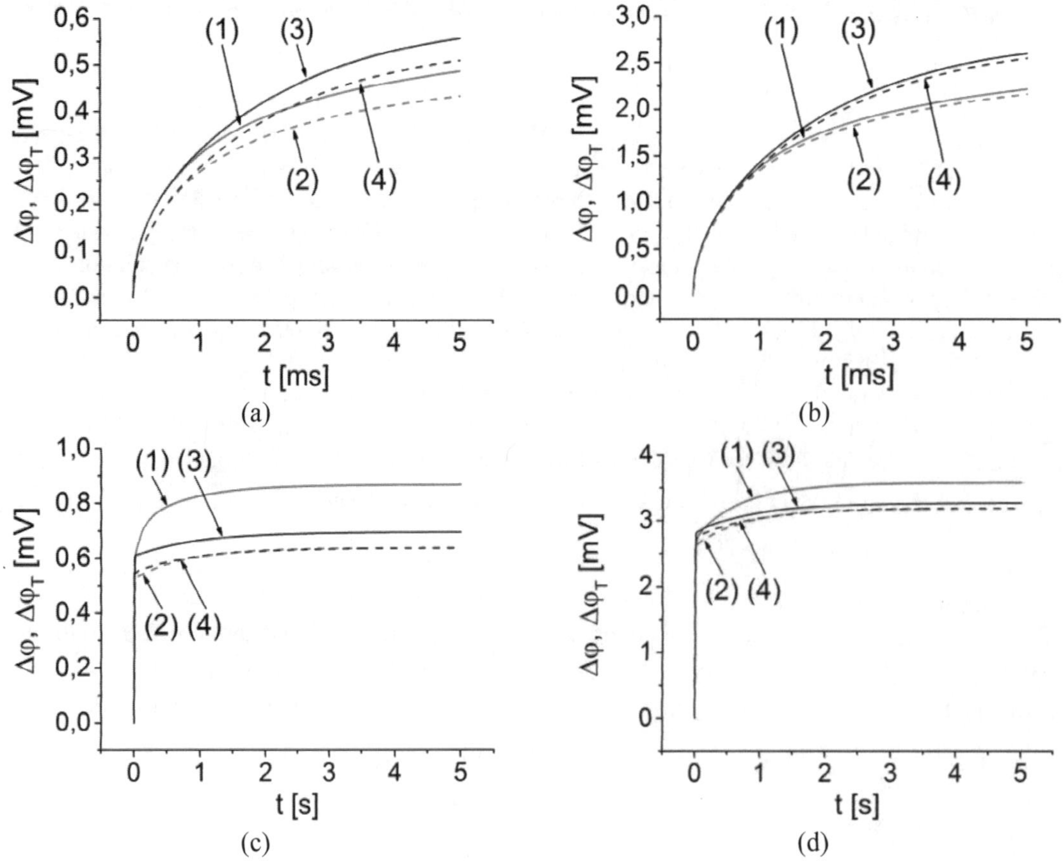

Figure 2. The voltage at the initial stage of heating (a, b) close to stationary thermal mode (c, d) for thermoelements with aspect ratio $l/h = 10$ (a, c) and $l/h = 50$ (b, d). Curves (1, 2) correspond to the mica substrate, and copper substrate (3, 4). Curves (1, 3) are calculated using the full model and simplified model (2, 4).

The influence of substrate thermal properties to the voltage is maintained at the steady-state thermal mode (figures (c) and (d)). In the case of mica substrate $\Delta\varphi$ (curves 1 and 3) and $\Delta\varphi_T$ (curves 2 and 4) are significantly different, but this difference is less for copper substrates. With the increase of the length of thermoelement, this difference decreases. It should also be noted a shorter time of the sensor signal output from the copper substrate to the quasi-stationary value.

Fig. 3 shows the ratio $\Delta\varphi_T/\Delta\varphi$ on a sufficiently large time when the thermal mode close to stationary. In the case of mica substrates, using of a simplified model causes to significant errors even for a fairly long thermoelements. For copper substrates, this approach can significantly improve the accuracy in the case of short thermoelements.

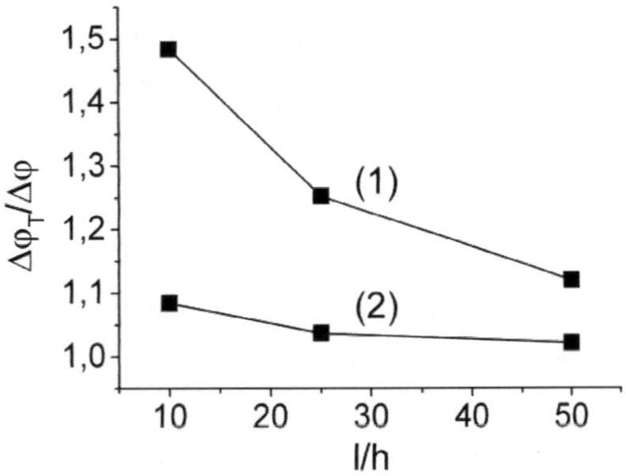

Figure 3. The ratio of the voltages, calculated according to the simplified model, to the value obtained for the full model for the different aspect ratios of the thermoelement l/h at stationary thermal mode. Curve 1 corresponds to the mica substrate and a curve 2 corresponds to the copper substrate.

Fig. 4 shows the temperature and the electric potential distribution in the case of thermoelement with aspect ratio $l/h = 10$ at stationary thermal mode. For the same heat flux q_h passes through the working surface of the sensor the temperature distribution is different. In the case of mica substrate, the isotherms have practically same slope relative to the working surface defined by the thermal conductivity anisotropy. The copper substrate significantly influences the temperature distribution in the thermoelement, and it becomes close to one-dimensional. In this case, the influence of the thermal conductivity anisotropy affects only near ends of thermoelement.

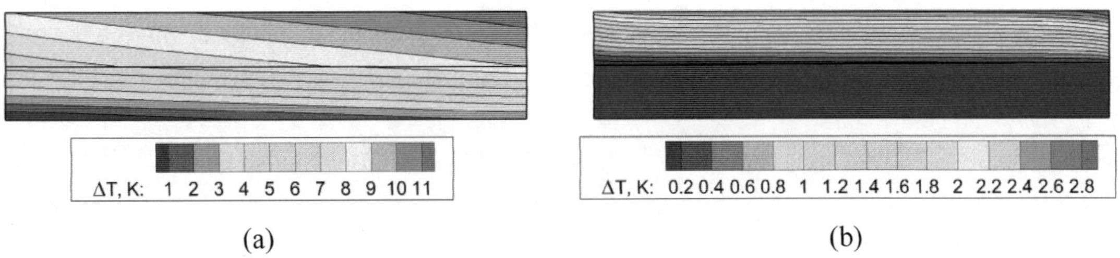

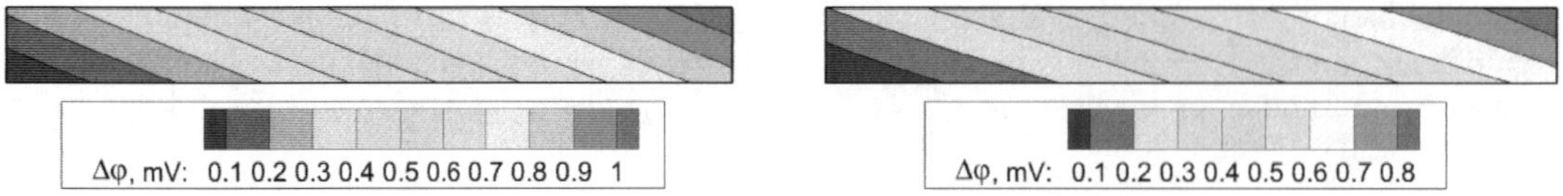

(c) (d)

Figure 4. Temperature in anisotropic thermoelement and substrate (a, b); electric potential in thermoelement (c, d) in the case of mica substrate (a, c) and copper substrate (b, d).

Conclusion

The studies show that at the initial stage of heating the simplified model allows to calculate the required temperature difference $\Delta T = T_h - T_0$ based on the electrical signal of heat sensor. When measuring pulsed heat fluxes with duration of $\sim 1\,ms$ it is better to use sensors with long thermoelements $l/h \approx 50$ as it increases the maximum time of applicability of the simplified model. Using the substrate with high thermal diffusivity (e.g. copper) allows to achieve close to one-dimensional temperature distribution in the thermoelement, and to apply the simplified model during the whole range of times. Also it increases the electrical signal of heat sensor at the same heat flux. The applicability of the simplified model means that there is geometric scalability effect in calculating the value of the sensor signal. It simplifies the preliminary calibration procedure. The calibration procedure should be carried out for the sensor with unit area, and then using a scale factor.

References

[1] Snarskii A.A., Pal'ti A.M., Ashcheulov A.A. Semiconductors (1997) 31: 1001.
[2] Rowe D.M. (ed.) Thermoelectrics Handbook: Macro to Nano (CRC, 2005).
[3] Sapozhnikov S.Z., Mityakov V.Y. & Mityakov A.V. High Temperature (2004) 42: 629.
[4] Mityakov A.V., Sapozhnikov S.Z. et al. Sensors and Actuators A: Physical (2012) 176: 1.
[5] Bobashev S.V., Mende N.P., Popov P.A. et al. Tech. Phys. Lett. (2009) 35: 214.
[6] Bobashev S.V., Golovachev Y.P., Kurbatov G.A. et al. Tech. Phys. (2009) 54: 33.
[7] Bobashev S.V., Mende N.P., Popov P.A. et al. Tech. Phys. (2010) 55: 1760.
[8] Reznikov B.I., Sakharov V.A. & Shteinberg A.S. Tech. Phys. Lett. (2008) 34: 193.

Photoionization microplasma sensor

A S Mustafaev[1], I V Rastvorova[1], S S Podenko[1] and A B Tsyganov[2]

[1] Saint-Petersburg Mining University, 2, 21-st Line, St. Petersburg, 199106 Russia
[2] Spectrum-Micro L.L.C., 29 Ligovskii pr., St. Petersburg, 191036 Russia

julia272727@mail.ru

Abstract. New developments in the physics of plasma are presented, specifically, research of completely new method of atoms' and molecules' detection in gaseous phase – collisional electron spectroscopy. As a result, the microplasma sensor for quality and quantity analysis of the gaseous mixture was created. It works in the discharge afterglow mode using He as a buffer gas. In addition, the modification of the sensor using resonance photon photoionization was developed. This consideration gives the opportunity for wide practical appliance as an individual gas analyzer for industrial and medical purposes.

1. Research method and its actuality

Method CES (collisional electron spectroscopy) is patented in Russia 2217739; the USA 7,309,992; Japan 4408810; China ZL200380106502.2; Germany 503 14 126.7; France EP1557667 and Great Britain EP1557667. This method allows to analyze gas mixtures at a high pressure up to atmospheric one without vacuum, using electron spectroscopy [1]. It is based on the fact, that only the value of characteristic electrons' energy needs to be measured, not the momentum vector value [2].

1.1. Review of traditional spectroscopy methods

Well known that to identify atoms and molecules in electron spectroscopy measurement of electrons energy released during ionization of atomic or molecular impurities A by particles B^* with definite energy E_p (photons, quick electrons, excited atoms, etc.) are taken.

$$A + B^* \rightarrow A^+ + B + e \tag{1}$$

After the collision the kinetic moment of appeared characteristic electrons is determined using various energy analyzers. Based on these results the energy value is calculated (m-particles' mass, v-particles' speed, p-particles kinetic moment).

$$E = \frac{mv^2}{2} = \frac{p^2}{2m} \tag{2}$$

Ionization potential E_i of atoms or molecules A is found by measured energy E_e (electrons' specter) and definite energy E_p of particles B^* and then these atoms or molecules are identified.

$$E_i = E_p - E_e \tag{3}$$

Consequently, to determine electron energy in the traditional spectroscopy methods it is necessary to use high vacuum and bulky energy analyzers to disperse electrons in 'kinetic moment - coordinate' space, for example, cylindrical or sphere mirror, electrostatic or magnetic lenses, time-of-flight tube and etc. Here, electron is flying out from the starting point with the definite momentum vector, then diverges to the analyzer field, and gets to the calculated according to the motion equations detector's

IOP Conf. Series: Journal of Physics: Conf. Series **929** (2017) 012105 doi:10.1088/1742-6596/929/1/012105

point. In reality, mentioned energy analyzers determine kinetic moment of the electron, based on which the starting kinetic energy is calculated.

Any collision of the identifying charged particle with atoms or molecules of residual gas leads to particle scattering and, so, to 'leaving the game' or wrong detection in unpredictable receiver point with the loss of resolution capability.

Additional technical and methodical problem is the sample's (which is under atmospheric pressure) atoms or molecules insertion into analyzer's space, where high vacuum is required.

This leads to the increase in mass, size, energy consumption and technical complexity of the traditional ways of gaseous mixtures' composition determination, which doesn't provide the solution for various important analytical challenges that modern science is facing.

1.2. Collisional electron spectroscopy

The basis for the proposed detection method is the fact that during one elastic collision of the electron with gas particles, it loses only small part of its initial kinetic energy. That's why, if during the movement toward the detector, electron goes under 100 elastic collisions, it completely 'loses' initial trajectory of its kinetic moment (like during Brownian movement), but at the same time its energy distortion will make relatively small part <1%.

This peculiarity of detection method allows to refuse from using electron energy analyzers with high vacuum and bulky construction for space electron dispersion, depending on its kinetic moment, and replace it with two plane parallel electrodes with pulse-actuated plasma between them. It significantly decreases the size and mass of the sensor, and as a consequence, its cost. Because of this, the size of the sensor CES is about 10*10*1 mm.

Determination of electron's energy in method CES is done by registration of the second derivative of electrons' current on the analyzing electrode, depending on counter potential.

Characteristic energy E_e of the electrons, which are formed during the ionization of the additives, according to (3) can lie in the range from 0 up to the energy of actuating particle, depending on the ionization potential of the additive. It is possible to assume that the highest limit for applicable energies of actuating particles should be equal to the ionization energy of buffer gas atom, because in case of use of actuating particles with higher energy, ionization of the main gas will start, which will cause troubles in determination of the additives. Therefore, measuring current-voltage characteristics in the range of applied voltage from 0 to the ionization potential of the main gas atom, it is possible to find the electron distribution speed function $F(v)$ (and also relative electron energy distribution $f(E_e)$) and the number of the electrons, which occur because of the particular additive ionization, using Druyvestein's formula for plane probe.

$$F\left(\sqrt{\frac{2eU}{m}}\right) = \frac{4m}{e^2 \cdot S \cdot N_e} \cdot U \cdot \frac{d^2 I_e}{dU^2} \qquad (4)$$

Here e – electron charge; U – voltage, applied to the electrodes; S – electrode area; I_e – electron current on plane cathode; N_e – electron concentration

Therefore, the second derivative of current-voltage measurement gives a curve, which contains peaks, each of which corresponds to the group of characteristic electrons produced by ionization of a particular additive. Energy of these electrons E_e and score of relative peaks in the range of applied voltage are determined by formula (3). Mentioned two-phased derivation can be done using different methods, including straight numerical derivation of the current-voltage measurement, using, if necessary, various ways of data smoothing. Different electronic circuits can be applied, which use modulation of applied voltage U with small ΔU with frequency ω and current registration on frequency ω for 1st derivative and 2ω for the 2nd.

2. Microplasma sensor CES

Micriplasma sensor is developed based on this method. It operates in the discharge afterglow mode. Electron energy spectrums of *He* as a buffer gas under different pressures and with various electrodes' distances are gotten [4].

Equipotentiality of the ionization camera's space, where ionization collisions occur, is provided by creation of afterglow plasma in the ionization camera between two plane parallel electrodes with the help of impulse source of energy (figure 1). During the afterglow of impulse discharge, after cooling of charged particles and when electro-neutral conditions are attained by plasma, almost equipotential space is leveling off in the camera between anode and cathode. At the same time, energy of the electrons, which occur as a result of additives' ionization, doesn't depend on coordinates of the point, where they occur.

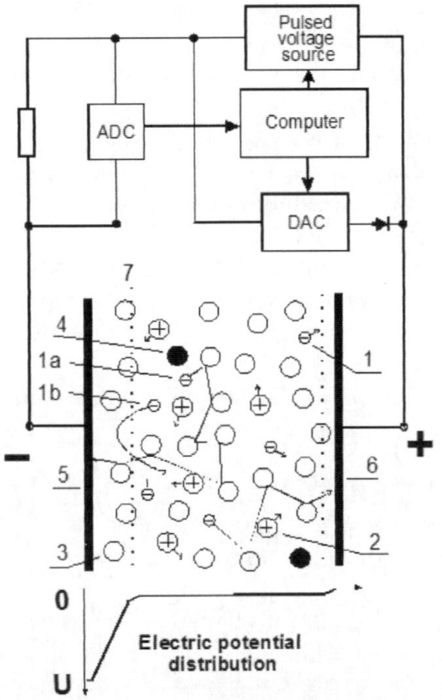

Figure 1. Scheme of microplasma CES-detector
1, 1a, 1b – characteristic electrons, 2 – additives' ions, 3 – main gas atoms, 4 – metastable atoms, 5,6 – plane parallel cathode and anode

The relation between concentrations of main gas and analyzing additives should be so, that major part of occurring metastable atoms is composed of buffer inert gas atoms. At the end of the discharge, quick 'decay lag' of all excited atoms will take place, and the afterglow plasma will mostly consist of metastable atoms of buffer gas with definite energy E_p. Helium atoms' energies are 19.8 and 20.6 eV, which is enough to ionize almost all known atoms and molecules.

To measure the spectrum of characteristic electrons probing pulse is applied on anode with dwell time after charge impulse. The voltage of probing pulse gradually changes with each charge impulse. The current on the cathode during probing pulse is measured, and as a result current-voltage characteristic is plotted in the range from 0 V to ionization limit of the main gas. Then the second derivative of *I-U* characteristic is taken and the spectrum of characteristic electrons, which occur in discharge afterglow as a result of analyzing atoms collision with metastable *He* atoms, is calculated. Ionization potentials of analyzing atoms is determined by electron energy spectrum and definite energy of ionization metastable atoms. Analyzing atoms are identified based on their ionization potential.

3. Photoionization microplasma sensor

Absolutely new construction of the VUV-photoionization CES detector was developed to use it as a portative gas analyzer for continuous personal biomedical diagnostics [5]. This way energy spectrum of electrons, occurred in the ionization with resonance photons, which wavelength is situated in the vacuum ultraviolet (VUV), is measured. Nowadays, micro plasma source of such photons on resonant line of *Kr* with energy of 10,6 eV is developed.

Electrons' energies analyzing is done in simple gas-filled block with additional grid between anode and cathode, where additives ionization is happening with a use of outside VUV-emission. In this case, full energy of the electrons, occurred in different parts of the grid, is determined by its characteristic kinetic energy. Gradually changing negative voltage is applied to the cathode, electron current is measured and *I-U* characteristic is plotted.

VUV-photoionization detectors CES have a big advantage in comparison with to all other detectors, where electron impact ionization is used. The reason for this is the fact, that only analyzing additives are ionized and detected using VUV-emission, main air components (oxygen, nitrogen) remain neutral, because their ionization potential is higher than VUV-photons' energy. This helps to decrease background signal, increase sensitivity and resolution capability.

Sensitivity limit of CES detectors is assessed by photoionization cross-section for detecting additives and VUV-photons' flow. According to the estimation, during the emission flow in order of 10^{14} photons/sec through photoionization zone and with sensitivity of recording equipment of current on the level of 10^{-12} A, the detection limit for molecules, which have a cross-section on resonant line of *Kr* or *Ar* is few digits in order of 10^{-19} cm^2, is about 0.1 ppm [6]. Work to further increase the sensitivity of the CES detectors by higher-aperture detector's construction and increased sensitivity of electrometer amplifiers, is in progress. The watt consumption may comprise less than 1W.

Numerical tests were done to assess the accuracy of the detection method [7].

The experimental model of photoionization sensor CES was developed and constructed (figure 2) [8].

Figure 2. The experimental model of photoionization sensor CES

Supposed photoionization sensor CES characteristics are shown in the Table 1.

Table 1. Supposed photoionization sensor CES characteristics

	Characteristics	Value	UM
1	Limit of the small additives detection	1,00	ppm
2	VUV-quantum energy	10,6	eV
3	Sensitivity of photocurrent, no less than	1,00	pA
4	Resolution on photoelectrons' energy	0,10	eV
5	Overall volume	100,00	cm3
6	Watt consumption, no more than	1,00	Watt
7	Maximum working temperature	50,00	°C
8	Minimum working temperature	-30,00	°C

Sensors, based on CES method, will be widely popular in industrial field, transportation, safety and ecological control systems, because of its small size and low cost. Nevertheless, it is possible to develop portative device for continuous personal biomedical diagnostics, because of its wide range of detectable molecules and high accuracy. Such sensor can become the base of the future preventive medicine [9].

This work was supported by the program Umnic-2015 (grant №7007SD/2015).

References

[1] Mustafaev A, Tsyganov A. 2011 Micro-plasma detector based on Collisional Electron Spectroscopy (CES) method for gas mixtures diagnostics *Contr. paper of 30th ICPIG* (Belfast, Northen Ireland, UK: Queen's University) B6

[2] Mustafaev A, Panasyuk G, Tsyganov A. 2013 Microplasma analyzer based on collisional electron spectroscopy (CES) *Contr. paper of 40th European Physics Society Conference on Plasma Physics* vol 4 (Helsinki, Finland) p 12-16

[3] Maksimova E, Rastvorova I 2014 The creation of "smart" microplasma devices for the composition detection and for of gas mixture's parameters control // *Materials of 52 International scientific student conference "Student and scientific-technical progress"* Physics of continuous medium (Novosibirsk: Novosibirsk state University) p 10

[4] Mustafaev A, Kudryavtsev A, Tsyganov et al 2012 Electron energy spectra in helium observed in a microplasma CES detector. *J. Technical physics* vol 10 p 1-6

[5] Mustafaev A, Macsimova E, Rastvorova I et al 2015 Smart Microplasma Analyzers for Continuous Personal Biomedical Diagnostics *The Proceedings of 2nd International Scientific Symposium "Sense. Enable. SPITSE."* (St. Petersburg: St. Petersburg Electrotechnical University "LETI") p 210-212

[6] Mustafaev A, Tsyganov A et al. 2013 On accuracy of CES microplasma and VUV photoionization spectra determination *Contr. paper of 55th Meeting of the APS Division of Plasma Physics* (Denver, CO, USA) p 387-392

[7] Rastvorova I, Khobnya K, Podenko S and Mustafaev A 2016 Microplasma photoionization sensor *Modern educational technology in the teaching of natural sciences and the humanities: Proceedings of the III International Scientific Conference* (St. Petersburg: National Mineral Resource University "Mining University") p 773-779

[8] Mustafaev A, Rastvorova I, Khobnya K and Podenko S 2016 Photoionization sensors for non-invasive medical diagnostics *69th Annual Gaseous Electronics Conference* vol 61 N 9 (Bochum, Germany) Abstract: MW6.00150

[9] Mustafaev A, Rastvorova I, Khobnya K and Podenko S 2016 Photoionization sensor CES for non-invasive medical diagnostics *58th Annual Meeting of the APS Division of Plasma Physics* vol 61 N18 (San Jose, CA) Abstract: YP10.00071

IOP Conf. Series: Journal of Physics: Conf. Series **929** (2017) 012106 doi:10.1088/1742-6596/929/1/012106

The axisymmetric oscillations of a cylindrical bubble in a liquid bounded volume with free deformable interface

A A Alabuzhev[1,2] and M I Kaysina[2]
[1]Institute of Continuous Media Mechanics UB RAS, Perm 614013, Russia
[2]Perm State University, Perm 614990, Russia

E-mail: alabuzhev@icmm.ru

Abstract. The eigen axisymmetric oscillations of a cylindrical gas bubble surrounded by an incompressible fluid with free deformable interface are considered. The bubble has an equilibrium cylindrical shape and is bounded axially by two parallel solid surfaces. Dynamics of contact lines is taken into account by an effective boundary condition: velocity of the contact line is assumed to be proportional to contact angle deviation from the equilibrium value. The equilibrium contact angle is right. Eigen frequency decreases with liquid outer free surface radius decreasing and increases with the radius-to-height ratio increasing. It's found that the eigen frequency can vanish in some wetting parameter interval for the volume mode of natural oscillations (which describes the radial compression of the bubble). The length of this interval increases with increasing ratio of the equilibrium bubble radius to the height. The eigen frequencies of other modes decrease with increasing Hocking's constant. The lowest natural frequency is observed for the freely sliding bubble.

1. Introduction

Motion of triple contact line dynamics have been examined in various formulations for many physical problems [1-4]. For very fast relaxation processes at the triple line the Hocking condition [5] is widely used [6-13]. This condition assumes a linear relationship between the velocity of the contact line motion and the contact angle (for the case of right equilibrium contact angle)

$$\frac{\partial \zeta^*}{\partial t^*} = \Lambda \vec{k} \cdot \vec{\nabla} \zeta^* , \tag{1}$$

where ζ^* is the deviation of the interface from the equilibrium position, $\vec{k}$ is the external normal to the solid surface, Λ is a phenomenological constant (the so-called wetting parameter or Hocking parameter) having the dimension of the velocity. There are two important limit of the boundary condition(1): (a) $\zeta^* = 0$ – the requirement of the fixed contact line (pinned-end edge condition), (b) $\vec{k} \cdot \vec{\nabla} \zeta^* = 0$ – the constant contact angle.

In the present article, we consider the axisymmetric oscillations of cylindrical bubble which surrounded by a liquid with free deformable interface. We apply condition (1) and continue the study of the cases with non-deformable interface [14]. The translation oscillations of this bubble were researched in [15]. The axisymmetric natural and forced oscillations of a compressible hemispherical bubble on a solid substrate are studied in [10]. The contact line dynamics is taken into account by application of the Hocking condition (1). Authors of [12] considered a sessile hemispherical bubble sitting on the transversally oscillating bottom of a deep liquid layer and focus on the interplay of the compressibility of the bubble and the contact angle hysteresis (by analogy [11] for hemispherical drop

[1] To whom any correspondence should be addressed.

of incompressible fluid). Authors of the above articles consider isothermal problem with assumption of adiabatic motionless bubble.

2. Problem formulation

By analogy [15] a gaseous bubble surrounded by a liquid with a free external surface are consider (see figure 1). The bubble is bounded by two parallel solid plates which separated by a distance h^*. In equilibrium, the bubble and fluid volume have circle cylindrical with a radius r_0^* and R_0^*, respectively, and contact angle is 90^0.

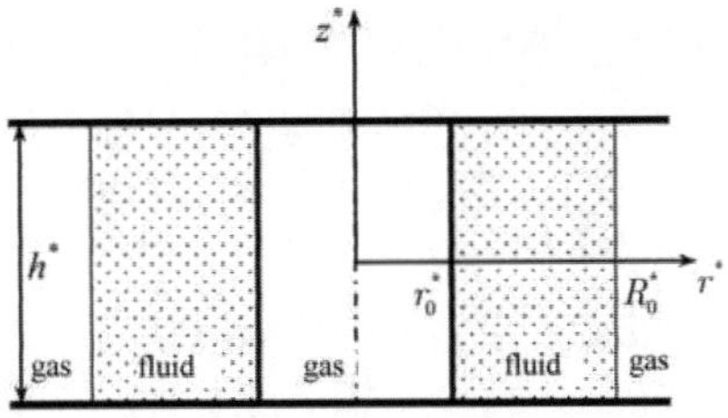

Figure 1. Problem geometry.

The oscillations amplitude A^* is small compared to the equilibrium radius r_0^*. The fluid motion is assumed to be incompressible: $\omega^* r_0^* << c$, where ω^* is fundamental oscillation frequency, c is the sound velocity. However ω^* is large enough for the viscosity could be ignored: $\delta = \sqrt{\nu/\omega^*} << r_0^*$ where δ is the boundary-layer thickness.

Owing to the problem symmetry, it is convenient to introduce cylindrical coordinates r^*, α, z^*. The azimuthal angle α is reckoned from the x axis, but α will be neglected the further. Let the lateral surface of the bubble and the external surface of liquid volume be described by the following equations, respectively

$$r^* = r_0^* + \zeta^*\left(z^*, t^*\right), \; r^* = R_0^* + \xi^*\left(z^*, t^*\right).$$

Following [10,14], we use $\sqrt{\rho_e^* r_0^{*3}/\sigma^*}$, r_0^*, h^*, A^*, $A^*\sigma^*/r_0^{*2}$, $A^*\sqrt{\sigma^*/\rho_e^* r_0^*}$ as the scales for the time, length, height, deviation of bubble surface and free surface from its equilibrium position, pressure, and velocity potential, respectively (σ^* is the surface tension and ρ_e^* is the liquid density). Thus, the dimensionless boundary value problem is determined by (intermediate steps can be found in [10,14])

$$p_e = -\varphi_t, \; \Delta\varphi = 0, \; p_i = -2n_p P_g^* r_0^* / \sigma^* \langle\zeta\rangle \equiv -P_0\langle\zeta\rangle, \tag{2}$$

$$\Delta = \frac{1}{r}\frac{\partial}{\partial r}\left(r\frac{\partial}{\partial r}\right) + b^2\frac{\partial^2}{\partial z^2},$$

$$r = 1: \; \zeta_t = \varphi_r, \; [p] = \zeta + b^2\zeta_{zz}, \tag{3}$$

$$r = R_0: \; \xi_t = \varphi_r, \; [p] = \xi + b^2\xi_{zz}, \tag{4}$$

$$z = \pm\frac{1}{2}: \; \varphi_z = 0, \tag{5}$$

$$r = 1, \; z = \pm\frac{1}{2}: \; \zeta_t = \pm\lambda\zeta_z, \tag{6}$$

$$r = R_0, \; z = \pm\frac{1}{2}: \; \xi_t = \pm\lambda\xi_z, \tag{7}$$

where p_e is the liquid pressure, φ is potential of liquid velocity, p_i is the gas pressure in the bubble, n_p is polytropic (e.g., adiabatic) exponent, P_g^* is dimension gas pressure in the bubble. The boundary-value problem (2)–(6) involves five parameters: the aspect ratio, the radius of free surface, the wetting parameter, the frequency and amplitude

$$b = r_0^*/h^*,\ R_0 = R_0^*/r_0^*,\ \lambda = \Lambda^*\Big/\sqrt{\rho_e^* r_0^*/\sigma^*},\ \omega = \omega^*\sqrt{\rho_e^* r_0^{*3}/\sigma^*},\ \varepsilon = A^*/r_0^*.$$

3. Natural axisymmetric oscillations of the bubble

In order to investigate the problem, it is convenient to begin with a consideration of the natural oscillations of a cylindrical bubble. By the evenness of the natural oscillation modes is meant the evenness of the functions under a change of sign of the axial coordinate z. In view of translational symmetry, the solution of the boundary value problem (2)–(7) without external force is written as

$$\varphi(r,z,t) = i\Bigg(\sum_{n=0}^{\infty}\left(a_n^{(1)}R_{1n}^i(r) + b_n^{(1)}R_{1mn}^e(r)\right)\sin\left((2n+1)\pi z\right) + \\ + \sum_{n=0}^{\infty}\left(a_n^{(2)}R_{2n}^i(r) + b_n^{(2)}R_{2n}^e(r)\right)\cos\left(2\pi nz\right)\Bigg)e^{i\omega t}, \tag{8}$$

$$\zeta(z,t) = \left(d_1\sin\left(\frac{z}{b}\right) + d_2\cos\left(\frac{z}{b}\right) + \sum_{n=0}^{\infty}h_n\sin\left((2n+1)\pi z\right) + \sum_{n=0}^{\infty}c_n\cos\left(2n\pi z\right)\right)e^{i\omega t}, \tag{9}$$

$$\xi(z,t) = \left(D_1\sin\left(\frac{z}{b}\right) + D_2\cos\left(\frac{z}{b}\right) + \sum_{n=0}^{\infty}H_n\sin\left((2n+1)\pi z\right) + \sum_{n=0}^{\infty}C_n\cos\left(2n\pi z\right)\right)e^{i\omega t}, \tag{10}$$

where Ω is eigen frequency, $R_{1n}^i(r) = \mathrm{I}_0((2n+1)\pi br)$, $R_{1n}^e(r) = \mathrm{K}_0((2n+1)\pi br)$, $R_{20}^i(r) = const$, $R_{2n}^i(r) = \mathrm{I}_0(2\pi nbr)$, $R_{2n}^e(r) = r^{-1}$, $R_{2n}^e(r) = \mathrm{K}_0(2\pi nbr)$, I_0 and K_0 are modified Bessel functions of the zero order.

Substituting solutions (8)–(10) into (2)–(7), we obtain a spectral-amplitude problem which eigenvalues are the values of the natural oscillation frequency Ω. From the solution of this problem it follows that the eigenvalues are found from the equations:

even modes

$$\frac{-1}{S_0+\omega^2}\left[\frac{f_0^2(P_0-1)}{R_0\left(S_0+\omega^2\right)}\frac{1}{\dfrac{\lambda}{i\omega b}\sin\dfrac{1}{2b} + \dfrac{f_0}{R_0\ln R_0\left(S_0+\omega^2\right)} + \displaystyle\sum_{n=1}^{\infty}\frac{f_n\Omega_{2n}^2(R_0)}{\omega^2-\Omega_{2n}^2(R_0)}} + f_0\left(\Omega_0^2(R_0)+\omega^2\right)\right] + \\ + \sum_{n=1}^{\infty}\frac{f_n\omega^2}{\omega^2-\Omega_{2n}^2(1)} + \cos\frac{1}{2b} = \frac{\lambda}{i\omega b}\sin\frac{1}{2b}, \tag{11}$$

$$\Omega_0^2(1) = \frac{P_0-1}{\ln R_0},\ \Omega_{2n}^2(1) = \frac{(4\pi^2n^2b^2-1)(R_{10n}'^e(R_0)R_{10n}'^i(1) - R_{10n}'^e(1)R_{10n}'^i(R_0))}{R_{10n}'^i(R_0)R_{10n}^e(1) - R_{10n}^i(1)R_{10n}'^e(R_0)},$$

$$\Omega_0^2(R_0) = \frac{P_1-1}{R_0\ln R_0},\ \Omega_{2n}^2(R_0) = \frac{(4\pi^2n^2b^2-1)(R_{10n}'^e(1)R_{10n}'^i(R_0) - R_{10n}'^e(R_0)R_{10n}'^i(1))}{R_{10n}'^e(1)R_{10n}^i(R_0) - R_{10n}^e(R_0)R_{10n}'^i(1)},$$

$$S_0 = \Omega_1^2(R_0) - \Omega_0^2(1),\ f_0 = 2b\sin\frac{1}{2b},\ f_n = \frac{4b}{4\pi^2n^2b^2-1}\sin\frac{1}{2b}.$$

Odd modes

$$\sum_{n=0}^{\infty}\frac{4b}{(2n+1)^2\pi^2b^2-1}\frac{\omega^2}{\omega^2-\Omega_{2n}^2(1)} - \mathrm{tg}\left(\frac{1}{2b}\right) = \frac{\lambda}{i\omega b}, \tag{12}$$

$$\sum_{n=0}^{\infty}\frac{4b}{(2n+1)^2\pi^2b^2-1}\cdot\frac{\Omega_{2n}^2(R_0)}{\omega^2-\Omega_{2n}^2(R_0)} = -\frac{\lambda}{i\omega b}, \tag{13}$$

$$\Omega_{1n}^2(1) = \frac{((2n+1)^2\pi^2b^2-1)(R_{10n}'^e(R_0)R_{10n}'^i(1) - R_{10n}'^e(1)R_{10n}'^i(R_0))}{R_{10n}^i(1)R_{10n}'^e(R_0) - R_{10n}'^i(R_0)R_{10n}^e(1)},$$

$$\Omega_{1n}^2(R_0) = \frac{((2n+1)^2\pi^2b^2-1)(R_{10n}'^e(1)R_{10n}'^i(R_0) - R_{10n}'^e(R_0)R_{10n}'^i(1))}{R_{10n}'^e(1)R_{10n}^i(R_0) - R_{10n}^e(R_0)R_{10n}'^i(1)}.$$

Here $\Omega_0^2(1)$, ,$\Omega_0^2(R_0)$ $\Omega_{1n}^2(1)$, $\Omega_{1n}^2(R_0)$, $\Omega_{2n}^2(1)$, $\Omega_{2n}^2(R_0)$ are the natural oscillation frequencies of the freely moving contact line. The complex algebraic equations (11)–(13) have complex solutions (were solved numerically with the usage of the two-dimensional secant method), this leads to oscillation damping due to the dissipation on the contact line.

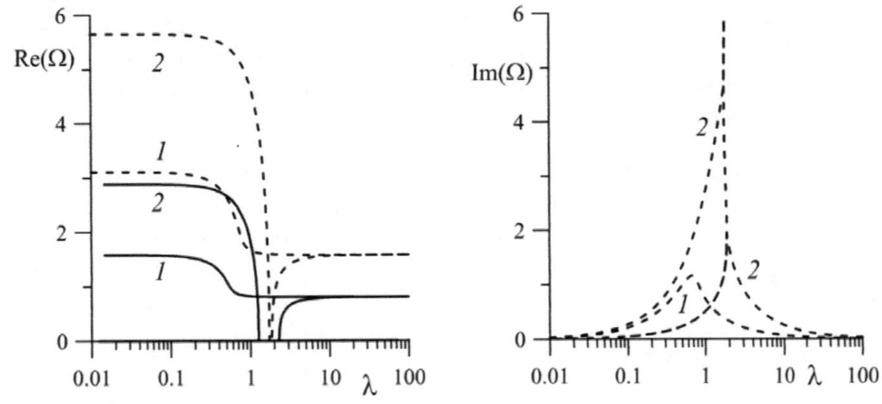

Figure 2. Frequency (a) and damping ratio (b) of volume natural oscillations vs wetting parameter λ for Ω_0 ($R_0 = 5$, $P_0 = 5$).
$b = 1$ – line 1, $b = 2$ – line 2, liquid – solid line, bubble – dashed line.

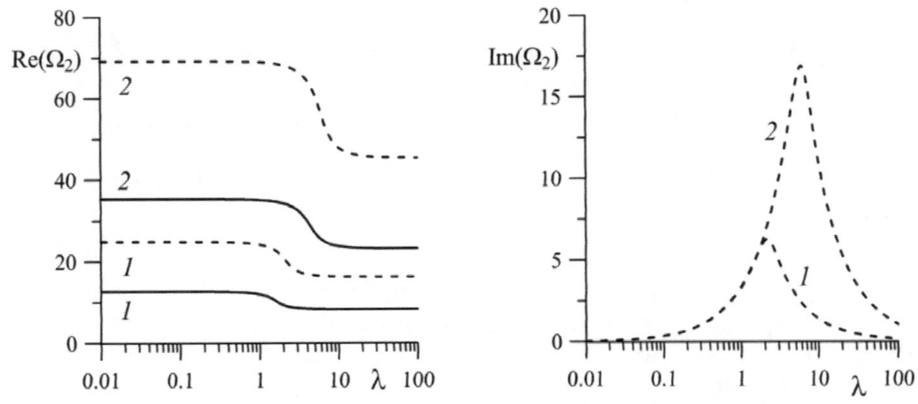

Figure 3. Frequency (a) and damping ratio (b) of shape natural oscillations vs wetting parameter λ for Ω_2 ($R_0 = 5$, $P_0 = 5$).
$b = 1$ – line 1, $b = 2$ – line 2, l liquid – solid line, bubble – dashed line.

Figures 2 and 3 show the real part of $\mathrm{Re}(\Omega)$ (oscillation frequency) and imaginary part $\mathrm{Im}(\Omega)$ (damping ratio) of the complex natural frequency Ω for the oscillation even modes Ω_0 (i.e., $k = 0$ is the wavenumber) and Ω_2. Even values of k correspond to the even modes (the solution of Eqs. (11)) and odd values of k correspond to the odd modes (the solution of Eqs. (12),(13)). The zero translational mode $k = 0$ describes volume (radial) oscillations. In a certain range of λ, the real part of the frequency $\mathrm{Re}(\Omega_0)$ can vanish, depending on the value of the geometrical parameter b (figure 2a). For small values of b, this range is absent. The vanishing of $\mathrm{Re}(\Omega_0)$ corresponds to the bifurcation of the branch of the increment $\mathrm{Im}(\Omega_0)$ (curve 2 in figure 2b). In this case dissipation is proportional to the length of the contact line, as it is the interaction of the contact line and the solid plate is a cause of dissipation. Therefore, growing of parameter b increases the length of the contact line at constant bubble volume, i.e. it increases the dissipation (also see [10,14]). Note, that the eigen frequencies of the bubble are higher than the eigen frequencies of the fluid volume. It is the result of the influence of the surface tension force on external interface.

For shape modes $k > 0$, the behavior resembles the case of the incompressible liquid. The frequency of shape oscillations decreases monotonically as λ increases (see figure 3a), the damping ratio has a maximum for a finite value of the capillary parameter and tends to zero as $\lambda \to 0$ and $\lambda \to \infty$ (figure 3b). However volume frequency can decreases monotonically as λ increases (see figure 4). It is possible when the complex frequencies of the two lowest modes become (see [10,14] for details).

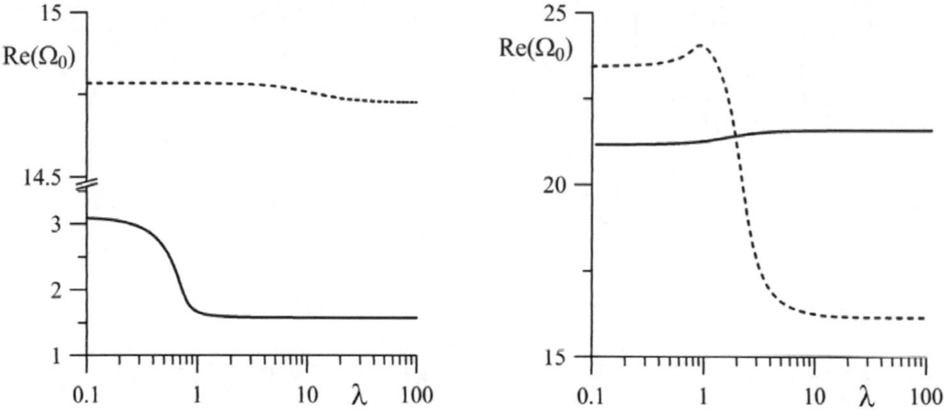

(a) (b)

Figure 4. Bubble frequency of volume natural oscillations vs wetting parameter λ for Ω_2 ($R_0 = 5$, $b = 1$).

(a) $P_0 = 5$ – solid line, $P_0 = 350$ – dashed; (b) $P_0 = 750$ – solid, $P_0 = 1000$ – dashed.

Figure 5 shows the real part of $\mathrm{Re}(\Omega)$ (oscillation frequency) and imaginary part $\mathrm{Im}(\Omega)$ (damping ratio) of the complex natural frequency Ω for the oscillation odd mode Ω_1. The real part of the frequency $\mathrm{Re}(\Omega_1)$ can vanish in a certain range of λ (figure 5a) also $\mathrm{Re}(\Omega_0)$ (figure 2a). But this range is absent for large values of b. The reason for the vanishing of the frequencies associated with such large dissipation (parameter λ is finite) that oscillations become impossible: dissipation is proportional to the area of the side surface for surface modes $k > 0$ of natural oscillations (than the surface is greater than the more energy is required for surface waves and the energy of bubble oscillations is proportional to the volume). Consequently, increase of geometrical parameter b is corresponds to a decrease side surface of the bubble at a constant bubble volume, i.e. dissipation is reducing for surface oscillations (see [14,16] for details).

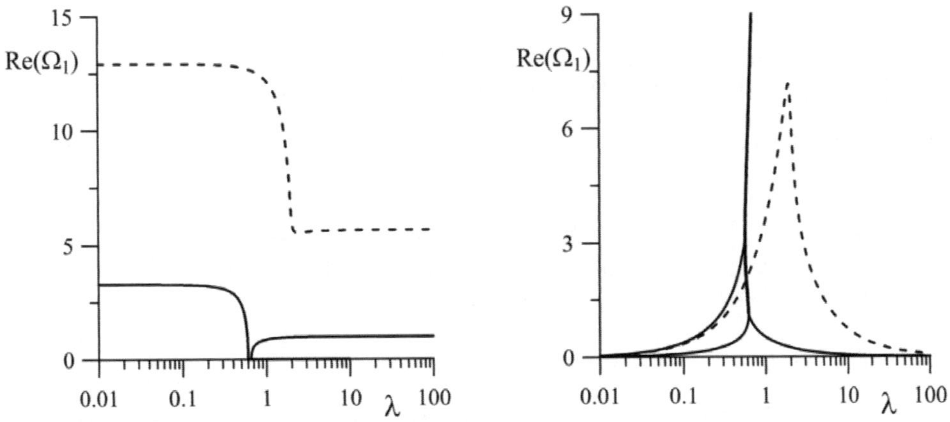

(a) (b)

Figure 5. Bubble frequency (a) and damping ratio (b) of shape natural oscillations vs wetting parameter λ for Ω_1 ($R_0 = 5$, $P_0 = 5$).
$b = 0.4$ – solid line, $b = 1$ – dashed line.

4. Conclusions

The eigen frequencies of the bubble are higher than the eigen frequencies of the fluid volume. Eigen frequency decreases with the decrease of free surface radius and increase with the bubble aspect ratio increasing. The radial frequency of the eigen oscillations increases at gas pressure P_0 growing. The shape frequencies are independent of the gas pressure inside the bubble.

The radial frequency of the eigen oscillations vanishes at a certain value of the Hocking parameter λ. This interval of values λ grows at increasing aspect ratio b. The first shape frequency also can vanish at some λ, but length of interval λ reduces at increasing b.

For $b < \pi^{-1}$ and a certain characteristic value of λ, the first shape frequency vanishes and the increment becomes negative, which corresponds to the occurrence of Rayleigh instability.

Acknowledgments
This work was supported by the Russian Foundation for Basic Research (project 14-01-96017-r-ural).

5. References

[1] Voinov O V 1976 *Fluid Dyn* **11** 714–721
[2] De Gennes P G 1985 *Rev. Mod. Phys.* **57** 827
[3] Bonn D, Eggers J, Indekeu J, Meunier J and Rolley E 2009 *Rev. Mod. Phys.* **81** 739–805
[4] Zhang L and Thiessen D B 2013 *J. Fluid Mech.* **719** 295–313
[5] Hocking L M 1987 *J. Fluid Mech.* **179** 253-66
[6] Borkar A and Tsamopoulus J 1991 *Phys. Fluids A* **3** 2866–74
[7] Perlin M, Schultz W W and Liu Z 2004 *Wave Motion* **40** 41–56
[8] Alabuzhev A A and Lyubimov D V 2005 *Fluid Dyn* **40** 183-192
[9] Alabuzhev A A and Lyubimov D V 2007 *J. Appl. Mech. Tech. Phys.* **48** 686–93
[10] Shklyaev S and Straube A V 2008 *Phys. Fluids* **20** 052102
[11] Fayzrakhmanova I S and Straube A V 2009 *Phys. Fluids* **21** 072104
[12] Fayzrakhmanova I S, Straube A V and Shklyaev S 2011 *Phys. Fluids* **23** 102105
[13] Alabuzhev A A and Lyubimov D V 2012 *J. Appl. Mech. Tech. Phys.* **53** 9–19
[14] Alabuzhev A A 2014 *Computational Continuum Mechanics* **7** 151–61 (in Russian)
[15] Alabuzhev A A and Kaysina M I 2016 *J. Phys.: Conf. Ser.* **681** 012043
[16] Alabuzhev A A 2016 *Computational Continuum Mechanics* **9** 316–30 (in Russian)

The oscillations of oblate drop under the influence of a alternating electric field

A A Alabuzhev[1,2] and M A Kashina[2]
[1]Institute of Continuous Media Mechanics UB RAS, Perm 614013, Russia
[2]Perm State University, Perm 614990, Russia

E-mail: alabuzhev@mail.ru

Abstract. The forced oscillations of incompressible fluid drop under the alternating electric field are considered. In equilibrium, the drop has the form of a rotational figure with arbitrary contact angle bounded axially parallel solid planes. The drop is surrounded by an incompressible fluid with another density. The external uniform electric field acts as an external force that causes motion of the contact line. In order to describe this contact line motion the modified Hocking boundary condition is applied: the velocity of the contact line is proportional to the deviation of the contact angle and the speed of the fast relaxation processes, which frequency is proportional to twice the frequency of the electric field. Using of this equation can qualitatively describe the experimental dependence of the contact angle vs voltage in contrast to the Young-Lippmann equation.

1. Introduction

Recent years have shown keen interest in the dynamics of droplets and bubbles on dielectric substrate in electric field (electrowetting-on-dielectric, EWOD) [1-4]. In the study of electrowetting one of the problems [5,6] is the lack of coordination of the experimental results with the predictions of the theoretical model using the Young-Lippmann equation [1,7,8] (figure 1):

$$\cos\vartheta = \cos\vartheta_0 + \frac{\varepsilon\varepsilon_0}{2d\sigma}V^2, \qquad (1)$$

where V is the value of the applied voltage, ϑ_0 is the contact angle without applied voltage – equilibrium contact angle, which is defined by well-known Young's equation, σ is the interfacial tension between the droplet and the surrounding fluid, d is the thickness of the dielectric film, ε_0 and ε are the permittivity of vacuum and the dielectric layer, respectively. The last term in the equation (1) is the modified term of Lippmann equation [9].

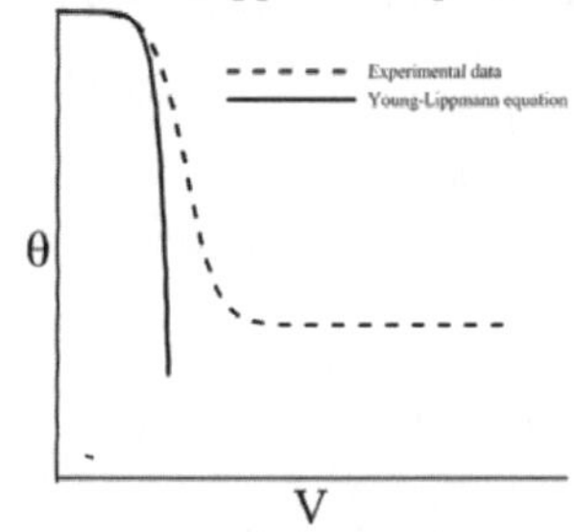

Figure 1. Contact angle θ as function of voltage V. The difference between the experimental data and the theoretical solution using the Young-Lippmann equation

In article [10] authors sought to develop a theory describing droplet oscillations induced by electrowetting and proposed another condition – modified Hocking boundary condition. Original

[1] To whom any correspondence should be addressed.

effective boundary condition has been proposed by Hocking for small oscillations of the contact line [11]. The velocity of the contact line is assumed to be proportional to the deviation of the contact angle from its equilibrium value (for simplicity, the equilibrium contact angle is considered to be $\pi/2$):

$$\frac{\partial \zeta^*}{\partial t} = \Lambda \vec{k} \cdot \vec{\nabla} \zeta^*, \tag{2}$$

where ζ^* is the deviation of the interface from the equilibrium position, $\vec{k}$ is the external normal to the solid surface, Λ is a phenomenological constant (the so-called wetting parameter or Hocking parameter), having the dimension of velocity. The equation (2) is well proven in the study of natural oscillations and fast relaxation processes [12-16].

In [10] we believed that the external electric field plays the role of source of motion and forces the contact angle to change in time. Hereafter, we assume that the external force is time periodic $\sim \cos \omega t$. For very fast relaxation processes at the triple line, the contact angle evolves according to the law $\sim \cos 2\omega t$. Taking into account the more general case when relaxation effects occur at timescales comparable with that of the external force, we can formulate an effective boundary condition to be fulfilled at the contact line

$$\frac{\partial \zeta^*}{\partial t^*} = \mp \Lambda^* \left(\frac{\partial \zeta^*}{\partial t^*} + A^* \cos(2\omega t) \right), \tag{3}$$

i.e. the velocity of the contact line is proportional to the deviation of the contact angle and the speed of the fast relaxation processes, which frequency is proportional to twice the frequency of the electric field.

In the present paper, we consider the behavior of a oblate drop sandwiched between two flat plates under AC. The drop has equilibrium form of rotational figure with arbitrary contact angle whereas cylindrical drop with right contact angle was considered in [10]. The natural oscillations of oblate drop were researched in [15].

2. Problem formulation

Following [10], we consider the dynamic behavior of an incompressible liquid drop of density ρ_i^* surrounded by other liquid of density ρ_e^*. The system is bounded by two parallel solid surfaces (see figure 2) separated by a distance h^*. The equilibrium contact angle ϑ_0 between the side surface of the drop and the solid surface is arbitrary. The external uniform alternating electric field acts as an external force that causes the contact line motion.

Let the surface of the drop be described by the equation $r^* = R^*(z) + \zeta^*(\alpha^*, z^*, t^*)$, where $R^*(z)$ is the equilibrium radius of the drop. We use $\sigma^{-1/2}\sqrt{(\rho_e^* + \rho_i^*)R_0^{*3}}$, $R_0^* = R(0)$, h^*, $\rho_e^* + \rho_i^*$, A^*, $A^*\sigma\left(R_0^*\right)^{-2}$, $A^*\sqrt{\sigma}\left(\left(\rho_e^* + \rho_i^*\right)R_0^{*3}\right)^{-1/2}$ as the scales for the time, length, height, density, deviation of drop surface from its equilibrium position, pressure, and velocity potential, respectively. The system is described by the following equations [10,15]:

$$p_j = -\rho_j \varphi_{jt}, \ \Delta\varphi_j = 0, \ j = i, e, \tag{4}$$

$$\Delta = \frac{1}{r}\frac{\partial}{\partial r}\left(r\frac{\partial}{\partial r}\right) + \frac{1}{r^2}\frac{\partial^2}{\partial \alpha^2} + b^2\frac{\partial^2}{\partial z^2},$$

$$r = R: \ [\varphi_r] = 0, \ \zeta_t = \varphi_r, \ [p] = \zeta + \zeta_{\alpha\alpha} + b^2\zeta_{zz}, \tag{5}$$

$$z = \pm\frac{1}{2}: \ \varphi_z = 0, \tag{6}$$

$$r = R, \ z = \pm\frac{1}{2}: \ \zeta_t = \mp\lambda\left(\zeta_z - \zeta_{0z} + a\cos(2\omega t)\right), \ \zeta_{0z} = \cot\vartheta_0 \tag{7}$$

where p is the fluid pressure, the square brackets denote the jump in the quantity at the interface between the external liquid and the drop. The boundary-value problem (5)–(8) involves six parameters: the aspect ratio, the dimensionless density, the wetting parameter, the AC frequency and amplitude

$$b = R_0 h^{-1},\ \rho_i = \rho_i^*\left(\rho_e^* + \rho_i^*\right)^{-1},\ \rho_e = \rho_e^*\left(\rho_e^* + \rho_i^*\right)^{-1},\ \lambda = \Lambda^*\sigma^{-1/2}\sqrt{\left(\rho_e^* + \rho_i^*\right)R_0^*},$$

$$\omega = \omega^*\sigma^{-1/2}\sqrt{\left(\rho_e^* + \rho_i^*\right)R_0^{*3}},\ a = A^*C\sigma^{-3/2}\sqrt{\left(\rho_e^* + \rho_i^*\right)R_0^{*3}}\Big/2.$$

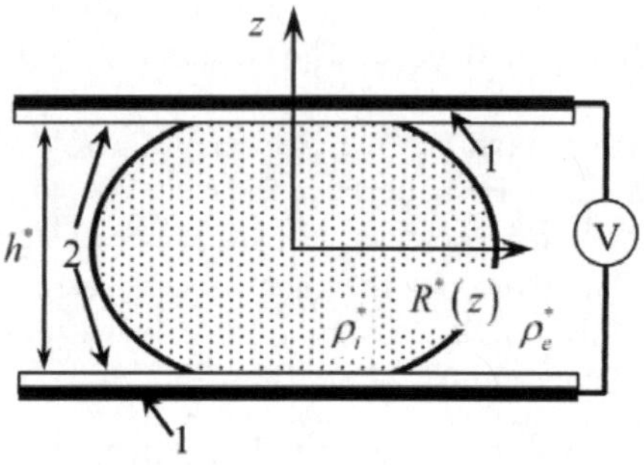

Figure 2. Problem geometry (1 – electrode, 2 – dielectric layer).

3. Uniform electric field

We consider the case of small distances between the solid planes. Assuming that $b \gg 1$, i.e. $\beta = b^{-1} \ll 1$, we expand the function $R(z)$ (equilibrium shape) in a series in the small parameter β: $R(z) = 1 + \beta R_1(z) + \ldots$ In order to investigate the problem it is convenient to begin with a consideration of the uniform electric field. Our interest deals with the axisymmetric oscillation modes, therefore system of the equations and boundary conditions (5)–(8) are axially symmetric.

The flow domain in the drop can be divided into a region having the shape of a cylinder with unit radius and a narrow transition region between the cylinder and the surrounding liquid. In the transition region, the velocity potential and the pressure of the liquids can vary significantly in the radial direction, so that it is necessary to introduce the expansion radial coordinate $r_1 = \left(r - R(z)\right)\beta^{-1}$. In view of axial symmetry, the solution of the Laplace equation (6) is written as

$$\varphi_i(r,z,t) = \mathrm{Re}\left(a_i r + \sum_{k=0}^{\infty} a_k \exp\left((2k+1)\pi b r_1\right)\sin\left((2k+1)\pi z\right)e^{i2\omega t}\right), \tag{8}$$

$$\varphi_e(r,z,t) = \mathrm{Re}\left(a_e r^{-1} + \sum_{k=0}^{\infty} b_k \exp\left(-(2k+1)\pi b r_1\right)\sin\left((2k+1)\pi z\right)e^{i2\omega t}\right), \tag{9}$$

The kinematic condition on the free surface (the second condition in (6)) gives the expression for the surface deviation

$$\zeta(z,t) = \mathrm{Re}\left(\left(dz + \sum_{k=0}^{\infty} c_k \cos\left((2k+)\pi z\right)\right)e^{i2\omega t}\right). \tag{10}$$

Substituting solutions (9)–(11) into (5)–(8) we obtain expressions for the unknown amplitudes a_k, b_k, c_k and d by Galerkin method .

The contact angle at the upper plate versus square root of amplitude a (i.e. proportional to AC potential V) is given in figure 3 for different values of the Hocking parameter λ and AC frequency (equilibrium contact angle $\vartheta_0 = \pi/2$). The dependences obtained in qualitative agreement with the experimental data. However the maximum deviation of contact angle tends to π/2, i.e. $\vartheta \to 0$ or $\vartheta \to \pi$ whereas contact angle is finite in experiments.

Different wetting situations are addresses by changing the values of the parameter λ and range between two limiting cases: i) for $\lambda = 0$, the contact line motion is completely uncoupled from the external field – the contact line remains fixed and the droplet overall is motionless; ii) the opposite limit, $\lambda \gg 1$, corresponds to infinitely fast relaxation of the contact angle, when the latter quickly adjusts to the slowly changing force. In both these situations, there is no dissipation at the contact line. Apparently the fluid viscosity must be taken into account. It is giving a finite amplitude for both limit cases of Hocking boundary especially in limiting case $\lambda \gg 1$.

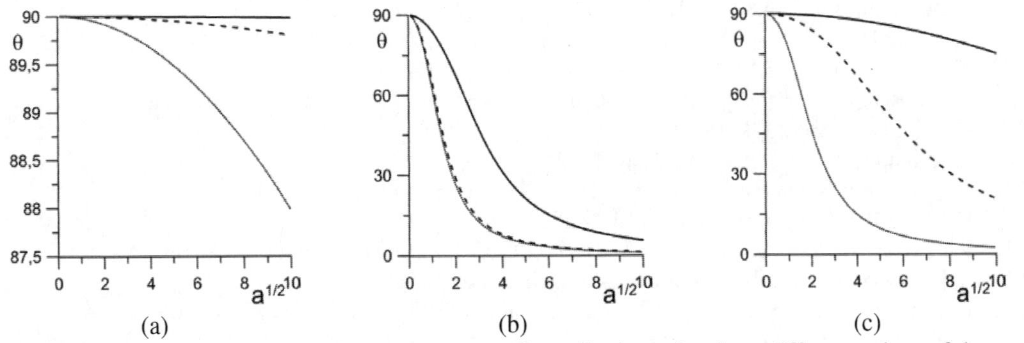

Figure 3. The contact angle ϑ vs square root of amplitude a for three different values of the Hocking parameter λ and of the oscillation frequency ω ($b=1$, $\rho_i=0.7$).
(a) $\omega=5$, (b) $\omega=10$, (a) $\omega=100$,
$\lambda=0.1$ (solid line), $\lambda=1$ (dashed line), $\lambda=10$ (dotted line).

Figure 4 shows the contact angle of highly-condensed drops as function of the square root of the amplitude for two values of the equilibrium contact angle ϑ_0. From these curves shows that for any values ϑ_0 curves behave the same way.

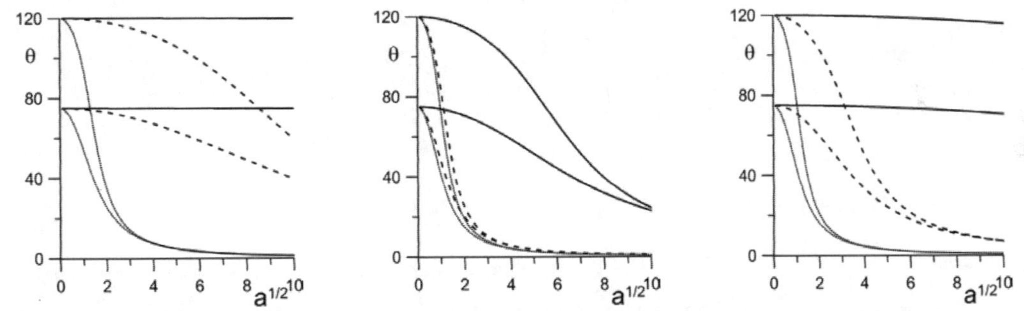

Figure 4. The contact angle ϑ vs square root of amplitude a for $\vartheta_0 = 75^{\circ}, 120^{o}$.
(a) $\omega=5$, (b) $\omega=10$, (a) $\omega=10$,
$\lambda=0.1$ (solid line), $\lambda=1$ (dashed line), $\lambda=10$ (dotted line).

4. Conclusions

The dependences of surface amplitude and contact angle from oscillation amplitude and frequency are obtained for Hocking parameter, geometric parameters and equilibrium contact angle. It is shown that The curve of contact angle ϑ vs square root of amplitude obtained in qualitative agreement with the experimental data. However the maximum deviation of contact angle tends to π/2, i.e. $\vartheta \to 0$ or $\vartheta \to \pi$ whereas contact angle is finite in experiments. Similar dependences are obtained for cylindrical drop ($\vartheta_0 = 90^{o}$) under nonuniform electric field [10]. Consequently, there are two ways of further development of problem solution: 1. fluid viscosity (see for example [13]), 2. contact angle hysteresis (see for example [14]). Both of these cases limit the amplitude of oscillation and may limit the change of the contact angle.

Acknowledgments
This work was supported by the Russian Science Foundation (project 14-21-00090).

IOP Conf. Series: Journal of Physics: Conf. Series **929** (2017) 012107 doi:10.1088/1742-6596/929/1/012107

5. References

[1] Mugele F and Baret J-C 2005 *J. Phys.: Condens. Matter.* **17** 705
[2] Mugele F 2009 *Soft Matter* **5** 3377–84
[3] Royal M W, Jokerst N M and Fair R B 2013 *IEEE Sensors Journal* **13** 4733-42
[4] Chen L and Bonaccurso E 2014 *Adv. Colloid Interface Sci.* **210** 2
[5] Klarman D, Andelman D and Urbakh M A 2011 *Langmuir* **27** 6031–41
[6] Zhao Y-P and Wang Y 2013 *Rev. Adhesion Adhesives* **1** 114
[7] Berge B 1993 *Comptes Rendus Acad. Sci. II* **317** 157
[8] Quilliet C and Berge B 2001 *Curr. Opin. Colloid Interface Sci.* **6** 34
[9] Lippmann G 1875 *Ann. Chim. Phys.* **5** 494–549
[10] Alabuzhev A A and Kashina M A 2016 *J. Phys.: Conf. Ser.* **681** 012042
[11] Hocking L M 1987 *J. Fluid Mech.* **179** 253-66
[12] Alabuzhev A A and Lyubimov D V 2007 *J. Appl. Mech. Tech. Phys.* **48** 686–93
[13] Shklyaev S and Straube A V 2008 *Phys. Fluids* **20** 052102
[14] Fayzrakhmanova I S, Straube A V and Shklyaev S 2011 *Phys. Fluids* **23** 102105
[15] Alabuzhev A A and Lyubimov D V 2012 *J. Appl. Mech. Tech. Phys.* **53** 9–19
[16] Alabuzhev A A and Kaysina M I 2016 *J. Phys.: Conf. Ser.* **681** 012043

IOP Conf. Series: Journal of Physics: Conf. Series **929** (2017) 012108 doi:10.1088/1742-6596/929/1/012108

A study of stationary states of a plasma diode in the presence of a transverse magnetic field.

V.I.Kuznetsov[1], A.B.Gerasimenko[1], S.Pramanik[2], N.Chakrabarti[2]

[1] Ioffe Institute, Russia

[2] Saha Institute of Nuclear Physics, India

E-mail: `gerasimenko.alexander@mail.ioffe.ru`

Abstract. In this paper we studied the stationary states of plasma diode in the presence of a transverse magnetic field. We considered regimes without and with electron turnaround by the magnetic field. Stationary solutions depending on the electric field strength at the emitter for different values of the electrode gap, the applied voltage and the magnetic field were found. The stability of solutions was studied.

1. Introduction

In recent years, due to the need to create effective environmentally friendly energy sources there is a significantly increased interest in the thermionic energy converters (TIC) [1, 2]. In addition to the experimental work, the research in the field of modelling of appropriate physical processes is actively conducting [3]. One of the important tasks is to study currents flowing in the TIC.

The new generation high-performance thermionic energy converter operates at high temperatures of emitter. Electrons in the inter-electrode gap move without collisions. Such a TIC can produce very strong current that creates a magnetic field. This field is perpendicular to the direction of electron current and can partially turnaround electrons. This leads to a decrease in the current passing. Instead of studying the influence of its own magnetic field, we can consider the problem of the influence of an external magnetic field on the currents of the diode. As shown earlier [4], this problem can be successfully solved by considering the Pierce diode, i.e., diode, in which a beam of electrons moves through a fixed ion background.

2. Problem statement

Consider the Pierce diode. Its scheme is shown in Fig. 1. There are two infinite parallel flat electrodes between which a voltage U is applied. An external magnetic field B is parallel to the electrode surface. A beam of electrons with a density n_b and velocity v_b emitted from the emitter ($z = 0$) and moves without collisions through a fixed ion background of constant density n_i towards the collector ($z = d$).

Let us define the neutralization coefficient as the ratio of the ion density to the beam density $\gamma = n_i/n_b$. We present the results for the case of $\gamma = 1$, because this case better corresponds to the TIC. However, the results for any other values of γ may also be obtained. Electrons moving in the diode, reach the collector but they can also be turned around by the magnetic field and return to the emitter. We suggest that in both cases they are completely absorbed by the electrodes. We believe that all quantities depend only on the z coordinate. In the case when

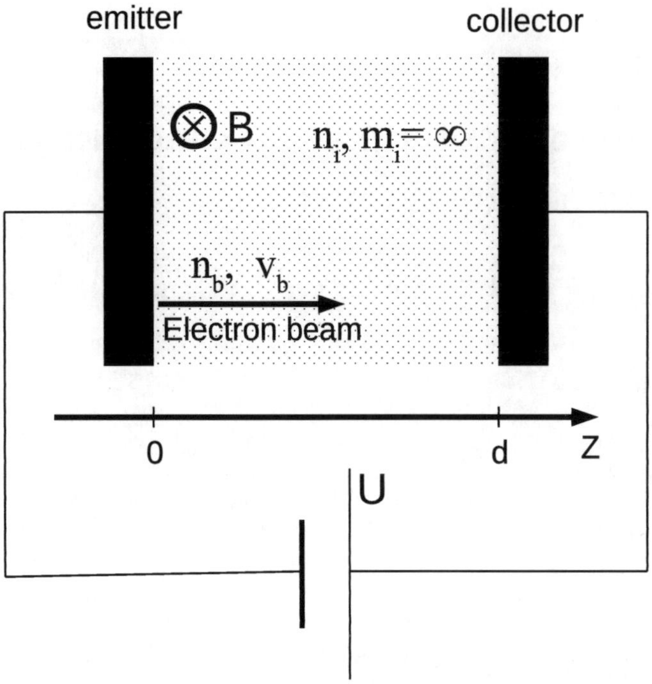

Figure 1. Diode scheme.

all electrons reach the collector the process is described by the following system of differential equations (which are continuity, momentum, and the Poisson's equations):

$$\frac{d}{dz}(nv_z) = 0,$$
$$v_z\frac{dv_z}{dz} = -\frac{e}{m}E - \omega v_x, \quad v_z\frac{dv_x}{dz} = \frac{eB}{m}v_z,$$
$$E = -\frac{d\varphi}{dz}, \quad \frac{dE}{dz} = -\frac{e}{\epsilon_0}(n-\gamma). \tag{1}$$

The Larmor frequency and radius are define as:

$$\omega = \frac{eB}{m} = \left(\frac{2e}{m}\right)\frac{V_b^{1/2}}{\lambda_L}[s^{-1}],\ \lambda_L = \frac{mv_b}{eB} \approx 0.3372\cdot 10^{-3}\frac{V_b^{1/2}}{B}[cm].$$

Here V_b is the beam accelerating voltage, e, m are charge and mass of the electron, ϵ_0 is the free-space permittivity.

It is convenient to switch to the dimensionless coordinate, time, speed, potential and the electric field: $(\zeta,\chi) = (z,x)/\lambda_D$, $\tau = t\omega_b$, $(u_\zeta, u_\chi) = (v_z, v_x)/\sqrt{2W_b/m}$, $\eta = e\varphi/(2W_b)$, $\varepsilon = eE\lambda_D/(2W_b)$, $\Omega = \omega/\omega_b$. Here $\lambda_D = [2\epsilon_0 W_b/(e^2 n_b)]^{1/2}$ is the beam Debye length, $W_b = mv_b^2/2$ is the kinetic energy of electrons at the emitter, $\omega_b = v_b/\lambda_D$ the characteristic frequency. The corresponding system of dimensionless equations takes the form:

$$\frac{d}{d\zeta}(nu_\zeta) = 0,$$
$$u_\zeta\frac{du_\zeta}{d\zeta} = -\varepsilon - \Omega u_\chi, \quad u_\zeta\frac{du_\chi}{d\zeta} = \Omega u_\zeta,$$

$$\varepsilon = -\frac{d\eta}{d\zeta}, \quad \frac{d\varepsilon}{d\zeta} = -n + \gamma. \tag{2}$$

The boundary conditions are of the form: density $n(\zeta = 0) = 1$, components of velocity $u_\zeta(\zeta = 0) = 1$, $u_\chi(\zeta = 0) = 0$, electric potential $\eta(\zeta = 0) = 0$, and electric field $\varepsilon(\zeta = 0) = \varepsilon_0$. The emitter electric field ε_0 will be used as a variable parameter for our following analysis.

3. Results

The resulting system of equations (2) is solved by introducing a Lagrangian coordinate τ and Lagrange transformation

$$\zeta = \int_0^\tau u_\zeta(\tau')d\tau'.$$

Thus, $u_\zeta d/d\zeta = d/d\tau$. Then equations (2) take the form

$$\begin{aligned} \frac{d}{d\tau}(nu_\zeta) &= 0, \\ \frac{du_\zeta}{d\tau} = -\varepsilon - \Omega u_\chi, &\quad \frac{du_\chi}{d\tau} = \Omega u_\zeta, \\ \frac{d\eta}{d\tau} = -u_\zeta \varepsilon, &\quad \frac{d\varepsilon}{d\tau} = -1 + \gamma u_\zeta. \end{aligned} \tag{3}$$

The following analytical solutions were obtained for this system:

$$\begin{aligned} u_\zeta(\tau) &= \frac{1}{\alpha^2} + \left(1 - \frac{1}{\alpha^2}\right)\cos(\alpha\,\tau) - \frac{\varepsilon_0}{\alpha}\sin(\alpha\,\tau). \\ \zeta(\tau) &= \frac{1}{\alpha^2}\tau + \frac{1}{\alpha}\left(1 - \frac{1}{\alpha^2}\right)\sin(\alpha\,\tau) + \frac{\varepsilon_0}{\alpha^2}\left[\cos(\alpha\,\tau) - 1\right]. \\ u_\chi(\tau) = \Omega\zeta(\tau) &= \frac{\Omega}{\alpha^2}\tau + \frac{\Omega}{\alpha}\left(1 - \frac{1}{\alpha^2}\right)\sin(\alpha\,\tau) + \frac{\varepsilon_0\Omega}{\alpha^2}\left[\cos(\alpha\,\tau) - 1\right]. \\ \varepsilon(\tau) = \varepsilon_0 - \tau + \gamma\zeta(\tau) &= \frac{\Omega^2}{\alpha^2}(\varepsilon_0 - \tau) + \frac{\gamma}{\alpha}\left(1 - \frac{1}{\alpha^2}\right)\sin(\alpha\,\tau) + \frac{\gamma\varepsilon_0}{\alpha^2}\cos(\alpha\,\tau). \\ \eta(\tau) &= \frac{1}{2}\left[u_\zeta^2(\tau) + \Omega^2\zeta^2(\tau) - 1\right], \end{aligned} \tag{4}$$

here the effective "frequency" $\alpha = \sqrt{\gamma + \Omega^2}$ is introduced. For every value of magnetic field Ω the family of solution of system (4) is obtained. Then the parameter ϵ_0 is varied. Calculations stopped when electron velocity u_ζ vanishes at the first time (it corresponds to the moment when electrons turnaround). The threshold value of ϵ_0 equals to $\pm\sqrt{2 - \alpha^2}$.

Examples of potential and velocity profiles are shown in Fig.2 and 3 respectively. As it can be seen from the figures, in weak magnetic fields, potential profile is an oscillating function, the same way as is the case in the absence of a magnetic field [5]. The oscillation period is equal to $2\pi(\gamma + \Omega^2)^{-3/2}$. Desired solutions correspond to points of intersection of potential curves with line $\eta = V$. The collector position corresponds to the values: $\tau = T$, $\zeta = \delta$, $\eta(\delta) = V$. Thus we have at the collector:

$$\begin{aligned} \delta &= \frac{1}{\alpha^2}T + \frac{1}{\alpha}\left(1 - \frac{1}{\alpha^2}\right)\sin(\alpha\,T) + \frac{\varepsilon_0}{\alpha^2}\left[\cos(\alpha\,T) - 1\right], \\ u_\zeta(T) &= \frac{1}{\alpha^2} + \left(1 - \frac{1}{\alpha^2}\right)\cos(\alpha\,T) - \frac{\varepsilon_0}{\alpha}\sin(\alpha\,T), \\ V &= \frac{1}{2}\left[u_\zeta^2(T) + \Omega^2\delta^2 - 1\right]. \end{aligned}$$

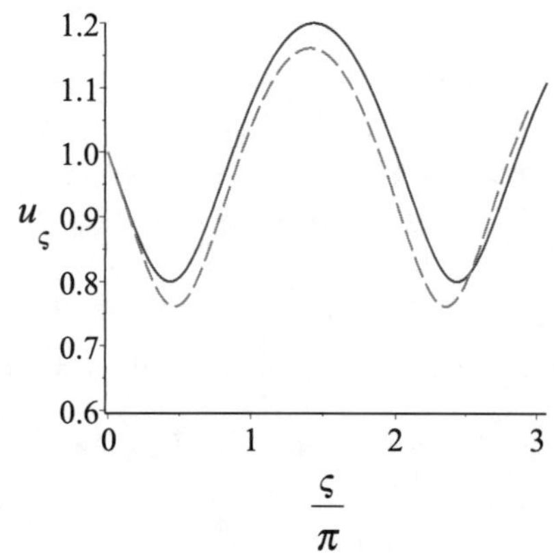

Figure 2. Velocity profiles. Solid line – $\Omega = 0$, dashed – $\Omega = 0.2$. $\gamma = 1$, $\varepsilon_0 = 0.2$.

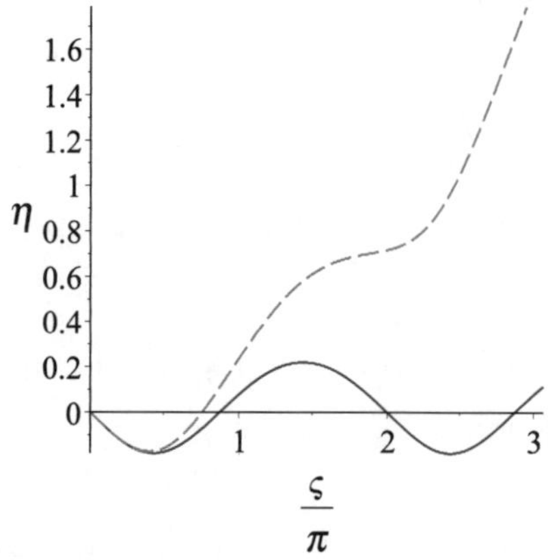

Figure 3. Potential profiles. Solid line – $\Omega = 0$, dashed – $\Omega = 0.2$. $\gamma = 1$, $\varepsilon_0 = 0.2$.

Here, T is the time-of-flight of an electron between electrodes. With an increase of magnitude of the magnetic field, the potential distribution becomes higher than $\eta = V$ and all solutions with $\delta > \delta^* = \pi(\gamma + \Omega^2)^{-3/2}$ disappear.

It is convenient to denote the solutions by the points on the plane (ε_0, δ), where δ is coordinate of the collector, and ε_0 is the electric field strength at the emitter. For fixed external parameters, solutions lie on the continuous curves. We call them branches of solutions. Several of these dependencies shown in Figure 4. Branches of solutions corresponding to $\delta < \delta^*$ are Bursian branches. The emitter electric field can take negative values unlike the Bursian diode, i. e. the diode without ion background. We see that Bursian branches could have several solutions. The right point on Bursian branch called space charge limit (SCL) point. It corresponds to maximum current that can go through diode in a certain regime.

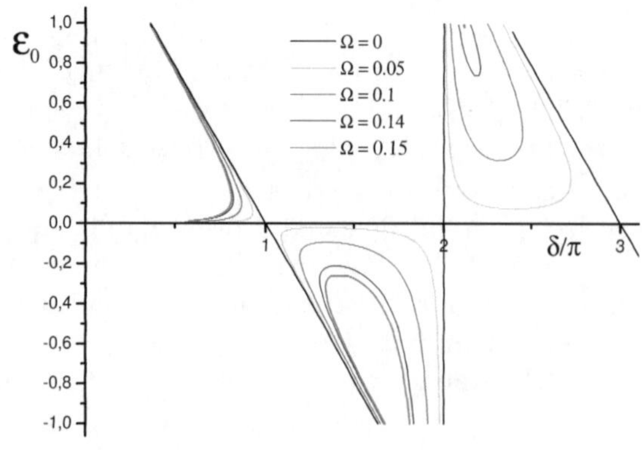

Figure 4. $\varepsilon - \delta$ diagram. $\gamma = 1$. No electron reflection.

For a complete consideration of the solutions we have to take into account a possible turnaround of electrons by the magnetic field. This allows us to supplement the $\varepsilon_0 - \delta$ diagram shown in Figure 4. We use method proposed in the paper [7]. Initially, we postulated that the electron beam is mono-energetic, so if at some point ζ_r the longitudinal electron velocity u_ζ vanishes then it should make all electrons turn back. Nevertheless, in real life electrons of the beam always have a finite (even if very small) velocity spread. In such a way we can assume that some electrons with a bit higher velocities could overcome zero velocity point and with lower velocities will be turned. In this case we can talk about beam "splitting". This situation could be described via introducing reflection coefficient r. This coefficient is the probability that an electron of the beam is turned around by the magnetic field. In such a way, there are direct and reverse flows in the region $0 < \zeta < \zeta_r$) and electron density there is $1/u_\zeta + r/|u_\zeta|$. To the right of ζ_r point the direct electron current will have a weight $1 - r$. All solutions with partial electron turn around could be derived by varying the r coefficient.

In this case, the system of equations takes the form:

$$
\begin{gathered}
nu_\zeta = H(\zeta;\zeta_r,r), \\
u_\zeta \frac{du_\zeta}{d\zeta} = -\varepsilon - \Omega u_\chi, \quad u_\zeta \frac{du_\chi}{d\zeta} = \Omega u_\zeta, \\
\varepsilon = -\frac{d\eta}{d\zeta}, \quad \frac{d\varepsilon}{d\zeta} = -n + \gamma.
\end{gathered} \tag{5}
$$

where $H(\zeta;\zeta_r,r) = [(1+r)\Theta(\zeta_r - \zeta) + (1-r)\Theta(\zeta - \zeta_r)]$, and Θ is the Heaviside function. Its solutions have the form

$$
\begin{gathered}
u_\zeta(\tau) = \frac{1+r}{\alpha^2} + \left(1 - \frac{1+r}{\alpha^2}\right)\cos(\alpha\,\tau) - \frac{\varepsilon_0}{\alpha}\sin(\alpha\,\tau), \\
\zeta(\tau) = \frac{1+r}{\alpha^2}\tau + \frac{1}{\alpha}\left(1 - \frac{1+r}{\alpha^2}\right)\sin(\alpha\,\tau) + \frac{\varepsilon_0}{\alpha^2}\left[\cos(\alpha\,\tau) - 1\right].
\end{gathered} \tag{6}
$$

for $\zeta < \zeta_r$ and

$$
\begin{gathered}
u_\zeta(\tau) = \frac{1-r}{\alpha^2}\left[1 - \cos\alpha(\tau - \tau_r)\right], \\
\zeta(\tau) = \zeta_r + \frac{1-r}{\alpha^2}(\tau - \tau_r) - \frac{1-r}{\alpha^3}\sin\alpha(\tau - \tau_r).
\end{gathered} \tag{7}
$$

for $\zeta > \zeta_r$. Here $\zeta_r = \alpha^{-2}[(1+r)\tau_r - \varepsilon_0]$. Depending on the values of γ and Ω, the function $\tau_r(\gamma,\Omega)$ reads for $\varepsilon_0 \geq 0$ as

$$
\begin{gathered}
\frac{1}{\alpha}\sin^{-1}\frac{\alpha\sqrt{2(1+r)-\alpha^2}}{1+r}, \quad if\ \Omega^2 \leq 1+r-\gamma, \\
\frac{1}{\alpha}\left(\pi - \sin^{-1}\frac{\alpha\sqrt{2(1+r)-\alpha^2}}{1+r}\right), \quad if\ \Omega^2 \geq 1+r-\gamma.
\end{gathered} \tag{8}
$$

and for $\varepsilon_0 < 0$ τ_r is equal to

$$
\begin{gathered}
\frac{1}{\alpha}\left(2\pi - \sin^{-1}\frac{\alpha\sqrt{2(1+r)-\alpha^2}}{1+r}\right), \quad if\ \Omega^2 \leq 1+r-\gamma, \\
\frac{1}{\alpha}\left(\pi + \sin^{-1}\frac{\alpha\sqrt{2(1+r)-\alpha^2}}{1+r}\right), \quad if\ \Omega^2 \geq 1+r-\gamma.
\end{gathered} \tag{9}
$$

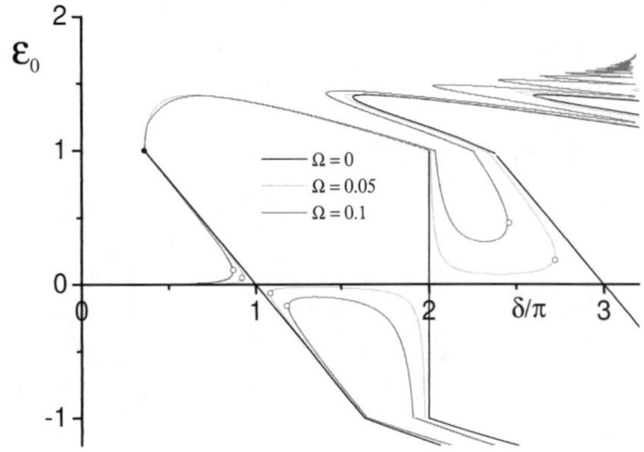

Figure 5. $\varepsilon - \delta$ diagram. $\gamma = 1$. Case with electron reflection.

For the potential distribution we have

$$\eta(\tau) = \frac{1}{2}\left[u_\zeta^2(\tau) + \Omega^2\zeta^2(\tau) - 1\right]. \tag{10}$$

Note that ε_0 connected with the parameter r by the formulae $\varepsilon_0 = \pm[2(1+r) - \alpha^2]^{1/2}$.

Solutions of the Eqs. (7)–(10) can also be plotted on $\varepsilon_0 - \delta$ plane (see Fig. 5). We can see from this Figure that each $\varepsilon_0 - \delta$–curve contains an oscillating segment as r tend to 1, i.e., it is many-valued over a certain range of δ's. It happens because for certain value of $r(\varepsilon_0)$, the electron velocity u_ζ could vanish several times in the region $\zeta > \zeta_r$ (see Eq. (8)). The Figure also shows that the amplitude of the oscillations diminishes with r for each zigzag segment. It is also found that the maximum width of the oscillating segment shrinks with the increase of Ω.

It can also be seen that along with the Bursian branches the non-Bursian branches arise. With an increase in the magnetic field these branches disappear gradually. It happens when the Larmor radius turns out to be about 10 times higher than the Debye length. There are two bifurcation points on the Bursian branches: the left one noted as BF (close circles on Fig. 5) and right one noted as SCL (open circles on Fig. 5).

We studied the stability of solutions upon small perturbations. The $\eta - \varepsilon$ diagram method [6] were used to investigate stability upon aperiodic perturbations. We have shown that solutions belonging to the Bursian family are stable with respect to aperiodic perturbations at $\varepsilon_0 < \varepsilon_{0,SCL}$. At the same time, those located between bifurcation points SCL and BF ($\varepsilon_{0,SCL} < \varepsilon_0 < \varepsilon_{0,BF}$) are unstable. Solutions are aperiodic stable when $\varepsilon_0 > \varepsilon_{0,BF}$ although they can be oscillatory unstable and it need further study.

Then we have studied the solutions belonging to the non-Bursian family. In the region of negative ε_0 values in Fig. 5 all solutions belonging to the leftmost branch which goes down from the SCL-type bifurcation point are aperiodic stable, and these belonging to the branch which goes to the right of this bifurcation point are unstable. In the region of positive ε_0 values the lower segment of the branch located to the left of the SCL-type bifurcation point contains the aperiodic stable solutions. The segment of the branch located between this SCL-type bifurcation point and the leftmost BF-type bifurcation point contains unstable solutions. The next segment of the branch which locates between two BF-type bifurcation points corresponds to aperiodic stable solutions and so on. We also studied stability of solutions corresponding to the zigzag part of branches (see Fig. 5). We found that each left-to-right segments of the zigzag is aperiodic stable and right-to-left segments are unstable.

In order to study the stability of the solutions for the regime without electron turning around we derived the dispersion equation. It has the following form:

$$F(\sigma;\delta,T) = -\exp(-\sigma T)\left[2\sigma\cos\alpha T + \frac{\sigma^2-\alpha^2}{\alpha}\sin\alpha T\right] + (\sigma^2+\alpha^2)^2\delta - (\sigma^2+\alpha^2)T + 2\sigma = 0. \tag{11}$$

Note that in Eq. (11), T and δ are derived from formulas (5). We found that at $\delta < \delta^+ \approx 3\pi/(\gamma+\Omega^2)^{3/2}$ all oscillatory eigen-modes have a negative growth rate, i. e. they are oscillatory stable whereas some solutions are aperiodic unstable. When $\delta > \delta^+$ a portion of solutions can be oscillatory unstable as well as in the absence of the magnetic field (see, e. g., [5]).

4. Conclusion

We explored in detail stationary states of the Pierce diode ($\gamma = 1$) in the presence of the magnetic field. Analytical expressions for electron trajectories, potential and electric field distributions were derived. We studied cases with and without electron turning by a magnetic field. The $\varepsilon_0 - \delta$ diagrams describing states of the diode were plotted for several most distinctive cases. Also we study stability of solutions upon small perturbations.

References

[1] Deng Y and Liu J 2009 Renew.Sustain.En., 2009, **1**, 052701
[2] Khalid K, Jien L, and Mohamed K 2016 IEEE Trans. Electr. Dev. 2016, **63**, 6
[3] Ito T and Cappelli M 2012 App.Phys.Lett. **101**, 213901
[4] Ender A, Kuhn S, Kuznetsov V 2006 Phys.Plasmas **13**, 113506
[5] Ender A, Kolinsky H, Kuznetsov V, Schamel H 2000 Phys. Rep. **328**, 1
[6] Kuznetsov V and Ender A 1977 Sov.Phys.Tech.Phys. **22**, 1295
[7] Pramanik S, Kuznetsov V, Chakrabarti N 2015 Phys. Plasmas **22**, 082103

Numerical simulation of multicellular natural convection in air-filled vertical cavities

A I Kunaeva and N G Ivanov

Department of Fluid Dynamics, Combustion and Heat Transfer,
Peter the Great St.Petersburg Polytechnic University,
Polytechnicheskaya str. 29, 195251, St.Petersburg, Russia

E-mail: nastya1kunaeva@gmail.com

Abstract. The paper deals with 2D laminar natural convection in vertical air-filled cavities of aspect ratio 20, 30 and 40 with differentially heated sidewalls. The airflow and heat transfer were simulated numerically with an in-house Navier-Stokes code SINF. The focus is on the appearance of stationary vortex structures, "cat's eyes", and their transition to unsteady regime in the Rayleigh number range from 4.8×10^3 to 1.3×10^4. The dependence of the predicted flow features and the local and integral heat transfer on the aspect ratio value is analysed.

1. Introduction

Transient natural convection in differentially heated infinite vertical fluid layers confined between two rigid boundaries is a classical problem in fluid dynamics considered first analytically by Gershuni [1]. Since that, numerous analytical, experimental and numerical contributions have been published considering convection of different fluids with wide range of the Prandtl number, and detailed reviews could be found in the books by Gershuni et al. [2] and by Lappa [3]. The instability scenario is as follows. The primary flow with one-dimensional odd cubic velocity and linear temperature profiles (the so called conduction regime) at a certain Rayleigh number is disturbed and a multicellular "cats-eye" steady-state pattern of two-dimensional (2D) transverse rolls is superimposed on the basic flow. At higher Ra the cells become unstable, and the oscillatory regime occurs.

For engineering applications, a slot of finite height is of interest, and a lot of work by different authors has been devoted to the study of this problem. In the applications related to building engineering, isotope separation, cooling of electronic devices etc., the slot aspect ratio A is large, and the multicellular "cats-eye" instability has been analysed by many authors. The paper by Vest and Arpaci [4] is an example of early experimental works. Natural convection of silicon oil with very high Pr for $A = 20$ was visualized there for Ra of 3.7×10^5. The experiment detected stable cats-eye pattern.

An example of a well-documented experimental study of natural convection in a vertical cavity is the paper by Wright at al. [5]. Natural convection of air ($Pr = 0.7$) in the cavity of $A = 40$ was studied using smoke patterns and interferometry. The range of the Rayleigh number considered ($Ra = 4.85\times10^3 \div 5.48\times10^4$) covered different flow regimes. Remarkably, secondary cells were not perfectly steady at any value of Ra considered.

Flow patterns of two-dimensional natural convection in a vertical air-filled tall cavity with differentially heated sidewalls were predicted numerically in [6]. For aspect ratios from 10 to 24, various cellular structures characterized by the number of secondary cells were clarified. The onset of the steady-state regime with the secondary cells was detected at the critical Ra of 8.5×10^3 for $A = 11.5$. It was found that with an increase of the aspect ratio the critical Ra decreases, and eventually becomes constant: for $A > 20$ $Ra_{crit} \approx 7\times10^3$.

Transition from unicellular to multicellular flow in differentially heated cavities with $A = 20$ was revisited in [7] where the transition between uni- and multicellular flows was explained by thermodynamic analysis. The detailed scenario of transition was proposed, and the value of $Ra_{crit} \approx 5.8\times10^3$ was reported. Numerical results for natural convection in vertical cavities filled with air were obtained also in [8] ($A = 12 \div 20$, the effect of layer inclination was also studied in the paper). It was reported that at $Ra = 1.1\times10^4$ the "cat's eye" instability arises for A > 12.

Recently, 2D laminar natural convective transient flow characteristics in a differentially heated air-filled tall cavity with gradual heating were investigated both experimentally and numerically in [9]. The results showed that for $Ra < 7.4\times10^3$ the flow pattern was unicellular, when Ra increases to about 6×10^3, the flow became multicellular, and for Ra $> 1.18\times10^4$ the flow pattern became unstable.

The current study considers the convective cells onset in air-filled vertical cavities for the same values of Ra that were investigated experimentally in [5]. Numerical simulation of 2D transient natural convection were performed, with the focus on the accurate resolution of the region near the end walls as the effect of the end walls could be important even for large A values.

2. Problem definition and computational aspects

Figure 1 shows the vertical cavity with the height H much higher than the distance between the vertical walls, L. The aspect ratio $A = H/L$ is varied, and three values of $A = 20$, 30, and 40 are considered in the current paper. The cavity is supposed to be long in the z-direction, so that 2D formulation is used.

The vertical plates are isothermal. The left plate is heated, with the temperature T_h, while the right plate is cooled, with the temperature T_c that is treated as the reference temperature. The reference temperature difference is defined as $\Delta T_{ref} = T_h - T_c$. The top and bottom boundaries are thermally insulated. All boundaries are solid walls with the no-slip boundary condition.

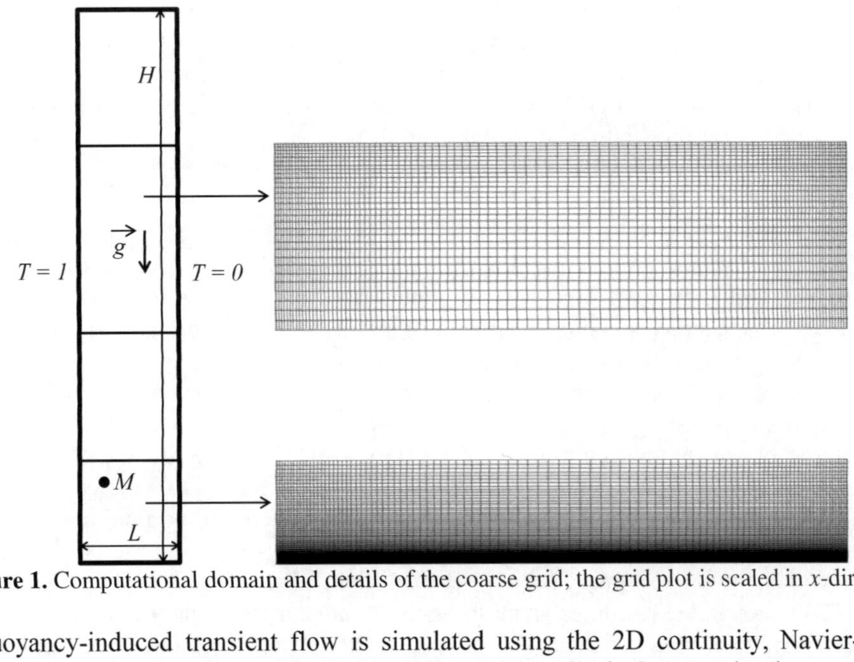

Figure 1. Computational domain and details of the coarse grid; the grid plot is scaled in x-direction

The buoyancy-induced transient flow is simulated using the 2D continuity, Navier-Stokes and energy equations. The incompressible fluid with constant physical properties is considered. The buoyancy effects are represented with the Boussinesq approximation. Assuming that the buoyancy velocity $V_b = (g\beta\Delta T_{\mathrm{ref}}L^3)^{1/2}$ is the velocity scale, here g is the gravity acceleration and β is the thermal expansion coefficient, and the ratio L/V_b is the time scale, the governing equations in the non-dimensional form are written as:

$$\frac{\partial u}{\partial x}+\frac{\partial v}{\partial y}=0$$

$$\frac{\partial u}{\partial t}+u\frac{\partial u}{\partial x}+v\frac{\partial u}{\partial y}=-\frac{\partial p}{\partial x}+\sqrt{Pr/Ra}\,\Delta u$$

$$\frac{\partial v}{\partial t}+u\frac{\partial v}{\partial x}+v\frac{\partial v}{\partial y}=-\frac{\partial p}{\partial y}+\sqrt{Pr/Ra}\,\Delta v+T \quad ,$$

$$\frac{\partial T}{\partial t}+u\frac{\partial T}{\partial x}+v\frac{\partial T}{\partial y}=\frac{1}{\sqrt{Pr\,Ra}}\Delta T$$

where u and v are the horizontal and vertical velocity components, respectively, p is the pressure and T is the temperature. Two non-dimensional parameters are included into the equations: the Rayleigh number, $Ra = g\beta\Delta T_{\mathrm{ref}}L^3/\nu a$, and the Prandtl number, $\mathrm{Pr} = \nu/a$, here ν is the kinematic viscosity and a is the thermal diffusivity. In the current study the Prandtl number value is fixed, $Pr = 0.7$, while the Rayleigh number is varied in the range from 4.8×10^3 to 1.3×10^4.

The computations were performed using the in-house 3D steady/unsteady Navier-Stokes code SINF (Supersonic to INcompressible Flows) being under development since 1993 [10-12]. The structured version of the code used in the current study operates with the second-order finite-volume spatial discretization using the cell-centered variable arrangement and body-fitted block-structured grids [10]. The artificial compressibility (AC) technique and/or the SIMPLEC method are used for the pressure-velocity coupling in the case of incompressible fluid flows (see [11] for details). The viscous fluxes are evaluated on the basis of central differences. Weighted corrections are calculated on the basis of the Rhie and Chow approach. Three-layer second-order scheme is implemented for physical time stepping. The code is parallelized based on the MPI standard and the domain decomposition strategy according to the grid block structure [12].

In the current contribution, the QUICK scheme was applied to compute the convective terms, and the coupled AC/SIMPLEC algorithm was activated (the AC-parameter value of seven was chosen, the Courant number was equal to five). A mesh dependency study was performed for the case of $A = 30$ using several successively refined grids, the meshes were clustered to the walls with special attention to the endwall regions, an illustration of a coarse mesh is given in figure 1. The solutions obtained with the mesh of 838×101 nodes (83,700 cells) were accepted as mesh-independent.

To simulate unsteady flow regimes, the time step of one time unit was set based on several tests with various time step values. Special attention was paid also on the effect of solution convergence at each time step, and after several additional tests it was concluded that 50 iterations per time step is enough to provide sufficient convergence. To control the convergence in case of a steady-state solution and local characteristics evolution in case of an unsteady solution, data from the monitoring point M ($x = 0.25$; $y = 5.06$) were used, see figure 1.

3. Results and discussion

Formation of the buoyancy-induced flow in the cavity with $A = 30$ is illustrated by the pathline patterns shown in figure 2. The figure gives the plots for the central part of the cavity, far from the endwalls, at six values of Ra. Note that the pathline plots illustrate the flow field qualitatively, as the pathlines are issued manually. At the lowest $Ra = 4850$, a steady-state solution is obtained, and the unicellular flow without any secondary cells is detected (figure 2a). A steady-state solution is obtained also at $Ra = 7300$, though the pathlines given in figure 2b are curved slightly that indicates formation of a weak secondary flow with narrow cells extended in the vertical direction. Note that experimental smoke patterns presented in [5] point to transition to secondary flow at $Ra \approx 6800$, and the secondary cells detected in [5] were always unsteady, with slow movement downward.

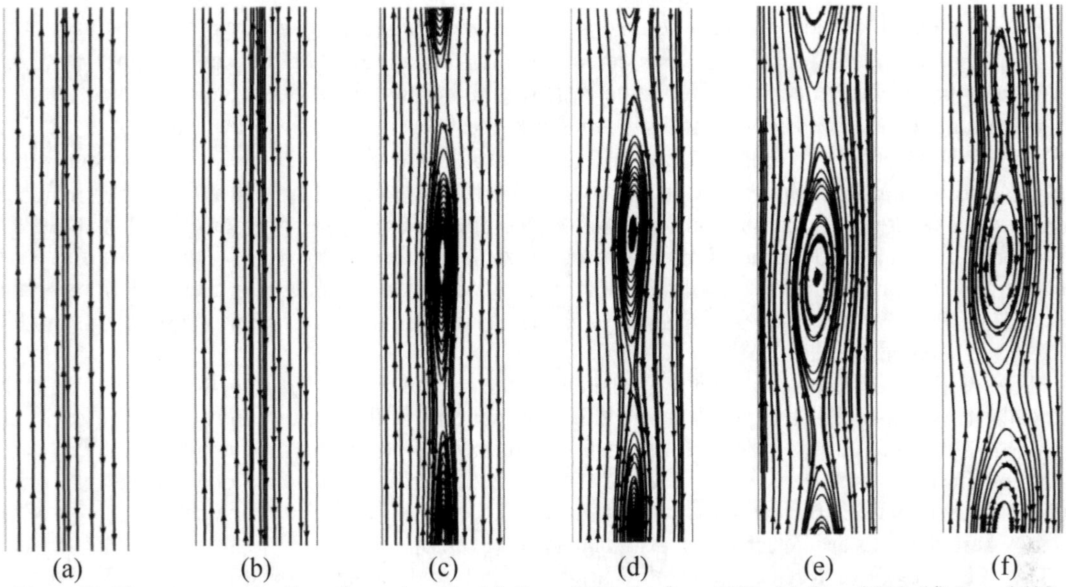

Figure 2. Flow patterns far from the end walls: (a) $Ra = 4850$; (b) $Ra = 7300$; (c) $Ra = 8600$; (d) $Ra = 9600$; (e) $Ra = 10500$ (f) $Ra = 12600$; $A = 30$

Pronounced secondary cells are visible in the plots corresponding to higher Ra values (figure 2c-f). The flow at $Ra \geq 8600$ is unsteady, and the plots give instantaneous flow fields. The width of the co-rotating cells increases with an increase in Ra, while the vertical length of the cells does not change much. The downward movement of the cells detected in all unsteady cases was very slow, e.g., at $Ra = 8600$ the period of high-amplitude velocity fluctuations at point M was about 270 time units. Instantaneous flow patterns at two successive time instants are given in figure 3 ($Ra = 8600$). Time interval between two plots is 105 time units. The distance travelled downward by the cell structure during this time interval is about one length unit.

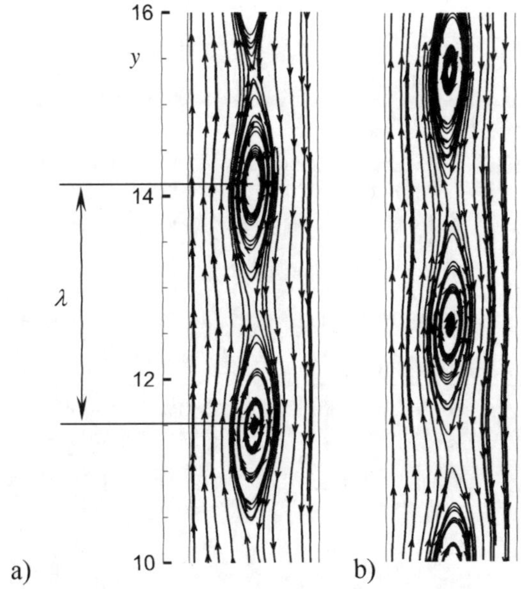

Figure 3. Successive instantaneous flow patterns, $A = 30$, $Ra = 8600$; time interval between (a) and (b) is 105 time units

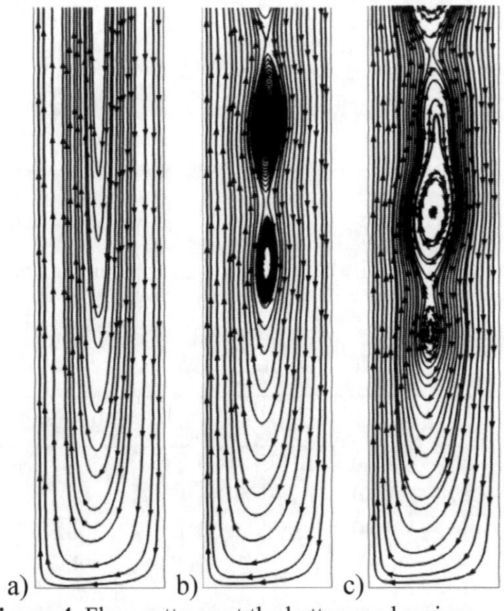

Figure 4. Flow patterns at the bottom end region: (a) $Ra = 7300$; (b) $Ra = 8600$; (c) $Ra = 12600$; $A = 30$

The wave length of the cells, λ, is defined as the distance between the centres of two neighbouring cells (see figure 3a), and for the case considered $\lambda \approx 2.7$. According to [9], the wave number $\alpha = 2\pi/\lambda$ can be also introduced to characterize the cells. It was found that the influence of the aspect ratio on the wave number value is significant. E.g., for the highest $Ra = 12600$ the wave number $\alpha = 2.71$ at $A = 20$, $\alpha = 2.33$ at $A = 30$, and $\alpha = 2.31$ at $A = 40$. At the lowest $A = 20$ the wave number values depend also on the Ra values: $\alpha = 2.63$ at $Ra = 8600$. Thus, at $A = 20$ the wave number increases slightly with the Rayleigh number growth, and the increment in the Ra range considered is about 3%. Stronger wave number dependence on Ra is reported in [9] for the cavity of $A = 16$, but the paper considered the wave number behavior at lower Ra values correspondent to the transitional range ($\alpha = 2.77$ at $Ra = 7153$). At higher aspect ratio values considered the wave number does not depend on Ra in the range considered.

The periodical structure of the cells immanent in the central part of the layer is destroyed near the endwalls. It is illustrated in figure 4 where the flow patterns at the bottom end region are given for three Ra values. In all three cases shown the flow patterns are skewed in the vicinity of the endwalls. Non-regular cells of smaller size are detected if the flow regime is unsteady. Note that the cells origin and unsteady behavior are very sensitive to the spatial resolution not only in the central region of the cavity where the cells are detected, but also near the endwalls.

The "cat's eye" convective cells play an important role in heat transfer. Figure 5 gives the distributions of the local Nusselt number, Nu, over the hot isothermal wall; $Nu = q_w L/k\Delta T_{ref}$, q_w is the local heat flux, k is the thermal conductivity. If there is only primary upward-downward flow ($Ra \leq 6800$), the heat transfer is dominated by conduction with $Nu = 1$ everywhere except near-endwall regions. As Ra increases, the convective cells form, and Nu should grow locally from unity. The local effects of convection are significant, especially at higher Ra: e.g., at $Ra = 12600$ the peak values of Nu exceed 1.7. The number of peaks and the distance between the neighboring maximum values in the Nu distribution correspond to the above discussion of A influence on the wave number.

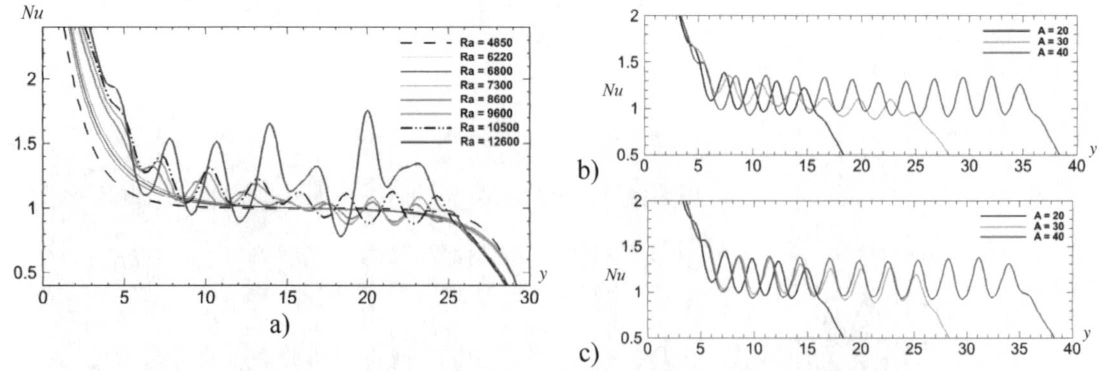

Figure 5. Local Nusselt number distributions over a vertical wall: (a) various Ra, $A = 30$; (b, c) various A, (b) $Ra = 8600$ and (c) $Ra = 10500$

Table. Computed <Nu> values for different A in comparison with correlation [13].

Ra	4850	6220	6800	7300	8600	9600	10500	12600
$\langle Nu \rangle$, $A = 20$	1.18	1.23	1.27	1.29	1.35	1.39	1.43	1.50
$\langle Nu \rangle$, $A = 30$	1.12	1.17	1.17	1.19	1.23	1.26	1.32	1.39
$\langle Nu \rangle$, $A = 40$	1.09	1.12	1.15	1.17	1.21	1.26	1.28	1.34
$\langle Nu \rangle$ [13]	1.052	1.094	1.113	1.133	1.194	1.250	1.301	1.402

The computed area-averaged Nusselt number values, $\langle Nu \rangle$, are compared in the table with the values estimated with the empirical correlation of Wright [13] valid for $A > 40$. As expected, the effect of the endwalls is pronounced in the transitional *Ra* range, $Ra \leq 10^4$: at lower *A* values the <*Nu*> values are higher, and even at $A = 40$ the computed values still exceed slightly the empirical data. As reported in [13], at $Ra > 10^4$ a new segment of correlation started that could be attributed to some 3D effects that start to develop. That could be the reason of slight underestimation of the *Nu* values at $Ra = 10500$ and $Ra = 12600$. It should be concluded that to predict accurately chaotic convective cell behavior at these *Ra* values 3D unsteady formulation is necessary.

4. Conclusions

Natural convection in vertical air-filled cavities of aspect ratio 20, 30 and 40 with differentially heated sidewalls was simulated numerically using an in-house Navier-Stokes code SINF. The focus is on the appearance of the convective vortex structures, "cat's eyes", and their transition to unsteady regime in the Rayleigh number range from 4.8×10^3 to 1.3×10^4.

It was proved that the cells origin and unsteady behavior are sensitive to the spatial resolution not only in the central region of the cavity where the cells are detected, but also near the endwalls. If sufficient spatial resolution in the vicinity of the endwalls is provided, the computed results correlate with the literature experimental data. It was found that the wave number of the convective cells and the integral Nusselt number values depend on the cavity aspect ratio up to $A = 40$.

References

[1] Gershuni G Z 1953 Stability of a plane convective flow of liquid *Zh. Tekh. Fiz.* **23** p 1838

[2] Gershuni G Z, Zhukhovitsky E M and Nepomniashchii A A 1989 *Ustojchivost' konvektivnykh techenij [Stability of convective flows]* (Moscow: Nauka) p 7

[3] Lappa M 2010 *Thermal convection: pattterns, evolution and stability* (Wiley) p 247

[4] Vest C M and Arpaci V S 1969 Stability of natural convection in a vertical slot *J. Fluid Mech.* **36** pp 1-15

[5] Wright J L, Jin H, Hollands K G T and Naylor D 2006 Flow visualization of natural convection in a tall, air-filled vertical cavity *Int. J. Heat and Mass Transfer* **49** pp 889–904

[6] Wakitani S 1998 Flow patterns of natural convection in an air-filled vertical cavity *Physics of Fluids* **10** pp 1924–28

[7] Pons M 2008 Transition from single to multi-cell natural convection of air in cavities with an aspect ratio of 20: a thermodynamic approach *Int. J. of Thermodynamics* **11** pp 71-79

[8] Baez E and Nicolas A 2013 From cat's eyes to multiple disjoint natural convection flow in tall tilted cavities: a direct primitive variables approach *Physics Letters A* **377** pp 2270-74

[9] Kolsi L, Ben Hamida M B, Hassen W, Hussein A K, Borjini M N, Sivasankaran S, Saha S C, Awad M M, Fathinia F and Ben Aissia H 2015 Experimental and numerical investigations of transient natural convection in differentially heated air-filled tall cavity *Am. J. of Modern Energy* **1** pp 30–43

[10] Smirnov E M 2000 Solving the full Navier-Stokes equations for very-long-duct flows using the artificial compressibility method *Proc. of ECCOMAS 2000 - European Congress on Computational Methods in Applied Sciences and Engineering (Barcelona, Spain, 11 - 14 September 2000)* pp 1–17

[11] Smirnov E M and Zaitsev D K 2004 Computations of internal flows using an artificial-compressibility solver enhanced with an elliptic pressure-correction procedure *Proc. of ECCOMAS 2004 - European Congress on Computational Methods in Applied Sciences and Engineering (Jyväskylä, Finland, 24 - 28 July 2004)* pp 1–13

[12] Zaitsev D K, Shchur N A and Smirnov E M 2009 Numerical simulation of 3D turbulent flows around bodies subjected to vortex-induced and forced vibration *Lecture Notes in Computational Science and Engineering* **67** pp 347–54

[13] Wright J L 1996 A correlation to quantify convective heat transfer between vertical window glazings *ASHRAE Trans.* **102** p 940-46

Evaluation of dispersion interaction in glyoxal/silica organomineral system

Y V Sokolova[1], A M Ayzenshtadt [1] and V V Strokova[2]

[1]Northern (Arctic) Federal University named after M.V. Lomonosov, Severnaya Dvina Emb. 17, Arkhangelsk, 163002 Russia
[2]Belgorod State Technological University named after V.G. Shoukhov, Kostyukov str. 46, Belgorod, 308012 Russia

E-mail: yu.sokolova@narfu.ru

Abstract. It is proposed to use an organomineral additive based on glyoxal and saponite-containing waste in order to stabilize the road base. Preliminary studies have shown that soil modification with the organomineral additive with a quantitative ratio of components of glyoxal – 0.52 % and saponite-containing material – 17 % of sand mass increases its specific cohesion by 50 times. However, it is necessary to study the mechanism of interaction between the additive components and soil to optimize the composition and technology of additive introduction into the soil. This paper examines dispersion interaction in glyoxal/silica organomineral system. The Hamaker constant was chosen as a main parameter to evaluate dispersion interaction. This constant is determined based on the principle of measuring the contact angle of wetting with service fluids and calculating the dispersion component of material surface tension. The Hamaker constant in the glyoxal/silica organomineral system was $0.25 \cdot 10^{-20} - 2.8 \cdot 10^{-20}$ J at different glyoxal contents, while its maximum value was observed at 0.52 % glyoxal content. The results compared with literature and previously obtained data demonstrated good reproducibility.

1. Introduction

Preliminary studies provided in the paper [1] proved the possibility to use the composite based on glyoxal and saponite-containing waste as an organomineral additive for sandy soil stabilization.

Glyoxal is a dialdehyde of oxalic acid. It is widely used in the industry not only due to the reactivity of its molecules, but also because of the high surface activity of this substance. The additive ensures uniform glyoxal polymerization throughout the soil with a sorbent included in its composition – a saponite-containing waste extracted from tailing dump slurry of industrial ore benefication at the diamond deposit named after M.V. Lomonosov.

The organomineral additive components introduced in the soil test samples with a quantitative ratio of glyoxal – 0.52 %, saponite-containing material – 17 % of sand mass increase specific cohesion of the composite by 50 times. The main interaction type of reaction system components is dispersion interaction on the phase boundary in the glyoxal/silica (sand) organomineral system. Currently, however, there is no quantitative estimation of this interaction. Meanwhile, solution of this problem will optimize the composition and technology of additive introduction into the soil.

The objective of studies presented in this paper is to estimate dispersion interaction in the glyoxal/silica (sand) organomineral system in terms of quantity.

IOP Conf. Series: Journal of Physics: Conf. Series **929** (2017) 012110 doi:10.1088/1742-6596/929/1/012110

The silica model used was a sedimentary rock – polymictic river sand from the Krasnoflotsky-Zapad deposit with 91 % of silicon dioxide [2].

The main parameter chosen for the study of dispersion interaction in the glyoxal/silica organomineral system was the Hamaker constant (*A*) which takes into account at least two components acting together – direct interparticle interaction between uniform particles and interface interaction on the solid-solution boundary. The papers [3-5] show that the Hamaker constant is functionally related to surface tension of the dispersion system generated on the phase boundary which characterizes the energy state of the unit of surface.

Surface tension between sand and glyoxal-based organic polymer particles was determined using the Owens, Wendt, Rabel and Kaelble (OWRK) method. The OWRK method rests on the fact that the surface free energy is a sum of polar and dispersive values and determined by linear regression built based on the results of measuring contact angles of wetting with service fluids containing known polar and dispersive components.

The polar component of the surface free energy is formed by means of electrostatic forces originated from polar groups of molecular structures that lost some atoms, and by means of chemical bonds forming electrically neutral dipoles. The dispersive component is defined by Van der Waals interaction formed by the molar internal energy in the material surface layer [3, 6, 7].

According to literature data, the Hamaker constant was already used to estimate dispersion interaction. Thus, [8] gives the following *A* values for polycrystalline quartz – $4.1 \cdot 10^{-20} - 4.5 \cdot 10^{-20}$ J.

2. Method

Sand was preliminarily washed free from clay intrusions and dried to its constant mass at 105±5 °C. The initial 40 % aqueous solution of glyoxal was diluted to obtain solutions at a concentration of 1, 2, 5, 10, 20 %. Test samples were made by mixing 10 ml of differently concentrated glyoxal solution with 20 g of sand. The resulting mixture was compacted using the PLG-20 press with a working force of 40 kN. Samples were 30 mm in diameter and 14 mm in thickness. Then samples were brought to their constant masses at 40 °C. The constant mass value was taken as a criterion determining the end of glyoxal polymerization.

The contact angle of wetting was determined at the OCA 35L unit. Service fluids were liquids with known polar and dispersive components – water, ethanol, ethylene glycol, decane.

The set of experiments consisting in three mandatory measurements with a maximum allowable divergence of 2-3 ° contact angle values was followed by calculating the dispersive component of surface tension and the Hamaker constant.

The Hamaker constant for the test material was calculated by the equation (1):

$$A = 24\pi l_0^2 \sigma_S^D \quad (1)$$

where l_0 – equilibrium distance between two interacting bodies (particles) which is taken as 0.165 nm for solids; σ_S^D – dispersive component of surface tension for the test material.

3. Results

Table 1 contains the surface tension of service fluids and average contact angles of wetting for the glyoxal/silica (sand) system at 0.52 % glyoxal content.

Figure 1 shows an example of functional dependence to determine the polar and dispersive components for the test material at 0.52 % glyoxal content.

Table 2 contains the polar and dispersive components, total surface tension for the glyoxal/silica (sand) system at different glyoxal weight percentages.

Figure 2 shows the dependence of the Hamaker constant for the glyoxal/silica system on glyoxal content.

Table 1. The surface tension of service fluids and average contact angles of wetting for the glyoxal/silica system at 0.52 % glyoxal content.

Service fluids	Surface tension, $\sigma_L \cdot 10^3$, N/m	Polar component of surface tension, $\sigma_L^P \cdot 10^3$, N/m	Dispersive component of surface tension, $\sigma_L^D \cdot 10^3$, N/m	Average contact angle of wetting, θ,°
water	72.8	51.0	21.8	74.5
ethanol	21.4	2.6	18.8	28.5
ethylene glycol	48.8	16.0	32.8	61.4
decane	23.8	0	23.8	18.8

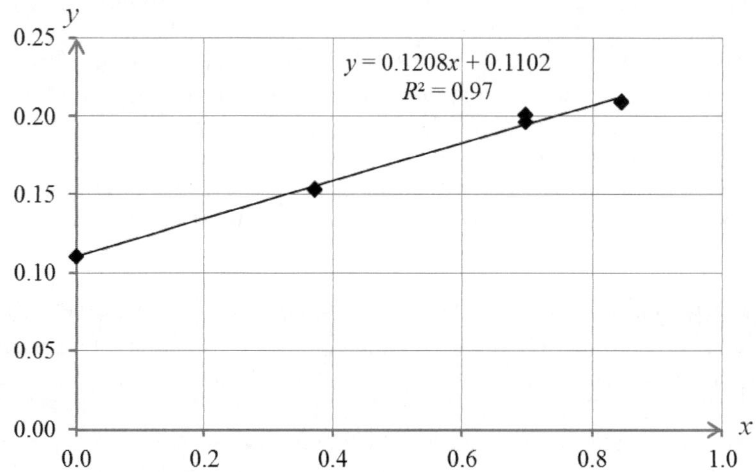

Figure 1. The functional dependence $y = f(x)$
for the glyoxal/silica system at 0.52 % glyoxal content.

$$\text{where } y = \frac{\sigma_L(\cos\theta + 1)}{2\sqrt{\sigma_L^D}}; \ x = \sqrt{\frac{\sigma_L^P}{\sigma_L^D}}$$

Table 2. The polar and dispersive components, total surface tension for the glyoxal/silica system at different glyoxal weight percentages.

Glyoxal mass content, N, % sand total amount	Polar component of surface tension, $\sigma_S^P \cdot 10^3$, N/m	Dispersive component of surface tension, $\sigma_S^D \cdot 10^3$, N/m	Total surface tension, $\sigma_S \cdot 10^3$, N/m
0.10	74.15	1.20	75.35
0.21	72.54	1.35	73.89
0.52	14.59	12.14	26.73
1.06	13.04	12.90	25.94
2.29	12.61	13.50	26.11

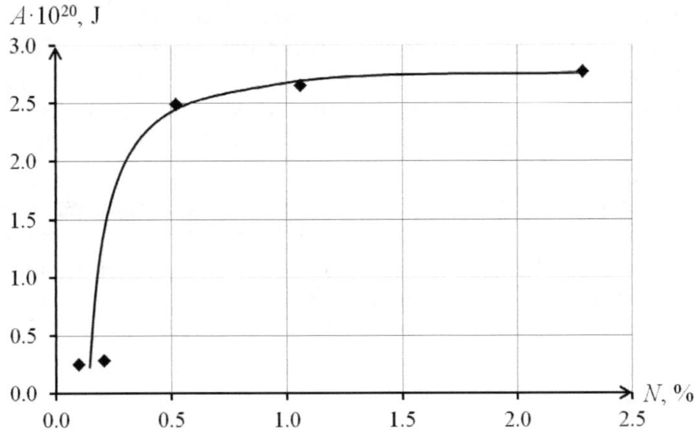

Figure 2. The dependence of the Hamaker constant for the glyoxal/silica system on glyoxal content.

4. Discussion

The Hamaker constant in the glyoxal/silica organomineral system at different glyoxal content was $0.25 \cdot 10^{-20} - 2.8 \cdot 10^{-20}$ J. The optimal glyoxal content was taken as 0.52 % of sand mass since such an additive content ensures maximum dispersion interaction in the system which does not considerably vary as glyoxal content increases. This fact is in good agreement with the previous studies aimed at determination of physical and mechanical properties of the soil with the organomineral additive.

In our opinion, the Hamaker constant values obtained compared with literature data demonstrated a satisfactory reproducibility. The lower content of this constant in the test sample was caused by inhomogeneity of its structure. Sorbent – saponite-containing material – introduced into the additive contributes to the uniform structure and higher soil strength.

The fact of strengthening Van der Waals interaction between mineral particles of ground in the presence of glyoxal, which was evaluated by Hamaker constant, in our opinion, is connected with forming strong polymer film of the organic component, substituting aqueous dispersion medium. This contributes to the creation of compact periodic colloidal structure (PCS) from monodispersed colloidal silica particles. PCS haves a quasicrystalline lattice. Therefore, the physical and mechanical properties of the resulting PCS will be determined by the forces of interaction between the surface of the dispersed particles and the properties of the organic polymer interlayer. The experiments showed to create the such interparticle interlayer in studied organomineral system, a sufficient amount of dialdehyde is 0.52 %.

At the same time, the test results indicate that the employed approach can be used to optimize the composition of the organomineral additive.

Acknowledgments

The research was carried out on the unique scientific equipment "Physical Chemistry of Surfaces of Nano-Dispersed Systems" with the financial support of the Russian Foundation for Basic Research, Project No. 16-33-50121.

References

[1] Gayda J, Ayzenshtadt A, Tutygin A and Frolova M 2016 *Procedia Engineering* vol 143 90–97
[2] Veshniakova L, Ayzenstadt A and Frolova M 2015 *Physics and Chemistry of Materials Treatment* **2** 68–72
[3] Volkov V A 2015 *Colloid chemistry. The surface phenomena and disperse systems* (St. Petersburg: Lan) p 659

[4] Opanasenko O N, Krutko N P, Zhigalova O L and Luksha O V 2013 *Proc. of the National Academy of Belarus Sciences. Series of Chemical Sciences* **1** 46–49
[5] Frolova M A, Tutygin A S, Ayzenshtadt A M, Lesovic V S, Makhova T A and Pospelova T A 2011 *Nanosystems: physics, chemistry, mathematics* **2** (4) 120–125
[6] Yakavets N, Krut'ko N and Opanasenko O 2012 *Collection of articles: Sviridov readings* **8** pp 253–260
[7] Dulina O A and Abramova A D 2015 *Collection of scientific works of the 2nd Int. Youth Competition: Youth in Science: New arguments* pp 17–20
[8] Guozhong Cao and Ying Wang 2012 *Nanostructures and nanomaterials. Synthesis, properties and applications* (Moscow: Scientific World) p 520

Sorption-desorption properties of saponite-containing material

M V Morozova[1], **M A Frolova**[1] **and T A Makhova**[1]

[1]Northern (Arctic) Federal University named after M.V. Lomonosov, Severnaya Dvina Emb. 17, Arkhangelsk, 163002 Russia

E-mail: m.morozova@narfu.ru

Abstract. The sorption mechanism of the mineral additive showed that self-saturation of the saponite-containing material with water vapors has a long-term nature (12 days). The nature of desorption differs from adsorption isotherm, i.e. the adsorption/desorption hysteresis is observed. Saponite-containing material samples studied after moisture desorption using infrared spectroscopy demonstrated that chemical compounds in the material have the ability to form calcium silicate hydrates when saturated with water. This fact along with the additive capability to control the water-cement ratio during concrete curing contributes to significantly better physical and chemical properties (strength, frost resistance) of the concrete composite.

1. Introduction

The efficiency of concrete curing is determined by the hydration of cement minerals forming a new colloidal dispersion phase (condensation reaction) of different basicity calcium silicate hydrates. However, condensation producing colloidally sized particles is only possible under certain conditions. Thus, the system must be oversaturated with the substance to enable its condensation. Nuclei of a new phase form at a certain degree of solution oversaturation. Therefore, formation of colloidal systems is controlled by the rate of nuclei formation and growing. In turn, the rate of new phase nuclei (calcium silicate hydrates) formation is proportional to the solution oversaturation degree. Consequently, water-cement (W/C) ratio regulation during cement curing by means of physisorption of water by the mineral additive can stabilize the degree of system oversaturation relative to new formation of hydrosilicates [1, 2]. This leads to the uniform distribution of cement mineral hydration products throughout the sample. The basic process, which enables W/C ratio regulation during concrete composite curing, is the water physisorption by the modifier.

It is known that bentonite clays are natural hydroalumosilicates and exhibit a number of properties typical of common clays (hydration swelling, dense gel formation to prevent moisture penetration, non-toxicity, chemical stability). A typical representative of such systems is a saponite-containing material (SCM) extracted from recycled water at mining plants [3-5]. Our preliminary studies conducted to investigate hydrophysical properties of this material identified the water-absorbing ability of the SCM. Moreover, the paper [6] demonstrates the possibility to use the saponite-containing material as a binder.

Therefore, the objective of research described in this paper is to characterize sorption-desorption properties of the saponite-containing material with respect to water and identify potential products of superfine mineral additive hydration.

2. Materials and methods

2.1. Materials

The tests used a composite sample of recycle water suspension with the saponite-containing material sampled at different heights of the tailing dump.

The SCM was extracted from the suspension by electrolytic coagulation based on transforming superfine particles to the state close to an isoelectric one.

2.2 Methods

The saponite-containing material extracted from recycle water was brought to the constant mass at 105 ^{0}C.

Dry milling at the Retsch PM100 planetary ball mill was used to reduce the material to the required particle size. Optimal dispersion parameters were selected for each test sample to ensure minimum particle sizes in samples and high reproducibility (at least three parallel tests). Rotor speed was 420 rpm. Dispersion was performed using tungsten carbide grinding balls (20 pcs).

Particle sizes were determined at the Delsa Nano zeta potential and submicron particle size analyzer by measuring dynamic and electrophoretic light scattering.

A superfine sample obtained was characterized by nitrogen sorption at the Autosorb-iQ-MP analyzer using the S_{sp} specific surface value.

The Zeiss Sigma VP scanning electronic microscope was used to take photographs of the superfine SCM (at 20,000X magnification with a resolution up to 1.3 nm) at the Arktika Common Use Center (Arktika CUC).

In addition, the Lab Center XRF-1800 unit was used for X-ray phase analysis at the Arktika CUC to determine mineral composition of the saponite-containing material by measuring the intensity of characteristic X-Rays (fluorescence) when exposing solid samples.

Sorption process was studied by placing a weight portion (100 g) of the SCM taken on the counterbalance to a desiccator above the distilled water layer at the constant ambient temperature 25±1 ^{0}C. Water absorption was characterized by changes in sample mass determined at specified time intervals. Measurement results were based on two parallel experiments.

For desorption the moisture-saturated test sample was placed in the desiccator above the calcium oxide layer, while the temperature remained constant (25±1 ^{0}C).

After sample desorption tests the SCM was placed in a drying box and brought to the constant mass at 105 ^{0}C, then infrared spectra of samples were recorded using the Vertex 70v FT-IR spectrometer. Spectra were recorded under the following conditions: range 4000-600 cm^{-1}, resolution 4 cm^{-1}.

3. Resuits and discussion

Dispersion produced the saponit-containing material samples with a particle size range of 1 μm to 400 nm and specific surface of 18610±10 m^2/kg to 50670±20 m^2/kg respectively (Table 1).

Table 1. Grinding time and characteristics of fractions of SCM samples.

Fraction, №	Samples dispersion time, min	Average particle size, nm	Specific surface area, S_{sp} m^2/kg
1	5	1113±23	18610±10
2	10	827±52	19810±10
3	30	764±43	21210±20
4	60	591±28	35280±30
5	90	445±40	50670±20
6	110	447±34	50672±20
7	120	517±22	30359±30

Relation between the SCM specific surface and dispersion time demonstrated that dispersion within over 90 min causes no S_{sp} increase. Therefore, the modifier of Size Fraction No. 5 (S_{sp} = 50670 m^2/kg) was used for further studies.

Figure 1 shows Size Fraction No. 5 indicating the uniformity of size distribution by their size in this range. Test data regarding particle sizes obtained using photon correlation spectroscopy are in good agreement with electron microscopy results (Figure 2). Moreover, electron microscopy showed that the saponite-containing material has a flaky structure with numerous voids able to be filled with water.

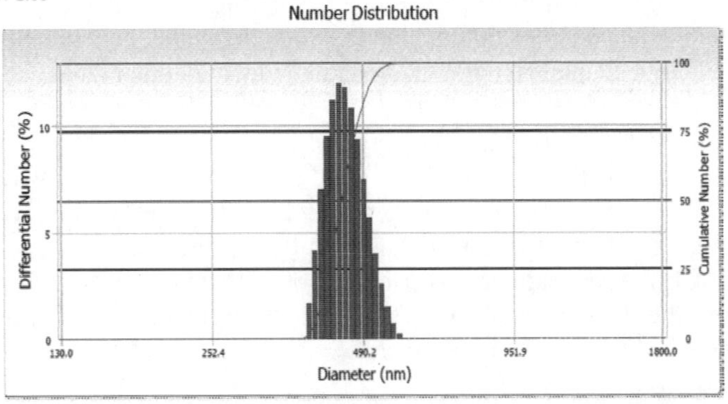

Figure 1. Dimensional distribution of particles in pre-production SCM sample of Size Fraction No. 5, obtained on Delsa Nano analyzer.

Figure 2. The microphoto of pre-production SCM sample (Size Fraction No. 5), obtained on raster electronic microscope Sigma VP.

The chemical analysis of the SCM sample showed that the major elements of the test sample (calculated as oxides) are SiO_2, MgO, Al_2O_3, Fe_2O_3, CaO and K_2O. The content of the above oxides is - 97.4% (Table 2)

Table 2. The composition of the SCM samples calculated as oxides.

Selected component	SCM content, %	Selected component	SCM content, %
SiO_2	51.746	SO_3	0.318
MgO	19.405	P_2O_5	0.660
Al_2O_3	9.971	Cr_2O_3	0.112
Fe_2O_3	10.406	ZnO	0.031
CaO	4.199	BaO	0.197
TiO_2	1.017	MnO	0.156
K_2O	1.693	SrO	0.044
CuO	0.045	other 0.001	

Figure 3 demonstrates test results regarding sorption properties of the saponite-containing material (Size Fraction No. 5) indicating that the process of solid phase saturation with water vapors has a long-term nature. The weight portion of the test material is fully saturated within 12 days from the test beginning. The sample mass is increased by 11% (saturation of the sample surface layer). The process of water desorption with the test sample has a nature other than the adsorption isotherm (Figure 3, Curve 2). Therefore, the test showed the presence of adsorption/desorption hysteresis in the SCM/water vapor system. It should be noted that the SCM weight after desorption differed from the initial sample weight by 8%.

Marked adsorptive hysteresis is characteristic of process of the polymolecular adsorption which is followed in parallel by the proceeding condensation of vapors of an adsorbtiv (water) in small pores of adsorbent (SCM). Therefore, it is necessary to consider two various mechanisms: polymolecular adsorption in direct and capillary evaporation in the return processes. Thus, different mechanisms in direct and return processes define the nature of the adsorptive hysteresis.

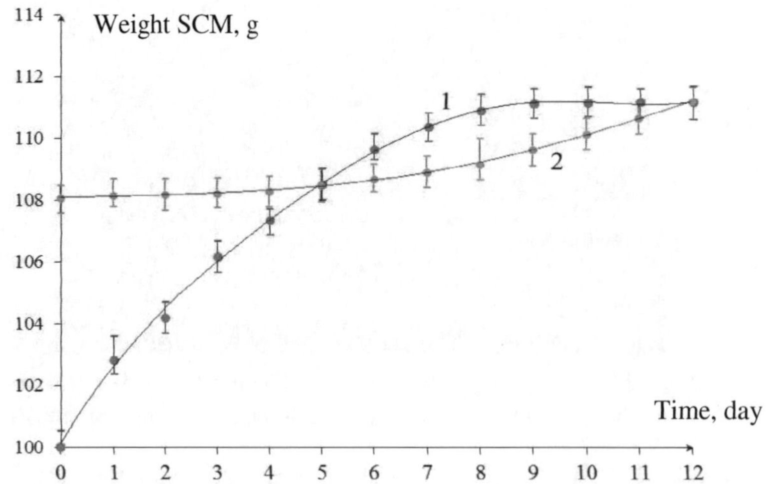

Figure 3. The dependence of SCM moisture absorption (S_{sp} = 50670±20 m^2/kg) on time: 1 - adsorption process; 2 - desorption process.

IR spectroscopy results (Figure 4) obtained after desorption and residual moisture removal from the SCM demonstrated the presence of new formations in the material – silicate groups (975 cm^{-1}).

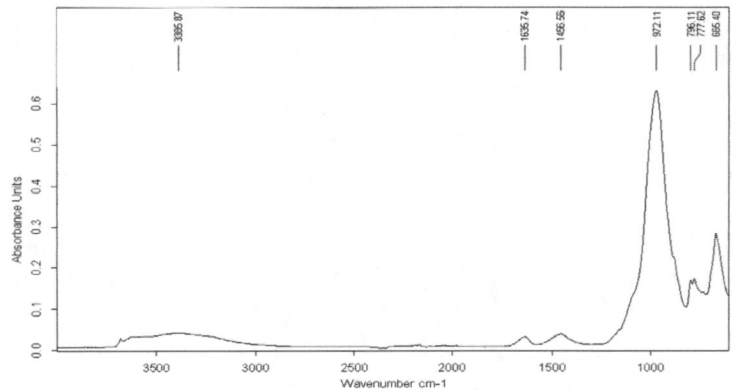

Figure 4. IR spectrum of the SCM sample.

The analysis of the peak form having a complex nature (with "shoulders") allows us to conclude that there are also hydrosilicates of different basicity. The lines of O-H groups are present in the 3600-3200 cm^{-1} and 1635 cm^{-1} areas that also indicates the presence of hydrosilicates.

4. Conclusion

Summarizing the obtained test results, we can conclude the following:

1. The process of water sorption with the superfine additive of the saponite-containing material proceeds spontaneously, while saturation of samples with water takes place within 12 days. Removal of moisture accumulated in the material is accompanied with adsorption/desorption hysteresis.

2. The study of saponite-containing material samples after moisture desorption by infrared spectroscopy demonstrated that chemical compounds in the material can form calcium silicate hydrates under test conditions.

5. Acknowledgements

The research was carried out on the unique scientific equipment "Physical Chemistry of Surfaces of Nano-Dispersed Systems".

References

[1] Morozova M V, Ayzenstadt A M and Tutygin A S 2013 *J. Industrial and Civil Engineering* **11** 29-31
[2] Basenov U M 2002 *The technology of concrete* (Moscow: DIA Press) p 500
[3] Tutygin A S, Aizenstadt M A, Aizenstadt A M and Makhova T A 2012 *J. Geoecology* **5** 379-83
[4] Apollonov V N, Verzhak V V, Garanin K V, Garanin V K, Kudryavtseva G P and Shlikov V G. *J. Geology and Exploration* **3** 20-37
[5] Morozova M V, Ayzenstadt A M, Frolova M A and Makhova T A 2015 *J. Academia. Architecture and Construction* **4** 137-41
[6] Drozdyk T A, Ayzenstadt A M, Tutygin A S and Frolova M A 2015 *J. Building materials* **5** 86-89